Lecture Notes in Computer Science 12813

Liesbeth De Mol · Andreas Weiermann ·
Florin Manea · David Fernández-Duque (Eds.)

Connecting
with Computability

17th Conference on Computability in Europe, CiE 2021
Virtual Event, Ghent, July 5–9, 2021
Proceedings

 Springer

Editors
Liesbeth De Mol
University of Lille
Lille, France

Florin Manea (iD)
University of Göttingen
Göttingen, Germany

Andreas Weiermann
Vakgroep Wiskunde
Ghent University
Ghent, Belgium

David Fernández-Duque
Ghent University
Ghent, Belgium

ISSN 0302-9743   ISSN 1611-3349   (electronic)
Lecture Notes in Computer Science
ISBN 978-3-030-80048-2   ISBN 978-3-030-80049-9   (eBook)
https://doi.org/10.1007/978-3-030-80049-9

LNCS Sublibrary: SL1 – Theoretical Computer Science and General Issues

This Springer imprint is published by the registered company Springer Nature Switzerland AG
The registered company address is: Gewerbestrasse 11, 6330 Cham, Switzerland

Preface

CiE 2021: Connecting with Computability, 5–9 July 2021, virtual conference hosted in Ghent, Belgium

Computability in Europe (CiE) is an annual conference organized under the auspices of the Association CiE (ACiE), a European association that brings together researchers from a broad variety of backgrounds (mathematics, computer science, logic, history, biology, philosophy, and physics, among others) connected to one another through their work on computability. The conference series has built up a strong tradition for developing a scientific program which is interdisciplinary and embracive at its core, bringing together all aspects of computability and foundations of computer science, as well as exploring the interplay of these theoretical areas with practical issues in both computer science and other disciplines. Its purpose is not only to report on ongoing research but also to broaden perspectives by engaging with the work of others from different backgrounds. As such, the conference has allowed participants to enlarge and transform our view on computability and its interface with other areas of knowledge.

Over the years the conference series has been organized in a spirit of open-mindedness and generosity, and also this year we have again aimed for this, despite the circumstances. The motto of CiE 2021 was Connecting with Computability, a clear acknowledgement of the connecting and interdisciplinary nature of the conference series, which is all the more important in a time when people are more disconnected from one another than ever due to the COVID-19 pandemic. It was organized virtually using the software gather.town as a means to create a virtual social environment with a nod to the host city of Ghent, with its characteristic castle and river as well as its vibrant social life.

CiE 2021 is the 17th conference in the series. Previous meetings have taken place in Amsterdam (2005), Swansea (2006), Siena (2007), Athens (2008), Heidelberg (2009), Ponta Delgada (2010), Sofia (2011), Cambridge (2012), Milan (2013), Budapest (2014), Bucharest (2015), Paris (2016), Turku (2017), Kiel (2018), Durham (2019), and, virtually, in Salerno (2020).

The conference series has become a major event and is the largest international conference that brings together researchers focusing on computability-related issues. The CiE conference series is coordinated by the ACiE conference Steering Committee consisting of Alessandra Carbone (Paris), Liesbeth De Mol (Lille), Gianluca Della

Vedova (Executive Officer, Milan), Mathieu Hoyrup (Nancy), Nataša Jonoska (Tampa), Benedikt Löwe (Amsterdam), Florin Manea (Chair, Göttingen), Klaus Meer (Cottbus), Russell Miller (New York), and Mariya Soskova (Wisconsin-Madison), along with ex-officio members Elvira Mayordomo (President of the Association, Zaragoza) and Dag Normann (Treasurer, Oslo).

Structure and Program of the Conference

The conference program is based on invited lectures and tutorials, and a set of special sessions on a variety of topics; there were also several contributed papers and informal abstract presentations. The Program Committee of CiE 2021 was chaired by Liesbeth De Mol (CNRS and Université de Lille, France) and Andreas Weiermann (Ghent University, Belgium). The committee, consisting of 33 members, selected the invited speakers, the tutorial speakers, and the special session organizers and coordinated the reviewing process and the selection of submitted contributions. The Program Committee selected for publication in this volume 28 of the 49 non-invited papers submitted to the conference. Each paper received at least three reviews by members of the Program Committee and their subreviewers. In addition to the contributed papers, this volume contains 18 invited papers and abstracts.

Invited Tutorials

Russell Miller (CUNY, USA), Computable procedures for field
Christine Tasson (Université de Paris, France), Probabilistic Programming and Computation

Invited Lectures

Laura Crosilla (University of Oslo, Norway), Cantor's paradise and the forbidden fruit
Markus Lohrey (Universität Siegen, Germany), Compression techniques in group theory
Joan Rand Moschovakis (Occidental College, USA), Minimum classical extensions of constructive theories
Joël Ouaknine (Max Planck Institute for Software Systems, Germany), Holonomic Techniques, Periods, and Decision Problems
Keita Yokoyama (Japan Advanced Institute of Science and Technology, Japan), Reverse mathematics and proof and model theory of arithmetic
Henry Yuen (Columbia University, USA), Einstein meets Turing: The Computability of Nonlocal Games

Special Sessions

Computational Geometry

Organizers: Maike Buchin (Ruhr-Universität Bochum, Germany) and Maarten Löffler (Utrecht University, The Netherlands)

Karl Bringmann (Saarland University, Germany), Conditional lower bounds for geometric problems.

Esther Ezra (Bar-Ilan University, Israel), On 3SUM-hard problems in the Decision Tree Model.

Tillmann Miltzow (Utrecht University, The Netherlands), Recent trends in geometric computation models and its relation to the existential theory of the reals.

Wolfgang Mulzer (Free University Berlin, Germany), The many computational models of computational geometry.

Classical Computability Theory: Open Problems and Solutions

Organizers: Noam Greenberg (Victoria University of Wellington, New Zealand) and Steffen Lempp (University of Wisconsin, USA)

Marat Faizrakhmanov (Kazan Federal University, Russia), Limitwise Monotonic Spectra and Their Generalizations

Andrea Sorbi (University of Siena, Italy), Effective inseparability and its applications

Liang Yu (Nanjing University, China), TD implies CCR

Ning Zhong (University of Cincinnati, USA), Computability of limit sets for two-dimensional flows

Proof Theory and Computation

Organizers: David Fernández-Duque (Ghent University, Belgium) and Juan Pablo Aguilera (Ghent University, Belgium)

Lorenzo Carlucci (University of Rome La Sapienza, Italy), Restrictions of Hindman's Theorem: an overview with questions

Leszek Kolodziejczyk (University of Warsaw, Poland), Reverse mathematics of combinatorial principles over a weak base theory

Francesca Poggiolesi (CNRS and Université Paris 1 Panthéon-Sorbonne, France), Defining Formal Explanation in Classical Logic by Substructural Derivability

Yue Yang (National University of Singapore, Singapore), Some results on Ramsey's theorems for trees

Quantum Computation and Information

Organizers: Harry Buhrman (Universiteit van Amsterdam, The Netherlands) and Frank Verstraete (Ghent University, Belgium)

Yfke Dulek (QuSoft and CWI, Netherlands). How to Verify a Quantum Computation
David Gross (University of Cologne, Germany), The axiomatic and the operational approaches to resource theories of magic do not coincide
David Pérez-Garciá (Universidad Complutense de Madrid, Spain), Uncomputability in quantum many body problems
Jens Eisert (Freie Universität Berlin, Germany), Undecidability in Quantum Physics

Church's Thesis in Constructive Mathematics (HaPoC Session)

Organizers: Marianna Antonutti-Marfori (Ludwig-Maximilians-Universität München, Germany) and Alberto Naibo (Université Paris 1 Panthéon-Sorbonne, France)

The HaPoC special session was part of a satellite workshop on the same topic. Other speakers invited to this workshop were Benno van den Berg, Takako Nemoto, Douglas Bridges, and Johanna Franklin.

Liron Cohen (Ben-Gurion University, Israel), Formally Computing with the Non-Computable
Angeliki Koutsoukou-Argyraki (University of Cambridge, UK) On preserving the computational content of mathematical proofs: toy examples for a formalising strategy
Máté Szabó (University of Oxford, UK), Péter on Church's Thesis, Constructivity and Computers
David Turner (University of Kent, UK), Constructive mathematics, Church's Thesis, and free choice sequences.

Computational Pangenomics

Organizers: Nadia Pisanti (University of Pisa, Italy) and Solon Pissis (CWI and Vrije Universiteit, The Netherlands)

Brona Brejova (Comenius University in Bratislava, Slovakia), Probabilistic models for k-mer frequencies
Rayan Chikhi (Pasteur Institute, France), A tale of optimizing the space taken by de Bruijn graphs
Francesca Ciccarelli (King's College London, UK), Gene deregulations driving cancer at single patient resolution
Benedict Paten (University of California, Santa Cruz, USA), Walk-preserving transformation of overlapped sequence graphs into blunt sequence graphs with GetBlunted

Women in Computability Workshop

Since CiE 2007, the Association CiE and the conference have built up a strong tradition of encouraging women to participate in computability-related research. In 2016, a Special Interest Group for Women in Computability was established. This year Mariya Soskova took the initiative of setting up an online mentorship programme for Women in Computability (https://www.acie.eu/women-in-computability-mentorship-programme). These kind of initiatives are anchored in the annual Women in Computability workshop, which was held virtually with the following speakers:

Laura Crosilla (University of Oslo, Norway)
Joan Rand Moschovakis (Occidental College, USA)
Christine Tasson (Université de Paris, France)

Organization and Acknowledgements

The CiE 2021 conference was organized by the Analysis, Logic and Discrete Mathematics group at the mathematics department of Ghent University, Belgium. It was chaired by David Fernández-Duque. We wish to thank all the other members of the Organizing Committee, without their help this conference would not have been possible.

We are very happy to acknowledge and thank the following for their basic financial support: the Fund for Scientific Research, Flanders (FWO), the National Centre for Research in Logic (CNRL-NCNL), Facultaire Commissie voor Wetenschappelijk Onderzoek (FCWO), and, finally, Springer.

The high quality of the conference was due to the careful work of the Program Committee, the special session organizers, and the external referees, and we are very grateful that they helped to create an exciting program for CiE 2021.

May 2021   Liesbeth De Mol
Ghent, Belgium   Andreas Weiermann
                                                        Florin Manea
                                          David Fernández-Duque

Organization

Program Committee

Marianna Antonutti Marfori	Ludwig Maximilian University of Munich, Germany
Nathalie Aubrun	ENS de Lyon, CNRS, Inria, and UCBL, France
Christel Baier	TU Dresden, Germany
Nikolay Bazhenov	Sobolev Institute of Mathematics, Russia
Arnold Beckmann	Swansea University, UK
David Belanger	Ghent University, Belgium
Marie-Pierre Béal	Université Gustave Eiffel, France
Joel Day	Loughborough University, UK
Liesbeth De Mol (Co-chair)	CNRS and Université de Lille, France
Carola Doerr	Sorbonne University and CNRS, France
Jérôme Durand-Lose	LIFO and Université d'Orléans, France
David Fernández-Duque	Ghent University, Belgium
Zuzana Hanikova	Institute of Computer Science, Czech Academy of Sciences, Czech Republic
Mathieu Hoyrup	Loria, France
Assia Mahboubi	Inria, France
Florin Manea	University of Göttingen, Germany
Irène Marcovici	Université de Lorraine, France
Klaus Meer	BTU Cottbus-Senftenberg, Germany
Ludovic Patey	Institut Camille Jordan, France
Cinzia Pizzi	University of Padova, Italy
Giuseppe Primiero	University of Milan, Italy
Simona Ronchi Della Rocca	Universita' di Torino, Italy
Svetlana Selivanova	KAIST, South Korea
Paul Shafer	University of Leeds, UK
Alexander Shen	LIRMM CNRS and Université de Montpellier, France
Alexandra Soskova	Sofia University, Bulgaria
Mariya Soskova	University of Wisconsin-Madison, USA
Frank Stephan	National University of Singapore, Singapore
Peter Van Emde Boas	Universiteit van Amsterdam, The Netherlands
Sergey Verlan	Université Paris-Est Créteil, France
Andreas Weiermann (Co-chair)	Ghent University, Belgium
Damien Woods	Maynooth University, Ireland

Organizing Committee

David Fernández-Duque (Chair)	Ghent University, Belgium
Juan Pablo Aguilera	Ghent University, Belgium
David Belanger	Ghent University, Belgium
Ana Borges	University of Barcelona, Spain
Liesbeth De Mol	CNRS and Université de Lille, France
Andreas Debrouwere	Ghent University, Belgium
Lorenz Demey	Catholic University of Leuven, Belgium
Oriola Gjetaj	Ghent University, Belgium
Eduardo Hermo-Reyes	University of Barcelona, Spain
Brett McLean	Ghent University, Belgium
Christian Michaux	University of Mons, Belgium
Fedor Pakhomov	Ghent University, Belgium
Konstantinos Papafilippou	Ghent University, Belgium
Pawl Pawlowsky	Ghent University, Belgium
Frederik Van De Putte	Ghent University, Belgium, Erasmus University of Rotterdam, The Netherlands
Peter Verdée	Catholic University of Leuven, Belgium
Andreas Weiermann	Ghent University, Belgium

Additional Reviewers

Aguilera, Juan
Anglès d'Auriac, Paul-Elliot
Baillot, Patrick
Berndt, Sebastian
Bienvenu, Laurent
Bournez, Olivier
Boyar, Joan
Buchin, Maike
Calvert, Wesley
Cardone, Felice
Carlucci, Lorenzo
Carton, Olivier
Dean, Walter
Dürr, Christoph
Epstein, Leah
Franklin, Johanna
Ganardi, Moses
Gao, Ziyuan
Gregoriades, Vassilios
Harizanov, Valentina

Hermo Reyes, Eduardo
Hofstra, Pieter
Kach, Asher
Kihara, Takayuki
Kohlenbach, Ulrich
Krejca, Martin S.
Kristiansen, Lars
Lechine, Ulysse
Lempp, Steffen
Loewe, Benedikt
Lubarsky, Robert
Martin, Eric
Melnikov, Alexander
Mercas, Robert
Minnes, Mia
Miquel, Alexandre
Miyabe, Kenshi
Monin, Benoît
Neumann, Eike
Neuwirth, Stefan

Normann, Dag
Ollinger, Nicolas
Pagani, Michele
Pakhomov, Fedor
Patey, Ludovic
Pisanti, Nadia
Pissis, Solon
Porter, Christopher
Pouly, Amaury
Péchoux, Romain
Quinon, Paula
Raskin, Mikhail
Richard, Gaétan

Sabili, Ammar Fathin
San Mauro, Luca
Sanders, Sam
Schmid, Markus L.
Schroeder, Matthias
Thies, Holger
Tirnauca, Cristina
Valiron, Benoît
Vatev, Stefan
Walsh, James
Weiss, Armin
Yokoyama, Keita
Zheng, Xizhong

Invited Talks

Don't Be Afraid to Burn Your Fingers on the Definition of the Real RAM

Tillmann Miltzow [ORCID]

Utrecht University, The Netherlands
t.miltzow@uu.nl

Abstract. We review the real RAM model of computation. The emphasis of this talk is its relation to the existential theory of the reals

Keywords: Real RAM · Computational geometry · Model of computation.

In Computational Geometry, we design and analyze algorithms for the real RAM model of computation. The real RAM is an abstraction of an ordinary computer. It consists of an array of registers to store the data, a central processing unit (CPU) to manipulate the data and a set of instructions for the CPU. The real RAM is capable to store and manipulate real numbers. It was originally defined by Shamos [3] in his PhD thesis. Unfortunately, the real RAM as defined by Shamos had some seriously unintended consequences. In short, it is possible to abuse the power of the real RAM in various ways by having access to the binary representation of real numbers [2]. Since then, researchers used *implicitly* a version of the real RAM that avoided access to the binary representation of real numbers. Recently, Erickson, van der Hoog, and Miltzow [1] made that new real RAM definition explicit. Furthermore, they gave some arguments why this new model may avoid previous pitfalls. In the talk, we will explain and repeat those arguments. Specifically, we will highlight the relation to the existential theory of the reals. It is conceivable that the new real RAM model has other weaknesses that will be discovered in the near or far future. In this case, the authors will have *burned their fingers*. Even if some new pitfalls and weaknesses will be found, we believe that this will not mean the end of Computational Geometry as we know it. This motivates the following hypothesis.

There is a rigorous definition of the real RAM model of computation for which the majority of algorithms and their analysis in Computational Geometry remain meaningful.

As the current model of the real RAM has not exposed any weaknesses within the last forty years, we may have confidence that it will not expose major weaknesses in the next forty years either.

Supported by NWO Veni grant EAGER.

References

1. Erickson, J., van der Hoog, I., Miltzow, T.: Smoothing the gap between NP and ER. In: 2020 IEEE 61st Annual Symposium on Foundations of Computer Science (FOCS), pp. 1022–1033. IEEE (2020)
2. Schönhage, A.: On the power of random access machines. In: Maurer, H.A. (ed.) ICALP 1979. LNCS, vol. 71, pp. 520–529. Springer, Heidelberg (1979). https://doi.org/10.1007/3-540-09510-1_42
3. Aichholzer, O., Cetina, M., Fabila-Monroy, R., Leaños, J., Salazar, G., Urrutia, J.: Convexifying monotone polygons while maintaining internal visibility. In: Márquez, A., Ramos, P., Urrutia, J. (eds.) EGC 2011. LNCS, vol. 7579, pp. 98–108. Springer, Heidelberg (2012). https://doi.org/10.1007/978-3-642-34191-5_9

The Many Computational Models of Computational Geometry

Wolfgang Mulzer[ID]

Institut für Informatik, Freie Universität Berlin, 14195 Berlin, Germany
mulzer@inf.fu-berlin.de

Abstract. I will present a short survey of the many different computational models that have been used in computational geometry over the last 50 years, both for describing geometric algorithms and for obtaining lower bounds.

Keywords: Real RAM · Word RAM · Algebraic decision tree · Pointer machine

Computational geometry is the subfield of theoretical computer science that is concerned with the design and analysis of algorithms that deal with geometric inputs, such as, e.g., points, lines, triangles, or circles. In many ways, computational geometry is very similar to the study of traditional combinatorial algorithms: we aim for methods that are provably correct on all inputs, we search for upper and lower bounds on the worst-case complexity of well-defined computational problems, and we consider traditional computational resources such as space, time, or randomness.

However, the geometric nature of the inputs presents an additional set of challenges over the combinatorial regime: geometric objects typically live in Euclidean spaces and are represented by arbitrary real numbers. Even if we choose to restrict our attention to inputs with integer coordinates, the need to compute angles and (sums of) distances arises often. High precision is necessary to evaluate geometric primitives accurately. Thus, in computational geometry, we must be more careful about our model of computation and which operations it allows. Over time, several such models have been proposed, both for describing upper and lower bounds for geometric algorithms.

The two most widespread models for algorithmic results are the *real RAM*, which allows operations on arbitrary real numbers, and the *word RAM*, which can handle only bit strings of a certain length. Lower bounds are typically proved in the *algebraic decision tree model*, which enhances traditional comparison-based decision trees by algebraic operations. For lower bounds in geometric data structures, the *pointer machine model* is used frequently.

In my talk, I will provide a survey of these models, their definitions, applications, advantages, and disadvantages.

Supported in part by ERC STG 757609.

Holonomic Techniques, Periods, and Decision Problems

Joël Ouaknine

Max Planck Institute for Software Systems, Saarland Informatics Campus,
Saarbrücken, Germany
joel@mpi-sws.org

Abstract. Holonomic techniques have deep roots going back to Wallis, Euler, and Gauss, and have evolved in modern times as an important subfield of computer algebra, thanks in large part to the work of Zeilberger and others over the past three decades. In this talk, I give an overview of the area, and in particular present a select survey of known and original results on decision problems for holonomic sequences and functions. I also discuss some surprising connections to the theory of periods and exponential periods, which are classical objects of study in algebraic geometry and number theory; in particular, I relate the decidability of certain decision problems for holonomic sequences to deep conjectures about periods and exponential periods, notably those due to Kontsevich and Zagier.

Contents

Searching for Applicable Versions of Computable Structures

P. E. Alaev[1] and V. L. Selivanov[2(✉)]

[1] S.L. Sobolev Institute of Mathematics SB RAS, Novosibirsk, Russia
alaev@math.nsc.ru
[2] A.P. Ershov Institute of Informatics Systems SB RAS and S.L. Sobolev Institute
of Mathematics SB RAS, Novosibirsk, Russia
vseliv@iis.nsk.su

Abstract. We systematise notions and techniques needed for development of the computable structure theory towards applications such as symbolic algorithms and on-line algorithms. On this way, we introduce some new notions and investigate their relation to the already existing ones. In particular, in the context of polynomial-time presentability such relation turns out to depend on complexity-theoretic conjectures like $P = NP$.

Keywords: Structure · Quotient-structure · Computable presentation · Grzegorczyk's presentation · Polynomial-time presentation

1 Introduction

"Structure" in this paper always means "countably infinite algebraic structure of a finite signature". Computable structure theory (CST) is a well established branch of computability theory [2,16]. The key notion here is that of a computably presentable structure, i.e. a structure isomorphic to a computable structure (a structure is computable if its universe is $\mathbb{N}$ and all signature functions and relations are computable). In the Russian literature, instead of the term "computably presentable structure" people often use the equivalent notion of a constructivizable structure defined in terms of numberings (see the next section for precise definitions). Using these notions, computability issues in algebra and model theory were thoroughly investigated.

Since CST is based on the general Turing computability and often uses the unbounded search through universes of structures, it is not well suited for computer implementations. In implementations, one has of course pay attention to the complexity of algorithms and of structure presentations. The complexity of

The work is supported by Mathematical Center in Akademgorodok under agreement No. 075-15-2019-1613 with the Ministry of Science and Higher Education of the Russian Federation.

L. De Mol et al. (Eds.): CiE 2021, LNCS 12813, pp. 1–11, 2021.
https://doi.org/10.1007/978-3-030-80049-9_1

structure presentations was pioneered by F. Cannonito [12] and studied in a more general setting by D. Cenzer, R. Downey, A. Nerode, and J. Remmel (see e.g., [13,14] and references therein) where, in particular, the notion of a polynomial-time presentable (P-presentable) structure was introduced and some interesting properties of such structures were established. This research was motivated by analogies with CST rather than by applications. A popular class of automatic structures (see e.g., [19] and references therein) has deep connections with computer science but the majority of structures of interest to mathematics fail to be automatic.

There exist active fields of applied computer science which use "feasible" structure presentations, often implicitly. An example is a well-developed theory of symbolic computations closely related to computer algebra (see e.g., [1,27,29]) which investigates, in particular, feasible algorithms over number fields and polynomial rings and implements them in computer systems which have important applications. Somehow related to this is a rather active research on complexity issues of finitely generated groups and of polynomial rings (see e.g., [22,23]). One usually studies "off-line" algorithms with a completely given finite input. However, many modern computational tasks need to be performed upon a constantly updated and potentially unbounded input, like in dealing with a huge constantly changing database where one cannot have knowledge of the whole of the database before acting. This leads to a new paradigm of so called "on-line" algorithms (see e.g., [6] and references therein).

Although the theories mentioned in the previous paragraph are clearly intimately related to CST, the two fields were developing apparently independently until very recently (at least, there were essentially no references and interactions between them). The situation changed with a series of papers (see [9,15,20] and references therein) promoting the development of feasible CST as a foundation for on-line computations. Independently, the study of P-presentable structures was pushed forward in [3–5], and in [7,8] where the relation of feasible CST to computer algebra is stressed. In [24–26] some applications of feasible CST to computable linear algebra and analysis were discovered and a bridge between symbolic and numeric computations was constructed.

Note that the term "feasible" is understood here in a very broad sense, including even primitive recursive (PR) structures which have some features making them closer to complexity theory than to general computability. A more precise term would be perhaps "subrecursive".

The aim of this paper is to systematise the sketched interaction of feasible CST to the aforementioned applied fields, and to discuss some related basic definitions and facts. On the way from CST to a more applicable CST some equivalent notions become non-equivalent, hence they have to be carefully analysed and tested on their relevance to the applied fields. On this way some new notions were already discovered, e.g., the notion of a PR-presentation without delay [20] and of a P-computable quotient-structure [7]. This paper is thus partially "ideological" but it also contains some new technical results.

The next section contains some definitions of feasible structure presentations and discussions of their relevance to the mentioned applied fields; in particular, we introduce $\mathcal{E}^n$-constructive structures, where $\mathcal{E}^n$ are the Grzegorczyk classes (see e.g., [28]). In Sect. 3 we show that $\mathcal{E}^n$-constructive structures share some properties of PR structures but also have specific features. In Sect. 4 we discuss P-constructive structures, in particular we show that the assertion "the P-presentable structures coincide with the P-constructivizable ones" is almost equivalent to the P = NP problem. We also show that any finitely generated substructure of a P-computable quotient-structure is isomorphic to a P-computable structure. The latter two results are probably the main technical contributions of this paper.

2 Structure Presentations

We start with recalling some basic notions of CST in the Russian terminology [21] based on numberings. Let $\nu : \mathbb{N} \to S$ be a numbering. A relation $P \subseteq S^n$ on S is ν-computable if the relation $P(\nu(k_1), \ldots, \nu(k_n))$ on $\mathbb{N}$ is computable. A function $f : S^n \to S$ is ν-computable if $f(\nu(k_1), \ldots, \nu(k_n)) = \nu g(k_1, \ldots, k_n)$ for some computable function $g : \mathbb{N}^n \to \mathbb{N}$.

Definition 1. *A structure* $\mathbb{A} = (A; \sigma)$ *of a finite signature* σ *is called constructivizable iff there is a numbering* α *of* A *such that all signature predicates and functions, and also the equality predicate, are* α-*computable. Such a numbering* α *is called a constructivization of* $\mathbb{A}$, *and the pair* $(\mathbb{A}, \alpha)$ *is called a constructive structure.*

The Mal'cev notion of a constructivizable structure is equivalent to the notion of a computably presentable structure popular in the western literature. We use both terminologies when we find it convenient. The feasible versions of constructivizable and of computably presentable structures are often not equivalent, as will be further discussed below. PR-versions of the notions defined above are obtained by changing "computable" to "PR" in the definitions above, and taking partial PR-constructivizations with PR-domains instead of the total constructivizations in Definition 1; see [21] for additional details. It is easy to see that a structure is PR-constructivizable iff it is PR-presentable, i.e. isomorphic to a structure with a PR universe A such that all signature functions are the restrictions of suitable PR functions to A and all signature relations are PR. This relates the Russian and western terminologies in the PR-case.

After a long break, PR-structures appeared as an intermediate stage of proving P-presentability of a structure (see e.g., [13,17]). After another break, PR-structures appeared as a promising candidate for capturing the on-line structures which give a practically relevant alternative to computable structures (see the discussion and additional references in [9]). Many results about on-line algorithms are focused on making algorithms PR as opposed to general Turing computable. The following notion from [9] is important for this approach: a structure is *fully PR-presentable (FPR-presentable, or PR-presentable without delay)* if it is isomorphic to a PR structure with universe $\mathbb{N}$. As shown in [20], there is a PR graph which is not FPR-presentable. In [26] a characterisation of FPR-structures in the Mal'cev terminology is given.

Note that some notions equivalent in CST become non-equivalent in the PR-version of CST. E.g., any computable structure whose universe is any infinite computable set, is computably isomorphic to a computable structure with universe $\mathbb{N}$ but, by the previous paragraph, the PR version of this fact fails. Thus, CST is much less relevant to on-line computations than its PR-version. Another example is the fact that, in CST, any computable structure is essentially equivalent to the computable relational structure where the signature functions are replaced by the relations corresponding to the graphs of the functions. This is not the case for the PR-version because any computable relational structure is isomorphic to a PR-structure (even to a P-computable structure), while this is not the case for structures of signatures with functional symbols [13]. Thus, functional symbols are more important for PR (and more feasible) structure theory than for CST.

The class of PR functions is naturally refined to the well known Grzegorczyk's hierarchy $\{\mathcal{E}^n\}_{n \geq 0}$; we refer to [28] for definitions. For $n \geqslant 2$, the Grzegorczyk classes $\mathcal{E}^n$ have many nice closure properties and characterisations which may be found e.g., in [28]. In particular, for any $n \geqslant 2$ the class $\mathcal{E}^n$ is closed under bounded minimization, and $\mathcal{E}^{n+1}$ coincides with the class of functions obtained from the simplest functions by substitutions and at most n applications of primitive recursion (see Theorem 2.18 and the corresponding references in [28]).

It is straightforward to adjust the definition of PR-constructivizable structure to Grzegorczyk's classes: a structure $\mathbb{A} = (A; \sigma)$ is $\mathcal{E}^n$-constructivizable, if there is a partial numbering α of A with an $\mathcal{E}^n$-domain such that all σ-functions and relations are $\mathcal{E}^n$-computable w.r.t. α; such α is called an $\mathcal{E}^n$-constructivization of $\mathbb{A}$, and the pair $(\mathbb{A}, \alpha)$ is called an $\mathcal{E}^n$-constructive structure. The western-style version looks as follows: a structure is $\mathcal{E}^n$-presentable if it is isomorphic to an $\mathcal{E}^n$-computable structure; a structure is $\mathcal{E}^n$-computable if its universe is an $\mathcal{E}^n$-set and all signature functions and relations are the restrictions of suitable $\mathcal{E}^n$-functions and relations. Similarly, a structure is $F\mathcal{E}^n$-presentable if it is isomorphic to an $\mathcal{E}^n$-computable structure with universe $\mathbb{N}$; this is the version of $\mathcal{E}^n$-computable structure "without delay".

For the usual "small" complexity classes like P or L, it is more natural to use words over a non-unary alphabet Σ as names for abstract objects instead of natural numbers in Definition 1. In this way we straightforwardly obtain e.g., the following polynomial-time versions of these definitions. By a *P-naming* we mean any function whose domain is a P-computable set of words over Σ. A P-naming μ is *P-reducible* to a P-naming ν (in symbols, $\mu \leqslant_P \nu$) if $\mu = \nu \circ f$ for a P-computable function $f : dom(\mu) \to dom(\nu)$, and μ, ν are *P-equivalent* (in symbols, $\mu \equiv_P \nu$) if $\mu \leqslant_P \nu$ and $\nu \leqslant_P \mu$. By a *P-naming of a set S* we mean a P-naming ν with $rng(\nu) = S$. A structure $(A; \sigma)$ is *P-constructivisable* if there is a P-naming α of A such that all the signature functions and predicates, as well as the equality predicate on A, are P-computable w.r.t. α. Such a P-naming is called a *P-constructivisation of $(A; \sigma)$*, and $((A; , \sigma), \alpha)$ is called a *P-constructive structure*.

The western-style version of P-constructive structures was introduced in [7] under the name "P-computable quotient-structures" and used to show P-equivalence of three presentations of the ordered field of algebraic reals popular in the literature on symbolic computations and computer algebra. This result could not be proved using only the more restricted notion of P-presentable structure from [13] (which is equivalent to P-constructivizability with a bijective P-naming) because some of those presentations essentially use non-bijective P-namings. Since we use the corresponding terminology in Sect. 4, we recall it here.

Let $A \subseteq \Sigma^*$ and let $E \subseteq A^2$ be an equivalence relation on A. With any such pair (A, E) we associate the quotient set $\bar{A} = A/E = \{[x]_E \mid x \in A\}$ consisting of the equivalence classes. We call the set $\bar{A}$ *a quotient set in* Σ^*. Clearly, any family $\bar{A}$ of nonempty pairwise disjoint subsets of Σ^* is a quotient set of the form A/E, and the pair (A, E) is uniquely determined by $\bar{A}$. Thus, we can sometimes identify $\bar{A}$ and (A, E).

A quotient-set $\bar{A}$ is P-*computable* if $\bar{A} = A/E$, where A and E are P-computable sets. Let $\bar{A} = A/E$ and $\bar{B} = B/F$ be P-computable quotient-sets. We call a map $f : \bar{A} \to \bar{B}$ P-*computable* if there is a P-computable function $f_0 : A \to B$ such that $f([x]) = [f_0(x)]$ for $x \in A$. Notice that in this case $(x, y) \in E$ implies $(f_0(x), f_0(y)) \in F$. Similarly we define the notion of a P-computable function $f : (\bar{A})^n \to \bar{B}$: for some P-computable function $f_0 : A^n \to B$ we have $f([x_1], \ldots, [x_n]) = [f_0(x_1, \ldots, x_n)]$.

We call a structure $\mathbb{A}$ *quotient-structure*, if its universe is a quotient-set. We say that a structure $\mathbb{A} = (\bar{A}, \sigma)$ is a P-*computable quotient-structure* if $\bar{A}$ is a P-computable quotient-set, and all signature operations and relations are P-computable on $\bar{A}$. Quotient-structures $\mathbb{A}$ and $\mathbb{B}$ are P-*computably isomorphic*, if there is an isomorphism $f : \mathbb{A} \to \mathbb{B}$ s.t. f and f^{-1} are P-computable functions. Identifying a set A with the quotient-set A/id_A, we can consider the usual structures as a particular case of quotient-structures. It is easy to see that, up to isomorphism (even up to P-isomorphism), the P-computable quotient-structures coincide with the P-constructivizable structures.

We conclude this section with mentioning some other notions of feasible structures. As for the class P, such notions are straightforwardly defined for any functional complexity class (more precise notations for such functional classes would be perhaps FP, FL and so on). These include, e.g., the classes L (logarithmic space), LSPACE (linear space), E (Kalmar's elementary functions), and PSPACE (polynomial space). For definitions of these and other complexity classes see e.g., [28]. It is known (see e.g., Theorem 10.16 in [28]) that $\mathrm{LSPACE} = \mathcal{E}^2$, $\mathrm{E} = \mathcal{E}^3$, and E coincides with the class of functions computed in deterministic tower-of-exponents time. To show the equalities, one identifies natural numbers with binary words via the dyadic bijection. Note that FPR-presentability is modified to any functional complexity class in a straightforward way.

"Small" complexity classes (like L, P or PSPACE) are very useful and popular but they are often not closed under basic mathematical constructions. "Large" complexity classes (like PR or $\mathcal{E}^n$ for $n \geqslant 3$) have much better closure properties but are not suitable for computer implementations. Although the upper complexity bounds for, say, PR-algorithms may be awfully large, such algorithms are much better than those in the general Turing computability where estimation of computation resources is impossible in principle. The relevance of presentability w.r.t. the "large" complexity classes stems from the fact that it is in some respects close to feasible presentability but technically easier, and the corresponding complexity classes have better closure properties.

3   Grzegorczyk's Structures

Here we discuss Grzegorczyk's versions of some facts known for PR-structures. First we show that the Russian-style notion of $\mathcal{E}^n$-constructivizable structure coincides with the western-style notion of $\mathcal{E}^n$-presentable structure.

Proposition 1. *For any* $n \geqslant 0$, *a structure* $\mathbb{A}$ *is* $\mathcal{E}^n$-*constructivizable iff it is* $\mathcal{E}^n$-*presentable.*

Proof. If φ is an isomorphism of $\mathbb{A}$ onto an $\mathcal{E}^n$-computable structure $\mathbb{B}$ then φ^{-1} is an $\mathcal{E}^n$-constructivization of $\mathbb{A}$ which proves the assertion from right to left.

Conversely, let $\alpha : D \to A$ be an $\mathcal{E}^n$-constructivization of $\mathbb{A}$, in particular, $D \in \mathcal{E}^n$ and $E = \{(x, y) \in D^2 \mid \alpha(x) = \alpha(y)\}$ is an $\mathcal{E}^n$-equivalence relation on D. Let $h(x) = 0$ for $x \in \mathbb{N} \setminus D$ and $h(x)$ be the smallest number in the equivalence class $[x]_E$ for $x \in D$. Let $sg(x) = 1 \Leftrightarrow \overline{sg}(x) = 0 \Leftrightarrow x > 0$ and $sg(x) = 0 \Leftrightarrow \overline{sg}(x) = 1 \Leftrightarrow x = 0$. Since the functions $sg(x), \overline{sg}(x), sg(x) \wedge sg(y)$, and $sg(x) \vee sg(y)$ are in $\mathcal{E}^0$, and the value $h(x)$ coincides with $\sum_{z=0}^{x} \prod_{t=0}^{z} \overline{sg}(E(x, t))$, we have $h \in \mathcal{E}^n$. Since $x \in h(D) \Leftrightarrow x \in D \wedge x = h(x)$, we have $B = h(D) \in \mathcal{E}^n$.

We interpret σ-symbols in B as follows. For a constant symbol $c \in \sigma$, let c^B be the smallest number in $\alpha^{-1}(c^{\mathbb{A}})$; then $c^B \in B$. For a k-ary predicate symbol $P \in \sigma$, $k \geqslant 1$, let $P^B(b_1, \ldots, b_k)$ iff $P^{\mathbb{A}}(\alpha(b_1), \ldots, \alpha(b_k))$; then P^B is an $\mathcal{E}^n$-relation on B. For a k-ary functional symbol $f \in \sigma$, $k \geqslant 1$, let $f^B(b_1, \ldots, b_k)$ be the smallest number in $\alpha^{-1}(f^{\mathbb{A}}(\alpha(b_1), \ldots, \alpha(b_k)))$; then $f^B : B^k \to B$ is the restriction of an $\mathcal{E}^n$-function to B. Thus, h induces an isomorphism of $\mathbb{A}$ onto the $\mathcal{E}^n$-computable structure $(B; \sigma^B)$. $\square$

A principal question is of course the question about non-collapse of the hierarchy of structures induced by the Grzegorczyk hierarchy, i.e., whether for any n there is an $\mathcal{E}^{n+1}$-computable structure which is not $\mathcal{E}^n$-presentable. We guess this is true for every $n \geqslant 2$, though do not yet have a proof at hand. But we can prove the following weaker version (see also [12] for similar facts, our proof here is much simpler).

Proposition 2. *For any* $n \geqslant 2$ *there is an* $\mathcal{E}^{n+2}$-*computable structure which is not* $\mathcal{E}^n$-*presentable.*

Proof. Let $\mathbb{A} = (\mathbb{N}; s, P)$ where P is a set in $\mathcal{E}^{n+2} \setminus \mathcal{E}^{n+1}$ (considered as a unary predicate on $\mathbb{N}$). Such P is known to exist (see e.g., Theorem 2.19(5) in

[28]). Then $\mathbb{A}$ is an $\mathcal{E}^{n+2}$-computable structure, so it remains to show that it is not $\mathcal{E}^n$-presentable. Suppose the contrary: $\mathbb{A}$ is isomorphic to an $\mathcal{E}^n$-computable structure $\mathbb{B} = (B; s^B, P^B)$ via isomorphism $\varphi : \mathbb{A} \to \mathbb{B}$, so P is an $\mathcal{E}^n$-set and s^B is the restriction to B of an $\mathcal{E}^n$-function. Let f be the iteration of s^B starting with $\varphi(0) \in B$, i.e. $f(0) = \varphi(0)$ and $f(x+1) = s^B(f(x))$. By Theorem 2.18 in [28] mentioned in the previous section, f is an $\mathcal{E}^{n+1}$-function. Since $x \in P \Leftrightarrow f(x) \in P^B$, we see that P is an $\mathcal{E}^{n+1}$-set. Contradiction. $\qquad\square$

We conclude this section with the Russian-style characterisation of the F$\mathcal{E}^n$-presentable structures. We call a numbering ν $\mathcal{E}^n$-*infinite* if there is an $\mathcal{E}^n$-function f such that $\nu(f(i)) \neq \nu(f(j))$ whenever $i \neq j$. The next result is a straightforward version of the corresponding fact for PR-presentations in [26].

Proposition 3. *For any $n \geqslant 3$ and any structure $\mathbb{B}$, $\mathbb{B}$ is F$\mathcal{E}^n$-presentable iff it has a total $\mathcal{E}^n$-constructivization β which is $\mathcal{E}^n$-infinite.*

Proof. We consider the less obvious direction. Let β be a total $\mathcal{E}^n$-constructivization of $\mathbb{B}$ which is $\mathcal{E}^n$-infinite via f. Define a function $g : \mathbb{N} \to \mathbb{N}$ as follows: $g(0) = 0$ and $g(n+1) = \mu x.\forall i \leqslant n(\beta(x) \neq \beta(g(i)))$. Since B is infinite, g is total and injective. Let $h(n) = max\{f(0), \dots, f(n)\}$. Then $g(n+1) = \mu x \leqslant h(n).\forall i \leqslant n(\beta(x) \neq \beta(g(i)))$ and h is $\mathcal{E}^n$, hence g is also $\mathcal{E}^n$ (by the well known closure properties of $\mathcal{E}^n$).

The numbering $\gamma = \beta \circ g$ is $\mathcal{E}^n$-reducible to β and injective. Conversely, $\beta \leqslant_{\mathcal{E}^n} \gamma$ via the $\mathcal{E}^n$ function $u(n) = \mu x \leqslant n.\beta(n) = \beta(x)$. Thus, γ is a bijective numbering of B $\mathcal{E}^n$-equivalent to β, so it is a bijective $\mathcal{E}^n$-constructivization of $\mathbb{B}$. Copying interpretations of signature symbols from $\mathbb{B}$ to $\mathbb{N}$ via γ^{-1} we obtain an $\mathcal{E}^n$-computable copy of $\mathbb{B}$ with universe $\mathbb{N}$. $\qquad\square$

As observed in [26], any PR-constructivization β of an associative commutative ring with 1 of characteristic 0 is PR-infinite (via any PR function f such that $\beta(f(i))$ coincide with the element $1 + \cdots + 1 \in B$ ($i+1$ summands)), hence any such PR-constructive ring is FPR-constructivizable. Unfortunately, the $\mathcal{E}^n$-version of this fact does not hold automatically, hence we currently do not have a rich class of natural F$\mathcal{E}^n$-presentable structures. Nevertheless, many structures popular in computer algebra (in particular, the ordered fields of rationals and algebraic reals, and the field of algebraic complex numbers), are F$\mathcal{E}^3$-presentable.

4 Polynomial Time Structures

Here we compare P-constructivizable and P-presentable structures.

Let Σ be a non-unary finite alphabet, $A \subseteq (\Sigma^*)^n$, $n \geqslant 1$. We say that the set A is in P, if it is P-computable, i.e. its characteristic function $\chi_A : (\Sigma^*)^n \to \{0, 1\}$ is P-computable. If $\bar{x} = x_1, \dots, x_n \in (\Sigma^*)^n$, then by $|\bar{x}|$ we denote $max_{i \leqslant n}\{|x_i|\}$. Suppose that $|\Sigma| \geqslant 2$. We say that A is in NP, if there is a set $B \subseteq (\Sigma^*)^{n+1}$ from P and a polynomial $p(u) \in \mathbb{Z}[u]$ such that, for all $x_1, \dots, x_n \in \Sigma^*$,

$$(x_1, \dots, x_n) \in A \Leftrightarrow \exists y \in \Sigma^* (x_1, \dots, x_n, y) \in B,$$

and $(\bar{x}, y) \in B$ implies $|y| \leqslant p(|\bar{x}|)$.

Proposition 4. *Let* $\mathbb{A} = (A/E, \sigma^{\bar{A}})$ *be a* P-*computable quotient-structure. Then the following are equivalent:*

a) $\mathbb{A}$ *is* P-*computably isomorphic to some* P-*computable structure* $\mathbb{B}$;

b) *there is a* P-*computable function* $\beta : A \to A$ *such that* $xE\beta(x)$ *and* $xEy \Rightarrow \beta(x) = \beta(y)$ *for* $x, y \in A$.

Proof. (a $\Rightarrow$ b) Let $f : \mathbb{A} \to \mathbb{B}$ be an isomorphism such that f and f^{-1} are P-computable functions. By the definition, there exist P-computable functions $g : A \to B$ and $h : B \to A$ such that $f([x]) = g(x)$ for $x \in A$ and $f^{-1}(y) = [h(y)]$ for $y \in B$. Let $\beta(x) = h(g(x))$. Then $[\beta(x)] = f^{-1}(f([x])) = [x]$ and $xE\beta(x)$ for $x \in A$. If $x, y \in A$ and xEy, then $[x] = [y]$ and $g(x) = f([x]) = f([y]) = g(y)$, hence $\beta(x) = \beta(y)$.

(b $\Rightarrow$ a) If $x \in A$, then $xE\beta(x)$, hence $\beta(x) = \beta(\beta(x))$. Let $B = \{x \in A \mid \beta(x) = x\}$. Then B is P-computable, $\beta : A \to B$ is surjective, and $\beta(x) = \beta(y) \Leftrightarrow xEy$ for $x, y \in A$. Define an interpretation of signature symbols in B as follows.

Let P be a predicate from σ. By the definition, there is a P-computable relation $P_0 \subseteq A^n$ such that $P^{\bar{A}}([x_1], \ldots, [x_n]) \Leftrightarrow P_0(x_1, \ldots, x_n)$ for $x_i \in A$, $i \leqslant n$. Let $P^{\bar{B}}(x_1, \ldots, x_n) \Leftrightarrow P_0(x_1, \ldots, x_n)$ for $x_i \in B$, $i \leqslant n$.

Let now f be a function from σ. There is a P-computable function $f_0 : A^n \to A$ such that $f^{\bar{A}}([x_1], \ldots, [x_n]) = [f_0(x_1, \ldots, x_n)]$. Let $f^{\bar{B}}(x_1, \ldots, x_n) = \beta(f_0(x_1, \ldots, x_n))$ for $x_i \in B$, $i \leqslant n$.

If the map $\beta' : A/E \to B$ is defined by $\beta'([x]) = \beta(x)$, then β' is a P-computable isomorphism from $\mathbb{A}$ onto $\mathbb{B}$. The inverse isomorphism is given by the function $\mathrm{id}_B : B \to A$. $\qquad \square$

This proposition shows that item a) reduces to the properties of $A \subseteq \Sigma^*$ and of the equivalence relation $E \subseteq A^2$. Function β with this property was considered in [10] for equivalence relations $E \subseteq (\Sigma^*)^2$. The question about the existence of such a function was called there the *normal form problem* because $\beta(x)$ may be considered as a unified normal form for all elements equivalent to x. In particular, it was shown that if P = NP then a P-computable function β with the specified property exists. Uniting this fact with some results in [18] and [11], we obtain the following discouraging result where Σ_n^p and Π_n^p denote levels of the polynomial-time hierarchy.

Theorem 1. a) *Suppose that* P = NP. *Then every* P-*computable quotient-structure* $\mathbb{A}$ *is* P-*computably isomorphic to some* P-*computable structure* $\mathbb{B}$.

b) *Suppose that* P $\neq$ NP *and, moreover,* $\Sigma_2^p \neq \Pi_2^p$. *Then there is a* P-*computable quotient-structure of the empty signature which is not* P-*computably isomorphic to a* P-*computable structure.*

Proof. a) The argument in [10] is not complicated, so we briefly sketch it for completeness. Define on Σ^* a natural order as follows: $x \leqslant y \Leftrightarrow |x| < |y|$ or ($|x| = |y|$ and $x \leqslant_l y$), where $x \leqslant_l y$ is the lexicographic order on Σ^*. Then $(\Sigma^*, \leqslant) \cong (\omega, \leqslant)$, i.e. the order is linear and well founded.

If E is a P-computable equivalence relation on A, where $A \subseteq \Sigma^*$ is a P-computable set then it can be extended to the relation $E_1 = E \cup \{(x, y) \mid$

$x, y \in \Sigma^* \setminus A\}$ on the whole Σ^*. Under the assumption $P = NP$, the function β that assigns to any $x \in \Sigma^*$ the smallest element of Σ^* equivalent to x, is P-computable. Indeed, the pair (x, y) is in the graph Γ_β of β, iff xE_1y, and $\forall y_1 \; [y_1 < y \rightarrow \neg(y_1 E_1 x)]$. The latter condition gives a set from co-NP, which coincides with P. If the graph Γ_β is P-computable then the first symbol $\beta(x)$ may be found by searching through $a \in \Sigma$ and checking whether there exists $y_1 \in \Sigma^*$ such that $(x, ay_1) \in \Gamma_\beta$. The latter is an NP-condition. Once we found the first symbol, we can in the same way find the second and so on. Since $|\beta(x)| \leqslant |x|$, the whole algorithm is polynomial.

b) We say that NP has the *shrinking* property if for any $A, B \in NP$ there are $A', B' \in NP$ such that $A' \subseteq A$, $B' \subseteq B$, $A' \cap B' = \emptyset$ and $A' \cup B' = A \cup B$. Of course, this definition may be formulated for any class of sets. By Theorem 2.9 in [18], the condition $\Sigma_2^p \neq \Pi_2^p$ implies that the shrinking property for NP fails.

In [11, Theorem 3] it was proved that if NP does not have the shrinking property then there is a p-computable equivalence relation $E \subseteq (\Sigma^*)^2$, for which there is no p-computable function β with the property in item b) of Proposition 4. Moreover, β can not be computed in polynomial time even by a non-deterministic Turing machine.   □

If we do not plan to solve the $P = NP$ problem, our best hope could be to prove that any P-computable quotient-structure of a finite signature is isomorphic to some P-computable structure. The next result shows that this holds in an important particular case.

Theorem 2. *Let* $\mathbb{A}$ *be a* P-*computable quotient-structure. Then every one of its finitely generated substructures is isomorphic to some* P-*computable structure.*

Proof. Let $\mathbb{A} = (A/E, \sigma^{\bar{A}})$ be a quotient-structure where $A \subseteq \Sigma^* \setminus \{\emptyset\}$ and $E \subseteq A^2$ are P-computable sets. If f is an n-ary function from σ then, by the definition, there is a P-computable function $f_0 : A^n \rightarrow A$ such that $f^{\bar{A}}([x_1], \ldots, [x_n]) = [f_0(x_1, \ldots, x_n)]$. If P is an n-ary predicate from σ then there is a P-computable relation $P_0 \subseteq A^n$ with $P^{\bar{A}}([x_1], \ldots, [x_n]) \Leftrightarrow P_0(x_1, \ldots, x_n)$. In this way we obtain a P-computable structure $\mathbb{A}_0$ with the universe A, the specified interpretations f_0 and P_0, and the congruence E such that $\mathbb{A}_0/E \cong \mathbb{A}$.

If t is a term then denote by $h(t)$ its *height* defined as follows:

a) if t is a constant or a variable then $h(t) = 0$;
b) if $t = f(t_1, \ldots, t_k)$ where t_i are terms for each $i \leqslant k$, then $h(t) = \max_{i \leqslant k}\{h(t_i)\} + 1$.

Let $\bar{e} = e_1, \ldots, e_n \in A$ and let $[e_1], \ldots, [e_n]$ be elements generating the substructure $\mathbb{A}_{\bar{e}}$ of $\mathbb{A}$. In [5] the following criterion is proved: a finitely generated structure $\mathbb{A}_{\bar{e}}$ has a P-computable presentation iff there is a constant $c \in \omega$ such that:

1) there is an algorithm that for given σ-terms $t_1(\bar{x}), t_2(\bar{x})$ checks whether $\mathbb{A}_{\bar{e}}$ satisfies the condition $t_1([e_1], \ldots, [e_n]) = t_2([e_1], \ldots, [e_n])$ in time $O(2^{2^{ch}})$, where $h = \max\{h(t_1), h(t_2)\}$;

2) for any k-ary predicate P from σ there is an algorithm that for given terms $t_1(\bar{x}), \ldots, t_k(\bar{x})$ of L checks whether $\mathbb{A}_{\bar{e}}$ satisfies the condition

$$P(t_1([e_1], \ldots, [e_n]), \ldots, t_k([e_1], \ldots, [e_n]))$$

in time $O(2^{2^{ch}})$, where $h = \max_{i \leqslant k}\{h(t_i)\}$.

We show that these conditions hold in our case. Let $t_1(x_1, \ldots, x_n)$ be a term of σ and $h = h(t)$. Then its value $t_1^{\bar{A}}([e_1], \ldots, [e_n])$ equals $[b_1]$, where $b_1 = t_1^{A_0}(e_1, \ldots, e_n)$ is the value in $\mathbb{A}_0$. As shown in [5, Theorem 1], in this case the computation of b_1 may be done in time $O(2^{2^{c_1 h}})$, where c_1 is a constant (depending on $\mathbb{A}_0$ and $\bar{e}$), and for the length $|b_1|$ a similar estimation holds.

To check the condition in 1), we have to compute $b_1 = t_1^{A_0}(\bar{e})$, $b_2 = t_2^{A_0}(\bar{e})$ and then check whether $[b_1] = [b_2]$, i.e. $b_1 E b_2$. The latter computation requires $O(\max\{|b_1|, |b_2|\}^p)$ steps where p is a constant. This estimation has the form $O(2^{2^{c_1 h + \log p}})$, equivalent to the estimation in 1).

The argument for 2) is similar. We compute $b_i = t_i^{A_0}(\bar{e})$ for $i \leqslant k$ in $O(2^{2^{c_1 h}})$ steps and then check the condition $P_0(b_1, \ldots, b_k)$ in time polynomial in $\max_{i \leqslant k}\{|b_i|\}$.   □

Concerning the P-version of on-line presentations, a natural candidate are the P-computable structures with universe Σ^*. Similar restrictive versions of P-presentable structures were discussed in [13,14] but no robust notion was identified so far. There are many other open questions related to this paper. E.g.: Is the ordered field $\mathbb{R}_{\mathrm{alg}}$ of algebraic reals FP-presentable? Are there P-computable real closed ordered fields not isomorphic to $\mathbb{R}_{\mathrm{alg}}$?

References

1. Akritas, A.G.: Elements of Computer Algebra with Applications. Wiley Interscience, New York (1989)
2. Ash, C., Knight, J.: Computable Structures and the Hyperarithmetical Hierarchy. Studies in Logic and the Foundations of Mathematics, vol. 144. North-Holland Publishing Co., Amsterdam (2000)
3. Alaev, P.E.: Existence and uniqueness of structures computable in polynomial time. Algebra Logic **55**(1), 72–76 (2016)
4. Alaev, P.E.: Structures computable in polynomial time. I. Algebra Logic **55**(6), 421–435 (2016)
5. Alaev, P.E.: Finitely generated structures computable in polynomial time. Algebra Logic **59**(3), 385–394 (2020). (in Russian, there is an English translation)
6. Albers, S.: Online algorithms. In: Goldin, D., Smolka, S.A., Wegner, P. (eds.) Interactive Computation, pp. 143–164. Springer, Heidelberg (2006). https://doi.org/10.1007/3-540-34874-3_7
7. Manea, F., Miller, R.G., Nowotka, D. (eds.): CiE 2018. LNCS, vol. 10936. Springer, Cham (2018). https://doi.org/10.1007/978-3-319-94418-0
8. Alaev, P.E., Selivanov, V.L.: Fields of algebraic numbers computable in polynomial time. I. Algebra Logic **58**(6), 673–705 (2019)

9. Bazhenov, N., Downey, R., Kalimullin, I., Melnikov, A.: Foundations of online structure theory. Bull. Symbolic Logic **25**(2), 141–181 (2019)
10. Blass, A., Gurevich, Yu.: Equivalence relations, invariants, and normal forms. SIAM J. Comput. **13**(4), 682–689 (1984)
11. Blass, A., Gurevich, Y.: Equivalence relations, invariants, and normal forms. II. In: Börger, E., Hasenjaeger, G., Rödding, D. (eds.) LaM 1983. LNCS, vol. 171, pp. 24–42. Springer, Heidelberg (1984). https://doi.org/10.1007/3-540-13331-3_31
12. Cannonito, F.B.: Hierarchies of computable groups and the word problem. J. Symbolic Logic **31**(3), 376–392 (1966)
13. Cenzer, D.A., Remmel, J.B.: Polynomial-time versus recursive models. Ann. Pure Appl. Logic **54**(1), 17–58 (1991)
14. Cenzer, D., Remmel, J.B.: Complexity theoretic model theory and algebra. In: Handbook of Recursive Mathematics, vol. 1 (1998)
15. Downey, R., Melnikov, A., Ng, K.M.: Foundations of on-line structure theory II: the operator approach. arxiv:2007.07401v1 (2020)
16. Ershov, Y.L., Goncharov, S.S.: Constructive Models. Plenum, New York (1999)
17. Grigorieff, S.: Every recursive linear ordering has a copy in DTIMESPACE(n, log(n)). J. Symb. Log. **55**(1), 260–276 (1990)
18. Glasser, C., Reitwiessner, C., Selivanov, V.: The shrinking property for NP and coNP. Theoret. Comput. Sci. **412**, 853–864 (2011)
19. Khoussainov B., Minnes M.: Three lectures on automatic structures. In: Logic Colloquium 2007. Lecture Notes in Logic, vol. 35, pp. 132–176. Association for Symbolic Logic, La Jolla, CA (2010)
20. Kalimullin, I., Melnikov, A., Ng, K.M.: Algebraic structures computable without delay. Theoret. Comput. Sci. **674**, 73–98 (2017)
21. Mal'cev A.I.: Constructive algebras. Uspechi mat. nauk **16**(3), 3–60 (1961). (in Russian, English translation). In: The Metamathematics of Algebraic Systems, North Holand, Amsterdam, pp. 148–214 (1971)
22. Mayr, E.W.: Some complexity results for polynomial ideals. J. Complex. **13**(3), 303–325 (1997)
23. Myasnikov, A., Roman'kov, V., Ushakov, A., Vershik, A.: The word and geodesic problems in free solvable groups. Trans. Amer. Math. Soc. **362**(9), 4655–4682 (2010)
24. Selivanova, S., Selivanov, V.: Computing solution operators of boundary-value problems for some linear hyperbolic systems of PDEs. Log. Meth. Comput. Sci. **13**(4:13), 1–31 (2017)
25. Selivanova, S.V., Selivanov, V.L.: Bit complexity of computing solutions for symmetric hyperbolic systems of PDEs (extended abstract). In: Manea, F., Miller, R.G., Nowotka, D. (eds.) CiE 2018. LNCS, vol. 10936, pp. 376–385. Springer, Cham (2018). https://doi.org/10.1007/978-3-319-94418-0_38
26. Selivanov V.L., Selivanova S.V.: Primitive recursive ordered fields and some applications. arxiv:2010.10189v1 (2020)
27. Winkler, F.: Polynomial Algorithms in Computer Algebra. Springer, Wien (1996). https://doi.org/10.1007/978-3-7091-6571-3
28. Wagner, K., Wechsung, G.: Computational Complexity. VEB, Berlin (1986)
29. Yap, C.K.: Fundamental Problems in Algorithmic Algebra. Oxford University Press (2000)

On Measure Quantifiers in First-Order Arithmetic

Melissa Antonelli$^{(\boxtimes)}$, Ugo Dal Lago, and Paolo Pistone

University of Bologna, Bologna, Italy
{melissa.antonelli2,ugo.dallago,paolo.pistone2}@unibo.it

Abstract. We study the logic obtained by endowing the language of first-order arithmetic with second-order measure quantifiers. This new kind of quantification allows us to express that the argument formula is true *in a certain portion* of all possible interpretations of the quantified variable. We show that first-order arithmetic with measure quantifiers is capable of formalizing simple results from probability theory and, most importantly, of representing every recursive random function. Moreover, we introduce a realizability interpretation of this logic in which programs have access to an oracle from the Cantor space.

Keywords: Probabilistic computation · Peano Arithmetic · Realizability

1 Introduction

The interactions between first-order arithmetic and the theory of computation are plentiful and deep. On the one side, proof systems for arithmetic can be used to prove termination of certain classes of algorithms [24], or to establish complexity bounds [5]. On the other, higher-order programming languages, such as typed λ-calculi, can be proved to capture the computational content of arithmetical proofs. These insights can be pushed further, giving rise to logical and type theories of various strengths. Remarkably, all the quoted research directions rely on the tight connection between the concepts of *totality* (of functions) and *termination* (of algorithms).

However, there is one side of the theory of computation which was only marginally touched by this fruitful interaction, that is, randomized computation. Probabilistic models have been widely investigated and are nowadays pervasive in many areas of computer science. The idea of relaxing the notion of algorithm to account for computations involving random decisions appeared early in the history of modern computability theory and studies on probabilistic computation have been developed since the 1950s and 1960s [17]. Today several formal models are available, such as probabilistic automata [20] probabilistic Turing machines (from now on, PTMs) [10,22], and probabilistic λ-calculi [21].

Supported by ERC CoG "DIAPASoN", GA 818616.

L. De Mol et al. (Eds.): CiE 2021, LNCS 12813, pp. 12–24, 2021.
https://doi.org/10.1007/978-3-030-80049-9_2

In probabilistic computation, behavioral properties, such as termination, have a *quantitative* nature: any computation terminates *with a given probability*. Can such quantitative properties be studied within a logical system? Of course, logical systems for set-theory and second-order logic can be expressive enough to represent measure theory [23] and, thus, are inherently capable of talking about randomized computations. Yet, what should one add to *first-order* arithmetic to make it capable of describing probabilistic computation?

In this paper we provide an answer to this question by introducing a somehow *minimal* extension of first-order Peano Arithmetic by means of measure quantifiers. We will call this system MQPA. Its language is obtained by enriching the language of PA with a special unary predicate, $\mathsf{FLIP}(\cdot)$, whose interpretation is an element of the Cantor space $\{0,1\}^{\mathbb{N}}$, and with measure-quantified formulas, such as $\mathbf{C}^{\frac{1}{2}}F$, which expresses the fact that F has probability $\geq \frac{1}{2}$ of being true (that is, the subset of $\{0,1\}^{\mathbb{N}}$ which makes A true has measure $\geq \frac{1}{2}$). The appeal to the Cantor space is essential here, since there is no *a priori* bound on the amount of random bits a given computation might need; at the same time, we show that it yields a very natural measure-theoretic semantics.

The rest of this paper is structured as follows. In Sect. 2 we introduce the syntax and semantics of MQPA. Section 3 shows that some non-trivial results in probability theory can be naturally expressed in MQPA. In Sect. 4 we establish our main result, that is, a representation theorem within MQPA for random functions computed by PTMs, which is the probabilistic analogous to Gödel's arithmetization theorem for recursive functions in PA [12]. Finally, in Sect. 5, a realizability interpretation for MQPA in terms of computable functions with oracles on the Cantor space is presented. For further details, an extended version of this work is available, see [1].

2   Measure-Quantified Peano Arithmetic

This section is devoted to the introduction of the syntax and semantics for formulas of MQPA. Before the actual presentation, we need some (very modest) preliminaries from measure theory.

Preliminaries. The standard model $(\mathbb{N}, +, \times)$ has nothing probabilistic in itself. Nevertheless, it can be naturally extended into a probability space: arithmetic being discrete, one may consider the underlying sample space as just $\mathbb{B}^{\mathbb{N}}$, namely the set of all infinite sequences of elements from $\mathbb{B} = \{0,1\}$. We will use metavariables, such as $\omega_1, \omega_2, \ldots$, for the elements of $\mathbb{B}^{\mathbb{N}}$. As it is known, there are standard ways of building a well behaved σ-algebra and a probability space on $\mathbb{B}^{\mathbb{N}}$, which we will briefly recall here. The subsets of $\mathbb{B}^{\mathbb{N}}$ of the form

$$\mathsf{C}_X = \{s \cdot \omega \mid s \in X \ \& \ \omega \in \mathbb{B}^{\mathbb{N}}\},$$

where $X \subseteq \mathbb{B}^n$ and $\cdot$ denotes sequence concatenation, are called n-*cylinders* [4]. Specifically, we are interested in Xs defined as follows: $X_n^b = \{s \cdot b \mid s \in \mathbb{B}^n \ \& \ b \in \mathbb{B}\} \subseteq \mathbb{B}^{n+1}$, with $n \in \mathbb{N}$. We will deal with cylinders of the form $\mathsf{C}_{X_n^1}$. We let $\mathscr{C}_n$

and $\mathscr{C}$ indicate the set of all n-cylinders and the corresponding algebra, made of the open sets of the natural topology on $\mathbb{B}^{\mathbb{N}}$. The smallest σ-algebra including $\mathscr{C}$, which is Borel, is indicated as $\sigma(\mathscr{C})$. There is a natural way of defining a probability measure $\mu_{\mathscr{C}}$ on $\mathscr{C}$, namely by assigning to C_X the measure $\frac{|X|}{2^n}$. There exist canonical ways to extend this to $\sigma(\mathscr{C})$. In doing so, the standard model $(\mathbb{N}, +, \times)$ can be generalized to $\mathscr{P} = (\mathbb{N}, +, \times, \sigma(\mathscr{C}), \mu_{\mathscr{C}})$, which will be our standard model for MQPA.[1] When interpreting sequences in $\mathbb{B}^{\mathbb{N}}$ as infinite supplies of random bits, the set of sequences such that the k-th coin flip's result is 1 (for any fixed k) is assigned measure $\frac{1}{2}$, meaning that each random bit is uniformly distributed and independent from the others.

Syntax. We now introduce the syntax of MQPA. Terms are defined as in classic first-order arithmetic. Instead, the formulas of MQPA are obtained by endowing the language of PA with *flipcoin formulas*, such as $\mathsf{FLIP}(t)$, and *measure-quantified formulas*, as for example $\mathbf{C}^{t/s}F$ and $\mathbf{D}^{t/s}F$. Specifically, $\mathsf{FLIP}(\cdot)$ is a special unary predicate with an intuitive computational meaning. It basically provides an infinite supply of independently and randomly distributed bits. Intuitively, given a closed term t, $\mathsf{FLIP}(t)$ holds if and only if the n-th tossing returns 1, where n denotes $t + 1$.

Definition 1 (Terms and Formulas of MQPA). *Let $\mathcal{G}$ be a denumerable set of ground variables, whose elements are indicated by metavariables such as x, y. The terms of MQPA, denoted by t, s, are defined as follows:*

$$t, s := x \mid 0 \mid \mathsf{S}(t) \mid t + s \mid t \times s. \tag{1}$$

The formulas of MQPA are defined by the following grammar:

$$F, G := \mathsf{FLIP}(t) \mid (t = s) \mid \neg F \mid F \vee G \mid F \wedge G \mid \exists x.F \mid \forall x.F \mid \mathbf{C}^{t/s}F \mid \mathbf{D}^{t/s}F. \tag{2}$$

Semantics. Given an environment $\xi : \mathcal{G} \to \mathbb{N}$, the interpretation $[\![t]\!]_\xi$ of a term t is defined as usual. Instead, the interpretation of formulas requires a little care, being it inherently *quantitative*: any formula F is associated with a *measurable set*, $[\![F]\!]_\xi \in \sigma(\mathscr{C})$ (similarly, for example, to [18]).

Definition 2 (Semantics for Formulas of MQPA). *Given a formula F and an environment ξ, the interpretation of F in ξ is the measurable set of sequences $[\![F]\!]_\xi \in \sigma(\mathscr{C})$ inductively defined as follows:*

---

[1] Here, we will focus on this structure as a "standard model" of MQPA, leaving the study of alternative models for future work.

$$[\![\mathsf{FLIP}(t)]\!]_\xi := \mathsf{C}_{X^1_{[\![t]\!]_\xi}}$$

$$[\![t = s]\!]_\xi := \begin{cases} \mathbb{B}^\mathbb{N} & \text{if } [\![t]\!]_\xi = [\![s]\!]_\xi \\ \emptyset & \text{otherwise} \end{cases}$$

$$[\![\neg G]\!]_\xi := \mathbb{B}^\mathbb{N} - [\![G]\!]_\xi$$

$$[\![G \vee H]\!]_\xi := [\![G]\!]_\xi \cup [\![H]\!]_\xi$$

$$[\![G \wedge H]\!]_\xi := [\![G]\!]_\xi \cap [\![H]\!]_\xi$$

$$[\![\exists x.G]\!]_\xi := \bigcup_{i \in \mathbb{N}} [\![G]\!]_{\xi\{x \leftarrow i\}}$$

$$[\![\forall x.G]\!]_\xi := \bigcap_{i \in \mathbb{N}} [\![G]\!]_{\xi\{x \leftarrow i\}}$$

$$[\![\mathbf{C}^{t/s} G]\!]_\xi := \begin{cases} \mathbb{B}^\mathbb{N} & \text{if } [\![s]\!]_\xi > 0 \text{ and } \mu_\mathscr{C}([\![G]\!]_\xi) \geq [\![t]\!]_\xi/[\![s]\!]_\xi \\ \emptyset & \text{otherwise} \end{cases}$$

$$[\![\mathbf{D}^{t/s} G]\!]_\xi := \begin{cases} \mathbb{B}^\mathbb{N} & \text{if } [\![s]\!]_\xi = 0 \text{ or } \mu_\mathscr{C}([\![G]\!]_\xi) < [\![t]\!]_\xi/[\![s]\!]_\xi \\ \emptyset & \text{otherwise} \end{cases}$$

The semantics is well-defined since the sets $[\![\mathsf{FLIP}(t)]\!]_\xi$ and $[\![t = s]\!]_\xi$ are measurable, and measurability is preserved by all the logical operators. It is not difficult to see that any n-cylinder can be captured as the interpretation of some MQPA formula. However, the language of MQPA allows us to express more and more complex measurable sets, as illustrated in the next sections.

A formula of MQPA, call it F, is *valid* if and only if for every ξ, $[\![F]\!]_\xi = \mathbb{B}^\mathbb{N}$. The notion of logical equivalence is defined in a standard way: two formulas of MQPA, call them F, G, are *logically equivalent* $F \equiv G$ if and only if for every ξ, $[\![F]\!]_\xi = [\![G]\!]_\xi$. Notably, the two measure quantifiers are inter-definable, since one has $[\![\mathbf{C}^{t/s} F]\!]_\xi = [\![\neg \mathbf{D}^{t/s} F]\!]_\xi$. The following examples illustrate the use of measure-quantifiers $\mathbf{C}^{t/s}$ and $\mathbf{D}^{t/s}$ and, in particular, the role of probabilities of the form $\frac{t}{s}$.

Example 1. The formula $F = \mathbf{C}^{1/1} \exists x.\mathsf{FLIP}(x)$ states that a true random bit will almost surely be met. It is valid, as the set of constantly 0 sequences forms a singleton, which has measure 0.

Example 2. The formula[2] $F = \forall x.\mathbf{C}^{1/2^x} \forall_{y \leq x}.\mathsf{FLIP}(y)$ states that the probability for the first x random bits to be true is at least $\frac{1}{2^x}$. This formula is valid too.

3   On the Expressive Power of MQPA

As anticipated, the language of MQPA allows us to express some elementary results from probability theory, and to check their validity in the structure $\mathscr{P}$. In this section we sketch a couple of examples.

---

[2] For the sake of readability, F has been written with a little abuse of notation the actual MQPA formula being $\forall x.\mathbf{C}^{1/z}(\mathrm{EXP}(z, x) \wedge \forall y.(\exists w.(y + w = x) \rightarrow \mathsf{FLIP}(y))$, where $\mathrm{EXP}(z, x)$ is an arithmetical formula expressing $z = 2^x$ and $\exists w.y + w = x$ expresses $y \leq x$.

The Infinite Monkey Theorem. Our first example is the so-called *infinite monkey theorem* (IMT). It is a classic result from probability theory stating that a monkey randomly typing on a keyboard has probability 1 of ending up writing the *Macbeth* (or any other fixed string), sooner or later. Let the formulas $F(x, y)$ and $G(x, y)$ of PA express, respectively, that "y is strictly smaller than the length of (the binary sequence coded by) x", and that "the $y+1$-th bit of x is 1". We can formalize IMT through the following formula:

$$F_{\text{IMT}} : \forall x. \mathbf{C}^{1/1} \forall y. \exists z. \forall w. F(x, w) \to (G(x, w) \leftrightarrow \text{FLIP}(y + z + w)). \tag{3}$$

Indeed, let x be a binary encoding of the *Macbeth*. The formula F_{IMT} says then that for all choice of start time y, there exists a time $y + z$ after which $\text{FLIP}(\cdot)$ will evolve exactly like x with probability 1.

How can we justify F_{IMT} using the semantics of MQPA? Let $\varphi(x, y, z, w)$ indicate the formula $F(x, w) \to (G(x, w) \leftrightarrow \text{FLIP}(y+z+w))$. We must show that for every natural number $n \in \mathbb{N}$, there exists a measurable set $S^n \subseteq \mathbb{B}^{\mathbb{N}}$ of measure 1 such that any sequence in S^n satisfies the formula $\forall y. \exists z. \forall w. \varphi(n, y, z, w)$. To prove this fact, we rely on a well-known result from measure theory, namely the *second Borel-Cantelli Lemma*:

Theorem 1 ([4], Thm. 4.4, p. 55). *If $(U_y)_{y \in \mathbb{N}}$ is a sequence of independent events in $\mathbb{B}^{\mathbb{N}}$, and $\sum_y^{\infty} \mu_{\mathscr{C}}(U_y)$ diverges, then $\mu_{\mathscr{C}} \left(\bigcap_y \bigcup_{z > y} U_z \right) = 1$.*

Let us fix $n \in \mathbb{N}$ and let $\ell(n)$ indicate the length of the binary string encoded by n. We suppose for simplicity that $\ell(n) > 0$ (as the case $\ell(n) = 0$ is trivial). We construct S^n in a few steps as follows:

- for all $p \in \mathbb{N}$, let U_p^n be the cylinder of sequences which, after p steps, agree with n; observe that the sequences in U_p^n satisfy the formula $\forall w. \varphi(n, p, 0, w)$;
- for all $p \in \mathbb{N}$, let $V_p^n = U_{p \cdot \ell(n)+1}^n$; observe that the sets V_p^n are pairwise independent and $\mu_{\mathscr{C}}(\sum_p^{\infty} V_p^n) = \infty$;
- for all $p \in \mathbb{N}$, let $S_p^n = \bigcup \{U_{p+q}^n \mid \exists_{s > p}. p + q = s \cdot \ell(n) + 1\}$. Observe that any sequence in S_p^n satisfies $\exists z. \forall w. \varphi(n, p, z, w)$; Moreover, one can check that $S_p^n = \bigcup_{q > p} V_q^n$;
- we finally let $S^n := \bigcap_p S_p^n$.

We now have that any sequence in S^n satisfies $\forall y. \exists z. \forall w. \varphi(n, y, z, w)$; furthermore, by Theorem 1, $\mu_{\mathscr{C}}(S^n) = \mu_{\mathscr{C}}(\bigcap_p \bigcup_{q > p} V_q^n) = 1$. Thus, for each choice of $n \in \mathbb{N}$, $\mu_{\mathscr{C}}(\llbracket \forall y. \exists z. \forall w. \varphi(x, p, z, w) \rrbracket_{\{x \leftarrow n\}}) \geq \mu_{\mathscr{C}}(S^n) \geq 1$, and we conclude that $\llbracket F_{\text{IMT}} \rrbracket_{\xi} = \mathbb{B}^{\mathbb{N}}$.

The Random Walk Theorem. A second example we consider is the *random walk theorem* (RW): any simple random walk over $\mathbb{Z}$ starting from 1 will pass through 1 infinitely many times with probability 1. More formally, any $\omega \in \mathbb{B}^{\mathbb{N}}$ induces a simple random walk starting from 1, by letting the n-th move be right if $\omega(n) = 1$ holds and left if $\omega(n) = 0$ holds. One has then:

Theorem 2 ([4], Thm. 8.3, p. 117). *Let $U_{ij}^{(n)} \subseteq \mathbb{B}^{\mathbb{N}}$ be the set of sequences for which the simple random walk starting from i leads to j in n steps. Then*
$$\mu_{\mathscr{C}} \left(\bigcap_x \bigcup_{y \geq x} U_{11}^{(y)} \right) = 1.$$

Similarly, the random predicate $\mathsf{FLIP}(n)$ induces a simple random walk starting from 1, by letting the n-th move be right if $\mathsf{FLIP}(n)$ holds and left if $\neg\mathsf{FLIP}(n)$ holds. To formalize RW in MQPA we make use two arithmetical formulas:

- $H(y, z)$ expresses that y is even and z is the code of a sequence of length $\frac{y}{2}$, such that for all $i, j < \frac{y}{2}$, $z_i < y$, and $z_i = z_j \Rightarrow i = j$ (that is, z codes a subset of $\{0, \ldots, y-1\}$ of cardinality $\frac{y}{2}$);
- $K(y, z, v) = H(y, z) \wedge \exists i. i < \frac{y}{2} \wedge z_i = v$.

The formula of MQPA expressing RW is as follows:

$$F_{\mathsf{RW}} : \mathbf{C}^{1/1} \forall x. \exists y. \exists z. y \geq x \wedge H(y, z) \wedge \forall v. \Big(v < y \rightarrow \big(K(y, z, v) \leftrightarrow \mathsf{FLIP}(v) \big) \Big). \quad (4)$$

F_{RW} basically says that for any fixed x, we can find $y \geq x$ and a subset z of $\{0, \ldots, y-1\}$ of cardinality $\frac{y}{2}$, containing all and only the values $v < y$ such that $\mathsf{FLIP}(v)$ holds (so that the number of $v < y$ such that $\mathsf{FLIP}(v)$ holds coincides with the number of $v < y$ such that $\neg\mathsf{FLIP}(v)$ holds). This is the case precisely when the simple random walk goes back to 1 after exactly y steps.

To show the validity of F_{RW} we can use the measurable set $S = \bigcap_n \bigcup_{p \geq n} U_{11}^{(p)}$. Let $\psi(y, z, v)$ be the formula $(v < y \rightarrow (K(y, z, v) \leftrightarrow \mathsf{FLIP}(v)))$. Observe that any sequence in $U_{11}^{(n)}$ satisfies the formula $\exists z. H(n, z) \wedge \forall v. \psi(y, z, v, w)$. Then, any sequence in S satisfies the formula $\forall x. \exists y. \exists z. y \geq x \wedge H(y, z) \wedge \forall v. \psi(y, z, v)$. Since, by Theorem 2, $\mu_{\mathscr{C}}(S) = 1$, we conclude that $\mu_{\mathscr{C}}(\llbracket F_{\mathsf{RW}} \rrbracket_\xi) \geq \mu_{\mathscr{C}}(S) \geq 1$, and thus that $\llbracket F_{\mathsf{RW}} \rrbracket_\xi = \mathbb{B}^{\mathbb{N}}$.

4   Arithmetization

It is a classical result in computability theory [12,24] that all computable functions are arithmetical, that is, for each partial recursive function $f : \mathbb{N}^m \rightharpoonup \mathbb{N}$ there is an arithmetical formula F_f, such that for every $n_1, \ldots, n_m, l \in \mathbb{N}$: $f(n_1, \ldots, n_m) = l \Leftrightarrow (\mathbb{N}, +, \times) \vDash F_f(n_1, \ldots, n_m, l)$. In this section we show that, by considering arithmetical formulas of MQPA, this fundamental result can be generalized to computable *random* functions.

Computability in Presence of Probabilistic Choice. Although standard computational models are built around determinism, from the 1950s on, models for *randomized* computation started to receive wide attention [10,17,22]. The first formal definitions of probabilistic Turing machines are due to Santos [22] and Gill [10,11]. Roughly, a PTM is an ordinary Turing machine (for short, TM) with the additional capability of making random decisions. Here, we consider the definition by Gill, in which the probabilistic choices performed by the machines are binary and fair.

Definition 3 (Probabilistic Turing Machine [10,11]). *A (one-tape) proba-bilistic Turing machine is a 5-tuple $(Q, \Sigma, \delta, q_0, Q_f)$, whose elements are defined as in a standard TM, except for the probabilistic transition function δ, which, given the current (non-final) state and symbol, specifies two equally-likely tran-sition steps.*

As any ordinary TM computes a partial function on natural numbers, PTM can be seen as computing a so-called *random function* [22, pp. 706–707]. Let $\mathbb{D}(\mathbb{N})$ indicate the set of *pseudo-distributions* on $\mathbb{N}$, i.e. of functions $f : \mathbb{N} \to \mathbb{R}_{[0,1]}$, such that $\sum_{n \in \mathbb{N}} f(n) \leq 1$. Given a PTM $\mathcal{M}$, a random function is a function $\langle \mathcal{M} \rangle : \mathbb{N} \to \mathbb{D}(\mathbb{N})$ which, for each natural number n, returns a pseudo-distribution supporting all the possible outcomes $\mathcal{M}$ produces when fed with (an encoding of) n in input, each with its own probability. As expected, the random function $f : \mathbb{N} \to \mathbb{D}(\mathbb{N})$ is said to be *computable* when there is a PTM $\mathcal{M}$, such that $\langle \mathcal{M} \rangle = f$.

Stating the Main Result. In order to generalize Gödel's arithmetization of par-tial recursive functions to the class of computable random functions, we first introduce the notion of arithmetical random function.

Definition 4 (Arithmetical Random Function) *. A random function $f : \mathbb{N}^m \to \mathbb{D}(\mathbb{N})$ is said to be* arithmetical *if and only if there is a formula of* MQPA, *call it F_f, with free variables $x_1, \ldots, x_m, y$, such that for every $n_1, \ldots, n_m, l \in \mathbb{N}$, it holds that:*

$$\mu_{\mathscr{C}}\big(\llbracket F_f(n_1, \ldots, n_m, l) \rrbracket\big) = f(n_1, \ldots, n_m)(l). \tag{5}$$

The *arithmetization theorem* below relates random functions and the formulas of MQPA, and is the main result of this paper.

Theorem 3. *All computable random functions are arithmetical.*

Actually, we establish a stronger fact. Let us call a formula A of MQPA Σ_1^0 if A is equivalent to a formula of the form $\exists x_1. \ldots .\exists x_n.B$, where B contains neither first-order or measure quantifiers. Then, Theorem 3 can be strengthened by saying that any computable random function is represented (in the sense of Definition 4) by a Σ_1^0-formula of MQPA. Moreover, we are confident that a sort of converse of this fact can be established, namely that for any Σ_1^0-formula $A(x_1, \ldots, x_m)$, there exists a computable random relation $r(x_1, \ldots, x_m)$ (i.e. a computable random function such that $r(x_1, \ldots, x_n)(i) = 0$, whenever $i \neq 0, 1$) such that $\mu_{\mathscr{C}}(\llbracket A(n_1, \ldots, n_m) \rrbracket) = r(n_1, \ldots, n_m)(0)$ and $\mu_{\mathscr{C}}(\llbracket \neg A(n_1, \ldots, n_m) \rrbracket) = r(n_1, \ldots, n_m)(1)$. However, we leave this fact and, more generally, the exploration of an *arithmetical hierarchy* of randomized sets and relations, to future work.

Given the conceptual distance existing between TMs and arithmetic, a direct proof of Theorem 3 would be cumbersome. It is thus convenient to look for an alternative route.

On Function Algebras. In [7], the class $\mathcal{PR}$ of *probabilistic* or *random recursive functions* is defined as a generalization of Church and Kleene's standard one [6, 14]. $\mathcal{PR}$ is characterized as the smallest class of functions, which (i) contains some basic random functions, and (ii) is closed under composition, primitive recursion and minimization. For all this to make sense, composition and primitive recursion are defined following the *monadic* structure of $\mathbb{D}(\cdot)$. In order to give a presentation as straightforward as possible, we preliminarily introduce the notion of *Kleisli extension* of a function with values in $\mathbb{D}(\mathbb{N})$.

Definition 5 (Kleisli Extension). *Given a k-ary function $f : X_1 \times \cdots \times X_{i-1} \times \mathbb{N} \times X_{i+1} \times \cdots \times X_k \to \mathbb{D}(\mathbb{N})$, its i-th Kleisli extension $f_i^{\mathbf{K}} : X_1 \times \cdots \times X_{i-1} \times \mathbb{D}(\mathbb{N}) \times X_{i+1} \times \cdots \times X_k \to \mathbb{D}(\mathbb{N})$ is defined as follows:*

$$f_i^{\mathbf{K}}(x_1, \ldots, x_{i-1}, d, x_{i+1}, \ldots, x_k)(n) = \sum_{j \in \mathbb{N}} d(j) \cdot f(x_1, \ldots, x_{i-1}, j, x_{i+1}, \ldots, x_k)(n).$$

The construction at the basis of the **K**-extension can be applied more than once. Specifically, given a function $f : \mathbb{N}^k \to \mathbb{D}(\mathbb{N})$, its *total* **K**-extension $f^{\mathbf{K}} : (\mathbb{D}(\mathbb{N}))^k \to \mathbb{D}(\mathbb{N})$ is defined as follows:

$$f^{\mathbf{K}}(d_1, \ldots, d_k)(n) = \sum_{i_1, \ldots, i_k \in \mathbb{N}} f(i_1, \ldots, i_k)(n) \cdot \prod_{1 \leq j \leq k} d_j(i_j).$$

We can now formally introduce the class $\mathcal{PR}$ as follows:

Definition 6 (The Class $\mathcal{PR}$ [7]). *The class of probabilistic recursive functions, $\mathcal{PR}$, is the smallest class of probabilistic functions containing:*

- *The zero function such that for every $x \in \mathbb{N}$, $z(x)(0) = 1$;*
- *The successor function such that for every $x \in \mathbb{N}$, $s(x)(x+1) = 1$;*
- *The projection function such that for $1 \leq m \leq n$, $\pi_m^n(x_1, \ldots, x_n)(x_m) = 1$;*
- *The fair coin function, $r : \mathbb{N} \to \mathbb{D}(\mathbb{N})$, defined as follows:*

$$r(x)(y) = \begin{cases} \frac{1}{2} & \text{if } y = x \\ \frac{1}{2} & \text{if } y = x + 1; \end{cases}$$

and closed under:

- Probabilistic composition. *Given $f : \mathbb{N}^n \to \mathbb{D}(\mathbb{N})$ and $g_1, \ldots, g_n : \mathbb{N}^k \to \mathbb{D}(\mathbb{N})$, their composition is defined as follows:*

$$(f \odot (g_1, \ldots, g_n))(\mathbf{x}) = f^{\mathbf{K}}(g_1(\mathbf{x}), \ldots, g_n(\mathbf{x}));$$

- Probabilistic primitive recursion. *Given $f : \mathbb{N}^k \to \mathbb{D}(\mathbb{N})$, and $g : \mathbb{N}^{k+2} \to \mathbb{D}(\mathbb{N})$, the function obtained from them by primitive recursion is as follows:*

$$h(\mathbf{x}, 0) = f(\mathbf{x}) \qquad h(\mathbf{x}, y+1) = g_{k+2}^{\mathbf{K}}(\mathbf{x}, y, h(\mathbf{x}, y));$$

- Probabilistic minimization. *Given $f : \mathbb{N}^{k+1} \to \mathbb{D}(\mathbb{N})$, the function obtained from it by minimization is defined as follows:*

$$\mu f(\mathbf{x})(y) = f(\mathbf{x}, y)(0) \cdot \left(\prod_{z < y} \left(\sum_{k > 0} f(\mathbf{x}, z)(k) \right) \right).$$

Proposition 1 ([7]). *$\mathcal{PR}$ coincides with the class of computable random functions.*

The class $\mathcal{PR}$ is still conceptually far from MQPA. In fact, while the latter has access to randomness in the form of a global supply of random bits, the former can fire random choices locally through a dedicated initial function. To bridge the gap between the two, we introduce a third characterization of computable random functions, which is better-suited for our purposes. The class of *oracle recursive functions*, $\mathcal{OR}$, is the smallest class of partial functions of the form $f : \mathbb{N}^m \times \mathbb{B}^{\mathbb{N}} \rightharpoonup \mathbb{N}$, which (i) contains the class of *oracle basic functions*, and (ii) is closed under composition, primitive recursion, and minimization. Remarkably, the only basic function depending on ω is the query function. All the closure schemes are independent from ω as well.

But in what sense do functions in $\mathcal{OR}$ represent random functions? In order to clarify the relationship between $\mathcal{OR}$ and $\mathcal{PR}$, we associate each $\mathcal{OR}$ function with a corresponding auxiliary function.

Definition 7 (Auxiliary Function). *Given an oracle function $f : \mathbb{N}^m \times \mathbb{B}^{\mathbb{N}} \to \mathbb{N}$, the corresponding auxiliary function, $f^* : \mathbb{N}^m \times \mathbb{N} \to \mathcal{P}(\mathbb{B}^{\mathbb{N}})$, is defined as follows: $f^*(x_1, \ldots, x_m, y) = \{\omega \mid f(x_1, \ldots, x_m, \omega) = y\}$.*

The following lemma ensures that the value of f^* is always a *measurable* set:

Lemma 1. *For every oracle recursive function $f \in \mathcal{OR}$, $f : \mathbb{N}^m \times \mathbb{B}^{\mathbb{N}} \to \mathbb{N}$, and natural numbers $x_1, \ldots, x_m, y \in \mathbb{N}$, the set $f^*(x_1, \ldots, x_m, y)$ is measurable.*

Thanks to Lemma 1, we can associate any oracle recursive function $f : \mathbb{N}^m \times \mathbb{B}^{\mathbb{N}} \to \mathbb{N}$ with a random function $f^\# : \mathbb{N}^m \to \mathbb{D}(\mathbb{N})$, defined as: $f^\#(x_1, \ldots, x_m)(y) = \mu_{\mathscr{C}}(f^*(x_1, \ldots, x_m, y))$. This defines a close correspondence between the classes $\mathcal{PR}$ and $\mathcal{OR}$.

Proposition 2. *For each $f \in \mathcal{PR}$, there is an oracle function $g \in \mathcal{OR}$, such that $f = g^\#$. Symmetrically, for any $f \in \mathcal{OR}$, there is a random function $g \in \mathcal{PR}$, such that $g = f^\#$.*

The Proof of the Main Result. The last ingredient to establish Theorem 3 is the following lemma, easily proved by induction on the structure of $\mathcal{OR}$ functions.

Lemma 2. *For every oracle function $f \in \mathcal{OR}$, the random function $f^\#$ is arithmetical.*

Since for both $\mathcal{OR}$ and MQPA the source of randomness consists in a denumerable amount of random bits, the proof of Lemma 2 is easy, and follows the standard induction of [12]. Theorem 3 comes out as a corollary of Lemma 2 above, together with Proposition 1: any computable random function is in $\mathcal{PR}$, by Proposition 1, and each $\mathcal{PR}$ function is arithmetical, by Lemma 2 and Proposition 2.

5   Realizability

In this section we sketch an extension of realizability, a well-known computational interpretation of Peano Arithmetics, to MQPA. The theory of realizability [26], which dates back to Kleene's 1945 paper [15], provides a strong connection between logic, computability, and programming language theory. The fundamental idea behind realizability is that from every proof of an arithmetical formula in HA or equivalently (via the Gödel-Gentzen translation) in PA, one can extract a program, called the *realizer* of the formula, which encodes the computational content of the proof. In Kreisel's *modified-realizability* [16] realizers are typed programs: any formula A of HA is associated with a type A^* and any proof of A yields a realizer of type A^*.

Our goal is to show that the modified-realizability interpretation of HA can be extended to the language MQPA. As we have not introduced a proof system for MQPA yet, we limit ourselves to establishing the soundness of modified-realizability with respect to the semantics of MQPA. Similarly to what happens with the class $\mathcal{OR}$, the fundamental intuition is that realizers correspond to programs which can query an *oracle* $\omega \in \mathbb{B}^{\mathbb{N}}$. For instance, a realizer of $\mathbf{C}^{t/s}A$ is a program which, for a randomly chosen oracle, yields a realizer of A with probability at least $[\![t]\!]_\xi / [\![s]\!]_\xi$.

Our starting point is a PCF-style language with oracles. The types of this language are generated by basic types nat, bool and the connectives $\to$ and $\times$. We let O := nat $\to$ bool indicate the type of *oracles*. For any type σ, we let $[\sigma]$ (resp. $[\sigma]_O$) indicate the set of closed terms of type σ (resp. of terms of type σ with a unique free variable o of type O). Moreover, for all $i \in \{0, 1\}$ (resp. $n \in \mathbb{N}$), we indicate as $\bar{i}$ (resp. $\bar{n}$) the associated normal form of type bool (resp. nat). For all term M and normal form N, we let $M \Downarrow N$ indicate that M converges to N. For any term $M \in [\sigma_O]$ and oracle $\omega \in \mathbb{B}^{\mathbb{N}}$, we let $M^\omega \in [\sigma]$ indicate the closed program in which any call to the variable o is answered by the oracle ω.

We consider the language of MQPA without negation and disjunction, enriched with implication $A \to B$. As is usually done in modified-realizability, we take $\neg A$ and $A \vee B$ as *defined* connectives, given by $A \to (0 = \mathsf{S}(0))$ and $\exists x.(x = 0 \to A) \wedge (x = \mathsf{S}(0) \to B)$, respectively. With any closed formula A of MQPA we associate a type A^* defined as follows:

$$
\begin{aligned}
\mathsf{FLIP}(t)^* &= \text{nat} & (\forall x.A)^* &= \text{nat} \to A^* \\
(t = u)^* &= \text{bool} & (\exists x.A)^* &= \text{nat} \times A^* \\
(A \wedge B)^* &= A^* \times B^* & (\mathbf{C}^{t/s}A)^* &= (\mathbf{D}^{t/s}A)^* = \text{O} \to A^* \\
(A \to B)^* &= A^* \to B^*
\end{aligned}
$$

We define by induction the realizability relation $M, \omega \Vdash A$ where $\omega \in \mathbb{B}^{\mathbb{N}}$ and, if $A = \mathbf{C}^{t/s}B$ or $A = \mathbf{D}^{t/s}B$, $M \in [\sigma]$, and otherwise $M \in [\sigma]_O$.

1. $M, \omega \Vdash \mathsf{FLIP}(t)$ iff $\omega([\![t]\!]) = 1$;
2. $M, \omega \Vdash t = s$ iff $[\![t]\!] = [\![s]\!]$;
3. $M, \omega \Vdash A_1 \wedge A_2$ iff $\pi_1(M), \omega \Vdash A_1$ and $\pi_2(M), \omega \Vdash A_2$;

4. $M, \omega \Vdash A \to B$ iff $\omega \in [\![A \to B]\!]$ and $P, \omega \Vdash A$ implies $MP, \omega \Vdash B$;
5. $M, \omega \Vdash \exists x.A$ iff $\pi_1(M^\omega) \Downarrow \overline{k}$ and $\pi_2(M), \omega \Vdash A(k/x)$;
6. $M, \omega \Vdash \forall x.A$ iff for all $k \in \mathbb{N}$, $M\overline{k}, \omega \Vdash A(k/x)$;
7. $M, \omega \Vdash \mathbf{C}^{t/s}A$ iff $[\![s]\!] > 0$ and $\mu_\mathscr{C}(\{\omega' \mid Mo, \omega' \Vdash A\}) \geq [\![t]\!]/[\![s]\!]$;
8. $M, \omega \Vdash \mathbf{D}^{t/s}A$ iff $\omega \in [\![\mathbf{D}^{t/s}A]\!]$, and $[\![s]\!] = 0$ or $\mu_\mathscr{C}(\{\omega' \mid Mo, \omega' \Vdash A\}) < [\![t]\!]/[\![s]\!]$.

Condition 7 is justified by the fact that for all term M and formula A, the set $\{\omega \mid M, \omega \Vdash A\}$ can be shown to be measurable. Conditions 5 and 9 include a semantic condition of the form $\omega \in [\![A]\!]$, which has no computational meaning. This condition is added in view of Theorem 4 below. In fact, also in standard realizability a similar semantic condition for implication is required to show that realizable formulas are true in the standard model, see [26].

Theorem 4 (Soundness). *For a closed formula A, if $M, \omega \Vdash A$, then $\omega \in [\![A]\!]$.*

For example, the term $M = \lambda o.\mathrm{fix}(\lambda f x.(\mathrm{iszero}(ox))(f(x+1))\langle x, x \rangle)\overline{0}$ realizes the valid formula $\mathbf{C}^{1/1}\exists x.\mathsf{FLIP}(x)$. Indeed, M looks for the first value k such that $o(k) = 1$ and returns the pair $\langle \overline{k}, \overline{k} \rangle$. Similarly, the program $\lambda x o y z.o(y)$, which checks whether the y-th bit of ω is true, realizes the formula $\forall x.\mathbf{C}^{1/2^x}\forall_{y \leq x}\mathsf{FLIP}(y)$. With the same intuition, one can imagine how a realizer M of the formula F_{IMT} can be constructed: given inputs x, o, y, M looks for the first k such that the finite sequence $o(y+k), o(y+k+1), \ldots, o(y+k+\ell(n))$ coincides with the string coded by x (where this last check can be encoded by a program $\lambda w.P(x, o, y, w)$), and returns the pair $\langle \overline{k}, \lambda w.P(x, 0, y, \overline{k}, w) \rangle$.

6   Conclusion

Future and Ongoing Work. This paper can be seen as "a first exploration" of MQPA, providing some preliminary results, but also leaving many problems and challenges open. The most compelling one is certainly that of defining a proof system for MQPA, perhaps inspired by realizability. Furthermore, our extension of PA is *minimal* by design. In particular, we confined our presentation to a unique predicate variable, $\mathsf{FLIP}(x)$. Yet, it is possible to consider a more general language with countably many predicate variables $\mathsf{FLIP}_a(x)$, and suitably-*named* quantifiers $\mathbf{C}_a^{t/s}$ and $\mathbf{D}_a^{t/s}$, as in [2]. We leave the exploration of this more sophisticated syntax to future work. Another intriguing line of work concerns the study of bounded versions of MQPA, which may suggest novel ways of capturing probabilistic complexity classes, different from those in the literature, e.g. [13].

Related Work. To the best of the authors' knowledge, the term "measure quantifier" was first introduced in 1979 to formalize the idea that a formula $F(x)$ is true *for almost all values of x* [19]. In the same years, similar measure quantifiers were investigated from a model-theoretic perspective by H. Friedman (see [25] for a survey). More recently, Mio et al. [18] studied the application of such quantifiers to define extensions of MSO. However, the main source of inspiration for our treatment of measure quantifiers comes from computational complexity, namely from Wagner's counting operator on classes of languages [27].[3] On the other hand, an extensive literature exists on formal methods to describe probabilistic reasoning (without any reference to arithmetic), in particular in the realm of modal logic [3,8]. Moreover, classes of probabilistic modal logics have been designed to model Markov chains and similar structure, e.g. in [9]. Universal modalities show affinities with counting quantifiers, which however focusses on *counting* satisfying valuations, rather than on identifying (sets of) worlds.

References

1. Antonelli, M., Dal Lago, U., Pistone, P.: On measure quantifiers in first-order Arithmetic (2021). https://arxiv.org/abs/2104.12124
2. Antonelli, M., Dal Lago, U., Pistone, P.: On counting propositional logic (2021). https://arxiv.org/abs/2103.12862
3. Bacchus, F.: Representing and Reasoning with Probabilistic Knowledge. MIT Press, Cambridge (1990)
4. Billingsley, P.: Probability and Measure. Wiley, New York (1995)
5. Buss, S.: Bounded Arithmetic. Ph.D. thesis, Princeton University (1986)
6. Church, A., Kleene, S.: Formal definitions in the theory of ordinal numbers. Fund. Math. **28**, 11–21 (1936)
7. Dal Lago, U., Gabbrielli, M., Zuppiroli, S.: Probabilistic recursion theory and implicit computational complexity. Sci. Ann. Comput. Sci. **24**(2), 177–216 (2014)
8. Fagin, R., Halpern, J., Megiddo, N.: A logic for reasoning about probabilities. Inf. Comput. **87**(1/2), 78–128 (1990)
9. Furber, R., Mardare, R., Mio, M.: Probabilistic logics based on Riesz spaces. LMCS **16**(1), (2020)
10. Gill, J.: Computational complexity of probabilistic Turing machines. In: STOC, pp. 91–95. ACM (1974)
11. Gill, J.: Computational complexity of probabilistic turing machines. J. Comput. **6**(4), 675–695 (1977)
12. Gödel, K.: Über formal unentscheidbare sätze der Principia Mathematica und verwandter systeme. Monatsch. Math. Phys. **38**, 173–178 (1931)
13. Jeřábek, E.: Approximate counting in bounded arithmetic. J. Symb. Log. **72**(3), 959–993 (2007)
14. Kleene, S.: General recursive functions of natural numbers. Math. Ann. **112**, 727–742 (1936)
15. Kleene, S.: On the interpretation of intuitionistic number theory. J. Symb. Log. **10**(4), 109–124 (1945)
16. Kreisel, G.: Gödel's interpretation of Heyting's arithmetic. In: Summaries of talks, Summer Institute for Symbolic Logic (1957)
17. de Leeuw, K., et al.: Computability by probabilistic machines. In: Press, P.U. (ed.) Automata Studies, pp. 183–212. No. 34, Shannon, C.E. and McCarthy, J. (1956)
18. Mio, M., Skrzypczak, M., Michalewski, H.: Monadic second order logic with measure and category quantifiers. LMCS **8**(2), (2012)

---

[3] For further details, see [2], where the model theory and proof theory of an extension of propositional logic with counting quantifiers is studied (in particular, the logic CPL$_0$ can be seen as a "finitary" fragment of MQPA).

19. Morgenstern, C.: The measure quantifier. J. Symb. Log. **44**, 1 (1979)
20. Rabin, M.O.: Probabilistic automata. Inf. Comput. **6**(3), 230–245 (1963)
21. Saheb-Djaromi, N.: Probabilistic LCF. In: MFCS, no. 64, pp. 442–452 (2021)
22. Santos, E.: Probabilistic Turing machines and computability. AMS **22**(3), 704–710 (1969)
23. Simpson, S.: Subsystems of Second Order Arithmetic. Cambridge Press, London (2009)
24. Sorensen, M., Urzyczyn, P.: Lectures on the Curry-Howard Isomorphism. Elsevier, Amsterdam (2006)
25. Steinhorn, C.I.: Borel structures and measure and category logics. Assoc. Symb. Log. **8**, 579–596 (1985)
26. Troelstra, A.: Realizability. In: Buss, S.R. (ed.) Handbook of Proof Theory, vol. 137, pp. 407–473. Elsevier (1998)
27. Wagner, K.: The complexity of combinatorial problems with succinct input representation. Acta Informatica **23**, 325–356 (1986)

Learning Languages with Decidable Hypotheses

Julian Berger, Maximilian Böther, Vanja Doskoč[⊠], Jonathan Gadea Harder,
Nicolas Klodt, Timo Kötzing, Winfried Lötzsch, Jannik Peters, Leon Schiller,
Lars Seifert, Armin Wells, and Simon Wietheger

Hasso Plattner Institute, University of Potsdam, Potsdam, Germany
{vanja.doskoc,timo.koetzing}@hpi.de
{julian.berger,maximilian.boether,jonathan.gadeaharder,nicolas.klodt,
winfried.loetzsch,jannik.peters,leon.schiller,lars.seifert,armin.wells,
simon.wietheger}@student.hpi.uni-potsdam.de

Abstract. In *language learning in the limit*, the most common type of hypothesis is to give an enumerator for a language, a W-index. These hypotheses have the drawback that even the membership problem is undecidable. In this paper, we use a different system which allows for naming arbitrary decidable languages, namely *programs for characteristic functions* (called C-indices). These indices have the drawback that it is now not decidable whether a given hypothesis is even a legal C-index.

In this first analysis of learning with C-indices, we give a structured account of the learning power of various restrictions employing C-indices, also when compared with W-indices. We establish a hierarchy of learning power depending on whether C-indices are required (a) on all outputs; (b) only on outputs relevant for the class to be learned or (c) only in the limit as final, correct hypotheses. We analyze all these questions also in relation to the mode of data presentation.

Finally, we also ask about the relation of semantic versus syntactic convergence and derive the map of pairwise relations for these two kinds of convergence coupled with various forms of data presentation.

1 Introduction

We are interested in the problem of algorithmically learning a description for a formal language (a computably enumerable subset of the set of natural numbers) when presented successively all and only the elements of that language; this is called *inductive inference*, a branch of (algorithmic) learning theory. For example, a learner h might be presented more and more even numbers. After each new number, h outputs a description for a language as its conjecture. The learner h might decide to output a program for the set of all multiples of 4, as long as all numbers presented are divisible by 4. Later, when h sees an even number not divisible by 4, it might change this guess to a program for the set of all multiples of 2.

This work was supported by DFG Grant Number KO 4635/1-1.

L. De Mol et al. (Eds.): CiE 2021, LNCS 12813, pp. 25–37, 2021.
https://doi.org/10.1007/978-3-030-80049-9_3

Many criteria for determining whether a learner h is *successful* on a language L have been proposed in the literature. Gold, in his seminal paper [10], gave a first, simple learning criterion, **TxtGEx**-*learning*[1], where a learner is *successful* if and only if, on every *text* for L (listing of all and only the elements of L) it eventually stops changing its conjectures, and its final conjecture is a correct description for the input language.

Trivially, each single, describable language L has a suitable constant function as a **TxtGEx**-learner (this learner constantly outputs a description for L). Thus, we are interested in analyzing for which *classes of languages* $\mathcal{L}$ is there a *single learner* h learning *each* member of $\mathcal{L}$. This framework is also known as *language learning in the limit* and has been studied extensively, using a wide range of learning criteria similar to **TxtGEx**-learning (see, for example, the textbook [11]).

In this paper, we put the focus on the possible descriptions for languages. Any computably enumerable language L has as possible descriptions any program enumerating all and only the elements of L, called a W-index (the language enumerated by program e is denoted by W_e). This system has various drawbacks; most importantly, the function which decides, given e and x, whether $x \in W_e$ is not computable. We propose to use different descriptors for languages: programs for characteristic functions (where such programs e describe the language C_e which it decides). Of course, only decidable languages have such a description, but now, given a program e for a characteristic function, $x \in C_e$ is decidable. Additionally to many questions that remain undecidable (for example, whether C-indices are for the same language or whether a C-index is for a finite language), it is not decidable whether a program e is indeed a program for a characteristic function. This leads to a new set of problems: Learners cannot be (algorithmically) checked whether their outputs are viable (in the sense of being programs for characteristic functions).

Based on this last observation, we study a range of different criteria which formalize what kind of behavior we expect from our learners. In the most relaxed setting, learners may output any number (for a program) they want, but in order to **Ex**-learn, they need to converge to a correct C-index; we denote this restriction with **Ex**$_C$. Requiring additionally to only use C-indices in order to successfully learn, we denote by **CIndEx**$_C$; requiring C-indices on *all* inputs (not just for successful learning, but also when seeing input from no target language whatsoever) we denote by $\tau(\mathbf{CInd})\mathbf{Ex}_C$. In particular, the last restriction requires the learner to be total; in order to distinguish whether the loss of learning power is due to the totality restriction or truly due to the additional requirement of outputting C-indices, we also study $\mathcal{R}\mathbf{CIndEx}_C$, that is, the requirement **CIndEx**$_C$ where additionally the learner is required to be total.

We note that $\tau(\mathbf{CInd})\mathbf{Ex}_C$ is similar to learning *indexable families*. Indexable families are classes of languages $\mathcal{L}$ such that there is an enumeration $(L_i)_{i \in \mathbb{N}}$ of all and only the elements of $\mathcal{L}$ for which the decision problem "$x \in L_i$" is decidable.

---

[1] **Txt** stands for learning from a *text* of positive examples; **G** for Gold, indicating full-information learning; **Ex** stands for *explanatory*.

Already for such classes of languages, we get a rich structure (see a survey of previous work [16]). For a learner h learning according to $\tau(\mathbf{CInd})\mathbf{Ex}_C$, we have that $L_x = C_{h(x)}$ gives an indexing of a family of languages, and h learns some subset thereof. We are specifically interested in the area between this setting and learning with W-indices ($\mathbf{Ex}_W$).

The criteria we analyze naturally interpolate between these two settings. We show that we have the following hierarchy: $\tau(\mathbf{CInd})\mathbf{Ex}_C$ allows for learning strictly fewer classes of languages than $\mathcal{R}\mathbf{CInd}\mathbf{Ex}_C$, which allow for learning the *same* classes as $\mathbf{CInd}\mathbf{Ex}_C$, which again are fewer than learnable by $\mathbf{Ex}_C$, which in turn renders fewer classes learnable than $\mathbf{Ex}_W$.

All these results hold for learning with full information. In order to study the dependence on the mode of information presentation, we also consider *partially set-driven* learners ($\mathbf{Psd}$, [2,19]), which only get the set of data presented so far and the iteration number as input; *set-driven* learners ($\mathbf{Sd}$, [20]), which get only the set of data presented so far; *iterative* learners ($\mathbf{It}$, [8,21]), which only get the new datum and their current hypothesis and, finally, *transductive* learners ($\mathbf{Td}$, [4,15]), which only get the current data. Note that transductive learners are mostly of interest as a proper restriction to all other modes of information presentation. In particular, we show that full-information learners can be turned into partially set-driven learners without loss of learning power and iterative learning is strictly less powerful than set-driven learning, in all settings.

Altogether we analyze 25 different criteria and show how each pair relates. All these results are summarized in Fig. 1(a) as one big map stating all pairwise relations of the learning criteria mentioned, giving 300 pairwise relations in one diagram, proven with 13 theorems in Sect. 3. Note that the results comparing learning criteria with W-indices were previously known, and some proofs could be extended to also cover learning with C-indices. For the proofs, please consider the full version of this paper [1].

In Sect. 4, we derive a similar map considering a possible relaxation on $\mathbf{Ex}_C$-learning: While $\mathbf{Ex}_C$ requires syntactic convergence to one single correct C-index, we consider *behaviorally correct* learning ($\mathbf{Bc}_C$, [6,17]) where the learner only has to semantically converge to correct C-indices (but may use infinitely many different such indices). We again consider the different modes of data presentation and determine all pairwise relations in Fig. 1(b). The proofs are again deferred to the full version [1].

2 Preliminaries

2.1 Mathematical Notations and Learning Criteria

In this section, we discuss the used notation as well as the system for learning criteria [15] we follow. Unintroduced notation follows the textbook [18].

With $\mathbb{N}$ we denote the set of all natural numbers, namely $\{0, 1, 2, \ldots\}$. We denote the subset and proper subset relation between two sets with $\subseteq$ and $\subsetneq$, respectively. We use $\emptyset$ and ε to denote the empty set and empty sequence, respectively. The set of all computable functions is denoted by $\mathcal{P}$, the subset of

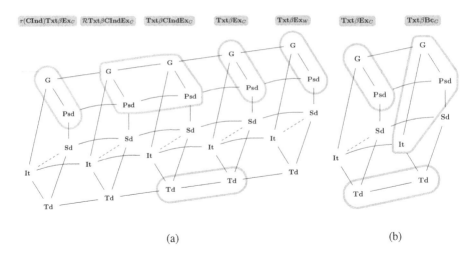

(a)   (b)

Fig. 1. Relation of **(a)** various requirements when to output characteristic indices and **(b)** various learning criteria, both paired with various memory restrictions β. Black solid respectively dashed lines imply trivial respectively non-trivial inclusions (bottom-to-top, left-to-right). Furthermore, greyly edged areas illustrate a collapse of the enclosed learning criteria and there are no further collapses.

all total computable functions by $\mathcal{R}$. If a function f is (not) defined on some argument $x \in \mathbb{N}$, we say that f converges (diverges) on x, denoting this fact with $f(x)\!\downarrow$ ($f(x)\!\uparrow$). We fix an effective numbering $\{\varphi_e\}_{e\in\mathbb{N}}$ of $\mathcal{P}$. For any $e \in \mathbb{N}$, we let W_e denote the domain of φ_e and call e a W-index of W_e. This set we call the e-th computably enumerable set. We call $e \in \mathbb{N}$ a C-index (characteristic index) if and only if φ_e is a total function such that for all $x \in \mathbb{N}$ we have $\varphi_e(x) \in \{0,1\}$. Furthermore, we let $C_e = \{x \in \mathbb{N}\}\varphi_e(x) = 1$. For a computably enumerable set L, if some $e \in \mathbb{N}$ is a C-Index with $C_e = L$, we write $\varphi_e = \chi_L$. Note that, if a set has a C-index, it is *recursive*. The set of all recursive sets is denoted by **REC**. For a finite set $D \subseteq \mathbb{N}$, we let $\mathrm{ind}(D)$ be a C-index for D. Note that $\mathrm{ind} \in \mathcal{R}$. Furthermore, we fix a Blum complexity measure $\varPhi$ associated with φ, that is, for all $e, x \in \mathbb{N}$, $\varPhi_e(x)$ is the number of steps the function φ_e takes on input x to converge [3]. The padding function $\mathrm{pad} \in \mathcal{R}$ is an injective function such that, for all $e, n \in \mathbb{N}$, we have $\varphi_e = \varphi_{\mathrm{pad}(e,n)}$. We use $\langle \cdot, \cdot \rangle$ as a computable, bijective function that codes a pair of natural numbers into a single one. We use π_1 and π_2 as computable decoding functions for the first and section component, i.e., for all $x, y \in \mathbb{N}$ we have $\pi_1(\langle x, y \rangle) = x$ and $\pi_2(\langle x, y \rangle) = y$.

We learn computably enumerable sets L, called *languages*. We fix a *pause* symbol $\#$, and let, for any set S, $S_\# := S \cup \{\#\}$. Information about languages is given from *text*, that is, total functions $T \colon \mathbb{N} \to \mathbb{N} \cup \{\#\}$. A text T is of a certain language L if its *content* is exactly L, that is, $\mathrm{content}(T) := \mathrm{range}(T) \setminus \{\#\}$ is exactly L. We denote the set of all texts as **Txt** and the set of all texts of a language L as **Txt**(L). For any $n \in \mathbb{N}$, we denote with $T[n]$ the initial sequence of the text T of length n, that is, $T[0] := \varepsilon$ and $T[n] := (T(0), \ldots, T(n-1))$. Given a language L and $t \in \mathbb{N}$, the set of sequences consisting of elements of $L \cup \{\#\}$ that are at most t long is denoted by $L_{\#}^{\leq t}$. Furthermore, we denote with $\mathbb{S}\mathrm{eq}$ all finite sequences over $\mathbb{N}_{\#}$ and define the *content* of such sequences analogous to the content of texts. The concatenation of two sequences $\sigma, \tau \in \mathbb{S}\mathrm{eq}$ is denoted by $\sigma\tau$ or, more emphasizing, $\sigma^\frown\tau$. Furthermore, we write $\subseteq$ for the *extension relation* on sequences and fix a order $\leq$ on $\mathbb{S}\mathrm{eq}$ interpreted as natural numbers.

Now, we formalize learning criteria using the following system [15]. A *learner* is a partial function $h \in \mathcal{P}$. An *interaction operator* β is an operator that takes a learner $h \in \mathcal{P}$ and a text $T \in \mathbf{Txt}$ as input and outputs a (possibly partial) function p. Intuitively, β defines which information is available to the learner for making its hypothesis. We consider *Gold-style* or *full-information* learning [10], denoted by $\mathbf{G}$, *partially set-driven* learning ($\mathbf{Psd}$, [2,19]), *set-driven* learning ($\mathbf{Sd}$, [20]), *iterative* learning ($\mathbf{It}$, [8,21]) and *transductive* learning ($\mathbf{Td}$, [4,15]). To define the latter formally, we introduce a symbol "?" for the learner to signalize that the information given is insufficient. Formally, for all learners $h \in \mathcal{P}$, texts $T \in \mathbf{Txt}$ and all $i \in \mathbb{N}$, define

$$\mathbf{G}(h, T)(i) = h(T[i]);$$
$$\mathbf{Psd}(h, T)(i) = h(\mathrm{content}(T[i]), i);$$
$$\mathbf{Sd}(h, T)(i) = h(\mathrm{content}(T[i]));$$
$$\mathbf{It}(h, T)(i) = \begin{cases} h(\varepsilon), & \text{if } i = 0; \\ h(\mathbf{It}(h, T)(i-1), T(i-1)), & \text{otherwise}; \end{cases}$$
$$\mathbf{Td}(h, T)(i) = \begin{cases} ?, & \text{if } i = 0; \\ \mathbf{Td}(h, T)(i-1), & \text{else, if } h(T(i-1)) = ?; \\ h(T(i-1)), & \text{otherwise}. \end{cases}$$

For any of the named interaction operators β, given a β-learner h, we let h^* (the *starred* learner) denote a $\mathbf{G}$-learner simulating h, i.e., for all $T \in \mathbf{Txt}$, we have $\beta(h, T) = \mathbf{G}(h^*, T)$. For example, let h be a $\mathbf{Sd}$-learner. Then, intuitively, h^* ignores all information but the content of the input, simulating h with this information, i.e., for all finite sequences σ, we have $h^*(\sigma) = h(\mathrm{content}(\sigma))$.

For a learner to successfully identify a language, we may oppose constraints on the hypotheses the learner makes. These are called *learning restrictions*. As a first, famous example, we required the learner to be *explanatory* [10], i.e., the learner must converge to a *single*, correct hypothesis for the target language. We hereby distinguish whether the final hypothesis is interpreted as a C-index ($\mathbf{Ex}_C$) or as a W-index ($\mathbf{Ex}_W$). Formally, for any sequence of hypotheses p and text $T \in \mathbf{Txt}$, we have

$$\mathbf{Ex}_C(p, T) \Leftrightarrow \exists n_0 \colon \forall n \geq n_0 \colon p(n) = p(n_0) \wedge \varphi_{p(n_0)} = \chi_{\mathrm{content}(T)};$$
$$\mathbf{Ex}_W(p, T) \Leftrightarrow \exists n_0 \colon \forall n \geq n_0 \colon p(n) = p(n_0) \wedge W_{p(n_0)} = \mathrm{content}(T).$$

We say that explanatory learning requires *syntactic* convergence. If there exists a C-index (or W-index) for a language, then there exist infinitely many. This motivates to not require syntactic but only *semantic* convergence, i.e., the learner may make mind changes, but it has to, eventually, only output correct hypotheses. This is called *behaviorally correct* learning ($\mathbf{Bc}_C$ or $\mathbf{Bc}_W$, [6,17]). Formally, let p be a sequence of hypotheses and let $T \in \mathbf{Txt}$, then

$$\mathbf{Bc}_C(p, T) \Leftrightarrow \exists n_0 \colon \forall n \geq n_0 \colon \varphi_{p(n)} = \chi_{\text{content}(T)};$$
$$\mathbf{Bc}_W(p, T) \Leftrightarrow \exists n_0 \colon \forall n \geq n_0 \colon W_{p(n)} = \text{content}(T).$$

In this paper, we consider learning with C-indices. It is, thus, natural to require the hypotheses to consist solely of C-indices, called *C-index learning*, and denoted by $\mathbf{CInd}$. Formally, for a sequence of hypotheses p and a text T, we have

$$\mathbf{CInd}(p, T) \Leftrightarrow \forall i, x \colon \varphi_{p(i)}(x) \in \{0, 1\}.$$

For two learning restrictions δ and δ', their combination is their intersection, denoted by their juxtaposition $\delta\delta'$. We let $\mathbf{T}$ denote the learning restriction that is always true, which is interpreted as the absence of a learning restriction.

A *learning criterion* is a tuple $(\alpha, \mathcal{C}, \beta, \delta)$, where $\mathcal{C}$ is the set of admissible learners, usually $\mathcal{P}$ or $\mathcal{R}$, β is an interaction operator and α and δ are learning restrictions. We denote this criterion with $\tau(\alpha)\mathcal{C}\mathbf{Txt}\beta\delta$, omitting $\mathcal{C}$ if $\mathcal{C} = \mathcal{P}$, and a learning restriction if it equals $\mathbf{T}$. We say that an admissible learner $h \in \mathcal{C}$ $\tau(\alpha)\mathcal{C}\mathbf{Txt}\beta\delta$-learns a language L if and only if, for arbitrary texts $T \in \mathbf{Txt}$, we have $\alpha(\beta(h, T), T)$ and for all texts $T \in \mathbf{Txt}(L)$ we have $\delta(\beta(h, T), T)$. The set of languages $\tau(\alpha)\mathcal{C}\mathbf{Txt}\beta\delta$-learned by $h \in \mathcal{C}$ is denoted by $\tau(\alpha)\mathcal{C}\mathbf{Txt}\beta\delta(h)$. With $[\tau(\alpha)\mathcal{C}\mathbf{Txt}\beta\delta]$ we denote the set of all classes $\tau(\alpha)\mathcal{C}\mathbf{Txt}\beta\delta$-learnable by some learner in $\mathcal{C}$. Moreover, to compare learning with W- and C-indices, these classes may only contain recursive languages, which we denote as $[\tau(\alpha)\mathcal{C}\mathbf{Txt}\beta\delta]_{\mathbf{REC}}$.

2.2   Normal Forms

When studying language learning in the limit, there are certain properties of learner that are useful, e.g., if we can assume a learner to be total. Cases where learners may be assumed total have been studied in the literature [13,14]. Importantly, this is the case for explanatory Gold-style learners obeying delayable learning restrictions and for behaviorally correct learners obeying delayable restrictions. Intuitively, a learning restriction is *delayable* if it allows hypotheses to be arbitrarily, but not indefinitely postponed without violating the restriction. Formally, a learning restriction δ is delayable, if and only if for all non-decreasing, unbounded functions $r \colon \mathbb{N} \to \mathbb{N}$, texts $T, T' \in \mathbf{Txt}$ and learning sequences p such that for all $n \in \mathbb{N}$, $\text{content}(T[r(n)]) \subseteq \text{content}(T'[n])$ and $\text{content}(T) = \text{content}(T')$, we have, if $\delta(p, T)$, then also $\delta(p \circ r, T')$. Note that $\mathbf{Ex}_W$, $\mathbf{Ex}_C$, $\mathbf{Bc}_W$, $\mathbf{Bc}_C$ and $\mathbf{CInd}$ are delayable restrictions.

Another useful notion are *locking sequences*. Intuitively, these contain enough information such that a learner, after seeing this information, converges correctly and does not change its mind anymore whatever additional information from the target language it is given. Formally, let L be a language and let $\sigma \in L_\#^*$. Given a **G**-learner $h \in \mathcal{P}$, σ is a *locking sequence* for h on L if and only if for all sequences $\tau \in L_\#^*$ we have $h(\sigma) = h(\sigma\tau)$ and $h(\sigma)$ is a correct hypothesis for L [2]. This concept can immediately be transferred to other interaction operators. Exemplary, given a **Sd**-learner h and a locking sequence σ of the starred learner h^*, we call the set content(σ) a *locking set*. Analogously, one transfers this definition to the other interaction operators. It shall not remain unmentioned that, when considering **Psd**-learners, we speak of *locking information*. In the case of **Bc$_W$**-learning we do not require the learner to syntactically converge. Therefore, we call a sequence $\sigma \in L_\#^*$ a **Bc$_W$**-*locking sequence* for a **G**-learner h on L if, for all sequences $\tau \in L_\#^*$, $h(\sigma\tau)$ is a correct hypothesis for L [11]. We omit the transfer to other interaction operators as it is immediate. It is an important observation that for any learner h and any language L it learns, there exists a (**Bc$_W$**-) locking sequence [2]. These notions and results directly transfer to **Ex$_C$**- and **Bc$_C$**-learning. When it is clear from the context, we omit the index.

3   Requiring *C*-Indices as Output

This section is dedicated to proving Fig. 1(a), giving all pairwise relations for the different settings of requiring C-indices for output in the various mentioned modes of data presentation. In general, we observe that the later we require C-indices, the more learning power the learner has. This holds except for transductive learners which converge to C-indices. We show that they are as powerful as **CInd**-transductive learners.

Although we learn classes of recursive languages, the requirement to converge to characteristic indices does heavily limit a learners capabilities. In the next theorem we show that even transductive learners which converge to W-indices can learn classes of languages which no Gold-style **Ex$_C$**-learner can learn. We exploit the fact that C-indices, even if only conjectured eventually, must contain both positive and negative information about the guess.

Theorem 1. *We have that* $[\mathbf{TxtTdEx}_W]_{\mathbf{REC}} \setminus [\mathbf{TxtGEx}_C]_{\mathbf{REC}} \neq \emptyset$.

Proof. We show this by using the Operator Recursion Theorem (**ORT**) to provide a separating class of languages. To this end, let h be the **Td**-learner with $h(\#) = ?$ and, for all $x, y \in \mathbb{N}$, let $h(\langle x, y \rangle) = x$. Let $\mathcal{L} = \mathbf{TxtTdEx}_W(h) \cap \mathbf{REC}$. Assume $\mathcal{L}$ can be learned by a **TxtGEx$_C$**-learner h'. We may assume $h' \in \mathcal{R}$ [13].

Then, by **ORT** there exist indices $e, p, q \in \mathbb{N}$ such that

$$L := W_e = \text{range}(\varphi_p);$$
$$\forall x: \tilde{T}(x) := \varphi_p(x) = \langle e, \varphi_q(\tilde{T}[x]) \rangle;$$
$$\varphi_q(\varepsilon) = 0;$$
$$\forall \sigma \neq \varepsilon: \bar{\sigma} = \min\{\sigma' \subseteq \sigma \mid \varphi_q(\sigma') = \varphi_q(\sigma)\};$$

$$\forall \sigma \neq \varepsilon: \varphi_q(\sigma) = \begin{cases} \varphi_q(\bar{\sigma}), & \text{if } \forall \sigma', \bar{\sigma} \subseteq \sigma' \subseteq \sigma: \Phi_{h'(\sigma')}(\langle e, \varphi_q(\bar{\sigma}) + 1 \rangle) > |\sigma|; \\ \varphi_q(\bar{\sigma}) + 1, & \text{else, for min. } \sigma' \text{ contradicting the previous case, if} \\ & \quad \varphi_{h'(\sigma')}(\langle e, \varphi_q(\bar{\sigma}) + 1 \rangle) = 0; \\ \varphi_q(\bar{\sigma}) + 2, & \text{otherwise.} \end{cases}$$

Here, Φ is a Blum complexity measure [3]. Intuitively, to define the next $\varphi_p(x)$, we add the same element to content($\tilde{T}$) until we know whether $\langle e, \tilde{T}[x] + 1 \rangle \in C_{h'(\bar{\sigma})}$ holds or not. Then, we add the element contradicting this outcome.

We first show that $L \in \mathcal{L}$ and afterwards that L cannot be learned by h'. To show the former, note that either L is finite or $\tilde{T}$ is a non-decreasing unbounded computable enumeration of L. Therefore, we have $L \in \mathbf{REC}$. We now prove that h learns L. Let $T \in \mathbf{Txt}(L)$. For all $n \in \mathbb{N}$ where $T(n)$ is not the pause symbol, we have $h(T(n)) = e$. With $n_0 \in \mathbb{N}$ being minimal such that $T(n_0) \neq \#$, we get for all $n \geq n_0$ that $\mathbf{Td}(h, T)(n) = e$. As e is a correct hypothesis, h learns L from T and thus we have that $L \in \mathbf{TxtTdEx}_W(h)$. Altogether, we get that $L \in \mathcal{L}$.

By assumption, h' learns L from the text $\tilde{T} \in \mathbf{Txt}(L)$. Therefore, there exists $n_0 \in \mathbb{N}$ such that, for all $n \geq n_0$,

$$h'(\tilde{T}[n]) = h'(\tilde{T}[n_0]) \text{ and } \chi_L = \varphi_{h'(\tilde{T}[n])},$$

that is, $h'(\tilde{T}[n])$ is a C-index for L. Now, as h' outputs C-indices when converging, there are $t, t' \geq n_0$ such that

$$\Phi_{h'(\tilde{T}[t'])}(\langle e, \varphi_q(\tilde{T}[n_0]) + 1 \rangle) \leq t.$$

Let t_0' and t_0 be the first such found. We show that $h'(\tilde{T}[t_0'])$ is no correct hypothesis of L by distinguishing the following cases.

1. Case: $\varphi_{h'(\tilde{T}[t_0'])}(\langle e, \varphi_q(\tilde{T}[n_0]) + 1 \rangle) = 0$. By definition of φ_q and by minimality of t_0', we have that $\langle e, \varphi_q(\tilde{T}[n_0]) + 1 \rangle \in L$, however, the hypothesis of $h'(\tilde{T}[t_0'])$ says differently, a contradiction.

2. Case: $\varphi_{h'(\tilde{T}[t_0'])}(\langle e, \varphi_q(\tilde{T}[n_0]) + 1 \rangle) = 1$. By definition of φ_q and by minimality of t_0', we have that $\langle e, \varphi_q(\tilde{T}[n_0]) + 1 \rangle \in L$, but $\langle e, \varphi_q(\tilde{T}[n_0]) + 1 \rangle \notin L$. However, the hypothesis of $h'(\tilde{T}[t_0'])$ conjectures the latter to be in L, a contradiction.

$\square$

Furthermore, the following known equalities from learning W-indices directly apply in the studied setting as well.

Theorem 2 ([12], [9,19]). *We have that*

$$[\mathbf{TxtItEx}_W]_{\mathbf{REC}} \subseteq [\mathbf{TxtSdEx}_W]_{\mathbf{REC}},$$
$$[\mathbf{TxtPsdEx}_W]_{\mathbf{REC}} = [\mathbf{TxtGEx}_W]_{\mathbf{REC}}.$$

The remaining separations we will show in a more general way, see Theorems 11 and 12. We generalize the latter result [9,19], namely that Gold-style learners may be assumed partially set-driven, to all considered cases. The idea here is to, just as in the $\mathbf{Ex}_W$-case, mimic the given learner and to search for minimal locking sequences. Incorporating the result that unrestricted Gold-style learners may be assumed total [13], we even get a stronger result.

Theorem 3. *For $\delta, \delta' \in \{\mathbf{CInd}, \mathbf{T}\}$, we have that*

$$[\tau(\delta)\mathbf{TxtG}\delta'\mathbf{Ex}_C]_{\mathbf{REC}} = [\tau(\delta)\mathcal{R}\mathbf{TxtPsd}\delta'\mathbf{Ex}_C]_{\mathbf{REC}}.$$

We also generalize the former result of Theorem 2 to hold in all considered cases. The same simulating argument (where one mimics the iterative learner on ascending text with a pause symbol between two elements) suffices regardless the exact setting.

Theorem 4. *Let $\delta, \delta' \in \{\mathbf{CInd}, \mathbf{T}\}$ and $C \in \{\mathcal{R}, \mathcal{P}\}$. Then, we have that*

$$[\tau(\delta')\mathcal{C}\mathbf{TxtIt}\delta\mathbf{Ex}_C]_{\mathbf{REC}} \subseteq [\tau(\delta')\mathcal{C}\mathbf{TxtSd}\delta\mathbf{Ex}_C]_{\mathbf{REC}}.$$

Interestingly, totality is not restrictive solely for Gold-style (and due to the equality also partially set-driven) learners. For the other considered learners with restricted memory, being total lessens the learning capabilities. This weakness results from the need to output some guess. A partial learner can await this guess and outperform it. This way, we obtain self-learning languages [5] to show the three following separations.

Theorem 5. *We have that* $[\mathcal{R}\mathbf{TxtSdCIndEx}_C]_{\mathbf{REC}} \subsetneq [\mathbf{TxtSdCIndEx}_C]_{\mathbf{REC}}.$

Theorem 6. *We have that* $[\mathcal{R}\mathbf{TxtItCIndEx}_C]_{\mathbf{REC}} \subsetneq [\mathbf{TxtItCIndEx}_C]_{\mathbf{REC}}.$

Theorem 7. *We have that* $[\mathcal{R}\mathbf{TxtTdCIndEx}_C]_{\mathbf{REC}} \subsetneq [\mathbf{TxtTdCIndEx}_C]_{\mathbf{REC}}.$

Next, we show the gradual decrease of learning power the more we require the learners to output characteristic indices. We have already seen in Theorem 1 that converging to C-indices lessens learning power. However, this allows for more learning power than outputting these indices during the whole learning process as shows the next theorem. The idea is that such learners have to be certain about their guesses as these are indices of characteristic functions. When constructing a separating class using self-learning languages [5], one forces the **CInd**-learner to output C-indices on certain languages to, then, contradict its choice there. This way, the $\mathbf{Ex}_C$-learner learns languages the **CInd**-learner cannot.

Theorem 8. *We have that* $[\mathbf{TxtItEx}_C]_{\mathbf{REC}} \setminus [\mathbf{TxtGCIndBc}_C]_{\mathbf{REC}} \neq \emptyset$.

Since languages which can be learned by iterative learners can also be learned by set-driven ones (see Theorem 4), this result suffices. Note that the idea above requires some knowledge on previous elements. Thus, it is no coincidence that this separation does not include transductive learners. Since these learners base their guesses on single elements, they cannot see how far in the learning process they are. Thus, they are forced to always output C-indices.

Theorem 9. *We have that* $[\mathbf{TxtTdCIndEx}_C]_{\mathbf{REC}} = [\mathbf{TxtTdEx}_C]_{\mathbf{REC}}$.

For the remainder of this section, we focus on learners which output characteristic indices on *arbitrary* input, that is, we focus on $\tau(\mathbf{CInd})$-learners. First, we show that the requirement of always outputting C-indices lessens a learners learning power, even when compared to total **CInd**-learners. To provide the separating class of self-learning languages, one again awaits the $\tau(\mathbf{CInd})$-learner's decision and then, based on these, learns languages this learner cannot.

Theorem 10. *We have* $[\mathcal{R}\mathbf{TxtTdCIndEx}_C]_{\mathbf{REC}} \setminus [\tau(\mathbf{CInd})\mathbf{TxtGBc}_C]_{\mathbf{REC}} \neq \emptyset$.

Proof. We prove the result by providing a separating class of languages. Let h be the **Td**-learner with $h(\#) = ?$ and, for all $x, y \in \mathbb{N}$, let $h(\langle x, y \rangle) = x$. By construction, h is total and computable. Let $\mathcal{L} = \mathcal{R}\mathbf{TxtTdCIndEx}_C(h) \cap \mathbf{REC}$. We show that there is no $\tau(\mathbf{CInd})\mathbf{TxtGBc}_C$-learner learning $\mathcal{L}$ by way of contradiction. Assume there is a $\tau(\mathbf{CInd})\mathbf{TxtGBc}_C$-learner h' which learns $\mathcal{L}$. With the Operator Recursion Theorem (**ORT**), there are $e, p \in \mathbb{N}$ such that for all $x \in \mathbb{N}$

$$L := \mathrm{range}(\varphi_p);$$

$$\varphi_e = \chi_L;$$

$$\tilde{T}(x) := \varphi_p(x) = \begin{cases} \langle e, 2x \rangle, & \text{if } \varphi_{h'(\varphi_p[x])}(\langle e, 2x \rangle) = 0; \\ \langle e, 2x + 1 \rangle, & \text{otherwise.} \end{cases}$$

Intuitively, for all x either $\varphi_p(x)$ is an element of L if it is not in the hypothesis of h' after seeing $\varphi_p[x]$, or there is an element in this hypothesis that is not in content($\tilde{T}$). As any hypothesis of h' is a C-index, we have that $\varphi_p \in \mathcal{R}$ and, as φ_p is strictly monotonically increasing, that L is decidable.

We now prove that $L \in \mathcal{L}$ and afterwards that L cannot be learned by h'. First, we need to prove that h learns L. Let $T \in \mathbf{Txt}(L)$. For all $n \in \mathbb{N}$ where $T(n)$ is not the pause symbol, we have $h(T(n)) = e$. Let $n_0 \in \mathbb{N}$ with $T(n_0) \neq \#$. Then, we have, for all $n \geq n_0$, that $\mathbf{Td}(h, T)(n) = e$ and, since e is a hypothesis of L, h learns L from T. Thus, we have that $L \in \mathcal{R}\mathbf{TxtTdCIndEx}_C(h) \cap \mathbf{REC}$.

By assumption, h' learns $\mathcal{L}$ and thus it also needs to learn L on text $\tilde{T}$. Hence, there is x_0 such that for all $x \geq x_0$ the hypothesis $h'(\tilde{T}[x]) = h'(\varphi_p[x])$ is a C-index for L. We now consider the following cases.

1. Case: $\varphi_{h'(\varphi_p[x])}(\langle e, 2x \rangle) = 0$. By construction, we have that $\tilde{T}(x) = \langle e, 2x \rangle$. Therefore, $\langle e, 2x \rangle \in L$, which contradicts $h'(\varphi_p[x])$ being a correct hypothesis.

2. Case: $\varphi_{h'(\varphi_p[x])}(\langle e, 2x \rangle) = 1$. By construction, we have that $\tilde{T}(x) \neq \langle e, 2x \rangle$ and thus, because $\tilde{T}$ is strictly monotonically increasing, $\langle e, 2x \rangle \notin L = \text{content}(\tilde{T})$. This, again, contradicts $h'(\varphi_p[x])$ being a correct hypothesis.

As in all cases $h'(\varphi_p[x])$ is a wrong hypothesis, h' cannot learn $\mathcal{L}$.   □

It remains to be shown that memory restrictions are severe for such learners as well. First, we show that partially set-driven learners are more powerful than set-driven ones. Just as originally witnessed by for W-indices [9,19], this is solely due to the lack of learning time. In the following theorem, we already separate from behaviorally correct learners, as we will need this stronger version later on.

Theorem 11. *We have that* $[\tau(\mathbf{CInd})\mathbf{TxtPsdEx}_C]_{\mathbf{REC}} \setminus [\mathbf{TxtSdBc}_W]_{\mathbf{REC}} \neq \emptyset$.

In turn, this lack of time is not as severe as lack of memory. The standard class (of recursive languages) to separate set-driven learners from iterative ones [11] can be transferred to the setting studied in this paper.

Theorem 12. *We have that* $[\tau(\mathbf{CInd})\mathbf{TxtSdEx}_C]_{\mathbf{REC}} \setminus [\mathbf{TxtItEx}_W]_{\mathbf{REC}} \neq \emptyset$.

Lastly, we show that transductive learners, having basically no memory, do severely lack learning power. As they have to infer their conjectures from single elements they, in fact, cannot even learn basic classes such as $\{\{0\}, \{1\}, \{0, 1\}\}$. The following result concludes the map shown in Fig. 1(a) and, therefore, also this section.

Theorem 13. *For* $\beta \in \{\mathbf{It}, \mathbf{Sd}\}$, *we have that*

$$[\tau(\mathbf{CInd})\mathbf{Txt}\beta\mathbf{Ex}_C]_{\mathbf{REC}} \setminus [\mathbf{TxtTdEx}_W]_{\mathbf{REC}} \neq \emptyset.$$

4 Syntactic Versus Semantic Convergence to C-indices

In this section, we investigate the effects on learners when we require them to converge to characteristic indices. We study both syntactically converging learners as well as semantically converging ones. In particular, we compare learners imposed with different well-studied memory restrictions.

Surprisingly, we observe that, although C-indices incorporate and, thus, require the learner to obtain more information during the learning process than W-indices, the relative relations of the considered restrictions remain the same. We start by gathering results which directly follow from the previous section.

Corollary 1. *We have that*

$$[\mathbf{TxtPsdEx}_C]_{\mathbf{REC}} = [\mathbf{TxtGEx}_C]_{\mathbf{REC}}, (\textit{Theorem 3}),$$
$$[\mathbf{TxtItEx}_C]_{\mathbf{REC}} \subseteq [\mathbf{TxtSdEx}_C]_{\mathbf{REC}}, (\textit{Theorem 4}),$$
$$[\mathbf{TxtGEx}_C]_{\mathbf{REC}} \setminus [\mathbf{TxtSdBc}_C]_{\mathbf{REC}} \neq \emptyset, (\textit{Theorem 11}),$$
$$[\mathbf{TxtSdEx}_C]_{\mathbf{REC}} \setminus [\mathbf{TxtItEx}_C]_{\mathbf{REC}} \neq \emptyset, (\textit{Theorem 12}),$$
$$[\mathbf{TxtItEx}_C]_{\mathbf{REC}} \setminus [\mathbf{TxtTdEx}_C]_{\mathbf{REC}} \neq \emptyset, (\textit{Theorem 13}).$$

We show the remaining results. First, we show that, just as for W-indices, behaviorally correct learners are more powerful than explanatory ones. We provide a separating class exploiting that explanatory learners must converge to a single, correct hypothesis. We collect elements on which mind changes are witnessed, while maintaining decidability of the obtained language.

Theorem 14. *We have that* $[\mathbf{TxtSdBc}_C]_{\mathbf{REC}} \setminus [\mathbf{TxtGEx}_C]_{\mathbf{REC}} \neq \emptyset$.

Next, we show that, just as for W-indices, a padding argument makes iterative behaviorally correct learners as powerful as Gold-style ones.

Theorem 15. *We have that* $[\mathbf{TxtItBc}_C]_{\mathbf{REC}} = [\mathbf{TxtGBc}_C]_{\mathbf{REC}}$.

We show that the classes of languages learnable by some behaviorally correct Gold-style (or, equivalently, iterative) learner, can also be learned by partially set-driven ones. We follow the proof which is given in a private communication with Sanjay Jain [7]. The idea there is to search for minimal **Bc**-locking sequences without directly mimicking the **G**-learner. We transfer this idea to hold when converging to C-indices as well. We remark that, while doing the necessary enumerations, one needs to make sure these are characteristic. One obtains this as the original learner eventually outputs characteristic indices.

Theorem 16. *We have that* $[\mathbf{TxtPsdBc}_C]_{\mathbf{REC}} = [\mathbf{TxtGBc}_C]_{\mathbf{REC}}$.

Lastly, we investigate transductive learners. Such learners base their hypotheses on a single element. Thus, one would expect them to benefit from dropping the requirement to converge to a *single* hypothesis. Interestingly, this does not hold true. This surprising fact originates from C-indices encoding characteristic functions. Thus, one can simply search for the minimal element on which no "?" is conjectured. The next result finalizes the map shown in Fig. 1(a) and, thus, this section.

Theorem 17. *We have that* $[\mathbf{TxtTdEx}_C]_{\mathbf{REC}} = [\mathbf{TxtTdBc}_C]_{\mathbf{REC}}$.

References

1. Berger, J., et al.: Learning languages with decidable hypotheses. CoRR (2020)
2. Blum, L., Blum, M.: Toward a mathematical theory of inductive inference. Inf. Control **28**, 125–155 (1975)
3. Blum, M.: A machine-independent theory of the complexity of recursive functions. J. ACM **14**, 322–336 (1967)
4. Carlucci, L., Case, J., Jain, S., Stephan, F.: Results on memory-limited U-shaped learning. Inf. Comput. **205**, 1551–1573 (2007)
5. Case, J., Kötzing, T.: Strongly non-U-shaped language learning results by general techniques. Inf. Comput. **251**, 1–15 (2016)
6. Case, J., Lynes, C.: Machine inductive inference and language identification. In: Nielsen, M., Schmidt, E.M. (eds.) ICALP 1982. LNCS, vol. 140, pp. 107–115. Springer, Heidelberg (1982). https://doi.org/10.1007/BFb0012761

7. Doskoč, V., Kötzing, T.: Cautious limit learning. In: Proceedings of the International Conference on Algorithmic Learning Theory (ALT) (2020)
8. Fulk, M.: A Study of Inductive Inference Machines. Ph.D. thesis (1985)
9. Fulk, M.A.: Prudence and other conditions on formal language learning. Inf. Comput. **85**, 1–11 (1990)
10. Gold, E.M.: Language identification in the limit. Inf. Control **10**, 447–474 (1967)
11. Jain, S., Osherson, D., Royer, J.S., Sharma, A.: Systems that Learn: An Introduction to Learning Theory. MIT Press, Cambridge, Second Edition (1999)
12. Kinber, E.B., Stephan, F.: Language learning from texts: Mindchanges, limited memory, and monotonicity. Inf. Comput. **123**, 224–241 (1995)
13. Kötzing, T., Palenta, R.: A map of update constraints in inductive inference. Theoret. Comput. Sci. **650**, 4–24 (2016)
14. Kötzing, T., Schirneck, M., Seidel, K.: Normal forms in semantic language identification. In: Proceedings of the International Conference on Algorithmic Learning Theory (ALT), pp. 76:493–76:516 (2017)
15. Kötzing, T.: Abstraction and Complexity in Computational Learning in the Limit. Ph.D. thesis, University of Delaware (2009)
16. Lange, S., Zeugmann, T., Zilles, S.: Learning indexed families of recursive languages from positive data: A survey. Theor. Comput. Sci. **397**, 194–232 (2008)
17. Osherson, D.N., Weinstein, S.: Criteria of language learning. Inf. Control **52**, 123–138 (1982)
18. Rogers, H., Jr.: Theory of Recursive Functions and Effective Computability. MIT Press, Cambridge (1987)
19. Schäfer-Richter, G.: Über Eingabeabhängigkeit und Komplexität von Inferenzstrategien. Ph.D. thesis, RWTH Aachen University, Germany (1984)
20. Wexler, K., Culicover, P.W.: Formal Principles of Language Acquisition. MIT Press, Cambridge (1980)
21. Wiehagen, R.: Limes-Erkennung rekursiver Funktionen durch spezielle Strategien. J. Inf. Proc. Cybern. **12**, 93–99 (1976)

Robust Online Algorithms for Dynamic Choosing Problems

Sebastian Berndt[1], Kilian Grage[2(✉)], Klaus Jansen[2], Lukas Johannsen[2], and Maria Kosche[3]

[1] University of Lübeck, 23562 Lübeck, Germany
s.berndt@uni-luebeck.de
[2] Kiel University, 24118 Kiel, Germany
{kig,kj}@informatik.uni-kiel.de
[3] Göttingen University, 37073 Göttingen, Germany
maria.kosche@cs.uni-goettingen.de

Abstract. Semi-online algorithms that are allowed to perform a bounded amount of repacking achieve guaranteed good worst-case behaviour in a more realistic setting. Most of the previous works focused on minimization problems that aim to minimize some costs. In this work, we study maximization problems that aim to maximize their profit.

We mostly focus on a class of problems that we call *choosing problems*, where a maximum profit subset of a set objects has to be maintained. Many known problems, such as KNAPSACK, MAXIMUMINDEPENDENTSET and variations of these, are part of this class. We present a framework for choosing problems that allows us to transfer offline α-approximation algorithms into $(\alpha - \epsilon)$-competitive semi-online algorithms with amortized migration $O(1/\epsilon)$. Moreover we complement these positive results with lower bounds that show that our results are tight in the sense that no amortized migration of $o(1/\epsilon)$ is possible.

Keywords: Online algorithms · Dynamic algorithms · Competitive ratio · Migration · Knapsack · Maximum independent set

1 Introduction

Optimization problems and how fast we can solve them optimally or approximatively have been a central topic in theoretical computer science. In this context, there is one major problem that is unique to applications: an unknown future. In the real world, we are often not given all the information in advance, as unforeseeable things like customers cancelling or new urgent customer requests can happen at any moment. The study of *online problems* addresses this kind of uncertainty in different variants. The classical model starts with an empty

Supported by DFG-Project JA 612 /19-1 and GIF-Project "Polynomial Migration for Online Scheduling". A full version of the paper is available at http://arxiv.org/abs/2104.09803.

instance and in subsequent time steps, new parts of the instance are added. In order to solve a problem, an online algorithm must generate a solution for every time step without knowing any information about future events.

In the strictest setting, the algorithm is not allowed to alter the solution generated in a previous step at all, so every mistake will carry weight into the future. As this is a very heavy restriction for online algorithms, there are also variants where the algorithm is allowed to change solutions to some degree. We cannot allow for an arbitrary number of changes as this only leads to the offline setting. We therefore consider the *migration model*, where every change in the instance, e. g., an added node to a graph or a new item, comes with a *migration potential*. Intuitively, this migration potential is linked to some size or weight which means objects that have a larger impact on the optimization criteria will yield larger migration potential allowing more change. Similarly, small objects will only allow for small changes of the solution.

We consider this migration setting in an amortized way by allowing the migration to accumulate over time. This way we allow our algorithm to generally handle newly arriving objects without changing the solution (apart from extending the solution with regards to the new objects) and at some later point of time we will repack the solution and use the sum of all migration potential of items that arrived up to that time. While there is usually a trade-off between maintaining a good solution and constraining the migration, in this paper, we will present a very simple framework that achieves results close to the best offline results for a large range of problems while maintaining an amortized migration factor of $O(1/\epsilon)$. In fact, we manage to keep solutions on par with the best offline algorithms except an additive ϵ-term. For many problems this framework even works when considering the problem variants where objects not only appear but are also removed from the instance. In addition to these positive results, we also show that a migration of only $\Omega(1/\epsilon)$ is needed even for relatively simple problems such as the SUBSETSUM problem.

2 Preliminaries

We are given some optimization problem Π_{off} consisting of a set of objects and consider the online version Π_{on}, where these objects arrive one by one over time (the *static* case). If arrived items can also be removed from the instance, we call this the *dynamic* case. For an instance $I \in \Pi_{\text{on}}$ and some time t we denote the instance at time t containing the first t objects by I_t. As discussed above, we allow a certain amount of repacking and thus, every object has an associated *migration potential*, which typically corresponds to its size or its weight. For the problems considered in this paper, the migration potential and cost is linked to the profits of the individual items being inserted into the problem instance.

The item arriving or departing at time t has migration potential $\Delta(I_t)$. Then, $\Delta(I_t \rightarrow I_{t'})$ denotes the migration potential that we received starting at time t until t' for $t \leq t'$. As the given instance is usually clear from the context, we simply this notation and write $\Delta_{t:t'} := \Delta(I_t \rightarrow I_{t'})$. The total migration potential up to some time t is thus given by $\Delta_{0:t}$.

We further assume that for every two feasible solutions S_t and $S'_{t'}$ at times t, t', we also have a necessary migration cost that we denote by $\phi(S_t \to S'_{t'})$, respectively. This resembles the costs to migrate from solution S_t to solution $S'_{t'}$. Note that the initial assignment of an item does not cost any migration. Therefore, $\phi(S_t \to S_{t+1})$ denotes the migration cost of changing solution S_t to solution S_{t+1} without accounting for the newly arrived item.

We say that an online algorithm has *amortized migration factor* γ if, for all time steps t, the sum of the migration costs is at most $\gamma \cdot \sum_{i=1}^{t} \Delta(I_t) = \gamma \cdot \Delta_{0:t}$, i.e. $\sum_{i=1}^{t} \phi(S_{i-1} \to S_i) \leq \gamma \cdot \Delta_{0:t}$, where S_i are the solutions produced by the algorithm with S_0 being the empty solution. We will sometimes make use of the amortized migration factor inside a time interval $t \to \cdots \to t'$, which we will denote by $\gamma_{t:t'}$. The notions of *migration* or *migration factor* are interchangeably used by us and always describe the amortized migration factor.

In this work, we only consider maximization problems. Hence, every solution S_t of an instance I_t has some profit $\text{PROFIT}(S_t)$ and $\text{OPT}(I_t)$ denotes the optimal profit of any solution to I_t. An algorithm that achieves competitive ratio $1 - \epsilon$ and has migration $f(1/\epsilon)$ for some function f is called *robust* [29].

2.1 KNAPSACK-Type Problems

The KNAPSACK problem is one of the classical maximization problems. In its most basic form, it considers a *capacity* $C \in \mathbb{N}$ and a finite set I of *items*, each of which is assigned a *weight* $w_i \in \mathbb{N}$ and a *profit* $p_i \in \mathbb{N}$. The objective is to find a subset $S \subseteq I$, interpreted as a packing of the figurative knapsack, with maximum profit $\text{PROFIT}(S) = \sum_{i \in S} p_i$ while the total weight $\text{WEIGHT}(S) = \sum_{i \in S} w_i$ does not exceed C. The special case in which all weights are equal to their respective profits is the SUBSETSUM problem. In this case, because weight and profit coincide, we will simply call both the *size* of an item. A natural generalization of KNAPSACK is to generalize capacity and weight vector to be d-dimensional vectors, i.e., $C, w_i \in \mathbb{N}^d$ for some $d \in \mathbb{N}$. The problem of finding a maximum profit packing fulfilling all d constraints is then known as the d-dimensional KNAPSACK problem.

In the MULTIPLEKNAPSACK problem, one is given an instance I consisting of items with assigned weights and profits, just like in the KNAPSACK problem. However, in contrast, we are given not one but m different knapsacks with respective capacities $C^{(1)}, \ldots, C^{(m)}$. The goal is to find m disjoint subsets $S^{(1)}, \ldots, S^{(m)}$ of I, such that the total profit $\sum_{j=1}^{m} \text{PROFIT}(S^{(j)})$ is maximized w.r.t. the capacity conditions

$$\text{WEIGHT}(S^{(j)}) \leq C^{(j)}, \quad \forall j = 1, \ldots, m.$$

Note that the KNAPSACK problem is a special case of the MULTIPLEKNAPSACK problem with $m = 1$. Another generalization of the standard KNAPSACK problem is 2DGEOKNAPSACK. This problem takes as input the width $W \in \mathbb{N}$ and height $H \in \mathbb{N}$ of the knapsack and a set I of axis-aligned rectangles $r \in I$ with widths $w_r \in \mathbb{N}$, heights $h_r \in \mathbb{N}$, and profits $p_r \in \mathbb{N}$. An optimal solution to this instance consists of a subset $S \subseteq I$ of the rectangles together with a non-overlapping axis-aligned packing of S inside the rectangular knapsack of size $W \times H$ such that PROFIT(S) is maximized.

2.2 Independent Set

Another classical optimization problem is the MAXIMUMINDEPENDENTSET problem. While it can be considered for different types of graphs, the most basic variant is defined on a graph $G = (V, E)$ with a set of nodes V and a set of corresponding edges E. A subset $S \subseteq V$ is called *independent* if for all $u, v \in S$ it holds that $(u, v) \notin E$, i.e., u and v are not neighbours. A maximal independent set is then an independent set that is no strict subset of another independent set. MAXIMUMINDEPENDENTSET is the problem of finding a maximum independent set for a given graph G. In the literature, MAXIMUMINDEPENDENTSET is, among others, studied in planar, perfect, or claw-free graphs. For the online variant, we usually assume that a node is added (or removed) to the instance in every time step along with its adjacent edges.

Closely related to the well-studied MAXIMUMINDEPENDENTSET problem is the MAXIMUMDISJOINTSET problem. For a given instance I that consists of items with a geometrical shape, the goal is to find a largest disjoint set which is a set of non-overlapping items. As we can convert an MDS instance to an MIS instance, we sometimes use MIS to also denote this problem. MDS is often considered limited to certain types of objects. These can be (unit-sized) disks, rectangles, polygons or other objects, and any d-dimensional generalization of them. We will also consider pseudo-disks, which are objects that pairwise intersect at most twice in an instance.

The standard MIS and MDS problems are both special cases of the generalized MAXWEIGHTINDEPENDENTSET or MAXWEIGHTDISJOINTSET, respectively, where each node i is assigned a profit value w_i. The goal for these problems is to find a maximum profit independent subset.

2.3 Our Results

Our main result is a framework that is strongly inspired by the approach of Berndt *et al.* [2]. For minimization problems, they proposed a framework using two known algorithms, one online and one offline algorithm, in order to solve a given problem. This approach behaves a bit differently for maximization problems in terms of the competitive ratio. While a respective α-approximation offline algorithm paired with a fitting β-competitive online algorithm yields an $\alpha + O(1)\beta\epsilon$ competitive-algorithm for minimization problems, we show that for maximization problems such fitting algorithms will result in a $\alpha \cdot \beta$-competitive algorithm. The general analysis on this framework had to be moved to the full version due to spacial constraints.

In this work we discuss a class of problems that is characterized by the common task of choosing a subset of objects with maximum profit fulfilling some secondary constraints. Many important problems like the above presented variants of KNAPSACK or MAXIMUMINDEPENDENTSET are covered by this class of problems. We show that for these kind of problems, which we will call *choosing problems* in the following, the framework can be simplified by completely removing the online algorithm.Using this simplified framework, we achieve $(1 - \epsilon)$-competitive algorithms for KNAPSACK even when generalized to arbitrary but fixed dimension d. In the 2DGEOKNAPSACK where we additionally interpret items as rectangles that need to be packed into a rectangular knapsack, we achieve a $(\frac{9}{17} - \epsilon)$-competitive ratio. We also consider problem variants outside of the class of choosing problems and show that the static cases of MAXIMUMINDEPENDENTSET for planar graphs with arriving edges and MULTIPLEKNAPSACK also admit robust approximation schemes. We complement these positive results by also proving lower bounds for the necessary migration showing that some of our results are indeed tight. We also give lower bounds for different variants, including starting with an adversarially chosen solution and different migration models. Due to page restrictions, parts of our detailed analysis and some proofs had to be omitted. We refer the reader to the full version. However, a summary of our results can be seen in the following table.

PROBLEM	Competitive Ratio	MF	Lower bound MF
KNAPSACK/SUBSETSUM (fixed d)	$(1 - \epsilon)$	$O(1/\epsilon)$	$\Omega(1/\epsilon)$
2DGEOKNAPSACK	$(9/17 - \epsilon)$	$O(1/\epsilon)$	Open[1]
MIS (unweighted planar)	$(1 - \epsilon)$	$O(1/\epsilon)$	Open
MIS (unweighted unit-disk)	$(1 - \epsilon)$	$O(1/\epsilon)$	Open
MIS (weighted disk-like objects[2])	$(1 - \epsilon)$	$O(1/\epsilon)$	$\Omega(1/\epsilon)$
MIS on pseudo-disks	$(1 - \epsilon)$	$O(1/\epsilon)$	Open
MULTIPLEKNAPSACK	$(1 - \epsilon)$	$O(1/\epsilon)$	$\Omega(1/\epsilon)$
MIS (unw. planar, edge arrival)	$(1 - \epsilon)$	$O(1/\epsilon)$	Open

[1]We prove a lower bound of $\Omega(1/\epsilon)$ for comp. ratio $(1 - \epsilon)$.
[2]Result applies to all objects that the PTAS from Erlebach *et al.* [16] can be used on, even in higher dimensions.

2.4   Related Work

Upper Bounds: The general idea of bounded migration was introduced by Sanders, Sivadasan, and Skutella [29]. They developed an $(1 + \epsilon)$-competitive algorithm with non-amortized migration factor $f(1/\epsilon)$ for the MAKESPANSCHEDULING problem. Gálvez *et al.* [19] showed two $(c + \epsilon)$-competitive algorithms with migration factor $(1/\epsilon)^{O(1)}$ for some constants c for the same problem. Skutella and Verschae [30] were able to transfer the $(1 + \epsilon)$-competitive algorithm also to the setting, where items depart. They also considered the MACHINECOVERING problem and obtained an $(1 + \epsilon)$-competitive algorithm with amortized migration factor $f(1/\epsilon)$. For BINPACKING, Epstein and Levin [12] presented a $(1 + \epsilon)$-competitive algorithm with non-amortized migration factor

$f(1/\epsilon)$. Jansen and Klein [24] were able to obtain a non-amortized migration factor of $(1/\epsilon)^{O(1)}$ for this problem, and Berndt *et al.* [4] showed that such a non-amortized migration factor is also possible for the scenario, where items can depart. Considering amortized migration, Feldkord *et al.* [17] presented a $(1 + \epsilon)$-competitive algorithm with migration factor $O(1/\epsilon)$ that also works for departing items. Epstein and Levin [13] investigated a multidimensional extension of BINPACKING problem, called HYPERCUBEPACKING where hypercubes are packed geometrically. They obtained an $(1 + \epsilon)$-competitive algorithm with worst-case migration factor $f(1/\epsilon)$. For the preemptive variant of MAKESPAN-SCHEDULING, Epstein and Levin [14] obtained an exact online algorithm with non-amortized migration factor $1 + 1/m$. Berndt *et al.* [3] studied the BINCOVERING problem with amortized migration factor and non-amortized migration factor and obtained $(1 + \epsilon)$-competitive algorithms and matching lower bounds, even if items can depart. Jansen *et al.* [25] developed a $(1 + \epsilon)$-competitive algorithm with amortized migration factor $f(1/\epsilon)$ for the STRIPPACKING problem. Finally, Berndt *et al.* [2] developed a framework similar to this work, but only for *minimization problems*, and showed that for a certain class of packing problems, any c-approximate algorithm can be combined with a suitable online algorithm to obtain a $(c+\epsilon)$-competitive algorithm with amortized migration factor $O(1/\epsilon)$.

Lower Bounds: Skutella and Verschae [30] showed that a non-amortized migration factor of $f(1/\epsilon)$ is not possible for any function f for MACHINECOVERING. Berndt *et al.* showed that a non-amortized migration factor of $\Omega(1/\epsilon)$ is needed [4] for BINPACKING, and Feldkord *et al.* [17] showed that this also holds for the amortized migration factor. Epstein and Levin [14] showed that exact algorithms for the makespan minimization problem on uniform machines and for identical machines in the restricted assignment setting have worst-case migration factor at least $\Omega(m)$.

Dynamic Algorithms: In the semi-online setting, there are several metrics that one tries to optimize. First of all, there is the competitive ratio measuring the quality of the solution. Second, in order to prevent that the online problem simply degrades to the offline problem, one needs to bound some resource. In the setting that we consider, we bound the amount of repacking possible, as such a repacking often comes with a high cost in practical applications. In an alternate approach, often called *dynamic algorithms*, we restrict the running time needed to update a solution, ideally to a sub-linear function (see e. g., the surveys [7,22]). Note that the amount of repacking used here can be arbitrarily high (using a suitable representation of the current solution). This setting has been also studied recently for KNAPSACK variants [6] and MAXIMUMINDEPENDENTSET variants [23]. There are also works aiming to combine both of the before mentioned approaches [21,27].

3   Upper Bounds for Choosing Problems

3.1   Framework for Choosing Problems

In this section, we will consider the aforementioned class of *choosing problems*. In general, a choosing problem is defined by a set of objects with some properties, and the objective is to select a subset of these objects with maximum profit, while potentially respecting some secondary constraints.

Definition 1. *Consider a problem Π where every instance $I \in \Pi$ is given by a set of objects, where each object $i \in I$ is assigned a fixed profit value p_i, and a set of feasible solutions $\mathrm{SOL}(I)$. We call Π a choosing problem if $\mathrm{SOL}(I) \subseteq \mathcal{P}(I)$, and the objective of some instance $I \in \Pi$ is to find a subset $S \in \mathrm{SOL}(I)$ while maximizing the total profit $\mathrm{PROFIT}(S) = \sum_{i \in S} p_i$. We further make the following two requirements for choosing problems that we will discuss in this paper:*

(i) For any feasible solution $S \in \mathrm{SOL}(I)$, we have that any subset $S' \subseteq S$ is also a feasible solution, i.e. $S' \in \mathrm{SOL}(I)$.

(ii) For any solution $S \in \mathrm{SOL}(I_t)$ for an instance I_t with respective follow-up instance $I_{t+1} = I_t \triangle \{i_{t+1}\}$, the solution $S' := S \backslash \{i_{t+1}\}$ stays feasible for I_{t+1}, i.e. $S' \in \mathrm{SOL}(I_{t+1})$.

For choosing problems, the profits of objects are their migration potential and costs; $\Delta(I_t) = p_i$, where i is the object added or removed at time t. Given two solutions $S_1, S_2 \subseteq I$, we further have that $\phi(S_1 \to S_2) = \mathrm{PROFIT}(S_1 \triangle S_2)$.

While we restrict the range of problems with these properties, it is necessary to do so since an adversary can enforce an unreasonably high migration factor for problems we excluded this way. In particular the first property guarantees that there is no low profit item with low migration potential added by the adversary which would allow for a completely new solution with high profit that we would need to switch to. The second property serves a similar purpose, as it prevents the adversary from adding any arbitrary items making our current solution infeasible.

We could approach choosing problems like Berndt *et al.* [2] and use a greedy online algorithm in conjunction with the best known offline algorithm. While this would also create good results, we want to show that an online algorithm is not even necessary. We propose instead the algorithm that computes an offline solution S with profit V and then waits until the total profit of items being added or removed from the instance exceeds ϵV. At this point we simply replace S with a new offline solution for the current instance. This algorithm already achieves a competitive ratio close to the approximation ratio of the offline algorithm.

Theorem 1. *Let A_{off} be an offline algorithm with an approximation ratio of α. Then the resulting framework using A_{off} is a $(1 - \epsilon) \cdot \alpha$-competitive algorithm requiring a migration factor of $O(1/(\epsilon \cdot \alpha))$.*

Proof. In the following, we refer to time steps where the offline algorithm is applied as repacking times. Note that for the start of any online instance and the first arriving items, we will just add the first eligible item with the offline algorithm when it arrives. This results in the first repacking time. Consider now any repacking time t with generated solution S_t, and let t' be the first point of time where $\Delta_{t:t'} > \epsilon\text{PROFIT}(S_t)$. First we will show that the migration factor of the applied framework is small. We do so by proving that the migration potential between two repacking times accommodates for the repacking at the later repacking time. Denote with $A_{t:t'}$ the total profit of items that arrived up until time t'.

We now have that

$$
\begin{aligned}
\frac{\phi(S_t \to S_{t'})}{\Delta_{t:t'}} &\leq \frac{\text{OPT}(I_t) + \text{OPT}(I_{t'})}{\Delta_{t:t'}} \leq \frac{\text{OPT}(I_t) + \text{OPT}(I_t) + A_{t:t'}}{\Delta_{t:t'}} \\
&\leq \frac{\text{OPT}(I_t) + \text{OPT}(I_t) + \Delta_{t:t'}}{\Delta_{t:t'}} \leq \frac{2 \cdot \text{OPT}(I_t)}{\Delta_{t:t'}} + 1 \leq \frac{2 \cdot \text{PROFIT}(S_t)}{\alpha\Delta_{t:t'}} + 1 \\
&< \frac{2 \cdot \text{PROFIT}(S_t)}{\alpha\epsilon\text{PROFIT}(S_t)} + 1 \in O(1/(\epsilon \cdot \alpha)).
\end{aligned}
$$

Now it is left to show that up until time $t' - 1$ we have maintained a competitive ratio of $(1 - \epsilon)\alpha$. Denote with $A_{t:t'-1}$ and $R_{t:t'-1}$ the total profit of items that arrived or departed up until time $t' - 1$, and note that by choice of t', we have that $A_{t:t'-1} + R_{t:t'-1} \leq \epsilon\text{PROFIT}(S_t)$. We now have

$$
\begin{aligned}
\text{PROFIT}(S_{t'-1}) &\geq \text{PROFIT}(S_t) - R_{t:t'-1} \\
&= \text{PROFIT}(S_t) + A_{t:t'-1} - A_{t:t'-1} - R_{t:t'-1} \geq \text{PROFIT}(S_t) + A_{t:t'-1} - \epsilon\text{PROFIT}(S_t) \\
&= (1 - \epsilon)\text{PROFIT}(S_t) + A_{t:t'-1} \geq (1 - \epsilon)\alpha\text{OPT}(S_t) + A_{t:t'-1} \\
&\geq (1 - \epsilon)\alpha(\text{OPT}(S_t) + A_{t:t'-1}) \geq (1 - \epsilon)\alpha(\text{OPT}(S_{t'-1})).
\end{aligned}
$$

Hence, our algorithm obtains a competitive ratio of $(1 - \epsilon)\alpha$.   □

3.2   Resulting Upper Bounds

The framework introduced can be applied to any choosing problem with some existing offline algorithm. We observe that by using Theorem 1 for problems that admit a PTAS, we get a respective robust PTAS, and for other problems, we achieve the ratio of any offline algorithm with small additional error. We note without proof that all problems mentioned in the following theorem are choosing problems as solutions are made up of sets of items or nodes and they stay feasible under the required circumstances. For a more detailed recap on these problems we refer to the full version.

Theorem 2. *For the following problems there exists an online algorithm with competitive ratio $1 - \epsilon$ and migration factor $O(1/\epsilon)$.*

- SUBSETSUM *and* KNAPSACK *using the FPTAS by Jin* [26]
- *d-dimensional* KNAPSACK *using the PTAS from Caprara* et al. [8]

- MAXIMUMINDEPENDENTSET *on unweighted planar graphs by using the PTAS by Baker* [1]
- MAXIMUMINDEPENDENTSET *on (weighted) d-dimensional disk-like objects with fixed d using the PTAS by Erlebach* et al. [16]
- MAXIMUMINDEPENDENTSET *on pseudo-disks using a PTAS by Chan and Har-Peled* [9]

Additionally, for the 2DGEOKNAPSACK *problem, there exists an online algorithm with a competitive ratio* $9/17 - \epsilon$ *and migration factor* $O(1/\epsilon)$ *by using the* $9/17$ *approximation algorithm from Gálvez* et al. [18]

Proof. We prove that a $(1 - \epsilon)$ approximation yields the desired result. The statement for 2DGEOKNAPSACK follows similarly. Note that using a $(1 - \epsilon)$ approximation with Theorem 1, we achieve an online algorithm with a competitive ratio of $(1 - \epsilon)^2 = 1 - 2\epsilon + \epsilon^2$. By applying the framework with $\epsilon' = 1/2\epsilon$, we achieve the desired PTAS quality. The required migration of our framework is bounded by $O(\frac{1}{\epsilon(1-\epsilon)}) = O(1/\epsilon)$. □

We also show how to apply our framework in a setting outside of choosing problems. We consider two problems still similar to the given context of KNAPSACK and MAXIMUMINDEPENDENTSET.

Theorem 3. *The static cases of* MAXIMUMINDEPENDENTSET *on weighted planar graphs when adding edges and* MULTIPLEKNAPSACK *both admit a robust PTAS with competitive ratio* $1 - \epsilon$ *and migration factor* $O(1/\epsilon)$.

3.3 Lower Bounds on Migration

We now want to complement these positive results by showing that the amount of migration these algorithms use, is also required. In the dynamic setting of online problems the adversary is quite powerful because removing objects only leaves the option to repack our solution while the range of possibilities has become smaller. This scenario is especially difficult when our current solution becomes inefficient and we might have to change the whole leftover solution in order to meet our competitive ratio. We want to create a scenario where the adversary forces our algorithm to switch between two solutions. For that we want to use two instances that we call *alternating instances*:

Definition 2. *Consider some dynamic online choosing problem* Π_{on}, *two instances* I_1 *and* I_2, *and some desired competitive ratio* $\beta < 1$. *We call the instances* I_1 *and* I_2 *alternating instances when there exists* $I_1' \subset I_1$, *such that the following properties hold:*

- *The solutions* $S_1 = I_1$ *and* $S_2 = I_2$ *are feasible and hence also* $S_1' = I_1'$ *is feasible, but any solution* S *with* $S \cap I_1 \neq \emptyset$ *and* $S \cap I_2 \neq \emptyset$ *is infeasible.*
- *We have that* $\mathrm{PROFIT}(I_1') < \beta\mathrm{PROFIT}(I_2) < \beta^2\mathrm{PROFIT}(I_1)$.

We will use these alternating instance and show how to gain lower bounds of migration in the following lemma.

Lemma 1. *Let Π_{on} be a dynamic online choosing problem and consider two alternating instances I_1, I_2 for a desired competitive ratio $\beta < 1$ with $I_1' \subseteq I_2$ fulfilling the requirements of Definition 2. When we additionally have that $\beta\text{PROFIT}(I_2) \in \Omega(\gamma)$ for some desired migration factor γ and $\text{PROFIT}(I_1) - \text{PROFIT}(I_1') \leq c$ for some c, then any β-competitive algorithm requires a migration factor of $\Omega(\frac{\gamma}{2(c+1)})$.*

Proof. Consider some β-competitive algorithm A and the following order of events: First add all items from I_1 and A will generate some approximate solution S_1 for I_1. Now, add also all items from I_2 and note that A will not change anything, since we have by definition that $\text{PROFIT}(I_2) < \beta\text{PROFIT}(I_1)$ and adding items from I_2 would make S_1 infeasible.

We now proceed to remove and add again all items of $I_1 \setminus I_1'$ and repeat this N times for some large $N \in \mathbb{N}$. By removing all these items the previous solution S_1 would consequently be reduced to a solution $S_1' \subseteq I_1'$, and as $\text{PROFIT}(I_1') < \beta\text{PROFIT}(I_2)$, our algorithm needs to change to a solution $S_2 \subseteq I_2$. When the items are added again, the algorithm switches from S_2 to a solution of I_1.

Let us now look at the necessary migration and the migration potential. The total migration potential we received is given through the arrival of all items and the repeated removal and re-adding of items and altogether we have migration potential of $\text{PROFIT}(I_1) + \text{PROFIT}(I_2) + N(\text{PROFIT}(I_1) - \text{PROFIT}(I_1')) = \text{PROFIT}(I_1) + \text{PROFIT}(I_2) + 2Nc$. The necessary migration results from the repacking of the solutions of I_1' and I_2. We have to note however, that the necessary migration for the solutions of I_1' might be small or even 0, when the approximate solution S_1 for I_1 does not use any items of I_1'. We know however that for I_2 we at least exchange a full approximate solution which yields a total necessary migration of at least $N\beta\text{PROFIT}(I_2)$. For the migration factor, we now have that:

$$\frac{N\beta\text{PROFIT}(I_2)}{\text{PROFIT}(I_1) + \text{PROFIT}(I_2) + 2Nc} \geq \frac{\beta\text{PROFIT}(I_2)}{2(c+1)}$$

when N is chosen large enough. $\qquad\square$

Using this approach we are actually able to prove matching lower bounds for the central choosing problems. For the proofs and an even more detailed analysis on lower bounds of migration we refer the reader to the full version.

Theorem 4. *For the online SUBSETSUM problem and (weighted) MAXIMUMIN-DEPENDENTSET problem, there is an instance such that the migration needed for a solution with value $(1 - \epsilon)\text{OPT}$ is $\Omega(1/\epsilon)$.*

4   Conclusion

In this paper, we present a general framework to transfer approximation algorithms for many maximization problems to the semi-online setting with bounded migration. Furthermore, we show that the algorithms constructed this way achieve optimal migration. We expect our framework to be also applicable to other problems such as 2DGEOKNAPSACK variants with more complex objects [20,28], 3DGEOKNAPSACK [11], DYNAMICMAPLABELING [5], THROUGHPUTSCHEDULING [10], or MAXEDGEDISJOINTPATHS [15].

References

1. Baker, B.S.: Approximation algorithms for NP-complete problems on planar graphs. In: 24th Annual Symposium on Foundations of Computer Science, pp. 265–273. IEEE (1983)
2. Berndt, S., Dreismann, V., Grage, K., Jansen, K., Knof, I.: Robust online algorithms for certain dynamic packing problems. In: Bampis, E., Megow, N. (eds.) WAOA 2019. LNCS, vol. 11926, pp. 43–59. Springer, Cham (2020). https://doi.org/10.1007/978-3-030-39479-0_4
3. Berndt, S., Epstein, L., Jansen, K., Levin, A., Maack, M., Rohwedder, L.: Online bin covering with limited migration. In: Bender, M.A., Svensson, O., Herman, G. (eds.) 27th Annual European Symposium on Algorithms, ESA 2019. LIPIcs, Munich/Garching, Germany, 9–11 September 2019, vol. 144, pp. 18:1–18:14. Schloss Dagstuhl - Leibniz-Zentrum für Informatik (2019)
4. Berndt, S., Jansen, K., Klein, K.-M.: Fully dynamic bin packing revisited. Math. Program., 109–155 (2018). https://doi.org/10.1007/s10107-018-1325-x
5. Bhore, S., Li, G., Nöllenburg, M.: An algorithmic study of fully dynamic independent sets for map labeling. In: ESA. LIPIcs, vol. 173, pp. 19:1–19:24. Schloss Dagstuhl - Leibniz-Zentrum für Informatik (2020)
6. Böhm, M., et al.: Fully dynamic algorithms for knapsack problems with polylogarithmic update time. CoRR arXiv:2007.08415 (2020)
7. Boria, N., Paschos, V.T.: A survey on combinatorial optimization in dynamic environments. RAIRO Oper. Res. **45**(3), 241–294 (2011)
8. Caprara, A., Kellerer, H., Pferschy, U., Pisinger, D.: Approximation algorithms for knapsack problems with cardinality constraints. Eur. J. Oper. Res. **123**(2), 333–345 (2000)
9. Chan, T.M., Har-Peled, S.: Approximation algorithms for maximum independent set of pseudo-disks. In: Proceedings of the 25th Annual Symposium on Computational Geometry, SCG 2009, pp. 333–340. Association for Computing Machinery (2009)
10. Cieliebak, M., Erlebach, T., Hennecke, F., Weber, B., Widmayer, P.: Scheduling with release times and deadlines on a minimum number of machines. In: Levy, J.-J., Mayr, E.W., Mitchell, J.C. (eds.) TCS 2004. IIFIP, vol. 155, pp. 209–222. Springer, Boston, MA (2004). https://doi.org/10.1007/1-4020-8141-3_18
11. Diedrich, F., Harren, R., Jansen, K., Thöle, R., Thomas, H.: Approximation algorithms for 3D orthogonal knapsack. J. Comput. Sci. Technol. **23**(5), 749–762 (2008)
12. Epstein, L., Levin, A.: A robust APTAS for the classical bin packing problem. Math. Program. **119**(1), 33–49 (2009)
13. Epstein, L., Levin, A.: Robust approximation schemes for cube packing. SIAM J. Optim. **23**(2), 1310–1343 (2013)
14. Epstein, L., Levin, A.: Robust algorithms for preemptive scheduling. Algorithmica **69**(1), 26–57 (2014)
15. Erlebach, T., Jansen, K.: The maximum edge-disjoint paths problem in bidirected trees. SIAM J. Discret. Math. **14**(3), 326–355 (2001)

16. Erlebach, T., Jansen, K., Seidel, E.: Polynomial-time approximation schemes for geometric intersection graphs. SIAM J. Comput. **34**(6), 1302–1323 (2005)
17. Feldkord, B., et al.: Fully-dynamic bin packing with little repacking. In: Proceedings of the ICALP, pp. 51:1–51:24 (2018)
18. Galvez, W., Grandoni, F., Heydrich, S., Ingala, S., Khan, A., Wiese, A.: Approximating geometric knapsack via l-packings. In: 2017 IEEE 58th Annual Symposium on Foundations of Computer Science (FOCS), Los Alamitos, CA, USA, pp. 260–271. IEEE Computer Society (October 2017)
19. Gálvez, W., Soto, J.A., Verschae, J.: Symmetry exploitation for online machine covering with bounded migration. In: Proceedings of the ESA, pp. 32:1–32:14 (2018)
20. Grandoni, F., Kratsch, S., Wiese, A.: Parameterized approximation schemes for independent set of rectangles and geometric knapsack. In: ESA. LIPIcs, vol. 144, pp. 53:1–53:16. Schloss Dagstuhl - Leibniz-Zentrum für Informatik (2019)
21. Gupta, A., Krishnaswamy, R., Kumar, A., Panigrahi, D.: Online and dynamic algorithms for set cover. In: STOC, pp. 537–550. ACM (2017)
22. Henzinger, M.: The state of the art in dynamic graph algorithms. In: Tjoa, A.M., Bellatreche, L., Biffl, S., van Leeuwen, J., Wiedermann, J. (eds.) SOFSEM 2018. LNCS, vol. 10706, pp. 40–44. Springer, Cham (2018). https://doi.org/10.1007/978-3-319-73117-9_3
23. Henzinger, M., Neumann, S., Wiese, A.: Dynamic approximate maximum independent set of intervals, hypercubes and hyperrectangles. In: Symposium on Computational Geometry. LIPIcs, vol. 164, pp. 51:1–51:14. Schloss Dagstuhl - Leibniz-Zentrum für Informatik (2020)
24. Jansen, K., Klein, K.: A robust AFPTAS for online bin packing with polynomial migration. In: Proceedings of the ICALP, pp. 589–600 (2013)
25. Jansen, K., Klein, K., Kosche, M., Ladewig, L.: Online strip packing with polynomial migration. In: Proceedings of the APPROX-RANDOM, pp. 13:1–13:18 (2017)
26. Jin, C.: An improved FPTAS for 0–1 knapsack. In: Baier, C., Chatzigiannakis, I., Flocchini, P., Leonardi, S. (eds.) 46th International Colloquium on Automata, Languages, and Programming, ICALP 2019. Leibniz International Proceedings in Informatics (LIPIcs), Dagstuhl, Germany, vol. 132, pp. 76:1–76:14. Schloss Dagstuhl-Leibniz-Zentrum fuer Informatik (2019)
27. Lacki, J., Ocwieja, J., Pilipczuk, M., Sankowski, P., Zych, A.: The power of dynamic distance oracles: efficient dynamic algorithms for the Steiner tree. In: STOC, pp. 11–20. ACM (2015)
28. Merino, A.I., Wiese, A.: On the two-dimensional knapsack problem for convex polygons. In: ICALP. LIPIcs, vol. 168, pp. 84:1–84:16. Schloss Dagstuhl - Leibniz-Zentrum für Informatik (2020)
29. Sanders, P., Sivadasan, N., Skutella, M.: Online scheduling with bounded migration. Math. Oper. Res. **34**(2), 481–498 (2009)
30. Skutella, M., Verschae, J.: Robust polynomial-time approximation schemes for parallel machine scheduling with job arrivals and departures. Math. Oper. Res. **41**(3), 991–1021 (2016)

On the Degrees of Constructively Immune Sets

Samuel D. Birns and Bjørn Kjos-Hanssen$^{(\boxtimes)}$ (iD)

University of Hawai'i at Mānoa, Honolulu, HI 96822, USA
{sbirns,bjoern.kjos-hanssen}@hawaii.edu
http://math.hawaii.edu/wordpress/bjoern/

Abstract. Xiang Li (1983) introduced what are now called constructively immune sets as an effective version of immunity. Such have been studied in relation to randomness and minimal indices, and we add another application area: numberings of the rationals. We also investigate the Turing degrees of constructively immune sets and the closely related Σ_1^0-dense sets of Ferbus-Zanda and Grigorieff (2008).

Keywords: Constructively immune · Turing degrees · Theory of numberings

1 Introduction

Effectively immune sets, introduced by Smullyan in 1964 [12], are well-known in computability as one of the incarnations of diagonal non-computability, first made famous by Arslanov's completeness criterion. A set $A \subseteq \omega$ is *effectively immune* if there is a computable function h such that $|W_e| \leq h(e)$ whenever $W_e \subseteq A$, where $\{W_e\}_{e \in \omega}$ is a standard enumeration of the computably enumerable (c.e.) sets.

There is a more obvious effectivization of immunity (the lack of infinite computable subsets), however: *constructive immunity*, introduced by Xiang Li [8] who actually (and inconveniently) called it "effective immunity".

Definition 1. *A set A is* constructively immune *if there exists a partial recursive ψ such that for all x, if W_x is infinite then $\psi(x) \downarrow$ and $\psi(x) \in W_x \setminus A$.*

The Turing degrees of constructively immune sets and the related Σ_1^0-dense sets have not been considered before in the literature, except that Xiang Li implicitly showed that they include all c.e. degrees. We prove in Sect. 3 that the Turing degrees of Σ_1^0-dense sets include all non-Δ_2^0 degrees, all high degrees, and all c.e. degrees. We do not know whether they include *all* Turing degrees.

The history of the study of constructive immunity seems to be easily summarized. After Xiang Li's 1983 paper, Odifreddi's 1989 textbook [9] included

This work was partially supported by a grant from the Simons Foundation (#704836 to Bjørn Kjos-Hanssen).

L. De Mol et al. (Eds.): CiE 2021, LNCS 12813, pp. 50–59, 2021.
https://doi.org/10.1007/978-3-030-80049-9_5

Li's results as exercises, and Calude's 1994 monograph [2] showed that the set $RAND_t^C = \{x : C(x) \geq |x| - t\}$ is constructively immune, where C is Kolmogorov complexity. Schafer 1997 [11] further developed an example involving minimal indices, and Brattka 2002 [1] gave one example in a more general setting than Cantor space. Finally in 2008 Ferbus-Zanda and Grigorieff proved an equivalence with constructive Σ_1^0-density.

Definition 2 (Ferbus-Zanda and Grigorieff [6]). *A set $A \subseteq \omega$ is Σ_1^0-dense if for every infinite c.e. set C, there exists an infinite c.e. set D such that $D \subseteq C$ and $D \subseteq A$.*

If there is a computable function $f : \omega \to \omega$ such that for each W_e, $W_{f(e)} \subseteq A \cap W_e$, and $W_{f(e)}$ is infinite if W_e is infinite, then A is constructively Σ_1^0-dense.

We should note that while the various flavors of immune sets are always infinite by definition, Ferbus-Zanda and Grigorieff do not require Σ_1^0-dense sets to be co-infinite.

The Σ_1^0-dense sets form a natural Π_4^0 class in 2^ω that coincides with the simple sets on Δ_2^0 but is prevalent (in fact exists in every Turing degree) outside of Δ_2^0 by Theorem 8 below.

2 Σ_1^0-density

To show that there exists a set that is Σ_1^0-dense, but not constructively so, we use Mathias forcing. A detailed treatment of the computability theory of Mathias forcing can be found in [3].

Definition 3. *A Mathias condition is a pair (d, E) where $d, E \subseteq \omega$, d is a finite set, E is an infinite computable set, and $\max(d) < \min(E)$. A condition (d_2, E_2) extends a condition (d_1, E_1) if*

- *$d_1 = d_2 \cap (\max d_1 + 1)$, i.e., d_1 is an initial segment of d_2,*
- *E_2 is a subset of E_1, and*
- *d_2 is contained in $d_1 \cup E_1$.*

A set A is Mathias generic *if it is generic for Mathias forcing.*

Theorem 1. *If A is Mathias generic, then*

1. $\omega \backslash A$ is Σ_1^0-dense.
2. $\omega \backslash A$ is not constructively Σ_1^0-dense.

Proof. 1. Let W_e be an infinite c.e. set. Let (d, E) be a Mathias condition.

Case (i): $E \cap W_e$ is finite. Then for any Mathias generic A extending the condition (d, E), $\omega \backslash A$ contains an infinite subset of W_e, in fact a set of the form $W_e \backslash F$ where F is finite.

Case (ii): $E \cap W_e$ is infinite. Then $E \cap W_e$ is c.e., hence has an infinite computable subset D. Write $D = D_1 \cup D_2$ where D_1, D_2 are disjoint infinite c.e. sets. The condition (d, D_1) extends (d, E) and forces a Mathias generic A

extending it to be such that $\omega \backslash A$ has an infinite subset in common with W_e, namely D_2.

We have shown that for each infinite c.e. set W_e, each Mathias condition has an extension forcing the statement that a Matias generic A satisfies

$$\omega \backslash A \text{ has an infinite c.e. subset in common with } W_e. \qquad (*)$$

Thus by standard forcing theory it follows that each Mathias generic satisfies $(*)$.

2. Let f be a computable function. It suffices to show that for each Mathias generic A, there exists an i such that W_i is infinite and $W_{f(i)}$ is either finite, or not a subset of W_i, or not a subset of $\overline{A}$. For this, as in (1) above it suffices to show that for each condition (d, D) there exists a condition (d', E') extending (d, E) and an i such that W_i is infinite and $W_{f(i)}$ is either finite, or not a subset of W_i, or not a subset of $\overline{A}$ for any A extending (d', D').

Let (d, E) be a Mathias condition and write $D = W_i$. If $W_{f(i)}$ is finite or not a subset of W_i then we are done. Otherwise there exists a condition (d', E') extending (d, E) such that $E' \cap W_{f(i)}$ is nonempty. This can be done by a finite extension (making only finitely many changes to the condition).

Theorem 2 ([6, Proposition 3.3]). *A set $Z \subseteq \omega$ is constructively immune if and only if it is infinite and $\omega \backslash Z$ is constructively Σ_1^0-dense.*

Since Ferbus-Zanda and Grigorieff's paper has not gone through peer review, we provide the proof.

Proof. $\Leftarrow$: Let the function g witness that $\omega \backslash Z$ is constructively Σ_1^0-dense. Define a partial recursive function φ by stipulating that $\varphi(i)$ is the first number in the enumeration of $W_{g(i)}$, if any.

$\Rightarrow$: Define a partial recursive function $\mu(i, n)$ by

- $\mu(i, 0) = \varphi(i)$;
- $\mu(i, n+1) = \varphi(i_n)$, where i_n is such that $W_{i_n} = W_i \backslash \{\mu(i, m) : m \leq n\}$.

Let g be total recursive so that $W_{g(i)} = \{\mu(i, m) : m \in \omega\}$. If W_i is infinite then all $\mu(i, m)$'s are defined and distinct and belong to $W_i \cap Z$. Thus, $W_{g(i)}$ is an infinite subset of $W_i \cap Z$.

Recall that a c.e. set is *simple* if it is co-immune.

Theorem 3 (Xiang Li [8]). *Let A be a set and let $\{\phi_x\}_{x \in \omega}$ be a standard enumeration of the partial computable functions.*

1. *If A is constructively immune then A is immune and $\overline{A}$ is not immune.*
2. *If A is simple then $\overline{A}$ is constructively immune.*
3. *$\{x : (\forall y)(\phi_x = \phi_y \rightarrow x \leq y)\}$ is constructively immune.*

2.1   Numberings

A *numbering* of a countable set $\mathcal{A}$ is an onto function $\nu : \omega \to \mathcal{A}$. The theory of numberings has a long history [5]. Numberings of the set of rational numbers $\mathbb{Q}$ provide an application area for Σ_1^0-density. Rosenstein [10, Sect. 16.2: Looking at $\mathbb{Q}$ effectively] discusses computable dense subsets of $\mathbb{Q}$. Here we are mainly concerned with noncomputable sets.

Proposition 1. *Let $A \subseteq \omega$. The following are equivalent:*

1. *$\nu(A)$ is dense for every injective computable numbering ν of $\mathbb{Q}$;*
2. *A is co-immune.*

Proof. (1) $\implies$ (2): We prove the contrapositive. Suppose $\overline{A}$ contains an infinite c.e. set W_e. Consider a computable numbering ν that maps W_e onto $[0, 1] \cap \mathbb{Q}$. Then $\nu(A)$ is disjoint from $[0, 1]$ and hence not dense.

(2) $\implies$ (1): We again prove the contrapositive. Assume that $\nu(A)$ is not dense for a certain computable ν. Let $\{x_n : n \in \omega\}$ be a converging infinite sequence of rationals disjoint from $\nu(A)$. Then $\{\nu^{-1}(x_n) : n \in \omega\}$ is an infinite c.e. subset of $\overline{A}$.

Definition 4. *A subset A of $\mathbb{Q}$ is* co-nowhere dense *if for each interval $[a, b] \subseteq \mathbb{Q}$, $[a', b'] \subseteq A$ for some $[a', b'] \subseteq [a, b]$.*

Proposition 2. *A set is co-nowhere dense under every numbering iff it is co-finite.*

Proof. Only the forward direction needs to be proven; the other direction is immediate. Let A be a co-infinite set, and define ν by letting ν map $\omega \backslash A$ onto $[0, 1]$. Then A is not co-nowhere dense.

Proposition 3. *A is infinite and non-immune iff there exists a computable numbering with respect to which A is co-nowhere dense.*

Proof. Let A be infinite and not immune. Thus, there is an infinite $W_e \subseteq A$ for some e. Let ν be a computable numbering that maps W_e onto $\mathbb{Q} \backslash \omega$. Then A is co-nowhere dense under ν.

Conversely, let A be co-nowhere dense under some computable numbering ν. Then $\nu^{-1}([a, b])$ is an infinite c.e. subset of A for some suitable a, b.

A set $D \subseteq \mathbb{Q}$ is *effectively dense* if there is a computable function $f(a, b)$ giving an element of $D \cap (a, b)$ for $a < b \in \mathbb{Q}$.

Proposition 4. *A set A is constructively Σ_1^0-dense iff it is effectively dense for all computable numberings.*

Proof. By Theorem 2, A is constructively Σ_1^0-dense iff it is infinite and $\omega \backslash A$ is constructively immune. Constructive immunity of $\omega \backslash A$ implies effective density of A since the witnessing function for constructive immunity can be used to witness effective density. For the converse we exploit the assumption that we get to choose a suitable ν.

Let A and B be sets, with B computable. We say that A is *co-immune within* B if there is no infinite computable subset of $A^c \cap B$. The following diagram includes some claims not proved in the paper, whose proof (or disproof) may be considered enjoyable exercises. The quantifiers $\exists \nu$, $\forall \nu$ range over computable numberings of $\mathbb{Q}$.

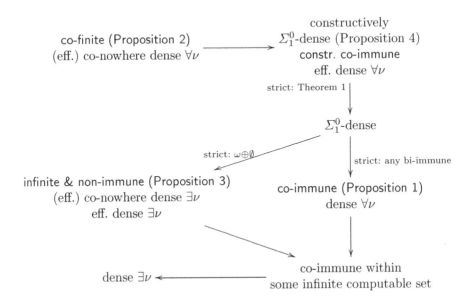

3   Prevalence of Σ_1^0-density

In this section we investigate the existence of Σ_1^0-density in the Turing degrees at large.

3.1   Closure Properties and Σ_1^0-density

Proposition 5. *1. The intersection of two Σ_1^0-dense sets is Σ_1^0-dense.*
2. The intersection of two constructively Σ_1^0-dense sets is constructively Σ_1^0-dense.

Proof. Let A and B be Σ_1^0-dense sets. Let W_e be an infinite c.e. set. Since A is Σ_1^0-dense, there exists an infinite c.e. set $W_d \subseteq A \cap W_e$. Since B is Σ_1^0-dense, there exists an infinite c.e. set $W_a \subseteq B \cap W_d$. Then $W_a \subseteq (A \cap B) \cap W_e$, as desired. This proves (1). To prove (2), let f and g witness the effective Σ_1^0-density of A and B, respectively. Given W_e, we have $W_{f(e)} \subseteq A \cap W_e$ and then

$$W_{g(f(e))} \subseteq B \cap W_{f(e)} \subseteq A \cap B \cap W_e.$$

In other words, $g \circ f$ witnesses the effective Σ_1^0-density of $A \cap B$.

Corollary 1. *Bi-Σ_1^0-dense sets do not exist.*

Proof. If A and A^c are both Σ_1^0-dense then by Proposition 5, $A \cap A^c$ is Σ_1^0-dense, which is a contradiction.

For sets A and B, $A \subseteq^* B$ means that $A \backslash B$ is a finite set.

Proposition 6. *1. If A is Σ_1^0-dense and $A \subseteq^* B$, then B is Σ_1^0-dense.*
2. If A is constructively Σ_1^0-dense and $A \subseteq^* B$, then B is constructively Σ_1^0-dense.

Proof. Let W_e be an infinite c.e. set. Since A is Σ_1^0-dense, there exists an infinite c.e. set W_d such that $W_d \subseteq A \cap W_e$. Let $W_c = W_d \backslash (A \backslash B)$. Since $A \backslash B$ is finite, W_c is an infinite c.e. set. Since $W_d \subseteq A$, we have $W_c = W_d \cap (B \cup A^c) = W_d \cap B$. Then, since $W_d \subseteq W_e$, we have $W_c \subseteq B \cap W_e$, and we conclude that B is Σ_1^0-dense. This proves (1). To prove (2), if f witnesses that A is constructively Σ_1^0-dense then a function g with $W_{g(e)} = W_{f(e)} \backslash (A \backslash B)$ witnesses that B is constructively Σ_1^0-dense.

Proposition 7. *Let B be a co-finite set. Then B is constructively Σ_1^0-dense.*

Proof. The set ω is constructively Σ_1^0-dense as witnessed by the identity function $f(e) = e$. Thus by Item 2 of Proposition 6, B is as well.

As usual we write $A \oplus B = \{2x \mid x \in A\} \cup \{2x + 1 \mid x \in B\}$.

Proposition 8. *1. If X_0 and X_1 are Σ_1^0-dense sets then so is $X_0 \oplus X_1$.*
2. If X_0 and X_1 are constructively Σ_1^0-dense sets then so is $X_0 \oplus X_1$.

Proof. Let $W_e = W_{c_0} \oplus W_{c_1}$ be an infinite c.e. set. For $i = 0, 1$, since X_i is Σ_1^0-dense there exists $W_{d_i} \subseteq X_i \cap W_{c_i}$ such that W_{d_i} is infinite if W_{c_i} is infinite. Then $W_{d_0} \oplus W_{d_1}$ is an infinite c.e. subset of $(X_0 \oplus X_1) \cap W_e$.

This proves (1). To prove (2), if d_i are now functions witnessing the effective Σ_1^0-density of X_i then $W_{d_i(c_i)} \subseteq X_i \cap W_{c_i}$, and $W_{d_0(c_0)} \oplus W_{d_1(c_1)}$ is an infinite c.e. subset of $(X_0 \oplus X_1) \cap W_e$. Thus a function g satisfying

$$W_{g(e)} = W_{d_0(c_0)} \oplus W_{d_1(c_1)},$$

where $W_e = W_{c_0} \oplus W_{c_1}$, witnesses the effective Σ_1^0-density of $X_0 \oplus X_1$.

Theorem 4. *There is no Σ_1^0-dense set A such that all Σ_1^0-dense sets B satisfy $A \subseteq^* B$.*

Proof. Suppose there is such a set A. Let W_d be an infinite computable subset of A. Let G be a Mathias generic with $G \cap W_d^c = \emptyset$, i.e., $G \subseteq W_d$. Then $B := G^c$ is Σ_1^0-dense by Theorem 1. Thus $A \cap G^c$ is also Σ_1^0-dense by Proposition 5. And $G \subseteq W_d \subseteq A$ and by assumption $A \subseteq^* G^c$ so we get $G \subseteq^* G^c$, a contradiction.

These results show that the Σ_1^0-dense sets under $\subseteq^*$ form a non-principal filter whose Turing degrees form a join semi-lattice.

Theorem 5. *Let A be a c.e. set. The following are equivalent:*

1. *A is co-infinite and constructively Σ_1^0-dense.*
2. *A is co-infinite and Σ_1^0-dense.*
3. *A is co-immune.*

Proof. $1 \implies 2 \implies 3$ is immediate from the definitions, and $3 \implies 1$ is immediate from Theorem 2 and Theorem 3.

Theorem 6. *Every c.e. Turing degree contains a constructively Σ_1^0-dense set.*

Proof. Let $\mathbf{a}$ be a c.e. degree. If $\mathbf{a} > \mathbf{0}$ then $\mathbf{a}$ contains a simple set A, see, e.g., [13], so Theorem 5 finishes this case. The degree $\mathbf{0}$ contains all the co-finite sets, which are constructively Σ_1^0-dense by Proposition 7.

3.2   Cofinality in the Turing Degrees of Constructive Σ_1^0-density

Definition 5. *For $k \geq 0$, let I_k be intervals of length $k+2$ such that $\min(I_0) = 0$ and $\max(I_k) + 1 = \min(I_{k+1})$.*

Let $V_e = \bigcup_{s \in \omega} V_{e,s}$ be a subset of W_e defined by the condition that $x \in I_k$ enters V_e at a stage s where x enters W_e if this makes $|V_{e,s} \cap I_k| \leq 1$, and for all $j > k$, $V_{j,s} \cap I_k = \emptyset$.

Lemma 1. *There exists a c.e., co-infinite, constructively Σ_1^0-dense, and effectively co-immune set.*

Proof. Let $A = \bigcup_{e \in \omega} V_e$. V_e is c.e. by construction, and if W_e is infinite, V_e is also infinite. So $V_e = W_{f(e)}$ is the set witnessing that A is constructively Σ_1^0-dense.

Moreover A is coinfinite since $|A \cap I_k| \leq k + 1 < k + 2 = |I_k|$ gives $I_k \not\subseteq A$ for each k and

$$|\omega \setminus A| = \left| \left(\bigcup_{k \in \omega} I_k \right) \setminus A \right| = \left| \bigcup_{k \in \omega} (I_k \setminus A) \right| = \sum_{k \in \omega} |I_k \setminus A| \geq \sum_{k \in \omega} 1 = \infty.$$

The set A is effectively co-immune because if W_e is disjoint from A then since as soon as a number in I_k for $k \geq e$ enters W_e then that number is put into A, $W_e \subseteq \bigcup_{k < e} I_k$ so $|W_e| \leq \sum_{k < e} (k + 2) = \sum_{k \leq e+1} k = \frac{(e+1)(e+2)}{2}$.

Theorem 7. *For each set R there exists a constructively Σ_1^0-dense, effectively co-immune set S with $R \leq_T S$.*

Proof. Let R be any set, which we may assume is co-infinite. Let A be as in the proof of Lemma 1. Let $S \supseteq A$ be defined by

$$S = A \cup \bigcup_{k \in R} I_k.$$

Since $A \subseteq S$ and S is co-infinite, S is constructively Σ_1^0-dense and effectively co-immune. Since $k \in R \iff I_k \subseteq S$, we have $R \leq_T S$.

3.3   Non-Δ_2^0 Degrees

Lemma 2. *Suppose that $T \subseteq 2^{<\omega}$ is a tree with only one infinite path. Then for each length n there exists a length $m > n$ such that exactly one string of length n has an extension of length m in T.*

Proof. Suppose not, i.e., there is a length n such that for all $m > n$ there are at least two strings σ_m, τ_m of length n with extensions of length m in T. By the pigeonhole principle there is a pair (σ, τ) that is a choice of (σ_m, τ_m) for infinitely many m. Then by compactness both σ and τ must be extendible to infinite paths of T.

Lemma 3. *Suppose that $T \subseteq 2^{<\omega}$ is a tree with only one infinite path A, and that T is a c.e. set of strings. Then A is Δ_2^0.*

Proof. By Lemma 2, for each length n there exists a length $m > n$ such that exactly one string of length n has an extension of length m in T. Using $0'$ as an oracle we can find that m and define $A \restriction n$ by looking for such a string. In fact, $T \leq_T 0'$ and so its unique path $A \leq_T 0'$ as well.

Theorem 8. *Given $A \in 2^\omega$, let $\hat{A} := \{\sigma \in 2^{<\omega} \mid \sigma \prec A\}$ be the set of finite prefixes of A. If A is not Δ_2^0 then $\hat{A}$ is co-Σ_1^0-dense.*

Proof. Let A^* be the complement of $\hat{A}$. Let $W_e \subseteq 2^{<\omega}$ be an infinite c.e. set of strings. Let T be the set of all prefixes of elements of W_e. Then T is an infinite tree, hence by compactness it has at least one infinite path. That is, there is at least one real B such that all its prefixes are in T.

Case 1: The only such real is $B = A$. Then by Lemma 3, A is Δ_2^0.

Case 2: There is a $B \neq A$ such that all its prefixes are in T. Let σ be a prefix of B that is not a prefix of A. Let $W_d = [\sigma] \cap W_e$. Since all prefixes of B are prefixes of elements of W_e, there are infinitely many extensions of σ that are prefixes of elements of W_e. Consequently W_d is infinite. Thus, W_d is our desired infinite subset of $A^* \cap W_e$.

3.4   High Degrees

Definition 6. *A set A is co-r-cohesive if its complement is r-cohesive. This means that for each computable (recursive) set W_d, either $W_d \subseteq^* A$ or $W_d^c \subseteq^* A$.*

Definition 7 (Odifreddi [9, Exercise III.4.8], Jockusch and Stephan [7]). *A set A is strongly hyperhyperimmune (s.h.h.i.) if for each computable $f : \omega \to \omega$ for which the sets $W_{f(e)}$ are disjoint, there is an e with $W_{f(e)} \subseteq \omega \setminus A$.*

A set A is strongly hyperimmune (s.h.i.) if for each computable $f : \omega \to \omega$ for which the sets $W_{f(e)}$ are disjoint and computable, with $\bigcup_{e \in \omega} W_{f(e)}$ also computable, there is an e with $W_{f(e)} \subseteq \omega \setminus A$.

Proposition 9. *Every s.h.i. set is co-Σ_1^0-dense.*

Proof. Let A be s.h.i. Let W_e be an infinite c.e. set. Let W_d be an infinite computable subset of W_e. Effectively decompose W_d into infinitely many disjoint infinite computable sets,

$$W_d = \bigcup_{i \in \omega} W_{g(d,i)}.$$

For instance, if $W_d = \{a_0 < a_1 < \ldots\}$ then we may let $W_{g(e,i)} = \{a_n : n = 2^i(2k+1), i \geq 0, k \geq 0\}$. Since A is s.h.i., there exists some i_e such that $W_{g(d,i_e)} \subseteq A^c$. The sets $W_{g(d,i_e)}$ witness that A^c is Σ_1^0-dense.

Clearly r-cohesive implies s.h.i., and s.h.h.i. implies s.h.i. It was shown by Jockusch and Stephan [7, Corollary 2.4] that the cohesive degrees coincide with the r–cohesive degrees and (Corollary 3.10) that the s.h.i. and s.h.h.i. degrees coincide.

Proposition 10. *Every high degree contains a Σ_1^0-dense set.*

Proof. Let $\mathbf{h}$ be a Turing degree. If $\mathbf{h} \not\leq \mathbf{0}'$, then $\mathbf{h}$ contains a Σ_1^0-dense set by Theorem 8.

If $\mathbf{h} \leq \mathbf{0}'$ and $\mathbf{h}$ is high then since the strongly hyperhyperimmune and cohesive degrees coincide, and are exactly the high degrees [4], $\mathbf{h}$ contains a strongly hyperimmune set. Hence by Theorem 9, $\mathbf{h}$ contains a Σ_1^0-dense set.

3.5   Progressive Approximations

Definition 8. *Let A be a Δ_2^0 set. A computable approximation $\{\sigma_t\}_{t \in \omega}$ of A, where each σ_t is a finite string and $\lim_{t \to \infty} \sigma_t = A$, is* progressive *if for each t,*

- *if $|\sigma_t| \leq |\sigma_{t-1}|$ then $\sigma_t \upharpoonright (|\sigma_t| - 1) = \sigma_{t-1} \upharpoonright (|\sigma_t| - 1)$ (the last bit of σ_t is the only difference with σ_{t-1});*
- *if $|\sigma_t| > |\sigma_{t-1}|$ then $\sigma_{t-1} \prec \sigma_t$; and*
- *if $\sigma_t \not\prec \sigma_s$ for some $s > t$ then $\sigma_t \not\prec \sigma_{s'}$ for all $s' \geq s$ (once an approximation looks wrong, it never looks right again).*

If A has a progressive approximation then we say that A is progressively approximable.

Note that a progressively approximable set must be h-c.e. where $h(n) = 2^n$.

Theorem 9. *Let A be a progressively approximable and noncomputable set. Let $\{\sigma_t\}_{t \in \omega}$ be a progressive approximation of A. Then $\{t : \sigma_t \prec A\}$ is constructively immune.*

Proof. Let W_e be an infinite c.e. set and let T be an infinite computable subset of W_e. Since A is noncomputable, we do not have $T \subseteq \{t : \sigma_t \prec A\}$. Since the approximation $\{\sigma_t\}_{t \in \omega}$ is progressive, once we observe a t for which $\sigma_t \not\prec \sigma_s$, for some $s > t$, then we know that $\sigma_t \not\prec A$. Then we define $\varphi(e) = t$, and φ witnesses that $\{t : \sigma_t \prec A\}$ is constructively immune.

A direction for future work may be to find new Turing degrees of progressively approximable sets.

References

1. Brattka, V.: Random numbers and an incomplete immune recursive set. In: Widmayer, P., Eidenbenz, S., Triguero, F., Morales, R., Conejo, R., Hennessy, M. (eds.) ICALP 2002. LNCS, vol. 2380, pp. 950–961. Springer, Heidelberg (2002). https://doi.org/10.1007/3-540-45465-9_81
2. Calude, C.: Information and Randomness: An Algorithmic Perspective. Monographs in Theoretical Computer Science. An EATCS Series. Springer, Berlin (1994). With forewords by Gregory J. Chaitin and Arto Salomaa. https://doi.org/10.1007/978-3-662-03049-3
3. Cholak, P.A., Dzhafarov, D.D., Hirst, J.L., Slaman, T.A.: Generics for computable Mathias forcing. Ann. Pure Appl. Logic **165**(9), 1418–1428 (2014)
4. Cooper, S.B.: Jump equivalence of the Δ_2^0 hyperhyperimmune sets. J. Symbolic Logic **37**, 598–600 (1972)
5. Ershov, Y.L.: Theory of numberings. In: Handbook of Computability Theory. Studies in Logic and the Foundations of Mathematics, vol. 140, pp. 473–503. North-Holland, Amsterdam (1999)
6. Ferbus-Zanda, M., Grigorieff, S.: Refinment of the "up to a constant" ordering using contructive co-immunity and alike. Application to the min/max hierarchy of Kolmogorov complexities. arXiv:0801.0350 (2008)
7. Jockusch, C., Stephan, F.: A cohesive set which is not high. Math. Logic Quart. **39**(4), 515–530 (1993)
8. Li, X.: Effective immune sets, program index sets and effectively simple sets—generalizations and applications of the recursion theorem. In: 1981 Southeast Asian Conference on Logic. Studies in Logic and the Foundations of Mathematics, Singapore, vol. 111, pp. 97–106. North-Holland, Amsterdam (1983)
9. Odifreddi, P.: Classical Recursion Theory: The Theory of Functions and Sets of Natural Numbers. Studies in Logic and the Foundations of Mathematics, vol. 125. North-Holland Publishing Co., Amsterdam (1989). With a foreword by G. E. Sacks
10. Rosenstein, J.G.: Linear Orderings. Pure and Applied Mathematics, vol. 98. Academic Press, Inc. [Harcourt Brace Jovanovich, Publishers], New York, London (1982)
11. Schaefer, M.: A guided tour of minimal indices and shortest descriptions. Arch. Math. Logic **37**(8), 521–548 (1998)
12. Smullyan, R.M.: Effectively simple sets. Proc. Amer. Math. Soc. **15**, 893–895 (1964)
13. Soare, R.: Recursively Enumerable Sets and Degrees. Springer, Heidelberg (1987)

Fine-Grained Complexity Theory: Conditional Lower Bounds for Computational Geometry

Karl Bringmann[✉]

Saarland University and Max Planck Institute for Informatics,
Saarland Informatics Campus, Saarbrücken, Germany
bringmann@cs.uni-saarland.de

Abstract. Fine-grained complexity theory is the area of theoretical computer science that proves conditional lower bounds based on the Strong Exponential Time Hypothesis and similar conjectures. This area has been thriving in the last decade, leading to conditionally best-possible algorithms for a wide variety of problems on graphs, strings, numbers etc. This article is an introduction to fine-grained lower bounds in computational geometry, with a focus on lower bounds for polynomial-time problems based on the Orthogonal Vectors Hypothesis. Specifically, we discuss conditional lower bounds for nearest neighbor search under the Euclidean distance and Fréchet distance.

Keywords: Fine-grained complexity · Computational geometry

1 Introduction

The term *fine-grained complexity theory* was coined in the last decade to describe the area of theoretical computer science that proves conditional lower bounds on the time complexity of algorithmic problems, assuming some hypothesis. The goal is to explain the computational complexity of many different problems based on a small number of core barriers. The general approach dates back to the introduction of 3SUM-hardness in '95 [25] (or even to the introduction of NP-hardness, depending on the interpretation). The last decade has seen several new hypotheses and a wealth of new techniques for proving conditional lower bounds, leading to a large body of literature on the topic, see the surveys [10,30]. In this article we give a self-contained introduction to recent fine-grained complexity results in the area of computational geometry. Instead of the most technically deep results, we focus on simple techniques that can be easily transferred to other problems. Moreover, we focus on lower bounds for polynomial-time problems.

This work is part of the project TIPEA that has received funding from the European Research Council (ERC) under the European Unions Horizon 2020 research and innovation programme (grant agreement No. 850979).

L. De Mol et al. (Eds.): CiE 2021, LNCS 12813, pp. 60–70, 2021.
https://doi.org/10.1007/978-3-030-80049-9_6

The basic setup of fine-grained lower bounds is similar to classic NP-hardness reductions: A fine-grained reduction from problem P to problem Q is an algorithm that given an instance I of size n for problem P computes in time $r(n)$ an equivalent instance J of size $s(n)$ for problem Q.[1] Thus, if there is an algorithm solving problem Q in time $T(n)$, by this reduction there is an algorithm solving problem P in time $r(n) + T(s(n))$. In particular, if $r(n) + T(s(n))$ is faster than the hypothesized optimal time complexity of problem P, then problem Q cannot be solved in time $T(n)$ assuming the hypothesis for P. We will see several concrete examples of this argumentation throughout this article.

1.1 Hardness Hypotheses

Let us discuss the three main hypotheses used in computational geometry.

3SUM Hypothesis. In the 3SUM problem, given n integers, we want to decide whether any three of them sum to 0. The 3SUM Hypothesis postulates that the classic $O(n^2)$-time algorithm for 3SUM cannot be improved to time $O(n^{2-\varepsilon})$ for any $\varepsilon > 0$. This hypothesis was introduced in '95 in a seminal work by Gajentaan and Overmars [25], making computational geometry a pioneer in fine-grained complexity theory. We refer to [30] for an overview of lower bounds based on the 3SUM Hypothesis; in this introduction we focus on other hypotheses.

Strong Exponential Time Hypothesis. The strongest new impulse for conditional lower bounds in the last two decades was the introduction of the Strong Exponential Time Hypothesis. This hypothesis concerns the fundamental k-SAT problem: Given a formula ϕ in conjunctive normal form of width k on n variables and m clauses, decide whether ϕ is satisfiable. Naively this problem can be solved in time $O(2^n m)$. Improved algorithms solve k-SAT in time $O(2^{(1-\varepsilon_k)n})$ for some constant $\varepsilon_k > 0$, but for all known algorithms the constant ε_k tends to 0 for $k \to \infty$. This lead Impagliazzo and Paturi [26] to postulate the following:

Hypothesis 1 (Strong Exponential Time Hypothesis – SETH). *For any* $\varepsilon > 0$, *there exists* $k \geq 3$ *such that* k-*SAT on formulas with* n *variables cannot be solved in time* $O(2^{(1-\varepsilon)n})$.

This has become the most standard hypothesis in fine-grained complexity theory [30], and it has been used to prove tight lower bounds for a wide variety of problems, see, e.g., [1,3,9,12,14–19,22,28,32].

---

[1] What we sketched here is a *many-one* reduction, since each instance of P is reduced to one instance of Q. One can also consider *Turing* reductions, where the reduction algorithm is allowed to make several calls to an oracle for Q. See [20, Definition 1] for the formal definition of (Turing-style) fine-grained reductions.

Orthogonal Vectors Hypothesis. In the Orthogonal Vectors problem (OV), given sets of Boolean vector $A, B \subseteq \{0,1\}^d$ of size n, we ask whether there exists a pair $(a, b) \in A \times B$ that is orthogonal, that is, $\langle a, b \rangle = \sum_{i=1}^{d} a_i \cdot b_i = 0$. Naively this problem can be solved in time $O(n^2 d)$. For small dimension $d = O(\log n)$ there are improved algorithms [4], but for $\omega(\log n) \leq d \leq n^{o(1)}$ no algorithm running in time $O(n^{2-\varepsilon})$ is known. This barrier is formalized as follows.

Hypothesis 2 (OV Hypothesis – OVH [31]). *For any $\varepsilon > 0$, OV cannot be solved in time $O(n^{2-\varepsilon}\text{poly}(d))$.*

Note that for $d = n^{\Omega(1)}$ we can naively solve OV in time $O(n^2 d) = \text{poly}(d) = O(n^{2-\varepsilon}\text{poly}(d))$, and thus OVH does not apply. Indeed, the hypothesis only asserts that there *exists* a dimension $d = d(n)$ such that OV cannot be solved in time $O(n^{2-\varepsilon}\text{poly}(d))$; this dimension d must be of the form $\omega(\log n) \leq d \leq n^{o(1)}$.

OVH has been used to prove tight conditional lower bounds for a wide range of problems, see, e.g., [1,9,12,14,16,17,19,32]. It is known that OVH is at least as believable as SETH, because SETH implies OVH [31].

In this article we focus on lower bounds based on OVH (since SETH implies OVH this also yields lower bounds based on SETH). Specifically, in Sect. 2 we consider nearest neighbor search, and in Sect. 3 we discuss curve similarity.

2   Nearest Neighbor Search

A fundamental problem of computer science is to compute the nearest neighbor of a point $q \in \mathbb{R}^d$ among a set of points $P \subset \mathbb{R}^d$, that is, to determine the point $p \in P$ minimizing the Euclidean distance $\|p - q\|$. This has an abundance of applications such as pattern recognition, spell checking, or coding theory. These applications often come in the form of a data structure problem, where we can first preprocess P to build a data structure that can then quickly answer nearest neighbor queries. Naively, a nearest neighbor query can be answered in time $O(nd)$, where n is the number of points in the data set P. Improved algorithms exist in small dimensions, for example k-d-trees have a worst-case query time of $O(d \cdot n^{1-1/d})$ [27]. However, already for a large constant dimension $d \geq 1/\varepsilon$ this query time is essentially linear, specifically it is $\Omega(n^{1-\varepsilon})$. We can thus ask:

Does high-dimensional nearest neighbor search require near-linear query time?

In the following we answer this question affirmatively assuming OVH. To connect nearest neighbor search to the OV problem we make use of the following embedding, which maps Boolean vectors to points in $\mathbb{R}^d$ such that from the points' Euclidean distance we can read off whether the vectors are orthogonal.

Lemma 1 (Embedding Orthogonality into Euclidean Distance). *There are functions $\mathcal{A}, \mathcal{B}: \{0,1\}^d \mapsto \mathbb{R}^d$ and a threshold τ such that $\langle a, b \rangle = 0$ if and only if $\|\mathcal{A}(a) - \mathcal{B}(b)\| \leq \tau$ for any $a, b \in \{0,1\}^d$. The functions $\mathcal{A}, \mathcal{B}$ and the threshold τ can be evaluated in time $O(d)$.*

Proof. For any $a \in \{0,1\}^d$ we construct $p := \mathcal{A}(a)$ by setting $p_i := 1 + 2a_i$ for any $1 \leq i \leq d$. Similarly, for any $b \in \{0,1\}^d$ we construct $q := \mathcal{B}(b)$ by setting $q_i := 2 - 2b_i$. Note that $|p_i - q_i| = |2(a_i + b_i) - 1|$, which evaluates to 3 if $a_i = b_i = 1$ and to 1 otherwise. Therefore, we obtain

$$\|p - q\| = \left(\sum_{i=1}^{d} |p_i - q_i|^2 \right)^{1/2} = \left(3^2 \cdot \langle a, b \rangle + 1^2 \cdot (d - \langle a, b \rangle) \right)^{1/2} = (d + 8\langle a, b \rangle)^{1/2}.$$

Setting $\tau := d^{1/2}$ yields $\|p - q\| = \|\mathcal{A}(a) - \mathcal{B}(b)\| \leq \tau$ if and only if $\langle a, b \rangle = 0$.

2.1 Bichromatic Closest Pair

We use the above embedding to prove a conditional lower bound for the Bichromatic Closest Pair problem, an offline variant of nearest neighbor search:

Problem 1 (Bichromatic Closest Pair). Given sets $P, Q \subset \mathbb{R}^d$ of size n, compute the pair $(p, q) \in P \times Q$ minimizing the Euclidean distance $\|p - q\|$.

Bichromatic Closest Pair cannot be solved in time $O(n^{2-\varepsilon}\text{poly}(d))$ under OVH:

Theorem 1 (Lower Bound for Bichromatic Closest Pair [6]). *For any $\varepsilon > 0$, Bichromatic Closest Pair cannot be solved in time $O(n^{2-\varepsilon}\text{poly}(d))$, unless OVH fails.*

Proof. We reduce from OV to Bichromatic Closest Pair using the embedding from Lemma 1. Given an OV instance (A, B) of size n in dimension d, we construct the point sets $P = \{\mathcal{A}(a) \mid a \in A\}$ and $Q := \{\mathcal{B}(b) \mid b \in B\}$. By Lemma 1, the bichromatic closest pair of (P, Q) has distance $\leq \tau$ if and only if there exists an orthogonal pair of vectors. Thus, a solution to the constructed Bichromatic Closest Pair instance solves the given OV instance. Since n and d do not change, the running time lower bound is immediate from OVH (Hypothesis 2).

2.2 Nearest Neighbor Data Structures

Now we consider the data structure version of nearest neighbor search.

Problem 2 (Nearest Neighbor Data Structure). In the preprocessing we are given a set $P \subset \mathbb{R}^d$ of size n and we build a data structure. The data structure allows to answer nearest neighbor queries: Given a point $q \in \mathbb{R}^d$, compute the point $p \in P$ minimizing the Euclidean distance $\|p - q\|$.

Observe that any nearest neighbor data structure also solves the Bichromatic Closest Pair Problem, by building the data structure for P and then querying every $q \in Q$. If the data structure has preprocessing time $T_P(n, d)$ and query time $T_Q(n, d)$, then this solves Bichromatic Closest Pair in time $T_P(n, d) + n \cdot T_Q(n, d)$. Theorem 1 thus implies that Bichromatic Closest Pair cannot be solved with preprocessing time $O(n^{2-\varepsilon}\text{poly}(d))$ and query time $O(n^{1-\varepsilon}\text{poly}(d))$:

Corollary 1 (Lower Bound for Nearest Neighbor Data Structures I).
For any $\varepsilon > 0$, there is no nearest neighbor data structure with preprocessing time $O(n^{2-\varepsilon}\mathrm{poly}(d))$ and query time $O(n^{1-\varepsilon}\mathrm{poly}(d))$, unless OVH fails.

It might seem natural that the preprocessing time is limited to $O(n^{2-\varepsilon})$, because from OVH we can prove only quadratic lower bounds. In the following we show that this intuition is wrong, and the above corollary can be improved to rule out *any polynomial* preprocessing time. To this end, we need an unbalanced version of OVH, which shows that the brute force enumeration of all $|A| \cdot |B|$ pairs of vectors is also necessary when $|A| = n^{\alpha} \ll n = |B|$. This tool was introduced in [3]; see also [14] for a proof that unbalanced OV and standard OV are equivalent.

Lemma 2 ([3]). *For any $\varepsilon, \alpha \in (0,1)$, OV on instances (A, B) with $|B| = n$ and $|A| = \Theta(n^{\alpha})$ cannot be solved in time $O(n^{1+\alpha-\varepsilon}\mathrm{poly}(d))$, unless OVH fails.*

Proof. Let (A', B') be a balanced instance of OV, that is, $|A'| = |B'| = n$. Split B' into $\Theta(n^{1-\alpha})$ sets $B'_1, \ldots, B'_\ell$ of size $\Theta(n^{\alpha})$. Run an unbalanced OV algorithm on each pair (A', B'_i), and note that from the results we can infer whether (A', B') contains an orthogonal pair of vectors. If each unbalanced instance can be solved in time $O(n^{1+\alpha-\varepsilon})$, then all $\Theta(n^{1-\alpha})$ unbalanced instances in total can be solved in time $O(n^{2-\varepsilon})$, contradicting OVH. $\square$

With this tool, we can rule out any polynomial preprocessing time $\mathrm{poly}(n, d)$ and query time $O(n^{1-\varepsilon}\mathrm{poly}(d))$ for nearest neighbor search:

Theorem 2 (Lower Bound for Nearest Neighbor Data Structures II).
For any $\varepsilon, \beta > 0$, there is no nearest neighbor data structure with preprocessing time $O(n^{\beta}\mathrm{poly}(d))$ and query time $O(n^{1-\varepsilon}\mathrm{poly}(d))$, unless OVH fails.

Proof. Fix $\varepsilon, \beta > 0$ and suppose nearest neighbor can be solved with preprocessing time $O(|P|^{\beta}\mathrm{poly}(d))$ and query time $O(|P|^{1-\varepsilon}\mathrm{poly}(d))$. Set $\alpha := 1/\beta$. Given an OV instance (A, B) with $|B| = n$ and $|A| = \Theta(n^{\alpha})$, we use the embedding from Lemma 1 to construct the sets $P := \{\mathcal{A}(a) \mid a \in A\}$ and $Q := \{\mathcal{B}(b) \mid b \in B\}$. We run the preprocessing of the nearest neighbor data structure on P; this takes time $O(|P|^{\beta}\mathrm{poly}(d)) = O((n^{\alpha})^{\beta}\mathrm{poly}(d)) = O(n\mathrm{poly}(d))$. Then we query the data structure for each $q \in Q$; over all $|Q|$ queries this takes total time

$$O(|Q| \cdot |P|^{1-\varepsilon}\mathrm{poly}(d)) = O(n^{1+\alpha\cdot(1-\varepsilon)}\mathrm{poly}(d)) = O(n^{1+\alpha-\varepsilon'}\mathrm{poly}(d)),$$

for $\varepsilon' := \alpha \cdot \varepsilon$. By Lemma 1, some query $q \in Q$ returns a point $p \in P$ within distance τ if and only if there exists an orthogonal pair of vectors in $A \times B$. We can thus solve unbalanced OV in time $O(n^{1+\alpha-\varepsilon'}\mathrm{poly}(d))$, contradicting Lemma 2. $\square$

We have thus shown that high-dimensional nearest neighbor search requires almost-linear query time, even if we allow any polynomial preprocessing time.

2.3   Further Results on Nearest Neighbor Search

Let us discuss some advanced research directions on nearest neighbor search. The proofs here are beyond the scope of this introduction to the topic.

- *Smaller Dimension:* The best known query time for nearest neighbor search is of the form $n^{1-\Theta(1/d)}$ [27], which is near-linear $n^{1-o(1)}$ for any unbounded dimension $d = \omega(1)$. Recall that OVH asserts hardness for some dimension $\omega(\log n) \leq d \leq n^{o(1)}$. A line of research has tried to close this gap [22,32]; the current record shows that Theorem 1 already holds in dimension $d = 2^{O(\log^* n)}$ [22]. It remains an important open problem to close the remaining gap and show hardness for any dimension $d = \omega(1)$.
- *Approximate Nearest Neighbor:* In many practical applications it suffices to compute nearest neighbors approximately. Note that the OV problem asks whether there is a pair of vectors with $\langle a, b \rangle = 0$ or whether all vectors have $\langle a, b \rangle \geq 1$. Inspecting the proof of Lemma 1, we see that it is hard to distinguish between Euclidean distance at most $d^{1/2}$ or at least $(d+8)^{1/2}$. This shows hardness of computing a $(d + 8)^{1/2}/d^{1/2} = 1 + \Theta(1/d)$ approximation for Bichromatic Closest Pair. A big leap forward was made by Rubinstein [28], who proved that Theorem 1 even holds for $(1 + \delta)$-approximation algorithms, where $\delta = \delta(\varepsilon)$ is some positive constant. See also [29] for more hardness of approximation results in fine-grained complexity theory.

3   Curve Similarity and the Fréchet Distance

We now turn to a different realm of applications. For our purposes, a *curve* is a sequence of points in the plane, that is, $\pi = (\pi_1, \ldots, \pi_n)$ with $\pi_i \in \mathbb{R}^2$. We call n the *length* of π. A typical task is to judge the similarity of two given curves. Several distance measures have been proposed for this task, but the most classical and most popular in computational geometry is the Fréchet distance[2]. For intuition, imagine a dog walking along curve π and its owner walking along curve σ, connected by a leash. They start at the respective startpoints and end at their endpoints, and at any point in time either the dog advances to the next vertex along its curve, or the owner advances, or they both advance together. The shortest possible leash length admitting such a traversal is called the Fréchet distance of π and σ.

Formally, for curves $\pi = (\pi_1, \ldots, \pi_n)$ and $\sigma = (\sigma_1, \ldots, \sigma_m)$, a *traversal* is a sequence $((i_1, j_1), \ldots, (i_T, j_T))$ such that $(i_1, j_1) = (1, 1)$, $(i_T, j_T) = (n, m)$, and for every $1 \leq t < T$ we have $(i_{t+1}, j_{t+1}) \in \{(i_t + 1, j_t), (i_t, j_t + 1), (i_t + 1, j_t + 1)\}$. The (discrete) Fréchet distance between π and σ is defined as

$$d_F(\pi, \sigma) = \min_{((i_1,j_1),\ldots,(i_T,j_T))} \max_{1 \leq t \leq T} \|\pi_{i_t} - \sigma_{j_t}\|,$$

where the minimum goes over all traversals of π and σ.

---

[2] For simplicity, we focus on the *discrete* Fréchet distance [24] instead of the slightly more standard *continuous* variant [7].

The Fréchet distance of two curves of length n can be computed in time $O(n^2)$, by a simple dynamic programming algorithm that computes the Fréchet distance of any prefix $(\pi_1, \ldots, \pi_i)$ of π and any prefix $(\sigma_1, \ldots, \sigma_j)$ of σ [24].

In the following, we first discuss the Fréchet distance from the viewpoint of nearest neighbor search, and then we elaborate on the problem of computing the Fréchet distance of two given curves.

3.1   Nearest Neighbor Search Under Fréchet Distance

We start with an embedding of vectors into curves, similar to Lemma 1.

Lemma 3 (Embedding Orthogonality into Fréchet Distance [9]). *There are functions $\mathcal{A}, \mathcal{B}$ mapping any $z \in \{0,1\}^d$ to a curve of length d in the plane, such that $\langle a, b \rangle = 0$ if and only if $d_F(\mathcal{A}(a), \mathcal{B}(b)) \leq 1$ for any $a, b \in \{0,1\}^d$. The functions $\mathcal{A}, \mathcal{B}$ can be evaluated in time $O(d)$.*

Proof. For any $a \in \{0,1\}^d$ we construct the curve $\pi := \mathcal{A}(a)$ by setting $\pi_i := (3i, 1 + 2a_i) \in \mathbb{R}^2$ for any $1 \leq i \leq d$. Similarly, for any $b \in \{0,1\}^d$ we construct $\sigma := \mathcal{B}(b)$ by setting $\sigma_i := (3i, 2 - 2b_i)$. Note that $\|\pi_i - \sigma_i\| = |2(a_i + b_i) - 1|$, which evaluates to 3 if $a_i = b_i = 1$ and to 1 otherwise. Moreover, for $i \neq j$ we have $\|\pi_i - \sigma_j\| \geq 3$.

Consider a traversal of π and σ. If at some point the dog advances but not the owner (or the owner advances but not the dog), we get a distance of the form $\|\pi_i - \sigma_j\|$ for $i \neq j$, and thus the leash length must be at least 3. In the remaining case, the dog and its owner always advance together, meaning that at time i the dog is at position π_i and the owner is at position σ_i. This traversal has distance $\max_{1 \leq i \leq d} \|\pi_i - \sigma_i\| = \max_{1 \leq i \leq d} |2(a_i + b_i) - 1|$, which is 1 if a, b are orthogonal, and 3 otherwise. Hence, $d_F(\pi, \sigma) \leq 1$ holds if and only if $\langle a, b \rangle = 0$.

Using this embedding, we can show lower bounds for nearest neighbor search among curves in the plane, analogously to the results for Euclidean nearest neighbor search from Sect. 2 (the same proofs work almost verbatim). Specifically, in the problem Bichromatic Closest Pair under Fréchet Distance we are given sets P, Q, each containing n curves of length d in the plane, and we want to compute the pair $(\pi, \sigma) \in P \times Q$ that minimizes the Fréchet distance $d_F(\pi, \sigma)$. Naively, this can be solved in time $O(n^2 d^2)$.

Theorem 3 (Lower Bound for Bichromatic Closest Pair under Fréchet Distance). *For any $\varepsilon > 0$, Bichromatic Closest Pair under Fréchet Distance cannot be solved in time $O(n^{2-\varepsilon} \mathrm{poly}(d))$, unless OVH fails.*

Similarly, in nearest neighbor data structures for the Fréchet distance we can preprocess a given set P consisting of n curves of length d in the plane, and then given a query curve σ of length d in the plane we want to find the curve $\pi \in P$ minimizing $d_F(\pi, \sigma)$.

Theorem 4 (Lower Bound for Nearest Neighbor Data Structures under Fréchet Distance). *For any $\varepsilon, \beta > 0$, there is no data structure for nearest neighbor search under Fréchet distance with preprocessing time $O(n^\beta \mathrm{poly}(d))$ and query time $O(n^{1-\varepsilon} \mathrm{poly}(d))$, unless OVH fails.*

3.2   Computing the Fréchet Distance

A classic dynamic programming algorithm computes the Fréchet distance between two curves of length n in time $O(n^2)$ [24]. A breakthrough result from '14 shows a tight lower bound ruling out time $O(n^{2-\varepsilon})$ under OVH [9]. This result paved the way for tight lower bounds for many other dynamic programming problems (mostly outside of computational geometry, see, e.g., [1,12]). Here we give a very brief sketch of this result.

Theorem 5 (Lower Bound for Fréchet Distance [9]). *For any $\varepsilon > 0$, the Fréchet distance cannot be computed in time $O(n^{2-\varepsilon})$.*

Proof Sketch. Given an OV instance (A, B) on n vectors in dimension d, we construct two curves π, σ of length $N = O(nd)$ such that $d_F(\pi, \sigma) \leq 1$ if and only if (A, B) contains an orthogonal pair. It then follows that if the Fréchet distance can be computed in time $O(N^{2-\varepsilon})$, then OV can be solved in time $O((nd)^{2-\varepsilon}) = O(n^{2-\varepsilon}\mathrm{poly}(d))$, contradicting OVH (Hypothesis 2).

To construct the curves π, σ, we start with *vector gadgets*. These gadgets are similar to the embedding in Lemma 3, but they are restricted to a much smaller region in space. Specifically, for each vector $a \in A$ we construct a curve $VG(a)$ as the sequence of points $((-1)^i \delta, 0.5 - (-1)^{a_i}\delta^2) \in \mathbb{R}^2$ for $1 \leq i \leq d$, where $\delta > 0$ is a small constant. Similarly, for each vector $b \in B$ we construct a curve $VG(b)$ as the sequence of points $((-1)^i \delta, -0.5 + (-1)^{b_i}\delta^2)$ for $1 \leq i \leq d$. Analogously to Lemma 3, we can show that $d_F(VG(a), VG(b)) \leq 1$ if and only if $\langle a, b \rangle = 0$.

The final and most complicated step of the reduction is the *OR gadget*. This gadget combines the curves $VG(a)$ for all $a \in A$ into one curve π, and similarly it combines the curves $VG(b)$ for all $b \in B$ into one curve σ, such that $d_F(\pi, \sigma) \leq 1$ if and only if there exist $a \in A, b \in B$ with $d_F(VG(a), VG(b)) \leq 1$. To this end, we introduce auxiliary points at the following positions:

$$s = (-0.5, 0), \ t = (0.5, 0), \ s^* = (-0.5, -1), \ t^* = (0.5, 1).$$

The final curve π repeats the pattern $(s, VG(a), t)$ for all $a \in A$. The final curve σ starts with s and s^*, then walks through all vector gadgets $VG(b)$, and ends with t^* and t. One can show that these curves satisfy $d_F(\pi, \sigma) \leq 1$ if and only if (A, B) contains an orthogonal pair, for details see [9].

3.3   Further Results on Fréchet Distance

- *Robustness:* For reductions to geometric problems a common concern is the precision needed to write down the constructed instances. The reductions shown in this article are very robust: they only require $O(\log d)$-bit coordinates, and some can even be made to work with $O(1)$-bit coordinates.
- *Hardness of Approximation:* Inspecting the proof of Lemma 3, we see that it is hard to distinguish Fréchet distance at most 1 or at least 3. Therefore, Theorems 3 and 4 even hold against multiplicative 2.999-approximation algorithms. For approximation algorithms we refer to [16,21].

- *One-dimensional Curves:* We showed hardness for curves in the plane. The same results hold for one-dimensional curves, of the form $\pi = (\pi_1, \ldots, \pi_d)$ with $\pi_i \in \mathbb{R}$, see [16,19].
- *Continuous and Weak Variants:* The same lower bounds as in Theorems 3, 4, and 5 also hold for other standard variants of the Fréchet distance [9,19].
- *Realistic Input Curves:* In order to avoid the quadratic worst-case complexity, geometers have studied several models of realistic input curves. For example, on so-called *c-packed* curves the Fréchet distance can be $(1+\varepsilon)$-approximated in time $\tilde{O}(cn/\sqrt{\varepsilon})$ [13,23], which matches a conditional lower bound [9].
- *Logfactor Improvements:* Lower bounds under OVH rule out polynomial improvements of the form $O(n^{2-\varepsilon})$. What about logfactor improvements? An algorithm running in time $O(n^2 \log \log n / \log n)$ is known [5]. Can we improve this to time $O(n^2 / \log^{100} n)$? Such an improvement was shown to be unlikely, as it would imply new circuit lower bounds [2].

4 More Fine-Grained Computational Geometry

In this article we focused on nearest neighbor search and the Fréchet distance. Further work on fine-grained complexity in computational geometry includes conditional lower bounds for a variant of Fréchet distance between k curves [18], the dynamic time warping distance [1,12], the Fréchet distance under translation [15] and Hausdorff distance under translation [17], curve simplification under Fréchet distance [11,18], and Maximum Weight Rectangle [8].

References

1. Abboud, A., Backurs, A., Vassilevska Williams, V.: Tight hardness results for LCS and other sequence similarity measures. In: FOCS, pp. 59–78. IEEE Computer Society (2015)
2. Abboud, A., Bringmann, K.: Tighter connections between formula-SAT and shaving logs. In: ICALP. LIPIcs, vol. 107, pp. 8:1–8:18 (2018)
3. Abboud, A., Vassilevska Williams, V.: Popular conjectures imply strong lower bounds for dynamic problems. In: FOCS, pp. 434–443. IEEE Computer Society (2014)
4. Abboud, A., Williams, R.R., Yu, H.: More applications of the polynomial method to algorithm design. In: SODA, pp. 218–230. SIAM (2015)
5. Agarwal, P.K., Avraham, R.B., Kaplan, H., Sharir, M.: Computing the discrete Fréchet distance in subquadratic time. SIAM J. Comput. **43**(2), 429–449 (2014)
6. Alman, J., Williams, R.: Probabilistic polynomials and Hamming nearest neighbors. In: FOCS, pp. 136–150. IEEE Computer Society (2015)
7. Alt, H., Godau, M.: Computing the Fréchet distance between two polygonal curves. Int. J. Comput. Geom. Appl. **5**, 75–91 (1995)
8. Backurs, A., Dikkala, N., Tzamos, C.: Tight hardness results for maximum weight rectangles. In: ICALP. LIPIcs, vol. 55, pp. 81:1–81:13 (2016)
9. Bringmann, K.: Why walking the dog takes time: Frechet distance has no strongly subquadratic algorithms unless SETH fails. In: FOCS, pp. 661–670. IEEE Computer Society (2014)

10. Bringmann, K.: Fine-grained complexity theory (tutorial). In: STACS. LIPIcs, vol. 126, pp. 4:1–4:7 (2019)
11. Bringmann, K., Chaudhury, B.R.: Polyline simplification has cubic complexity. In: Symposium on Computational Geometry. LIPIcs, vol. 129, pp. 18:1–18:16 (2019)
12. Bringmann, K., Künnemann, M.: Quadratic conditional lower bounds for string problems and dynamic time warping. In: FOCS, pp. 79–97. IEEE Computer Society (2015)
13. Bringmann, K., Künnemann, M.: Improved approximation for Fréchet distance on c-packed curves matching conditional lower bounds. Int. J. Comput. Geom. Appl. **27**(1–2), 85–120 (2017)
14. Bringmann, K., Künnemann, M.: Multivariate fine-grained complexity of longest common subsequence. In: SODA, pp. 1216–1235. SIAM (2018)
15. Bringmann, K., Künnemann, M., Nusser, A.: Fréchet distance under translation: conditional hardness and an algorithm via offline dynamic grid reachability. In: SODA, pp. 2902–2921. SIAM (2019)
16. Bringmann, K., Mulzer, W.: Approximability of the discrete Fréchet distance. J. Comput. Geom. **7**(2), 46–76 (2016)
17. Bringmann, K., Nusser, A.: Translating Hausdorff is hard: fine-grained lower bounds for Hausdorff distance under translation. In: Symposium on Computational Geometry (2021, to appear)
18. Buchin, K., Buchin, M., Konzack, M., Mulzer, W., Schulz, A.: Fine-grained analysis of problems on curves. In: EuroCG, Lugano, Switzerland (2016)
19. Buchin, K., Ophelders, T., Speckmann, B.: SETH says: weak Fréchet distance is faster, but only if it is continuous and in one dimension. In: SODA, pp. 2887–2901. SIAM (2019)
20. Carmosino, M.L., Gao, J., Impagliazzo, R., Mihajlin, I., Paturi, R., Schneider, S.: Nondeterministic extensions of the strong exponential time hypothesis and consequences for non-reducibility. In: ITCS, pp. 261–270. ACM (2016)
21. Chan, T.M., Rahmati, Z.: An improved approximation algorithm for the discrete Fréchet distance. Inf. Process. Lett. **138**, 72–74 (2018)
22. Chen, L.: On the hardness of approximate and exact (bichromatic) maximum inner product. Theor. Comput. **16**, 1–50 (2020)
23. Driemel, A., Har-Peled, S., Wenk, C.: Approximating the Fréchet distance for realistic curves in near linear time. Discret. Comput. Geom. **48**(1), 94–127 (2012)
24. Eiter, T., Mannila, H.: Computing Discrete Fréchet Distance. Technical report, CD-TR 94/64, Christian Doppler Laboratory (1994)
25. Gajentaan, A., Overmars, M.H.: On a class of $O(n^2)$ problems in computational geometry. Comput. Geom. **5**, 165–185 (1995)
26. Impagliazzo, R., Paturi, R.: On the complexity of k-SAT. J. Comput. Syst. Sci. **62**(2), 367–375 (2001)
27. Lee, D.T., Wong, C.K.: Worst-case analysis for region and partial region searches in multidimensional binary search trees and balanced quad trees. Acta Informatica **9**, 23–29 (1977)
28. Rubinstein, A.: Hardness of approximate nearest neighbor search. In: STOC, pp. 1260–1268. ACM (2018)
29. Rubinstein, A., Vassilevska Williams, V.: SETH vs approximation. SIGACT News **50**(4), 57–76 (2019)
30. Vassilevska Williams, V.: On some fine-grained questions in algorithms and complexity. In: Proceedings of the ICM, vol. 3, pp. 3431–3472. World Scientific (2018)

31. Williams, R.: A new algorithm for optimal 2-constraint satisfaction and its implications. Theor. Comput. Sci. **348**(2–3), 357–365 (2005)
32. Williams, R.: On the difference between closest, furthest, and orthogonal pairs: nearly-linear vs. barely-subquadratic complexity. In: SODA, pp. 1207–1215. SIAM (2018)

The Lost Melody Theorem for Infinite Time Blum-Shub-Smale Machines

Merlin Carl[✉]

Europa-Universität Flensburg, Auf dem Campus 1b, 24943 Flensburg, Germany
merlin.carl@uni-flensburg.de
https://www.uni-flensburg.de/mathematik/wer-wir-sind/mitarbeiterinnen-
und-mitarbeiter/dr-merlin-carl/

Abstract. We consider recognizability for Infinite Time Blum-Shub-Smale machines, a model of infinitary computability introduced by Koepke and Seyfferth. In particular, we show that the lost melody theorem (originally proved for ITTMs by Hamkins and Lewis), i.e. the existence of non-computable, but recognizable real numbers, holds for ITBMs, that ITBM-recognizable real numbers are hyperarithmetic and that both ITBM-recognizable and ITBM-unrecognizable real numbers appear at every level of the constructible hierarchy below $L_{\omega_1^{CK}}$ above ω^ω.

Keywords: Infinite Time Blum-Shub-Smale Machines · Recognizability · Ordinal computability · Admissibility

1 Introduction

In ordinal computability, a considerable variety of machine models of infinitary computability was defined, including Infinite Time Turing Machines (ITTMs), (weak) Infinite Time Register Machines (ITRMs), Ordinal Turing Machines (OTMs), Ordinal Register Machines (ORMs) and Infinite Time Blum-Shub-Smale Machines (ITBMs) etc. For each of these models, a real number (more generally, a set of ordinals) x is called "recognizable" if and only if there is a program P such that, when executed on a machine of the type under consideration, P halts on every input y with output 0 or 1 and outputs 1 if and only if $x = y$. The term was originally defined for ITTMs in Hamkins and Lewis [9], where the most prominent statement about ITTM-recognizability was proved, namely the existence of real numbers that are ITTM-recognizable, but not ITTM-computable, so called "lost melodies".

Later on, recognizability was also studied for other machine models. The lost melody theorem was shown to also hold for ITRMs (see [6]; see [4] and [3] for a detailed study of ITRM-recognizability) and OTMs with parameters (where computability amounts to constructibility, while recognizability takes us up to

L. De Mol et al. (Eds.): CiE 2021, LNCS 12813, pp. 71–81, 2021.
https://doi.org/10.1007/978-3-030-80049-9_7

M_1, the canonical inner model for a Woodin cardinal, see [7]). On the other hand, it fails for OTMs without parameters and weak ITRMs[1], see [5].

Infinite Time Blum-Shub-Smale machines, introduced by Koepke and Seyfferth in [14] are register models of infinitary computability that compute with real numbers rather than ordinals as their register contents. ITBMs are known to compute exactly the real numbers in L_{ω^ω} by Welch [19] and Koepke and Morozov [13]. Moreover, it is known from Koepke and Seyfferth [14] (Theorem 1) that an ITBM-program with n nodes either halts in $< \omega^{n+1}$ many steps or not at all. So far, recognizability for ITBMs was not considered. Indeed, as ITBMs are extremely weak in comparison with the other models mentioned above, many of the usual methods for studying recognizability are not available in this setting.

In this paper, we close this gap by (i) showing that the lost melody theorem holds for ITBMs and in particular the ITBM-recognizability of the ITBM-halting number, (ii) showing $L_{\omega_1^{CK}}$ to be the minimal L-level containing all ITBM-recognizable real numbers and (iii) that both new ITBM-recognizable and new ITBM-unrecognizable real numbers appear at every level after ω^ω below ω_1^{CK}.

Most arguments in this paper are variants of the corresponding arguments used in the investigation of register models of ordinal computability, specifically Infinite Time Register Machines (ITRMs, see Koepke and Miller [12]) and weak Infinite Time Register Machines (now called wITRMs, see Koepke [11]). However, due to the weakness of ITBMs, considerable adaptations are required. In this respect, ITBMs turn out to be a kind of mixture between these two machine types with respect to recognizability: Like ITTMs and ITRMs but other than wITRMs, they have lost melodies, even though they are too weak to check whether a given real number codes a well-ordering (which is crucial in the constructions for ITRMs and ITTMs). The real number coding the ITBM-halting problem is ITBM-recognizable, which is also true for ITRMs, but fails for ITTMs. The distribution of the ITBM-recognizable real numbers in Gödel's constructible hierarchy L is different for ITBMs than for all other machine types considered so far: From ω^ω up to ω_1^{CK}, new unrecognizable and new recognizable real numbers occur at every level, while for ITTMs and ITRMs, there are "gaps" in the set of levels at which new recognizable real numbers are constructed.[2]

An ordinal α is called an "index" if and only if $L_{\alpha+1} \setminus L_\alpha$ contains a real number. By standard fine-structure (see, e.g., Jensen [10]), $L_{\alpha+1}$ contains a bijection $f : \omega \to L_\alpha$ when α is an index. Moreover, by Theorem 1 of Boolos and Putnam [2], if α is an index, then $L_{\alpha+1}$ contains an "arithmetical copy" of L_α, i.e., a real number coding L_α. It is known – see, e.g., [15] – that every infinite α below ω_1^{CK} is an index (in fact, by [15], a non-index level L_α is a model of ZF$^-$, so that the first non-index level appears way above ω_1^{CK}); we will freely

---

[1] "Weak ITRMs", also known as "unresetting ITRM", differ from ITRMs in that a computation in which the inferior limit of the sequence of contents of some register is infinite at some limit time, the computation is undefined, while for ITRMs, the content of such a register is just reset to 0; they were defined by Koepke in [11].

[2] For ITRMs, this is proved in [4]; for ITTMs, it is known that there are, e.g., no recognizable real numbers in $L_\Sigma \setminus L_\lambda$, see, e.g., [8], Theorem 4.2.6.

use this fact below. Below, unless indicated otherwise, p will denote Cantor's pairing function.

1.1   Infinite Time Blum-Shub-Smale Machines

Infinite Time Blum-Shub-Smale machines were introduced in Koepke and Seyfferth ([14]) and then studied further in Koepke and Morozov [13] and Welch [19]. We briefly recall the definitions and results required for this article.

Like a Blum-Shub-Smale machine (BSSM), an ITBM has finitely many registers, each of which can store a single real number. An ITBM-program is just an ordinary Blum-Shub-Smale-machine program, i.e., a finite, numerated list of commands for applying a rational functions to the contents of some registers and (i) replacing the content of some register with the result or (ii) jumping to some other program line, depending on whether the value of the function is positive or not; this latter kind of command is called a "node". At successor times, an ITBM works like a BSSM, while at limit levels, the active program line is the inferior limit of the sequence of earlier program lines and the content of each register R is the Cauchy limit of the sequence of earlier contents of R, provided this sequence converges; if this sequence does not converge for some register, the computation is undefined.

We fix a natural enumeration $(P_i : i \in \omega)$ of the ITBM-programs. For an ITBM-program P and a real number x, we write P^x for the computation of P that starts with x in the first register.

Definition 1. *A real number x is ITBM-computable if and only if there is an ITBM-program P that starts with 0s in all of its registers and halts with x in its first register.*

We say that a real number x is ITBM-recognizable if and only if there is an ITBM-program P such that, for all real numbers y, P^y halts with output 1 if and only if $y = x$ and otherwise, P^y halts with output 0.

We summarize the relevant results about ITBMs in the following theorem.

Theorem 1. *(i) (Koepke, Seyfferth, [14]) If P is an ITBM-program using $n \in \omega$ many nodes and x is a real number, then P^x halts in $< \omega^{n+1}$ many steps or it does not halt at all. In particular, any ITBM-program P^x either halts in $< \omega^\omega$ many steps or not at all. An ordinal α is ITBM-clockable if and only if $\alpha < \omega^\omega$.*

(ii) (Koepke, Morozov [13], Welch [19]) A real number x is ITBM-computable from the real input y if and only if $x \in L_{\omega^\omega}[y]$. In particular, x is ITBM-computable if and only if $x \in L_{\omega^\omega}$.

As a consequence of (i), it is possible to decide, for every ITBM-program P, the set $\{x \subseteq \omega : P^x \text{ halts}\}$ on an ITBM: Namely, if P uses n nodes, simply run P^x for ω^{n+1} many steps and see whether it has halted up to this point (clearly, any ITBM-program that does this will use more than n nodes). Thus, if a partial function $f : \mathbb{R} \to \mathbb{R}$ is ITBM-computable, there is also a total ITBM-computable function $\hat{f} : \mathbb{R} \to \mathbb{R}$ such that $\hat{f}(x) = f(x)$ whenever $f(x)$ is defined and otherwise $\hat{f}(x) = 0$. These properties of ITBMs will be freely used below.

2   The Lost Melody Theorem for ITBMs

In this section, we will show that there is a real number x which is ITBM-recognizable, but not ITBM-computable.

Let x be a real number with the following properties:

1. There is a bijection $f : \omega \to L_{\omega^\omega}$ such that $x = \{p(i,j) : f(i) \in f(j)\}$ and f is such that $f[\{2i : i \in \omega\}] = \omega^\omega$, so that ordinals are coded exactly by the even numbers. We fix f from now on.
2. $x \in L_{\omega^\omega} + 1$. In particular, x is definable over L_{ω^ω}, and in fact definable without parameters (by fine-structure). Let ϕ_x be an $\in$-formula such that $x = \{i \in L_{\omega^\omega} : L_{\omega^\omega} \models \phi_x(i)\}$.
3. The real number $c := \{p(i,j) : p(2i, 2j) \in x\}$ (which, by definition, is a code of ω^ω) is recursive.

Lemma 1. *Let $c \subseteq \omega$ be such that, for some ordinal α and some bijection $f : \omega \to \alpha$, we have $c = \{p(i,j) : i, j \in \omega \wedge f(i) \in f(j)\}$. Then $f \in L_{\alpha+1}[c]$. In particular, if c is recursive and $\alpha > \omega + 2$, then $f \in L_{\alpha+1}$.*

Proof. We need to show that f is definable over $L_\alpha[c]$. First suppose that α is a limit ordinal. Then f is defined as follows. For $i \in \omega$, we have $f(i) = \beta$ if and only if there is a sequence $(a_\iota : \iota \le \beta)$ of natural numbers with the following properties:

1. For all $k \in \{a_\iota : \iota \le \beta\} =: A$, and all $j \in \omega$, if $p(j,k) \in c$, then $j \in A$
2. for all $\iota, \xi \le \beta$, we have $\iota < \xi$ if and only if $p(a_\iota, a_\xi) \in c$
3. $a_\beta = i$

When α is a limit ordinal, these sequences will be contained in L_α, so the above provides a definition of f over L_α. When α is a successor ordinal, the above works up to the last limit ordinal before α and then the remaining values of f can be defined separately explicitly; we skip the details of this case.

The second claim now follows from the first as a recursive real number c is contained in $L_{\omega+1}$, so that $L_{\alpha+1}[c] = L_{\alpha+1}$ when $\alpha > \omega + 2$. □

Lemma 2. *There is a real number x satisfying (1)-(3) above.*

Proof. It is clear that the Skolem hull of the empty set in L_{ω^ω} is equal to L_{ω^ω}. By standard fine-structure (see [10]), this implies that $L_{\omega^\omega + 1}$ contains a bijection $g : \omega \to L_{\omega^\omega}$.

Moreover, as $\omega^\omega < \omega_1^{CK}$, there is a recursive code c for ω^ω. Using Lemma 1, a function $h : \omega \to \omega^\omega$ such that $c = \{p(i,j) : h(i) \in h(j)\}$ is definable over L_{ω^ω}.

Now define $f : \omega \to L_{\omega^\omega}$ by letting, for $i \in \omega$, $f(2i) = h(i)$ and letting $f(2i+1)$ be the g-image of the ith natural number whose g-image is not an ordinal. Since g is definable over L_{ω^ω}, so is f.

Now let $x := \{p(i,j) : i, j \in \omega \wedge f(i) \in f(j)\}$. Then x is definable over L_{ω^ω} and by definition as desired.

We now show that x is a lost melody for ITBMs.

Lemma 3. *(Truth predicate evaluation) Given a real number y coding the structure (Y, R) (with Y a set, $R \subseteq Y \times Y$ a binary relation on Y, $g : \omega \to Y$ a bijection and $y = \{p(i, j) : (f(i), f(j)) \in Y\}$) there is an ITBM-program P_{truth} that computes the truth predicate over (Y, R) (i.e., for each $\in$-formula ϕ and all $i_1, ..., i_n \in \omega$, P_{truth} will decide, on inputs y and $(\phi, (i_1, ..., i_n))$, whether or not $(Y, R) \models \phi(g(i_1), ..., g(i_n))$).*

Proof. By Proposition 2.7 of Koepke and Morozov [13], there is an ITBM-program P such that, for each input $y \subseteq \omega$, P^y computes the (classical) Turing-jump of y. By the iteration lemma in [12], there is also an ITBM-program H that computes the ω-th iteration $y^{(\omega)}$ of the Turing-jump of y. But it is clear that the truth predicate for (Y, R) is recursive in $y^{(\omega)}$, and a fortiori ITBM-computable.

Corollary 1. *(Identification of natural numbers) There is a program that identifies the natural numbers coding natural numbers in a real code r for a structure (A, R). Moreover, there is a program P_{id} that, for each natural number k, identifies the natural number i that codes k in the sense of r, provided such i exists.*

Proof. The first part is an immediate consequence of the last lemma.

For the second part, note that there is a recursive function that maps each $k \in \omega$ to an $\in$-formula ψ_k such that $\psi_k(x)$ holds if and only if $x = k$.[3] But then, searching for the natural number coding k is an easy application of the last lemma: Just check successively, for each $i \in \omega$, whether $\psi_k(g(i))$ holds in (A, R) and output the first $i \in \omega$ for which it is true.

Lemma 4. *x is not ITBM-computable.*

Proof. Since ITBM-halting times are bounded by ω^ω, L_{ω^ω} contains all halting ITBM-computations. Thus, the statement that the ith ITBM-program P_i halts is Σ_1 over L_{ω^ω}. By bounded truth predicate evaluation, the set H of $i \in \omega$ for which P_i halts - i.e., the ITBM halting set - is thus ITBM-computable from x. By Koepke and Morozov [13] (transitivity lemma), H would thus be ITBM-computable if x was ITBM-computable. Thus, x is not ITBM-computable.

Lemma 5. *x is ITBM-recognizable.*

Proof. Let the input y be given. First, use truth predicate evaluation to check whether y codes a model of the sentence "I am an L-level" from Boolos [1], Theorem 1'. If not, output 0.

If yes, check whether, in y, $i \in \omega$ codes an ordinal if and only if i is even. This can be determined by computing $y^{(\omega)}$ in which the set s of natural numbers coding ordinals in the sense of y is recursive, and then checking whether the Turing program that (in the oracle s) runs through ω and halts once it has found an odd number in s or an even number not in s halts.

---

[3] For example, we can take $\psi_0(x)$ to be $\forall y \in x(y \neq y)$ and then let $\psi_{k+1}(x)$ be $\forall y(y \in x \leftrightarrow (\exists z(\psi_k(z) \wedge y \in z) \vee \psi_k(y)))$.

If not, return 0. Otherwise, continue.

Check whether $\{p(i,j) : p(2i,2j) \in y\} = c$. This is possible as c is recursive, so we can simply compute c and compare it to $\{p(i,j) : p(2i,2j) \in y\}$. If not, return 0. Otherwise, we know that y codes L_{ω^ω}, and we only need to check whether it is the "right" code. To do this, we continue as follows:

Using bounded truth predicate evaluation and identification of natural numbers, compute the set s of natural numbers i such that $L_{\omega^\omega} \models \phi_x(i)$. At this point, we know that $s = x$. Now simply compare s to y. If they are equal, output 1, otherwise output 0.

We note for later use that the proof of Lemma 5 shows more:

Corollary 2. *Let $\alpha < \omega_1^{CK}$, so that α is an index. Then L_α has an ITBM-recognizable code c. In fact, c can be taken to be contained in $L_{\alpha+1}$.*

Proof. Since $\alpha < \omega_1^{CK}$, there is a recursive real number r that codes α. But then, since $L_{\alpha+1}$ contains a bijection $f : \omega \to L_\alpha$, there is, as in Lemma 2, a code c for L_α that is (i) contained in $L_{\alpha+1}$, (ii) codes ordinals by even numbers and such that (iii) $\{p(i,j) : i, j \in \omega \wedge p(2i,2j) \in c\} = r$. Now c is recognizable as in the proof of Lemma 5.

Thus, we obtain:

Theorem 2. *There is a lost melody for Infinite Time Blum-Shub-Smale machines.*

It is known from [3] that the set of indices of halting ITRM-programs is ITRM-recognizable, while the set of indices of halting ITTM-programs is not ITTM-recognizable. Here, we show that ITBMs resemble ITRMs in this respect: Namely, define H to be the set of natural numbers i such that P_i halts. It is not hard to see, (though a bit cumbersome)[4] that a code c for L_{ω^ω} is ITBM-computable from H, say by the program P. Now, to identify whether a given real number x is equal to H, first check, using the bounded halting problem solver, whether P^x will halt. If not, output 0 and halt. If yes, let y be the output of P^x and check, as in the proof of Lemma 5, whether y is a code for L_{ω^ω}. If not, output 0 and halt. Otherwise, use y to compute, again as in the proof of Lemma 5, the set H, (which is Σ_1 over L_{ω^ω}) and compare x to H. Thus, we get:

---

[4] Here is a sketch for the construction: Use p to split ω into ω many disjoint portions of the form $\{p(k,i) : i \in \omega\}$. For $i \in \omega$, let f_0 be the $<_L$-minimal bijection $f_0 : \omega \to L_\omega$ and, for $i > 0$, let f_i be the $<_L$-minimal bijection $f_i : \omega \to L_{\omega^{i+1}} \setminus L_{\omega^i}$; for $k \in \omega$, let $F_k := \bigcup_{i<k} f_i$. Then let $c_k := \{p(i,j) : F_k(i) \in F_k(j)\}$ and $c := \bigcup_{k \in \omega} c_k$. Thus, c is a code for L_{ω^ω}. To compute c, it suffices to compute f_k and c_k, uniformly in k. For this, run through the ITBM-programs and use H to identify the first program Q that computes a code d for $L_{\omega^{k+1}}$. Using H, we can actually obtain d by considering, for each $i \in \omega$, the program Q_i' which, for $i \in \omega$, halts when $Q(i)$ halts with output 0 and loops otherwise and using H to check whether Q_i' halts (note that Q_i' is recursive in Q and i). From d, one can compute f_0, ..., f_{k-1}, and hence also F_k, using truth predicate evaluation (since natural numbers are definable without parameters); again using truth predicate evaluation, one obtains c_k.

Theorem 3. *The real number H coding the halting problem for ITBMs is ITBM-recognizable.*

3 The Distribution of ITBM-Recognizable Real Numbers

Where do ITBM-recognizable real numbers occur in L? This question was studied in detail in [4] and [3] for the case of ITRMs and in [5] for wITRMs, where it turned out that the wITRM-recognizable real numbers coincide with the wITRM-computable real numbers (i.e., there are no lost melodies for wITRMs), which are known from Koepke [11] to coincide with the hyperarithmetical real numbers. By an adaptation of the proof in [5], we obtain:

Lemma 6. *Let $x \subseteq \omega$ be ITBM-recognizable. Then $x \in L_{\omega_1^{CK}}$.*

Proof. Let $x \subseteq \omega$ be ITBM-recognizable, and let P be an ITBM-program that recognizes x. It follows that there is an ordinal $\gamma < \omega^\omega$ such that P^x halts in exactly γ many steps. As a snapshot of an ITBM can easily be encoded as a real number, the same holds for a γ-sequence of such snapshots.

Now, the statement "There is a real number g such that g codes an ITBM-computation of length γ by P in the oracle y that halts with output 1" is Σ_1^1; thus, the set of such y is Σ_1^1, and, in particular, $\{x\}$ is Σ_1^1. By Kreisel's basis theorem (see, e.g., [18], Theorem 7.2), it follows that $x \in L_{\omega_1^{CK}}$.

Lemma 7. *For every $\alpha < \omega_1^{CK}$, there is an ITBM-recognizable real number x such that $x \notin L_\alpha$. More specifically, if $\alpha < \omega_1^{CK}$, then $L_{\alpha+1} \setminus L_\alpha$ contains an ITBM-recognizable real number.*

Proof. The first claim clearly follows from the second one. We thus show the second claim. Let $\alpha < \omega_1^{CK}$, so that α is an index. If $\alpha < \omega^\omega$, every real number in $L_{\alpha+1} \setminus L_\alpha$ is ITBM-computable and thus ITBM-recognizable. So suppose that $\alpha \geq \omega^\omega$.

By Corollary 2, $L_{\alpha+1}$ contains an ITBM-recognizable code c for L_α. It thus suffices to see that $c \notin L_\alpha$. But it is clear that, as c codes all real numbers contained in L_α, we can define by diagonalization from c a real number not contained in L_α. For $c \in L_\alpha$, that real number would then be contained in L_α as well (see, e.g., [2], Theorem 2), a contradiction.

Thus x is as desired.

In combination with Lemma 6 above, this shows that new ITBM-recognizable real numbers appear wherever they can, i.e., at every L-level $< \omega_1^{CK}$. This is in contrast both with ITRMs and ITTMs, for which there are "gaps" in the set of constructible levels at which new recognizable ordinals appear, see, e.g., [4] or [8].

4   Non-recognizability with and Without Resource Bounds

Given the results of the preceding section that the ITBM-recognizable real numbers appear cofinally in $L_{\omega_1^{CK}}$, it becomes natural to ask whether the same happens for ITBM-nonrecognizability. (Note that, for weak ITRMs, the set of recognizable real numbers coincides with $\mathbb{R} \cap L_{\omega_1^{CK}}$.) Moreover, since ITBMs increase in computational strength the more computational nodes are allowed in the program (so that, in particular, there is no universal ITBM, see Koepke and Morozov [13]), one may wonder whether the same happens for their recognizability strength (which is the case for ITRMs when one increases the number of registers, see [3]).

In this section, we will show that (i) non-ITBM-recognizable real numbers appear cofinally often in $L_{\omega_1^{CK}}$ and (ii) for every $n \in \omega$, there is a real number x that is ITBM-recognizable (in fact ITBM-computable), but not ITBM-recognizable by a program with $\leq n$ many nodes.

The proof idea for both results is to consider real numbers that are Cohen-generic over sufficiently high L-levels below $L_{\omega_1^{CK}}$.[5] However, since we are working below the first admissible ordinal, the amount of set theory available in the relevant ground models is very small. Fortunately, forcing over the very weak set theory PROVI has been worked out by Mathias in [16] and Bowler and Mathias in [17]. The results from these papers that will be relevant below are summarized in the following lemma:

Lemma 8. *(Mathias and Bowler)*
(i) [17] L_α is provident, i.e., a model of PROVI, if and only if α is indecomposable. In particular, L_{ω^ι} is provident for all ordinals $\iota > 0$.
(ii) [16], Theorem 4.17, the forcing theorem for Δ_0-formulas over provident sets] If L_α is provident, then the forcing theorem for Δ_0-formulas (and forcings contained in L_α) holds for L_α. In particular, if G is Cohen-generic over L_α and $\phi(\dot{a}_1, ..., \dot{a}_k)$ is Δ_0 (where $\dot{a}_1, ..., \dot{a}_n$ are names for Cohen forcing in L_α) and G is a Cohen-generic filter over L_α (i.e., G intersects every dense subset of Cohen forcing that is contained in L_α) then $L_\alpha[G] \models \phi(\dot{a}_1^G, ..., \dot{a}_k^G)$ if and only if there is $p \in G$ such that $p \Vdash \phi(\dot{a}_1, ..., \dot{a}_k)$.

Lemma 9. *If $n \in \omega$ and $x \subseteq \omega$ is Cohen-generic over $L_{\omega^{n+1}}$, then x is not ITBM-recognizable by an ITBM-program using $< n$ many nodes. In particular, if x is Cohen-generic over L_{ω^ω}, then x is not ITBM-recognizable.*

Proof. Suppose that x is Cohen-generic over $L_{\omega^{n+1}}$ and ITBM-recognizable by the program P which uses $< n$ nodes. By Theorem 1, since P^x halts, P^x will run for less than ω^n many steps. Consequently, the halting computation of P^x with output 1 will be contained in $L_{\omega^n}[x]$. Thus $L_{\omega^{n+1}}[x]$ believes that $L_{\omega^n}[x]$ contains a halting computation of P in the oracle x with output 1, which is a Δ_0-formula. Let $\dot{A}$ be a name for $L_{\omega^n}[x]$ and let $\dot{x}$ be a name for x. By the

---
[5] The same approach was used in [3] to obtain non-recognizables for ITRMs.

forcing theorem for Δ_0-formulas over provident sets, there is a condition $p \subseteq x$ which forces that $\dot{A}$ contains a halting computation of P in the oracle $\dot{x}$ with output 1. Now let y be a real number that is Cohen-generic over $L_{\omega^{n+1}}$, extends p and is different from x. Then p will also force that P^y halts with output 1, contradicting the assumption that P recognizes x.

If x is Cohen-generic over L_{ω^ω}, it is in particular Cohen-generic over L_{ω^n} for every $n \in \omega$, thus not recognizable by an ITBM-program with any number of nodes, and thus not ITBM-recognizable.

Theorem 4. *(i) For each $n \in \omega$, there is a real number x that is ITBM-computable (and thus ITBM-recognizable), but not ITBM-recognizable by a program with $< n$ many nodes.*

(ii) For each $\alpha < \omega_1^{CK}$, there is an ITBM-nonrecognizable real number in $L_{\omega_1^{CK}} \setminus L_\alpha$. In fact, if $\alpha > \omega^\omega$ is an index, then $L_{\alpha+1} \setminus L_\alpha$ contains a non-ITBM-recognizable real number. Thus, below ω_1^{CK}, new non-ITBM-recognizable real numbers appear at every level after ω^ω.

Proof. (i) For $n \in \omega$, let $x \in L_{\omega^{n+1}+1}$ be Cohen-generic over $L_{\omega^{n+1}}$ (the existence of such an x is proved in part (ii) below). By Lemma 8, $L_{\omega^{n+1}}$ is provident. However, as $x \in L_{\omega^\omega}$, it is ITBM-computable and thus ITBM-recognizable, but not by a program with $< n$ many nodes by Lemma 9.

(ii) By Lemma 9, it suffices to show that $L_{\alpha+1} \setminus L_\alpha$ contains a Cohen-generic real number over L_{ω^ω} whenever $\alpha \geq \omega^\omega$ is an index.

We will show that $L_{\alpha+1}$ in fact contains a real number that is Cohen-generic over L_α, which will in most cases be much more than demanded. By fine-structure, a surjection from ω to L_α is definable over L_α; a fortiori, there is a surjection f from ω to the dense subsets of Cohen-forcing in L_α definable over L_α, say by the formula $\phi_f(x, y, q)$, q a finite tuple of elements of L_α. Now define $x : \omega \to 2$ by letting $x(i) = b$ if and only if there is a finite sequence $(p_j : j \leq k)$ of Cohen-conditions such that $p_0 = \emptyset$ and, for all $j < k$, p_{j+1} is the $<_L$-minimal element of $f(j)$ that extends p_j and such that $p_k(i) = b$. By definition of x, it is Cohen-generic and definable over L_α.

Note that the programs used in the proof of Corollary 2 can all be taken to use the same number of nodes, so that there is a fixed number n such that the real numbers that are ITBM-recognizable by programs with n nodes are already cofinal in $L_{\omega_1^{CK}}$.

This leaves us with the question whether, for any $n \in \omega$, there is a non-ITBM-computable real number that is ITBM-recognizable, but not by a program with n nodes. This is indeed the case.

Theorem 5. *Let $n \in \omega$. Then there are cofinally in $L_{\omega_1^{CK}}$ many real numbers x that are ITBM-recognizable, but not with $\leq n$ nodes. In particular, there are cofinally in $L_{\omega_1^{CK}}$ many ITBM-lost melodies that are not ITBM-recognizable with $\leq n$ nodes.*

Proof. We will show that there is an ITBM-computable injection $f : \mathbb{R} \to \mathbb{R}$ that has an ITBM-computable (partial) inverse function g and maps each real number x to a real number $\tilde{x}$ that is Cohen-generic over $L_{\omega^{n+1}}$. Once this is done, the result can be seen as follows: Let P_g be an ITBM-program that computes g. Given $\alpha < \omega_1^{CK}$, pick a real number c that is ITBM-recognizable, but not contained in L_α (the existence of such real numbers was proved above); we can assume without loss of generality that $\alpha > \omega^\omega$. Let P be an ITBM-program for recognizing c. We claim that $\tilde{c}$ is ITBM-recognizable, but not with $\leq n$ nodes. The latter claim follows since $\tilde{c}$ is by definition Cohen-generic over $L_{\omega^{n+1}}$. To recognize $\tilde{c}$, first check whether $g(\tilde{c})$ is defined. Note that this can be done by clocking $P_g^{\tilde{c}}$ for ω^{m+1} many steps, where m is the number of nodes used in P_g. If not, halt with output 0. Otherwise, compute $g(\tilde{c})$ and return the output of $P^{g(\tilde{c})}$. Since g is injective, this will give the right result.

The encoding f works as follows: Let $\mathcal{D} = (D_i : i \in \omega)$ be an ITBM-computable enumeration of the dense subsets of Cohen-forcing contained in $L_{\omega^{n+1}}$, encoded in some natural way as a real number d. Define a sequence $(\tilde{c}_i : i \in \omega)$ by letting $\tilde{c}_0 = \emptyset$ and $\tilde{c}_{i+1}$ the lexically minimal element e_i of D_i that extends $\tilde{c}_i$ when $i \in c$ and otherwise $\tilde{c}_{i+1}$ is the lexically first element of D_i that properly extends $\tilde{c}_i$ and is incompatible with e_i. Then let $\tilde{c} := \bigcup_{i \in \omega} \tilde{c}_i$. It is now easy to see that there the function $h : \mathbb{R} \times \omega \to \mathbb{R}$ that maps (x, i) to the ith bit of $\tilde{x}$ is actually recursive in d.

Similarly, we can recursively reconstruct x from $\tilde{x}$ and d: Namely, given x, i and d, compute a sufficiently long initial segment of $(\tilde{x}_i : i \in \omega)$ such that the last element fixes the ith bit. To compute $\tilde{x}_{i+1}$ from $\tilde{x}_i$, exhaustively search (in lexicographic order) through all finite partial functions from ω to 2 that extend $\tilde{x}_i$, find the lexically minimal element e_i that properly extends $\tilde{x}_i$ and see whether x extends e_i. If that is the case, then $i \in x$, otherwise, we have $i \notin x$. Now, if we had $\tilde{c} \in L_\alpha$, then $d \in L_{\omega^\omega} \subseteq L_\alpha$, combined with c being recursive in d and $\tilde{c}$, would imply $c \in L_\alpha$, a contradiction; thus, we have $\tilde{c} \notin L_\alpha$.

The second claim now follows, as the ITBM-computable real numbers belong to L_{ω^ω}.

Acknowledgements. We thank our three anonymous referees for various helpful suggestions for improving the presentation of this paper.

References

1. Boolos, G.: On the semantics of the constructible levels. Math. Logic Q. **16**, 139–148 (1970)
2. Boolos, G., Putnam, H.: Degrees of unsolvability of constructible sets of integers. J. Symb. Log. **33**, 497–513 (1968)
3. Carl, M.: Optimal results on ITRM-recognizability. arXiv: Logic (2013)
4. Carl, M.: The distribution of ITRM-recognizable reals. Ann. Pure Appl. Log. **165**, 1403–1417 (2014)
5. Carl, M.: The lost melody phenomenon, pp. 49–70. arXiv: Logic (2014)

6. Carl, M., Fischbach, T., Koepke, P., Miller, R., Nasfi, M., Weckbecker, G.: The basic theory of infinite time register machines. Arch. Math. Logic **49**, 249–273 (2010)
7. Carl, M., Schlicht, P., Welch, P.: Recognizable sets and woodin cardinals: computation beyond the constructible universe. Ann. Pure Appl. Log. **169**, 312–332 (2018)
8. Carl, M.: Ordinal Computability. An Introduction to Infinitary Machines. De Gruyter, Berlin, Boston (2019). https://doi.org/10.1515/9783110496154
9. Hamkins, J., Lewis, A.: Infinite time turing machines. J. Symb. Logic **65**, 567–604 (2000)
10. Jensen, R.: The fine structure of the constructible hierarchy. Ann. Math. Logic **4**, 229–308 (1972)
11. Koepke, P.: Infinite time register machines. In: Beckmann, A., Berger, U., Löwe, B., Tucker, J.V. (eds.) CiE 2006. LNCS, vol. 3988, pp. 257–266. Springer, Heidelberg (2006). https://doi.org/10.1007/11780342_27
12. Koepke, P., Miller, R.: An Enhanced Theory of Infinite Time Register Machines. In: Beckmann, A., Dimitracopoulos, C., Löwe, B. (eds.) CiE 2008. LNCS, vol. 5028, pp. 306–315. Springer, Heidelberg (2008). https://doi.org/10.1007/978-3-540-69407-6_34
13. Koepke, P., Morozov, A.: The computational power of infinite time blum-shub-smale machines. Algebra Logic **56**, 37–62 (2017). https://doi.org/10.1007/s10469-017-9425-x
14. Koepke, P., Seyfferth, B.: Towards a theory of infinite time blum-shub-smale machines. In: Cooper, S.B., Dawar, A., Löwe, B. (eds.) CiE 2012. LNCS, vol. 7318, pp. 405–415. Springer, Heidelberg (2012). https://doi.org/10.1007/978-3-642-30870-3_41
15. Marek, W., Srebrny, M.: Gaps in the constructible universe. Ann. Math. Logic **6**, 359–394 (1974)
16. Mathias, A.: Provident sets and rudimentary set forcing. Fundamenta Mathematicae **230**, 99–148 (2015)
17. Mathias, A.R.D., Bowler, N.J.: Rudimentary recursion, gentle functions and provident sets. Notre Dame J. Formal Logic **56**(1), 3–60 (2015)
18. Sacks, G.E.: Higher Recursion Theory. Perspectives in Logic. Cambridge University Press, Cambridge (2017). https://doi.org/10.1017/9781316717301
19. Welch, P.D.: Discrete transfinite computation. In: Sommaruga, G., Strahm, T. (eds.) Turing's Revolution, pp. 161–185. Springer, Cham (2015). https://doi.org/10.1007/978-3-319-22156-4_6

Randomising Realizability

Merlin Carl[1], Lorenzo Galeotti[2(✉)], and Robert Passmann[3,4]

[1] Europa-Universität Flensburg, 24943 Flensburg, Germany
[2] Amsterdam University College, Postbus 94160,
1090 GD Amsterdam, The Netherlands
l.galeotti@uva.nl
[3] Institute for Logic, Language and Computation, Faculty of Science, University
of Amsterdam, P.O. Box 94242, 1090 GE Amsterdam, The Netherlands
[4] St John's College, University of Cambridge, Cambridge CB2 1TP, England

Abstract. We consider a randomised version of Kleene's realizability interpretation of intuitionistic arithmetic in which computability is replaced with randomised computability with positive probability. In particular, we show that (i) the set of randomly realizable statements is closed under intuitionistic first-order logic, but (ii) different from the set of realizable statements, that (iii) "realizability with probability 1" is the same as realizability and (iv) that the axioms of bounded Heyting's arithmetic are randomly realizable, but some instances of the full induction scheme fail to be randomly realizable.

1 Introduction

Have you met skeptical Steve? Being even more skeptical than most mathematicians, he only believes what he actually sees. To convince him that there is an x such that A, you have to give him an example, together with evidence that A holds for that example. To convince him that $A \to B$, you have to show him a *method* for turning evidence of A into evidence of B, and so on. Given that Steve is "a man provided with paper, pencil, and rubber, and subject to strict discipline" [9], we can read "method" as "Turing program", which leads us to Kleene's realizability interpretation of intuitionistic logic [4].

Steve has a younger brother, pragmatical Per. Like Steve, Per is equipped with paper and pencil; however, he also has a coin on his desk, which he is allowed to throw from time to time while performing computations. By his pragmatical nature, he does not require being successful at obtaining evidence for a given proposition A every time he gives it a try; he is quite happy when it works with probability $(1 - \frac{1}{10^{100}})$ or so, which makes it highly unlikely to ever fail in his lifetime.

The authors would like to thank Rosalie Iemhoff and Jaap van Oosten for discussions about the material included in this paper.

L. De Mol et al. (Eds.): CiE 2021, LNCS 12813, pp. 82–93, 2021.
https://doi.org/10.1007/978-3-030-80049-9_8

Per wonders whether his pragmatism is more powerful than Steve's method. After all, he knows about Sacks's theorem[1] [1, Corollary 8.12.2] that every function $f : \omega \to \omega$ that is computable using coin throws with positive probability is recursive. Can he find evidence for some claims where Steve fails? He also notices that turning such "probabilistic evidence" for A into "probabilistic evidence" for B is a job considerably different (and potentially harder) than turning evidence for A into evidence for B. Could it be that there are propositions whose truth Steve can see, but Per cannot? Although Per is skeptical, e.g., of the law of the excluded middle just like Steve, he is quite fond of the deduction rules of intuitionistic logic; thus, he wonders whether the set of statements for which he can obtain his "highly probable evidence" is closed under these.

Steve is unhappy with his brother's sloppiness. After all, even probability $(1-\frac{1}{10^{100}})$ leaves a nonzero, albeit small, chance of getting things wrong. He might consider changing his mind if that chance was brought down to 0 by strengthening Per's definition, demanding that the "probabilistic evidence" works with probability 1. However, he is only willing to give up absolute security if that leads to evidence for more statements. Thus, he asks whether "probability 1 evidence" is the same as "evidence".

These and other questions will be considered in this paper. To begin with, we will model Per's attitude formally, which gives us the concepts of μ-*realizability* and *almost sure realizability*. We will then show the following: There are statements that are μ-realizable, but not realizable (Theorem 13). The set of μ-realizable statements are closed under deduction in intuitionistic predicate calculus (Theorem 15); in a certain sense to be specified below, the law or excluded middle fails for μ-realizability (Lemma 14). The axioms of Heyting arithmetic except for the induction schema are μ-realized (Theorem 16); and there are instances of the induction schema that are not μ-realized (Theorem 17). Almost sure realizability is the same as realizability (Theorem 22).

2   Preliminaries

Realizability is one of the most common semantic tools for the study of constructive theories and was introduced by Kleene in his seminal 1945 paper [4]. In this work, Kleene connected intuitionistic arithmetic—nowadays called *Heyting arithmetic*—and recursive functions. The essential idea is that a statement is true if and only if there is a recursive function witnessing its truth. For more details on realizability, see also Troelstra's 344 [8], and see van Oosten's paper [7] for an excellent historical survey of realizability. In particular, see [8, Definition 3.2.2] for a definition of realizability in terms of recursive functions. In what follows, we denote this realizability relation by "$\Vdash$", and we will call it "Kleene realizability."

As mentioned in the introduction, we want to give pragmatic Per the ability to throw coins while he tries to prove the truth of a statement. We will imple-

---

[1] Sacks's theorem is the corollary of an earlier result by de Leeuw, Moore, Shannon, and Shapiro, see, e.g., [1, Theorem 8.12.1].

ment this coin throwing by allowing Per to access an infinite binary sequence. Therefore, we will make use of the Lebesgue measure on Cantor space 2^ω. For a full definition, see Kanamori's section on 'Measure and Category' [2, Chapter 0]. We denote the Lebesgue measure by μ. Recall that a set A is Lebesgue measurable if and only if there is a Borel set B such that the symmetric difference of A and B is null. Given an element u of Cantor space we will denote by $N_{u \restriction n}$ the basic clopen set $\{v \in 2^\omega \,;\, u \restriction n \subset v\}$ where as usual $u \restriction n$ is the prefix of u of length n, and $u \restriction n \subset v$ if $u \restriction n$ is a prefix of v. We recall that given a binary sequence s of length n, we have that N_s is measurable and $\mu(N_s) = \frac{1}{2^n}$.

We fix a computable enumeration $(p_n)_{n \in \mathbb{N}}$ of programs. Given a program p that uses an oracle and an element $u \in 2^\omega$ of Cantor space we will denote by p^u the program p where the oracle tape contains u at the beginning of the computation. Moreover, given $n \in \mathbb{N}$ we will denote by $p(n)$ the program that for every oracle $u \in 2^\omega$ returns $p^u(n)$.

A sentence in the language of arithmetic is said to be Δ_0 if it does not contain unbounded quantifiers. We will say that a sentence is *pretty* Σ_1 if it is Δ_0 or of the form $Q_0 Q_1 \ldots Q_n \psi$ where ψ is Δ_0 and Q_i is either an existential quantifier or a bounded universal quantifier for every $0 \leq i \leq n$. Similarly, we will say that a sentence is *universal* Π_1 if it is Δ_0 or of the form $Q_0 Q_1 \ldots Q_n \psi$ where ψ is Δ_0 and Q_i is a universal quantifier for every $0 \leq i \leq n$.

Throughout this paper, we fix codings for formulas and programs. In order to simplify notation, we will use φ to refer to both the formula and its code, and similar for programs p. We end this section with some lemmas on Kleene realizability of pretty Σ_1 and universal Π_1 formulas.

Lemma 1. *There is a program p that for every pretty Σ_1 sentence φ does the following: If φ is true then $p(\varphi)$ halts and outputs a realizer of φ and otherwise it diverges.*

Lemma 2. *There is a program p that for every universal Π_1 sentence φ does the following: If φ is true then $p(\varphi)$ halts and outputs a realizer of φ (we do not specify a behaviour otherwise).*

Lemma 3. *A pretty Σ_1 sentence in the language of arithmetic is realized if and only if it is true. The same result holds for universal Π_1 sentences.*

3   Random Realizability

In this section we will introduce the notion of μ-realizability and prove its basic properties. As mentioned before, we will modify realizability in order to use realizers that can access an element of Cantor space. Then we will say that a sentence is randomly realized if for a non-null set of oracles in Cantor space the program does realize the sentence. Formally we define μ-realizability as follows:

Definition 4 (μ-realizability). *We define two realizability relations by mutual recursion: $\Vdash_O$ whose domain consists of pairs (u, p) of an oracle tape $u \in 2^\omega$ and* *a program p (potentially appealing to an oracle tape); and $\Vdash_\mu$ whose domain consists of programs p (potentially appealing to an oracle tape). The range of both relations are sentences φ in the language of arithmetic. We define:*

1. $(p, u) \not\Vdash_O \bot$,
2. $(p, u) \Vdash_O n = m$ *if and only if $n = m$,*
3. $(p, u) \Vdash_O \varphi \wedge \psi$ *if and only if $(p^u(0), u) \Vdash_O \varphi$ and $(p^u(1), u) \Vdash_O \psi$,*
4. $(p, u) \Vdash_O \varphi \vee \psi$ *if and only if we have $p^u(0) = 0$ and $(p^u(1), u) \Vdash_O \varphi$ or $p^u(0) = 1$ and $(p^u(1), u) \Vdash_O \psi$,*
5. $(p, u) \Vdash_O \varphi \to \psi$ *if and only if for all s such that $s \Vdash_\mu \varphi$, we have that $p^u(s) \Vdash_\mu \psi$,*
6. $(p, u) \Vdash_O \exists \mathbf{x} \varphi$ *if and only if $(p^u(0), u) \Vdash_O \varphi(p^u(1))$,*
7. $(p, u) \Vdash_O \forall \mathbf{x} \varphi$ *if and only if for all $n \in \omega$ we have $(p^u(n), u) \Vdash_O \varphi(n)$.*

For every program p that uses an oracle and every sentence φ in the language of arithmetic, we will denote by $C_{p,\varphi}$ the set: $\{u \in 2^\omega \,;\, (p, u) \Vdash_O \varphi\}$. Let φ be a sentence in the language of arithmetic, r be a positive real number, and p be a program using an oracle. We define $p \Vdash_\mu \varphi \geq r$ as follows: $p \Vdash_\mu \varphi \geq r$ if and only if $\mu(C_{p,\varphi}) \geq r$. In this case we will say that p randomly realizes (or μ-realizes) φ with probability at least r.

We will say that φ is randomly realizable (or μ-realizable) with probability at least r if and only if there is p such that $p \Vdash_\mu \varphi \geq r$. Moreover, we write $p \Vdash_\mu \varphi$ and say that p randomly realizes (or μ-realizes) φ if and only if $p \Vdash_\mu \varphi \geq r$ for some $r > 0$. Finally, we will say that φ is randomly realizable (or μ-realizable) if $\sup\{\mu(C_{p,\varphi}) \,;\, p \Vdash_\mu \varphi\} = 1$.

Note that $\Vdash_\mu$ is well-defined in virtue of the following lemma.

Lemma 5. *For all programs p and sentences φ the set $C_{p,\varphi}$ is Borel. In particular $C_{p,\varphi}$ is measurable.*

Proof. The proof is an induction on the complexity of φ. All the cases except implication follow directly from the closure properties of the pointclass of Borel sets, see, e.g., [5, Theorem 1C.2]. Let us just prove the implication case. Let $\varphi \equiv \psi_0 \to \psi_1$ and p be a program. For every program s let A_s be 2^ω if $s \not\Vdash_\mu \psi_0$ and $C_{p(s),\psi_1}$, otherwise. Then $C_{p,\varphi} = \bigcap_{s \in \mathbb{N}} A_s$. By inductive hypothesis we have that A_s is Borel for every s so $C_{p,\varphi}$ is a countable intersection of Borel sets, which is Borel. □

Why does the definition of μ-realizability have to be so complicated? A simpler, natural attempt would be the following: Consider the notion of Oracle-realizability $\Vdash_{Or}$ which is defined just like Kleene realizability but allowing the realizers to access a fixed oracle, i.e., $\Vdash_{Or}$ is defined like $\Vdash_O$ except for case 5., which would be modified as follows:

5'. $(p, u) \Vdash_{Or} \varphi \to \psi$ if and only if for all s such that $(s, u) \Vdash_{Or} \varphi$, we have that $p^u(s) \Vdash_{Or} \psi$.

On the basis of $\Vdash_{\mathrm{Or}}$, we could then define an alternative notion of μ-realizability as follows:

$$p \Vdash'_\mu \varphi \geq r \Leftrightarrow \mu(\{u \,;\, (p, u) \Vdash_{\mathrm{Or}} \varphi\}) \geq r$$

Unfortunately, it turns out that this relation is not closed under modus ponens and the $\forall$-GEN rule of predicate logic, i.e., the rule that asserts that if x is not free in ψ then from $\psi \to \varphi$ we can infer $\psi \to (\forall x \; \varphi)$. We will discuss some other natural approaches in Sect. 6. For now, we begin our study of μ-realizability by showing that the set of μ-realized sentences of arithmetic is consistent.

Lemma 6. *Let φ be a sentence in the language of arithmetic. Then φ is μ-realized if and only if $\neg\varphi$ is not μ-realized.*

Proof. Assume that both $p \Vdash_\mu \varphi$ and $q \Vdash_\mu \neg\varphi$. Suppose that $u \in C_{q,\neg\varphi}$. Then $q^u(p)$ is a realizer of $\bot$, a contradiction. $\qquad\square$

The following lemma has a crucial role in the theory of μ-realizability.

Lemma 7 (Push Up Lemma). *Let φ be a sentence in the language of first order arithmetic and $0 < r \leq r' < 1$ be positive real numbers. Then φ is randomly realizable with probability at least r if and only if φ is randomly realizable with probability at least r'.*

Proof. The right-to-left direction is trivial. For the left-to-right direction, let φ be randomly realizable with probability at least r. We will show that φ is randomly realized with probability at least r'. Let p be a program such that $\mu(C_{p,\varphi}) \geq r > 0$. By the Lebesgue Density Theorem [3, Exercise 17.9] there are $u \in 2^\omega$ and $n \in \omega$ such that $\frac{\mu(C_{p,\varphi} \cap N_{u\upharpoonright n})}{\mu(N_{u\upharpoonright n})} > r'$. Now, let p' be the program that given an oracle runs p with oracle $(u \upharpoonright n) \circ u$. Note that $\mu(C_{p',\varphi}) = \frac{\mu(C_{p,\varphi} \cap N_{u\upharpoonright n})}{\mu(N_{u\upharpoonright n})} > r'$. Finally, it follows trivially by the definition that p' randomly realizable with probability at least r' as desired. $\qquad\square$

The Push Up Lemma allows to simplify the definition of μ-realizability as follows.

Corollary 8. *A sentence φ in the language of arithmetic is μ-realizable if and only if there is a non-zero $r \leq 1$ such that φ is μ-realizable with probability at least r.*

We conclude this section with some basic interactions between μ-realizability and the logical operators.

Corollary 9. *Let ψ_0 and ψ_1 be sentences and let φ be a formula. Then for every p the following hold:*

1. *$p \Vdash_\mu \psi_0 \wedge \psi_1$ if and only if there are s and q such that $s \Vdash_\mu \psi_0$ and $q \Vdash_\mu \psi_1$.*
2. *$p \Vdash_\mu \psi_0 \vee \psi_1$ if and only if there is q such that $q \Vdash_\mu \psi_0$ or $q \Vdash_\mu \psi_1$.*
3. *If $p \Vdash_\mu \psi_0 \to \psi_1$ then $p(s) \Vdash_\mu \psi_1$ for all s such that $s \Vdash_\mu \psi_0$.*
4. *If $p \Vdash_\mu \exists x\varphi$ then there is $n \in \mathbb{N}$ such that $p(0) \Vdash_\mu \varphi(n)$.*

5. If $p \Vdash_\mu \forall x \varphi$ then for all $n \in \mathbb{N}$ we have $p(n) \Vdash_\mu \varphi(n)$.

Corollary 10. *Let ψ_0 and ψ_1 be sentences and let φ be a formula. Then for every p the following hold:*

1. $p \Vdash_\mu \psi_0 \wedge \psi_1 \geq 1$ *if and only if $p(0) \Vdash_\mu \psi_0 \geq 1$ and $p(1) \Vdash_\mu \psi_1 \geq 1$.*
2. *If $p(1) \Vdash_\mu \psi_0 \geq 1$ or $p(1) \Vdash_\mu \psi_1 \geq 1$ then $p \Vdash_\mu \psi_0 \vee \psi_1 \geq 1$.*
3. *If $p(s) \Vdash_\mu \psi_1 \geq 1$ for all s such that $s \Vdash_\mu \psi_0$ then $p \Vdash_\mu \psi_0 \to \psi_1 \geq 1$.*
4. *If there is $n \in \mathbb{N}$ such that $p(n) \Vdash_\mu \varphi(n) \geq 1$ then $p \Vdash_\mu \exists x \varphi \geq 1$.*
5. *If all $n \in \mathbb{N}$ we have $p(n) \Vdash_\mu \varphi(n) \geq 1$ then $p \Vdash_\mu \forall x \varphi \geq 1$.*

4   Kleene Realizability and Random Realizability

In this section, we will study the relationship between Kleene and random realizability. In particular we will show that the two notion do not coincide.

We start by proving that Kleene realizability and μ-realizability agree on pretty Σ_1 sentences and that therefore μ-realizability for pretty Σ_1 sentences coincides with arithmetic truth.

Theorem 11. *Let φ be a pretty Σ_1 sentence in the language of arithmetic. Then, there are two computable functions P_μ and P_μ^{-1} such that for every p, (i) $p \Vdash \varphi$ implies $P_\mu(p, \varphi) \Vdash_\mu \varphi \geq 1$, and (ii) $p \Vdash_\mu \varphi$ implies $P_\mu^{-1}(p, \varphi) \Vdash \varphi$. Therefore a pretty Σ_1 formula is true if and only if it is μ-realized. The same result holds for universal Π_1 sentences.*

Proof. The case for Δ_0 formulas is straightforward. We show how to extend to (1) pretty Σ_1 formulas, and (2) universal Π_1 formulas.

(1) Let $\varphi \equiv \exists x \psi$, where ψ is pretty Σ_1. First assume that $p \Vdash_\mu \exists x \psi$. Then, by Corollary 9 there must be $n \in \mathbb{N}$ such that $p(0) \Vdash_\mu \psi(n)$. Therefore, by inductive hypothesis $\psi(n)$ is realized. Let $P_\mu^{-1}(p, \varphi)$ be the program that does the following: run in parallel all the instances of the program of Lemma 1 with input $\psi(n)$ with $n \in \mathbb{N}$. By inductive hypothesis, one of these instances must halt. Let $i \in \mathbb{N}$ be the first for which the $\psi(i)$ instance halts. Then on input 0, the program returns $P_\mu^{-1}(p(0), \psi(i))$ and for 1, the program returns i. By inductive hypothesis, the program halts and returns a realizer of φ, as desired.

Now assume that $p \Vdash \exists x \psi$. Then $p(0) \Vdash \psi(p(1))$. Let $f(p, \varphi)$ be the program that returns $P_\mu(p(0), \psi(p(1)))$ if the input is 0 and $p(1)$ if the input is 1. By inductive hypothesis $P_\mu(p(0), \psi(p(1))) \Vdash_\mu \psi(p(1)) \geq 1$. But then by Corollary 10 since for all u we have $P_\mu(p, \varphi)^u(0) = f(p(0), \psi(p(1)))$ and $P_\mu(p, \varphi)^u(0) = p(1)$, we have that $P_\mu(p, \varphi) \Vdash_\mu \exists x \psi \geq 1$ as desired.

(2) Let $\varphi \equiv \forall x \psi$ where ψ is universal Π_1. First assume that $p \Vdash_\mu \forall x \psi$. Let $P_\mu^{-1}(p, \varphi)$ be the program that for all n runs $P_\mu(P_\mu^{-1}(p(n), \psi(n)))$. By Corollary 9 and the inductive hypothesis $P_\mu^{-1}(p(n), \psi(n))$ is a realizer of $\psi(n)$. Therefore $P_\mu^{-1}(p, \varphi)$ is a realizer of $\forall x \psi$ as desired.

Now assume that $p \Vdash \forall x \psi$. Let $P_\mu(p, \varphi)$ be the program that for every n and for every oracle returns $P_\mu(p(n), \psi(n))$. Once more by inductive hypothesis for all n $\mu(C_{P_\mu(p(n),\psi(n)),\psi(n)}) = 1$ but then $\mu(\bigcap_{n\in\mathbb{N}} C_{P_\mu(p(n),\psi(n)),\psi(n)}) = 1$ and $P_\mu(p,\varphi) \Vdash_\mu \varphi \geq 1$ as desired.

The second part of the statement follows from Lemma 3.   □

Corollary 12. *Let φ be any false pretty Σ_1 sentence in the language of arithmetic. Then $(p,u) \Vdash_O \varphi \to \bot$ and $p \Vdash_\mu \varphi \to \bot \geq 1$ for every p and u. The same holds for universal Π_1 formulas.*

We are now ready to prove the main result of this section, namely that μ-realizability and Kleene realizability do not coincide. This result is surprising given that by Sacks's theorem [1, Corollary 8.12.2] functions that are computable with a non-null set of oracles are computable by a classical Turing machine.

Theorem 13. *There is a sentence φ in the language of arithmetic that is randomly realizable but not realizable.*

Proof. Let $\psi(k)$ be the sentence "There is n such that for all ℓ the execution of $p_k(k)$ does not stop in at most ℓ steps or $p_k(k) \neq n$", and φ be the sentence "For all k, $\psi(k)$ or $p_k(k) \neq n$". We will show that φ is not Kleene realizable but that it is μ-realizable.

A Kleene realizer for φ would be a program that computes a total function that, for every code k of a program, returns a natural number which is not the output of $p_k(k)$. By diagonalization, such a program cannot exists: If p_k was a code for such a program, then for every $n \in \omega$ we would have that $p_k(k) = n \Leftrightarrow p_k(k) \neq n$.

Now we want to show that φ is randomly realizable. Fix any realizer s. Let p be the program that given an oracle $u \in 2^\omega$, a natural number k, and $i \in \{0,1\}$ does the following:[2] Let $p^u(k)(0) = u{\restriction}(k+1)$ and $p^u(k)(1) = p'$ where p' is the program that ignores the oracle and does the following:

On input ℓ, p' checks whether $p_k(k)$ stops in ℓ steps. If not, then $p'(\ell)$ is the code of a program that returns 0 on input 0 and s on input 1. Otherwise $p'(\ell)$ is the code of a program that returns 1 on input 0 and on input 1 looks for an μ-realizer of the Δ_0 formula expressing the fact that "$p_k(k) \neq u{\restriction}(k+1)$" by running the algorithms in Lemma 3 and Theorem 11.

Now, for every $k \in \omega$ and $u \in 2^\omega$ we have that $p^u(k)(1) = u{\restriction}(k+1)$ and $p^u(k)(0) = p'$. There are two cases.

If $p_k(k)$ does not halt, then $p'(\ell)(0) = 0$ and $p'(\ell)(1) = s$ for every ℓ. Moreover, by Corollary 12 $(s,u) \Vdash_O$ "$p_k(k)$ does not halt in ℓ steps" and therefore $(p^u(k),u) \Vdash_O \psi(k)$.

If $p_k(k)$ halts, then let ℓ be such that $p_k(k)$ halts in at most ℓ steps. Then, $p'(\ell)(0) = 1$. Moreover, note that if the output of $p_k(k)$ is not the same as the first k bits of the oracle then $(p'(\ell)^u(1),u) \Vdash_O u{\restriction}(k+1) \neq p_k(k)$.

We only need to show that $\mu(C_{p,\varphi}) > 0$. To see this, it is enough to note that the set of u such that $p_k(k) \neq u{\restriction}(k+1)$ has measure $\geq 1 - \frac{1}{2^{(k+1)}}$. Therefore, $\mu(C_{p,\varphi}) = \prod_{k\in\mathbb{N}}(1 - \frac{1}{2^{(k+1)}}) > 0$ as desired.   □

---

[2] Here, we do not distinguish between the finite sequence $u{\restriction}(k+1)$ and the natural number coding it.

5   Soundness and Arithmetic

In this section, we study the logic and arithmetic of μ-realizability. We first observe that some instances of the Law of Excluded Middle are not μ-realizable.

Lemma 14. *There is φ such that $\forall x(\varphi(x) \lor \neg\varphi(x))$ is not μ-realizable.*

Proof. Let $\varphi(x)$ be the formula expressing the fact that the program $p_x(x)$ halts. Assume that $\forall x(\varphi(x) \lor \neg\varphi(x))$ is randomly realized. Then, there is a program p such that $p \Vdash_\mu \forall x(\varphi(x) \lor \neg\varphi(x))$. Therefore, p computes the halting problem for a set of oracles of measure > 0. But this directly contradicts Sacks's theorem [1, Corollary 8.12.2]. ☐

Theorem 15 (Soundness). *The set of μ-realizable statements is closed under the rules of intuitionistic first-order calculus.*

It is a classical result that the axioms of Heyting Aritmetic are realizable, see [6, Theorem 1]. However, only a fragment of HA is μ-realizable. Let HA$^-$ denote the axioms of Peano arithmetic without the induction schema. As usual, *Heyting arithmetic* HA is the theory obtained from adding the induction schema to HA$^-$. We say that a set of formulas Γ is μ-realized if φ is μ-realized for all $\varphi \in \Gamma$.

Since all the axioms except for the induction schema are universal Π_1 statements, it follows by Theorem 11 that the axioms of HA$^-$ are all μ-realized.

Theorem 16. *The set HA$^-$ is μ-realized.*

Contrary to Kleene realizability, the induction schema is not μ-realizable. A consequence of the following theorem is that the negation of some instances of the induction schema are μ-realized.

Theorem 17. *Some instances of the induction schema are not μ-realized.*

Proof. Let $\varphi(x)$ be the formula expressing the fact that "Every program with code $i < x$ halts or does not halt". By the proof of Lemma 14, φ is not μ-realizable.

On the other hand, a μ-realizer $p(n)$ for $\varphi(n)$ is given by a program that does the following: for every $i < n$, p returns a program that if the ith element of the oracle is 1 returns 1 on input 0 and any number on input 1. While if the ith element of the oracle is 0 the program returns 0 on input 0 and on input 1 starts building a realizer of "the program i halts" using the algorithm in Lemma 1; if it finds one, it runs the algorithm in Theorem 11 to compute the desired μ-realizer.

It is not hard to see that the algorithm works with probability $\frac{1}{2^n}$. Thus, to realize the implication $\varphi(n) \rightarrow \varphi(n+1)$, we can ignore the μ-realizer for $\varphi(n)$ and just output $p(n)$. So the premise of the instance of the induction schema is μ-realized, while the conclusion is not. ☐

Note that the proof of Theorem 17 heavily relies on the fact that the definition of μ-realizability does not require any relationship between the measures of the set of oracles realizing the antecedent of an implication and the set of oracles realizing the consequent. We think that a modification of this definition could lead to a notion of probabilistic realizability that realizes the induction schema.

Even though the axiom schema of induction is not μ-realizable, one can prove that all Δ_0-instances of the schema are realizable. Indeed, by Theorem 11 and the fact that if φ is a Δ_0 formula then $\forall x \varphi(x, \bar{y})$ is a universal Π_1 formula, we have the following:

Corollary 18. *The set* HA$^-$ *together with the induction schema restricted to* Δ_0 *formulas is μ-realizable.*

6   Big Realizability

When we defined μ-realizability in Sect. 3, we explained that a *prima facie* more natural definition would not work because it is not sound for intuitionistic logic. There are yet some other potential definitions that we will consider in this section. These notions of realizability arise from notions of *big sets of oracles* on Cantor space. The idea is that a statement is realised whenever it is realised with a 'big' set of oracles. More specifically, we will consider "almost sure realizability," "comeagre realizability," "interval-free realizability," and "positive measure realizability." It will turn out, however, that these notions are not so interesting: The first three are equivalent to Kleene realizability, and the fourth notion coincides with arithmetic truth. We begin with the following general definition.

Definition 19. *Let $\mathcal{F}$ be a family of subsets of Cantor space 2^ω. We then define $\mathcal{F}$-realizability recursively as follows:*

1. $p \Vdash_{\mathcal{F}} \bot$ *never,*
2. $p \Vdash_{\mathcal{F}} n = m$ *if and only if $n = m$,*
3. $p \Vdash_{\mathcal{F}} \psi_0 \wedge \psi_1$ *if and only if $p(i) \Vdash_{\mathcal{F}} \psi_i$ for $i < 2$,*
4. $p \Vdash_{\mathcal{F}} \psi_0 \vee \psi_1$ *if and only if there is some $O \in \mathcal{F}$ and some $i < 2$ such that for every $u \in O$, we have $p^u(0) = i$ and $p^u(1) \Vdash_{\mathcal{F}} \psi_i$,*
5. $p \Vdash_{\mathcal{F}} \varphi \rightarrow \psi$ *if and only if there is a set $O \in \mathcal{F}$, such that for every $u \in O$ and $s \Vdash_{\mathcal{F}} \varphi$, we have $p^u(s) \Vdash_{\mathcal{F}} \psi$,*
6. $p \Vdash_{\mathcal{F}} \exists x \varphi$ *if and only if there is some $O \in \mathcal{F}$ and $n \in \omega$ such that for all $u \in O$ we have $p^u(0) = n$ and $p^u(1) \Vdash \varphi(n)$,*
7. $p \Vdash_{\mathcal{F}} \forall x \varphi$ *if and only if there is a set $O \in \mathcal{F}$, such that for every $u \in O$ and $n \in \mathbb{N}$ we have $p^u(n) \Vdash_{\mathcal{F}} \varphi(n)$.*

From this definition, we derive the following notions of realizability: Let $\mathcal{F}_{\text{if}}$ be the family of *co-interval-free* subsets of the Cantor space, i.e. $X \in \mathcal{F}_{\text{cif}}$ if and only if $X \in 2^\omega$ and there is no open interval I such that $I \subseteq 2^\omega \setminus X$, and $\Vdash_{\text{cif}}$ denotes $\mathcal{F}_{\text{cif}}$-realizability. Let $\mathcal{C}$ be the family of comeagre subsets of the Cantor space, then let $\Vdash_{\mathcal{C}}$ denote $\mathcal{C}$-realizability. Let $\mathcal{F}_{=1}$ be the family of subsets of the

Cantor space that are of measure 1, and let $\Vdash_{=1}$ denote $\mathcal{F}_{=1}$-realizability. Let $\mathcal{F}_{>0}$ be the family of subsets of the Cantor space of positive measure, and $\Vdash_{>0}$ denotes $\mathcal{F}_{>0}$-realizability. As before, we will write $\Vdash_{\mathcal{F}} \varphi$ if and only if there is some realizer p such that $p \Vdash_{\mathcal{F}} \varphi$.

In what follows we will make use of the *bounded exhaustive search with* $p(n)$, i.e., the following procedure. Given a program p (and possibly some input n), do the following successively for all $k \in \omega$. Enumerate all 0-1-strings of length k. For each of these strings s, do the following: Run $p^s(n)$ for k many steps. If the computation does not halt within that time (which implies in particular that at most the first k many bits of the oracle were requested), continue with the next s (if there is one, otherwise with $(k+1)$). If the computation halts with output x within that time, then the search terminates with output x.

A similar procedure is used in [1, Theorem 8.12.1, Corollary 8.12.2] and its crucial property is the following:

Lemma 20. *Let $G \subseteq \omega$, $n \in \omega$ and let p be a program. Suppose that there is a set $S \subseteq 2^\omega$ such that $2^\omega \setminus S$ is interval-free and $p^u(n)$ terminates for all $u \in S$ with output $k \in G$. Then the bounded exhaustive search with $p(n)$ will terminate with output $k \in G$.*

Proof. Note that for every n and $u \in S$ we have that $p^u(n)$ terminates with output in G. So, there is a finite initial segment s of u such that $p^s(n)$ terminates with output $p^u(n)$. So, the bounded exhaustive search will halt.

Now, note that if the search halts on the string s with output $k \in \omega$, but $k \notin G$, then $p^x(n) \downarrow k$ for all $u \in N_s$. But then, $N_s \subseteq 2^\omega \setminus S$ which contradicts the fact that $2^\omega \setminus S$ is interval free. $\square$

Lemma 21. *Let $X \subseteq 2^\omega$ be a subset of Cantor space. If $\mu(X) = 0$ or X is meagre, then X is interval-free.*

Theorem 22. *Let $\mathcal{F}$ be a family of subsets of Cantor space such that every $X \in \mathcal{F}$ is co-interval-free. There are programs $P_{\mathcal{F}}$, "realisability to $\mathcal{F}$-realisability," and $P_{\mathcal{F}}^{-1}$, "$\mathcal{F}$-realisability to realisability," such that the following hold for all statements φ:*

1. *If $r \Vdash \varphi$, then $P_{\mathcal{F}}(r, \varphi) \Vdash_{\mathcal{F}} \varphi$.*
2. *If $r \Vdash_{\mathcal{F}} \varphi$, then $P_{\mathcal{F}}^{-1}(r, \varphi) \Vdash \varphi$.*

Consequently, φ is realisable if and only if it is $\mathcal{F}$-realisable.

Proof. Only the cases for $\exists$, $\forall$ and $\vee$ make use of bounded exhaustive search. The other cases are straightforward. The case for $\vee$ is essentially a special case of the $\exists$-case.

(1) φ is $\exists x \psi(x)$. Let $p \Vdash \exists x \psi(x)$. Then $p(0) = n$ and $p(1) \Vdash \psi(n)$. By induction hypothesis, it follows that $P_{\mathcal{F}}(p(1), \psi) \Vdash_{\mathcal{F}} \psi(n)$. So let $P_{\mathcal{F}}(p, \varphi)$ be the program that output n on input 0, and $P_{\mathcal{F}}(p(1), \psi)$ on input 1. Then, $P_{\mathcal{F}}(p, \varphi) \Vdash_{\mathcal{F}} \varphi$.

Conversely, let $p \Vdash_{\mathcal{F}} \exists x \psi(x)$. Then there is some $O \in \mathcal{F}$ and $n \in \omega$ such that $p^u(0) = n$ and $p^u(1) \Vdash_{\mathcal{F}} \psi(n)$. By induction hypothesis, $P_{\mathcal{F}}^{-1}(p^u(1), \psi) \Vdash \psi(n)$.

Define $P_{\mathcal{F}}^{-1}(p, \varphi)$ to be the following program: First, start a bounded exhaustive search with $p(0)$. By Lemma 20 this search must terminate with output n. Return n on input 0, and return $P_{\mathcal{F}}^{-1}(p^u(1), \psi)$ on input 1. Then $P_{\mathcal{F}}^{-1}(p, \varphi) \Vdash \exists x \psi(x)$.

(2) φ is $\forall x \psi(x)$. Let $p \Vdash \forall x \psi(x)$. Then $p(n) \Vdash \psi(n)$ for every $n \in \omega$. Let $P_{\mathcal{F}}(p, \varphi)$ be the program that, given $n \in \omega$, returns $P_{\mathcal{F}}(p(n), \psi)$. With the induction hypothesis, it follows that $P_{\mathcal{F}}(p, \varphi) \Vdash_{\mathcal{F}} \varphi$.

Conversely, let $p \Vdash_{\mathcal{F}} \forall x \psi(x)$. Then there is some $O \in \mathcal{F}$ such that for every $u \in O$ and $n \in \mathbb{N}$ we have that $p^u(n) \Vdash \psi(n)$. Define $P_{\mathcal{F}}^{-1}(p, \varphi)$ to be the following program: Start a bounded exhaustive search with $p(n)$. By Lemma 20, this search will terminate with $p' \Vdash_{\mathcal{F}} \psi(n)$. Then return $P_{\mathcal{F}}^{-1}(p', \psi)$, which, by induction hypothesis, is a realizer of $\psi(n)$. Hence, $P_{\mathcal{F}}^{-1}(p, \varphi) \Vdash \psi$.   □

As a consequence, we derive the corollary that realisability, co-interval-free realisability, comeagre realisability and measure-1 realisability coincide.

Corollary 23. *We have that* $\Vdash \, = \, \Vdash_{\mathrm{cif}} \, = \, \Vdash_{\mathcal{C}} \, = \, \Vdash_{=1}$.

On the other hand, positive-measure-realizability coincides with arithmetical truth.

Theorem 24. *Let* φ *be a formula. Then* $\Vdash_{>0} \varphi$ *if and only if* φ *is an arithmetic truth.*

Proof. This is shown by induction on φ.   □

The following corollary is an immediate consequence.

Corollary 25. *Every instance of the law of excluded middle is* $\mathcal{F}_{>0}$-*realised. Also, the halting problem is* $\mathcal{F}_{>0}$-*realised. Hence,* $\Vdash_{>0}$ *and* $\Vdash$ *are not the same.*

References

1. Downey, R., Hirschfeldt, D.: Algorithmic Randomness and Complexity. Theory and Applications of Computability. Springer, New York (2010). https://doi.org/10.1007/978-0-387-68441-3
2. Kanamori, A.: The Higher Infinite: Large Cardinals in Set Theory from Their Beginnings. Springer Monographs in Mathematics. Springer, Berlin (2008). https://doi.org/10.1007/978-3-540-88867-3
3. Kechris, A.: Classical Descriptive Set Theory. Graduate Texts in Mathematics, vol. 156. Springer, New York (2012). https://doi.org/10.1007/978-1-4612-4190-4
4. Kleene, S.C.: On the interpretation of intuitionistic number theory. J. Symb. Logic **10**, 109–124 (1945)
5. Moschovakis, Y.: Descriptive Set Theory. Studies in Logic and the Foundations of Mathematics, vol. 100. Elsevier, Amsterdam (1987)
6. Nelson, D.: Recursive functions and intuitionistic number theory. Trans. Am. Math. Soc. **61**(2), 307–368 (1947)
7. van Oosten, J.: Realizability: a historical essay. Math. Struct. Comput. Sci. **12**(3), 239–263 (2002)

8. Troelstra, A.S. (ed.): Metamathematical Investigation of Intuitionistic Arithmetic and Analysis. Lecture Notes in Mathematics, Vol. 344. Springer-Verlag, Berlin, New York (1973). https://doi.org/10.1007/BFb0066739

9. Turing, A.M.: 'Intelligent machinery', national physical laboratory report. In: Meltzer, B., Michie, D. (eds.) Machine Intelligence 5 (1969). Edinburgh University Press, Edinburgh (1948)

Restrictions of Hindman's Theorem: An Overview

Lorenzo Carlucci[✉]

Department of Computer Science, Sapienza University, Rome, Italy
carlucci@di.uniroma1.it

Abstract. I give a short overview on bounds on the logical strength and effective content of restrictions of Hindman's Theorem based on the family of sums for which monochromaticity is required, highlighting a number of questions I find interesting.

1 Introduction

The Finite Sums Theorem by Neil Hindman [14] (HT) is a celebrated result in Ramsey Theory stating that for every finite coloring of the positive integers there exists an infinite set such that all the finite non-empty sums of distinct elements from it have the same color. Thirty years ago Blass, Hirst and Simpson [3] proved that *all* X-computable instances of HT have *some* solutions computable in $X^{(\omega+1)}$ and that for *some* X-computable instance of HT *all* solutions compute X'. In terms of Reverse Mathematics: ACA_0^+ proves HT and HT proves ACA_0 over RCA_0. For the latter implication two colors are sufficient. Closing the gap between the upper and lower bound is one of the major open problems in Reverse Mathematics (see, e.g., [19]).

Hindman, Leader and Strauss posed the following question some twenty years ago: Is there a proof that whenever $\mathbf{N}$ is finitely coloured there is a sequence $x_1, x_2, \ldots$ such that all x_i and all $x_i + x_j$ ($i \neq j$) have the same colour, that does not also prove the Finite Sums Theorem? (Question 12 in [15]). This question can be profitably translated and made precise in the setting of Reverse Mathematics: does HT restricted to sums of at most two terms imply the full version over, say, RCA_0? Around the same years, Blass [2] advocated the study of restrictions of HT in which a bound is put on the length (i.e., number of distinct terms) of sums for which monochromaticity is required, asking whether the complexity of HT grows as a function of the length of sums.

The above questions stirred some research attention in recent years [4–7,9, 12,16]. Although no major advance has been achieved regarding the strength of HT, some interesting results concerning the logical and effective strength of many natural restrictions have been obtained. The insights obtained might be of help for solving the main problem and also suggest interesting new problems.

The general picture is the following: any choice of a family $\mathcal{F}$ of finite sums of distinct elements gives rise to a restriction of Hindmans' Theorem, in which the monochromaticity is only required for sums specified by $\mathcal{F}$. How does the

© Springer Nature Switzerland AG 2021
L. De Mol et al. (Eds.): CiE 2021, LNCS 12813, pp. 94–105, 2021.
https://doi.org/10.1007/978-3-030-80049-9_9

choice of $\mathcal{F}$ impact on the logical and effective strength of the corresponding restriction of Hindman's Theorem?

Implications from Ramsey-type theorems to Hindman-type theorems and viceversa are of crucial interest in the perspective of improving upper and lower bounds on HT. In particular this is true if the implications can be witnessed by natural combinatorial reductions, or, more formally, by one of the many notions of *effective reduction* that have been defined and investigated intensively in recent years, such as Weirauch, (strong) computable reductions etc. (see [10]).

This paper is a short, non exhaustive, overview of some of these developments. I focus on results obtained by reasonably simple proofs, basically of two types: coding of the Halting Set and reduction proofs using simple functionals for transforming instances/solutions of one problem to instances/solutions of another problem. Despite the simplicity of the constructions, these techniques gives most of what we know about the relations between Hindman- and Ramsey-type theorems, probably indicating that only the surface of the combinatorics of Hindman's Theorem has been scratched.

2 Hindman's Theorem(s) and the Apartness Condition

Below I introduce a notation for restrictions of Hindman's Theorem based on the type of sums for which monochromaticity is required. Let us use $\mathbf{N}$ to denote the positive integers. Let $\mathcal{F}$ be a family of finite subsets of $\mathbf{N}$. If $X \subseteq \mathbf{N}$ is an infinite set and $x_1 < x_2 < x_3 < \ldots$ are its elements, we denote by $\mathrm{FS}^{\mathcal{F}}(X)$ the set of all sums of the form $\sum_{i \in F} x_i$ for $F \in \mathcal{F}$. For $\mathcal{F} = [\mathbf{N}]^{<\omega}$ we just write $\mathrm{FS}(H)$.

Definition 1 (Hindman's Theorem for sums in $\mathcal{F}$). *Let $\mathcal{F}$ be a family of finite subsets of $\mathbf{N}$, and k a positive integer. $\mathrm{HT}_k^{\mathcal{F}}$ is the following principle: For every coloring $f : \mathbf{N} \to k$ there exists an infinite set $H \subseteq \mathbf{N}$ such that $\mathrm{FS}^{\mathcal{F}}(H)$ is monochromatic for f.*

In this notation Hindman's Theorem is $\mathrm{HT}_k^{[\mathbf{N}]^{<\omega}}$, which we denote HT_k. For any Hindman-type principle P_k indexed by the number of colors, we denote $\forall k \mathsf{P}_k$ by P. The best bounds on the strength of HT have been proved by Blass, Hirst and Simpson some thirty years ago.

Theorem 1 (Blass, Hirst, Simpson, [3]). HT *is provable in* ACA_0^+ *and* HT_2 *implies* ACA_0 *over* RCA_0.

The upper bound comes from an analysis of Hindman's original proof, while the lower bound is based on a clever coloring allowing coding of the Turing Jump.

The following restrictions of HT received some attention in recent years. Let $n \geq 1$. $\mathrm{HT}^{[\mathbf{N}]^n}$ is Hindman's Theorem for sums of exactly n terms, commonly denoted by $\mathrm{HT}^{=n}$. For $n \geq 1$, $\mathrm{HT}^{[\mathbf{N}]^{\leq n}}$ is Hindman's Theorem for sums of at most n terms, commonly denoted by $\mathrm{HT}^{\leq n}$. We will also deal with $\mathrm{HT}^{\mathcal{F}}$ for the following choices $\mathcal{F}$: non-empty intervals of $\mathbf{N}$, exactly ω-large subsets of $\mathbf{N}$, indices of finite rooted branches in the full binary tree, etc.

We are interested in how the strength of the principles $\mathsf{HT}_k^{\mathcal{F}}$ varies as a function of $\mathcal{F}$. An ambitious goal is that of identifying a relation $\prec$ on families of subsets of $\mathbf{N}$ such that $\mathcal{F} \prec \mathcal{G}$ if and only if $\mathsf{HT}^{\mathcal{F}} \leq \mathsf{HT}^{\mathcal{G}}$, where $\leq$ denotes one of the meaningful logico-computational relations between principles, e.g., computable reducibility, implication over RCA_0, etc.

One background intuition that is put to test in the study of restrictions of HT is whether the hidden universal quantifier over all (lengths of) finite sums in HT could behave similarly to the quantifier over all dimensions in Ramsey's Theorem, possibly yielding an ACA_0' lower bound for HT. In this respect, uniform implications from Hindman restrictions to Ramsey's Theorem are of great interest.

Hindman's Finite Sums Theorem has a well-known equivalent formulation in terms of unions, which we state in its general form.

Definition 2 (Hindman's Finite Unions Theorem for unions in $\mathcal{F}$). $\mathsf{FUT}_k^{\mathcal{F}}$ is the following assertion: For every coloring $f : [\mathbf{N}]^{<\omega} \to k$ there exists an infinite sequence $(S_i)_{i \in \mathbf{N}}$ of finite subsets of $\mathbf{N}$ such that

1. f is constant on all finite unions $\bigcup_{i \in F} S_i$ for $F \in \mathcal{F}$, and
2. For all $i < j$, $max(S_i) < min(S_j)$.

The second condition is called the *block condition*. Hirst [16] showed that dropping it yields a principle equivalent to $B\Sigma_2^0$. A corresponding property – called *apartness* in [5] – is crucial in Hindman's original proof of the Finite Sums Theorem and in all proofs of strong lower bounds on $\mathsf{HT}^{\mathcal{F}}$s in [3,5,12]. Fix $b \geq 2$. For $n \in \mathbf{N}$ we denote by $\lambda_b(n)$ the least exponent of n written in base b and by $\mu_b(n)$ the largest exponent of n written in base b. We drop the subscript when clear from context. As we will see, the choice of the base is mostly irrelevant.

Definition 3 (Apartness Condition). *Fix $b \geq 2$. A set $X \subseteq \mathbf{N}$ satisfies the b-apartness condition (or is b-apart) if for all $x, x' \in X$, if $x < x'$ then $\mu_b(x) < \lambda_b(x')$.*

For a Hindman-type principle P, let "P *with b-apartness*" denote the corresponding version in which the solution has to satisfy the b-apartness condition. We state the following Proposition in a somewhat vague form.

Proposition 1 (Folklore, see [6]). *For many choices of $\mathcal{F}$, for all $b \geq 2$, $\mathsf{FUT}_k^{\mathcal{F}}$ is equivalent to $\mathsf{HT}_k^{\mathcal{F}}$ with b-apartness over RCA_0.*

Proof. The direction from the union version to the sum version is given by the following simple functional Φ from functions of type $[\mathbf{N}]^{<\omega} \to k$ to functions of type $\mathbf{N} \to k$:
$$\Phi(c)(b^{e_1} + \cdots + b^{e_\ell}) := c(\{e_1, \ldots, e_\ell\}).$$

For the other implication we use the following simple functional Φ from functions of type $\mathbf{N} \to k$ to functions of type $[\mathbf{N}]^{<\omega} \to k$:
$$\Phi(d)\{s_1, \ldots, s_p\} := d(b^{s_1} + \cdots + b^{s_p}).$$

$\square$

In Hindman's original proof 2-apartness is ensured by a simple counting argument and computable thinning out process, under the assumption that we have a solution to the full Finite Sums Theorem (Lemma 2.2 in [14]).

Proposition 2. *Over* RCA$_0$, HT *is equivalent to* HT *with apartness.*

Hindman's proof for this equivalence strongly relies on monochromaticity of *arbitrary finite sums* and therefore cannot be immediately adapted to generic restrictions HT$^{\mathcal{F}}$. As we observe below, it is significantly easier to prove lower bounds on P with b-apartness than on P in all the cases we consider.

3   Sums of at Most n Elements

The principles HT$_k^{\leq n}$, Hindman's Theorem for sums of *at most n* terms, are arguably the most natural restrictions of HT.

By Proposition 1 the HT$_k^{\leq n}$s *with apartness* should be considered as the natural restrictions of HT to sums of at most n terms, in that they are strongly equivalent to FUT$_k^{\leq n}$, i.e. FUT$_k$ restricted to unions of at most n terms.

As the next lemma shows, apartness can be ensured for HT$_k^{\leq n}$s at the cost of doubling the colors. The idea of the proof is from the first part of the proof of Theorem 3.1 in [12], with some needed adjustments.

Lemma 1. *For all $n \geq 2$, for all $d \geq 1$, HT$_{2d}^{\leq n}$ implies HT$_d^{\leq n}$ with apartness over* RCA$_0$.

Proof. We show 3-apartness for technical convenience. Let $f : \mathbf{N} \to d$ be given. Define $g : \mathbf{N} \to 2d$ as follows. $g(n) := f(n)$ if $i(n) = 1$, and $d + f(n)$ if $i(n) = 2$. where $i(n)$ is the coefficient of the smallest term in the base-3 representation of n. Let H be an infinite set such that FS$^{\leq n}(H)$ is homogeneous for g of color k. For $h, h' \in$ FS$^{\leq n}(H)$ we have $i(h) = i(h')$. Then we claim that for each $m \geq 0$ there is at most one $h \in H$ such that $\lambda(h) = m$. Suppose otherwise, by way of contradiction, as witnessed by $h, h' \in H$. Then $i(h) = i(h')$ and $\lambda(h) = \lambda(h')$. Therefore $i(h + h') \neq i(h)$, but $h + h' \in$ FS$^{\leq n}(H)$. Contradiction. Therefore we can computably obtain a 3-apart infinite subset of H.   □

The above is not known to be optimal. My former student Daniele Tavernelli obtained the following slight refinement in his MSc. Thesis.

Lemma 2 (D. Tavernelli [21]). *For all $k, n \geq 2$ such that $2k + 1$ is prime,* HT$_k^{\leq \max(n,k)}$ *implies* HT$_{k-1}^{\leq n}$ *with apartness over* RCA$_0$.

Question 1. For which values of n, k, p, q, HT$_k^{\leq n}$ implies HT$_q^{\leq p}$ with apartness by an effective reduction?

The recent interest in restrictions of Hindman's Theorem was first revived, to my best knowledge, by Dzhafarov et al. [12]. The authors establish the interesting result that the ACA$_0$ lower bound on HT is already true of HT$^{\leq 3}$.

Theorem 2 (Dzhafarov, Jockusch, Solomon and Westrick, [12]). *Over* RCA$_0$, HT$_3^{\leq 3}$ *implies* ACA$_0$.

Essentially the proof first shows the bound for HT$_2^3$ with 3-apartness, then invokes Lemma 1 and finally manages to get rid of one more color. The same coloring as in Blass-Hirst-Simpsons' proof [3] is used.

Blass [2] states that inspection of the proof of the ACA$_0$-lower bound for HT in [3] shows that this bound is true for the restriction of the Finite Unions Theorem to unions of at most *two sets*, yet he observes that things might be different for restrictions of the Finite *Sums* Theorem, as HT$_k^{\leq n}$ (Remark 12 of [2]). The proofs of Theorem 1 and of Theorem 2 require sums of length three. However, Blass' claim is vindicated by the next Theorem, whose proof requires a different coloring from [3].

Theorem 3 (C., Kołodziejczyk, Lepore, Zdanowski [6]). *Over* RCA$_0$, HT$_4^{\leq 2}$ *implies* ACA$_0$.

Theorem 3 is better explained by isolating the role of apartness in the proof: by Lemma 1 the following result is sufficient.

Theorem 4 (C., Kołodziejczyk, Lepore, Zdanowski, [6]). *For any fixed* $b \geq 2$, HT$_2^{\leq 2}$ *with b-apartness implies* ACA$_0$ *over* RCA$_0$.

Proof. We write the proof for $b = 2$. Assume HT$_2^{\leq 2}$ with 2-apartness and consider $f \colon \mathbf{N} \to \mathbf{N}$. We have to prove that the range of f exists.

For a number n, written as $2^{n_0} + \cdots + 2^{n_r}$ in base 2, with $n_0 < \cdots < n_r$, we call $j \in \{0, \ldots, r\}$ *important in* n if some value of $f \restriction [n_{j-1}, n_j)$ is below n_0 (here $n_{-1} = 0$). The coloring $c_{\mathrm{imp}} \colon \mathbf{N} \to 2$ is defined as the parity the cardinality of the set of numbers important in n. Let H be a solution to HT$_2^{\leq 2}$ with 2-apartness for c_{imp}. We claim that for each $n \in H$ and each $x < \lambda(n)$, $x \in \mathrm{rg}(f)$ if and only if $x \in \mathrm{rg}(f \restriction \mu(n))$; which gives an algorithm for $\mathrm{rg}(f)$.

To prove the claim, consider $n \in H$ and assume that there is some element below $n_0 = \lambda(n)$ in $\mathrm{rg}(f) \setminus \mathrm{rg}(f \restriction \mu(n))$. By the consequence of Σ_1^0-induction known as *strong Σ_1^0-collection* (see Exercise II.3.14 in [20], Thm I.2.23), there is a number ℓ such that for any $x < \lambda(n)$, $x \in \mathrm{rg}(f)$ if and only if $x \in \mathrm{rg}(f \restriction \ell)$. By 2-apartness, there is $m \in H$ with $\lambda(m) \geq \ell > \mu(n)$. Write $n + m$ in base 2 notation,

$$n + m = 2^{n_0} + \cdots + 2^{n_r} + 2^{n_{r+1}} + \cdots + 2^{n_s},$$

where $n_0 = \lambda(n) = \lambda(n + m)$, $n_r = \mu(n)$, and $n_{r+1} = \lambda(m)$. Clearly, $j \leq s$ is important in $n + m$ if and only if either $j \leq r$ and j is important in n or $j = r + 1$. Hence $c_{\mathrm{imp}}(n) \neq c_{\mathrm{imp}}(n+m)$, a contradiction to the monochromaticity of FS$^{\leq 2}(H)$. □

Anglès-D'Auriac, Monin and Patey studied the limits of the coloring c_{imp} and of the coloring used in [3,12] (based on the notion of very short gap) showing that they cannot be used to improve the lower bound on HT.

Proposition 3 (Anglès-D'Auriac, Monin, Patey, [1]). *The coloring c_{imp} admits an arithmetical monochromatic solution.*

Theorem 3 improves on Theorem 2 with respect to length of sums (≤ 2 vs ≤ 3), although it uses 4 rather than 3 colours. The impact of colours on Hindman Theorem(s) has not been studied yet (a few results are in [21]).

Question 2. Does $\mathsf{HT}_2^{\leq 2}$ with apartness imply ACA_0? More generally, what is the impact of the number of colors on $\mathsf{HT}_k^{\mathcal{F}}$?

It is easy to observe that, for Hindman-type theorems, one color can be used to rule out a single (definable) infinite sum-free set from the solution space.

4   Sums of Exactly n Elements

By analogy with $\mathsf{HT}_k^{\leq n}$, Dzhafarov et al. [12] considered restrictions of HT_k to sums of *exactly n* distinct terms, denoted $\mathsf{HT}_k^{=n}$.

As for the $\mathsf{HT}_k^{\leq n}$s, we have that $\mathsf{FUT}_k^{=n}$ and $\mathsf{HT}_k^{=n}$ with b-apartness are equivalent, for any $t \geq 2$ by Proposition 1; thus we consider the principles $\mathsf{HT}_k^{=n}$ with apartness as the natural restrictions of Hindman's Theorem to sums of exactly n elements.

In striking contrast to the $\mathsf{HT}_k^{\leq n}$s, the $\mathsf{HT}_k^{=n}$s have an easy proof in ACA_0: the following functional Φ from functions of type $\mathbf{N} \to k$ to functions of type $[\mathbf{N}]^n \to k$ reduces $\mathsf{HT}_k^{=n}$ to RT_k^n:

$$\Phi(c)(x_1, \ldots, x_n) := c(x_1 + \cdots + x_n).$$

An application of RT_k^n relativized to an apart set also guarantees $\mathsf{HT}_k^{=n}$ with apartness. The following is the strongest lower bound known on these principles.

Theorem 5 (C., Kołodziejczyk, Lepore, Zdanowski, [6]). *Over RCA_0, $\mathsf{HT}_2^{=3}$ with apartness (eq., $\mathsf{FUT}_2^{=3}$) implies ACA_0.*

The proof is an extension of the proof of Theorem 3 using important numbers.

Question 3. Does $\mathsf{HT}_2^{=3}$ with apartness imply RT_2^3 by an effective reduction?

The principle $\mathsf{HT}_2^{=2}$ has an interesting strict connection with a Ramsey-type Principle, the so-called Increasing Polarized Ramsey's Theorem IPT_2^2 introduced by Dzhafarov and Hirst in [11]. IPT_2^2 can be pictured as Ramsey's Theorem for pairs (RT_2^2) in which one asks for a monochromatic bipartite "skew" graph (i.e., with only forward edges) rather than for a monochromatic clique: a solution to an instance $c : [\mathbf{N}]^2 \to 2$ of IPT_2^2 is a pair of infinite sets (H_1, H_2) such that all pairs $(x, y) \in H_1 \times H_2$ with $x < y$ have the same colour. It turns out that $\mathsf{HT}_2^{=2}$ with apartness implies IPT_2^2 by a computable reduction. The simple functional from functions of type $[\mathbf{N}]^2 \to 2$ to functions of type $\mathbf{N} \to 2$ used in the proof is essentially the following Φ:

$$\Phi(f)(n) := f(\lambda(n), \mu(n)),$$

which can be used to reduce IPT_k^2 to a number of other Hindman-type theorems (see *infra*; it was originally applied by the author to $\mathsf{HT}_4^{\leq 2}$).

Theorem 6 (C., Kołodziejczyk, Lepore, Zdanowski, [6]). *Over* RCA_0, $\mathsf{HT}_2^{=2}$ *with 2-apartness implies* IPT_2^2. *The implication is witnessed by a computable reduction.*

Proof. Let $f : [\mathbf{N}]^2 \to 2$ be given. Define $g : \mathbf{N} \to 2$ as follows. $g(n) := 0$ if $n = 2^m$, $f(\lambda(n), \mu(n))$ if $n \neq 2^m$. Note that g is well-defined since $\lambda(n) < \mu(n)$ if n is not a power of 2. Let $H = \{h_1, h_2, \dots\}_<$ witness $\mathsf{HT}_2^{=2}$ with 2-apartness for g. Let the color be $k < 2$. Let $H_1 := \{\lambda(h_{2i-1}) : i \in \mathbf{N}\}$ and $H_2 := \{\mu(h_{2i}) : i \in \mathbf{N}\}$. We claim that (H_1, H_2) is increasing p-homogeneous for f. First observe that $\lambda(h_1) < \lambda(h_3) < \lambda(h_5) < \dots$ and $\mu(h_2) < \mu(h_4) < \mu(h_6) < \dots$ by 2-apartness. Then we claim that $f(x_1, x_2) = k$ for every increasing pair $(x_1, x_2) \in H_1 \times H_2$. Note that $(x_1, x_2) = (\lambda(h_i), \mu(h_j))$ for some $i < j$ (the case $i = j$ is impossible by construction of H_1 and H_2). Then we have

$$k = g(h_i + h_j) = f(\lambda(h_i + h_j), \mu(h_i + h_j)) = f(\lambda(h_i), \mu(h_j)) = f(x_1, x_2),$$

since $\mathsf{FS}^{=2}(H)$ is monochromatic for g with color k. □

It is natural to ask for a reversal (the analogue is open for RT_2^2 vs IPT_2^2).

Question 4. Does IPT_2^2 imply $\mathsf{HT}_2^{=2}$ (with apartness)?

The relation between Hindman Theorems and Increasing Polarized Theorems is intriguing. The natural version of an "Increasing Polarized" $\mathsf{HT}_2^{=2}$, IPHT_2^2 is the only known restriction of HT known to be equivalent to a Ramsey-type principle: IPT_2^2 and IPHT_2^2 are strongly computably interreducible (see [6]).

We know that $\mathsf{HT}_2^{=3}$ with apartness (as well as $\mathsf{HT}_3^{\leq 3}$ and $\mathsf{HT}_4^{\leq 2}$) implies IPT_2^3 over RCA_0, and similarly for all dimensions $n \geq 3$. Contrary to the proof of Theorem 4, one can hope to lift the proof of Theorem 6 to higher dimensions. The answer might imply playing around with adequate additive representations of the integers. A uniform reduction would show a ACA_0'-lower bound on HT.

Question 5. Can the implication from $\mathsf{HT}_k^{=n}$ (or $\mathsf{HT}_k^{\leq n}$) to IPT_ℓ^m be witnessed by an effective reduction for some $n \geq m \geq 3$, $\ell, k \geq 1$?

$\mathsf{HT}_2^{=2}$ *and Lovász' Local Lemma* Csima et al. [9] recently investigated lower bound on $\mathsf{HT}_2^{=2}$ without apartness. They obtained the following interesting result, where RRT_2^2 denotes the Rainbow Ramsey Theorem for pairs.

Theorem 7 (Csima et al., [9]). *Over* RCA_0, $\mathsf{HT}_2^{=2}$ *implies* RRT_2^2.

The proof of Theorem 7 inaugurates the use of tools from the Probabilistic Method from Combinatorics to the study of the logical and effective content of combinatorial theorems: the lower bound on $\mathsf{HT}_2^{=2}$ is obtained by applying an effective version of the famous Lovász' Local Lemma, which allows to give a computable instance of $HT_2^{=2}$ with no computable solutions. The argument can be strengthened to obtain an implication between $HT_2^{=2}$ and $2 - DNC$ over RCA_0. A result of Miller gives the missing link to RRT_2^2.

Inspection of the proof shows that the same bound works for other restrictions of HT, such as $IPHT_2^2$ defined *infra*.

Interestingly, the lower bound on $HT_2^{=2}$ without apartness is significantly weaker than the corresponding IPT_2^2-lower bound on $HT_2^{=2}$ *with apartness* from Theorem 4.

Question 6. Does $HT_2^{=2}$ imply $HT_2^{=2}$ with apartness?

5   Weak Yet Strong Hindman Theorems

An important obstacle in the study of Hindman's Theorem(s) is that no other proof of $HT_2^{\leq 2}$ is known except for the proof(s) of full HT. This makes the following question natural: are there restrictions of HT that admit a better upper bound than HT while still preserving non-trivial lower bounds?

Results presented above show that, for example, $HT_2^{=3}$ with apartness is equivalent to ACA_0; similarly, $HT_2^{=2}$ with apartness is between RT_2^2 and IPT_2^2. The $HT^{=n}$ principles, however, might not tell the whole story: after all, one of the distinctive features of Hindman's Theorem is that it guarantees *simultaneous* monochromaticity of sums of different length and structure.

In the perspective of understanding the limits of provability of restrictions of Hindman's Theorem in various systems of arithmetic it is interesting to investigate other versions with similar properties. While the $HT^{=n}$s are interesting for isolating the *weakest/simplest* form of a restriction of HT that satisfies some upper and lower bound, it is of interest to test the limits of provability of restrictions of $HT^{\mathcal{F}}$ by looking for the *strongest/most complex* restriction that admits a proof in a given system while retaining a strong lower bound. Restrictions of this kind, dubbed "weak yet strong principles" in [5], are known at the level of ACA_0, RT_2^2, $B\Sigma_2^0$ and $I\Sigma_2^0$. Interestingly, their study pre-dated the previously shown results on $HT^{=n}$s.

Equivalents of ACA_0. In [5], an infinite family of natural restrictions of Hindman's Theorem each equivalent to ACA_0 is isolated, The following definition collects some examples based, respectively, on well-known theorems of Schur, Brauer and Van Der Waerden from finite combinatorics.

Definition 4 (C., [5]).

1. *Schur type: Let* HT_k^{Sch} *denote the following statement: Whenever* **N** *is colored in k colors there is an infinite set H and there exist positive integers a, b such that* $FS^{\{a,b,a+b\}}(H)$ *is monochromatic.*
2. *Brauer type: Let* $HT_k^{Bra(\ell)}$ *denote the following statement: Whenever* **N** *is 2-colored there is an infinite set H and there exist positive integers a, b such that* $FS^{\{a,b,a+b,a+2b,\dots,a+\ell b\}}(H)$ *is monochromatic.*

3. *Van Der Waerden type:* Let $\mathsf{HT}_k^{VdW(\ell)}$ denote the following statement: Whenever $\mathbf{N}$ is colored in k colors there is an infinite set H and positive integers a, b such that $\mathrm{FS}^{\{a,a+b,a+2b,\ldots,a+(\ell-1)b\}}(H)$ is monochromatic.

All these versions can be proved in ACA_0 using the same simple scheme.

Theorem 8 (C., [5]). *For each k, ℓ in $\mathbf{N}$, HT_k^{Sch}, $\mathsf{HT}_k^{Br(\ell)}$ and $\mathsf{HT}_k^{WdV(\ell)}$ with apartness are equivalent to ACA_0 over RCA_0.*

Proof. We detail the proof for HT_2^{Sch}. Let $c : \mathbf{N} \to 2$. Let $k = 6 = R_2(3)$, i.e., the Ramsey number ensuring a monochromatic triangle in any coloring of $[1, R_2(3)]$ with 2 colors. Let H_0 be an infinite (computable) set satisfying the Apartness Condition. Let $H_1 \subseteq H_0$ be an infinite homogeneous set for c, witnessing RT_2^1 relative to H_0. Let $f_2 : [\mathbf{N}]^2 \to 2$ be defined as $f(x, y) = c(x+y)$. Let $H_2 \subseteq H_1$ be an infinite homogeneous set for f_2, witnessing RT_2^2 relative to H_1. Let $f_3 : [\mathbf{N}]^2 \to 2$ be defined as $f(x, y, z) = c(x+y+z)$. Let $H_3 \subseteq H_2$ be an infinite homogeneous set for f_3, witnessing RT_2^3 relative to H_2.

We continue in this fashion for k steps. This determines a finite chain of infinite sets $H_0 \supseteq H_1 \supseteq H_2 \supseteq \cdots \supseteq H_k$. Each H_i satisfies the Apartness Condition. Furthermore, for each $i \in [1, k]$, $\mathrm{FS}^{=i}(H_j)$ is monochromatic under c for all $j \in [i, k]$. Also, $\mathrm{FS}^{=i}(H_k)$ is monochromatic for each $i \in [1, k]$. Let c_i be the color of $\mathrm{FS}^{=i}(H_k)$ under c. The construction induces a coloring $C : [1, k] \to 2$, setting $C(i) = c_i$. In general, if k is large enough, C will enjoy some regularity. Since $k = R_3(2)$, by Schur's Theorem there exists $a, b > 0$ in $[1, k]$ such that $\{a, b, a + b\} \subseteq [1, k]$ is monochromatic for C. Let $i < 2$ be the color. Then $\mathrm{FS}^{\{a,b,a+b\}}(H_k)$ is monochromatic of color i for the original coloring c.

The lower bound is obtained by a straightforward adaptation of the proof of Theorem 4.

$\square$

Question 7. Can the equivalences between HT_k^{Sch}, $\mathsf{HT}_k^{Bra(\ell)}$, $\mathsf{HT}_k^{VdW(\ell)}$ with apartness and RT_k^n with $n \geq 3$ be witnessed by effective reductions?

Adjacent Hindman Theorem. All restrictions discussed so far differ from HT in that they only require monochromaticity for sums of bounded length. Can we isolate a family $\mathcal{F}$ of finite subsets of $\mathbf{N}$ such that $\mathcal{F}$ contains arbitrary length sums and $\mathsf{HT}^{\mathcal{F}}$ has some interesting upper and lower bounds? The family $\mathcal{F}_{int}$ of finite intervals of $\mathbf{N}$ answers the question in the positive. We call $\mathsf{HT}_k^{\mathcal{F}_{int}}$ the Adjacent Hindman Theorem, denoted by AHT_k. The Adjacent Hindman Principles guarantee joint monochromaticity for sums of *arbitrary length* yet *severely* constrain the way the terms of these sums are chosen.

Perhaps surprisingly, it is easy to establish an upper bound on AHT with apartness. This should be contrasted with the case of $\mathsf{HT}^{\leq 2}$, for which no upper bound other than ACA_0^+ is currently known.

In fact, the Adjacent Hindman's Theorem was inspired by the proof below: it is obtained by applying RT_2^2 to the output of the functional Φ from functions of type $\mathbf{N} \to 2$ to functions of type $[\mathbf{N}]^2 \to 2$:

$$\Phi(c)(i, j) := c(2^{i+1} + \cdots + 2^{j-1} + 2^j).$$

Proposition 4 (C., [4]). *Over* RCA_0, RT^2_2 *implies* AHT_2 *with apartness and the latter implies* IPT^2_2. *Both implications are witnessed by computable reductions.*

Proof. Fix a coloring $c : \mathbf{N} \to 2$. This induces a coloring f of $[\mathbf{N}]^2$ in 2 colors by setting $f(i, j) := c(2^{i+1} + \cdots + 2^{j-1} + 2^j)$. By RT^2_2 let $J = \{j_1 < j_2 < \ldots\}_<$ be an infinite homogeneous set for f, of color $i < 2$. Consider the set

$$H = \{(2^{j_1+1} + \cdots + 2^{j_2}), (2^{j_2+1} + \cdots + 2^{j_3}), \ldots, (2^{j_n+1} + \cdots + 2^{j_{n+1}}), \ldots\}.$$

It is easy to verify that H satisfies AHT for c.

The implication from AHT_2 to IPT^2_2 is an easy adaptation of Theorem 4. $\square$

The above proof is uniform in k (thus $\mathsf{RT}^2 \to \mathsf{AHT}$ with apartness $\to \mathsf{IPT}^2$).

Question 8. Does AHT_2 imply RT^2_2? Does IPT^2_2 imply AHT_2?

Hindman on Branches of the Binary Tree and Σ^0_2-Induction. Inspired by a proof in Hirst [17] showing that Hindman's Theorem implies the so-called Pigeonhole Principle for Trees (TT^1), my former student Daniele Tavernelli and I recently investigated a restriction of HT in which monochromaticity is required only for sums whose terms correspond to finite rooted branches of the full binary tree [7]. Let btnodes denote the family of finite subsets of $\mathbf{N}$ that corresponds to labelings of the full binary tree in a breadth-first left-to-right visit, i.e.,

$$\text{btnodes} = \{\{1\}, \{1, 2\}, \{1, 3\}, \{1, 2, 4\}, \{1, 2, 5\}, \{1, 3, 6\}, \{1, 3, 7\}, \ldots\}.$$

$\mathsf{HT}^{\text{btnodes}}_k$ is the corresponding restriction of HT_k. When apartness is imposed, this principle turns out to be equivalent to Σ^0_2-induction over RCA_0. Interestingly the lower bound proof is analogous to the ACA_0-lower bound proof of Theorem 4 using "important numbers".

Theorem 9 (C., Tavernelli, [7]). $\mathsf{HT}^{\text{btnodes}}$ *with apartness is equivalent to* Σ^0_2-*induction over* RCA_0.

The study of this principle goes through an $I\Sigma^0_2$-equivalent version of the Pigeonhole Principle for Trees (TT^1) with an added condition on the solution tree inherited from the apartness condition.

Hindman's Theorem for Exactly Large Sums. A finite set $S \subseteq \mathbf{N}$ is *exactly large*, or $!\omega$-*large*, if $|S| = \min(S) + 1$. Let $[X]^{!\omega}$ denote the set of exactly large subsets of $X \subseteq \mathbf{N}$. Ramsey's Theorem for exactly large sets ($\mathsf{RT}^{!\omega}_2$) asserts that every 2-coloring f of the exactly large subsets of an infinite set $X \subseteq \mathbf{N}$ admits an infinite set $H \subseteq X$ such that f is constant on $[H]^{!\omega}$. It was studied in [8] and there proved equivalent to ACA^+_0. Thus, by Theorem 1, $\mathsf{RT}^{!\omega}_2$ implies HT, and the following question is of interest.

Question 9. Does $\mathsf{RT}^{!\omega}_2$ imply HT by an effective reduction?

As any Ramsey-type theorem, $\mathsf{RT}_2^{!\omega}$ naturally implies some restriction of HT: the functional Φ from functions of type $\mathbf{N} \to 2$ to functions of type $[\mathbf{N}]^{!\omega} \to 2$ given by $\Phi(c)(S) := c(\sum S)$, yields the principle $\mathsf{HT}_2^{[\mathbf{N}]^{!\omega}}$, Hindman's Theorem for exactly large sums, also with apartness. For lower bounds we have the following Proposition.

Proposition 5 (C., Kołodziejczyk, Lepore, Zdanowski, [6]). $\mathsf{HT}_2^{[\mathbf{N}]^{!\omega}}$ *with apartness implies* ACA_0 *over* RCA_0 *and computably reduces* IPT_2^2.

Question 10. Does $\mathsf{HT}_2^{[\mathbf{N}]^{!\omega}}$ imply RT_2^3 by an effective reduction?

First-Order Versions and Iterated Largeness Analysis. It is natural to investigate first-order consequences of Hindman's Theorem in the style of Paris-Harrington as well as to inquire into an ordinal or iterated-largeness analysis of Hindman's Theorem in the hope of getting unprovability in Peano Arithmetic or subsystems thereof. An ordinal/iterated-largeness analysis has been attempted in [1] and, within a different framework, in [22]; yet no improvement on the upper bound on HT has been obtained so far. It might be interesting to look at what this approach can say about restrictions of HT, e.g., $\mathsf{HT}^{\leq 2}$. Regarding first-order miniaturizations of HT, a recent result by Mohsenipour and Shelah [18] shows that the finite form of Hindman's Theorem with a largeness condition on the solution yields a principle with primitive recursive upper bounds. This does not rule out the possibility of extracting stronger finitary principles from HT.

Question 11. Is there a miniaturization of HT unprovable in Peano Arithmetic?

Other Types of Restrictions. Restricting the type of monochromatic sums is only one of the possible restrictions of HT we can consider. Other natural forms restrict the type of colorings or put further constraints on where the solution set lies. Both types of restrictions have not been investigated so far in the context of Reverse Mathematics. The second type has to be handled with care. It is easy to come up with Hindman-type theorems with an ACA_0'-lower bound. For example, let's call a positive integer n *exactly-large* if the set of its exponents in base 2 is exactly large. Let A_k be the following principle: For every coloring c of the positive integers $\mathbf{N}$ in 2 colors there exists an infinite $H \subseteq \mathbf{N}$ such that each $n \in H$ is exactly large, H is apart, and $\mathsf{FS}^{=k}(H)$ is monochromatic. It is not hard to show that RCA_0 proves: For all k, A_k with apartness is equivalent to RT_2^k. Yet $\forall k A_k$ might *not* be a consequence of HT.

References

1. Anglès-D'Auriac, P.E.: Infinite Computations in Algorithmic Randomness and Reverse Mathematics. Université Paris-Est. Ph.D. Thesis (2020)
2. Blass, A.: Some questions arising from Hindman's theorem. Scientiae Mathematicae Japonicae **62**, 331–334 (2005)

3. Blass, A.R., Hirst, J.L., Simpson, S.G.: Logical analysis of some theorems of combinatorics and topological dynamics. In: Logic and Combinatorics. Contemporary Mathematics, vol. 65, pp. 125–156. American Mathematical Society, Providence, RI (1987)
4. Carlucci, L.: A weak variant of Hindman's theorem stronger than Hilbert's theorem. Arch. Math. Logic **57**, 381–389 (2018)
5. Carlucci, L.: Weak yet strong restrictions of Hindman's finite sums theorem. Proc. Am. Math. Soc. **146**, 819–829 (2018)
6. Carlucci, L., Kołodziejczyk, L.A., Lepore, F., Zdanowski, K.: New bounds on the strength of some restrictions of Hindman's theorem. Computability **9**, 139–153 (2020)
7. Carlucci, L., Tavernelli, D.: Hindman's theorem for sums along the full binary tree, Σ_2^0-induction and the Pigeonhole Principle for trees. Accepted for publication in Archive for Mathematical Logic
8. Carlucci, L., Zdanowski, K.: The strength of Ramsey's theorem for coloring relatively large sets. J. Symb. Logic **79**(1), 89–102 (2014)
9. Csima, B.F., Dzhafarov, D.D., Hirschfeldt, D.R., Jockusch, C.G., Jr., Solomon, R., Westrick, L.B.: The reverse mathematics of Hindman's theorem for sums of exactly two elements. Computability **8**, 253–263 (2019)
10. Dorais, F.G., Dzhafarov, D., Hirst, J.L., Mileti, J.P., Paul, P.S.: On uniform relationships between combinatorial problems. Trans. Am. Math. Soc. **368**(2), 1321–1359 (2016)
11. Dzhafarov, D.D., Hirst, J.L.: The polarized Ramsey's theorem. Arch. Math. Logic **48**(2), 141–157 (2011)
12. Dzhafarov, D.D., Jockusch, C.G., Solomon, R., Westrick, L.B.: Effectiveness of Hindman's theorem for bounded sums. In: Day, A., Fellows, M., Greenberg, N., Khoussainov, B., Melnikov, A., Rosamond, F. (eds.) Computability and Complexity. LNCS, vol. 10010, pp. 134–142. Springer, Cham (2017). https://doi.org/10.1007/978-3-319-50062-1_11
13. Hindman, N.: The existence of certain ultrafilters on N and a conjecture of Graham and Rothschild. Proc. Am. Math. Soc. **36**(2), 341–346 (1972)
14. Hindman, N.: Finite sums from sequences within cells of a partition of N. J. Comb. Theor. Ser. A **17**, 1–11 (1974)
15. Hindman, N., Leader, I., Strauss, D.: Open problems in partition regularity. Comb. Probab. Comput. **12**, 571–583 (2003)
16. Hirst, J.: Hilbert vs Hindman. Arch. Math. Logic **51**(1–2), 123–125 (2012)
17. Hirst, J.L.: Disguising induction: proofs of the pigeonhole principle for trees. Found. Adventures Tribut. **22**, 113–123 (2014). College Publications, London
18. Mohsenipour, S., Shelah, S.: On finitary Hindman numbers. Combinatorica **39**(5), 1185–1189 (2019). https://doi.org/10.1007/s00493-019-4002-7
19. Montalbán, A.: Open questions in reverse mathematics. Bull. Symb. Logic **17**(3), 431–454 (2011)
20. Simpson, S.: Subsystems of Second Order Arithmetic, 2nd edn. Cambridge University Press, Cambridge. Association for Symbolic Logic, New York (2009)
21. Tavernelli, D.: On the strength of restrictions of Hindman's theorem. MSc. Thesis, Sapienza University of Roma (2018)
22. Beiglböck, M., Towsner, H.: Transfinite approximation of Hindman's theorem. Isr. J. Math. **191**, 41–59 (2012)

Complexity and Categoricity of Injection Structures Induced by Finite State Transducers

Richard Krogman and Douglas Cenzer[✉]

University of Florida, Gainesville, FL 32611, USA
{richard.krogman,cenzer}@ufl.edu

Abstract. An injection structure $\mathcal{A} = (A, f)$ is a set A together with a one-place one-to-one function f. $\mathcal{A}$ is a Finite State Transducer (abbreviated FST) injection structure if A is a regular set, that is, the set of words accepted by some finite automaton, and f is realized by a deterministic finite-state transducer. We study the complexity of the character of an FST injection structure. We also examine the effective categoricity of such structures.

Keywords: Computability theory · Injection structures · Automatic structures · Finite state automata · Finite state transducers

1 Introduction and Preliminaries

The main goal of this paper is to study the complexity and categoricity of automatic injection structures. An injection structure $\mathcal{A} = (A, f)$ is a set A together with a one-place one-to-one function f. $\mathcal{A}$ is a Finite State Transducer (abbreviated FST) injection structure if A is a regular set, that is, the set of words accepted by some finite automaton, and f is realized by a finite-state transducer. In a recent paper [4], Buss, Cenzer, Minnes and Remmel developed the study of FST injection structures. It was shown that the model checking problem for FST injection structures is undecidable, contrasting with the fact that the model checking problem for automatic relational structures is decidable. They also explored which isomorphisms types of injection structures can be realized by FST injections, in particular characterizing the isomorphism types that can be realized by FST injection structures over a unary alphabet. They showed that any FST injection structure is isomorphic to an FST injection structure over a binary alphabet, and gave a number of results about the possible isomorphism types of FST injection structures over an arbitrary alphabet.

There has been considerable work on automatic structures for languages that contain only relation symbols. A structure, $\mathcal{A} = (A; R_0, \ldots, R_m)$, is *automatic* if its domain A and all its basic relations $R_0, \ldots, R_m$ are recognized by finite automata. Independently, Hodgson [9] and later Khoussainov and Nerode [12] proved that for any given automatic structure there is an algorithm that solves

L. De Mol et al. (Eds.): CiE 2021, LNCS 12813, pp. 106–119, 2021.
https://doi.org/10.1007/978-3-030-80049-9_10

the model checking problem for the first-order logic in the language of the structure. In particular, the first-order theory of the structure is decidable. The following result will be needed.

Theorem 1. *There is an algorithm that take an automatic structure $\mathcal{A}$ and a formula $\phi(\boldsymbol{x})$ in the language of $\mathcal{A}$, and produces an automaton which recognizes $\{\boldsymbol{a} : A \models \phi(\boldsymbol{a})\}$.*

For some classes of automatic structures, there are characterization theorems that have direct algorithmic implications. For example, in [8], Delhommé proved that automatic well-ordered sets all have order type strictly less than ω^ω. Using this characterization, [14] gives an algorithm which decides the isomorphism problem for automatic well-ordered sets. Another characterization theorem of this ilk is that automatic Boolean algebras are exactly those that are finite products of the Boolean algebra of finite and co-finite subsets of ω [13]. Again, this result can be used to show that the isomorphism problem for automatic Boolean algebras is decidable.

Another body of work is devoted to the interaction between the representation of an automatic structure and the complexity of the model checking problem. In particular, every automatic structure has a presentation over a binary alphabet but there are automatic structures which do not have presentations over a unary (one letter) alphabet. Moreover, for automatic structures with unary presentations, the monadic second-order theory (not just the first-order theory) is decidable. There are also feasible time bounds on deciding the first-order theories of automatic structures over the unary alphabet [3,10].

Automatic equivalence structures were studied in a recent paper [7] by Cenzer et al. They compared and contrasted these automatic structures with computable equivalence structures. Equivalence structures $\mathcal{A}$ may be characterized by their characters $\chi_\mathcal{A}$ which encodes the number of equivalence classes of any given size. The characters of computably categorical, Δ_2^0 categorical but not computably categorical, or Δ_3^0 categorical but not Δ_2^0 categorical have been determined. It was shown in [7] that every computably categorical equivalence structure has an automatic copy, but not every Δ_2^0 categorical structure has an automatic copy. They constructed an automatic equivalence structure which is Δ_2^0 categorical but not computably categorical and another automatic equivalence structure which is not Δ_2^0 categorical. It was observed that the theory of an automatic equivalence structure is decidable and hence the character of any automatic equivalence structure is computable. On the other hand, there is a computable character which is not the character of any automatic equivalence structure. It was shown that any two automatic equivalence structures which are isomorphic are in fact computably isomorphic. Moreover, for certain characters, there is always a double exponential time isomorphism between two automatic equivalence structures with that character.

In this paper, we restrict our attention to *injection structures*. We begin by fixing notation and terminology. Let $\mathbb{N} = \{0, 1, 2, \ldots\}$ denote the natural numbers and $\mathbb{Z} = \{0, \pm 1, \pm 2, \ldots\}$ denote the integers. Let ω denote the order type of $\mathbb{N}$ under the usual ordering and ζ denote the order type of $\mathbb{Z}$ under the usual ordering. Let ϵ denote the empty word and for any word $w = w_1 \ldots w_n$, let $|w| = n$ denote the length of w. For any finite nonempty set Σ, let Σ^* denote the set of all words over the alphabet Σ, let $\Sigma^+ = \Sigma^* \setminus \{\epsilon\}$. For any $n \in \mathbb{N}$, let $\Sigma^n = \{w \in \Sigma^* : |w| = n\}$ and let $\Sigma^{\leq n} = \{w \in \Sigma^* : |w| \leq n\}$.

An injection is a one-place one-to-one function and an injection structure $\mathcal{A} = (A, f)$ consists of a set A and an injection $f : A \to A$. Given $a \in A$, the orbit $\mathcal{O}_f(a)$ of a under f is

$$\mathcal{O}_f(a) = \{b \in A : (\exists n \in \mathbb{N})(f^n(a) = b \ \vee \ f^n(b) = a)\}.$$

We define the *character* $\chi(\mathcal{A})$ of an injection structure $\mathcal{A} = (A, f)$ by

$$\chi(\mathcal{A}) = \{(n, k) : \mathcal{A} \text{ has at least n orbits of size k}\}$$

An orbit of finite size $k \in \mathbb{N}$ will be a k-cycle of the form $\mathcal{O}_f(a) = \{f^i(a) : 0 \leq i \leq k - 1\}$ where $f^k(a) = a$. Infinite orbits can have two forms. One is of type ζ, which are of the form $\mathcal{O}_f(a) = \{f^n(a) : n \in \mathbb{Z}\}$ in which every element is in the range of f and $f^{-n}(a)$ for $n > 0$ refers to the unique element $b \in A$ with $f^n(b) = a$. The other is of type ω, which have the form $\mathcal{O}_f(a) = \{f^n(a) : n \in \mathbb{N}\}$ for some $a \notin ran(f)$ which serves as the initial element. It is easy to see that the character of an injection structure plus the information about the number of ζ-orbits and ω-orbits completely characterizes its isomorphism type.

The algorithmic properties of injection structures were studied by Cenzer, Harizanov and Remmel [5] and [6]. They characterized computably categorical injection structures, and showed that they are all relatively computably categorical. Among other things, they proved that a computable injection structure $\mathcal{A}$ is computably categorical if and only if it has finitely many infinite orbits. They also characterized Δ^0_2-categorical injection structures as those with finitely many orbits of type ω, or with finitely many orbits of type ζ. They showed that they coincide with the relatively Δ^0_2-categorical structures. Finally, they proved that every computable injection structure is relatively Δ^0_3-categorical. They also showed that the character of any computable injection structure is a c.e. set and that any c.e. character may be realized by a computable injection structure.

A deterministic finite automaton (DFA) is specified by the tuple $M = (Q, \iota, \Sigma, \delta, F)$ where Q is the finite set of states, ι is the initial state, Σ is the input alphabet, $\delta : Q \times \Sigma \to Q$ is the (possibly partial) transition function, and $F \subseteq Q$ is the set of final, or accepting states. The transition function may be extended to $\delta : Q \times \Sigma^* \to \Sigma^*$ by recursion on the length of a word. For $q \in Q$, $\delta(q, \epsilon) = q$, and for $w \in \Sigma^*$ and a letter $a \in \Sigma$, $\delta(q, wa) = \delta(\delta(q, w), a)$. Then $\delta(\iota, w)$ represents the final state M reaches when scanning through w while transitioning states according to δ, starting at ι. A DFA M *accepts* a string w if $\delta(\iota, w) \in F$. The set $L(M) \subseteq \Sigma^*$ of strings accepted by M is the language *recognized* by M. A language $L \subseteq \Sigma^*$ is said to be *regular* or *automatic* if it is accepted by some DFA. To recognize a relation $R \subseteq \Sigma^* \times \Sigma^*$, we use a two-tape synchronous DFA, where the transition function $\delta : Q \times \Sigma \cup \{\Box\} \times \Sigma \cup \{\Box\} \to Q$ and $\Box$ denotes a blank square. The blank square is needed in the case that one input is longer than the other. Then M halts after reaching the end of the longer word. Automatic relations and structures have been studied by Khoussainov, Liu, Minnes, Nies, Rubin, Stephan and others [10–16].

A deterministic finite-state transducer (DFST) is specified by the tuple $M = (Q, \iota, \Sigma, \Gamma, \delta, \tau)$ where Q is the finite set of states, ι is the initial state, Σ is the input alphabet, Γ is the output alphabet, $\delta : Q \times \Sigma \to Q$ is the (possibly partial) transition function, and $\tau : Q \times \Sigma \to \Gamma^*$ is the (possibly partial) output function. Here δ extends as with an automata, and τ is extended to $\tau : Q \times \Sigma^* \to \Gamma^*$ as follows: for $q \in Q$, $\tau(q, \epsilon) = \epsilon$, and $\tau(q, wa) = \tau(q, w)\tau(\delta(q, w), a)$ for word $w \in \Sigma^*$ and letter $a \in \Sigma$. A DFST M defines a (possibly partial) function, $f_M : \Sigma^* \to \Gamma^*$ with $f_M(w) = \tau(\iota, w)$. We say that the DFST M *realizes*, *computes*, or *generates* a function f on a set $U \subseteq \Sigma^*$ if $f_M \restriction U = f$. We will exclusively work with deterministic automata and transducers so we will drop the moniker of determinism in the succeeding discussion. We occasionally use $\oplus$ to denote concatenation in operator notation when convenient.

Definition 2. *An injection structure* $\mathcal{A} = (A, f)$ *consists of a set A together with a one-to-one mapping $f : A \to A$. $\mathcal{A}$ is an FST injection structure if A is a regular set of words in a language Σ^* and f is realized by a DFST.*

It is possible to combine the underlying automaton that accepts the domain, and the transducer that computes the function as shown in [4]. Hence, we suppose from here on, without loss of generality, that (A, f) is computed by $T = (Q, \iota, \delta, \tau, F)$ where (Q, ι, δ, F) accepts A, and (Q, ι, δ, τ) computes f.

Since the isomorphism class of an injection structure is determined by its character, and the number of orbits of type ω and ζ, we seek to characterize those types which have presentations using transducers.

The main goal of this paper is explore the complexity of automatic injection structures and isomorphisms between two such structures. Results will sometimes depend on factors such as (i) the underlying alphabet Σ, (ii) the number of states in of the underlying transducer $T = (Q, \iota, \Sigma, \Gamma, \delta, \tau)$ for f, and (iii) the nature of the output function τ of the transducer T. For example, we say that a transducer $T = (Q, \iota, \Sigma, \Gamma, \delta, \tau)$ has ϵ-*outputs* if there is state $q \in Q$ and $a \in \Sigma$ such that $\tau(q, a) = \epsilon$. We say T is *length preserving* if $\tau(q, a) \in \Gamma$ for all $q \in Q$ and $a \in \Sigma$. Thus when a length preserving transducer reads a symbol $a \in \Sigma$ in any state q, it outputs a single letter in Γ. (A, f) has *bounded growth* if there is a constant c such that, for all $w \in A$, $|w| - c \leq |f(w)| \leq |w| + c$. We say that (A, f) has *full domain* if $\Sigma^+ \subseteq A$. Recall that the *length-lexicographic* order on Σ^* is defined by $u <_{lex} v$ if and only if $|u| < |v|$ or $|u| = |v| = n$ and $u <_n v$, where $<_n$ is the lexicographic order on Σ^n.

The outline of this paper is as follows. In Sect. 2, we construct some additional FST injection structures not found in the previous paper [4]. We characterize those FST injection structures (A, f) for which the graph G_f is automatic, in terms of the notion of bounded growth. In Sect. 3, we examine the complexity of the character of a FST injection structure. In particular, it is shown that the character of a graph relational FST injection structure is decidable in exponential time and for unary structures, in linear time. In Sect. 4, we show that isomorphic unary FST injection structures are exponential time isomorphic, and furthermore, for graph relational structures, they are quadratic time isomorphic. It is shown that not all isomorphic pairs of FST injection structures are computably isomorphic. It is shown that two isomorphic graph relational injection structures with full universe $\{0,1\}^*$ are double exponential time isomorphic.

For the analysis of relational injection structures, we will need to consider the problem of counting the number of elements of a regular language of a certain bounded length. The following well-known lemma provides an upper bound on the search space for the first few elements of a regular set. The proof will be in the full paper.

Lemma 1. *Let* $M = (Q, \iota, \delta, F)$ *be a* Σ-*DFA with* s *states.*

1. $L(M) \neq \emptyset$ *if and only if* $L(M)$ *contains a string of length* $\leq s$.
2. $L(M)$ *is infinite if and only if* $L(M)$ *contains a string of length* $> s$.
3. *If* $L(M) = \{u_0 <_{lex} u_1 <_{lex} \cdots\}$, *then* $|u_0| \leq s$ *and for each* i, $|u_{i+1}| \leq |u_i| + s$.
4. *For any* n, $|L(M)| \geq n$ *if and only if either* $L(M)$ *is infinite or* $card(L(M) \cap \Sigma^{\leq s}) \geq n$.

Using this lemma, we can see how many steps it takes to decide whether $card(L(M)) \geq n$.

Proposition 1. *For any DFA* M *with* s *states,* $\{(n, m) : card(L(M) \cap \Sigma^m) \geq n\}$ *can be decided in time* $\leq ms^3n$, *we can decide* $\{(n, m) : card(L(M) \cap \Sigma^{\leq m}) \geq n\}$ *in* $\leq ms^3n$ *steps, and we can decide whether* $\{n : card(L(M)) \geq n\}$ *in* $\leq 2s^4n$ *steps.*

Proof. This proof is relegated to the full paper.

2   *FST* Injection Structures

A variety of FST injection structures were constructed in [4]. For example, there are FST injection structures consisting of any number of orbits of type ζ. There is a FST injection structure consisting of exactly one cycle of length k for each finite k. For each finite $\ell > 1$, there is an FST injection structure with exactly one cycle of length ℓ^i for all finite i.

Proposition 2. *For any two FST injection structures* $\mathcal{A} = (A, f)$ *and* $\mathcal{B} = (B, g)$, *there is an injection structure* $\mathcal{C}$ *consisting of a copy of* $\mathcal{A}$ *together with a copy of* $\mathcal{B}$.

Proof. Let $\mathcal{A}$ be given by the FST T_A and $\mathcal{B}$ by the FST T_B. Then the structure $\mathcal{C} = (C, h)$ consisting of the join of $\mathcal{A}$ and $\mathcal{B}$ may be defined by letting $h(0x) = 0f(x)$ and $h(1x) = 1g(x)$ for each $x \in \{0,1\}^*$, and letting $C = \{0x : x \in A\} \cup \{1x : x \in B\}$. The injection structure $\mathcal{C}$ may be defined by the FST T_C as follows. On initial input 0, T_C outputs 0 and hands over the control to T_A, and on initial input 1, T_C outputs 1 and hands control to T_B. Then $\phi(a) = 0a$ and $\psi(b) = 1b$ are injective homomorphisms that map A and B to their respective copies in C.

Corollary 1. *For each $m, n \in \mathbb{N}$, not both equal to 0, there is an FST injection structure consisting of exactly m orbits of type ω and n orbits of type ζ.*

Proposition 3. *For any FST injection structure $\mathcal{A}$, there is an FST injection structure $\mathcal{A}_\omega$ which consists of infinitely many copies of $\mathcal{A}$.*

Proof. Let the structure $\mathcal{A} = (A, f)$ be given by the *FST* T with initial state q_0. Then the structure $\mathcal{A}_\omega = (B, g)$ consisting of infinitely many copies of $\mathcal{A}$ may be defined by letting $g(0^m 1x) = 0^m 1 f(x)$ for each $m \in \mathbb{N}$ and $x \in \{0, 1\}^*$, and letting $B = \{0^m 1x : m \in \omega, x \in A\}$. The injection structure $\mathcal{A}_\omega$ may be realized by the *FST* T_ω, with initial state q_ω, as follows. In state q_ω, there are two possibilities. Upon reading a 0, T_ω outputs 0 and remains in state q_ω; upon reading a 1, T_ω outputs a 1 and transitions to state q_0. Having reached state q_0, T_ω simply follows the transitions and outputs of T. For each m, $A_m = \{0^m 1x : x \in A\}$ is a distinct sub injection structure isomorphic to A through $\phi_m(a) = 0^m 1a$.

Theorem 3. *(a) There is an FST injection structure consisting of infinitely many orbits of size k, for each $k \in \mathbb{N}$.*
(b) There is an FST injection structure consisting of infinitely many orbits of types ζ.
(c) There is an FST injection structure consisting of infinitely many orbits of each finite size, as well as any predetermined amount (finite or infinite) of orbits of types ω and of type ζ.

Proof. (a) It was shown in Theorem 8 from [4] that there is an *FST* injection structure consisting of exactly one class of size k for each finite k. Then the desired *FST* may be obtained by applying Proposition 3.

(b) This may be shown by applying Proposition 3 to the *FST* from [4] with one orbit of type ζ.

(c) This now follows by applying Proposition 2 to the structures obtained in the above results.

Now we can improve a result from [4], as follows.

Theorem 4. *There is a computable injection structure $\mathcal{A}$ such that $Inf^{\mathcal{A}} = \{a \in A : card(\mathcal{O}_f(a)) = \omega\}$ is Σ_1^0 complete. Furthermore, $\mathcal{A}$ has infinitely many infinite orbits and may be assumed to have infinitely many finite orbits of all sizes.*

Proof. The first part is proved in [4]. For the second part simply use Proposition 2 to adjoin the *FST* structure consisting of infinitely many classes of size k for each finite k.

Graph Relational FST Injection Structures

It is natural to ask whether there is a difference between FST injection structures and automatic structures $\mathcal{A} = (A, G_f)$ where G_f is the graph of the function f, i.e., the set of all pairs $(a, f(a))$ for $a \in f$. Since an automatic structure

$\mathcal{A} = (A, G_f)$ has a decidable theory, it would follow that if G_f is recognizable by a DFA, then (A, f) would also have a decidable theory. However, it was observed in [4] that there are simple FST's with no ϵ-outputs whose graph is not recognizable by a 2-tape synchronous automaton, for example where the injection maps 1^i to 1^{2i}. Thus we have the following.

Proposition 4. *There exists an FST computable injection structure which is not graph automatic.*

Proposition 5. *There exists a graph automatic injection structure which is not FST computable.*

Proof. Observe that an FST injection structure must satisfy $f(u) \sqsubseteq f(v)$ for $u \sqsubseteq v$ where $\sqsubseteq$ denotes the prefix/initial segment relation. Consider $(1^*, f)$ where $f = \{(1^{2n+1}, 1^{2n-2}) : n \in \mathbb{N}\} \cup \{(1^{2n}, 1^{2n+1}) : n \in \mathbb{N}\}$. Then G_f is automatic, but we have $1^{2n} \sqsubseteq 1^{2n+1}$ and yet $f(1^{2n}) = 1^{2n+1} \not\sqsubseteq 1^{2n-2} = f(1^{2n+1})$.

On the other hand, it is easy to see that if $T = (Q, \iota, \Sigma, \Gamma, \delta, \tau)$ is length preserving, i.e. $\tau(q, a) \in \Gamma$ for all $q \in Q$ and $a \in \Sigma$, then we can use the transition diagram for δ and τ to construct of DFA which accepts the graph $(w, \tau(w))$ for all w accepted by the DFA $(Q, \iota, \Sigma, \delta)$. There is a stronger result.

Some definitions are needed. For any state q and any string y, we can extend the transition function δ and the output function τ to strings by letting $\delta(q, y)$ and $\tau(q, y)$ be the state and output obtained by starting in state q at the beginning of y and applying the *FST* T to the string y. For a string $w = a_0 a_1 \dots a_k \neq \varepsilon$, we will let $w^- = a_1 a_2 \dots a_k$.

Definition 1. *Let T be a FST which generates the injection structure (A, f).*

1. *A state cycle is a pair $(q, y) \in Q \times \Sigma^*$ such that $|y| > 0$ and $\delta(q, y) = q$. A word $w \in \Sigma^*$ is said to contain a state cycle if $w = xyz$ with $(\delta(\iota, x), y)$ a state cycle.*
2. *T has bounded growth if there is a constant c such that for any $w \in A$, $|w| - c \leq |f(w)| \leq |w| + c$; T has strongly bounded growth if there is a constant c such that for any state q and any $w \in \Sigma^*$, $|w| - c \leq |\tau(q, w)| \leq |w| + c$, provided that there is a computation of T on a string $w \in A$ in which T reads the input w starting in state q.*
3. *T length-preserves state cycles from A if for any $w \in A$ with $w = xyz$ and $(\delta(\iota, x), y)$ a state cycle, then we have $|\tau(\delta(\iota, x), y)| = |y|$.*
4. *(A, f) is graph relational if G_f is recognized by a two tape DFA.*

Proposition 6. *If T is an FST with s states that generates (A, f), then the following are equivalent:*

1. *T length-preserves state cycles from A;*
2. *T has strongly bounded growth;*
3. *(A, f) is graph relational and, furthermore, if T has s states, and $|T(x)| \leq |x| + c$ for all inputs x, then there is a DFA M with $\leq 2s|\Sigma|^{c+1}$ states which recognizes G_f;*
4. *The function f has bounded growth.*

Proof. A DFA to recognize the graph will read both words and verify if the second coordinate comprises of the images of the symbols in the first coordinate under τ. It would have to remember what it has seen in the first coordinate, though the condition of bounded growth means that the DFA need only remember a bounded amount of information using its states. Details are left to the full paper.

Note that in a unary alphabet, we may identify the string 1^n with the number n, and accordingly identify the isomorphic prefix relation $\sqsubseteq$ and the usual ordering on $\mathbb{N}$, $\leq$. The possible types of FST injection structures over a unary alphabet were characterized in [4] as follows.

Theorem 5. *(BCMR2017) An FST injection structure over $\{1\}^*$ is either finite in which case it consists of cycles of length 1, or infinite, in which case it may have any of the following types:*

1. *A structure consisting of infinitely many cycles of length 1;*
2. *For each finite $m \geq 0$ and each $n \geq 1$, a structure with m orbits of length 1 and n orbits of type ω;*
3. *For each finite m, a structure with m orbits of length 1 and infinitely many orbits of type ω;*

Here is an improvement for graph relational structures.

Proposition 7. *Unary graph relational FST injection structures have exactly the following two types:*

1. *A structure consisting of infinitely many cycles of length 1;*
2. *For each finite m and each $n \geq 1$, a structure with m orbits of length 1 and n orbits of type ω*

Proof. Examples of these types constructed in [4] have bounded growth. Now suppose by way of contradiction that (A, f) has bounded growth and has infinitely many orbits of type ω. For simplicity we may assume that there are no orbits of type 1. Let us identify 1^n with n and fix b so that $f(n) \leq n + b$ for all n. Consider $b + 1$ distinct orbits with initial elements $a_0, a_1, \ldots, a_b$ and let $a \geq max\{a_0, a_1, \ldots, a_b\}$. Since the distance between x and $f(x)$ is always $\leq b$, it follows that each infinite orbit must have an element in the interval $[a, a + b]$. But this interval contains only b elements, contradicting the assumption that the orbits are distinct.

Corollary 2. *The first-order theory of each unary injection structure is decidable.*

3   The Character of FST Injection Structures

In the previous section, a condition was proven to indicate when a particular FST injection structure is graph relational.

Proposition 8. *The character of a graph relational FST injection structure is computable.*

Proof. By Theorem 1, the theory of an automatic structure is decidable. Fix $(n, k) \in \mathbb{N}^2$. Let

$$\alpha_k(x) = (f^k(x) = x) \wedge (\bigwedge_{1 \leq i \leq k-1} f^i(x) \neq x)$$

so that $|\mathcal{O}_f(a)| = k$ if and only if $\alpha_k(a)$ holds. Similarly, consider

$$\beta_k(x, y) = \bigwedge_{0 \leq i \leq k-1} f^i(x) \neq y$$

so that $\mathcal{O}_f(a) \cap \mathcal{O}_f(b) = \emptyset$ if and only if $\beta_k(a, b)$ holds. Then for

$$\phi_{n,k} = \exists^n \boldsymbol{x} \, [\bigwedge_{0 \leq i \leq n-1} \alpha_k(x_i)] \wedge [\bigwedge_{0 \leq i < j \leq n-1} \beta_k(x_i, x_j)],$$

we have $\chi(A) = \{(n, k) : A \models \phi_{n,k}\}$. Hence, $\chi(A)$ is decidable. It follows from [5] that the theory is also decidable. $\qquad \square$

For automatic structures, the character is less complicated. The study of these characters makes use of the sets $I_k := \{a : card(\mathcal{O}_f(a)) = k\}$, for each fixed $k \in \mathbb{N}$. The next result follows from Theorem 1, together with the proof of Proposition 8.

Proposition 9. *If A is FST and graph relational, then for each $k \in \mathbb{N}$, I_k is computable.*

Since every automatic set is decidable in linear time, we have the following.

Corollary 3. *For each $k \in \mathbb{N}$, $I_k \subseteq A$ is decidable in linear time. If $\chi(A)$ is bounded, then there is an algorithm that computes, given $a \in A$, the size of $\mathcal{O}_f(a)$.*

Some of these results can be improved.

Proposition 10. *Let (A, f) be an FST injection structure.*

1. *There is a constant c such that $\{(a, k) \in A \times \mathbb{N} : card(\mathcal{O}_f(a)) = k\}$ is computable in time $O(c^k \cdot |a|)$.*
2. *If (A, f) is graph relational, then there is a constant c such that $\{(a, k) \in A \times \mathbb{N} : card(\mathcal{O}_f(a)) = k\}$ is computable in time $O(\lceil k^2 \rceil c + k|a|)$.*

Proof. 1. Let $T = (Q, \iota, \Sigma, \Gamma, \delta, \tau)$ be a FST which computes the injection f and let $c = max\{|\tau(q, a)| : (q, a) \in Q \times \Sigma\}$, say that $c \geq 2$. For $q_i = \delta(\iota, a_{\restriction a})$ It follows that $|f(a)| = |\bigoplus_{i \in |a|} \tau(q_i, a_i)| \leq c \cdot |a|$ and thus by induction $|f^i(a)| \leq c^i \cdot |a|$ for each i. Then for each $i < k$, the computation of $f^{(i)}(a)$ takes $\leq c^i \cdot |a|$ steps in the computation by the FST. Given $(a, k) \in A \times \mathbb{N}$, in order to verify if $card(\mathcal{O}_f(a)) = k$ or not we must compute $f^i(a)$. To check whether $f^{(k)}(a) = a$ takes an additional $|a| + 1$ steps. It follows that the overall time required to see whether $card(\mathcal{O}_f(a)) = k$ is $\leq c^k \cdot |a|$, we must compute each $f^i(a)$ for $1 \leq i \leq k$ and verify that $f^i(a) \neq a$ if $i \neq k$ and $f^k(a) = a$. The computation of $f^i(a)$ takes $\leq c^i |a|$ and at each i, the comparison of $f^i(a)$ and a takes $\leq |a| + 1$. Hence, the entire procedure takes $\leq c^k |a| + c^{k-1}|a| + \cdots + |a| + k|a| + k$ steps, which is $O(c^k|a|)$.

2. Now suppose that G_f is automatic. Then by Proposition 6, there is a constant c such that $|f(a)| \leq |a| + c$ for all inputs a. Note then that for all $1 \leq i \leq k$, we have that $|f(a)| \leq |a| + ic$. To compute each of $f^i(a)$ then takes $\leq |a| + ic$ steps. Following the same process used in 1, we have that $card(\mathcal{O}_f(a)) = k$ can be determined in $\leq (|a| + c) + (|a| + 2c) + \cdots + (|a| + kc) + k|a| + k = 2k|a| + k + (1/2)k(k + 1)c$ steps, which is $O(\lceil k^2 \rceil c + k|a|)$.

Corollary 4. *For any FST injection structure (A, f) and any fixed k, $\{a : card(\mathcal{O}_f(a)) = k\}$ is computable in linear time.*

Analysis of the size of the orbit of x under an injection computed by a FST requires looking at the iterated composition $f^{(k)}$ of f. So the following result from [1] is needed.

Theorem 6. *Let T and U be finite state transducers that compute $f : \Sigma^* \to \Gamma^*$ and $g : \Gamma^* \to \Delta^*$, respectively. Let T have m states and let U have n states. Then there is a finite state transducer V that computes $g \circ f : \Sigma^* \to \Delta^*$ which has $m \cdot n$ states.*

Proof. Say that $T = (Q, \iota, \delta, \tau)$ and $U = (R, \kappa, \zeta, \xi)$. Consider $V = (S, \sigma, \mu, \lambda)$ as follows. In practice, to compute the composition $g \circ f$, we would like a scratch tape to write down $T(u)$ and then compute $V(T(u))$. Here when T outputs a string rather than a bit, we will let λ perform several steps in the computation of U. Take $S = Q \times R$ and $\sigma = (\iota, \kappa)$. Here is how to determine $\mu((q, r), a)$ and $\lambda((q, r), a)$. First let $w = \tau(q, a) = b_0 b_1 \ldots b_k$. In the case that $w = \varepsilon$, let $\mu((q, r), a) = (\delta(q, a), r)$ and let $\lambda((q, r), a) = \varepsilon$. Otherwise the definition proceeds in stages. Let $r_1 = \zeta(r, b_0)$ and let $w_1 = \xi(r, b_0)$ and for each $i \leq k$, let $r_{i+1} = \zeta(r_i, b_i)$ and let $w_{i+1} = \xi(r_i, b_i)$. Then $\mu((q, r), a) = r_{k+1}$ and $\lambda((q, r), a)) = w_1 w_2 \ldots w_{k+1}$. We note that all of these calculations are simply to define the FST V and are being done during a computation $V(w)$.

The following corollary follows by induction from the preceding theorem.

Corollary 5. *If (A, f) is an FST structure computed by T with s states and $k \in \mathbb{N}$, then there exists an FST that computes (A, f^k) with s^k states.*

Finally we look at the complexity of the character $\chi(\mathcal{A})$.

Theorem 7. *Let $\mathcal{A} = (A, E)$ be a graph relational FST injection structure. Then there is an algorithm which decides whether $(n, k) \in \chi(\mathcal{A})$ in time exponential in $max\{n, k\}$. If $\mathcal{A}$ is a unary structure, then $\chi(\mathcal{A})$ is decidable in linear time.*

Proof. Let $\mathcal{A}$ be a relational FST injection structure. Given finite n and $k > 0$, we need to decide whether $\mathcal{A}$ has at least n distinct orbits of size k. It follows from Corollary 5 that there is an FST that computes $f^{(k)}$ with s^k states. Then by Proposition 6, there is a DFA M with $\leq 2s^k|\Sigma|^{c+1}$ states which recognizes the graph G_{f^k}. This is easily modified to a DFA M_k which recognizes $\{x : f^{(k)}(x) = x\}$ with $\leq 2s^k|\Sigma|^{c+1}$ states. Now $\mathcal{A}$ has at least n distinct orbits of size k if and only if there are at least kn elements belonging to orbits of size k and thus accepted by M_k. It now follows from Proposition 1 that this may be decided in time $\leq 2(2s^k|\Sigma|^{c+1})^4 n$ steps, which is of order $s^{4k}|n|$, and thus in exponential time in $max\{k, n\}$.

Unary injection structures are trivial and hence have linear time characters.

4   Isomorphisms Between FST Injection Structures

In this section, we analyze the complexity of isomorphisms between FST injection structures.

Definition 8. *A Scott family for a structure $\mathcal{A}$ is a countable family Φ of $L_{\omega_1\omega}$ formulas, possibly with finitely many fixed parameters from A, such that (i) Each finite tuple in $\mathcal{A}$ satisfies some $\psi \in \Phi$; and (ii) If $\overrightarrow{a}$, $\overrightarrow{b}$ are tuples in $\mathcal{A}$, of the same length, satisfying the same formula in Φ, then there is an automorphism of $\mathcal{A}$ that maps $\overrightarrow{a}$ to $\overrightarrow{b}$.*

Scott families were used by Ash, Knight, Manasse, and Slaman [2] to characterize the relatively Δ_n^0-categorical structures. The following result for automatic relational structures was obtained in [4].

Theorem 9. *(BCMR2017) Let $\mathcal{A}$ be an automatic structure for which there is a c.e. set of formulas Φ with fixed parameters $\boldsymbol{d}$ such that (i) For any tuple $\boldsymbol{a}$ in $\mathcal{A}$, there is some $\phi \in \Phi$ such that $\mathcal{A} \models \phi(\boldsymbol{d}, \boldsymbol{a})$; and (ii) if $\boldsymbol{a}$ and $\boldsymbol{b}$ are tuples satisfying the same formula $\phi \in \Phi$, then $(\mathcal{A}, \boldsymbol{d}, \boldsymbol{a}) \cong (\mathcal{B}, \boldsymbol{d}, \boldsymbol{b})$.*
Then any two automatic copies of $\mathcal{A}$ are computably isomorphic.

This result can be applied to graph relational FST injection structures as follows.

Theorem 10. *If two graph relational FST injection structures with bounded character are isomorphic, then they are computably isomorphic.*

Proof. The desired formulas $\Phi(\boldsymbol{x})$ for a tuple $\boldsymbol{a}$ simply give the size of the orbit of a_i for each i and state whether $f^{(t)}(a_i) = a_j$ for each i, j, t. We have seen above in Proposition 9 that, for each finite k, there is a formula $\Phi_k(x)$ so that $(A, f) \models \Phi_k(x)$ if and only if $|\mathcal{O}(x)| = k$. To define elements belonging to an infinite orbit, just let k be the maximum size of any finite orbit. Then $\mathcal{O}_f(a)$ is infinite if and only if $\forall i \leq k(\neg f^{(i)}(a) = a)$.

The remainder of the section is devoted to improving this result.

Lemma 2. *For any unary FST injection structure (A, f) with domain $A = \{a_0 < a_1 < \cdots\}$, the function f is monotone strictly increasing. Furthermore, for all i, $f(a_i) \geq i$.*

Proof. Let $a, b \in A$ with $a < b$. Then as noted before, $f(a) \leq f(b)$ since f is FST computable. However, $f(a) \neq f(b)$ since f is an injection. Hence, $f(a) < f(b)$.

For the second part, let $A = \{a_0 < a_1 < \cdots\}$. We proceed by induction on i. Since $f(a_0) \in A$, we have $a_0 \leq f(a_0)$. Suppose by induction that $a_i \leq f(a_i)$. By monotonicity, $f(a_i) \leq f(a_{i+1})$ and in fact $f(a_i) < f(a_{i+1})$ since f is one-to-one. It follows that $a_{i+1} \leq f(a_{i+1})$.

Here is the general result for the complexity of isomorphisms of unary FST injection structures.

Theorem 11. *Let $\mathcal{A}$ and $\mathcal{B}$ be two isomorphic FST injection structures with universe a subset of $\{1\}^*$.*

1. *If $\mathcal{A}$ and $\mathcal{B}$ are graph relational, then $\mathcal{A}$ and $\mathcal{B}$ are quadratic time isomorphic.*
2. *In general, $\mathcal{A}$ and $\mathcal{B}$ are exponential time isomorphic.*

Proof. The core of the argument is for the infinite orbits. Given specific orbits $\{a, f(a), f(f(a)), \dots\}$ from $\mathcal{A}$ and $\{b, f(b), f(f(b)), \dots\}$ from $\mathcal{B}$, and an element $x \in \mathcal{O}_f(a)$, we can compute j such that $x = f^{(j)}(a)$ and then compute $g^{(j)}(b)$ all in quadratic time. When there are infinitely many orbits with initial elements $\{a_0, a_1, \dots\}$ and $\{b_0, b_1, \dots\}$, additional effort is required to compute from x the value i so that $x \in \mathcal{O}_f(a_i)$.

Details are left to the full paper.

The case of *FST* injection structures with binary universes is rather more complicated.

Proposition 11. *There are two isomorphic FST injection structures which are not computably isomorphic.*

Proof. First recall from Theorem 4 the *FST* injection structure $\mathcal{A}$ with infinitely many orbits of type ω together with infinitely many orbits of each finite size, such that $Inf^{\mathcal{A}}$ is c.e. complete. Then consider the copy $\mathcal{B}$ of this structure obtained by using Proposition 2 to join one *FST* structure consisting of infinitely many orbits of each finite size to another *FST* structure consisting of infinitely many orbits of type ω. It follows from the proof of Proposition 2 that $Inf^{\mathcal{B}}$ will be a regular set and therefore computable. Hence there can be no computable isomorphism between $\mathcal{A}$ and $\mathcal{B}$.

Thus we shall have only partial success in trying to show that two isomorphic FST structures must be computably isomorphic. In particular, we will restrict the discussion to structures with full domain $\{0,1\}^*$.

Lemma 3. *For any binary FST injection structure (A, f), the function f is monotone strictly increasing. Furthermore, if $A = \{0,1\}^*$, then for all w, $|f(w)| \geq |w|$.*

Proof. Let (A, f) be a FST injection structure. Suppose $u \sqsubset v$. It is immediate from the definition of a finite state transducer that $f(u) \sqsubset f(v)$. Now suppose that $A = \{0,1\}^*$. Since f is an injection, $f(x^\frown i) \neq f(x)$ so that $|f(x^\frown i)| > f(x)$ and it follows by induction that $|f(x)| \geq |x|$ for all x. Note that this means that there are no ϵ-outputs from any reachable state.

This leads to two results about full domain structures.

Proposition 12. *Let $\mathcal{A} = (\{0,1\}^*, f)$ be an FST injection structure.*

1. *$\mathcal{A}$ has no orbits of type ζ.*
2. *If $\mathcal{O}_f(a)$ is finite, then for all i, $|f^{(i)}(a)| = |a|$.*

Lemma 3 and Proposition 12 may be used to prove the following result with an argument similar to Theorem 11. Details are left to the full paper.

Theorem 12. *Let $\mathcal{A}$ and $\mathcal{B}$ be two isomorphic FST injection structures with full universe $A = \{0,1\}^* = B$, with bounded character, and with finitely many infinite orbits. If $\mathcal{A}$ and $\mathcal{B}$ are graph relational, then $\mathcal{A}$ and $\mathcal{B}$ are isomorphic in double exponential time.*

5   Conclusions and Further Research

Results of this paper extend the known family of FST injection structures from [4]. FST injection structures (A, f), for which the graph G_f is automatic, are characterized in terms of the notion of bounded growth. The character of a graph relational FST injection structure is decidable in exponential time and for unary structures, in linear time. Isomorphic unary FST injection structures are exponential time isomorphic, and furthermore, for graph relational structures, they are quadratic time isomorphic. Not all isomorphic pairs of FST injection structures are computably isomorphic. Any two isomorphic graph relational injection structures with full universe $\{0,1\}^*$ are double exponential time isomorphic.

Ongoing research continues on characterizing the FST injection structures. We have obtained partial results on the Isomorphism Problem for unary FST injection structures and will continue to study this. An important question remains whether every FST injection structure (A, f) have a decidable theory.

References

1. Aho, A.V., Hopcroft, J.E., Ullman, J.D.: A general theory of translation. Math. Syst. Theor. **3**, 193–221 (1969)
2. Ash, C., Knight, J., Manasse, M., Slaman, T.: Generic copies of countable structures. Ann. Pure Appl. Logic **42**, 195–205 (1989)
3. Blumensath, A.: Automatic structures. Diploma Thesis, RWTH Aachen (1999)
4. Buss, S., Cenzer, D., Minnes, M., Remmel, J.B.: Injection structures specified by finite state transducers. In: Day, A., Fellows, M., Greenberg, N., Khoussainov, B., Melnikov, A., Rosamond, F. (eds.) Computability and Complexity. LNCS, vol. 10010, pp. 394–417. Springer, Cham (2017). https://doi.org/10.1007/978-3-319-50062-1_24
5. Cenzer, D., Harizanov, V., Remmel, J.B.: Σ_1^0 and Π_1^0 structures. Ann. Pure Appl. Logic **162**, 490–503 (2011)
6. Cenzer, D., Harizanov, V., Remmel, J.B.: Computability theoretic properties of injection structures. Algebra Logic **53**, 39–69 (2014)
7. Cenzer, D., Remmel, J.B., Carson, J.: Effective categoricity of automatic equivalence and nested equivalence structures. Theor. Comput. Syst. **64**, 1110–1139 (2020)
8. Delhommé, C.: Automaticité des ordinaux et des graphes homogènes. C. R. Math. Acad. Sci. Paris **339**(1), 5–10 (2004)
9. Hodgson, B.R.: On direct products of automaton decidable theories. Theoret. Comput. Sci. **19**(3), 331–335 (1982)
10. Khoussainov, B., Liu, J., Minnes, M.: Unary automatic graphs: an algorithmic perspective. Math. Struct. Comput. Sci. **19**(1), 133–152 (2009)
11. Khoussainov, B., Minnes, M.: Model-theoretic complexity of automatic structures. Ann. Pure Appl. Logic **161**(3), 416–426 (2009)
12. Khoussainov, B., Nerode, A.: Automatic presentations of structures. In: Leivant, D. (ed.) LCC 1994. LNCS, vol. 960, pp. 367–392. Springer, Heidelberg (1995). https://doi.org/10.1007/3-540-60178-3_93
13. Khoussainov, B., Nies, A., Rubin, S., Stephan, F.: Automatic structures: richness and limitations. In: Logic Methods in Computer Science (Special issue: Conference "Logic in Computer Science 2004") 2:2, 18 (2004)
14. Khoussainov, B., Rubin, S., Stephan, F.: On automatic partial orders. In: Proceedings of the LICS 2003, pp. 168–177 (2003)
15. Liu, J., Minnes, M.: Deciding the isomorphism problem in classes of unary automatic structures. Theoret. Comput. Sci. **412**(18), 1705–1717 (2011)
16. Liu, J., Minnes, M.: Analysing complexity in classes of unary automatic structures. In: Dediu, A.H., Ionescu, A.M., Martín-Vide, C. (eds.) LATA 2009. LNCS, vol. 5457, pp. 518–529. Springer, Heidelberg (2009). https://doi.org/10.1007/978-3-642-00982-2_44

A Tale of Optimizing the Space Taken by de Bruijn Graphs

Rayan Chikhi[✉]

Institut Pasteur, CNRS, Paris, France
rayan.chikhi@pasteur.fr

Abstract. In the last decade in bioinformatics, many computational works have studied a graph data structure used to represent genomic data, the de Bruijn graph. It is closely tied to the problem of genome assembly, i.e. the reconstruction of an organism's chromosomes using a large collection of overlapping short fragments. We start by highlighting this connection, noting that assembling genomes is a computationnally intensive task, and then focus our attention on the reduction of the space taken by de Bruijn graph data structures. This extended abstract is a retrospective centered around my own previous work in this area. It complements a recent review [10] by providing a less technical and more introductory exposition of a selection of concepts.

Keywords: Bioinformatics · Data structures · de Bruijn graphs

1 Context

Let us travel back in time in 2008, when the problem of reconstructing genomes using DNA sequencing, termed *de novo* genome assembly, had a rebirth as an active area of research within many computational research groups. Back when I started a PhD in bioinformatics in 2008, my advisor Dominique Lavenier told me about a relatively new problem consisting in reconstructing genomes using DNA sequencing, termed *de novo* genome assembly. It was a somewhat "fresh" problem at the time: many genomes were already assembled, e.g. the Human Genome Project completed at the beginning of the 2000s, yet performing the assembly of any organism was just starting to be within reach for most biological labs. The vast majority of organisms did not have their genomes assembled (and as of today: they still do not). So the challenge was to create software that any individual lab could use, not just large organizations. The main type of data at the time were *short reads*, i.e. fragments of around 100 nucleotides, meaning that only a tiny fraction of a genome could be read contiguously at a time. (Genomes of viruses are in the order of thousands of nucleotides, but for most other organisms they range from millions to billions.) By repeatedly sequencing fragments from random locations, reads would significantly overlap which makes genome reconstruction possible. Short reads were produced mainly

© Springer Nature Switzerland AG 2021
L. De Mol et al. (Eds.): CiE 2021, LNCS 12813, pp. 120–134, 2021.
https://doi.org/10.1007/978-3-030-80049-9_11

from the company Illumina, still a market leader on DNA sequencing today; some of the previous technologies (e.g. 454) were on their way out.

The EULER-SR assembler was one of the first specialized genome assembly software for short reads, and it came out in 2008. It achieved an assembly of a bacterium (*E. coli*) in 199 pieces [9]. This means that the genome was near-completely reconstructed, yet in a fragmented way to due ambiguities. This may seem unremarkable by current standards, as nowadays we can reconstruct nearly all bacteria in a single piece per chromosome. Yet the task was fundamentally hindered by the length of the short reads. Still, at the time it was clear that the next frontier would be to assemble larger genomes, e.g. animals or plants, even if the final assembly would still be largely fragmented.

The widely-used Velvet [43], ABySS [41] and SOAPdenovo [24] assemblers appared in the following two years. And indeed, the first two were able to assemble a human genome using a cluster or a large-memory single machine. These assemblers were all based on a certain representation of the input data (de Bruijn graph), that we will explain in the next section. These graphs come from mathematics and had not yet been widely used outside of some networking applications. This was before the era of more advanced assemblers (IDBA/SPAdes [3,37]); early assemblers only constructed a single graph, as opposed to iterating over multiple graphs with different parameters.

Even so, the construction of a large de Bruijn graph was the most computation-intensive step of genome assembly at the time. This should come as no surprise, as 1) this was the first period in history when one had to construct large de Bruijn graphs in any domain; there existed no previous litterature describing how to do it efficiently, and no software library. 2) It did not matter so much if construction was slow or memory-intensive, as long as some large-memory machine managed to run it. 3) The volume of input data was really large by historical standards: in the order of a hundred of gigabytes in compressed form. Yet, as genome assembly later became a routine task, along with the advent of huge instances such as metagenomics (the analysis of multiple genomes at once), the efficient construction de Bruijn graphs naturally became a critical aspect of genome assembly. It also turns out that de Bruijn graphs would be useful for other biological sequence analysis tasks, such as the sequencing of RNA [35], the compression of genomic data [20], and the detection or representation of variations across a single or multiple genomes [16].

The goal of this extended abstract is to retrace some of the steps that the community and I took towards achieving space-efficient representations of de Bruijn graphs, starting from the initial attempts in the first assemblers, making a detour through theoretical lower bounds, and finishing with current advances and some perspectives.

2 Problem Formulation

Let us introduce some of the concepts. A DNA sequence is seen as a string over four possible characters (A,C,T,G). A k-mer is a portion of DNA sequence of length k, e.g. ACT is a 3-mer. The de Bruijn graph is a directed graph where nodes are k-mers, and edges are the exact suffix-prefix overlaps of length $k - 1$ between two nodes; e.g. ACT→CTA or AAA→AAT, but ACT and AAA are not connected by an edge. See Fig. 1 for another example. Note that in practice, k is typically greater than 20. A de Bruijn graph is constructed by inserting all the possible k-mers present in an input dataset. If the same k-mer is seen multiple times, all of its occurrences are associated to the same node.

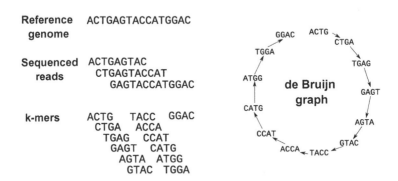

Fig. 1. Left panel: example of a toy reference genome sequence, a set of 3 sequenced reads, and the corresponding 4-mers extracted from the reads. Right panel: the de Bruijn graph constructed these reads with $k = 4$ and drawn using a circular layout.

The scientific question we will be interested in can be informally stated as follows: given a set of nodes of the de Bruijn graph, stored on disk, construct an in-memory representation[1] that supports a reasonable subset of standard graph operations, e.g. determine all the neighbors of a node, determine if some putative node is present or absent, etc. The representation should take as little memory as possible, and answer queries reasonably fast, although as we will see next, the main limiting factor here is typically not the query time but the representation size.

Note that prior to circa 2012, the problem as stated above was not recognized as its own area of investigation within bioinformatics nor computer science. Arguably it became one when several data structures were published as standalone articles [5,13,14].

3   Caveats

We will focus here on only a selection of major milestones, where space usage was reduced, ignoring other features such as query times. The presentation will also

---

[1] Such a representation is also often called a *data structure*, and the abstract model that encompasses all the data structures supporting the same operations is called an *abstract data type*.

sacrifice some technical accuracy in favor of accessibility. For a more complete and technical exposition, please refer to this review [10].

Note that genome assembly cannot be reduced to the representation of the de Bruijn graph. In fact, many older tools even used different paradigms [32]. Among those which do use a de Bruijn graph, they implement many steps before (e.g. error correction) and after (e.g. graph cleaning) the construction of the graph that crucially affect results quality. However, for the sake of keeping the story coherent, we will set aside this broader environment to focus solely on the efficiency of graph representation.

4   The Early Days

The early assembly programs from the 2008–2010 era did not particularly aim to optimize the space usage of de Bruijn graphs. Therefore, their memory usage may be seen as wasteful by current standards, yet they laid the bases for future progress.

The EULER-SR assembler reported building the graph using what they describe as "an efficient hashing structure" which was then transformed into a sorted list of vertices, queried using binary search. Notably, k-mers were represented explicitly as strings.[2]

Similarly the Velvet assembler, published the same year, used a hash table to record for each k-mer "the ID of the first read encountered containing that k-mer and the position of its occurrence within that read". It is natural to want to keep track of where each k-mer is coming from, however as we will see next, storing this information in the graph is prohibitively expensive. The authors note: "The main bottleneck, in terms of time and memory, is the graph construction. The initial graph of the *Streptococcus* reads needs 2.0 [gigabytes] of RAM." Given that the *Streptococcus* genome is 2 million nucleotides in length, and under the assumption that there were roughly 10x more erroneous k-mers than correct ones, we infer that the de Bruijn graph representation of Velvet required in the order of 100 bytes per k-mer.

The SOAPdenovo assembler followed the Velvet assembler strategy, except that its authors realized that one could achieve the nearly identical (or even better) results quality despite discarding a lot of space-intensive information in the hash table (read locations and paired-end information). Its graph representation required 120 GB of memory for storing 5 billion nodes of a human genome [32], i.e. around 24 bytes per k-mer. This prowess demonstrated that the quality of genome assembly was not sacrificed when trimming down the de Bruijn graph data structure. There existed some minimal set of supported operations that would make a de Bruijn graph fit for purpose, although this set was not described at the time. As long as a data structure would support all these

---

[2] The total space usage of the graph was reported to be $O(L) * (v + k)$ bytes, where L is the genome size, k is the k-mer length and v is the memory allocated per vertex, reported to be 40 bytes.

features, then computer scientists would be free to optimize it as much as they could.

The Meraculous assembler, published in 2011, took a radically different approach by storing the de Bruijn graph using collision-free hashing. Its representation only supports the lookup of the next nucleotide following a k-mer (i.e. the out-neighbor of a node), where k-mers having multiple out-neighbors were previously discarded during a pre-processing step. As in other assemblers, there are further steps taken to attempt to "fill the gaps" between the discarded k-mers and to orient the assembled fragments, yet these are outside our current scope. The Meraculous de Bruijn graph structure does not support enumeration of vertices. Despite the apparent minimality in terms of supported operations, it appeared to be sufficient for enabling genome assembly. While this technique was not further re-used by other assembly tools, we revisited it 6 years later to develop the general-purpose *minimal perfect hashing* library BBHash [25].

5   The Birth of a Line of Research

The years 2011–2012 saw a remarkable amount of independent contributions proposing new ways to represent the de Bruijn graph in a space-efficient manner. In retrospect, the field was ripe for such contributions as there was an important problem to be solved (genome assembly of human genomes was taking a prohibitive amount of memory), which was well-defined computationally[3], and there were no previous "clever" solutions apart from using off-the-shelf data structures.

To my knowledge, the first article on this topic was published in 2011 by T. Conway and A. Bromage [14]. They describe an encoding of the de Bruijn graph using an existing state-of-the-art efficient encoding of bit arrays. Furthermore, they also show that their representation is 'optimal', in the following sense: information theory dictates that any other exact de Bruijn graph representation will have to use as many bits per k-mer in the worst case. The key words of the previouse sentence are "exact" and "worst case", and we will revisit this statement later in this document. But for now, it is sufficient to note that to this date, the Conway-Bromage data structure is provably optimal. Then, does this mark the end of the line of research on the representation of de Bruijn graphs?

Not quite, despite the lower bound argument being convincing. We will briefly expose it here. Observe that to represent a de Bruijn graph, one only needs to represent its vertices[4]. The edges are indeed implicit in the representation, as

---

[3] At least implicitly, as to my knowledge, it has not been explicitly formulated as an open question in an article.

[4] We omit a technicality here that will only be of interest to specialists. We only consider node-centric de Bruijn graphs. For edge-centric de Bruijn graphs, the argument stated in this paragraph does not apply. Yet, edge-centric graphs are tightly related to node-centric ones and in practice, using one definition or the other does not matter.

one could determine all the neighbors of a certain k-mer by querying for the presence of all the potential k-mers shifted to the left or to the right.

Then, a bijection is established between the set of all possible sets of n vertices, and the set of all possible binary vectors having n ones and $4^k - n$ zeros. The bijection is actually rather straightforward: each k-mer is directly encoded as an integer in base 4 (see Fig. 2 middle panel), and a bit vector has a **1** at position i if and only if i is the encoding of a k-mer that belongs to the set of graph vertices. Since the number of possible bit vectors is classically known, one deduces that to represent a de Bruijn graph for a certain parameter k having n vertices, one must uses in the worst case as many bits as the logarithm of the number of possible bit vectors of size 4^k that have n ones.

6   Beating the Lower Bound (by Inexactness)

As it turns out, this lower bound did not discourage researchers from proposing data structures that exhibited even lower space usages in practice than those dictated by the bound. One such data structure is the encoding of a de Bruijn graph using a Bloom filter by Pell *et al.* [36] (Fig. 2). By inserting all the vertices of the graph inside a probabilitic membership data structure (here, the Bloom filter), it is possible to represent a set of k-mers approximatively. The trade-off is then that the graph is not exactly represented, yet the space usage is an order of magnitude lower than the one dictated by the Conway-Bromage lower bound: around 4 bits per k-mer. Pell *et al.* showed that despite having many false positive nodes resulting from the approximate representation, it was still possible to perform useful analysis on the graph – not quite genome assembly, but another related task (read partitioning).

Following this, G. Rizk and I proposed to extend this representation to make it exact within a certain setting, and perform genome assembly [13]. Due to the lower bound, any attempt at removing all the false positives of the Bloom filter would result in a data structure that would necessarily be at least as large as the one from Conway-Bromage. The key insight was to realize that only a fraction of the false positives of the Bloom filter mattered: those which were neighbors of a true positive vertex. By explicitly storing them in a "blacklist" hash table, our data structure managed to represent a graph in typically ≈ 13 bits per k-mer[5], and the neighbor query operation would be exact. This "beats" the Conway-Bromage bound by a factor of roughly 2x. The caveat is that the representation is not exact everywhere, but as long as a user traverses the graph from a true positive vertex, then the representation would act identically to the exact one. We implemented this data structure inside a genome assembler, Minia [13].

Another instance of an inexact de Bruijn graph representation is the *sparse de Bruijn graph* [42], which is a de Bruijn graph that skips g intermediate k-mers, providing roughly a $1/g$ space saving, where g was set to 16. Finally, along the same line of thought the A-Bruijn graph formalism [26] selects an arbitrary set

---

[5] Which would later be improved by Sahlikov & Kucherov to ≈ 8.5 bits per k-mer, using cascading Bloom filters [40].

Exact encoding of the de Bruijn graph using a hash table

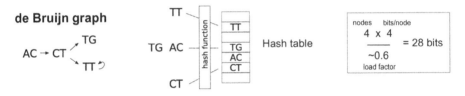

Exact encoding of the de Bruijn graph using a bit vector

Approximate encoding of the de Bruijn graph using a Bloom filter

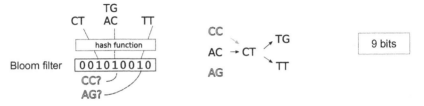

Fig. 2. Example of a de Bruijn graph (top left panel) and three possible encodings. Top right panel: hash table, each node is inserted at a position given by a random hash function. The collision between TG and AC is resolved using linear probing, i.e. by inserting AC at the next free slot in the table. The load factor is number of occupied cells over total cells. Middle panel: bit vector, storing each node converted into an integer using the classical binary encoding of characters A,C,G,T=00b,01b,10b,11b, where b indicates that the number is written in binary. Bottom panel: Bloom filter, where each node is inserted at a position given by a random hash function. Two false positives nodes (CC, AG) are shown in red. They arise because the hash function causes collisions between any possibly existing node and true nodes. (Color figure online)

of strings, and creates an edge when two strings appear consecutively in at least one read. This concepts generalizes de Bruijn graphs; yet A-Bruijn graphs may contain one or several orders of magnitude less nodes than de Bruijn graphs. There are some potentially interesting parallels between A-Bruijn graphs and sparse de Bruijn graphs, yet to the best of my knowledge they have not been explored (Fig. 3).

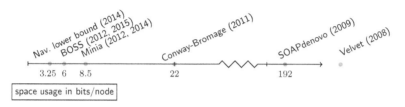

Fig. 3. Space taken by various representations of de Bruijn graphs, in bits per node. "Conway-Bromage" is both the Conway-Bromage exact lower bound and its matching upper bound. "Nav. lower bound" is the navigational lower bound for general de Bruijn graphs from [11]. BOSS is the flavor of [23].

7   Beating the Lower Bound (by Instance Specificity)

Independently of Minia, and presented at the same session of the WABI conference in 2012, the BOSS data structure proposes a completely different yet exact de Bruijn graph representation [5]. It uses a variant of the Burrows-Wheeler transform specifically tailored for k-mers. A complete description of the BOSS structure, or even the Burrows-Wheeler transform, would be beyond the technical level of this document, and can be found in [1]. Intuitively, the Burrows-Wheeler transform [7] is a permutation of the characters of a string that facilitates substring search and compression. BOSS extends this concept by storing a permutation of the last characters of each k-mer[6] together with a bit array. The result is a data structure that supports efficient membership queries and neighborhood traversal of the graph, all in around 6 bits per k-mer in practice. While in the first few years the construction of this structure was relatively impractical, recent improvements lifted those limitations, allowing to process even terabases of input data [22].

Taking a step back, BOSS is an exact representation that appears to somehow beat the Conway-Bromage lower bound. How is this even possible? While this aspect was not discussed in nearly all of the publications related to BOSS, it turns out that BOSS has been mainly applied to k-mer sets that have a so-called *spectrum-like property* [10], i.e. where all the k-mers originate from some underlying long strings. Should BOSS be applied to an arbitrary set of k-mers, its space usage would mechanically be raised to match or exceed the Conway-Bromage lower bound; yet, this fact has to my knowledge never been properly tested in practice.

Regardless, the spectrum-like property and the effectiveness of BOSS are important insights: a data structure may do better than the worst-case lower bound while still remaining exact, when it is restricted to a certain class of inputs that matter in practice. Then, a natural next question arises: what would be a more realistic lower bound for representing 'practical' de Bruijn graphs, i.e. those having spectrum-like property?

---

[6] along with some additional artificial k-mers to "pad" those which do not have a large enough neighborhood.

Several collaborators and I addressed this question in [11], where we formulated several concepts. First, we defined a *navigational* data structure as one that enables navigation in the graph but does not necessarily support membership queries. We showed that navigational data structures for general de Bruijn graphs require at least 3.24 bits per k-mer in the worst case. When restricted to the family of linear de Bruijn graphs (i.e. graphs where all nodes have a single in-neighbor and/or single out-neighbor), then a lower bound for navigational data structures is 2 bits per k-mer. This last lower bound is tight, as representing the linear de Bruijn graph using the Burrows-Wheeler transform (or its optimized flavor, FM-index [17]) yields also a data structure that is asymptotically close to 2 bits per k-mer. In [11], we also proposed a new data structure for de Bruijn graphs having the spectrum-like property, using the Burrows-Wheeler transform, and showed that it takes $2 + (k + 2)c/n$ bits per k-mer, where c is essentially the maximal number of k-mer-disjoint strings the k-mers could have been generated from[7].

Lastly, one may also wonder how a de Bruijn graph could be further compressed, e.g. to be stored on disk. Supposedly such a compressed representation would be even smaller than the previously mentioned data structures. The trade-off is the inability to perform fast queries. Two independent works [6,38], one on *simplitigs* and the other on *spectrum-preserving string sets* which I was associated with, proposed to store non-overlapping paths of the compacted de Bruijn graph as sequences, and store them in compressed form on disk. Despite the representation apparently storing an incorrect representation of the graph, due to paths being constructed by choosing edges arbitrarily, one may observe that the original graph can be reconstructed losslessly from its path representation. Such a disk representation achieved a space very close to the 2 bits per k-mer lower bound: 4.1 bits per k-mer for a whole human genome read dataset, and 2.7 bits per k-mer for a human metagenome.

8   Construction Algorithms

An apparté will be made in this section, where we will briefly mention the data structure construction algorithms. One typical pre-processing step commonly done prior to creating a de Bruijn graph data structure is k-mer counting. This step takes the input sequencing data and yields the set of distinct k-mers present in the input along with their abundances. It essentially constructs the nodes of the de Bruijn graph.

During the development of Minia, we had ran into an issue. The graph representation was so succinct that other steps of the genome assembly pipeline acted as bottlenecks, including k-mer counting. At the time, the most efficient k-mer counter was Jellyfish [28], which used a custom thread-safe hash table optimized specifically to store k-mers. Yet, Jellyfish would have used much more memory than Minia. We therefore set out to design a low-memory k-mer counting tool

---

[7] For specialists, c is the number of unitigs.

that would use the disk to alleviate memory usage (DSK [39]). This strategy was also used by other popular k-mer counting tools, e.g. KMC [15].

The problem of k-mer counting is fascinating in its simplicity but also difficult to engineer correctly, given that large volumes of input sequences need to be processed with high CPU utilization, low memory usage, and bandwidth-limited disk accesses. A relatively current review is [27]. After k-mer counting, nearly all of the data structures presented above have their own, customized construction algorithms. As such, there does not exist an 'universal' construction algorithm for de Bruijn graph that would then be slightly adapted to derive a particular data structure.

However, several recent data structures (the one presented in [11], Puffer-fish [2], BLight [30]) require as input a common object: the compacted de Bruijn graph. It is obtained from a classical de Bruijn graph by transforming each non-branching path into a single node, similarly to suffix tries are transformed into suffix trees by collapsing paths of vertices having one child. However, this is a circular situation: in order to construct an efficient representation of the classical de Bruijn graph, one must have already constructed a compacted de Bruijn graph, which itself is obtained from the classical de Bruijn graph. In order to break this circularity, My colleagues and I proposed an efficient construction algorithm for the compacted de Bruijn graph [11], which uses a constant amount of memory. It was further extended to make use of multiple threads efficiently [12].

9   Current State of the Art

Since the influx of de Bruijn graph data structures in 2012, several more have been published in the recent years. As it turns out, many of them are based on minimal perfect hashing. It is a variation of a hash table which does not store its keys, yet still manages to resolves collisions. Minimal perfect hashing is unable to confirm if an arbitrary key is present or absent in the structure, however for any key that was inserted during its construction, it returns an exact answer. This makes the structure highly space-efficient. Along with colleagues, we proposed a fast parallel C++ library for constructing minimal perfect hashes (BBHash [25]) which has been engineered to scale significantly better than other existing implementations at the time.

Among current de Bruijn graph data structures, I will briefly highlight Puffer-fish [2] and Blight [30] which are both based on the compacted de Bruijn graph, and queries are supported by an additional minimal perfect hashing structure that quickly locates positions within nodes of the compacted graph. The Bloom Filter trie [19] and Bifrost [18] structures both use Bloom filters in addition with other auxiliary data structures to keep the representation exact, yet even brushing their algorithmic details would be too technical for this survey. Counting quotient filters [34] improve upon Bloom filters by also storing the number of occurrences of each node in the input data. Belazzougui *et al.* proposed a navigational data structure based on minimal perfect hashing, with a clever addition of a tree data structure to restore membership queries. For more details on all of these data structures, see [10].

In a way, one might acknowledge that "the dust has settled" in the landscape of de Bruijn graph data structures. The bioinformatics algorithms community has attempted for several years to come up with solutions that combine low space usage, fast query speed and a reasonable set of features. The outcome is a set of current data structures that achieve reasonable trade-offs, with space close to the known lower bounds. As a result, de Bruijn graphs are no longer a bottleneck in genome assembly, partly also due to decreasing RAM costs (Fig. 4).

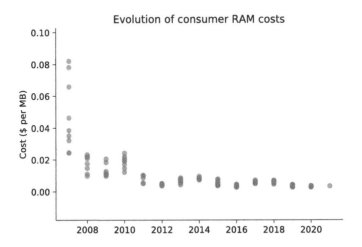

Fig. 4. Consumer RAM costs from the 2007–2021 period. Each dot is a retail DIMM product sold on the current year. Source: https://jcmit.net/memoryprice.htm

Then, is this the end of this line of research? Not quite, as the natural next frontier is the representation of multiple genomes within a generalization of the de Bruijn graph.

10   Colored de Bruijn Graphs

As a coincidence of dates (or perhaps not), 2012 was not only the year where many seminal data structures for de Bruijn graphs were proposed, but also the year when the term *colored de Bruijn graph* was coined (in [21]), which will pave the way to the next type of contributions that we will mention here. Colored de Bruijn graphs generalize de Bruijn graphs to multiple samples. When faced with multiple samples, a classical de Bruijn graph would bundle them together and consider the union of all samples as a single "mega-sample". Colored de Bruijn graph also do that, but they add additional information associated to the nodes so that one can tell the origin of each node across samples. Naturally, speaking in terms of lower bounds, storing such a graph for multiple samples should require strictly more space than storing the graph of any subset of samples.

Several data structures have been proposed to store colored de Bruijn graph, the first of which was based on an efficient hash table [21], then later using the Burrows-Wheeler transform [33]. More recently my colleagues and I proposed the REINDEER structure, based on compacted de Bruijn graphs and minimal perfect hashing [29], with the distinctive feature of not only storing the presence/absence of nodes, but also the approximate frequency of each node within each sample.

To the best of my knowledge, there has been no attempt made at formulating space lower bounds for colored de Bruijn graphs. One may obtain one through an immediate application of existing lower bounds to the union graph disregarding color information. However, this would be a loose bound as much of the difficulty of storing colored graphs lies in the color information.

11   Wrap-Up and Open Questions

As we reviewed above, many data structures have been proposed to store de Bruijn graphs, achieving several order of magnitudes improvement in space usage compared to using off-the-shelf data structures for graph storage. From this perspective, the theoretical study of data structures along with their practical implementations has been successful at providing performance gains for widely-used software tools (e.g. [3,23]). Looking back, the improvements have mainly be due to two realizations. 1) Data structure exactness can be sacrified yet still provide exact results in a certain frame of operations. 2) The theoretical worst-case analysis of data structures inadequately applies to practical instances. The latter realization is the topic of an upcoming article from Medvedev [31], critically reflecting on the analysis of bioinformatics algorithms more broadly.

Several topics were not covered in this document, to keep it simple. One is double-strandedness, which forces all the data structures mentioned above to consider that a k-mer and its reverse-complement should be the same object; this adds theoretical and especially practical complications, yet does not fundamentally change the exposition of the data structures. An additional one is the use of multiple k values. Nowadays genome assembly tools on short reads typically construct multiple de Bruijn graphs iteratively. This is a somewhat orthogonal matter as presented here, given that each individual graph is represented using one of the techniques above. We note however that some works have attempted to unify multiple graphs into one [4,8]. Another consideration is how to store the number of times each k-mer is seen in the input. All these considerations are discussed in more details in [10].

We summarize here a few open questions:

1. Can compressed representations e.g. spectrum-preserving string sets be made efficiently queryable? This would lead to even more compressed de Bruijn graphs.
2. What would be a space lower bound for exactly representing a colored de Bruijn graph of n samples, each sample i having D_i distinct k-mers?
3. A matching upper bound of the above.

4. How to efficiently represent not only the presence/absence of a node but also its abundance in colored de Bruijn graphs (improving upon REINDEER [29]).

References

1. Alipanahi, B., Kuhnle, A., Puglisi, S.J., Salmela, L., Boucher, C.: Succinct Dynamic de Bruijn Graphs. Bioinformatics (2020). https://academic.oup.com/bioinformatics/advance-article/doi/10.1093/bioinformatics/btaa546/5848003
2. Almodaresi, F., Sarkar, H., Srivastava, A., Patro, R.: A space and time-efficient index for the compacted colored de Bruijn graph. Bioinformatics **34**(13), i169–i177 (2018)
3. Bankevich, A., et al.: SPAdes: a new genome assembly algorithm and its applications to single-cell sequencing. J. Comput. Biol. **19**(5), 455–477 (2012)
4. Boucher, C., Bowe, A., Gagie, T., Puglisi, S.J., Sadakane, K.: Variable-Order de Bruijn graphs. In: 2015 Data Compression Conference, pp. 383–392 (2015)
5. Bowe, A., Onodera, T., Sadakane, K., Shibuya, T.: Succinct de Bruijn graphs. In: Raphael, B., Tang, J. (eds.) WABI 2012. LNCS, vol. 7534, pp. 225–235. Springer, Heidelberg (2012). https://doi.org/10.1007/978-3-642-33122-0_18
6. Břinda, K., Baym, M., Kucherov, G.: Simplitigs as an efficient and scalable representation of de Bruijn graphs. Genome. Biol. **22**, 96 (2021). https://doi.org/10.1186/s13059-021-02297-z
7. Burrows, M., Wheeler, D.: A block-sorting lossless data compression algorithm. Report 124, Digital Systems Research Center, Palo Alto, CA, USA (May 1994)
8. Cazaux, B., Rivals, E.: Hierarchical overlap graph. Inf. Proc. Lett. **155**, 105862 (2020)
9. Chaisson, M.J., Pevzner, P.A.: Short read fragment assembly of bacterial genomes. Genome Res. **18**(2), 324–330 (2008)
10. Chikhi, R., Holub, J., Medvedev, P.: Data structures to represent sets of k-long DNA sequences. arXiv preprint arXiv:1903.12312 (2019)
11. Chikhi, R., Limasset, A., Jackman, S., Simpson, J.T., Medvedev, P.: On the representation of de Bruijn graphs. In: Sharan, R. (ed.) RECOMB 2014. LNCS, vol. 8394, pp. 35–55. Springer, Cham (2014). https://doi.org/10.1007/978-3-319-05269-4_4
12. Chikhi, R., Limasset, A., Medvedev, P.: Compacting de Bruijn graphs from sequencing data quickly and in low memory. Bioinformatics **32**(12), i201–i208 (2016)
13. Chikhi, R., Rizk, G.: Space-efficient and exact de Bruijn graph representation based on a bloom filter. Algorithms Mol. Biol. **8**(1), 1–9 (2013)
14. Conway, T.C., Bromage, A.J.: Succinct data structures for assembling large genomes. Bioinformatics **27**(4), 479–486 (2011)
15. Deorowicz, S., Debudaj-Grabysz, A., Grabowski, S.: Disk-based k-mer counting on a PC. BMC Bioinf. **14**(1), 1–12 (2013)
16. Eizenga, J.M., et al.: Pangenome graphs. Ann. Rev. Genomics Hum. Genet. **21**(1), 139–162 (2020). PMID: 32453966
17. Ferragina, P., Manzini, G.: Opportunistic data structures with applications. In: Proceedings 41st Annual Symposium on Foundations of Computer Science, pp. 390–398. IEEE (2000)
18. Holley, G., Melsted, P.: Bifrost: highly parallel construction and indexing of colored and compacted de Bruijn graphs. Genome Biol. **21**(1), 1–20 (2020)

19. Holley, G., Wittler, R., Stoye, J.: Bloom filter trie: an alignment-free and reference-free data structure for pan-genome storage. Algorithms Mol. Biol. **11**(1), 1–9 (2016)
20. Holley, G., Wittler, R., Stoye, J., Hach, F.: Dynamic alignment-free and reference-free read compression. In: Sahinalp, S.C. (ed.) RECOMB 2017. LNCS, vol. 10229, pp. 50–65. Springer, Cham (2017). https://doi.org/10.1007/978-3-319-56970-3_4
21. Iqbal, Z., Caccamo, M., Turner, I., Flicek, P., McVean, G.: De novo assembly and genotyping of variants using colored de Bruijn graphs. Nature Genet. **44**(2), 226–232 (2012)
22. Karasikov, M.: Indexing and analysing nucleotide archives at petabase-scale. bioRxiv (2020)
23. Li, D., Liu, C.-M., Luo, R., Sadakane, K., Lam, T.-W.: MEGAHIT: an ultra-fast single-node solution for large and complex metagenomics assembly via succinct de Bruijn graph. Bioinformatics **31**(10), 1674–1676 (2015)
24. Li, R., et al.: De novo assembly of human genomes with massively parallel short read sequencing. Genome Res. **20**(2), 265–272 (2010)
25. Limasset, A., Rizk, G., Chikhi, R., Peterlongo, P.: Fast and scalable minimal perfect hashing for massive key sets. arXiv preprint arXiv:1702.03154 (2017)
26. Lin, Y., Yuan, J., Kolmogorov, M., Shen, M.W., Chaisson, M., Pevzner, P.A.: Assembly of long error-prone reads using de bruijn graphs. Proc. Natl. Acad. Sci. **113**(52), E8396–E8405 (2016)
27. Manekar, S.C., Sathe, S.R.: A benchmark study of k-mer counting methods for high-throughput sequencing. GigaScience **7**(12), 125 (2018)
28. Marçais, G., Kingsford, C.: A fast, lock-free approach for efficient parallel counting of occurrences of k-mers. Bioinformatics **27**(6), 764–770 (2011)
29. Marchet, C., Iqbal, Z., Gautheret, D., Salson, M., Chikhi, R.: REINDEER: efficient indexing of k-mer presence and abundance in sequencing datasets. Bioinformatics **36**(Supplement_1):i177–i185 (2020)
30. Marchet, C., Kerbiriou, M., Limasset, A.: Blight: Efficient exact associative structure for k-mers. bioRxiv (2020)
31. Medvedev, P.: The theoretical analysis of sequencing bioinformatic algorithms. in preparation (2020)
32. Miller, J.R., Koren, S., Sutton, G.: Assembly algorithms for next-generation sequencing data. Genomics **95**(6), 315–327 (2010)
33. Muggli, M.D.: Succinct colored de Bruijn graphs. Bioinformatics **33**(20), 3181–3187 (2017)
34. Pandey, P., Bender, M.A., Johnson, R., Patro, R.: A general-purpose counting filter: Making every bit count. In: Proceedings of the 2017 ACM International Conference on Management of Data, pp. 775–787 (2017)
35. Patro, R., Duggal, G., Love, M.I., Irizarry, R.A., Kingsford, C.: Salmon provides fast and bias-aware quantification of transcript expression. Nat. Methods **14**(4), 417–419 (2017)
36. Pell, J.: Scaling metagenome sequence assembly with probabilistic de Bruijn graphs. Proc. Natl. Acad. Sci. **109**(33), 13272–13277 (2012)
37. Peng, Yu., Leung, H.C.M., Yiu, S.M., Chin, F.Y.L.: IDBA – a practical iterative de Bruijn graph de novo assembler. In: Berger, B. (ed.) RECOMB 2010. LNCS, vol. 6044, pp. 426–440. Springer, Heidelberg (2010). https://doi.org/10.1007/978-3-642-12683-3_28
38. Rahman, A., Chikhi, R., Medvedev, P.: Disk Compression of k-mer Sets. In: 20th International Workshop on Algorithms in Bioinformatics (WABI 2020). Schloss Dagstuhl-Leibniz-Zentrum für Informatik (2020)

39. Rizk, G., Lavenier, D., Chikhi, R.: DSK: k-mer counting with very low memory usage. Bioinformatics **29**(5), 652–653 (2013)
40. Salikhov, K., Sacomoto, G., Kucherov, G.: Using cascading Bloom filters to improve the memory usage for de Brujin graphs. Algorithms Mol. Biol. **9**(1), 1–10 (2014)
41. Simpson, J.T., Wong, K., Jackman, S.D., Schein, J.E., Jones, S.J., Birol, I.: ABySS: a parallel assembler for short read sequence data. Genome Res. **19**(6), 1117–1123 (2009)
42. Ye, C., Ma, Z.S., Cannon, C.H., Pop, M., Douglas, W.Y.: Exploiting sparseness in de novo genome assembly. In: BMC bioinformatics, vol. 13, pp. 1–8 (2012) BioMed Central
43. Zerbino, D.R., Birney, E.: Velvet: algorithms for de novo short read assembly using de Bruijn graphs. Genome Res. **18**(5), 821–829 (2008)

Formally Computing with
the Non-computable

Liron Cohen[(✉)]

Department of Computer Science, Ben-Gurion University, Be'er Sheva, Israel
cliron@bgu.ac.il

Abstract. Church–Turing computability, which is the standard notion of computation, is based on functions for which there is an effective method for constructing their values. However, intuitionistic mathematics, as conceived by Brouwer, extends the notion of effective algorithmic constructions by also admitting constructions corresponding to human experiences of mathematical truths, which are based on temporal intuitions. In particular, the key notion of infinitely proceeding sequences of freely chosen objects, known as *free choice sequences*, regards functions as being constructed over time. This paper describes how free choice sequences can be embedded in an implemented formal framework, namely the constructive type theory of the Nuprl proof assistant. Some broader implications of supporting such an extended notion of computability in a formal system are then discussed, focusing on formal verification and constructive mathematics.

1 Introduction

Church–Turing computability is the standard notion of computation. It defines the computable functions as those for which there is an effective method for obtaining the values of the function. Turing used the term 'purely mechanical', whereas Church used 'effectively calculable':

> "define the notion ... of an effectively calculable function of positive integers by identifying it with the notion of a recursive function of positive integers (or of a λ-definable function of positive integers)." [13]

Intuitionistic mathematics, which originated in the ideas of L.E.J. Brouwer, extends the Church–Turing notion of computability by putting forward novel forms of computation, namely the bar induction principle and the continuity principle. *Bar induction* is a strong intuitionistic induction principle which is equivalent to the classical principle of transfinite induction [34][1], while the *continuity principle for numbers* states that all functions from $\mathbb{N} \to \mathbb{N}$ to $\mathbb{N}$ are continuous. Brouwer used the bar induction principle to derive the fan theorem, which was used in turn, together with the continuity principle for numbers,

---

[1] Variants of bar induction were shown to be compatible with constructive type theory, and used to enhance the logical functionality implemented by proof assistants [29,32].

© Springer Nature Switzerland AG 2021
L. De Mol et al. (Eds.): CiE 2021, LNCS 12813, pp. 135–145, 2021.
https://doi.org/10.1007/978-3-030-80049-9_12

to derive the *uniform continuity principle* [12, Thm 3]. The uniform continuity principle states that every continuous function on a closed interval of the reals into the reals is *uniformly continuous* and has a supremum. Historians of mathematics consider that "in just ten lines a revolution was launched" [40].

But to obtain these foundational principles, the standard function space had to be expanded to include non-recursive functions. For this, Brouwer proposed accepting non-lawlike computations, and thus he introduced the bold notion of *choice sequences*. Choice sequences are fundamental objects that are at the core of intuitionistic mathematics. They are never-finished sequences of objects created *over time* by continuously picking elements from a previously well-defined collection, e.g., the natural numbers.[2] Choice sequences can be *lawlike*, in the sense that they are determined by an algorithm (i.e., standard computable functions), or *lawless* (i.e., *free*), in the sense that they are not subject to any law (e.g., generated by throwing dice), or a combination of both. Free choice sequences are described as

> "new mathematical entities ... in the form of infinitely proceeding sequences, whose terms are chosen more or less freely from mathematical entities previously acquired..." [11]

While this notion clearly steps out of the realm of sequences constructed by an algorithm, there is a mental conception of how to create such sequences: the ideal mathematician, or creative subject , can simply pick elements as time proceeds. Brouwer used the concept of choice sequences to develop a novel theory of the continuum, defining real numbers as choice sequences of nested rational intervals.

The foundations of Brouwer's intuitionistic mathematics have been widely studied [21,26,34,35,38,41]. These works have examined Brouwer's ideas from a theoretical, foundational point of view. However, the focus of this paper is the study of intuitionistic mathematics and its extended notion of computability in a formal setting, namely that of a proof assistant. We show that while choice sequences, and in particular *free* choice sequences, are considered noncomputable in the traditional sense, they can be integrated into a mechanized system and used in computations. This will not only extend the standard notion of computation in theory, but will in practice provide us with a mechanized system in which such forms of computation are supported and utilized, which, in turn, will enable the exploration of the wider implications of the resulting computational theory.

Currently, the standard Church–Turing notion of computability is the one that underlies the computational theories invoked by standard constructive type theories, which in turn are the basis of extant proof assistants such as Nuprl [2, 14], Coq [17], and Agda [1].[3] Thus, for example, the elements of the function

---

[2] For simplicity, throughout this paper we focus on choice sequences of natural numbers.

[3] For a survey of the status of Church Thesis in type-theory-based proof assistants see [20].

type $\mathbb{N} \to T$ are taken to be the effective (computable) functions from the type of the natural numbers, $\mathbb{N}$, to the type T. The integration of the notion of choice sequences into the constructive type theory would entail, among other things, that choice sequences whose elements are chosen from T become first-class citizens of the function type $\mathbb{N} \to T$.

While many of Brouwer's intuitionistic principles and theorems have already been implemented in the Nuprl proof assistant [28–30], the constructive type theory underlying Nuprl has only recently been extended to fully integrate the notion of choice sequences. This extension required major modifications to the type theory, starting from the structure of the underlying library of definitions and lemmas, and working up to the semantics of the type system. The current paper presents the main components of that extension, with full details available in [6,7]. We further discuss the wider implications of the resulting mechanized theory, especially with respect to formal verification and constructive mathematics.

2   Integrating Choice Sequences into a Proof Assistant

The Nuprl proof assistant [2,14] implements a type theory called constructive type theory, which is a dependent type theory in the spirit of Martin-Löf's extensional theory [25], based on an untyped functional programming language. It has a rich type theory including equality types, W-types, quotient types, set types, union and (dependent) intersection types, partial equivalence relation (PER) types, approximation and computational equivalence types, and partial types. This section demonstrates how the constructive type theory implemented by the Nuprl proof assistant can be consistently extended to an intuitionistic type theory, that is, a type theory that supports Brouwer's intuitionistic principles. In particular, we focus on the integration of Brouwer's broader sense of computability through an embedding of choice sequences [6,7]. This extended theory provides a formal account of the notion of choice sequences driven by the design constraints of their implementation in a theorem prover.[4]

The Nuprl proof assistant can be (very roughly) described as consisting of the following components. Underlying the whole system is the library, which stores all the definitions and proofs the system currently holds. The computation system encapsulates the operational semantics of the system. The type system defines the type constructors, the behaviors of types and their associated equalities, based on the semantics of types employed. Then there is a set of axioms and inference rules for manipulating the terms and types of the system.[5] Fully integrating choice sequences into Nuprl required a major overhaul to all of the aforementioned components, as we will describe below.

---

[4] The extended framework described was formalized in Coq's formalization of Nuprl's constructive type theory [3,27].

[5] This simplified description omits many components of the system which are not relevant to the current paper.

2.1 Storing Choice Sequences in the Library

Choice sequences are implemented as a new type of entry in the digital library of facts and definitions underlying Nuprl, which holds a (finite) list of terms. Thus, the library is used as a *state* in which we store the choices of values that have been made for a particular choice sequence at a given point in time ([31] provides details on the treatment of choice sequences in the library). We utilize the library to perform what is known as memoization, a programming language method originally designed to improve efficiency. In this scheme, we allow the values of a choice sequence to be chosen freely, but once the fifth element of the sequence has been chosen to be 7, say, we store that in the library, and from that point on we return it whenever the input is 5. Thus, the Nuprl library can be extended in two orthogonal directions: by adding more entries to the library, or by adding more values to a specific choice sequence entry. The fact that the library can always be extended allows choice sequences to be represented as finite at any given point in time (i.e., the state of the library), but as infinitely proceeding as the library extends over time. This corresponds to Brouwer's notion of a choice sequence progressing over *time*, as implemented by progressing over *library extensions*.

Concretely, each choice sequence entry in the library has a name, taken from a nominal set of atoms, and it also comes equipped with a restriction[6], in the form of a predicate, which specifies which sequence extensions are valid (e.g., the restriction '$\lambda n, t.$ if $n < 10$ then true else $2 \leq t$' forces the choices starting from position 10 to be greater than or equal to 2). The restriction constitutes a proof obligation that has to be enforced when adding more values to a choice sequence. Using the restriction mechanism we can also define the lawlike choice sequences, by simply posing their generating rule as the restriction.

Since choice sequences are open-ended objects, it may be the case that, to prove a theorem or carry out a computation, the value of a choice sequence at a certain point may need to be known, but at that stage it has yet to be defined. There are different implementation approaches in such cases. In the intuitionistic theory of choice sequences, a reasonable answer is to 'wait until the creative subject picks enough values in the sequence' (consistent with thinking about a choice sequence as the advancement of knowledge over time). This suggests one possible implementation: the system can print out a message to the user asking for more values until there is sufficient data. Another possibility is to have the system automatically fill in values up to the desired place in the sequence, using some number generator. The generator could be random or not, or even probabilistic. The current implementation in Nuprl takes the first approach, but can be combined with the second approach if needed.

2.2 Extending the Computation System

Nuprl's programming language is an untyped (à la Curry), lazy λ-calculus with pairs, injections, a fixpoint operator, etc. For efficiency, integers are primitive and

---

[6] See, e.g., [35,37,38] for discussions on the various types of restrictions.

$$T \in \texttt{Type} \quad ::= \mathbb{N} \mid \mathbb{U}_i \mid \boldsymbol{\Pi} x{:}t.t \mid \boldsymbol{\Sigma} x{:}t.t \mid \{x : t \mid t\} \mid t = t \in t \mid t{+}t \mid \ldots$$
$$\mid \texttt{Free} \quad (\text{choice sequence type})$$
$$v \in \texttt{Value} ::= T \mid \star \mid \underline{n} \mid \lambda x.t \mid \langle t, t \rangle \mid \texttt{inl}(t) \mid \texttt{inr}(t) \mid \ldots$$
$$\mid \eta \quad (\text{choice sequence name})$$
$$t \in \texttt{Term} \quad ::= x \mid v \mid t\ t \mid \texttt{fix}(t) \mid \texttt{let } x := t \texttt{ in } t \mid \texttt{case } t \texttt{ of } \texttt{inl}(x) \Rightarrow t \mid \texttt{inr}(y) \Rightarrow t$$
$$\mid \texttt{if } t{=}t \texttt{ then } t \texttt{ else } t \mid \ldots$$

$(\lambda x.t_1)\ t_2$	$\longmapsto_{lib}$	$t_1[x \backslash t_2]$	$\texttt{let } x_1, x_2 = \langle t_1, t_2 \rangle \texttt{ in } t$	$\longmapsto_{lib} \quad t[x_1 \backslash t_1; x_2 \backslash t_2]$
$\texttt{fix}(v)$	$\longmapsto_{lib}$	$v\ \texttt{fix}(v)$	$\ldots$	
$\eta(i)$	$\longmapsto_{lib}$	$\eta[i]$	, if $\eta[i]$ is defined in lib	

Fig. 1. Extended syntax (top) and operational semantics (bottom) (Color figure online)

Nuprl provides operations on integers as well as comparison operators. Nuprl's computation system also had to be revised to support choice sequences and, in particular, to make explicit the tight dependency on the library. Figure 1 presents a subset of Nuprl's extended syntax and small-step operational semantics, where the additional components related to choice sequences are highlighted in blue.

Choice sequences are incorporated as values of the form η, and the new type $\texttt{Free}$ is the type of choice sequences. The operational semantics is then extended so that all small-step reduction rules are parameterized by a library, *lib*. In particular, an application of the form $\eta(i)$ reduces to $\eta[i]$ if $0 \le i$ and the ith value in the choice sequence named η is available in the current library, in which case $\eta[i]$ returns that value; otherwise it is left undefined.

2.3 Possible Library Semantics

The introduction of choice sequences entails a radical shift in our understanding of mathematical truth. The meaning of proposition $P(\eta)$ mentioning a choice sequence η may not be determined by our current knowledge of η, and so mathematical truth is no longer a *timeless* concept. Instead, truth now depends on current knowledge of η and the possible ways that η may be extended in the future. To support this, the semantics of the Nuprl system was turned into a possible-world-style semantics [18,24], in which the possible worlds correspond to extensions of the library (thus providing a computational interpretation of the possible-world semantics in terms of libraries). In any particular state of the library the semantics is induced by Nuprl's standard realizability semantics.

Nonetheless, the standard Kripke semantics, in which a statement is true in a library only if it is true in *all possible extensions* of the current library, is insufficient to support choice sequences. To demonstrate the problem, consider the claim "there is some value in a given place of a choice sequence" (e.g., formally, $\exists x.\eta(100) = x$). This should be a valid statement in the theory of choice sequences, based on their "infinitely proceeding" nature. However, if in the current stage of the library the choice sequence a has only three values, this will be false under the Kripke-like semantics, since there are extensions of the library in which the 100th value is yet to be filled in. Thus, to support the evolving nature of choice sequences the possible-world semantics has to be more subtle in its treatment of possible extensions.

Two different possible-world semantics that depend on the current Nuprl library lib and its possible extensions, $lib \mapsto lib'$, have been considered [6,7]. The two semantics are especially well-suited to model choice sequences because in both, expressions only need to "eventually" compute to values, which is compatible with the "eventual" nature of choice sequences that are only partially available at a given time, with the promise that they can always be extended in the future. The first semantics is a *Beth-style semantics* [5,18], where $P(\eta)$ is true in library lib when, roughly speaking, there is a bar for lib (i.e., a collection of libraries such that each path in the tree of library extensions that goes through lib intersects it) in which $P(\eta)$ is true [6]. This is equivalent to saying that there is a proof of $P(\eta)$ by bar induction on the tree of possible extensions of lib. Another semantics is a variant of the Beth-style semantics called *open bar* semantics [7]. In the open bar semantics, $P(\eta)$ is true in library lib when, for each extension lib' of lib, P holds for some extension of lib'. The open bar semantics enables a more general bar induction argument and hence validates some classical principles (see Sect. 3).

2.4   Extending the Type System

These new semantics entail new interpretations of Nuprl's type system, in which types are interpreted as PERs on closed terms. The resulting type systems satisfy all the standard properties (e.g., transitivity and symmetry), but also two additional properties that are unique to such a possible-world interpretation: monotonicity and locality. Monotonicity ensures that true facts remain true in the future, and locality allows one to deduce a fact about the current library if it is true in a bar of that library. While monotonicity is a general feature of possible-world semantics (including Kripke semantics), locality is a distinctive feature of Beth-like models.

3   The Resulting Theories of Choice Sequences

The two semantics induce two theories: the one based on the Beth-style semantics is called BITT, for 'Brouwerian Intuitionistic Type Theory', and the one based on the open bar semantics is called OTT. Both theories fully embed choice sequences as first-class citizens, in the sense that choice sequences inhabit the extended function type $\mathbb{N} \to \mathbb{N}$ (also called the Baire space, $\mathcal{B}$). That is, in both BITT and OTT the following holds: $\eta \in \texttt{Free} \to \eta \in \mathcal{B}$.

3.1   Axioms for Choice Sequences

Both BITT and OTT validate (variants) of the following key properties governing choice sequences that have been suggested in the literature (see, e.g., [23,39]). In what follows we write $\mathcal{B}_n$ for $\mathbb{N}_n \to \mathbb{N}$, where $\mathbb{N}_n = \{k : \mathbb{N} \mid k < n\}$.

Density Axiom $\Pi n{:}\mathbb{N}.\Pi f{:}\mathcal{B}_n.\Sigma\alpha{:}\mathsf{Free}.f = \alpha \in \mathcal{B}_n$
Discreteness Axiom $\Pi\alpha,\beta{:}\mathsf{Free}.(\alpha{=}\beta \in \mathcal{B})+(\neg\alpha{=}\beta \in \mathcal{B})$
Open Data Axiom $\Pi\alpha{:}\mathsf{Free}.P(\alpha) \to \Sigma n{:}\mathbb{N}.\Pi\beta{:}\mathsf{Free}.(\alpha{=}\beta \in \mathcal{B}_n \to P(\beta))$

The Density Axiom intuitively states that, for any finite list of values, there is a choice sequence that extends it. In BITT, proving its validity required an additional machinery of name spaces for choice sequences (see [6] for full details). In OTT, however, such machinery is not necessary for validating the variant of the statement in which the existential quantifier is 'squashed'. The squashing mechanism erases the evidence that a type is inhabited by squashing it down to a single constant inhabitant using set types: $\downarrow T = \{x : \mathsf{True} \mid T\}$ [14, p. 60]. Intuitively, a squashed existential quantifier asserts the existence of an object without specifying how it can be computed. The Discreteness Axiom states that intensional equality over choice sequences is decidable, and it is easily validated since choice sequences are identified by their names in the library, which are unique.

The Open Data Axiom, roughly speaking, states that if a property (with certain side-conditions essentially ensuring that α is the only free choice sequence in $P(\alpha)$) holds for a free choice sequence, then there is a finite initial segment of that sequence such that this property holds for all free choice sequences with the same initial segment. Since the axiom does not provide information on a specific choice sequence, but rather on the collection of all sequences determined by an initial segment, it constitutes a continuity principle in a sense. The non-squashed continuity principle is, however, incompatible with Nuprl (following similar arguments to those in [19,22,33]). Nonetheless, in OTT two squashed variants of the Open Data Axiom have been validated. The key observation is that when the Σ type is $\downarrow$-squashed, there is no need to provide a witness for the modulus of continuity of P at α. Instead, one can simply find a suitable meta-theoretical number in the proof of its validity, without having to provide an expression from the object theory that computes that number.

3.2   Classical Axioms

One main difference between BITT and OTT relates to compatibility with classical logic. BITT is incompatible with classical logic in the sense that it validates the negation of many classically valid principles. In particular, it proves the negation of the $\downarrow$-squashed law of excluded middle, $\neg\Pi P.\downarrow(P{+}\neg P)$, the negation of Markov's principle (a principle of constructive recursive mathematics [9, ch. 3]), and the negation of the independence of premise axiom (a controversial axiom which is classically true but generally not accepted by constructivists, which was used by Gödel in his famous Dialectica interpretation [4]). Proofs of these negated properties follow similar arguments such as in [15,16]. For example, notice that in order to prove the validity of the $\downarrow$-squashed law of excluded middle, we would have to prove in the metatheory that for all propositions, there exists a bar of the current library such that either the proposition is true at the bar, or that it is false in all extensions of the bar. To prove its negation we show that neither option is valid anymore because choice sequences can always evolve differently when multiple choices are possible. The open bar semantics invoked by OTT, on the other hand, is based on a more relaxed notion of time that is flexible enough to be compatible with classical reasoning. In particular, it enables the validation of the $\downarrow$-squashed law of excluded middle.

4   Implications of the Formalization of Choice Sequences

The integration of Brouwer's ex ed notion of computation into a mechanized proof assistant is not only important from a foundational standpoint, but also has interesting consequences and practical applications. This section informally and briefly discusses such implications in two main fields, namely formal verification and constructive mathematics.

4.1   Formal Verification

Leveraging the foundational, novel computational capabilities that go beyond the Church–Turing notion of computation has the potential to facilitate significant advances in the *internal* verification of complex systems, e.g., distributed protocols. The dominant approach in verification of such systems is *external*; that is, one develops a model of the system and then proves that the system behaves correctly according to its desired specification, assuming this model is correct. While this strategy is extremely flexible (a model can describe any kind of computational system), it has the major disadvantage that the model may be incorrect. Extending the computation system with Brouwer's broader notion of computation enables an *internal* approach in such verifications. This is because the embedding of the notion of choice sequences within the computational system provides a means to internally formalize non-deterministic behaviors. In large distributed systems, a lot of information is, de facto, 'lawless', in the sense that it is unpredictable. It is far too expansive (source and money-wise) to keep track of all the computation steps, and the environment cannot be controlled, and so in practice there is no way to determine, e.g., the order of messages sent. Therefore, one can use standard computable functions to model the processes of a distributed system, and free choice sequences to model sensors (or unpredictable environmental inputs).

4.2   Intuitionistic Mathematics

Standard mathematical discourse is based on classical mathematics, and thus standard textbooks in mathematics contain, e.g., proofs by contradiction or by cases, as well as impredicative structures. Constructive mathematics (e.g., [8]), on the other hand, does not allow "non-constructive" methods of formal proof, and in particular rejects the law of excluded middle. Because of this restriction, the practice of constructive mathematics is often quite remote from the (classical) standard practice of mathematics, and proofs tend to require more elaborate arguments. For example, without some version of the compactness theorem (which, classically, requires the axiom of choice), point-wise versions of continuity and the derivative are of no use and the more complicated notions of uniform continuity and a uniform version of the derivative must be used.

Most works in constructive mathematics adopt E. Bishop's approach [8,10] and remain agnostic towards the fundamental intuitionistic principles such as choice sequences, bar induction and the uniform continuity principle.[7] But these intuitionistic principles, which go beyond constructive mathematics, have the potential to simplify the practice of mathematical theories. For one, the uniform continuity principle obviates the need for the compactness theorem, thus making intuitionistic calculus more elegant than constructive calculus, because restrictions on key theorems can be eliminated. For example, in intuitionistic mathematics we can again use the point-wise versions of continuity and the derivative in a manner similar to the way they are employed in classical mathematics. Thus, Brouwer's intuitionistic mathematics has the computational advantages of constructive mathematics, while at the same time enabling proofs that resemble those of classical mathematics to a greater extent than constructive ones.

The computational account of choice sequences in Nuprl also provides a natural framework for the formalization of the Brouwerian, choice-sequence-based constructive real numbers, and, in turn, the development of the corresponding real analysis, and the exploration of its computational benefits. Brouwer used choice sequences to define the constructive real numbers as sequences of nested rational intervals. The standard formalization of the reals, even in classical or constructive mathematics, is also achieved via converging sequences. Nonetheless, there are two major differences between the standard formalizations and the intuitionistic (Brouwerian) formalization. First, in the intuitionistic formalization, a real number *is* the choice sequence itself, as opposed to it being the limit point (i.e., equivalence class). Second, the notion of what these sequences can be incorporates, in the intuitionistic setting, the *free* choice sequences.

Acknowledgments. The author thanks Vincent Rahli, Robert Constable and Mark Bickford as the framework described in the paper is based on a joint ongoing work with them.

References

1. Agda wiki. http://wiki.portal.chalmers.se/agda/pmwiki.php
2. Allen, S.F., et al.: Innovations in Computational Type Theory using Nuprl. J. Appl. Logic **4**(4), 428–469 (2006)
3. Anand, A., Rahli, V.: Towards a formally verified proof assistant. In: Klein, G., Gamboa, R. (eds.) ITP 2014. LNCS, vol. 8558, pp. 27–44. Springer, Cham (2014). https://doi.org/10.1007/978-3-319-08970-6_3
4. Avigad, J., Feferman, S.: Gödel's functional ("Dialectica") interpretation. Handb. Proof Theor. **137**, 337–405 (1998)

---

[7] Notable exceptions include, e.g., [36,42].

5. Beth, E.W.: Semantic construction of intuitionistic logic. J. Symbolic Logic **22**(4), 363–365 (1957)
6. Bickford, M., Cohen, L., Constable, R.L., Rahli, V.: Computability beyond church-turing via choice sequences. In: Proceedings of the 33rd Annual ACM/IEEE Symposium on Logic in Computer Science, LICS 2018, pp. 245–254 (2018)
7. Bickford, M., Cohen, L., Constable, R.L., Rahli, V.: Open Bar - a Brouwerian intuitionistic logic with a pinch of excluded middle. In: Baier, C., Goubault-Larrecq, J., (eds.), 29th EACSL Annual Conference on Computer Science Logic (CSL), vol. 183, LIPIcs, pp. 11:1–11:23 (2021)
8. Bishop, E., Bridges, D.: Constructive Analysis. GW, vol. 279. Springer, Heidelberg (1985). https://doi.org/10.1007/978-3-642-61667-9
9. Bridges, D., Richman, F.: Varieties of Constructive Mathematics. London Mathematical Society Lecture Notes Series, Cambridge University Press (1987)
10. Bridges, D., Richman, F.: Varieties of Constructive Mathematics. Cambridge University Press, Cambridge (1988)
11. Brouwer, L.E.J.: Begründung der mengenlehre unabhängig vom logischen satz vom ausgeschlossen dritten. zweiter teil: Theorie der punkmengen. Koninklijke Nederlandse Akademie van Wetenschappen te Amsterdam **12**(7), (1919). Reprinted in Brouwer, L.E.J., Collected Works, Volume I: Philosophy and Foundations of Mathematics, edited by Heyting, A., North-Holland Publishing Co., Amsterdam, pp. 191–221 (1975)
12. Brouwer, L.E.J.: From frege to Gödel: A Source Book in Mathematical Logic, 1879–1931, chapter On the Domains of Definition of Functions (1927)
13. Church, A.: An unsolvable problem of elementary number theory. Am. J. Math. **58**(2), 345–363 (1936)
14. Constable, R.L. et al.: Implementing Mathematics with the Nuprl Proof Development System. Prentice-Hall Inc, Hoboken (1986)
15. Coquand, T., Mannaa, B.: The independence of markov's principle in type theory. In: Kesner, D., Pientka, B., (eds.) FSCD 2016, vol. 52 of LIPIcs, pp. 17:1–17:18. Schloss Dagstuhl - Leibniz-Zentrum fuer Informatik (2016)
16. Coquand, T., Mannaa, B., Ruch, F.: Stack semantics of type theory. In: 2017 32nd Annual ACM/IEEE Symposium on Logic in Computer Science (LICS), pp. 1–11 (2017)
17. The Coq Proof Assistant. http://coq.inria.fr/
18. Dyson, V.H., Kreisel, G.: Analysis of Beth's Semantic Construction of Intuitionistic Logic. Stanford University (1961)
19. Escardó, M.H., Xu, C.: The inconsistency of a Brouwerian continuity principle with the curry-howard interpretation. In: 13th International Conference on Typed Lambda Calculi and Applications (TLCA), pp. 153–164 (2015)
20. Forster, Y.: Church's thesis and related axioms in Coq's type theory. In: Baier, C., Goubault-Larrecq, J., (eds.) 29th EACSL Annual Conference on Computer Science Logic (CSL 2021), vol. 183 of Leibniz International Proceedings in Informatics (LIPIcs), pp. 21:1–21:19, Dagstuhl, Germany (2021). Schloss Dagstuhl-Leibniz-Zentrum für Informatik
21. Kleene, S.C., Vesley, R.E.: The Foundations of Intuitionistic Mathematics, Especially in Relation to Recursive Functions. North-Holland Publishing Company, Amsterdam (1965)
22. Kreisel, G.: On weak completeness of intuitionistic predicate logic. J. Symb. Log. **27**(2), 139–158 (1962)
23. Kreisel, G.: Lawless sequences of natural numbers. Compositio Mathematica **20**, 222–248 (1968)

24. Kripke, S.A.: Semantical considerations on modal logic. Acta Philosophica Fennica **16**(1963), 83–94 (1963)
25. Martin-Löf, P.: Constructive mathematics and computer programming. In: Proceedings of the Sixth International Congress for Logic, Methodology, and Philosophy of Science, pp. 153–175. Amsterdam, North Holland (1982)
26. Moschovakis, J.R.: An intuitionistic theory of lawlike, choice and lawless sequences. In: Logic Colloquium'90: ASL Summer Meeting in Helsinki, pp. 191–209. Association for Symbolic Logic (1993)
27. Rahli, V., Bickford, M.: A nominal exploration of intuitionism. In: Proceedings of the 5th ACM SIGPLAN Conference on Certified Programs and Proofs, CPP, pp. 130–141, p. 2016. New York (2016)
28. Rahli, V., Bickford, M.: Validating Brouwer's continuity principle for numbers using named exceptions. Math. Struct. Comput. Sci. **28**(6), 942–990 (2018)
29. Rahli, V., Bickford, M., Cohen, L., Constable, R.L.: Bar induction is compatible with constructive type theory. J. ACM **66**(2), 13:1–13:35 (2019)
30. Rahli, V., Bickford, M., Constable, R. L.: Bar induction: The Good, the Bad, and the Ugly. In: 2017 32nd Annual ACM/IEEE Symposium on Logic in Computer Science (LICS), pp. 1–12 (2017)
31. Rahli, V., Cohen, L., Bickford, M.: A verified theorem prover backend supported by a monotonic library. In: Barthe, G., Sutcliffe, G., Veanes, M., (eds.) LPAR-22. 22nd International Conference on Logic for Programming, Artificial Intelligence and Reasoning, vol. 57 of EPiC Series in Computing, pp. 564–582 (2018)
32. Rathjen, M.: A note on bar induction in constructive set theory. Math. Logic Q. **52**(3), 253–258 (2006)
33. Troelstra, A.S.: A note on non-extensional operations in connection with continuity and Recursiveness. Indagationes Mathematicae **39**(5), 455–462 (1977)
34. Troelstra, A.S.: Choice Sequences: a Chapter of Intuitionistic Mathematics. Clarendon Press, Oxford (1977)
35. Troelstra, A.S.: Choice sequences and informal rigour. Synthese **62**(2), 217–227 (1985)
36. Troelstra, A.S., van Dalen, D.: Constructivism in Mathematics, An Introduction, vol. I. II. North-Holland, Amsterdam (1988)
37. van Atten, M.: On Brouwer. Cengage Learning, Wadsworth Philosophers (2004)
38. van Atten, M., van Dalen, D.: Arguments for the continuity principle. Bull. Symbolic Logic **8**(3), 329–347 (2002)
39. van Dalen, D.: An interpretation of intuitionistic analysis. Ann. Math. Logic **13**(1), 1–43 (1978)
40. van Dalen, D.: L.E.J. Brouwer: Topologist, Intuitionist, Philosopher: How Mathematics is Rooted in Life. Springer, New York (2013) https://doi.org/10.1007/978-1-4471-4616-2
41. Veldman, W.: Understanding and using brouwer's continuity principle. In: Schuster, P., Berger, U., Osswald, H. (eds.) Reuniting the Antipodes - Constructive and Nonstandard Views of the Continuum. Synthese Library (Studies in Epistemology, Logic, Methodology, and Philosophy of Science), vol. 306, pp. 285–302. Springer, Dordrecht (2001). https://doi.org/10.1007/978-94-015-9757-9_24
42. Veldman, W.: Some applications of brouwer's thesis on bars. In: Atten, M., Boldini, P., Bourdeau, M., Heinzmann, G. (eds.) One Hundred Years of Intuitionism (1907–2007). Publications of the Henri Poincaré Archives, pp. 326–340. Berkhäuser, Berlin (2008) https://doi.org/10.1007/978-3-7643-8653-5_20

Mapping Monotonic Restrictions in Inductive Inference

Vanja Doskoč[(⊠)] and Timo Kötzing

Hasso Plattner Institute, University of Potsdam, Potsdam, Germany
{vanja.doskoc,timo.koetzing}@hpi.de

Abstract. In *inductive inference* we investigate computable devices (learners) learning formal languages. In this work, we focus on *monotonic* learners which, despite their natural motivation, exhibit peculiar behaviour. A recent study analysed the learning capabilities of *strongly monotone* learners in various settings. The therein unveiled differences between *explanatory* (syntactically converging) and *behaviourally correct* (semantically converging) such learners motivate our studies of *monotone* learners in the same settings.

While the structure of the pairwise relations for monotone explanatory learning is similar to the strongly monotone case (and for similar reasons), for behaviourally correct learning a very different picture emerges. In the latter setup, we provide a *self-learning* class of languages showing that monotone learners, as opposed to their strongly monotone counterpart, do heavily rely on the order in which the information is given, an unusual result for behaviourally correct learners.

1 Introduction

Algorithmically learning a formal language from a growing but finite amount of its positive information is referred to as *inductive inference* or *language learning in the limit*. For example, a learner h (a computable device) might be presented more and more data from a formal language (a computably enumerable subset of the natural numbers), say, the set of all odd prime numbers $\mathbb{P}_o$. With each new element presented, h outputs a description for a formal language as its guess. As such, the learner may decide to conjecture a code for the set of all odd numbers $\mathbb{N}_o$. With more data given, the learner may infer some structure and finally decide to output a program for the set $\mathbb{P}_o$. If h does not change its mind any more, we say that h learned the language $\mathbb{P}_o$ correctly.

Originally introduced by Gold [9], such learning is referred to as *explanatory learning*, as the learner eventually provides a syntactically fixed explanation of the language. We denote such learning by **TxtGEx**, where **Txt** indicates that the information is given from text, **G** stands for *Gold-style* or *full-information* learning and, lastly, **Ex** refers to explanatory learning. Since a single language can be learned by a learner which always guesses one and the same code for this language, we study classes of languages which can be **TxtGEx**-learned by a single learner and denote the set of all such classes with [**TxtGEx**]. We refer to this set as the *learning power* of **TxtGEx**-learners.

This work was supported by DFG Grant Number KO 4635/1-1.

L. De Mol et al. (Eds.): CiE 2021, LNCS 12813, pp. 146–157, 2021.
https://doi.org/10.1007/978-3-030-80049-9_13

Picking up the initial example, we observe that the learner h outputs a code for $\mathbb{N}_o$ overgeneralizing the target language $\mathbb{P}_o$ before outputting a correct code. The question arises whether such overgeneralizations are necessary in order to obtain full learning power? Various restrictions mimicking overgeneralizations have been investigated in the literature and show such a behaviour to be crucial. A prominent example are *monotonic* learners [11, 23], where the hypotheses must show a monotone behaviour. In the strongest from, the hypotheses of *strongly monotone* (**SMon**) learners must form ascending chains. In a less restrictive form, only the *correctly* inferred elements, that is, elements that belong to the target language, in the hypotheses of *monotone* learners (**Mon**) need to form ascending chains.

A recent study of strongly monotone learners under various additional restrictions provided a full overview of the pairwise relations between these [13]. The studied restrictions affect the data given to the learners as well as the learners themselves. In particular, the learners may be given solely the set of elements to infer their hypotheses from, referred to as *set-driven* (**Sd**, [22]) learning, or may additionally be given an iteration-counter, called *partially set-driven* or *rearrangement-independent* (**Psd**, [2, 21]) learning. When learning indexed families of recursive languages [1] rather than classes of recursively enumerable languages, monotonic learners have been studied under similar restrictions [16–18]. Directly affecting the learner are requirements such as them being *total* (denoted using the prefix $\mathcal{R}$) or them being monotone on arbitrary information (denoted by the prefix $\tau(\mathbf{Mon})$).

Comparing all the possible pairwise combinations, Kötzing and Schirneck [13] show that Gold-style strongly monotone learners may be assumed so on *arbitrary* information. Besides that, they provide self-learning classes of languages [4] to show that all other combinations separate from each other. Contrasting this are their findings when studying *behaviourally correct* learners (**Bc**, [5, 19]), which need to provide a *semantic* explanation (rather than a syntactic one) in the limit. Behaviourally correct strongly monotone learners turn out to be equally powerful, regardless the considered restriction on the given data (that is, whether the learner has full information, is partially set-driven or set-driven) or learner itself (that is, whether it is partial, total or required to be strongly monotone on arbitrary input).

These interesting findings motivate the present study. In Sect. 3.1, we study monotonic explanatory learners. In particular, we observe that the overall behaviour of monotone learners resembles the one of strongly monotone learners. This similarity culminates in Theorem 3, where we prove learners which are monotone on arbitrary input, so called *globally* monotone learners, to be equal to globally strongly monotone ones. We additionally observe that most proof strategies used to separate the diverse strongly monotone learning paradigms [13] can be carried over to fit monotone learners. While these transitions are often non-trivial, they do indicate a deep similarity between these two restrictions. We provide all the necessary comparisons in Sect. 3.1 and depict the overall picture in a lucid map, see Fig. 1(a). Please consider the full version [6] for the proofs.

In Sect. 3.2, we transfer the problem of finding the pairwise relations to behaviourally correct monotonic learners and discover an unexpected result. In Theorem 7, we provide a self-learning class of languages [4] using the Operator Recursion Theorem [3] showing that Gold-style monotone learners are strictly more

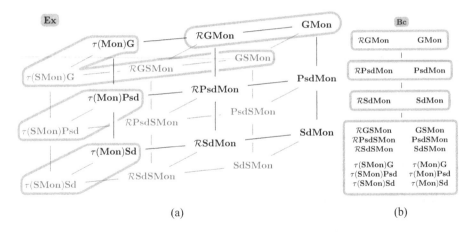

(a)   (b)

Fig. 1. Relation of various monotonic learning restrictions in the (a) explanatory (**Ex**) and (b) behaviourally correct (**Bc**) case. We omit mentioning **Txt** to favour readability. Solid lines imply trivial inclusions (bottom-to-top, left-to-right). Greyly edged areas illustrate a collapse of the enclosed learning criteria. There are no further collapses.

powerful than their partially set-driven counterpart. This is particularly surprising as usually behaviourally correct learners cope rather well with such memory restrictions [7, 13]. This marks the most important and surprising insight of this work. We provide the necessary results in Sect. 3.2 and collect our findings in the lucid Fig. 1(b).

2 Preliminaries

2.1 Language Learning in the Limit

In this section, we discuss the notation used and the system for learning criteria which we follow [15]. Notation which is not introduced follows the textbook [20].

Starting with the mathematical notation, we use $\subsetneq$ and $\subseteq$ to denote the proper subset and subset relation between sets, respectively. We denote with $\mathbb{N} = \{0, 1, 2, \dots\}$ the set of all natural numbers. With $\emptyset$ and ε we denote the *empty set* and *empty string*, respectively. Furthermore, we let $\mathcal{P}$ and $\mathcal{R}$ be the set of all partial and total computable functions $p\colon \mathbb{N} \to \mathbb{N}$, respectively. We fix an effective numbering $\{\varphi_e\}_{e\in\mathbb{N}}$ of $\mathcal{P}$ and denote with $W_e = \mathrm{dom}(\varphi_e)$ the e-th computably enumerable set. This way, we interpret the natural number e as an *index* or *hypothesis* for the set W_e. Regarding important computable functions, we fix with $\langle ., .\rangle$ a computable coding function. We use π_1 and π_2 to recover the first and second component, respectively. Furthermore, we write pad for an injective computable function such that, for all $e, k \in \mathbb{N}$, we have $W_e = W_{\mathrm{pad}(e,k)}$. We use unpad_1 and unpad_2 to compute the first and second component of $\mathrm{pad}(.,.)$, respectively. Note that both functions can be extended iteratively to more coordinates. Lastly, we let ind compute an index for any given finite set.

We aim to learn *languages*, that is, recursively enumerable sets $L \subseteq \mathbb{N}$. These will be learned by *learners* which are partial computable functions. By $\#$ we denote the *pause symbol* and for any set S we denote $S_\# := S \cup \{\#\}$. Furthermore, a *text* is a total function $T \colon \mathbb{N} \to \mathbb{N} \cup \{\#\}$, the collection of all texts we denote with $\mathbf{Txt}$. For any text or sequence T, we let $\mathrm{content}(T) := \mathrm{range}(T) \setminus \{\#\}$ be the *content* of T. A text of a language L is such that $\mathrm{content}(T) = L$, the collection of all texts of L we denote with $\mathbf{Txt}(L)$. For $n \in \mathbb{N}$, we denote by $T[n]$ the initial sequence of T of length n, that is, $T[0] := \varepsilon$ and $T[n] := (T(0), T(1), \ldots, T(n-1))$. For a set S, we call the text where all elements of S are presented in strictly increasing order (followed by infinitely many pause symbols if S is finite) the *canonical text of S*. Furthermore, we call the sequence of all elements of S presented in strictly ascending order the *canonical sequences of S*. On finite sequences we use $\subseteq$ to denote the *extension relation* and $\leq$ to denote the order on sequences interpreted as natural numbers. Furthermore, for tuples of finite sets and numbers (D, t) and (D', t'), we define the order $\preceq$ such that $(D, t) \preceq (D', t')$ if and only if $t \leq t'$ and there exists a text T such that $D = \mathrm{content}(T[t])$ and $D' = \mathrm{content}(T[t'])$. In addition, given two sequences σ and τ we write $\sigma^\frown \tau$ to denote the concatenation of these. Occasionally, we omit writing $^\frown$ for readability.

We formalise learning criteria using the following system [15]. An *interaction operator* β is given a learner $h \in \mathcal{P}$ and a text $T \in \mathbf{Txt}$ and outputs a (partial) function p. Intuitively, β provides the information for the learner to make its guesses. We consider the interaction operators $\mathbf{G}$ for *Gold-style* or *full-information* learning [9], $\mathbf{Psd}$ for *partially set-driven* or *rearrangement-independent* learning [2,21] and $\mathbf{Sd}$ for *set-driven* learning [22]. Define, for any $i \in \mathbb{N}$,

$$\mathbf{G}(h, T)(i) := h(T[i]),$$
$$\mathbf{Psd}(h, T)(i) := h(\mathrm{content}(T[i]), i),$$
$$\mathbf{Sd}(h, T)(i) := h(\mathrm{content}(T[i])).$$

Intuitively, Gold-style learners have full information on the elements presented to them. Partially set-driven learners, however, base their guesses on the total amount of elements presented and the content thereof. Lastly, set-driven learners only base their conjectures on the content given to them. Furthermore, for any β-learner h, we write h^* for its starred learner, that is, the $\mathbf{G}$-learner which simulates h. For example, if $\beta = \mathbf{Sd}$, then, for any sequence σ, $h^*(\sigma) = h(\mathrm{content}(\sigma))$.

When it comes to learning, we can distinguish between various criteria for successful learning. The first such criterion is *explanatory* learning ($\mathbf{Ex}$, [9]). Here, a learner is expected to converge to a single, correct hypothesis in order to learn a language. This can be loosened to require the learner to converge semantically, that is, from some point onwards it must output correct hypotheses which may change syntactically [5,19]. This is referred to as *behaviourally correct* learning ($\mathbf{Bc}$). Formally, a *learning restriction* δ is a predicate on a total learning sequence p, that is, a total function, and a text $T \in \mathbf{Txt}$. For the mentioned criteria we have

$$\mathbf{Ex}(p, T) :\Leftrightarrow \exists n_0 \forall n \geq n_0 \colon p(n) = p(n_0) \wedge W_{p(n_0)} = \mathrm{content}(T),$$
$$\mathbf{Bc}(p, T) :\Leftrightarrow \exists n_0 \forall n \geq n_0 \colon W_{p(n)} = \mathrm{content}(T).$$

These success criteria can be expanded in order to model natural learning restrictions. Our focus lies on monotonic learners [11,23]. *Strongly monotone* learning ($\mathbf{SMon}$)

forms the basis. Here, the learner may never discard elements which were once present in its previous hypotheses. This restrictive criterion can be loosened to hold only on the elements of the target language, that is, the learner may never discard such elements from the language which it already proposed in previous hypotheses. This is referred to as *monotone* learning (**Mon**). This is formalized as

$$\mathbf{SMon}(p,T) :\Leftrightarrow \forall n, m : n \leq m \Rightarrow W_{p(n)} \subseteq W_{p(m)},$$

$$\mathbf{Mon}(p,T) :\Leftrightarrow \forall n, m : n \leq m \Rightarrow W_{p(n)} \cap \mathrm{content}(T) \subseteq W_{p(m)} \cap \mathrm{content}(T).$$

Given two restrictions δ and δ', we denote their combination, that is, their intersection, with $\delta\delta'$. Finally, $\mathbf{T}$, the always true predicate, denotes the absence of a restriction.

Now, a *learning criterion* is a tuple $(\alpha, \mathcal{C}, \beta, \delta)$, where $\mathcal{C}$ is a set of admissible learners, typically $\mathcal{P}$ or $\mathcal{R}$, β is an interaction operator and α and δ are learning restrictions. We denote this learning criterion as $\tau(\alpha)\mathcal{C}\mathbf{Txt}\beta\delta$. In the case of $\mathcal{C} = \mathcal{P}$, $\alpha = \mathbf{T}$ or $\delta = \mathbf{T}$ we omit writing the respective symbol. Now, an admissible learner $h \in \mathcal{C}$ $\tau(\alpha)\mathcal{C}\mathbf{Txt}\beta\delta$-learns a language L if and only if on *arbitrary* text $T \in \mathbf{Txt}$ we have $\alpha(\beta(h,T),T)$ and on texts of the target language $T \in \mathbf{Txt}(L)$ we have $\delta(\beta(h,T),T)$. With $\tau(\alpha)\mathcal{C}\mathbf{Txt}\beta\delta(h)$ we denote the class of languages $\tau(\alpha)\mathcal{C}\mathbf{Txt}\beta\delta$-learned by h and with $[\tau(\alpha)\mathcal{C}\mathbf{Txt}\beta\delta]$ we denote the set containing, for all $h' \in \mathcal{C}$, all classes $\tau(\alpha)\mathcal{C}\mathbf{Txt}\beta\delta(h')$. Note that restrictions which hold globally (that is, on arbitrary text) are denoted using $\tau(.)$.

2.2   Normal Forms in Inductive Inference

The introduced learning restrictions all fall into the scope of delayable restrictions. Informally, the hypotheses of a delayable restriction may be postponed arbitrarily but not indefinitely. Formally, we call a learning restriction δ *delayable* if and only if for all texts T and T' with $\mathrm{content}(T) = \mathrm{content}(T')$, all learning sequences p and all total, unbounded non-decreasing functions r, we have that if $\delta(p,T)$ and, for all n, $\mathrm{content}(T[r(n)]) \subseteq \mathrm{content}(T'[n])$, then $\delta(p \circ r, T')$. Furthermore, we call a restriction *semantic* if and only if for any learning sequences p and p' and any text T, we have that if $\delta(p,T)$ and, for all n, $W_{p(n)} = W_{p'(n)}$, then $\delta(p',T)$. Intuitively, a restriction is semantic if any hypothesis could be replaced by a semantically equivalent one without violating the learning restriction. Note that all mentioned restrictions are delayable and all except for **Ex** are semantic. In particular, one can provide general results when talking about delayable or semantic restrictions.

Theorem 1. *([12,14]). For any interaction operator β, delayable restriction δ and semantic restriction δ', we have $[\mathcal{R}\mathbf{Txt}\mathbf{G}\delta] = [\mathbf{Txt}\mathbf{G}\delta]$ and $[\mathcal{R}\mathbf{Txt}\beta\delta'] = [\mathbf{Txt}\beta\delta']$.*

3   Studying Monotone Learning Restrictions

We investigate monotone learners imposed with various restrictions and compare them to their strongly monotone counterpart. We split this study into two parts, first studying explanatory learners in Sect. 3.1 and then behaviourally correct ones in Sect. 3.2.

Before we dive into the respective part, we note that it is a well-established fact that strongly monotone learners are significantly weaker than their monotone counterpart. In particular, the class $\mathcal{L} = \{2\mathbb{N}\} \cup \{\{0, 2, 4, \ldots, 2k, 2k + 1\} \mid k \in \mathbb{N}\}$ is learnable by a **TxtSdMonEx**-learner, however, any **TxtGSMonBc**-learner fails to do so. We remark that the separating class can also be learned by a total monotone learner.

Theorem 2. *We have* $[\mathcal{R}\mathbf{TxtSdMonEx}] \setminus [\mathbf{TxtGSMonBc}] \neq \emptyset$.

Despite this fundamental separation, we observe similarities between monotone and strongly monotone explanatory learners. These similarities are not only reflected by the overall pairwise relation of the different settings, but also by the techniques used to obtain these relations. The main difficulty thereby is to reason why the elements used to contradict strongly monotone learning suddenly are part of a learnable language and, thus, also contradict monotone learning. Furthermore, in order to show strong results, all of these adaptations have to be done while maintaining the original learnability by some strongly monotone learner.

These similarities culminate in Theorem 3, where we show globally monotone learners to be equally powerful as globally strongly monotone ones. This result also holds true when requiring semantic convergence. However, as monotone learners may discard elements from their guesses, the strategy of keeping all once suggested elements regardless of the order (as for strongly monotone learners [13]) is not fruitful for monotone learners. On the contrary, we show that such an equality cannot be obtained. In particular, in Theorem 7 we show that partially set-driven learners are strictly less powerful than their Gold-style counterpart, an unusual result as we discuss in Sect. 3.2.

3.1 Explanatory Monotone Learning

In this section, we investigate monotone learners when requiring syntactic convergence and also compare them to their strongly monotone counterpart. Building on the thorough discussion of strongly monotone learners [13], we show that the general behaviour of both types of learners is alike. This can be seen, firstly, in the resulting overall picture of the pairwise relations and, secondly, in the way these results are obtained.

Our first result is already a good indication towards how similar these restrictions are. We show that requiring both restrictions to hold globally results in equal learning power. To motivate the idea, note that monotone learners exhibit a strongly monotone behaviour on target languages. If now the learner is required to be monotone on any possible set, as required by global restrictions, it is already globally strongly monotone. Note that this equality, in fact, holds on the level of the restrictions itself.

Theorem 3. *For all restrictions δ and all interaction operators β we have*

$$[\tau(\mathbf{SMon})\mathbf{Txt}\beta\delta] = [\tau(\mathbf{Mon})\mathbf{Txt}\beta\delta].$$

Proof. The inclusion $[\tau(\mathbf{SMon})\mathbf{Txt}\beta\delta] \subseteq [\tau(\mathbf{Mon})\mathbf{Txt}\beta\delta]$ is immediate. For the other inclusion, let h^* be a $\tau(\mathbf{Mon})\mathbf{Txt}\beta\delta$-learner in its starred form. Assume that h^*

is not $\tau(\mathbf{SMon})$. Then, there exists some text T, $i < j$ and x such that $x \in W_{h^*(T[i])} \setminus W_{h^*(T[j])}$. Considering the text $T' := T[j]^\frown x^\frown T(j)^\frown T(j+1)^\frown \cdots$, we have

$$x \in W_{h^*(T[i])} \cap \mathrm{content}(T') \setminus W_{h^*(T[j])} \cap \mathrm{content}(T').$$

Thus, h^* is not $\tau(\mathbf{Mon})$ on text T', a contradiction.   □

In particular, this implies that all separations and equalities known for globally strongly monotone learners also hold for globally monotone ones. Most notably, Gold-style globally monotone learners are strictly less powerful than their total counterpart.

Gold-style monotone learners, being delayable, can be assumed *total* without loss of learning power [12]. We show that these learners are more powerful than their partially set-driven counterpart. In particular, we show that even strongly monotone Gold-style learners are more powerful than any partially set-driven monotone learner. We do so by learning a class of languages on which the learner, in order to discard certain elements, needs to know the order the information appeared in. This, no partially set-driven monotone learner can do.

Theorem 4. *We have* $[\mathbf{TxtGSMonEx}] \setminus [\mathbf{TxtPsdMonEx}] \neq \emptyset$.

Next, we show that a partial learner, even sustaining a severe memory restriction and expected to be strongly monotone, is still more powerful than any total monotone partially set-driven learner. When constructing a separating class of languages, the partial learner simply awaits the guess of the total learner to, then, learn a different language.

Theorem 5. *We have* $[\mathbf{TxtSdSMonEx}] \setminus [\mathcal{R}\mathbf{TxtPsdMonEx}] \neq \emptyset$.

Proof. We adapt the proof of the separation from total $\mathbf{SMon}$-learners [13, Thm. 11] as follows. Let $h \in \mathcal{P}$ be the following learner. With p_0 being such that $W_{p_0} = \emptyset$, let for each finite set $D \subseteq \mathbb{N}$

$$h(D) = \begin{cases} p_0, & \text{if } D = \emptyset, \\ \mathrm{ind}(D), & \text{else, if } |D| = 1, \\ \uparrow, & \text{else, if } \exists x \in D\colon \varphi_x(0)\uparrow \vee \mathrm{unpad}_2(\varphi_x(0)) \notin \{1,2\}, \\ e, & \text{else, if } \forall x \in D\colon \mathrm{unpad}_1(\varphi_x(0)) = e, \\ e', & \text{else, if} \\ & \quad \big(\exists y \forall x \in D\colon \mathrm{unpad}_2(\varphi_x(0)) = 1 \Rightarrow \mathrm{unpad}_1(\varphi_x(0)) = y\big)\wedge \\ & \quad \wedge \big(\forall x \in D\colon \mathrm{unpad}_2(\varphi_x(0)) = 2 \Rightarrow \mathrm{unpad}_1(\varphi_x(0)) = e'\big), \\ \uparrow, & \text{otherwise.} \end{cases}$$

The intuition is the following. While no elements are presented, h conjectures (a code for) the empty set. Once, a single element is presented, h suggests (a code for) that singleton. Thus, h learns all singletons. Given more elements, h either outputs the first coordinate of the elements (if they all coincide), or another code if there are different second coordinates. In case of equal second coordinates but different first coordinates, h is undefined.

Let $\mathcal{L} = \mathbf{TxtSdSMonEx}(h)$. Assume there exists a $\mathcal{R}\mathbf{TxtPsdMonEx}$-learner h' which learns $\mathcal{L}$, that is, $\mathcal{L} \subseteq \mathcal{R}\mathbf{TxtPsdMonEx}(h')$. Since h learns all singletons, so does h'. Thus, there is a total, strictly monotone function $t \in \mathcal{R}$ such that $t(0) > 0$ and for each x

$$x \in W_{h'(\{x\}, t(x))}. \tag{1}$$

With **ORT** ([3]), we get a total recursive predicate $P \in \mathcal{R}$, a strictly monotone increasing $a \in \mathcal{R}$ and indices $e, e' \in \mathbb{N}$ such that for all $i \in \mathbb{N}$, using $\tilde{t}(i) := \sum_{j=0}^{i} t(a(j)) + j$ as abbreviation,

$$P(i) \Leftrightarrow h'(\text{content}(a[i]), \tilde{t}(i)) \neq h'(\text{content}(a[i+1]), \tilde{t}(i) + 1),$$
$$W_e = \{a(i) \mid \forall j \le i \colon P(j)\},$$
$$W_{e'} = \{a(i) \mid \forall j < i \colon P(j)\},$$
$$\varphi_{a(i)}(0) = \begin{cases} \text{pad}(e, 1), & \text{if } P(i), \\ \text{pad}(e', 2), & \text{otherwise.} \end{cases}$$

We show that W_e and $W_{e'}$ are in $\mathcal{L}$.

1. Case: W_e is infinite. This means for all i we have $P(i)$. Thus, $W_e = W_{e'}$. Thus, it suffices to show $W_e \in \mathcal{L}$. Let $T \in \mathbf{Txt}(W_e)$. For $n > 0$, let $D_n := \text{content}(T[n])$. As long as $D_n = \emptyset$, we have $h(D_n) = p_0$, i.e. a code for the empty set. When $|D_n| = 1$, we have $h(D_n) = \text{ind}(D_n)$, a code for the singleton D_n. Once D_n contains more than one element, $h(D_n)$ starts unpadding. As, for all i, $\varphi_{a(i)}(0) = \text{pad}(e, 1)$, we have $\text{unpad}_1(\{\varphi_x(0) \mid x \in D_n\}) = \{e\}$. Thus, h is strongly monotone and will output e correctly.

2. Case: W_e is finite. Let k be such that $W_e = \{a(j) \mid j < k\}$ and $W_{e'} = \{a(j) \mid j < k + 1\}$. Again, as long as no elements or only one element is shown, h will output a code for the empty, respectively singleton set. As $W_e \subseteq W_{e'}$ and $\text{unpad}_1(\{\varphi_x(0) \mid x \in W_e\}) = \{e\}$, h will output e as long as it sees only elements from W_e. Once it sees $a(k) \in W_{e'}$, it correctly changes its mind to e'. This maintains strong monotonicity and is the correct behaviour.

Thus, $W_e, W_{e'} \in \mathcal{L}$. We show that h' cannot learn both simultaneously.

1. Case: W_e is infinite. On the text $a(0)^{t(a(0))} a(1)^{t(a(1))+1} a(2)^{t(a(2))+2} \ldots$ of W_e, the learner h' makes infinitely many mind changes. Thus, it cannot learn W_e, a contradiction.

2. Case: W_e is finite. Let k be minimal such that $\neg P(k)$, and thus $W_e = \text{content}(a[k])$ and $W_{e'} = \text{content}(a[k+1])$. By Condition (1) and monotonicity of h' on $W_{e'}$ we have $a(k) \in W_{h'(\text{content}(a[k+1]), \tilde{t}(k)+1)}$, as $a(k)^{\tilde{t}(k)} \frown a[k]$ is a sequence of elements in $W_{e'}$ and $a(k) \in W_{e'}$. Since $\neg P(k)$, we get $h'(\text{content}(a[k]), \tilde{t}(k)) = h'(\text{content}(a[k+1]), \tilde{t}(k) + 1)$ and, thus, $a(k) \in W_{h'(\text{content}(a[k]), t(a(k))+k)}$. For each $t \ge \tilde{t}(k)$, we have that $(\text{content}(a[k]), t)$ is an initial sequence for some text of $W_{e'}$, and thus, by monotonicity of h' we get $a(k) \in W_{h'(\text{content}(a[k]), t)}$. As $a(k) \notin W_e = \text{content}(a[k])$, h' cannot identify W_e, a contradiction. $\qquad\square$

To complete Fig. 1(a), it remains to be shown that globally strongly monotone partially set-driven learners are more powerful than their monotone set-driven counterpart. The separation from strongly monotone set-driven learners has already been shown [13]. We provide a self-learning class [4] to show that globally strongly monotone partially set-driven learners outperform *unrestricted* set-driven learners. This result emphasises the weakness of set-driven learners which results from a lack of "learning time" [8].

We note that, when studying learners which may be undefined even on input belonging to a target language, a similar class is used to separate strongly monotone Gold-style learners from total set-driven learners [10].

Theorem 6. *We have* $[\tau(\mathbf{SMon})\mathbf{TxtPsdEx}] \setminus [\mathbf{TxtSdEx}] \neq \emptyset$.

3.2 Behaviourally Correct Monotone Learning

In this section we consider an analogous question: How do monotone and strongly monotone learners interact when requiring semantic convergence? By Theorem 3 and the findings of Kötzing and Schirneck [13], we already have that globally monotone set-driven (and even Gold-style) learners are as powerful as strongly monotone Gold-style learners. The mentioned learners are, due to Theorem 2, less powerful than total set-driven monotone ones. This, in particular, implies that a "complete collapse" of the learning considered criteria as for strongly monotone learners [13] is impossible. As partially set-driven monotone (explanatory) learners are more powerful than set-driven behaviourally correct ones [14], only one question remains, namely, whether Gold-style **Mon**-learners may be separated from partially set-driven **Mon**-learners? Studies of various other restrictions [7, 13], show that behaviourally correct partially set-driven learners are often as powerful as their respective Gold-style counterpart.

Surprisingly, for monotone behaviourally correct learners, such an equality does *not* hold, as we show with the next result. The idea is to construct a class of languages where the learner must keep track of the order the elements were presented in, in order to safely discard them at a later point in learning-time. To obtain this result, we apply the technique of self-learning classes [4] using the Operator Recursion Theorem [3]. Note that this result already completes Fig. 1(b), as monotone **Bc**-learners may be assumed total [14].

Theorem 7. *We have* $[\mathbf{TxtGMonEx}] \setminus [\mathbf{TxtPsdMonBc}] \neq \emptyset$.

Proof. We provide a class witnessing the separation using self-learning classes [4, Thm. 3.6]. Consider the learner which for a finite sequence σ is defined as

$$h(\sigma) = \begin{cases} \mathrm{ind}(\emptyset), & \text{if } \mathrm{content}(\sigma) = \emptyset, \\ \varphi_{\max(\mathrm{content}(\sigma))}(\sigma), & \text{otherwise.} \end{cases}$$

Let $\mathcal{L} = \mathbf{TxtGMonEx}(h)$. Assume there exists a **TxtPsdMonBc**-learner h' which learns $\mathcal{L}$, that is, $\mathcal{L} \subseteq \mathbf{TxtPsdMonBc}(h')$. By the Operator Recursion Theorem (**ORT**, [3]), there exists a family of strictly monotone increasing, total computable functions $(a_j)_{j \in \mathbb{N}}$ with pairwise disjoint range, a total computable function $f \in \mathcal{R}$, an index

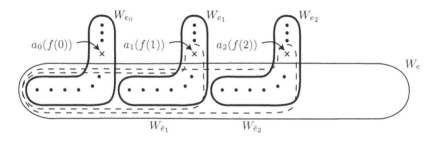

Fig. 2. A depiction of the class $\mathcal{L}'$. Given j, the dashed line depicts the set $W_{\hat{e}_j}$ and the cross indicates the element $a_j(f(j))$.

$e \in \mathbb{N}$ and two families of indices $(e_j)_{j \in \mathbb{N}}$, $(\hat{e}_k)_{k \in \mathbb{N}}$ such that for all finite sequences σ, where $\operatorname{first}(\sigma)$ is the first non-pause element in the sequence σ, we have

$$\varphi_{a_j(i)}(\sigma) = \begin{cases} e_j, & \text{if } \operatorname{content}(\sigma) \subseteq \operatorname{range}(a_j), \\ \hat{e}_k, & \text{else, if } \exists k \colon a_k(f(k)) \in \operatorname{content}(\sigma) \vee \\ & \qquad \exists k \colon \operatorname{first}(\sigma) \in \operatorname{range}(a_k) \wedge \\ & \qquad \wedge \max\{j \mid \operatorname{content}(\sigma) \cap \operatorname{range}(a_j) \neq \emptyset\} = k, \\ e, & \text{otherwise.} \end{cases}$$

$$f(j) = \text{first } i \text{ found such that } a_j(i) \in W_{h'(\operatorname{content}(a_j[i]),i)},$$

$$W_{e_j} = \operatorname{range}(a_j),$$

$$W_{\hat{e}_k} = \bigcup_{j' \leq k} \operatorname{content}(a_{j'}[f(j')]) \cup \{a_k(f(k))\},$$

$$W_e = \bigcup_j \operatorname{content}(a_j[f(j)]).$$

Let $\mathcal{L}' = \{W_{e_j} \mid j \in \mathbb{N}\} \cup \{W_{\hat{e}_k} \mid k > 0\} \cup \{W_e\}$. Figure 2 shows a depiction of the class $\mathcal{L}'$. We show that $\mathcal{L}'$ can be learned by h, but not by h'. The intuition is the following. For some j, as long as only elements from W_{e_j} are presented, h will suggest e_j as its hypothesis. Thus, h' needs to learn W_{e_j} as well and eventually overgeneralize, that is, at some point i we have $\operatorname{content}(a_j[i]) \subsetneq W_{h'(\operatorname{content}(a_j[i]),i)}$. The function $f(j)$ finds such i. Once the overgeneralization happens, the text presents, for $j' \neq j$, elements from $\operatorname{range}(a_{j'})$. Knowing the order in which the elements were presented, the learner h now either keeps or discards the element $a_j(f(j))$ in its next hypothesis depending whether $j' < j$ or $j < j'$, respectively. If $j' < j$, h needs to keep $a_j(f(j))$ in its hypothesis as it still may be presented the set $W_{\hat{e}_j}$. Otherwise, it suggests the set W_e, only changing its mind if it sees, for appropriate $i \in \mathbb{N}$, an element of the form $a_i(f(i))$. Then, h is certain to be presented $W_{\hat{e}_i}$. So the full-information learner h can deal with this new information and preserve monotonicity, while h' cannot, as it does not know which information came first.

We proceed with the formal proof that h **TxtGMonEx**-learns $\mathcal{L}'$. Let $L' \in \mathcal{L}'$ and $T' \in \mathbf{Txt}(L')$. We first show the **Ex**-convergence and the monotonicity afterwards. For the former, we distinguish the following cases.

1. Case: For some j, we have $L' = W_{e_j}$. Let n_0 be such that content$(T'[n]) \neq \emptyset$. Then, for $n \geq n_0$, there exists some i such that $a_j(i) = \max(\text{content}(T'[n]))$. Thus,

$$h(T'[n]) = \varphi_{\max(\text{content}(T'[n]))}(T'[n]) = \varphi_{a_j(i)}(T'[n]) = e_j.$$

   Hence, h learns W_{e_j} correctly.

2. Case: We have $L' = W_e$. Let $n_0 \in \mathbb{N}$ be the minimal and let $k_0 \in \mathbb{N}$ be such that content$(T'[n_0]) \neq \emptyset$ and first$(T'[n_0]) \in \text{range}(a_{k_0})$. Let $n_1 \geq n_0$ be minimal such that there exists $k > k_0$ such that content$(T'[n_1])$ also contains elements from content(a_k). Then, for $n > n_1$ we have that $h(T'[n]) = e$, as there exists no j with $a_j(f(j)) \in \text{content}(T')$ and also $\max\{j \mid \text{content}(T'[n]) \cap \text{range}(a_j) \neq \emptyset\} \neq k_0$. Thus, h learns W_e correctly.

3. Case: For some $k > 0$ we have $L' = W_{\hat{e}_k}$. In this case, there exists n_0 such that, for some $k' < k$, $\text{range}(a_{k'}) \cap \text{content}(T'[n_0]) \neq \emptyset$ and $a_k(f(k)) \in \text{content}(T'[n_0])$. Then, for $n \geq n_0$, we have $h(T'[n]) = \hat{e}_k$. Therefore, h learns $W_{\hat{e}_k}$ correctly.

We show that the learning is monotone. Let $n \in \mathbb{N}$. As long as content$(T'[n])$ is empty, h returns $\text{ind}(\emptyset)$. Once content$(T'[n])$ is not empty anymore and as long as content$(T'[n])$ only contains elements from, for some j, $\text{range}(a_j)$, the learner h outputs (a code for) the set W_{e_j}. Note that j is the index of the element first$(T'[n])$, that is, first$(T'[n]) \in \text{range}(a_j)$. If ever, for some later n, content$(T'[n]) \setminus \text{range}(a_j) \neq \emptyset$, then h only changes its mind if there exists $k > j$ such that content$(T'[n]) \cap \text{range}(a_k) \neq \emptyset$ (note that in case $j < k$, h does not change its mind). Depending on whether $a_k(f(k)) \in \text{content}(T'[n])$ or not, h changes its mind to (a code of) either $W_{\hat{e}_k}$ or W_e, respectively. In the former case, the learner h is surely presented the set $W_{\hat{e}_k}$, making this mind change monotone. In the latter case, no element of $W_{e_j} \setminus \text{content}(a_j[f(j)])$ is contained the target language. These are exactly the elements h discards from its hypothesis, keeping a monotone behaviour. The learner only changes its mind again if it witnesses, for some $k' \geq k$, the element $a_{k'}(f(k'))$. It will then output (a code of) the set $W_{\hat{e}_{k'}}$. This is, again, monotonic behaviour, as h is sure to be presented the set $W_{\hat{e}_{k'}}$. Altogether, h is monotone on any text of L'.

Thus, h identifies all languages in $\mathcal{L}'$ correctly. Now, we show that h' cannot do so too. We do so by providing a text of W_e where h' makes infinitely many wrong guesses. To that end, consider the text T of W_e given as $a_0[f(0)]a_1[f(1)]a_2[f(2)]\ldots$ For $j > 0$, since $a_j(f(j)) \in W_{h'(\text{content}(a_j[f(j)]),f(j))}$, we have

$$a_j(f(j)) \in W_{h'(\text{content}(T[\sum_{m \leq j} f(m)]),\sum_{m \leq j} f(m))},$$

as $T[\sum_{m \leq j} f(m)]$ is an initial sequence for a text for $W_{\hat{e}_j}$. But, since $a_j(f(j)) \notin W_e$, h' makes infinitely many incorrect conjectures and thus does not identify W_e on the text T correctly, a contradiction. $\qquad\square$

Acknowledgements. We would like to thank the anonymous reviewers for their helpful suggestions and comments. We believe that their feedback helped improve this work.

References

1. Angluin, D.: Inductive inference of formal languages from positive data. Inf. Control **45**, 117–135 (1980)
2. Blum, L., Blum, M.: Toward a mathematical theory of inductive inference. Inf. Control **28**, 125–155 (1975)
3. Case, J.: Periodicity in generations of automata. Math. Syst. Theor. **8**, 15–32 (1974)
4. Case, J., Kötzing, T.: Strongly non-U-shaped language learning results by general techniques. Inf. Comput. **251**, 1–15 (2016)
5. Case, J., Lynes, C.: Machine inductive inference and language identification. In: Nielsen, M., Schmidt, E.M. (eds.) ICALP 1982. LNCS, vol. 140, pp. 107–115. Springer, Heidelberg (1982). https://doi.org/10.1007/BFb0012761
6. Doskoč, V., Kötzing, T.: Mapping monotonic restrictions in inductive inference. CoRR (2020)
7. Doskoč, V., Kötzing, T.: Cautious limit learning. In: Proceedings of the International Conference on Algorithmic Learning Theory (ALT) (2020)
8. Fulk, M.A.: Prudence and other conditions on formal language learning. Inf. Comput. **85**, 1–11 (1990)
9. Gold, E.M.: Language identification in the limit. Inf. Control **10**, 447–474 (1967)
10. Jain, S.: Strong monotonic and set-driven inductive inference. J. Exp. Theor. Artif. Intell. **9**, 137–143 (1997)
11. Jantke, K.: Monotonic and non-monotonic inductive inference. New Gener. Comput. **8**, 349–360 (1991)
12. Kötzing, T., Palenta, R.: A map of update constraints in inductive inference. Theor. Comput. Sci. **650**, 4–24 (2016)
13. Kötzing, T., Schirneck, M.: Towards an atlas of computational learning theory. In: Proceedings of the Symposium on Theoretical Aspects of Computer Science (STACS), pp. 47:1–47:13 (2016)
14. Kötzing, T., Schirneck, M., Seidel, K.: Normal forms in semantic language identification. In: Proceedings of the International Conference on Algorithmic Learning Theory (ALT), pp. 76:493–76:516 (2017)
15. Kötzing, T.: Abstraction and Complexity in Computational Learning in the Limit. Ph.D. thesis, University of Delaware (2009)
16. Lange, S., Zeugmann, T.: Monotonic versus non-monotonic language learning. In: Nonmonotonic and Inductive Logic, pp. 254–269 (1993)
17. Lange, S., Zeugmann, T.: Set-driven and rearrangement-independent learning of recursive languages. Math. Syst. Theor. **29**, 599–634 (1996)
18. Lange, S., Zeugmann, T., Kapur, S.: Monotonic and dual monotonic language learning. Theor. Comput. Sci. **155**, 365–410 (1996)
19. Osherson, D.N., Weinstein, S.: Criteria of language learning. Inf. Control **52**, 123–138 (1982)
20. Rogers Jr., H.: Theory of Recursive Functions and Effective Computability. Reprinted by MIT Press, Cambridge (MA) (1987)
21. Schäfer-Richter, G.: Über Eingabeabhängigkeit und Komplexität von Inferenzstrategien. Ph.D. thesis, RWTH Aachen University, Germany (1984)
22. Wexler, K., Culicover, P.W.: Formal Principles of Language Acquisition. MIT Press, Cambridge (MA) (1980)
23. Wiehagen, R.: A thesis in inductive inference. In: Nonmonotonic and Inductive Logic, pp. 184–207 (1991)

Normal Forms for Semantically Witness-Based Learners in Inductive Inference

Vanja Doskoč[(✉)] and Timo Kötzing

Hasso Plattner Institute, University of Potsdam, Potsdam, Germany
{vanja.doskoc,timo.koetzing}@hpi.de

Abstract. In *inductive inference*, we study learners (computable devices) inferring formal languages. In particular, we consider *semantically witness-based* learners, that is, learners which are required to *justify* each of their semantic mind changes. This natural requirement deserves special attention as it is a specialization of various important learning paradigms. As such, it has already proven to be fruitful for gaining knowledge about other types of restrictions.

In this paper, we provide a thorough analysis of semantically converging, semantically witness-based learners, obtaining *normal forms* for them. Most notably, we show that *set-driven globally* semantically witness-based learners are equally powerful as their *Gold-style semantically conservative* counterpart. Such results are key to understanding the, yet undiscovered, mutual relation between various important learning paradigms of semantically converging learners.

1 Introduction

Computably learning formal languages from a growing but finite amount of information thereof is referred to as *inductive inference* or *language learning in the limit* [5], a branch of (algorithmic) learning theory. Here, a learner h (a computable device) is successively presented all and only the information from a formal language L (a computably enumerable subset of the natural numbers). We call such a list of elements of L a *text* of L. With every new datum, the learner h makes a guess (a description for a c.e. set) about which language it believes to be presented. Once these guesses converge to a single, correct hypothesis explaining the language, the learner successfully *learned* the language L on this text. We say that h *learns* L if it learns L on every text of L.

We refer to this as *explanatory learning* as the learner, in the limit, provides an explanation of the presented language. If we drop the requirement to converge to a *single* correct hypothesis and allow the learner to oscillate between arbitrarily many correct ones, we refer to this as *behaviourally correct* learning [3,13] and denote it as[1] **TxtGBc**. Since a learner which always guesses the same language can learn this very language, we study classes of languages which can be **TxtGBc**-learned by a single

---

[1] Here, **Txt** indicates that the information is given from text, **G** stands for *Gold-style* learning, where the learner has *full information* on the elements presented to make its guess, and, lastly, **Bc** refers to behaviourally correct learning.

This work was supported by DFG Grant Number KO 4635/1-1.

© Springer Nature Switzerland AG 2021
L. De Mol et al. (Eds.): CiE 2021, LNCS 12813, pp. 158–168, 2021.
https://doi.org/10.1007/978-3-030-80049-9_14

learner. We denote the set of all such classes with $[\mathbf{TxtGBc}]$, which we refer to as the *learning power* of $\mathbf{TxtGBc}$-learners.

Additional restrictions, modelling various learning strategies, may be imposed on the learner. By studying these we discover how seemingly intuitive strategies affect the learning power. For example, it may seem evident to only make semantic mind changes when seeing a new datum *justifying* this mind change. However, it is known that following such a strategy, referred to as *semantically witness-based* learning ($\mathbf{SemWb}$, [9, 10]), severely lessens the obtainable learning power.

Besides being intuitive yet restrictive, this restriction proved to be important in the literature. Together with *target-cautious* learning [8], where learners may not overgeneralize the target language, this paradigm encloses various important learning restrictions. Exemplary for explanatory learning, in settings where (syntactically) witness-based learning, as *specialization* or lower bound, and target-cautious learning, as *generalization* or upper bound, permit equivalent learning power, the three enclosed but seemingly incomparable restrictions, namely *conservativeness* [1], *weak monotonicity* [7, 17] and *cautiousness* [12], are equivalent as well [9].

The still undiscovered mutual relation between the mentioned restrictions in the behaviourally correct setting makes it worthwhile to study semantically witness-based learning in this setting as well. The previous literature indicates analogous equalities to be possible. Particularly, semantically witness-based learners and, a generalization thereof, *semantically conservative* learners ($\mathbf{SemConv}$, [10]), which keep their guesses while they are consistent with the data given, are shown to be equally powerful [10]. This equality holds true regardless of the amount of information given, particularly, it holds true for both Gold-style and *set-driven* learners ($\mathbf{Sd}$, [16]), which base their hypotheses solely on the set of elements given. We enhance the analogy by showing that the learners perform equally well regardless of the amount of information given, drawing parallels to target-cautious learning, where Gold-style and set-driven learners are also equally powerful [4].

The latter result already provides a powerful *normal form*. It states that semantically witness-based learners do not need to know the order and amount of the information given. This significantly extends the result [10] that such set-driven learners are as powerful as *partially set-driven* ones [2, 15]. Note that the latter learners base their hypotheses on the content and amount of information given, but have no access to the order in which the information came. Another normal form shows that witness-based learners display such behaviour also *globally*, that is, on arbitrary text. This means that the learners always display a "decent" behaviour regardless whether the information given belongs to a language they actually learn. Lastly, semantically witness-based and semantically conservative learning is interchangeable also when required globally.

This paper is structured as follows. In Sect. 2 we introduce all necessary notation and preliminary results. In Sect. 3, we show that three normal forms can be assumed *simultaneously*. Our main result (Theorem 1) states that semantically conservative $\mathbf{G}$-learners may be assumed (a) globally (b) semantically witness-based and (c) set-driven.

2   Language Learning in the Limit

In this section we introduce notation and preliminary results used throughout this paper. Thereby, we consider basic computability theory as known [14]. We start with the mathematical notation and use $\subsetneq$ and $\subseteq$ to denote the proper subset and subset relation between sets, respectively. We denote the set of all natural numbers as $\mathbb{N} = \{0, 1, 2, \ldots\}$ and the empty set as $\emptyset$. Furthermore, we let $\mathcal{P}$ and $\mathcal{R}$ be the set of all partial and total computable functions $p : \mathbb{N} \to \mathbb{N}$. Next, we fix an effective numbering $\{\varphi_e\}_{e \in \mathbb{N}}$ of all partial computable functions and denote the e-th computably enumerable set as $W_e = \mathrm{dom}(\varphi_e)$ and interpret the number e as an *index* or *hypothesis* thereof.

We learn recursively enumerable sets $L \subseteq \mathbb{N}$, called *languages*, using *learners*, that is, partial computable functions. By $\#$ we denote the *pause symbol* and for any set S we denote $S_\# := S \cup \{\#\}$. Then, a *text* is a total function $T : \mathbb{N} \to \mathbb{N} \cup \{\#\}$ and the collection of all texts is denoted as **Txt**. For any text (or sequence) T we define the *content* of T as $\mathrm{content}(T) := \mathrm{range}(T) \setminus \{\#\}$. Here, range denotes the image of a function. A text of a language L is such that $\mathrm{content}(T) = L$. We denote the collection of all texts of L as $\mathbf{Txt}(L)$. Additionally, for $n \in \mathbb{N}$, we denote by $T[n]$ the initial sequence of T of length n, that is, $T[0] := \varepsilon$ (the empty string) and $T[n] := (T(0), T(1), \ldots, T(n-1))$. For a set S, we call the sequence (text) where all elements of S are presented in strictly increasing order without interruptions (followed by infinitely many pause symbols if S is finite) the *canonical sequence (text) of S*. On finite sequences we use $\subseteq$ to denote the *extension relation*. Given two sequences σ and τ we write $\sigma^\frown \tau$ or (if readability permits) $\sigma \tau$ to denote the concatenation of these.

We use the following system to formalize learning criteria [11]. An *interaction operator* β takes a learner $h \in \mathcal{P}$ and a text $T \in \mathbf{Txt}$ as argument and outputs a possibly partial function p. Intuitively, β provides the information for the learner to make its guesses. We consider the interaction operators $\mathbf{G}$ for *Gold-style* or *full-information* learning [5] and $\mathbf{Sd}$ for *set-driven* learning [16]. Define, for any $i \in \mathbb{N}$,

$$\mathbf{G}(h, T)(i) := h(T[i]),$$
$$\mathbf{Sd}(h, T)(i) := h(\mathrm{content}(T[i])).$$

Intuitively, a Gold-style learner has full information on the elements presented, while a set-driven learner bases its guesses solely on the content, that is, set of elements, given.

Given a learning task, we can distinguish between various criteria for successful learning. A first such criterion is *explanatory* learning (**Ex**, [5]), where the learner is expected to converge to a single, correct hypothesis in order to learn a language. Allowing the learner to oscillate between arbitrarily many semantically correct, but possibly syntactically different hypotheses we get *behaviourally correct* learning (**Bc**, [3,13]). Formally, a *learning restriction* δ is a predicate on a total learning sequence p, that is, a total function, and a text $T \in \mathbf{Txt}$. For the mentioned criteria we have

$$\mathbf{Ex}(p, T) :\Leftrightarrow \exists n_0 \forall n \geq n_0 : p(n) = p(n_0) \wedge W_{p(n_0)} = \mathrm{content}(T),$$
$$\mathbf{Bc}(p, T) :\Leftrightarrow \exists n_0 \forall n \geq n_0 : W_{p(n)} = \mathrm{content}(T).$$

We impose restrictions on the learners. In particular, we focus on *semantically witness-based* learning (**SemWb**, [9,10]), where the learners need to *justify* each of their semantic mind changes. A generalization thereof is *semantically conservative* learning (**SemConv**, [1]). Here, the learners may not change their mind while their hypotheses are consistent with the information given. A hypothesis is consistent if it contains all the information it is based on and if we require the learners to output consistent hypotheses we speak of *consistent* learning (**Cons**, [1]). Formally, we define the restrictions as

$$\mathbf{SemWb}(p, T) :\Leftrightarrow \forall n, m\colon (\exists k\colon n \leq k \leq m \wedge W_{p(n)} \neq W_{p(k)}) \Rightarrow$$
$$\Rightarrow (\text{content}(T[m]) \cap W_{p(m)}) \backslash W_{p(n)} \neq \emptyset,$$
$$\mathbf{Cons}(p, T) :\Leftrightarrow \forall n\colon \text{content}(T[n]) \subseteq W_{h(T[n])},$$
$$\mathbf{SemConv}(p, T) :\Leftrightarrow \forall n, m\colon \big(n < m \wedge \text{content}(T[m]) \subseteq W_{p(n)}\big) \Rightarrow$$
$$\Rightarrow W_{p(n)} = W_{p(m)}.$$

Given two restrictions δ and δ', the juxtaposition $\delta\delta'$ denotes their combination, that is, intersection. Finally, the always true predicate **T** denotes the absence of a restriction.

Now, a *learning criterion* is a tuple $(\alpha, \mathcal{C}, \beta, \delta)$, where $\mathcal{C}$ is a set of admissible learners, typically $\mathcal{P}$ or $\mathcal{R}$, β is an interaction operator and α and δ are learning restrictions. We denote this learning criterion as $\tau(\alpha)\mathcal{C}\mathbf{Txt}\beta\delta$. In the case of $\mathcal{C} = \mathcal{P}$, $\alpha = \mathbf{T}$ or $\delta = \mathbf{T}$ we omit writing the respective symbol. For an admissible learner $h \in \mathcal{C}$ we say that h $\tau(\alpha)\mathcal{C}\mathbf{Txt}\beta\delta$-learns a language L if and only if on arbitrary text $T \in \mathbf{Txt}$ we have $\alpha(\beta(h, T), T)$ and on texts of the target language $T \in \mathbf{Txt}(L)$ we have $\delta(\beta(h, T), T)$. With $\tau(\alpha)\mathcal{C}\mathbf{Txt}\beta\delta(h)$ we denote the class of languages $\tau(\alpha)\mathcal{C}\mathbf{Txt}\beta\delta$-learned by h and the set of all such classes we denote with $[\tau(\alpha)\mathcal{C}\mathbf{Txt}\beta\delta]$.

Lastly, we discuss **Bc**-locking sequences, the semantic counterpart to locking sequences [2]. Intuitively, a **Bc**-locking sequence is a sequence where the learner correctly identifies the target language and does not make a semantic mind change anymore regardless what information of the language it is presented. Formally, given a language L and a **G**-learner h, a sequence $\sigma \in L_{\#}^*$ is called a **Bc**-*locking sequence* for h on L if and only if for every sequence $\tau \in L_{\#}^*$ we have that $W_{h(\sigma\tau)} = L$ [6]. When talking about **Sd**-learners, we call a finite set D a **Bc**-*locking set* of L if and only if for all D', with $D \subseteq D' \subseteq L$, we have $W_{h(D')} = L$.

While for each **G**-learner h there exists a **Bc**-locking sequence on every language it learns [2], not every text may contain an initial segment which is a **Bc**-locking sequence. Learners which do have a **Bc**-locking sequence on every text of a language they learn are called *strongly* **Bc**-*locking* [8]. Formally, a learner is *strongly* **Bc**-*locking* if on every language L it learns and on every text $T \in \mathbf{Txt}(L)$ there exists n such that $T[n]$ is a **Bc**-locking sequence for h on L. The transition to set-driven learners is immediate.

3   Semantic Witness-Based Learning

In this section, we provide a normal form for semantically witness-based learners, namely that $\tau(\mathbf{SemWb})\mathbf{TxtSdBc}$-learners are as powerful as $\mathbf{TxtGSemConvBc}$ ones (Theorem 1). We prove this normal form stepwise. We start by showing that each $\mathbf{TxtGSemConvBc}$-learner may be assumed semantically conservative on arbitrary text (Theorem 2). Afterwards, we prove that such learners base their guesses solely on the content given (Theorem 3). Lastly, we observe that they remain equally powerful when being globally semantically witness-based (Theorem 4).

Theorem 1. *We have that* $[\tau(\mathbf{SemWb})\mathbf{TxtSdBc}] = [\mathbf{TxtGSemConvBc}]$.

We make a **TxtGSemConvBc**-learner h *globally* semantically conservative first.

Theorem 2. *We have that* $[\tau(\mathbf{SemConv})\mathbf{TxtGBc}] = [\mathbf{TxtGSemConvBc}]$.

Proof. The inclusion $[\tau(\mathbf{SemConv})\mathbf{TxtGBc}] \subseteq [\mathbf{TxtGSemConvBc}]$ is immediate. For the other, let h be a consistent learner [10] and $\mathcal{L} = \mathbf{TxtGSemConvBc}(h)$. We provide a learner h' which $\tau(\mathbf{SemConv})\mathbf{TxtGBc}$-learns $\mathcal{L}$.

We do so with the help of an auxiliary $\tau(\mathbf{SemConv})\mathbf{TxtGBc}$-learner $\hat{h}$, which only operates on sequences without repetitions or pause symbols. For convenience, we subsume these using the term *duplicates*. When h' is given a sequence with duplicates, say $(7, 1, 5, 1, 4, \#, 3, 1)$, it mimics $\hat{h}$ given the same sequence without duplicates, that is, $h'(7, 1, 5, 1, 4, \#, 3, 1) = \hat{h}(7, 1, 5, 4, 3)$. First, note that this mapping of sequences preserves the $\subseteq$-relation on sequences, thus making h' also a $\tau(\mathbf{SemConv})$-learner. Furthermore, it suffices to focus on sequences without duplicates since consistent, semantically conservative learners cannot change their mind when presented a datum they have already witnessed (or a pause symbol). Thus, $\hat{h}$ will be presented sufficient information for the learning task, which then again is transferred to h'. With this in mind, we only consider **sequences without duplicates**, that is, without repetitions or pause symbols, for the entirety of this proof. Sequences where duplicates may potentially still occur (for example when looking at the initial sequence of a text) are also replaced as described above. To ease notation, given a set A, we write $\mathbb{S}(A)$ for the subset of A^* where the sequences do not contain duplicates. Now, we define the auxiliary learner $\hat{h}$.

Algorithm 1: The auxiliary $\tau(\mathbf{SemConv})$-learner $\hat{h}$.

Parameter: **TxtGSemConv**-learner h.
Input: Finite sequence $\sigma \in \mathbb{S}(\mathbb{N})$.
Semantic Output: $W_{\hat{h}(\sigma)} = \bigcup_{t \in \mathbb{N}} E_t$.
Initialization: $t' \leftarrow 0$, $E_0 \leftarrow \text{content}(\sigma)$ and, for all $t > 0$, $E_t \leftarrow \emptyset$.

1 **for** $t = 0$ **to** ∞ **do**
2   **if** $\exists \sigma' \subsetneq \sigma : \text{content}(\sigma) \subseteq W^t_{\hat{h}(\sigma')}$ **then**
3         $\Sigma'_t \leftarrow \{\sigma' \subsetneq \sigma \mid \text{content}(\sigma) \subseteq W^t_{\hat{h}(\sigma')}\}$
4         $E_{t+1} \leftarrow E_t \cup \bigcup_{\sigma' \in \Sigma'_t} W^t_{\hat{h}(\sigma')}$
5   **else if** $\forall \sigma' \subsetneq \sigma : \text{content}(\sigma) \not\subseteq W^t_{\hat{h}(\sigma')}$ **then**
6         $S(\sigma, t') \leftarrow \mathbb{S}\left(W^{t'}_{\hat{h}(\sigma)} \setminus \text{content}(\sigma)\right)$
7   **if** $\forall \tau \in S(\sigma, t') : \bigcup_{\tau' \in S(\sigma, t')} W^{t'}_{\hat{h}(\sigma\tau')} \subseteq W^t_{\hat{h}(\sigma\tau)}$ **then**
8             $E_{t+1} \leftarrow E_t \cup W^{t'}_{\hat{h}(\sigma)}$
9             $t' \leftarrow t' + 1$
10   **else**
11       $E_{t+1} \leftarrow E_t$

Consider the learner $\hat{h}$ as in Algorithm 1 with parameter h. Given some input σ, the intuition is the following. Once $\hat{h}$, on any previous sequence σ', is consistent with the currently given information content(σ), the learner only enumerates the same as such hypotheses (lines 2 to 4). While no such hypothesis is found, $\hat{h}$ does a forward search (lines 5 to 9) and only enumerates elements if all visible future hypotheses also witness these elements. As already discussed, $\hat{h}$ operates only on sequences without repetitions or pause symbols, thus making it possible to check *all* necessary future hypotheses.

First we show that for any $L \in \mathcal{L}$ and any $T \in \mathbf{Txt}(L)$ we have, for $n \in \mathbb{N}$,

$$W_{\hat{h}(T[n])} \subseteq W_{h(T[n])}. \tag{1}$$

Note that, while the (infinite) text T may contain duplicates, the (finite) sequence $T[n]$ does not by our assumption. Now, we show Eq. (1) by induction on n. The case $n = 0$ follows immediately. Assume Eq. (1) holds up to n. As content$(T[n+1]) \subseteq W_{h(T[n+1])}$ by consistency of h and as, for $n' \le n$, $W_{h(T[n'])} = W_{h(T[n+1])}$ whenever content$(T[n+1]) \subseteq W_{h(T[n'])}$, we get

$$W_{\hat{h}(T[n+1])} \subseteq \bigcup_{\substack{n' \le n, \\ \text{content}(T[n+1]) \subseteq W_{\hat{h}(T[n'])}}} W_{\hat{h}(T[n'])} \cup W_{h(T[n+1])} \subseteq W_{h(T[n+1])}.$$

The first inclusion follows as the big union contains all previous hypotheses found in the first if-clause (lines 2 to 4) and as $W_{h(T[n+1])}$ contains all elements possibly enumerated by the second if-clause (lines 5 to 9). Note that the latter also contains content$(T[n+1])$, thus covering the initialization. The second inclusion follows by the induction hypothesis and semantic conservativeness of h.

We continue by showing that $\hat{h}$ $\mathbf{TxtGBc}$-learns $\mathcal{L}$. To that end, let $L \in \mathcal{L}$ and $T \in \mathbf{Txt}(L)$. We distinguish the following two cases.

1. Case: L is finite. Then there exists n_0 with content$(T[n_0]) = L$. Let $n \ge n_0$. By $\mathbf{SemConv}$ and consistency of h, we have $L = W_{h(T[n])}$. By Eq. (1), we have $W_{h(T[n])} \supseteq W_{\hat{h}(T[n])}$ and, by consistency of $\hat{h}$, $W_{\hat{h}(T[n])} \supseteq$ content$(T[n]) = L$. Altogether we have $W_{\hat{h}(T[n])} = L$ as required.
2. Case: L is infinite. Let n_0 be minimal such that $W_{h(T[n_0])} = L$. Then, as h is semantically conservative, $T[n_0]$ is a $\mathbf{Bc}$-locking sequence for h on L and we have

$$\forall i < n_0 \colon \text{content}(T[n_0]) \not\subseteq W_{h(T[i])}.$$

Thus, elements enumerated by $W_{\hat{h}(T[n_0])}$ cannot be enumerated by the first if-clause (lines 2 to 4) but only by the second one (lines 5 to 9). We show $W_{\hat{h}(T[n_0])} = L$. The $\subseteq$-direction follows immediately from Eq. (1). For the other direction, let t' be the current step of enumeration. As $T[n_0]$ is a $\mathbf{Bc}$-locking sequence, we have, for all $\tau \in S(T[n_0], t') = \mathbb{S}\left(W_{h(T[n_0])}^{t'} \setminus \text{content}(T[n_0])\right)$,

$$\bigcup_{\tau' \in S(T[n_0], t')} W_{h(T[n_0]) \frown \tau'}^{t'} \subseteq W_{h(T[n_0] \frown \tau)} = L.$$

Thus, at some step t, $E_{t+1} \leftarrow W_{h(T[n_0])}^{t'}$ and, then, the enumeration continues with $t' \leftarrow t' + 1$. In the end we have $L \subseteq W_{\hat{h}(T[n_0])}$ and, altogether, $L = W_{\hat{h}(T[n_0])}$.

We now show that, for any $n > n_0$, $L = W_{\hat{h}(T[n])}$ holds. Note that at some point content$(T[n]) \subseteq W_{\hat{h}(T[n_0])}$ will be witnessed. Thus, $W_{\hat{h}(T[n])}$ will enumerate the same as $W_{\hat{h}(T[n_0])} = L$, and it follows that $L \subseteq W_{\hat{h}(T[n])}$. By Eq. (1), $W_{\hat{h}(T[n])}$ will not enumerate more than $W_{h(T[n])} = L$, that is, $W_{\hat{h}(T[n])} \subseteq W_{h(T[n])} = L$, concluding this part of the proof.

It remains to be shown that $\hat{h}$ is **SemConv** on arbitrary text $T \in \mathbf{Txt}$. The problem is that when a previous hypothesis becomes consistent with information currently given, the learner may have already enumerated incomparable data in its current hypothesis. This is prevented by closely monitoring the time of enumeration, namely by waiting until the enumerated data will certainly not cause such problems. We prove that $\hat{h}$ is $\tau(\mathbf{SemConv})$ formally. Let $n < n'$ be such that content$(T[n']) \subseteq W_{\hat{h}(T[n])}$. We show that $W_{\hat{h}(T[n])} = W_{\hat{h}(T[n'])}$ by separately looking at each inclusion.

$\subseteq$: The inclusion $W_{\hat{h}(T[n])} \subseteq W_{\hat{h}(T[n'])}$ follows immediately since by assumption content$(T[n']) \subseteq W_{\hat{h}(T[n])}$, meaning that at some point the first if-clause (lines 2 and 4) will find $T[n]$ as a candidate and then $W_{\hat{h}(T[n'])}$ will enumerate $W_{\hat{h}(T[n])}$.

$\supseteq$: Assume there exists $x \in W_{\hat{h}(T[n'])} \setminus W_{\hat{h}(T[n])}$. Let x be the first such enumerated and let t_x be the step of enumeration with respect to $h(T[n'])$, that is, $x \in W_{h(T[n'])}^{t_x}$ but $x \notin W_{h(T[n'])}^{t_x - 1}$. Furthermore, let t_{content} be the step where content$(T[n']) \subseteq W_{\hat{h}(T[n])}$ is witnessed for the first time. Now, by the definition of $\hat{h}$, we have

$$W_{\hat{h}(T[n'])} \subseteq W_{h(T[n'])}^{t_{\text{content}} - 1} \cup W_{\hat{h}(T[n])},$$

as $W_{\hat{h}(T[n'])}$ enumerates at most $W_{h(T[n'])}^{t_{\text{content}} - 1}$ until it sees the consistent prior hypothesis, namely $\hat{h}(T[n])$. This happens exactly at step $t_{\text{content}} - 1$, at which $W_{\hat{h}(T[n'])}$ stops enumerating elements from $W_{h(T[n'])}^{t_{\text{content}} - 1}$ and continues to follow $W_{\hat{h}(T[n])}$. Now, observe that $t_x < t_{\text{content}}$ since $x \in W_{\hat{h}(T[n'])}$ but $x \notin W_{\hat{h}(T[n])}$. But then, with $S(T[n], t_{\text{content}}) = \mathbb{S}\left(W_{h(T[n])}^{t_{\text{content}}} \setminus \text{content}(T[n]) \right)$,

$$x \in \bigcup_{\tau' \in S(T[n], t_{\text{content}})} W_{h(T[n] \frown \tau')}^{t_{\text{content}}} \subseteq W_{\hat{h}(T[n])},$$

which must be witnessed in order for $W_{\hat{h}(T[n])}$ to enumerate content$(T[n'])$ via the second if-clause (lines 5 to 9), that is, to get content$(T[n']) \subseteq W_{\hat{h}(T[n])}$. This contradicts $x \notin W_{\hat{h}(T[n])}$, concluding the proof. $\qquad\square$

This result proves that h may be assumed semantically conservative on arbitrary text. Next, we show that h does not rely on the order or amount of information given.

Theorem 3. *We have that* $[\tau(\mathbf{SemConv})\mathbf{TxtSdBc}] = [\tau(\mathbf{SemConv})\mathbf{TxtGBc}]$.

Proof. Let h be a learner and $\mathcal{L} = \tau(\mathbf{SemConv})\mathbf{TxtGBc}(h)$. We may assume h to be globally consistent [10]. We provide a learner h' which $\tau(\mathbf{SemConv})\mathbf{TxtSdBc}$-learns $\mathcal{L}$. To that end, we introduce the following auxiliary notation used throughout this proof. For each finite set $D \subseteq \mathbb{N}$ and each $x \in \mathbb{N}$, let $d := \max(D)$, σ_D be the canonical sequence of D and $D_{<x} := \{y \in D \mid y < x\}$. Note that the definition of $D_{<x}$ can be extended to $\leq, >$ and $\mathbf{Geq}$ as well as infinite sets in a natural way. Now, let h' be such that, for each finite set D,

$$W_{h'(D)} = D \cup \left(W_{h(\sigma_D)} \right)_{>d} \cup \left\{ x \in \left(W_{h(\sigma_D)} \right)_{<d} : D \cup \{x\} \subseteq W_{h(\sigma_{(D_{<x})})} \right\}.$$

Intuitively, $h'(D)$ simulates h assuming it got the information in the canonical order, that is, $h'(D)$ simulates $h(\sigma_D)$. All elements $x \in W_{h(\sigma_D)}$ such that $x > d$ can be enumerated, as any later, consistent hypothesis will do so as well. If $x < d$, then we check whether the learner h given the canonical sequence up to x is consistent with $D \cup \{x\}$, that is, whether $D \cup \{x\} \subseteq W_{h(\sigma_{(D_{<x})})}$. If so, we enumerate x as it will be done by the previous hypotheses as well. Note that, for each finite $D \subseteq \mathbb{N}$, we have

$$W_{h'(D)} \subseteq W_{h(\sigma_D)}. \tag{2}$$

We proceed by proving that h' $\tau(\mathbf{SemConv})\mathbf{TxtSdBc}$-learns $\mathcal{L}$. First, we show the $\mathbf{TxtSdBc}$-convergence. The idea here is to find a $\mathbf{Bc}$-locking sequence of the canonical text. Doing so ensures that even if elements are shown out of order they will be enumerated as h will not make a mind change and thus the consistency condition will be observed. To that end, let $L \in \mathcal{L}$. We distinguish whether L is finite or not.

1. Case: L is finite. We show that $W_{h'(L)} = L$. By definition of h', we have $L \subseteq W_{h'(L)}$. For the other inclusion, note that as h is consistent and semantically conservative (which in particular implies it being target-cautious), we have that $W_{h(\sigma_L)} = L$. Then, by Eq. (2), we have $W_{h'(L)} \subseteq W_{h(\sigma_L)} = L$, concluding this case.

2. Case: L is infinite. Let T_c be the canonical text of L, and let σ_0 be a $\mathbf{Bc}$-locking sequence for h on T_c. Such a $\mathbf{Bc}$-locking sequence exists, as h is strongly $\mathbf{Bc}$-locking [10, Thm. 7]. Let $D_0 := \mathrm{content}(\sigma_0)$. For any input $D \subseteq L$ such that $D \supseteq D_0$, we show that $W_{h'(D)} = L$. By Eq. (2), we get $W_{h'(D)} \subseteq W_{h(\sigma_D)} = L$. To show $L \subseteq W_{h'(D)}$, let $x \in L$. We distinguish the relative position of x and d.

$x > d$: In this case we have $x \in W_{h'(D)}$ by definition of h'.

$x \leq d$: In this case either $x \in D$ and we immediately get $x \in W_{h'(D)}$, or we have to check whether $D \cup \{x\} \subseteq W_{h(\sigma_{(D_{<x})})}$. Since σ_0 is an initial segment of the canonical text of L, it holds that $x > \max(\mathrm{content}(\sigma_0))$ and, thus, we get $\sigma_0 \subseteq \sigma_{(D_{<x})}$. Now $W_{h(\sigma_{(D_{<x})})} = L$, meaning that $D \cup \{x\} \subseteq W_{h(\sigma_{(D_{<x})})}$ will be observed at some point in the computation. Thus, $x \in W_{h'(D)}$.

Altogether, we get $W_{h'(D)} = L$ and thus $\mathbf{TxtSdBc}$-convergence. It remains to be shown that h' is $\tau(\mathbf{SemConv})$. Let $D' \subseteq D''$ and $D'' \subseteq W_{h'(D')}$. The trick here is that upon checking for consistency with elements shown out of order, the learner has to check the same, minimal sequence regardless whether the input is D' or D''. We proceed with the formal proof. Therefore, we expand the initially introduced notation of this proof. For any $x \in \mathbb{N}$ define $\sigma' := \sigma_{D'}$, $d' := \max(D')$ and $\sigma'_{<x} := \sigma_{(D'_{<x})}$. Analogously, we use σ'', d'' and $\sigma''_{<x}$ when D'' is the underlying set. First, we show that

$W_{h(\sigma')} = W_{h(\sigma'')}$. Since $W_{h'(D')}$ enumerates D'', that is, $D'' \subseteq W_{h'(D')}$, we have for all $y \in (D'' \setminus D')_{<d'}$ that $D' \cup \{y\} \subseteq W_{h(\sigma'_{<y})}$ by definition of h'. Thus, we have

$$W_{h(\sigma'_{<y})} = W_{h(\sigma')}. \tag{3}$$

Note that, if $(D'' \setminus D')_{<d'} = \emptyset$, then $\sigma'_{<d'+1} = \sigma'$. Thus, Eq. (3) also holds for

$$m := \begin{cases} \min(D''_{<d'} \setminus D'), & \text{if } D''_{<d'} \setminus D' \neq \emptyset, \\ d'+1, & \text{otherwise.} \end{cases}$$

Furthermore, it holds true that for any $x \leq m$ we have

$$\sigma'_{<x} = \sigma''_{<x}. \tag{4}$$

By Eqs. (2) and (3), we have $D'' \subseteq W_{h'(D')} \subseteq W_{h(\sigma')} = W_{h(\sigma'_{<m})}$. As, by Eq. (4), $\sigma'_{<m} = \sigma''_{<m} \subseteq \sigma''$ and h is $\tau(\mathbf{SemConv})$, we get

$$W_{h(\sigma')} = W_{h(\sigma'')}. \tag{5}$$

We conclude the proof by showing that $W_{h'(D')} = W_{h'(D'')}$. We check each direction separately by checking every possible position of an element, which is a candidate for enumeration, relative to the given information D' and D''.

$\supseteq$: Let $x \in W_{h'(D'')}$. For $x \in D''$ we have $x \in W_{h'(D')}$ by assumption. Otherwise, by Eqs. (2) and (5), we get $x \in W_{h(\sigma')}$. Thus, x will be considered in the enumeration of $W_{h'(D')}$. We distinguish the relation between x and d'.

$x > d'$: In this case $x \in (W_{h(\sigma')})_{>d'} \subseteq W_{h'(D')}$.

$x < d'$: As $d' \leq d''$ and since x is enumerated into $W_{h'(D'')}$, we have $D'' \cup \{x\} \subseteq W_{h(\sigma''_{<x})}$. We, again, distinguish the relative position of x and m and get

$$x < m: \quad D' \cup \{x\} \subseteq D'' \cup \{x\} \subseteq W_{h(\sigma''_{<x})} \overset{(5)}{=} W_{h(\sigma'_{<x})},$$

$$m < x < d': \quad D' \cup \{x\} \subseteq D'' \cup \{x\} \subseteq W_{h(\sigma''_{<x})} \overset{(*)}{=} W_{h(\sigma'')} \overset{(3)}{=} W_{h(\sigma')} \overset{(3)}{=}$$

$$\overset{(3)}{=} W_{h(\sigma'_{<m})} \overset{(*)}{=} W_{h(\sigma'_{<x})}.$$

We use h being $\tau(\mathbf{SemConv})$ in the steps marked by $(*)$. Thus, $x \in W_{h'(D')}$.

$\subseteq$: Let $x \in W_{h'(D')}$. For $x \in D''$ we have $x \in W_{h'(D'')}$ by definition of h'. Otherwise,

$$x \in D'' \cup \{x\} \subseteq W_{h'(D')} \subseteq W_{h(\sigma')} \overset{(5)}{=} W_{h(\sigma'')}.$$

Thus, x will be considered in the enumeration of $W_{h'(D'')}$. We now distinguish between the possible relation of x and d''.

$x > d''$: In this case $x \in W_{h'(D'')}$ by definition of h'.

$x < d''$: We show that $D'' \cup \{x\} \subseteq W_{h(\sigma''_{<x})}$ and, thus, x is enumerated by $W_{h'(D'')}$.

$$x < m : D'' \cup \{x\} \subseteq W_{h(\sigma'_{<x})} \overset{(4)}{=} W_{h(\sigma''_{<x})},$$

$$m < x < d' : D'' \cup \{x\} \subseteq W_{h(\sigma'_{<x})} \overset{(*)}{=} W_{h(\sigma'_{<m})} \overset{(4)}{=} W_{h(\sigma''_{<m})} \overset{(*)}{=} W_{h(\sigma''_{<x})},$$

$$d' < x < d'' : D'' \cup \{x\} \subseteq W_{h(\sigma')} = W_{h(\sigma'_{<m})} = W_{h(\sigma''_{<m})} \overset{(4)}{=} W_{h(\sigma''_{<x})}.$$

We use h being $\tau(\mathbf{SemConv})$ in the steps marked by $(*)$. In the end, $x \in W_{h'(D'')}$. $\square$

Hence, we may assume h to be $\tau(\mathbf{SemConv})\mathbf{TxtSdBc}$. Lastly, we observe that h may even be assumed globally semantically witness-based. This concludes the proof of Theorem 1 and, thus, also this section.

Theorem 4. *We have that* $[\tau(\mathbf{SemWb})\mathbf{TxtSdBc}] = [\tau(\mathbf{SemConv})\mathbf{TxtSdBc}]$.

Proof. Let $\delta \in \{\mathbf{SemWb}, \mathbf{SemConv}\}$. Since δ-learners may be assumed to be consistent [10, Thm. 8], which also holds true when the restrictions are required globally, we have $[\tau(\mathbf{Cons}\delta)\mathbf{TxtSdBc}] = [\tau(\delta)\mathbf{TxtSdBc}]$. Since $\mathbf{Cons} \cap \mathbf{SemWb} = \mathbf{Cons} \cap \mathbf{SemConv}$ [10, Lem. 11], the theorem holds. $\square$

References

1. Angluin, D.: Inductive inference of formal languages from positive data. Inf. Control **45**, 117–135 (1980)
2. Blum, L., Blum, M.: Toward a mathematical theory of inductive inference. Inf. Control **28**, 125–155 (1975)
3. Case, J., Lynes, C.: Machine inductive inference and language identification. In: Nielsen, M., Schmidt, E.M. (eds.) ICALP 1982. LNCS, vol. 140, pp. 107–115. Springer, Heidelberg (1982). https://doi.org/10.1007/BFb0012761
4. Doskoč, V., Kötzing, T.: Cautious limit learning. In: Proceedings of the International Conference on Algorithmic Learning Theory (ALT) (2020)
5. Gold, E.M.: Language identification in the limit. Inf. Control **10**, 447–474 (1967)
6. Jain, S., Osherson, D., Royer, J.S., Sharma, A.: Systems that Learn: An Introduction to Learning Theory, 2nd edn. MIT Press, Cambridge (1999)
7. Jantke, K.: Monotonic and non-monotonic inductive inference. New Gener. Comput. **8**, 349–360 (1991)
8. Kötzing, T., Palenta, R.: A map of update constraints in inductive inference. Theor. Comput. Sci. **650**, 4–24 (2016)
9. Kötzing, T., Schirneck, M.: Towards an atlas of computational learning theory. In: Proceedings of the Symposium on Theoretical Aspects of Computer Science (STACS), pp. 47:1–47:13 (2016)
10. Kötzing, T., Schirneck, M., Seidel, K.: Normal forms in semantic language identification. In: Proceedings of the International Conference on Algorithmic Learning Theory (ALT), pp. 76:493–76:516 (2017)
11. Kötzing, T.: Abstraction and Complexity in Computational Learning in the Limit. Ph.D. thesis, University of Delaware (2009)
12. Osherson, D.N., Stob, M., Weinstein, S.: Learning strategies. Inf. Control **53**, 32–51 (1982)

13. Osherson, D.N., Weinstein, S.: Criteria of language learning. Inf. Control **52**, 123–138 (1982)
14. Rogers Jr., H.: Theory of Recursive Functions and Effective Computability. Reprinted by MIT Press, Cambridge (1987)
15. Schäfer-Richter, G.: Über Eingabeabhängigkeit und Komplexität von Inferenzstrategien. Ph.D. thesis, RWTH Aachen University, Germany (1984)
16. Wexler, K., Culicover, P.W.: Formal Principles of Language Acquisition. MIT Press, Cambridge (1980)
17. Wiehagen, R.: A thesis in inductive inference. In: Nonmonotonic and Inductive Logic, pp. 184–207 (1991)

Walk-Preserving Transformation of Overlapped Sequence Graphs into Blunt Sequence Graphs with GetBlunted

Jordan M. Eizenga⬥, Ryan Lorig-Roach⬥, Melissa M. Meredith⬥, and Benedict Paten$^{(\boxtimes)}$⬥

University of California Santa Cruz Genomics Institute, 1156 High Street, Santa Cruz, CA 95064, USA
bpaten@ucsc.edu

Abstract. Sequence graphs have emerged as an important tool in two distinct areas of computational genomics: genome assembly and pangenomics. However, despite this shared basis, subtly different graph formalisms have hindered the flow of methodological advances from pangenomics into genome assembly. In genome assembly, edges typically indicate overlaps between sequences, with the overlapping sequence expressed redundantly on both nodes. In pangenomics, edges indicate adjacency between sequences with no overlap—often called *blunt* adjacencies. Algorithms and software developed for blunt sequence graphs often do not generalize to overlapped sequence graphs. This effectively silos pangenomics methods that could otherwise benefit genome assembly. In this paper, we attempt to dismantle this silo. We have developed an algorithm that transforms an overlapped sequence graph into a blunt sequence graph that preserves walks from the original graph. Moreover, the algorithm accomplishes this while also eliminating most of the redundant representation of sequence in the overlap graph. The algorithm is available as a software tool, GetBlunted, which uses little enough time and memory to virtually guarantee that it will not be a bottleneck in any genome assembly pipeline.

Keywords: Genome assembly · Graph genome · Pangenomics

1 Introduction

Genome assembly is the process of determining a sample's full genome sequence from the error-prone, fragmentary sequences produced by DNA sequencing technologies. Sequence graphs have a long history of use in this field [16,17,20]. In these graphs, nodes are labeled with sequences derived from sequencing data, and edges indicate overlaps between observed sequences, which may in turn indicate adjacency in the sample's genome (Fig. 1A). The sample genome then

J. M. Eizenga and R. Lorig-Roach—Contributed equally.

L. De Mol et al. (Eds.): CiE 2021, LNCS 12813, pp. 169–177, 2021.
https://doi.org/10.1007/978-3-030-80049-9_15

corresponds to some walk through graph. There are several specific sequence graph articulations in wide use, including de Bruijn graphs, overlap graphs, and string graphs. They each present computational and informational trade-offs that make them better suited to certain configurations of sequencing technologies and genome complexity.

The common topological features of genome assembly graphs are driven primarily by the repetitiveness of the underlying genomes. In many species, a large fraction of the genome consists of repeats (for instance, more than 50% of the human genome [11]). Because all copies of a repeat are highly similar to each other, the corresponding nodes in the sequence graph frequently overlap each other. In contrast, the unique regions of the genome have few erroneous overlaps. These two factors tend to create graphs that consist of long non-branching paths (corresponding to the unique regions), which meet in a densely tangled core with a complicated topology (corresponding to the repeats).

Recently, sequence graphs have also emerged into prominence in the growing field of pangenomics, which seeks to analyze the full genomes of many individuals from the same species [4]. In pangenomics, sequence graphs are used to represent genomic variation between individual haplotypes. Sequences in the graph furcate and rejoin around sites of variation so that each individual genome corresponds to a walk through the graph (Fig. 1B). The growth of pangenomics has fueled major advances in both formal algorithms research [12,21] and practical genomics tools [10,22].

Pangenome graphs have much simpler topologies than genome assembly graphs. Having fuller knowledge of the constituent genomes makes it possible to distinguish different copies of a repeat. Thus, pangenome graphs tend to be mostly non-branching, much like the portions of assembly graphs that correspond to unique sequences in the genome. Moreover, most of the branching in pangenome graphs consists of localized bubble-like motifs. In contrast to assembly graphs, pangenome graphs have few if any cycles.

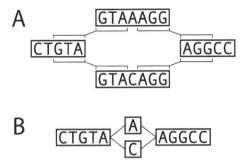

Fig. 1. A: An overlapped sequence graph. **B**: A blunt sequence graph.

Intuitively, the shared basis in sequence graphs should permit the advances in pangenomics to spill over into genome assembly. However, such cross-pollination is stymied by a small difference in the graph formalisms. The edges in assembly graphs indicate sequence overlaps, which are necessary because of the uncertain adjacencies in the underlying genome. In pangenome graphs, the underlying genomes are known, and the edges are *blunt* in that they indicate direct adjacency with no overlap. Blunt sequence graphs can be trivially converted into overlap graphs (with overlaps of length 0), but the reverse requires nontrivial merging operations between the overlapping sequences. As a result, methods have remained siloed within pangenomics despite potential uses in genome assembly.

In this work, we present a method to transform an overlapped sequence graph into a blunt sequence graph. We state the formal guarantees of our formulation and discuss their computational complexity. We then present an algorithm and compare its results to similar methods.

2   Problem Statement

In transforming an overlapped sequence graph to a blunt one, we seek to provide two guarantees:

1. All walks in the overlapped graph are preserved in the blunt graph.
2. Every walk in the blunt graph corresponds to some walk in the overlapped graph.

These two properties prohibit the intuitive solution of transitively merging all overlapped sequences. Doing so can result in walks that are not present in the overlapped graph, because walks can transition between nodes that are not connected by an edge via the transitively merged sequences (Fig. 2). Because overlapped sequences cannot be fully merged, it is necessary to retain multiple copies of some sequences in the blunt graph. However, excessive duplication can create problems for downstream analysis, for instance by increasing alignment uncertainty. Thus, we add one further criterion to the above formulation:

3. Minimize the amount of duplicated sequence.

3   Notation

An overlapped sequence graph consists of a set of sequences S and a set of overlaps $O \subset (S \times \{+, -\} \times S \times \{+, -\})$. In this notation, the symbols $+$ and $-$ indicate whether the overlap involves a prefix or suffix (collectively *affix*) of the sequence. This makes the overlapped graph a bidirected graph.

In a bidirected graph, a *walk* consists of a sequence of nodes $s_1 s_2 \ldots s_N$, $s_i \in S$ such that 1) each pair of subsequent nodes is connected by an overlap and 2) if s_{i-1} and s_i are connected by an overlap on s_i's prefix, then s_i and s_{i+1} are a connected by an overlap on s_i's suffix (or vice versa). In the case that a walk traverses a node $s \in S$ from suffix to prefix, we interpret the sequence as

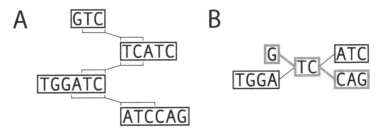

Fig. 2. A: An overlapped sequence graph, and **B**: the blunt sequence graph that results from transitively merging its overlaps. The highlighted walk in the blunt graph does not correspond to any walk in the original overlapped graph.

its *reverse complement*, which is the sequence of the antiparallel strand of the DNA molecule.

Finally, an *adjacency component* is a collection of affixes (in $S \times \{+, -\}$) that can reach each other via a sequence of adjacent overlaps in O (Fig. 3). This sequence need not form a valid bidirected walk.

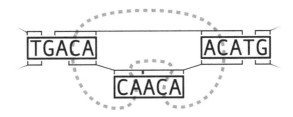

Fig. 3. An adjacency component in a larger sequence graph. Each of the indicated affixes can reach the others by a sequence of overlaps.

4   Methods

To minimize the amount of duplicated sequence, overlapped sequences must be merged. However, we have already mentioned that our criteria prohibit transitively merging all overlaps. We must then minimize the total number of groups within which overlaps are merged transitively, which coincides with the number of times the sequences need to be duplicated.

Consider a group of overlaps that contains $(s_1, s_2, +, -)$ and $(t_1, t_2, +, -)$. For merging to not introduce any walks that are not in the overlapped graph, the overlaps $(s_1, t_2, +, -)$ and $(t_1, s_2, +, -)$ must also be overlaps in O. Extending this logic, the entire group of overlaps must be contained within a *biclique* subgraph of the adjacency component: two sets of affixes B_1 and B_2 such that every affix in B_1 is connected to every affix in B_2 by an overlap. Thus, we can minimize the number of duplicated sequences by minimizing the number of bicliques needed to cover every overlap edge.

The problem of covering edges with the minimum number of bicliques is known as *biclique cover* (Fig. 4), and it is known to be NP-hard [19]. However, there are domain-specific features of overlapped sequence graphs that often make it tractable to solve large portions of the graph optimally.

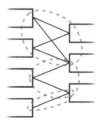

Fig. 4. A biclique cover of an adjacency component with three bicliques. A biclique cover of an adjacency component with three bicliques.

First, many adjacency components are bipartite. Consider the case that an adjacency component is not bipartite, in which case there is cycle of overlaps between affixes with odd parity. Each overlap indicates high sequence similarity, so an odd cycle means that each sequence is similar to itself, reverse complemented an odd number of times. Such sequences are called DNA palindromes, and they do exist in nature. However, they comprise a small fraction of most real genomes.

Second, most adjacency components are *domino-free*. This property refers to the absence of a particular induced subgraph, the *domino* (Fig. 5). A sufficient condition to prohibit dominoes is for overlapping to be a transitive property. That is, whenever sequence s_1 overlaps sequences t_1 and t_2, and sequence s_2 overlaps t_1, then s_2 also overlaps t_2. In reality, this is not always the case. However, it is very often the case, since overlaps indicate sequence similarity, and similarity is approximately transitive.

These features guided the design of the following algorithm. If an adjacency component is bipartite and domino-free, we compute the biclique cover in polynomial time with the algorithm of Amilhastre, Vilaren, and Janssen [1]. When an adjacency component is bipartite but not domino-free, we instead use the dual graph reduction algorithm of Ene, et al. [7], followed by their lattice-based

Fig. 5. The domino graph. If either of the dotted edges are present, the induced subgraph is not a domino.

post-processing if the algorithm does not identify the optimal solution. Finally, if an adjacency component is not bipartite, we first reduce it to the bipartite case by computing an approximate solution to the maximum bipartite subgraph problem using the algorithm of Bylka, Idzik, and Tuza [3]. The maximum bipartite subgraph problem is equivalent to max cut, which is also NP-hard [13]. This process is repeated recursively on the edges that are not included in the bipartite subgraph.

The amount of duplicated sequence is also affected by the manner in which sequences are merged among the overlaps of a biclique. To minimize duplicated sequence, we must maximize matches in the alignment between the overlapped sequences. This is the multiple sequence alignment problem, which is NP-hard. We use the partial order alignment algorithm to approximate the optimal multiple sequence alignment [14]. Partial order alignment also has the advantage that the alignment is expressed as a blunt sequence graph, which can be directly incorporated in the full blunt graph.

5   Implementation

We have implemented the algorithm described here as a genomics tool called Get-Blunted. GetBlunted takes as input a GFA file (a common interchange format for sequence graphs [15]) and outputs a GFA containing a blunt graph. In addition, it provides a translation table from sequences in the output to sequences in the input, which can be used to translate analyses performed on the blunt graph into analyses on the overlapped graph. The implementation is written entirely in C++, and it use several auxiliary libraries: GFAKludge is used for manipulating GFA files [5], libbdsg is used to represent sequence graphs [6], and SPOA is used for partial order alignment [24].

6   Results

We compared the performance of GetBlunted to two other tools that transform overlapped sequence graphs into blunt graphs: the gimbricate/seqwish [8,9] pipeline and Stark [18]. These are, to our knowledge, the only other such tools besides GetBlunted. However, they are not completely comparable. Neither tool provides the guarantees that GetBlunted does for preserving the walk space of the graph. In addition, Stark only works with de Bruijn graphs, a restricted subset of overlap graphs in which all overlaps are exact matches of a uniform length.

We profiled speed and memory usage on three assembly graphs. The first two are assembly graphs produced by the Shasta assembler [23] for the haploid human cell line CHM13 and for human sample HG002. Both of these were built using Oxford Nanopore reads[1]. The last graph is a de Bruijn graph of Pacific

---

[1] Publicly available at https://s3-us-west-2.amazonaws.com/miten-hg002/index.html?prefix=guppy_3.6.0/.

Biosciences HiFi reads of an *Escherichia coli* strain (SRR10382245), which was constructed using jumboDB [2].

All of the bluntifying tools were run on a single core of a c5.9xlarge AWS instance with an Intel Xeon Scalable Processor. Memory usage and compute time were measured with the Unix time tool. The results of the profiling are presented in Table 1. GetBlunted is over 1000 times faster than and comparably memory-intensive to the gimbricate/seqwish pipeline. For de Bruijn graphs, Stark is faster than either tool, although this performance comes at the cost of limited generality.

Table 1. Table of speed and memory usage of bluntifing tools run on a single core of an AWS server.

Assembly	Bluntification tool	Run time (min)	RAM (GB)
HG002 Shasta	GetBlunted	0.35	9
	Gimbricate/seqwish	917.5	6
CHM13 Shasta	GetBlunted	0.38	4
	Gimbricate/seqwish	314.6	6
E. coli de Bruijn	GetBlunted	8.36	26
	Gimbricate/seqwish	10.74	4
	Stark	0.65	3

7   Discussion

In this work, we described an algorithm and software tool, GetBlunted, which transforms overlapped sequence graphs into blunt sequence graphs. This provides a route for sequence graph methods developed for pangenomics to be applied to sequence graphs in genome assembly. In both fields, walks through the sequence graph are of primary importance. In genome assembly, some walk through the graph corresponds to the sample genome. In pangenomics, the genomes used to construct the pangenome each correspond to a walk through the graph. GetBlunted provides attractive guarantees that it faithfully preserves the walk space of the input while also producing parsimonious output. Other comparable methods either do not provide these guarantees or only provide them in limited cases. In addition, GetBlunted is (except in the case of de Bruijn graphs) faster than alternatives that do not provide these guarantees, and it has resource requirements that are easily achievable in any computational environment that is used for genome assembly. In the future, GetBlunted could serve as an step in genome assembly pipelines to improve the quality of their overlap graphs. It could also facilitate direct analyses of assembly graphs in metagenomics applications.

References

1. Amilhastre, J., Vilarem, M.C., Janssen, P.: Complexity of minimum biclique cover and minimum biclique decomposition for bipartite domino-free graphs. Discret. Appl. Math. **86**(2–3), 125–144 (1998)
2. Bankevich, A., Bzikadze, A., Kolmogorov, M., Pevzner, P.A.: Assembling Long Accurate Reads Using de Bruijn Graphs. bioRxiv p. 2020.12.10.420448 (December 2020). https://doi.org/10.1101/2020.12.10.420448, https://www.biorxiv.org/content/10.1101/2020.12.10.420448v1, publisher: Cold Spring Harbor Laboratory Section: New Results
3. Bylka, S., Idzik, A., Tuza, Z.: Maximum cuts: improvements and local algorithmic analogues of the Edwards-Erdos inequality. Discret. Math. **194**(1–3), 39–58 (1999)
4. Computational Pan-Genomics Consortium: Computational pan-genomics: status, promises and challenges. Brief. Bioinform. **19**(1), 118–135 (2018)
5. Dawson, E.T., Durbin, R.: GFAKluge: A C++ library and command line utilities for the graphical fragment assembly formats. J. Open Source Softw. **4**(33), 1083 (2019)
6. Eizenga, J.M., et al.: Efficient dynamic variation graphs. Bioinformatics **36**(21), 5139–5144 (2020)
7. Ene, A., Horne, W., Milosavljevic, N., Rao, P., Schreiber, R., Tarjan, R.E.: Fast exact and heuristic methods for role minimization problems. In: Proceedings of the 13th ACM Symposium on Access Control Models and Technologies, pp. 1–10 (2008)
8. Garrison, E.: ekg/gimbricate. https://github.com/ekg/gimbricate (October 2020)
9. Garrison, E.: ekg/seqwish. https://github.com/ekg/seqwish (February 2021)
10. Garrison, E., et al.: Variation graph toolkit improves read mapping by representing genetic variation in the reference. Nat. Biotechnol. **36**(9), 875–879 (2018)
11. Haubold, B., Wiehe, T.: How repetitive are genomes? BMC Bioinform. **7**(1), 1–10 (2006)
12. Jain, C., Zhang, H., Gao, Y., Aluru, S.: On the complexity of sequence to graph alignment. bioRxiv (January 2019). https://doi.org/10.1101/522912
13. Karp, R.M.: Reducibility among combinatorial problems. In: Miller R.E., Thatcher J.W., Bohlinger J.D. (eds.) Complexity of Computer Computations, pp. 85–103. Springer, Boston (1972) https://doi.org/10.1007/978-1-4684-2001-2_9
14. Lee, C., Grasso, C., Sharlow, M.F.: Multiple sequence alignment using partial order graphs. Bioinformatics **18**(3), 452–464 (2002)
15. Li, H., et al.: GFA specification (2013). https://github.com/GFA-spec/GFA-spec
16. Myers, E.W.: Toward simplifying and accurately formulating fragment assembly. J. Comput. Biol. **2**(2), 275–290 (1995)
17. Myers, E.W.: The fragment assembly string graph. Bioinformatics **21**(suppl 2), ii79–ii85 (2005)
18. Nikaein, H.: hnikaein/stark (January 2021). https://github.com/hnikaein/stark
19. Orlin, J., et al.: Contentment in graph theory: covering graphs with cliques. In: Indagationes Mathematicae (Proceedings), vol. 80, pp. 406–424. North-Holland (1977)
20. Pevzner, P.A., Tang, H., Waterman, M.S.: An Eulerian path approach to DNA fragment assembly. Proc. Natl. Acad. Sci. **98**(17), 9748–9753 (2001)
21. Rautiainen, M., Marschall, T.: Aligning sequences to general graphs in O(V+ mE) time. bioRxiv p. 216127 (2017)

22. Rautiainen, M., Marschall, T.: GraphAligner: rapid and versatile sequence-to-graph alignment. Genome Biol. **21**(1), 1–28 (2020)
23. Shafin, K., et al.: Nanopore sequencing and the Shasta toolkit enable efficient de novo assembly of eleven human genomes. Nature Biotechnology **38**(9), 1044–1053 (2020). https://doi.org/10.1038/s41587-020-0503-6, https://www.nature.com/articles/s41587-020-0503-6. number: 9 Publisher: Nature Publishing Group
24. Vaser, R., Sović, I., Nagarajan, N., Šikić, M.: Fast and accurate de novo genome assembly from long uncorrected reads. Genome Res. **27**(5), 737–746 (2017)

On 3SUM-hard Problems in the Decision Tree Model

Esther Ezra[✉][iD]

School of Computer Science, Bar Ilan University, Ramat Gan, Israel
ezraest@cs.biu.ac.il

Abstract. We describe subquadratic algorithms, in the algebraic decision-tree model of computation, for detecting whether there exists a triple of points, belonging to three respective sets A, B, and C of points in the plane, that satisfy a pair of polynomial equations. In particular, this has an application to detect collinearity among three sets A, B, C of n points each, in the complex plane, when each of the sets A, B, C lies on some constant-degree algebraic curve. In another development, we present a subquadratic algorithm, in the algebraic decision-tree model, for the following problem: Given a pair of sets A, B each consisting of n pairwise disjoint line segments in the plane, and a third set C of arbitrary line segments in the plane, determine whether $A \times B \times C$ contains a triple of concurrent segments. This is one of four 3SUM-hard geometric problems recently studied by Chan (2020). The results reported in this extended abstract are based on the recent studies of the author with Aronov and Sharir (2020, 2021).

Keywords: 3SUM-hard problems · Algebraic decision tree model · Collinearity testing · Segment concurrency

1 Introduction

In theoretical computer science, analysis of algorithms in non-uniform models is often applied when one is interested in optimizing the number of certain operations, while the remaining operations performed by the algorithm are disregarded in the complexity analysis.

A central problem, which received considerable attention due to its relation to conditional lower bounds on the complexity of fundamental geometric questions, is 3SUM, namely, given a set of n real numbers, decide whether there is a triple of them that sums to 0. We sometimes refer to the trichromatic version of 3SUM, where we are given three sets A, B, C of n real numbers each, and the question is to determine whether there is a triple $(a, b, c) \in A \times B \times C$ with $a + b + c = 0$.

There is a large family of geometric problems that are known to be 3SUM-hard, in the sense that 3SUM can be reduced to them. In fact, Gajentaan and

Work partially supported by NSF CAREER under grant CCF:AF-1553354 and by Grant 824/17 from the Israel Science Foundation.

L. De Mol et al. (Eds.): CiE 2021, LNCS 12813, pp. 178–188, 2021.
https://doi.org/10.1007/978-3-030-80049-9_16

Overmars [15], who introduced this concept, initially coined these problems "n^2-hard," as it was strongly believed that 3SUM cannot be solved in sub-quadratic time (it can easily be solved in $O(n^2)$ time). Moreover, Erickson [11] showed a matching quadratic lower bound in the *3-linear decision-tree model* (see also [4]),namely, in the *linear decision-tree model* the only operation allowed to directly access the input data is a sign test of a linear function of the input numbers; these are the operations counted by the model, while all the remaining operations are free. In the *3-linear* model, the linear functions are restricted to three parameters, in particular, Erickson [11] used comparisons of the form: "Is $a + b$ greater/smaller/equal to c?"

As a consequence, it was conjectured that there is no subquadratic solution. However, this prevailing conjecture was refuted by Grønlund and Pettie [16], who showed an upper bound close to $O(n^{3/2})$ in the *4-linear* decision-tree model (using the so-called *Fredman's trick* - see Sect. 2 for this notion in geometric problems), and a slightly subquadratic algorithm in the RAM model. This pioneering work raised the interest of researchers from the entire theoretical computer science community. Soon afterwards, the more general question of k-SUM (deciding whether there is a k-tuple of the input numbers which sum to 0) was studied by Cardinal *et al.* [8], Kane *et al.* [19] and by the author and collaborators [13, 14]. The main idea is to reduce the problem to *point location* in high-dimensional "hyperplane arrangements," where the challenge is to bring down the query time, so that its dependence on the dimension d is polynomial rather than exponential (see also [20, 21]). In fact, the work of the author in [13] presents a point-location mechanism for arbitrary hyperplanes in the RAM model (whereas the work in [14] considered this problem only in the linear decision-tree model). Very recently Hopkins *et al.* [18] showed an improvement for such a mechanism, yielding an almost optimal bound in the linear decision-tree model. The approach in [18, 19] uses the concept of *inference* adapted from active learning. *Inference dimension* refers to the situation where one considers linear comparisons on a set of items, and the quantity to be measured is, roughly speaking, what is the smallest number of comparisons from which one can deduce the result of another comparison. A key property in the analysis in [19], which eventually led to the near-linear bound for the k-SUM problem, is that the inference dimension is only linear in n in this case.

The seminal work of Gajentaan and Overmars [15] presents a fairly long list of geometric 3SUM-hard problems in two and three dimensions (they all have a simple reduction from 3SUM), such as: *(i) Collinearity testing*: Given a set of points in $\mathbb{R}^2$, do they contain a collinear triple? *(ii) Covering by strips*: Given strips in the plane, does their union cover the unit square? *(iii) Separator problems*: Given a set of non-intersecting axis-parallel line segments in $\mathbb{R}^2$, is there a line separating them into two non-empty sets? Among geometric 3SUM-hard problems, collinearity testing is perhaps the most fundamental one, since many such problems are intrinsically "collinearity-hard". In this extended abstract we refer to the trichromatic version of collinearity testing, which is described next.

In the sequel we focus on two main developments concerning collinearity testing in the *algebraic decision tree model*, that is, each comparison is a sign test of some constant-degree polynomial in the coordinates of a constant number of input points. In the first we consider the trichromatic problem in the complex plane, where the input consists of three sets of points A, B, C, each lying on some constant-degree algebraic (complex) curve. We in fact present a more general scheme, for detecting whether there exists a triple of points, belonging to three respective sets A, B, and C of points in the plane, that satisfy two polynomial equations, and show a subquadratic bound in the algebraic decision tree model. These results are briefly described in Sect. 2; we refer the reader to [5] for further details.[1]

In the second development we study the *segment concurrency* problem in the algebraic decision tree model: Given a pair of sets A, B each consisting of n pairwise disjoint line segments in the plane, and a third set C of arbitrary line segments in the plane, determine whether $A \times B \times C$ contains a triple of concurrent segments. When A, B, C are three sets of lines in the plane (rather than line segments as above), the problem is exactly the dual to collinearity testing, which is 3SUM-hard. In fact, our restricted setting is still 3SUM-hard, and is among four such problems studied by Chan [9], all of which can be reduced to the problem of *triangle intersection counting*: That is, given two sets A and B, each consisting of n pairwise disjoint line segments in the plane, and a set C of n triangles in the plane, count, for each triangle $\Delta \in C$, the number of intersection points between the segments of A and those of B that lie inside Δ. The other two problems are: *(i) Intersection of three polygons:* Given three simple n-gons A, B, C in the plane, determine whether $A \cap B \cap C$ is nonempty. *(ii) Coverage by three polygons:* Given three simple n-gons A, B, C in the plane, determine whether $A \cup B \cup C$ covers a given triangle Δ_0. The property that these problems are 3SUM-hard, as well as the reduction to the problem of triangle intersection counting, are described in [9].

In Sect. 3 we briefly describe a subquadratic solution to the segment concurrency problem in the algebraic decision tree model. This is part of the work in progress of the author with Aronov and Sharir [6] where they presented a subquadratic solution to the triangle intersection counting problem in the algebraic decision tree model.

2   Testing a Pair of Polynomial Equations and Collinearity Testing in the Complex Plane

For simplicity of presentation, we present a solution sketch for the following problem: We are given three sets A, B, and C, each consisting of n points in the plane, and we seek a triple $(a, b, c) \in A \times B \times C$ that satisfies two polynomial

---

[1] The work in [5] also shows how to detect whether A, B, C satisfy a single polynomial equation under the condition that two of the sets lie on two respective one-dimensional curves and the third is placed arbitrarily in the plane. We do not report this particular development in this extended abstract.

equations. We assume that they are of the form $c_1 = F(a, b)$, $c_2 = G(a, b)$, for $c = (c_1, c_2)$, where F and G are constant-degree 4-variate polynomials with *good fibers*, in the following sense: For any pair of real numbers κ_1, κ_2, the two-dimensional surface $\pi_{(\kappa_1, \kappa_2)} := \{(a, b) \in \mathbb{R}^4 \mid F(a, b) = \kappa_1, \ G(a, b) = \kappa_2\}$ has good fibers[2].

Polynomial Partitioning. Our analysis relies on planar polynomial partitioning and on properties of Cartesian products of pairs of them. For a polynomial $f \colon \mathbb{R}^d \to \mathbb{R}$, for any $d \geq 2$, the *zero set* of f is $Z(f) := \{x \in \mathbb{R}^d \mid f(x) = 0\}$. We refer to an open connected component of $\mathbb{R}^d \setminus Z(f)$ as a *cell*. A fundamental result of Guth and Katz [17] is:

Proposition 1 (Polynomial partitioning [17]). *Let P be a finite set of points in $\mathbb{R}^d$, for any $d \geq 2$. For any real parameter D with $1 \leq D \leq |P|^{1/d}$, there exists a real d-variate polynomial f of degree $O(D)$ such that $\mathbb{R}^d \setminus Z(f)$ has $O(D^d)$ cells, each containing at most $|P|/D^d$ points of P.*

Agarwal *et al.* [3] presented an algorithm that computes a polynomial partitioning in expected $O(n\mathrm{poly}(D))$ time, whose degree is within a constant factor of that stated in Proposition 1.

We now proceed as follows. We apply a "block" partitioning for the points in A and in B, using Cartesian products of pairs polynomial partitions, based on the analysis of Solymosi and de Zeeuw [22]. Roughly speaking, we fix a parameter $g \ll n$ (whose value will be set later), and use Proposition 1 in order to form a polynomial partitioning of degree $D = O(\sqrt{n/g})$ for each of the sets A, B. Each connected component in the partition of A (resp., B) contains at most g point of A (resp.,., B). We then form the Cartesian product of the partitioning of A and B. Let ζ denote a cell in this partition, this cell is the Cartesian product $\tau \times \tau'$ of a cell τ from the partition of A and a cell τ' from the partition of B. Since $|A \cap \tau|, |B \cap \tau'| \leq g$, we have that ζ contains at most g^2 points of $A \times B$, and the overall number of cells in this partition is $O((n/g)^{2+\varepsilon})$, for any prescribed $\varepsilon > 0$.[3]

Put $A_\tau := A \cap \tau$ and $B_{\tau'} := B \cap \tau'$. The high-level idea of the algorithm is to sort lexicographically each of the sets $H_{\tau, \tau'} := \{(F(a, b), G(a, b)) \mid (a, b) \in A_\tau \times B_{\tau'}\}$, over all pairs of cells (τ, τ'). We then search with each $c = (c_1, c_2) \in C$ through the sorted lists of those sets $H_{\tau, \tau'}$ that might contain (c_1, c_2). We show that each $c \in C$ has to be searched for in only a small number of sets. Typically to this kind of problems [16], sorting the sets explicitly is too expensive. We overcome this issue by considering the problem in the algebraic decision-tree model, and by using an algebraic variant of *Fredman's trick* (extending those used in [7,16]).

---

[2] A two-dimensional algebraic surface S in $\mathbb{R}^4$ has *good fibers* if, for every point $p \in \mathbb{R}^2$, the fibers $(\{p\} \times \mathbb{R}^2) \cap S$ and $(\mathbb{R}^2 \times \{p\}) \cap S$ are finite.

[3] The number of cells is in fact $O(n/g)^2)$, but the analysis in [5] uses *hierarchical polynomial partitioning* in order to speed up computation, which slightly increases the number of cells to $O((n/g)^{2+\varepsilon})$. We skip this variant in this extended abstract.

Implicit Sorting and Batched Point Location. Consider the step of sorting $\{F(a,b) \mid (a,b) \in A_\tau \times B_{\tau'}\}$ (the sorting of the values $G(a,b)$ is done in a secondary round and is treated analogously). It has to perform various comparisons of pairs of values $F(a,b)$ and $F(a',b')$, for $a,a' \in A_\tau$, $b,b' \in B_{\tau'}$. We consider $A_\tau \times A_\tau$ as a set of g^2 points in $\mathbb{R}^4$, and associate, with each pair $(b,b') \in B_{\tau'} \times B_{\tau'}$, the 3-surface $\sigma_{b,b'} = \{(a,a') \in \mathbb{R}^4 \mid F(a,b) = F(a',b')\}$. Let $\Sigma_{\tau'}$ denote the set of these surfaces. The arrangement $\mathcal{A}(\Sigma_{\tau'})$ has the property that each of its cells ζ has a fixed sign pattern with respect to all these surfaces. That is, each comparison of $F(a,b)$ with $F(a',b')$, for any (a,b), $(a',b') \in A_\tau \times B_{\tau'}$, has a fixed outcome for all points $(a,a') \in \zeta$ (for a fixed pair b,b'). In other words, if we locate the points of $A_\tau \times A_\tau$ in $\mathcal{A}(\Sigma_{\tau'})$, we have available the outcome of all the comparisons needed to sort the set $\{F(a,b) \mid (a,b) \in A_\tau \times B_{\tau'}\}$

Following the above steps is still too expensive, and takes $\Omega(n^2)$ steps (in the algebraic decision-tree model) if implemented naïvely. We circumvent this issue, in the algebraic decision-tree model, by forming the unions $P := \bigcup_\tau A_\tau \times A_\tau$, and $\Sigma := \bigcup_{\tau'} \Sigma_{\tau'}$; we have $|P|, |\Sigma| = O(g^2 \cdot (n/g)^{1+\varepsilon}) = O(n^{1+\varepsilon} g^{1-\varepsilon})$. By locating each point of P in $\mathcal{A}(\Sigma)$, we get all the signs that are needed to sort all the sets $\{F(a,b) \mid (a,b) \in A_\tau \times B_{\tau'}\}$, over all pairs τ, τ' of cells, and then the actual sorting costs nothing in algebraic decision tree model. In a main step in [5] we show:

Lemma 1. *One can complete the above sorting step, in the algebraic decision tree model, in $O\left((ng)^{8/5+\varepsilon}\right)$ randomized expected time, for any prescribed $\varepsilon > 0$, where the constant of proportionality depends on ε and the degree of F and G.*

Searching with the Points of C. We next search the structure with every $c = (c_1, c_2) \in C$. We only want to visit subproblems (τ, τ') where there might exist $a \in \tau$ and $b \in \tau'$, such that $F(a,b) = c_1$ and $G(a,b) = c_2$. To find these cells, and to bound their number, we consider the two-dimensional surface $\pi_{c=(c_1,c_2)} := \{(a,b) \in \mathbb{R}^4 \mid F(a,b) = c_1, G(a,b) = c_2\}$, and our goal is to enumerate the cells $\tau \times \tau'$ in the polynomial partition of $A \times B$ crossed by π_c. By assumption, π_c has good fibers, so, by [5, Theorem 3.2], it crosses only $O((n/g)^{1+\varepsilon})$ cells $\tau \times \tau'$, and we can compute them in time $O((n/g)^{1+\varepsilon})$, for any $\varepsilon > 0$ (see [5] for these details).

Summing over all the n possible values of c, the number of crossings between the surfaces π_c and the cells $\tau \times \tau'$ is $O(n^{2+\varepsilon}/g^{1+\varepsilon})$, for any $\varepsilon > 0$. Thus computing all such surface-cell crossings, over all $c \in C$, costs $O(n^{2+\varepsilon}/g^{1+\varepsilon})$ time. The cost of searching with any specific c, in the structure of a cell $\tau \times \tau'$ crossed by π_c, is $O(\log g)$ (it is simply a binary search over the sorted lists). Hence the overall cost of searching with the elements of C through the structure is (with a slightly larger ε) $O\left(\dfrac{n^{2+\varepsilon}}{g^{1+\varepsilon}}\right)$.

The Overall Algorithm. Combining the above costs we get total expected running time of $O\left((ng)^{8/5+\varepsilon} + \dfrac{n^{2+\varepsilon}}{g^{1+\varepsilon}}\right)$. We now choose $g = n^{2/13}$, and obtain expected running time of $O\left(n^{24/13+\varepsilon}\right)$, where the implied constant of proportionality depends on the degrees of F and G and on ε. In summary, we have shown:

Theorem 1 (Aronov et al. [5]). *Let A, B, C be three n-point sets in the plane, and let F, G be a pair of constant-degree 4-variate polynomials with good fibers (in the sense defined at the beginning of this section). Then one can test, in the algebraic decision-tree model, whether there exists a triple $a \in A$, $b \in B$, $c = (c_1, c_2) \in C$, such that $c_1 = F(a, b)$ and $c_2 = G(a, b)$, using only $O\left(n^{24/13+\varepsilon}\right)$ polynomial sign tests (in expectation), for any $\varepsilon > 0$.*

Collinearity Testing in the Complex Plane. In a further development in the Arxiv version of [5], the author with Aronov and Sharir extend Theorem 1 and show that one can test in the same asymptotic time bound (in the algebraic decision tree model) whether there exists a triple $a \in A$, $b \in B$, $c \in C$, such that $F(a, b, c) = 0$, $G(a, b, c) = 0$, where now F, G are algebraic functions that satisfy some mild assumptions. Generally speaking, this involves a non-trivial and technical procedure for sorting roots of polynomials in the algebraic decision tree model.

We next apply this property to the problem of *collinearity testing* in the complex plane, for the following setup. The sets A, B, C are now sets of points in the complex plane $\mathbb{C}^2$, each consisting of n points and lying on a constant-degree algebraic curve, and we wish to determine whether $A \times B \times C$ contains a collinear triple. For simplicity of exposition, we assume that the curves γ_A, γ_B, and γ_C that contain, respectively, A, B, and C are represented parametrically by equations of the form $(w, z) = (f_A(t), g_A(t))$, $(w, z) = (f_B(t), g_B(t))$, and $(w, z) = (f_C(t), g_C(t))$, where t is a complex parameter and f_A, g_A, f_B, g_B, f_C, and g_C are constant-degree univariate (complex) polynomials.

With this parameterization the points of A, B, C can be represented as points in the real plane (representing the complex numbers t), and the complex polynomial whose *vanishing* (that is, once it is compared to zero) asserts collinearity of a triple $a = (z_a, w_a)$, $b = (z_b, w_b)$, $c = (z_c, w_c)$, is

$$H(t_a, t_b, t_c) := \begin{vmatrix} 1 & f_A(t_a) & g_A(t_a) \\ 1 & f_B(t_b) & g_B(t_b) \\ 1 & f_C(t_c) & g_C(t_c) \end{vmatrix}, \tag{1}$$

where t_a, t_b, t_c are the parameters that specify a, b, c, respectively, so its real and imaginary components form a pair of real polynomial equations. This is the role assignment of the above polynomials F and G. In summary this shows (once again, refer to the Arxiv version of [5]):

Corollary 1 (Aronov et al. [5]). *Let A, B, C be n-point sets in the complex zw-plane, so that A (resp., B, C) lies on a curve γ_A (resp., γ_B, γ_C) represented by parametric equations of the form $(z, w) = (f_A(t), g_A(t))$ (resp., $(z, w) = (f_B(t), g_B(t))$, $(z, w) = (f_C(t), g_C(t))$), where f_A, g_A, f_B, g_B, f_C,*

g_C are constant-degree univariate complex polynomials. Then one can determine, in the algebraic decision-tree model, whether there exists a collinear triple $(a, b, c) \in A \times B \times C$, with $O\left(n^{24/13+\varepsilon}\right)$ real polynomial sign tests, in expectation, for any $\varepsilon > 0$.

3 Segment Concurrency

We now sketch a subquadratic algorithm, in the algebraic decision tree model, for the segment concurrency problem. We use the notation of Sect. 1.

The Decomposition. Fix a parameter $g \ll n$, put $r := n/g$. We construct a $(1/r)$-cutting Ξ_A for the segments of A, that is, a partition of the plane into interior-disjoint simplices, the interior of each of which meets at most $n/r = g$ segments of A. Since the segments are pairwise disjoint, we can construct Ξ_A so that it consists of only $O(r)$ trapezoids, each of which is crossed by at most g segments of A. The construction time, in the real-RAM model, is $O(n \log r) = O(n \log n)$; see [10, Theorem 1] for these details. We apply a similar construction for B, and let Ξ_B denote the resulting cutting, with similar properties.

We next overlay Ξ_A with Ξ_B, to obtain a decomposition Ξ of the plane into $O(r^2)$ convex polygons of constant complexity. Each cell σ of Ξ is identified by a pair (τ, τ'), where τ and τ' are the respective cells (simplices) of Ξ_A and Ξ_B whose intersection is σ. Each cell σ of Ξ is crossed by at most $n/r = g$ segments of A and by at most $n/r = g$ segments of B.

Classifying the Segments in a Cell. Let $\sigma = (\tau, \tau')$ be a cell of Ξ. Call a segment e of A *long* (resp., *short*) within σ if e crosses σ and neither of its endpoints lies in σ (resp., at least one endpoint lies in σ). We apply analogous definitions to the segments of B and to the segments of C.

The high-level structure of the algorithm proceeds as follows. We construct Ξ_A and Ξ_B. For each simplex τ of Ξ_A (resp., τ' of Ξ_B), we compute its *conflict list* A_τ (resp., $B_{\tau'}$), which is the set of all segments of A that cross τ (resp., segments of B that cross τ'). We then form the overlay Ξ, and for each of its cells $\sigma = (\tau, \tau')$, we compute the set A_σ of the segments of A_τ that cross σ, and the set B_σ of the segments of $B_{\tau'}$ that cross σ. We partition A_σ into the subsets of long and short segments (within σ), respectively, and apply an analogous partition to B_σ. We also trace each segment $c \in C$ through the cells of Ξ that it crosses, and form, for each cell σ of the overlay, the list C_σ of segments of C that cross σ, partitioned into the subsets of long and short segments. Using standard properties of $(1/r)$-cuttings, the overall complexity of these sets (of both long and short segments), over all cells σ, is $O(r^2 \cdot n/r) = O(nr) = O(n^2/g)$, and the additional overall cost of constructing them is $O(n^2 \log n/g)$.

In this extended abstract we only present the main steps of the analysis for triples in $A \times B \times C$ all of whose segments are long in a cell σ, the remaining cases (that is, when at least one of these segments is short in σ) are fairly standard and can be handled by the theory of *range search* [1]; these steps are presented [6].

Point Location in Planar Arrangements. Passing to the dual plane, the segments of A (resp., B, C), or rather the lines containing them, are mapped to points. We denote the set of points dual to the segments of A (resp., B, C) as A^* (resp., B^*, C^*). Our goal is to determine whether there exists a collinear triple $(a^*, b^*, c^*) \in A^* \times B^* \times C^*$, so that the corresponding primal segments a, b, c intersect each other (necessarily at the same common point).

We follow the scheme of Sect. 2, let $F(\xi, \eta, \zeta)$ be the quadratic 6-variate polynomial whose vanishing expresses collinearity of the three points $\xi, \eta, \zeta \in \mathbb{R}^2$. Ignoring C for the time being, we preprocess A and B into a data structure that we will then search with the points dual to the (lines containing the) segments of C. For each $a \in A$, $b \in B$, we define the line

$$\gamma_{a,b} = \{\zeta \in \mathbb{R}^2 \mid F(a^*, b^*, \zeta) = 0\},$$

which is the line passing through a^* and b^*. Let Γ_0 denote the set of these $O(n^2)$ lines. Our goal is to determine whether any point $c^* \in C^*$ lies on any of the lines $\gamma_{a,b}$, and then also make sure that the corresponding primal segments a, b, c intersect each other (it is possible that a, b, c are long in a cell σ but intersect outside σ; this scenario is handled using range search, as shown in [6]). This requires to preprocess the arrangement $\mathcal{A}(\Gamma_0)$ into a point-location data structure, and then search that structure with each $c^* \in C^*$.

Fredman's Trick and Batched Point Location. As above, a naïve implementation of this approach would be way too expensive. Instead, we return to the partitions Ξ_A, Ξ_B and Ξ, and iterate over all cells $\sigma = (\tau, \tau')$ of Ξ, defining

$$\Gamma_\sigma = \{\gamma_{a,b} \mid (a, b) \in A_\sigma \times B_\sigma\}.$$

In principle, we want to construct the separate arrangements $\mathcal{A}(\Gamma_\sigma)$, over the cells σ, preprocess each of them into a point-location structure, and search with each $c^* \in C^*$ in the structures that correspond to the cells of Ξ that c crosses. This is also too expensive if implemented naïvely. We circumvent it, using Fredman's trick, as follows.

Consider the step of constructing $\mathcal{A}(\Gamma_\sigma)$ for some fixed cell σ. We observe that it suffices to construct and sort the vertices of $\mathcal{A}(\Gamma_\sigma)$ in the x-direction, and also to sort the lines of Γ_σ at $x = -\infty$ (see [5,6] for details). The rest of the construction, including the preprocessing of the arrangement into an efficient point-location data structure, costs nothing in the algebraic decision-tree model, as it is based on a sweeping procedure on the lines of Γ_σ, and all the input-dependent data that the sweep requires has already been computed by the steps just mentioned, so the sweep does not have to access A or B explicitly; Searching the structure with any $c^* \in C^*_\sigma$ (that is, applying a point location query) takes $O(\log g)$ time, since Γ_σ consists of only g^2 lines.

Consider then the step of sorting the vertices of $\mathcal{A}(\Gamma_\sigma)$. In this step we need to compare the x-coordinates of pairs of these vertices. In general, such a comparison involves four pairs $(a_i, b_i) \in A_\sigma \times B_\sigma$, $i = 1, \ldots, 4$, where γ_{a_1, b_1} and γ_{a_2, b_2} intersect at one of the vertices and γ_{a_3, b_3} and γ_{a_4, b_4} intersect at the other vertex. Roughly speaking, such a comparison can be expressed as testing the sign of some constant-degree 16-variate polynomial $G(a_1, a_2, a_3, a_4; b_1, b_2, b_3, b_4)$.

We now apply Fredman's trick. We fix a cell τ of Ξ_A. For each quadruple $(a_1, a_2, a_3, a_4) \in A_\tau^4$, we define the surface

$$\psi_{a_1, a_2, a_3, a_4} = \{(b_1, b_2, b_3, b_4) \in \mathbb{R}^8 \mid G(a_1, a_2, a_3, a_4; b_1, b_2, b_3, b_4) = 0\},$$

and denote by Ψ the collection of these surfaces, over all cells τ. We have $|\Psi| = O((n/g) \cdot g^4) = O(ng^3)$. Similarly, we let P denote the set of all quadruples (b_1, b_2, b_3, b_4), for $b_1, b_2, b_3, b_4 \in B_{\tau'}^4$, over all cells τ' of Ξ_B. We have $|P| = O(ng^3)$.

We apply a batched point location procedure to the points of P and the surfaces of Ψ. The output of this procedure tells us the sign of $G(a_1, a_2, a_3, a_4; b_1, b_2, b_3, b_4)$, for every pair of quadruples $(a_1, a_2, a_3, a_4) \in A_\tau^4$, $(b_1, b_2, b_3, b_4) \in B_{\tau'}^4$, over all pairs of cells $(\tau, \tau') \in \Xi_A \times \Xi_B$, and these signs allow us to sort the vertices of the arrangements $\mathcal{A}(\Gamma_\sigma)$, for all cells σ of Ξ, at no extra cost in our model. In a main technical step in [6], based on the recent multilevel polynomial partitioning technique of Agarwal *et al.* [2, Corollary 4.8], we show:

Lemma 2. *One can complete the above sorting step, in the algebraic decision tree model, in $O\left((ng^3)^{16/9+\varepsilon}\right)$ randomized expected time, for any prescribed $\varepsilon > 0$, where the constant of proportionality depends on ε.*

Searching with the Elements of C and Wrapping Up. We now need to search the structures computed at the preceding phase with the dual points of C^*. Each such point c^* comes from a segment $c \in C$, which crosses only $O(r) = O(n/g)$ cells of Ξ (once again, this follows from standard properties of $(1/r)$-cuttings). For each of these cells σ, we need to locate c^* in the arrangement $\mathcal{A}(\Gamma_\sigma)$. There are $O(nr) = O(n^2/g)$ such segment-cell crossings, and each resulting search takes $O(\log g)$ time, using a suitable point-location data structure for each such arrangement, for a total of $O\left(\dfrac{n^2 \log g}{g}\right)$ time. The cost of the overall algorithm is thus

$$O\left((ng^3)^{16/9+\varepsilon} + \frac{n^2 \log g}{g}\right).$$

We (nearly) balance this bound by taking $g = n^{2/57}$, and conclude that the cost of this procedure, in the algebraic decision-tree model, is $O(n^{112/57+\varepsilon})$, for any $\varepsilon > 0$. In summary, we have shown:

Theorem 2. *Given three sets A, B, C, each consisting of n line segments in the plane, where the segments of A, B are pairwise disjoint, one can determine whether there exists a concurrent triple of segments $(a, b, c) \in A \times B \times C$, in the algebraic decision-tree model, with $O(n^{112/57+\varepsilon})$ polynomial sign tests (in expectation), for any $\varepsilon > 0$.*

Discussion and Open Problems. In spite of the progress in the study of collinearity testing in the algebraic decision tree model [5–7], as well as in the RAM model [7,9], the *unrestricted* setting of collinearity testing (that is, where A, B, C are collections of arbitrary points in the plane) has still remained elusive in both models of computation. Moreover, we are not aware of subquadratic solutions in either model even for the case where one set of points is one-dimensional and the other two are unrestricted.

Given the subquadratic solutions for 3SUM [16,18] in both models of computation, one may hope that this should also be the case for collinearity testing, as well as other geometric 3SUM-hard problems (see problems (i)–(iii) in Sect. 1). Unfortunately, we are still facing a gap between 3SUM and its geometric counterparts. Perhaps, this is because the sign test required for 3SUM is a linear function of the input, whereas other geometric 3SUM-hard problems do not have this property. In particular, for collinearity testing, the basic operation one needs to apply is orientation testing, which corresponds to a quadratic inequality in the input point coordinates, and thus does not benefit from the linear structural properties of 3SUM.

References

1. Agarwal, P.K.: Simplex range searching and its variants: a review. In: Loebl, M., Nešetřil, J., Thomas, R. (eds.) A Journey Through Discrete Mathematics, pp. 1–30. Springer, Cham (2017). https://doi.org/10.1007/978-3-319-44479-6_1
2. Agarwal, P.K., Aronov, B., Ezra, E., Zahl, J.: An efficient algorithm for generalized polynomial partitioning and its applications. In: Proceedings of 35th Symposium on Computational Geometry, pp. 5:1–5:14 (2020). arXiv:1812.10269
3. Agarwal, P.K., Matoušek, J., Sharir, M.: On range searching with semialgebraic sets II. SIAM J. Comput. **42**, 2039–2062 (2013)
4. Ailon, N., Chazelle, B.: Lower bounds for linear degeneracy testing. J. ACM **52**(2), 157–171 (2005)
5. Aronov, B., Ezra, E., Sharir, M.: Testing polynomials for vanishing on Cartesian products of planar point sets, In: Proceedings of 36th Symposium on Computational Geometry, pp. 8:1–8:14 (2020). arXiv:2003.09533
6. Aronov, B., Ezra, E., Sharir, M.: Subquadratic algorithms for some 3SUM-Hard geometric problems in the algebraic decision-tree model, Manuscript (2021)
7. Barba, L., Cardinal, J., Iacono, J., Langerman, S., Ooms, A., Solomon, N.: Subquadratic algorithms for algebraic 3SUM, Discrete Comput. Geom. **61**, 698–734 (2019). Also in Proceedings 33rd International Symposium on Computational Geometry, pp. 13:1–13:15 (2017)
8. Cardinal, J., Iacono, J., Ooms, A.: Solving k-SUM using few linear queries, In: Proceedings of 24th European Symposium on Algorithms, pp. 25:1–25:17 (2016)
9. Chan, T.M.: More logarithmic-factor speedups for 3SUM, (median,+)-convolution, and some geometric 3SUM-hard problems, ACM Trans. Algorithms **16**, 7:1–7:23 (2020)
10. de Berg, M., Schwarzkopf, O.: Cuttings and applications. Int. J. Comput. Geometry Appl. **5**, 343–355 (1995)
11. Erickson, J.: Lower bounds for linear satisfiability problems. Chicago. J. Theoret. Comput. Sci. **8**, 388–395 (1997)

12. Erickson, J., Seidel, R.: Better lower bounds on detecting affine and spherical degeneracies. Discrete Comput. Geom. **13**(1), 41–57 (1995). https://doi.org/10.1007/BF02574027
13. Ezra, E., Har-Peled, S., Kaplan, H., Sharir, M.: Decomposing arrangements of hyperplanes: VC-dimension, combinatorial dimension, and point location. Discrete Comput. Geom **64**(1), 109–173 (2020)
14. Ezra, E., Sharir, M.: A nearly quadratic bound for point-location in hyperplane arrangements, in the linear decision tree model. Discrete Comput. Geom. **61**(4), 735–755 (2018). https://doi.org/10.1007/s00454-018-0043-8
15. Gajentaan, A., Overmars, M.H.: On a class of $O(n^2)$ problems in computational geometry. Comput. Geom. Theory Appl. **5**, 165–185 (1995)
16. Grønlund, A., Pettie, S.: Threesomes, degenerates, and love triangles, J. ACM **65** 22:1–22:25 (2018). Also in Proceedings of 55th Annul Symposium on Foundations of of Computer Science, pp. 621–630 (2014)
17. Guth, L., Katz, N.H.: On the Erdős distinct distances problem in the plane, Annals Math. **181**, 155–190 (2015). arXiv:1011.4105
18. Hopkins, M., Kane, D.M., Lovett, S., Mahajan, G.: Point location and active learning: learning halfspaces almost optimally. In: Proceedings of 61st IEEE Annual Symposium on Foundations of Computer Science, (FOCS) (2020)
19. Kane, D.M., Lovett, S., Moran, S.: Near-optimal linear decision trees for k-SUM and related problems, J. ACM **66**, 16:1–16:18 (2019). Also in Proceedings of 50th Annul ACM Symposium on Theory Computational, pp. 554–563 (2018). arXiv:1705.01720
20. Meiser, S.: Point location in arrangements of hyperplanes. Inf. Comput. **106**(2), 286–303 (1993)
21. Meyer auf der Heide, F.: A polynomial linear search algorithm for the n-dimensional knapsack problem. J. ACM **31**, 668–676 (1984)
22. Solymosi, J., de Zeeuw, F.: Incidence bounds for complex algebraic curves on cartesian products. In: Ambrus, G., Bárány, I., Böröczky, K.J., Fejes Tóth, G., Pach, J. (eds.) New Trends in Intuitive Geometry. BSMS, vol. 27, pp. 385–405. Springer, Heidelberg (2018). https://doi.org/10.1007/978-3-662-57413-3_16

Limitwise Monotonic Spectra and Their Generalizations

Marat Faizrahmanov$^{(\boxtimes)}$ [iD]

Kazan (Volga Region) Federal University, Volga Region Scientific-Educational Centre of Mathematics, 18 Kremlyovskaya Street, Kazan 420008, Russian Federation

Abstract. The current work studies the limitwise monotonic spectra introduced by Downey, Kach and Turetsky [6]. In the first part of the paper, we study the limitwise monotonic spectra of subsets and sequences of subsets of $\mathbb{N}$. In particular, we study their measure-theoretical and topological properties. Then we generalize them to the spectra of subsets and sequences of subsets of $\mathbb{R}$ and provide some new degree spectra of structures.

Keywords: Limitwise monotonic function · Limitwise monotonic set · Limitwise monotonic spectrum · Degree spectrum

1 Introduction

This paper is motivated by research on spectra of structures from various classes given by algebraic or model-theoretic properties. Examples of such classes of structures include abelian groups, equivalence structures and graphs coding countable families. Let us recall how to construct a graph from a given countable family of subsets of $\mathbb{N}$. To code a particular set $A \subseteq \mathbb{N}$, we let $\mathfrak{H}(A)$ be the "daisy graph" starting with a central vertex v and adding a loop from v to itself of length $n + 3$ for each $n \in A$. The "bouquet graph" $\mathfrak{H}(\mathcal{F})$ of a family of sets $\mathcal{F}$ consists of infinitely many disjoint copies of the daisy graph $\mathfrak{H}(A)$ for each $A \in \mathcal{F}$ (see [1,17]). Then it is easy to see that a Turing degree $\mathbf{a}$ can enumerate a family $\mathcal{F}$ if and only if it computes a copy of $\mathfrak{H}(\mathcal{F})$.

Given a countable structure $\mathfrak{A}$, we define the degree spectrum of $\mathfrak{A}$ to be

$$\mathrm{Spec}(\mathfrak{A}) = \{X \subseteq \mathbb{N} : \exists \mathfrak{B} \cong \mathfrak{A} \, [\mathfrak{B} \leqslant_T X]\},$$

where we identify $\mathfrak{B}$ with its atomic diagram. Since $\mathrm{Spec}(\mathfrak{A})$ is degree-invariant, we often replace $\mathrm{Spec}(\mathfrak{A})$ by the collection of Turing degrees of elements of $\mathrm{Spec}(\mathfrak{A})$. The notion of degree spectrum was introduced by Richter [24]. It is motivated, in particular, by the fact that intuitively, the isomorphism type of a

The work is supported by the Russian Science Foundation (grant no. 18-11-00028) and performed under the development program of Volga Region Mathematical Center (agreement no. 075-02-2020-1478).

L. De Mol et al. (Eds.): CiE 2021, LNCS 12813, pp. 189–198, 2021.
https://doi.org/10.1007/978-3-030-80049-9_17

structure $\mathfrak{A}$ captures the computability-theoretic properties of $\mathrm{Spec}(\mathfrak{A})$. In this way, classes of degrees which cannot be captured by any single countable set (as they may not have least elements), are nonetheless captured by a single countable structure. For example, Slaman [25] and Wehner [29] showed that the collection of nonzero Turing degrees is a degree spectrum, and so there is a structure which captures the property of being non-computable.

There are many examples of classes of Turing degrees that are spectra of graphs of the form $\mathfrak{H}(\mathcal{F})$, where $\mathcal{F}$ is a countable family. For instance, the previous collection of nonzero [25, 29] degrees, the collections of hyperimmune [3], non-superlow [12], non-jump traceable [4] degrees form spectra of such graphs. Much less is known about the spectra of natural algebras and models, such as abelian groups, boolean algebras, linear orders, etc. For example, by the well-known result of Downey and Jockush [5] each boolean algebra $\mathfrak{B}$ such that $\mathrm{Spec}(\mathfrak{B})$ contains some low degree (see Knight and Stob [19] for the case of low_4 degree) has a computable copy. For the case of linear orders, it is known, for example, that for each integer $n \geqslant 2$ there is an order $\mathfrak{L}_n$ with $\mathrm{Spec}(\mathfrak{L}_n)$ consisting of all non-low_n degrees (see [8]). Now consider the case of spectra of abelian groups. So, Melnikov found some specific examples of degree spectra of torsion-free abelian groups. Namely, in [20] he proved that for every computable ordinal β of the form $\delta + 2n + 1 > 1$, where δ is zero or is a limit ordinal and $n \in \mathbb{N}$, there exists a torsion-free abelian group with the degree spectrum consisting of all non-low_β degrees. Kalimullin, Khoussainov and Melnikov [15] showed that there is an abelian p-group (and equivalence structure) $\mathfrak{A}$ such that $\mathrm{Spec}(\mathfrak{A})$ contains a Δ_2^0-degree $\mathbf{a}$ if and only if $\mathbf{a} > \mathbf{0}$. They also constructed a torsion abelian group $\mathfrak{G}$ such that (a) $\mathfrak{G}$ has no computable copy and (b) $\mathfrak{G}$ has an $\mathbf{a}$-computable copy for every hyperimmune degree $\mathbf{a}$. The proofs of the last two results essentially use the concept of limitwise monotonic sets and functions. The goal of this paper is to further study their spectral capabilities. So, using a generalization of the notion of limitwise monotonicity, it was possible to obtain a series of new examples of degree spectra (see Theorem 8).

It should be noted that algebraic structures constructed using limitwise monotonic tools as in Corollary 1 and Theorem 3 are not universal in the sense of the paper of Hirschfeldt, Khoussainov, Shore, and Slinko [10] as, for example, graphs, rings (with zero-divisors), 2-step nilpotent groups, etc. Therefore, their degree spectra should be studied using specific techniques.

2   Limitwise Monotonic Sets and Sequences of Integers

We begin by introducing some requisite terminology. The following definition was introduced by Khisamiev.

Definition 1. *A function $F : \mathbb{N} \to \mathbb{N}$ is limitwise (X-limitwise) monotonic if there is a total computable (X-computable) function $f : \mathbb{N} \times \mathbb{N} \to \mathbb{N}$ satisfying $f(x, s) \leqslant f(x, s + 1)$ such that $F(x) = \lim_s f(x, s)$ exists for all x.*

A set S is *limitwise (X-limitwise) monotonic* if there is a limitwise (X-limitwise) monotonic function F whose range is S. The study of limitwise monotonic sets is motivated mainly by our desire to investigate the degree spectra of specific structures, such as torsion abelian groups, equivalence structures, and graphs coding countable families. It is not hard to see that a set S is limitwise monotonic if and only if the family

$$I(S) = \{\{y \leqslant x\} : x \in S\}$$

is c.e. For abelian groups and equivalence structures, we recall the following two well-known theorems.

Theorem 1 (Khisamiev [16]). *For an infinite set S ($0 \notin S$) and prime p the group $A_p(S) = \bigoplus_{n \in S} \mathbb{Z}_{p^n}$ has a computable copy iff S is limitwise monotonic.*

Theorem 2 (Calvert, Cenzer, Harizanov, Morozov [2]). *For an infinite set $S = \{c_0, c_1, \dots\} \subseteq \mathbb{N}$ the equivalence structure $\mathfrak{E}(S)$ in which all equivalence classes are finite and have distinct cardinalities $c_0, c_1, \dots$ has a computable copy iff S is limitwise monotonic.*

Definition 2 (Downey, Kach, Turetsky [6]). *If S is any nonempty set, define the limitwise monotonic spectrum of S to be the set*

$$\mathrm{LmSp}(S) = \{X : S \text{ is } X\text{-limitwise monotonic}\}.$$

By relativization of the Theorems 1 and 2 we have the following corollary.

Corollary 1. *Let S ($0 \notin S$) be an infinite set. Then we have the following statements.*

1. $\mathrm{Spec}(A_p(S)) = \mathrm{LmSp}(S)$.
2. $\mathrm{Spec}(\mathfrak{E}(S)) = \mathrm{LmSp}(S)$.
3. $\mathrm{Spec}(\mathfrak{H}(I(S))) = \mathrm{LmSp}(S)$.

Definition 3 (Kalimullin, Khoussainov, Melnikov [15]). *A sequence of sets $\mathcal{S} = \{S_n\}_{n \in \mathbb{N}}$ is limitwise (X-limitwise) monotonic if there is a total computable (X-computable) function $f : \mathbb{N}^3 \to \mathbb{N}$ such that*

- $f(n, x, s) \leqslant f(n, x, s+1)$ *for all n, x, s;*
- $\lim_s f(n, x, s)$ *exists for all n, x;*
- S_n *is range of the function $x \mapsto \lim_s f(n, x, s)$ for every n.*

As in the case of sets, we define the limitwise monotonic spectrum of a sequence $\mathcal{S} = \{S_n\}_{n \in \mathbb{N}}$ by letting $\mathrm{LmSp}(\mathcal{S}) = \{X : \mathcal{S} \text{ is } X\text{-limitwise monotonic}\}$.

Let $p_0, p_1, \dots$ be the sequence of prime numbers listed in increasing order.

Theorem 3 (Kalimullin, Khoussainov, Melnikov [15]). *A sequence $\mathcal{S} = \{S_n\}_{n \in \mathbb{N}}$ ($0 \notin \bigcup_n S_n$) is limitwise monotonic iff the abelian group $G(\mathcal{S}) = \bigoplus_n A_{p_n}(S_n)$ has a computable copy.*

It was shown in [15] that the Slaman-Wehner Theorem [25,29] is not true for the limitwise monotonic spectra of sequences of sets. Moreover, the complement of $\mathrm{LmSp}(\mathcal{S})$ of any non-limitwise monotonic sequence $\mathcal{S}$ has continuum cardinality. In contrast, Wallbaum [28] had shown that there is a Δ_2^0-set which is not limitwise monotonic but is limitwise monotonic relative to each member of a co-null class (class of measure one) in the Cantor space. Now we provide a natural co-null class that can be contained in non-trivial limitwise monotonic spectra. Recall that a set A is said to be 1-*random* (see [23]) if there is no uniformly c.e. sequence $\{G_m\}_{m \in \mathbb{N}}$ of open sets in the Cantor space $2^{\mathbb{N}}$ such that $\forall m \, [\lambda G_m \leqslant 2^{-m}]$ and $A \in \bigcap_m G_m$, where λ is the uniform probability measure on the Cantor space. A set a is called 2-*random* if A is 1-random relative to $\emptyset'$.

Theorem 4 (Kalimullin, Faizrahmanov [13]). *There is a Δ_2^0-set which is not limitwise monotonic but is limitwise monotonic relative to each 2-random set.*

Since there exist Π_1^0-classes that contain only 1-random members, we can not replace 2-random sets by 1-random ones in Theorem 4 due to the following theorem.

Theorem 5 (Kalimullin, Faizrahmanov [13]). *Let $\mathcal{P}$ be a nonempty Π_1^0-class. Then there is no set S such that S is not limitwise monotonic and $\mathcal{P} \subseteq \mathrm{LmSp}(S)$.*

Among classes which are not cocountable, in addition to the notion of measure, we can appeal to the notion of category to obtain notion of largeness, namely being co-meagre. The largeness notions given by category and measure are not compatible: there is a meagre co-null class, and also a null co-meagre class. Greenberg, Montalbán and Slaman have shown [9] that this incompatibility is reflected in degree spectra; namely that there is a null and co-meagre degree spectrum, and a meagre and co-null spectrum. Let us show that one of these incompatibilities can also be obtained for the limitwise monotonic spectra.

Theorem 6 (Kalimullin, Faizrahmanov [13]). *(a) If $S \in \Sigma_2^0$ then $\mathrm{LmSp}(\mathcal{S})$ is co-meagre. (b) There is a set $S \in \Delta_2^0$ such that $\mathrm{LmSp}(\mathcal{S})$ is null.*

Sketch of Proof. (a) Let $S \in \Sigma_2^0$. As it was shown in [13], there is an increasing function $f \leqslant_T \emptyset'$ such that each X which computes some function g not dominated by f is an element of $\mathrm{LmSp}(S)$. For any integer i let $\mathcal{L}_i$ be the class consisting of all infinite sets $X = \{x_0 < x_1 < \dots\}$ such that $x_k \leqslant f(k)$ for each $k \geqslant i$. It is not hard to show that the class $\mathcal{L} = \bigcup_i \mathcal{L}_i \cup \{F \subseteq \mathbb{N} : F \text{ is finite}\}$ is meagre and S is X-limitwise monotonic for each $X \notin \mathcal{L}$. Therefore, $\mathrm{LmSp}(S)$ is co-meagre.

(b) Let $\{\Xi_n\}_{n \in \mathbb{N}}$ be a Gödel enumeration of all Turing functionals $\Xi : 2^{\mathbb{N}} \times \mathbb{N}^2 \to \mathbb{N}$ such that

- $\Xi(X \upharpoonright s; x, s) \downarrow$;
- $\Xi(X; x, s) \leqslant \min\{s, \Xi(X; x, s+1)\}$,

for all X, x, s. Let

$$\Xi_n^*(X, x) = \begin{cases} \lim_s \Xi_n(X; x, s), & \text{if } \lim_s \Xi_n(X; x, s) < \infty, \\ \text{undefined}, & \text{otherwise.} \end{cases}$$

Then any infinite set $S \in \Sigma_2^0(X)$ is X-limitwise monotonic iff there is an n such that

- the function $\Xi_n^*(X)$ is total and its range is a subset of S;
- $\Xi_n^*(X; x) > x$ for each x.

Now we define the following predicates on $\mathbb{N}^3 \times \mathbb{Q}^+$:

$$P(n, x, y, \delta) \Leftrightarrow \lambda\{X : \exists s \,[\Xi_n(X; x, s) \geqslant y]\} > \delta,$$

$$Q(n, x, y, \delta) \Leftrightarrow \lambda\{X : \Xi_n^*(X; x) = y\} > \delta.$$

It has been shown in [13] that $P \in \Sigma_1^0$ and $Q \in \Sigma_2^0$. Using this bounds we can construct a set $S \in \Delta_2^0$ satisfying the requirements

$$\mathcal{R}_e : \lambda\{X : \Xi_e^*(X) \text{ is total} \,\&\, \forall x \,[\Xi_e^*(X; x) > x] \,\&\, \text{rng}\,\Xi_e^*(X) \subseteq S\} \leqslant \frac{1}{2}$$

for each e. By the Lebesgue Density Theorem we will have that $\lambda\text{LmSp}(S) = 0$. The strategy for meeting the requirements $\mathcal{R}_e$ is similar to the standard strategy for meeting the requirements of non-limitwise monotonicity (for details, see Khoussainov, Nies, Shore [18]).

It should be noted that if $\text{LmSp}(\mathcal{S})$ is co-null then it is also co-meagre. Indeed, in this case the class $\{X : S \in \Sigma_2^0(X)\}$ is co-null. Then, by Stillwell's result [27] we have $S \in \Sigma_2^0$. Therefore, $\text{LmSp}(\mathcal{S})$ is co-meagre.

3   Limitwise Monotonic Sets and Sequences of Reals

The notions of limitwise monotonic sets and sequences can be extended to subsets and sequences of subsets of reals.

Definition 4. *A countable set $S \subseteq \mathbb{R}$ is limitwise (X-limitwise) monotonic [7] if there exists a computable (X-computable) function $f : \mathbb{N}^2 \to \mathbb{Q}$ satisfying the following conditions:*

- *$f(x, s) \leqslant f(x, s + 1)$ for all x, s;*
- *$\lim_s f(x, s)$ exists for all x;*
- *S is range of the function $x \mapsto \lim_s f(x, s)$.*

Definition 5. *A sequence $\mathcal{S} = \{S_n\}_{n \in \mathbb{N}}$ of countable subsets of reals is said to be limitwise (X-limitwise) monotonic if there exists a computable (X-computable) function $f : \mathbb{N}^3 \to \mathbb{Q}$ such that:*

- $f(n, x, s) \leqslant f(n, x, s + 1)$ *for all* n, x, s;
- $\lim_s f(n, x, s)$ *exists for all* n, x;
- S_n *is range of the function* $x \mapsto \lim_s f(n, x, s)$ *for every* n.

If $\mathcal{S} = \{S_n\}_{n \in \mathbb{N}}$ is any sequence of nonempty countable subsets of reals, define the *limitwise monotonic spectrum* of $\mathcal{S}$ to be the set

$$\mathrm{LmSp}(\mathcal{S}) = \{X : \mathcal{S} \text{ is } X\text{-limitwise monotonic}\}.$$

It is not hard to see that

$$\mathrm{LmSp}(\mathcal{S}) = \mathrm{Spec}(\mathfrak{H}(I(\mathcal{S}))),$$

where $I(\mathcal{S}) = \{\{n\} \oplus \{q \in \mathbb{Q} : q < \alpha\} : n \in \mathbb{N}, \ \alpha \in S_n\}$.

Using the concept of limitwise monotonicity on reals, we can establish the existence of Slaman-Wehner family that has no uniform enumeration in every non-computable set.

Theorem 7 (Faizrahmanov, Kalimullin [7]). *There is a set* $S \subseteq \mathbb{Q}$ *such that* S *is* X-*limitwise monotonic iff* X *is non-computable. Moreover, there is no uniform limitwise monotonic approximation of* S *for all non-computable* X, *i.e., there is no Turing functional* Φ *such that for every non-computable* X *we have*

1. $\Phi^X(x, s) \leqslant \Phi^X(x, s + 1)$;
2. $\lim_s \Phi^X(x, s) < \infty$;
3. S *is range of the function* $x \mapsto \lim_s \Phi^X(x, s)$.

Let us provide some other degree spectra using the concept of limitwise monotonicity. For this purpose, we recall the definitions of c.e. traces and jump traceable sets.

Definition 6. *A uniformly c.e. sequence of nonempty sets* $\mathcal{R} = \{T_n\}_{n \in \mathbb{N}}$ *is a c.e. trace (see Ishmukhametov [11]) if there exists a computable function* h *such that* $|T_n| \leqslant h(n)$ *for each* n. *We say that* $\mathcal{R}$ *is a trace for the partial function* ψ *if* $\psi(n) \downarrow \Rightarrow \psi(n) \in T_n$ *for each* n. *A set* A *is jump traceable (see Nies [23]) if there is a c.e. trace for the partial function* $J^A(e) = \Phi_e^A(e)$.

We also need the following propositions.

Proposition 1. *Let* $\{S_n\}_{n \in \mathbb{N}}$ *be a uniformly* Σ_2^0-*sequence of subsets of* $\mathbb{Q}$. *If there exists a limitwise monotonic sequence* $\{L_n\}_{n \in \mathbb{N}}$ *of subsets of* $\mathbb{Q}$ *such that* $L_n \subseteq S_n$ *and* $\sup L_n = \sup S_n$ *for every* n, *then the sequence* $\{S_n\}_{n \in \mathbb{N}}$ *is limitwise monotonic.*

Proof. The proof follows from the uniformity of the similar Proposition 1.4 [7] for the subsets of $\mathbb{Q}$.

Let $\{U_n\}_{n \in \mathbb{N}}$ and $\{\Theta_n\}_{n \in \mathbb{N}}$ be Gödel enumerations of all c.e. operators from $2^{\mathbb{N}}$ to $2^{\mathbb{Q}}$ and all Turing functionals from $2^{\mathbb{N}} \times \mathbb{N}$ to $\mathbb{Q}$, respectively. The use-function of the functional Θ_n is denoted by θ_n.

Proposition 2. *Let $C \subseteq 2^{\mathbb{N}}$ be a class of sets such that for all $Y_0, Y_1 \in C$ there is a $Z \in C$ such that $Y_0, Y_1 \leqslant_T Z$, and let $X \nleqslant_T Y$ for every $Y \in C$. Then there exists a computable function g such that for all rationals $a < b$ we have $a < \sup U^X_{g(a,b)} < b$ and the real $\sup U^X_{g(a,b)}$ is not left Y-c.e. for every $Y \in C$.*

Proof. Let $\alpha_0 = \sum_{x \in X} 2^{-x}$ and $\alpha_1 = \sum_{x \notin X} 2^{-x}$. It is not hard to see that each of α_0 and α_1 it Turing equivalent to X. If there exist $Y_0, Y_1 \in C$ such that α_i is left Y_i-c.e., $i = 0, 1$, then $X \leqslant_T Z$ for some $Z \in C$ with $Y_0, Y_1 \leqslant_T Z$. This is a contradiction. So one of $\alpha \in \{\alpha_0, \alpha_1\}$ is not left Y-c.e. for every $Y \in C$. It remains to fix a computable function g such that $U^X_{g(a,b)} = \{q \in \mathbb{Q} : q < a + \frac{(b-a)\cdot\alpha}{2}\}$ for all rationals a and b.

Theorem 8. *Let $\{A_m\}_{m \in \mathbb{N}}$ be a uniformly c.e. sequence of jump traceable sets such that for all m_0, m_1 there is an n with $A_{m_0}, A_{m_1} \leqslant_T A_n$. Then there exists a sequence of subsets of rationals $\mathcal{J} = \{J_i\}_{i \in \mathbb{N}}$ such that*

$$\mathrm{LmSp}(\mathcal{J}) = \{X : \forall m \, [X \nleqslant_T A_m]\}.$$

Proof. Let us fix a uniformly c.e. double sequence $\{T^n_x\}_{n,x \in \mathbb{N}}$ satisfying the following conditions:

- for every n the sequence $\mathcal{R}^n = \{T^n_x\}_{x \in \mathbb{N}}$ is a c.e. trace;
- for every c.e. trace $\mathcal{R}$ there is an n with $\mathcal{R}^n = \mathcal{R}$.

For example, the double sequence $\{T^n_x\}_{n,x \in \mathbb{N}}$ can be defined as following:

$$T^n_{x,0} = \emptyset, \ n, x \in \mathbb{N},$$

$$T^{c(e,b)}_{x,s+1} = \begin{cases} W_{\varphi_{e,s}(x),s}, & \text{if } \forall y \leqslant x \, [\varphi_{e,s}(y) \downarrow \& \varphi_{b,s}(y) \downarrow \& |W_{\varphi_e(y),s}| \leqslant \varphi_b(y)], \\ T^{c(e,b)}_{x,s}, & \text{otherwise,} \end{cases}$$

where $e, b, x, s \in \mathbb{N}$ and c is the Cantor pairing function. Then let $T^n_x = \bigcup_s T^n_{x,s}$, $n, x \in \mathbb{N}$.

Now we define an auxiliary Turing functional Γ. For all integers m, s, k, z let $\Gamma^{A_m,0}$ be nowhere defined and

$$\Gamma^{A_m,s+1}(c(k,z)) = \begin{cases} s, & \text{if } \theta^{A_m,s+1}_k(z) \neq \theta^{A_m,s}_k(z), \\ \Gamma^{A_m,s}(c(k,z)), & \text{otherwise.} \end{cases}$$

Using the functional Γ, we define the following modifications of the functionals $\{\Theta_k\}_{k \in \mathbb{N}}$:

$$\widehat{\Theta}^{A_m,s}_{k,n}(z) = \begin{cases} \Theta^{A_m,s}_k(z), & \text{if } \Gamma^{A_m,s}(c(k,z)) \downarrow \in T^n_{c(e,z),s}, \\ \text{undefined}, & \text{otherwise,} \end{cases}$$

for all $m, s, k, n, z \in \mathbb{N}$. Note that if $\widehat{\Theta}^{A_m}_{k,n}(z) \uparrow$ then there exists an s such that $\widehat{\Theta}^{A_m,t}_{k,n}(z) \uparrow$ for every $t \geqslant s$. Moreover, if $\mathcal{R}^n$ is a trace for $\Gamma^{A_m,s}$ then

$$\widehat{\Theta}^{A_m}_{k,n} = \Theta^{A_m}_k.$$

Now we define the required sequence $\mathcal{J}$ by letting

$$J_{c(m,n)} = \mathbb{Q} \setminus \{k + \sup_z \widehat{\Theta}_{k,n}^{A_m}(z) : k \in \mathbb{N}\}$$

for all m, n, where

$$\sup_z \widehat{\Theta}_{k,n}^{A_m}(z) \stackrel{\text{def}}{=} \sup\{\widehat{\Theta}_{k,n}^{A_m}(z) : z \in \mathbb{N} \text{ and } \forall y \leqslant z \, [\widehat{\Theta}_{k,n}^{A_m}(z) \downarrow]\}.$$

Let us show that $\mathcal{J}$ is X-limitwise monotonic iff $X \not\leqslant_T A_m$ for every m. Fix an arbitrary set X with $\forall m \, [X \not\leqslant_T A_m]$. By Proposition 1, it is suffice to define an X-computable function $f : \mathbb{N}^4 \to \mathbb{Q}$ such that

$$\max_s f(m, n, k, s) \in J_{c(m,n)} \text{ and } \max_s f(m, n, k, s) \geqslant k$$

for all m, n, k. Let g be the computable function from Proposition 2 with $\mathcal{C} = \{A_m : m \in \mathbb{N}\}$. Choose arbitrary integers m, k and n. Since $\sup U_{g(k,k+1)}^X$ is not left A_m-c.e., there exists an s_0 such that for all $s \geqslant s_0$ and $l \leqslant k$ one of the following inequalities hold:

$$\max U_{g(k,k+1),s}^X < l + \limsup_t \sup_z \widehat{\Theta}_{l,n}^{A_{m,t}}(z) - \frac{1}{s},$$

$$l + \limsup_t \sup_z \widehat{\Theta}_{l,n}^{A_{m,t}}(z) < \max U_{g(k,k+1),s}^X - \frac{1}{s}.$$

So we can define $f(m, n, k, 0) = k$ and

$$f(m, n, k, s+1) = \begin{cases} \max U_{g(k,k+1),s}^X, & \text{if there exists an } l \leqslant k \text{ such that,} \\ & (f(m,n,k,s) \geqslant l + \sup_z \widehat{\Theta}_{l,n}^{A_{m,s}}(z) - \frac{1}{s}) \, \& , \\ & (l + \sup_z \widehat{\Theta}_{l,n}^{A_{m,s}}(z) \geqslant f(m,n,k,s) - \frac{1}{s}), \\ f(m, n, k, s), & \text{otherwise,} \end{cases}$$

for all m, n, k, s.

Conversely, assume that $\mathcal{J}$ is A_m-limitwise monotonic for some m. Choose an n such that $\mathcal{R}^n$ is a trace for Γ^{A_m}. Since the set $J_{c(m,n)}$ is also A_m-limitwise monotonic, there exists an A_m-computable function $h : \mathbb{N} \times \mathbb{N} \to \mathbb{Q}$ such that $h(k, 0) > k$ and $\sup_z h(k, z) \in J_{c(m,n)}$ for every k. Fix a computable function r such that

$$\Theta_{r(k)}^{A_m}(z) = h(k, z) - k.$$

Let p be a fixed point of the function r. Then

$$\sup_z h(p, z) = p + \sup_z \Theta_{r(p)}^{A_m}(z) = p + \sup_z \Theta_p^{A_m}(z) = p + \sup_z \widehat{\Theta}_{p,n}^{A_m}(z) \in J_{c(m,n)}.$$

This contradiction completes the proof of the theorem.

Recall that a set A is K-*trivial* if there is a constant b such that $K(A \upharpoonright n) \leqslant K(n) + b$ for each n, where K is the prefix-free Kolmogorov complexity.

Corollary 2. *There exists a sequence of countable sets of rationals* $\mathcal{K} = \{K_i\}_{i \in \mathbb{N}}$ *with* $\mathrm{LmSp}(\mathcal{K}) = \{X : X$ *is not* K*-trivial*$\}$.

Proof. Since the index set of all K-trivial c.e. sets is Σ_3^0 (see [23]) and each finite set is K-trivial, by [30] (see also Theorem 3.2 [26]) there is a uniformly c.e. sequence $\{A_m\}_{m \in \mathbb{N}}$ of all K-trivial c.e. sets. By [22], each K-trivial set is super-low and hence [21], each K-trivial c.e. set is jump traceable. Note that the K-trivial Turing degrees form the ideal generated by the K-trivial c.e. degrees [23]. Thus, it remains to apply Theorem 8 to the sequence $\{A_m\}_{m \in \mathbb{N}}$.

Note that a structure whose spectrum consists of all non-K-trivial degrees has also been constructed in [14].

Acknowledgments. I am very grateful to the referees for the careful reading of the paper and for their comments and detailed suggestions, which helped me to improve the manuscript considerably.

References

1. Ash, C.J., Knight, J.F.: Computable structures and the hyperarithmetical hierarchy, volume 144 of Studies in Logic and the Foundations of Mathematics. North-Holland Publishing Co., Amsterdam (2000)
2. Calvert, W., Cenzer, D., Harizanov, V., Morozov, A.: Effective categoricity of equivalence structures. Ann. Pure Appl. Logic **141**, 61–78 (2006)
3. Csima, B., Kalimullin, I.: Degree spectra and immunity properties. Math. Logic Q. **56**(1), 67–77 (2010)
4. Diamondstone, D., Greenberg, N., Turetsky, D.: Natural large degree spectra. Computability **2**(1), 1–8 (2013)
5. Downey, R., Jockusch, C.G.: Every low boolean algebra is isomorphic to a recursive one. Proc. Am. Math. Soc. **122**(3), 871–880 (1994)
6. Downey, R., Kach, A., Turetsky, D.: Limitwise monotonic functions and their applications. In: Proceedings of the Eleventh Annual Asian Logic Conference, World Scientific, Hackensack, pp. 59–85, NJ (2012)
7. Faizrahmanov, M., Kalimullin, I.: Limitwise monotonic sets of reals. Math. Logic Q. **61**(3), 224–229 (2015)
8. Frolov, A., Kalimullin, I., Harizanov, V., Kudinov, O., Miller, R.: Spectra of high$_n$ and non-low$_n$ degrees. J. Logic Comput. **22**(4), 755–777 (2012)
9. Greenberg, N., Montalbán, A., Slaman, T.: Relative to any non-hyperarithmetic set. J. Math. Logic **13**, 1250007 (2013)
10. Hirschfeldt, D.R., Khoussainov, B.M., Shore, R.A., Slinko, A.M.: Degree spectra and computable dimensions in algebraic structures. Ann. Pure Appl. Logic **115**(1–3), 71–113 (2002)
11. Ishmukhametov, S.: Weak recursive degrees and a problem of Spector. Recursion Theor. Complexity (Kazan, 1997) **2**, 81–87 (1999)
12. Kalimullin, I.: Spectra of degrees of some structures. Algebra Logic **46**, 399–408 (2007)
13. Kalimullin, I., Faizrakhmanov, M.: Limitwise monotonic spectra of Σ_2^0-sets. Uchenye Zapiski Kazanskogo Universiteta. Seriya Fiziko-Matematicheskie Nauki (Russian) **154**(2), 107–116 (2012)

14. Kalimullin, I., Faizrakhmanov, M.: Degrees of enumerations of countable Wehner-like families. In: Proceedings of the Seminar on Algebra and Mathematical Logic of the Kazan (Volga Region) Federal University, Itogi Nauki i Tekhniki. Ser. Sovrem. Mat. Pril. Temat. Obz. (Russian), vol. 157, pp. 59–69, VINITI, Moscow (2018)

15. Kalimullin, I., Khoussainov, B., Melnikov, A.: Limitwise monotonic sequences and degree spectra of structures. Proc. Am. Math. Soc. **141**(9), 3275–3289 (2013)

16. Khisamiev, N.: Constructive abelian groups. In: Handbook of Recursive Mathematics, Vol. 2, volume 139 of Studies in Logic and the Foundations of Mathematics, pp. 1177–1231, North-Holland, Amsterdam (1998)

17. Khoussainov, B.: Strongly effective unars and nonautoequivalent constructivizations. In: Some problems in differential equations and discrete mathematics (Russian), pp. 33–44. Novosibirsk. Gos. Univ., Novosibirsk (1986)

18. Khoussainov, B., Nies, A., Shore, R.: Computable models of theories with few models. Notre Dame J. Formal Logic **38**(2), 165–178 (1997)

19. Knight, J.F., Stob, M.: Computable boolean algebras. J. Symbolic Logic **65**(4), 1605–1623 (2000)

20. Melnikov, A.G.: New degree spectra of abelian groups. Notre Dame J. Formal Logic **58**(4), 507–525 (2017)

21. Nies, A.: Reals which compute little. In: Chatzidakis, Z, Koepke, P., Pohlers, W., (eds.) Proceedings of Logic Colloquium 2002, Lecture Notes in Logic, vol. 27, pp. 261–275 (2002)

22. Nies, A.: Lowness properties and randomness. Adv. Math. **197**(1), 274–305 (2005)

23. Nies, A.: Computability and randomness. Oxford Logic Guides, vol. 51. Oxford University Press, Oxford (2009)

24. Richter, L.J.: Degrees of unsolvability of models, Ph.D. Thesis, University of Illinois at Urbana-Champaign (1977)

25. Slaman, T.: Relative to any nonrecursive set. Proc. Am. Math. Soc. **126**(7), 2117–2122 (1998)

26. Soare, R.I.: Recursively Enumerable Sets and Degrees. A Study of Computable Functions and Computably Generated Sets. Perspectives in Mathematical Logic. Springer, Berlin (1987)

27. Stillwell, J.: Decidability of the almost all theory of degrees. J. Symbolic Logic **37**(3), 501–506 (1972)

28. Wallbaum, J.: Computability of algebraic structures. Ph.D. Thesis, University of Notre Dame (2010)

29. Wehner, S.: Enumerations, countable structures and Turing degrees. Proc. Am. Math. Soc. **126**(7), 2131–2139 (1998)

30. Yates, C.E.M.: On the degrees of index sets. II. Trans. Am. Math. Soc. **35**, 249–266 (1969)

On False Heine/Borel Compactness
Principles in Proof Mining

Fernando Ferreira[(⊠)] [ID]

Faculdade de Ciências, Universidade de Lisboa, Lisbon, Portugal
fjferreira@fc.ul.pt

Abstract. The use of *certain* false Heine/Borel compactness principles
is justified in source theories of proof mining. The justification rests
on the metatheorems of the theory of proof mining. Ulrich Kohlenbach
recently produced a counterexample showing that the metatheorems do
not apply unrestrictedly to Heine-Borel compactness principles. In this
short note, we present a simpler counterexample than Kohlenbach's,
showing that the metatheorems can fail because the source theory is
already inconsistent.

Keywords: Proof mining · Heine/Borel compactness · Bounded
collection · Uniform boundedness · Bounded functional interpretation

1

Ten years ago, the strong convergence theorem [1] of Felix Browder was analyzed
using proof mining methods by Ulrich Kohlenbach in [6]. Browder's proof uses
a sequential weak compactness argument. Surprisingly, the last part of Kohlenbach's analysis discloses that what is left of sequential weak compactness to
analyze is trivial. Recently, [4] proposed an explanation for this phenomenon.
It argued that the proof being analyzed can be seen as using a (countable)
Heine/Borel compactness argument, instead of the sequential weak compactness argument. This modified argument is explicitly given in Section 2 of [4] and
applies a Heine/Borel compactness principle to open sets of the form

$$\left\{ y \in X : \|U(y) - y\| > \frac{1}{m+1} \right\} \text{ or } \left\{ y \in X : \langle w, x - y \rangle < \frac{1}{k+1} \right\},$$

where X is a real Hilbert space with inner product $\langle \cdot, \cdot \rangle$, U is a nonexpansive
mapping of X into itself, x and w are elements of X, and k and m are natural
numbers. If a countable collection of open sets of this form covers a bounded
closed convex subset of X, then it has a finite subcover. This is a consequence
of the countable Heine/Borel covering principle CHBC:

$$\forall x \in C \, \exists n \in \mathbb{N} \, (x \in \Omega_n) \rightarrow \exists n \in \mathbb{N} \, \forall x \in C \, \exists m \leq n \, (x \in \Omega_m),$$

We acknowledge the support of Fundação para a Ciência e Tecnologia by way of the
grant UIDB/04561/2020 given to the research center CMAFcIO.

L. De Mol et al. (Eds.): CiE 2021, LNCS 12813, pp. 199–203, 2021.
https://doi.org/10.1007/978-3-030-80049-9_18

where C is a bounded closed set of the Hilbert space X and $(\Omega_n)_{n\in\mathbb{N}}$ is a family of open sets (for the norm topology). As it is well-known, in infinite dimensional Hilbert spaces, CHBC fails (even for convex C).

The concrete use of CHBC in Browder's modified proof is justified because of certain metatheorems of proof mining: see, for instance, the Main Theorem of [2] (the original, but different, metatheorems are due to Kohlenbach and are discussed in the book referred below). The justification rests on the fact that the Heine/Borel principles used are forms of bounded collection. Truth be told, the natural statement of the Main Theorem would be in the form of a conservation result, but the formulation given in [2] is good enough for applications. The monotone functional interpretation of Kohlenbach can also justify the concrete use of CHBC in Browder's result, with so-called uniform boundedness principles $\exists\text{-UB}^X$ in place of the bounded collection principles (see Sections 7 and 8 of Chapter 17 of the book [5] for a reference on these matters).

The crux of the matter lies in the fact that it is not enough that the open sets Ω_n used in CHBC be merely topologically open: they must also be *logically open*. They must be defined by Σ-formulas. These formulas are of the form $\exists\underline{r}\,B(\underline{r})$, with B a bounded formula (possibly with parameters) and $\exists\underline{r}$ a string of existential quantifiers. A particular (and familiar) case is when the string of existential quantifiers of the Σ-formulas consists of number quantifiers. The bounded formulas here at play include the usual bounded formulas of first-order arithmetic but go beyond these (since they allow so-called bounded quantifications in finite types). Of course, the open sets of the first paragraph above are given by Σ-formulas and are, therefore, logically open.

It is important to delineate the scope of central theorems (in this case, of proof mining). Very recently, Kohlenbach presented in [7] an example showing that things can go wrong when the open sets are not logically open. In the next section, we give an elementary example of this phenomenon and show that not only the metatheorem is no longer true in this case, but that the theory itself, with CHBC, is contradictory.

In Reverse Mathematics, membership in open sets is *defined* by Σ-formulas with existential number quantifiers (see Definition II.5.6 of the book [9], but also the recent [8] for a discussion and critique). The theory is indeed much smoother with the open sets being, by *fiat*, logically open. It is true that, in the real world, every open set of the real line (in general, every open set of a separable metric space) is open in the sense of Reverse Mathematics provided that we accept a suitable second-order parameter in the Σ-definition. Formally, this second-order parameter is obtained using comprehension. The catch is that, from the point of view of a weak subsystem of arithmetic with restricted comprehension, topologically open sets need not be logically open.

2

We work with the theory of real inner product spaces (a.k.a. real pre-Hilbert spaces). It is a classical finite type theory with an abstract type X for a real pre-Hilbert space. The theory can be set up with a universal axiomatization of the real pre-Hilbert space as in Section 3 of Chapter 17 of [5] – indirectly, via an axiomatization of normed spaces with the parallelogram law – or directly as hinted in Sect. 5.4 of [2]. We add to our language an infinite number of constants e_n of type X (indexed by natural numbers n) axiomatized by the universal sentences $\langle e_i, e_j \rangle = 0$, for $i \neq j$, and $\|e_i\| = 1$. We show that adding the CHBC principle to this theory is contradictory.

Fix $x \in X$ such that $\|x\| \leq 1$. Our theory shows that

$$\forall n \in \mathbb{N} \left(\sum_{j=0}^{n} |\langle x, e_j \rangle|^2 \leq 1 \right).$$

It is now easy to see, by contradiction, that there is $m_0 \in \mathbb{N}$ such that

$$\sum_{j=m_0+1}^{n} |\langle x, e_j \rangle|^2 \leq \frac{1}{16},$$

for all $n > m_0$. In particular, $|\langle x, e_n \rangle| \leq \frac{1}{4}$, for all $n > m_0$.

Let $S(w)$ be the formula $\forall n \in \mathbb{N} (|\langle w, e_n \rangle| \leq \frac{1}{4})$. We claim that the point $p := x - \sum_{j=0}^{m_0} \langle x, e_j \rangle e_j$ is such that $S(p)$. To see this, take a natural number n. Of course, $\langle p, e_n \rangle = \langle x, e_n \rangle - \sum_{j=0}^{m_0} \langle x, e_j \rangle \langle e_j, e_n \rangle$. If $n \leq m_0$, we get $\langle p, e_n \rangle = \langle x, e_n \rangle - \langle x, e_n \rangle = 0$. If $n > m_0$, we have $\langle p, e_n \rangle = \langle x, e_n \rangle$ and, therefore, $|\langle p, e_n \rangle| \leq \frac{1}{4}$.

Let $B := \{x \in X : \|x\| \leq 1\}$. We have showed the following

$$\forall x \in B \, \exists w \in X \, \exists n \in \mathbb{N} \left(S(w) \wedge \left\| (x - w) - \sum_{j=0}^{n} \langle x, e_j \rangle e_j \right\| = 0 \right).$$

A fortiori, we have $\forall x \in B \, \exists n \in \mathbb{N} \, (x \in \Omega_n)$, where

$$\Omega_n = \left\{ x \in X : \exists w \in X \left(S(w) \wedge \left\| (x - w) - \sum_{j=0}^{n} \langle x, e_j \rangle e_j \right\| < \frac{1}{2} \right) \right\}.$$

It is not difficult to argue that each Ω_n is open for the norm topology. Therefore, if we apply the CHBC principle, we conclude that there is n_0 such that

$$\forall x \in B \, \exists m \leq n_0 \, \exists w \in X \left(S(w) \wedge \left\| (x - w) - \sum_{j=0}^{m} \langle x, e_j \rangle e_j \right\| < \frac{1}{2} \right).$$

In particular, for $x := e_{n_0+1}$,

$$\exists m \leq n_0 \, \exists w \in X \left(S(w) \wedge \left\| (e_{n_0+1} - w) - \sum_{j=0}^{m} \langle e_{n_0+1}, e_j \rangle e_j \right\| < \frac{1}{2} \right).$$

Therefore, for a certain $\tilde{w} \in X$ with $S(\tilde{w})$, one has $\|e_{n_0+1} - \tilde{w}\| < \frac{1}{2}$. Hence,

$$1 - 2\langle \tilde{w}, e_{n_0+1} \rangle + \|\tilde{w}\|^2 < \frac{1}{4},$$

and we get the following contradiction:

$$1 \leq 1 + \|\tilde{w}\|^2 < 2\langle \tilde{w}, e_{n_0+1} \rangle + \frac{1}{4} \leq \frac{2}{4} + \frac{1}{4} = \frac{3}{4}.$$

3

The problem with the definition of the Ω_ns in the above section is not the existential quantifier $\exists w \in X$ but rather the condition $S(w)$. This problem would disappear had we enough comprehension to form the set $S := \{w \in X : S(w)\}$ internally (i.e., the theory shows that there is an element of type $X \to 0$ which is the characteristic function of S). In this case, we could bring the sets Ω_n into syntactic Σ-form. Therefore, the example of this section shows that the kind of comprehension needed to obtain S is contradictory (in the presence of bounded collection). Note that the logical form of the formula $S(w)$ is relatively simple: it is a universal statement with respect to numbers. So, it is contradictory to have comprehension for the elements w of the abstract space even for relatively simple conditions. This kind of failure of comprehension is already implicit in a very elementary counterexample of Kohlenbach given in remark 17.103 of [5]. There is an analogous failure of comprehension for type 1 objects: this is kind of folklore, but was explicitly discussed in Section 8 of [3]. It is nevertheless a consequence of deep work of Clifford Spector in [10] that unrestricted comprehension for *numbers* is admissible in proof mining studies.

That a condition can be put into syntactic Σ-form depends on how much comprehension one has in the theory. The notion of logically open set is not a semantic notion, it is rather proof-theoretic. However, some conditions are unmistakably in Σ-form. For instance, conditions (on w) of the form $\|w - a\| < \frac{1}{k+1}$, where a is an element of the Hilbert space and $k \in \mathbb{N}$. It could perhaps cross the mind of the reader that, with open sets given by conditions of this sort, then CHBC is true in every Hilbert space. This is obviously not the case. We can recycle an example of Kohlenbach in Section 8.2 of Chapter 17 of [5] to show this. Take X a separable infinite dimensional (real) Hilbert space. Let $(a_n)_{n \in \mathbb{N}}$ be an enumeration of a dense subset of X. Fix a natural number k. It is clear that X (and, hence, the closed unit ball B) is covered by the open sets $\Omega_n := \{w \in X : \|w - a_n\| < \frac{1}{k+1}\}$. By CHBC applied to this covering, B has a finite subcovering. We have therefore shown that the unit closed ball is totally bounded (and, hence, compact). As it is well known, this is only true in finite dimensional spaces.

4

The contradiction of Sect. 2 shows – in particular – the main result of [7], namely that the metatheorems of proof mining do not apply to the CHBC principle (as opposed to the bounded collection principles or to $\exists-\mathrm{UB}^X$). Our theory above was necessarily of an infinite dimensional space, given that the CHBC principle is true in finite dimensional Hilbert spaces. If we do not impose infinite dimensionality, we cannot get a contradiction but, with CHBC in place, we still get the failure of the proof mining metatheorems. Actually, the example of Sect. 2 can be slightly modified to show just this. As before, we work in our theory of real pre-Hilbert real spaces with a language that has an infinite number of constants e_n. Our new axiomatization has now just the universal sentences $\langle e_i, e_j \rangle = 0$, for $i \neq j$. Of course, this theory T has finite dimensional models.

The argument of Sect. 2 shows that $\forall n\,(\|e_n\| = 1)$ leads to absurdity in the presence of the principle CHBC. So, in the presence of this principle, the theory T proves $\exists n\,(\|e_n\| \neq 1)$. This sentence has the right logical form for the application of the proof mining metatheorems. Therefore, there would be an absolute constant n_0 such that, in every (real) Hilbert space satisfying T, $\exists n \leq n_0\,(\|e_n\| \neq 1)$. This is false in many models of T (both of finite and infinite dimension).

References

1. Browder, F.E.: Convergence of approximants to fixed points of nonexpansive non-linear mappings in Banach spaces. Arch. Rational Mech. Anal. **24**, 82–90 (1967)
2. Engrácia, P., Ferreira, F.: Bounded functional interpretation with an abstract type. In: Rezuş, A. (ed.) Contemporary Logic and Computing. Landscapes in Logic, vol. 1, pp. 87–112. College Publications, London (2020)
3. Ferreira, F.: A most artistic package of a jumble of ideas. Dialectica **62**, 205–222 (2008). Special Issue: Gödel's dialectica interpretation. Guest editor: Thomas Strahm
4. Ferreira, F., Leustean, L., Pinto, P.: On the removal of weak compactness arguments in proof mining. Adv. Math. **354**, 106728 (2019)
5. Kohlenbach, U.: Applied Proof Theory: Proof Interpretations and their Use in Mathematics. Springer Monographs in Mathematics, Springer, Berlin (2008). https://doi.org/10.1007/978-3-540-77533-1
6. Kohlenbach, U.: On quantitative versions of theorems due to F. E. Browder and R. Wittmann. Adv. Math. **226**, 2764–2795 (2011)
7. Kohlenbach, U.: Proof-theoretic uniform boundedness and bounded collection principles and countable Heine-Borel compactness. Arch. Math. Logic. https://doi.org/10.1007/s00153-021-00771-w
8. Normann, D., Sanders, S.: Open sets in computability theory and reverse mathematics. J. Logic Comput. **30**(8), 1639–1679 (2020)
9. Simpson, S.G.: Subsystems of Second Order Arithmetic. Perspectives in Mathematical Logic, Springer, Heidelberg (1999)
10. Spector, C.: Provably recursive functionals of analysis: a consistency proof of analysis by an extension of principles in current intuitionistic mathematics. In: Dekker, F.D.E. (ed.) Recursive Function Theory: Proceedings of Symposia in Pure Mathematics, vol. 5, pP. 1–27. American Mathematical Society, Providence (1962)

Placing Green Bridges Optimally, with a Multivariate Analysis

Till Fluschnik[ID] and Leon Kellerhals[(✉)][ID]

Faculty IV, Algorithmics and Computational Complexity,
Technische Universität Berlin, Berlin, Germany
{till.fluschnik,leon.kellerhals}@tu-berlin.de

Abstract. We study the problem of placing wildlife crossings, such as green bridges, over human-made obstacles to challenge habitat fragmentation. The main task herein is, given a graph describing habitats or routes of wildlife animals and possibilities of building green bridges, to find a low-cost placement of green bridges that connects the habitats. We develop three problem models for this task, which model different ways of how animals roam their habitats. We settle the classical complexity and parameterized complexity (regarding the number of green bridges and the number of habitats) of the three problems.

Keywords: Wildlife crossings · Computational complexity ·
Computational sustainability · Parameterized algorithmics · Connected
subgraphs

1 Introduction

Sustainability is an enormous concern impacting today's politics, economy, and industry. Accordingly, sustainability sciences are well-established by now. Yet, the *interdisciplinary* scientific field "computational sustainability" [8], which connects practical and theoretical computer science with sustainability sciences, is quite young. For instance, the Institute for Computational Sustainability at Cornell University was founded in 2008, the 1st International Conference on Computational Sustainability (CompSust '09) took place in 2009, and special tracks on computational sustainability and AI were established in 2011 (AAAI) and 2013 (IJCAI). This work contributes to computational sustainability: We model problems of elaborately placing wildlife crossings and give complexity-theoretical and algorithmic analyses for each. Wildlife crossings are constructions (mostly bridges or tunnels) that allow wildlife animals to safely cross human-made transportation lines (mostly roads). We will refer to wildlife crossings as *green bridges*.

Huijser *et al.* [10] give an extensive report on wildlife-vehicle collisions. They identify several endangered animal species suffering from high road mortality

T. Fluschnik—Supported by DFG, project TORE (NI 369/18).

L. De Mol et al. (Eds.): CiE 2021, LNCS 12813, pp. 204–216, 2021.
https://doi.org/10.1007/978-3-030-80049-9_19

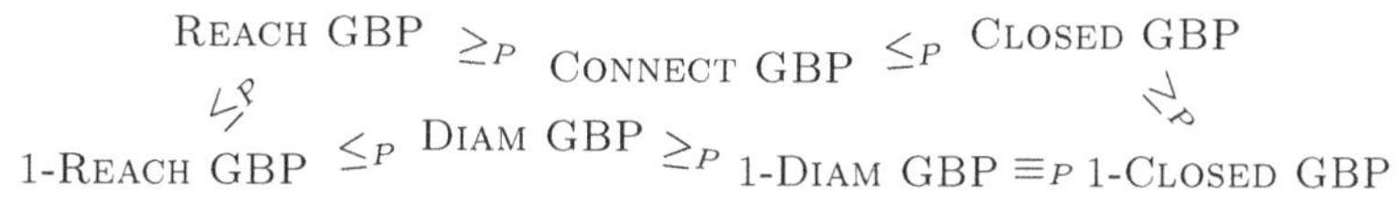

$$\text{REACH GBP} \;\;\geq_P\;\; \text{CONNECT GBP} \;\;\leq_P\;\; \text{CLOSED GBP}$$

$$\text{1-REACH GBP} \;\;\leq_P\;\; \text{DIAM GBP} \;\;\geq_P\;\; \text{1-DIAM GBP} \;\equiv_P\; \text{1-CLOSED GBP}$$

Fig. 1. Polynomial-time many-one reducibility directly derived from problem definitions.

and estimate the annual cost associated with wildlife-vehicle collisions to be about 8 billion US dollars. Wildlife fencing with wildlife crossings can reduce collisions by over 80% [10], enables populations to sustain [17], and is therewith among the most cost-effective [9]. The implementation, though, is an important problem: "*The location, type, and dimensions of wildlife crossing structures must be carefully planned with regard to the species and surrounding landscape. [...] In addition, different species use different habitats, influencing their movements and where they want to cross the road.*" [10, p. 16] It is further pointed out that data about wildlife habitats is basic for mitigation plans, yet challenging to obtain [18]. In this work, our main problem is placing green bridges at low cost and under several variants of habitat-connectivity requirements, thereby inherently modeling different availabilities of data on habitats. The problem is hence the following: *Given a graph describing habitats of wildlife animals and possibilities of building green bridges, find a low-cost placement of green bridges that sufficiently connects habitats.* In particular, we comparatively study in terms of computational complexity and parameterized algorithmics the following three different (families of) decision problems.

Π GREEN BRIDGES PLACEMENT (Π GBP)

Input: An undirected graph $G = (V, E)$, a set $\mathcal{H} = \{V_1, \ldots, V_r\}$ of habitats where $V_i \subseteq V$ for all $i \in \{1, \ldots, r\}$, and $k \in \mathbb{N}_0$.

Question: Is there an edge set $F \subseteq E$ with $|F| \leq k$ such that for every $i \in \{1, \ldots, r\}$, it holds that

$\Pi \equiv d$-REACH:	$G[F]^d[V_i]$ is connected?	(Problem 2)	(Sect. 2)
$\Pi \equiv d$-CLOSED:	$G[F]^d[V_i]$ is a clique?	(Problem 3)	(Sect. 3)
$\Pi \equiv d$-DIAM(ETER):	$\mathrm{diam}(G[F][V_i]) \leq d$?	(Problem 4)	(Sect. 4)

In words: d-REACH GBP seeks to connect each habitat such that every patch has some other patch at short distance. d-CLOSED GBP seeks to connect each habitat such that any two habitat's patches are at short distance. Finally, d-DIAM GBP seeks to connect each habitat such that the habitat forms a connected component of low diameter. Figure 1 depicts a relationship between the problems in terms of Karp reductions.

Our Contributions. Our results are summarized in Table 1. We settle the classical complexity and parameterized complexity (regarding the number k of green bridges and the number r of habitats) of the three problems. While d-REACH GBP is (surprisingly) already NP-hard for $d = 1$ on planar graphs or graphs of maximum degree $\Delta = 4$, d-CLOSED GBP and d-DIAM GBP become NP-hard for $d \geq 2$, but admit an $(r+\Delta)^{\mathcal{O}(1)}$-sized problem kernel and thus are linear-time

Table 1. Overview of our results. NP-c., P, and K stand for NP-complete, "polynomial-size", and "problem kernel", respectively. [a](even on planar graphs or if $\Delta = 4$) [b](even on bipartite graphs with $\Delta = 4$ or graphs of diameter four.) [c](even if $r = 1$ or if $r = 2$ and $\Delta = 4$) [d](even on bipartite graphs of diameter three and $r = 1$, *but* linear-time solvable when $r + \Delta$ is constant) [e](admits a linear-size problem kernel if Δ is constant) [f](linear-time solvable when $r+\Delta$ is constant) [†](no polynomial problem kernel unless NP $\subseteq$ coNP / poly, but an $\mathcal{O}(k^3)$-vertex kernel on planar graphs)

Problem (Π GBP)		Comput. Complex.	Parameterized Algorithmics			Ref.
			k	r	$k + r$	
d-REACH (Sect.2)	$d = 1$	NP-c. [a]	$O(k4^k)$ K	*Open*	$O(rk + k^2)$ PK	(Sect.2.1)
	$d = 2$	NP-c.[b]	$O(2^{4k})$ K[†]	p-NP-h. [c]	FPT[†]	(Sect.2.2)
	$d \geq 3$	NP-c	XP, W[1]-h	p-NP-h	XP, W[1]-h	(Sect.2.3)
d-CLOSED (Sect.3)	$d = 1$	Lin. time	—	—	—	(Sect.3)
	$d = 2$	NP-c. [d]	$O(2^{4k})$ K[†]	p-NP-h.[e]	FPT[†]	(Sect.3.1)
	$d \geq 3$	NP-c	XP, W[1]-h	p-NP-h.[e]	XP, W[1]-h	(Sect.3.2)
d-DIAM (Sect.4)	$d = 1$	Lin. time	—	—	—	(Sect.4)
	$d = 2$	NP-c. [f]	$2k$-vertex K	p-NP-h	$O(rk + k^2)$ PK	(Sect.4.1)
	$d \geq 3$	NP-c	$2k$-vertex K	p-NP-h	$O(rk + k^2)$ PK	(Sect.4.1)

solvable if $r + \Delta$ is constant. Except for 1-REACH GBP, we proved all variants to be para-NP-hard regarding r. d-REACH GBP and d-CLOSED GBP are fixed-parameter tractable regarding k when $d \leq 2$, but become W[1]-hard (yet XP) regarding k and $k + r$ when $d > 2$. Additionally, we prove that d-REACH GBP admits an rd-approximation in $\mathcal{O}(mn + rnd)$ time.

Further Related Work. Our problems deal with finding (small) spanning connected subgraphs obeying some (connectivity) constraints, and thus can be seen as network design problems [11]. Most related to our problems are Steiner multigraph problems [16] with an algorithmic study [7,12] (also in the context of wildlife corridor construction). Requiring small diameter appears also in the context of spanning trees [15] and Steiner forests [4]. A weighted version of 4-DIAM GBP is proven to be NP-hard with two different weights [14]. As to wildlife crossing placement, models and approaches different to ours are studied [5,13].

Connecting Habitats Arbitrarily. The following obvious model *just* requires that each habitat is connected.

Problem 1. CONNECTED GREEN BRIDGES PLACEMENT (CONNECT GBP)
Input: An undirected graph $G = (V, E)$, a set $\mathcal{H} = \{V_1, \ldots, V_r\}$ of habitats where $V_i \subseteq V$ for all $i \in \{1, \ldots, r\}$, and an integer $k \in \mathbb{N}_0$.
Question: Is there a subset $F \subseteq E$ with $|F| \leq k$ such that for every $i \in \{1, \ldots, r\}$ it holds that in $G[F]$ exists a connected component containing V_i?

CONNECT GBP with edge costs is also known as STEINER FOREST [7] and generalizes the well-known NP-hard STEINER TREE problem. Gassner [7] proved

STEINER FOREST to be NP-hard even if every so-called terminal net contains two vertices, if the graph is planar and has treewidth three, and if there are two different edge costs, each being upper-bounded linearly in the instance size. It follows that CONNECT GBP is also NP-hard in this case. Bateni et al. [1] proved that STEINER FOREST is polynomial-time solvable on treewidth-two graphs and admits approximation schemes on planar and bounded-treewidth graphs.

From a modeling perspective, CONNECT GBP allows for solutions in which habitats are arbitrarily scattered and the habitat's patches are far away from another; thus animals may need to take long walks through areas outside of their habitats. With our models we avoid solutions with this property.

Preliminaries. Let $\mathbb{N}$ ($\mathbb{N}_0$) be the natural numbers without (with) zero. Let $G = (V, E)$ be an undirected graph with vertex set V and edge set $E \subseteq \binom{V}{2}$. We also denote by $V(G)$ and $E(G)$ the vertices and edges of G, respectively. For $F \subseteq E$ let $V(F) := \{v \in V \mid \exists e \in F : v \in e\}$ and $G[F] := (V(F), F)$. A path P is a graph with $V(P) := \{v_1, \ldots, v_n\}$ and $E(P) := \{\{v_i, v_{i+1}\} \mid 1 \leq i < n\}$. The length of the path P is $|E(P)|$. The distance $\text{dist}_G(v, w)$ between vertices $v, w \in V(G)$ is the length of the shortest path between v and w in G. The diameter $\text{diam}(G)$ is the greatest distance between any two vertices in G. For $p \in \mathbb{N}$, the graph G^p is the p-th power of G containing the vertex set V and edge set $\{\{v, w\} \in \binom{V}{2} \mid \text{dist}_G(v, w) \leq p\}$. We remark that the graph operations introduced above can be stacked. For example, let $F \subseteq E$, let $d \in \mathbb{N}$ and let $V' \subseteq V$. Then $G[F]^d[V'] = ((G[F])^d)[V']$. Let $N_G(v) := \{w \in V \mid \{v, w\} \in E\}$ be the (open) neighborhood of v, and $N_G[v] := N_G(v) \cup \{v\}$ be the closed neighborhood of v. For $p \in \mathbb{N}$, let $N_G^p(v) := \{w \in V \mid \{v, w\} \in E(G^p)\}$ be the (open) p-neighborhood of v, and $N_G^p[v] := N_G^p(v) \cup \{v\}$ be the closed p-neighborhood of v. Two vertices $v, w \in V$ are called twins if $N_G(v) = N_G(w)$. The (vertex) degree $\deg_G(v) := |N_G(v)|$ of v is the number if its neighbors. The maximum degree $\Delta(G) := \max_{v \in V} \deg_G(v)$ is the maximum over all (vertex) degrees.

We use basic definitions from parameterized algorithmics [2].

2 Connecting Habitats with a Patch at Short Reach

The following problem ensures that any habitat patch can reach the other patches via patches of the same habitat and short strolls over "foreign" ground.

Problem 2. d-REACH GREEN BRIDGES PLACEMENT (d-REACH GBP)

Input: An undirected graph $G = (V, E)$, a set $\mathcal{H} = \{V_1, \ldots, V_r\}$ of habitats where $V_i \subseteq V$ for all $i \in \{1, \ldots, r\}$, and an integer $k \in \mathbb{N}_0$.
Question: Is there a subset $F \subseteq E$ with $|F| \leq k$ such that for every $i \in \{1, \ldots, r\}$ it holds that $G[F]^d[V_i]$ is connected?

Theorem 1. d-REACH GBP *is (i) if $d = 1$, NP-hard even on planar graphs or graphs with maximum degree four; (ii) if $d = 2$, NP-hard even on graphs*

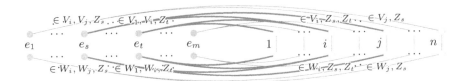

Fig. 2. Illustration to Construction 1 for 1-REACH GBP. Here, e.g., $e_s = \{i,j\}$ and $e_t = \{1,i\}$. Every solution (if existent) contains all red-colored edges. (Color figure online)

with maximum degree four and $r = 2$ or graphs with diameter four and $r = 1$, and in FPT *regarding k; (iii) if $d \geq 3$, NP-hard and* W[1]-*hard regarding $k + r$. Moreover, d-*REACH GBP *admits an rd-approximation of the minimum number of green bridges in $\mathcal{O}(mn + rnd)$ time.*

The approximation algorithm computes for every habitat V_i a spanning tree in $G^d[V_i]$, and adds the edges of the corresponding paths to the solution F. Each of the spanning trees then is a d-approximation for just the one habitat, hence the union of the spanning trees is an rd-approximation for all habitats.[1]

2.1   When a Next Habitat Is Directly Reachable ($d = 1$)

Setting $d = 1$ may reflect perfect knowledge about the habitats. In this case, we want that in $G[F]$, each habitat V_i forms a connected component.

Proposition 1. 1-REACH GBP *is NP-hard even on graphs of maximum degree four and on series-parallel graphs.*

We remark that series-parallel graphs are also planar. We present the construction for showing hardness for graphs of degree at most four. To this end, we reduce from the NP-hard [6] 3-REGULAR VERTEX COVER problem, where given a 3-regular (all vertices are of degree 3) graph $G = (V, E)$ and an integer $k \in \mathbb{N}$, the question is whether there is a subset $V' \subseteq V$ with $|V'| \leq k$ such that $G[V \setminus V']$ contains no edge.

Construction 1. For an instance $\mathcal{I} = (G, k)$ of 3-REGULAR VERTEX COVER with $G = (V, E)$, $V = \{1, \ldots, n\}$, and $E = \{e_1, \ldots, e_m\}$, construct an instance $\mathcal{I}' := (G', \mathcal{H}, k')$ where $\mathcal{H} := \mathcal{V} \cup \mathcal{W} \cup \mathcal{Z}$, $\mathcal{V} := \{V_1, \ldots, V_n\}$, $\mathcal{W} := \{W_1, \ldots, W_n\}$, $\mathcal{Z} := \{Z_1, \ldots, Z_m\}$, and $k' := 4m + k$, as follows (see Fig. 2 for an illustration). Construct vertex sets $V_E := \{x_i, y_i \mid e_i \in E\}$ and $V_G := \{v_i, w_i \mid i \in V\}$. Next, construct edge sets $E^* := \bigcup_{i \in V}\{\{v_i, x_j\}, \{w_i, y_j\} \mid i \in e_j\}$ and $E' := \{\{v_i, w_i\} \mid i \in V\} \cup E^*$. Finally, construct habitats $V_i := \{v_i\} \cup \bigcup_{i \in e_j}\{x_j\}$ and $W_i := \{w_i\} \cup \bigcup_{i \in e_j}\{y_j\}$ for every $i \in \{1, \ldots, n\}$, and $Z_j := \{x_j, y_j\} \cup \bigcup_{i \in e_j}\{v_i, w_i\}$ for every $j \in \{1, \ldots, m\}$.   ◇

To obtain NP-hardness for series-parallel graphs, replace E^* by two stars.

---

[1] All omitted proofs can be found in the full version of this paper, available on arXiv:
https://arxiv.org/abs/2102.04539.

2.2   One Hop Between Habitat Patches ($d = 2$)

We prove that 2-REACH GBP is already NP-complete even if there are two habitats and the graph has maximum degree four, or if there is only one habitat.

Proposition 2. *d-REACH GBP with $d \geq 2$ is NP-complete even if (i) $r = 2$ and $\Delta \leq 4$ or (ii) $r = 1$ and the input graph has diameter $2d$.*

Proposition 2 leaves k unbounded. This leads to the following.

Parameterizing with k. We show that 2-REACH GBP admits a problem kernel of size exponential in k.

Proposition 3. *2-REACH GBP admits a problem kernel with at most $2k + \binom{2k}{k}$ vertices, at most $\binom{2k}{2} + k\binom{2k}{k}$ edges, and at most 2^{2k} habitats.*

Let $\bar{V} := V \setminus \bigcup_{V' \in \mathcal{H}} V'$ for graph $G = (V, E)$ and habitat set $\mathcal{H} = \{V_1, \ldots, V_r\}$. The following reduction rules are immediate.

Reduction Rule 1. (i) If $|V_i| = 1$ for some i, delete V_i. (ii) If a vertex in $\bar{V}$ is of degree at most one, delete it. (iii) If there is an $i \in \{1, \ldots, r\}$ with $|V_i| > 1$ and an $v \in V_i$ of degree zero, return a trivial no-instance. (iv) If there is $v \in V \setminus \bar{V}$ of degree at most one, delete it (also from $V_1, \ldots, V_r$), and set $k := k - 1$.

Clearly, k edges can connect at most $2k$ vertices; thus we obtain the following.

Reduction Rule 2. If $|V \setminus \bar{V}| > 2k$, then return a trivial no-instance.

So we have at most $2k$ vertices in habitats. Next, we upper-bound the number of non-habitat vertices. No minimal solution has edges between two such vertices.

Reduction Rule 3. If there is an edge $e \in E$ with $e \subseteq \bar{V}$, then delete e.

Moreover, no minimum solution connects through non-habitat twins.

Reduction Rule 4. If $N(v) \subseteq N(w)$ for distinct $v, w \in \bar{V}$, then delete v.

We still need to bound the number of vertices in $\bar{V}$. For a $2n$-element set S let $\mathcal{F} \subseteq 2^S$ be a family of subsets such that for every $A, B \in \mathcal{F}$ we have $A \not\subseteq B$. Then $|\mathcal{F}| \leq \binom{2n}{n}$ by Sperner's Theorem. Hence, after applying the reduction rules, we get an instance with at most $2k + \binom{2k}{k}$ vertices and $\binom{2k}{2} + 2k\binom{2k}{k}$ edges.

Reduction Rule 5. If $V_i = V_j$ for distinct $i, j \in \{1, \ldots, r\}$, then delete V_j.

It follows that we can safely assume that $r \leq 2^{2k}$. Thus, Proposition 3 follows. Unfortunately, improving the problem kernel to polynomial-size appears unlikely, which can be shown by a linear parametric transformation from SET COVER.

Proposition 4. *Unless NP $\subseteq$ coNP / poly, d-REACH GBP for $d \geq 2$ admits no problem kernel of size $k^{\mathcal{O}(1)}$, even if $r \geq 1$ is constant.*

Proposition 4 holds for general graphs. In fact, for planar graphs, the above reduction rules allow for an $\mathcal{O}(k^3)$-vertex kernel.

Proposition 5. *2-REACH GBP on planar graphs admits a problem kernel with $\mathcal{O}(k^3)$ vertices and edges and at most 2^{2k} habitats.*

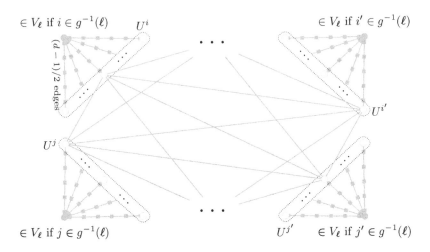

Fig. 3. Illustration to Construction 2 for d-REACH GBP for $d \geq 3$.

2.3 At Least Two Hops Between Habitat Patches ($d \geq 3$)

If the data is more sparse, that is, the observed habitat patches are scattered, then the problem becomes significantly harder to solve.

Proposition 6. d-REACH GBP *with* $d \geq 3$ *is* NP-*complete and* W[1]-*hard when parameterized by* $k + r$.

We give the construction for odd d. Adapting it for even d is straightforward.

Construction 2. Let G with $G = (U^1, \ldots, U^k, E)$ be an instance of MULTI-COLORED CLIQUE where $G[U^i]$ forms an independent set for every $i \in \{1, \ldots, k\}$. Assume without loss of generality that $U^i = \{u^i_1, \ldots, u^i_{|U^i|}\}$. Construct the instance $(G', \{V_1, \ldots, V_{\binom{k}{2}}\}, k')$ with $k := \frac{(d-1)}{2}k + \binom{k}{2}$ as follows (see Fig. 3).

Let $g \colon \left(\begin{smallmatrix} \{1,\ldots,k\} \\ 2 \end{smallmatrix}\right) \to \{1, \ldots, \binom{k}{2}\}$ be a bijective function. Let G' be initially G. For each $i \in \{1, \ldots, k\}$, add a vertex v_i to G', add v_i to each habitat V_ℓ with $i \in g^{-1}(\ell)$, and connect v_i with u^i_j for each $j \in \{1, \ldots, |U^i|\}$ via a path with $\frac{d-1}{2}$ edges, where v_i and u^i_j are the endpoints of the path.   ◇

3 Connecting Habitats at Short Pairwise Distance

In the next problem, we require short pairwise reachability.

Problem 3. d-CLOSED GREEN BRIDGES PLACEMENT (d-CLOSED GBP)

Input: An undirected graph $G = (V, E)$, a set $\mathcal{H} = \{V_1, \ldots, V_r\}$ of habitats where $V_i \subseteq V$ for all $i \in \{1, \ldots, r\}$, and $k \in \mathbb{N}_0$.

Question: Is there a subset $F \subseteq E$ with $|F| \leq k$ such that for every $i \in \{1, \ldots, r\}$ it holds that $G[F]^d[V_i]$ is a clique?

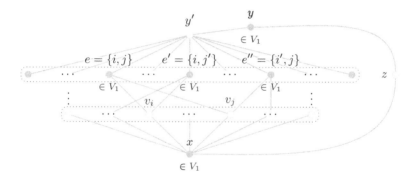

Fig. 4. Illustration to Construction 3 for 2-CLOSED GBP.

If $G[F]^d[V_i]$ is a clique, then $\text{dist}_{G[F]}(v, w) \leq d$ for all $v, w \in V_i$. Note that 2-CLOSED GBP is an unweighted variant of the 2NET problem [3].

Theorem 2. d-CLOSED GBP *is, (i) if* $d = 1$, *solvable in time linear to the input size; (ii) if* $d = 2$, *NP-hard even on bipartite graphs of diameter three and* $r = 1$, *in* FPT *regarding* k, *and in* FPT *regarding* $r + \Delta$; *(iii) if* $d \geq 3$, *NP-hard and* W[1]*-hard regarding* k *even if* $r = 1$.

For $d = 1$, the problem is solvable in linear time: Check whether each habitat induces a clique. If so, check if the union of the cliques is small enough.

Observation 1. 1-CLOSED GBP *is solvable in linear time.*

3.1   When Each Part Is Just Two Steps Away ($d = 2$)

For $d = 2$, d-CLOSED GBP becomes NP-hard already on quite restrictive inputs.

Proposition 7. 2-CLOSED GBP *is NP-complete, even if* $r = 1$ *and the input graph is bipartite and of diameter three.*

Construction 3. Let $\mathcal{I} = (G, k)$ with $G = (V, E)$ be an instance of VERTEX COVER, and assume without loss of generality that $V = \{1, \ldots, n\}$. Construct an instance of 2-CLOSED GBP with graph $G' = (V', E')$, habitat V_1, and integer $k' := 2|E| + k + 3$ as follows (see Fig. 4 for an illustration). To construct G' and V_1, add the vertex set $V_E := \{v_e \mid e \in E\}$, add two vertices y' and y, and make y' adjacent with y and all vertices in V_E. Add a vertex x and introduce a path of length two from x to y. Add the vertex set $V_G := \{v_i \mid i \in V\}$ and add the edge set $E_G := \bigcup_{i \in V}\{\{v_i, v_e\} \mid i \in e\}$. Finally, set $V_1 := V_E \cup \{x, y\}$.   ◇

Graphs of Constant Maximum Degree. 2-REACH GBP is NP-hard if the number r of habitats and the maximum degree Δ are constant (Proposition 2). 2-CLOSED GBP is linear-time solvable in this case:

Proposition 8. d-CLOSED GBP *admits an $\mathcal{O}(r\Delta(\Delta-1)^{3d/2})$-sized problem kernel computable in $\mathcal{O}(r(n+m))$ time.*

Proof. Let $\mathcal{I} = (G, \mathcal{H}, k)$ be an instance of d-CLOSED GBP. For every $i \in \{1, \ldots, r\}$, fix a vertex $u_i \in V_i$. We assume that we have $V_i \subseteq N_G^d[u_i]$ for all $i \in \{1, \ldots, r\}$, otherwise $\mathcal{I}$ is a no-instance. Now let $W_i = N_G^{\lceil 3d/2\rceil}[u_i]$ and let $G' := G[\bigcup_{i=1}^r W_i]$. Note that G' contains at most $r\Delta(\Delta-1)^{\lceil 3d/2\rceil}$ vertices and can be computed by r breadth-first searches. We claim that G' contains every path of length at most d between every two vertices $v, w \in V_i$, for every $i \in \{1, \ldots, r\}$. Recall that an edge set $F \subseteq E$ is a solution if and only if for every $i \in \{1, \ldots, r\}$ and for every $v, w \in V_i$, $G[F]$ contains a path of length at most d from v to w. As by our claim G' contains any such path, that $\mathcal{I}$ is a yes-instance if and only if $\mathcal{I}' := (G', \mathcal{H}, k)$ is a yes-instance (since $V_i \subseteq V(G')$ for every $i \in \{1, \ldots, r\}$).

Assuming that $V_i \subseteq N_G^d[u_i]$, $G[W_i]$ contains all paths of length at most d between u_i and any $v \in V_i$. So let $v, w \in V_i$ be two vertices, both distinct from u_i. As $v, w \in N_G^d[u_i]$ and $W_i = N_G^{\lceil 3d/2\rceil}[u_i]$, the subgraph $G[W_i]$ contains all vertices in $N_G^{\lceil d/2\rceil}[v]$ and $N_G^{\lceil d/2\rceil}[w]$. Consider now a path of length at most d between v and w. Suppose it contains a vertex $x \in V(G) \setminus (N_G^{\lceil d/2\rceil}[v] \cup N_G^{\lceil d/2\rceil}[w])$. Then $\operatorname{dist}_G(v, x) + \operatorname{dist}_G(w, x) > 2\lceil d/2\rceil \geq d$, a contradiction to x being on a path from v to w of length at most d. The claim follows.

Parameterizing with k. All the reduction rules that worked for 2-REACH GBP also work for 2-CLOSED GBP. It thus follows that 2-CLOSED GBP admits a problem kernel of size exponential in k, and hence, 2-CLOSED GBP is FPT. As with 2-REACH GBP, the problem kernel presumably cannot be much improved. For this, we combine the constructions for Propositions 4 & 7.

Corollary 1. *2-CLOSED GBP admits a problem kernel of size exponential in k and, unless $\mathrm{NP} \subseteq \mathrm{coNP}\,/\,\mathrm{poly}$, none of size polynomial in k, even if $r = 1$.*

3.2 When Reaching Each Part Is a Voyage ($d \geq 3$)

For $d \geq 3$, the problem is W[1]-hard regarding the number k of green bridges, even for one habitat. The reduction is similar to the one for Proposition 6.

Proposition 9. *d-CLOSED GBP with $d \geq 3$ is NP-complete and W[1]-hard when parameterized by k, even if $r = 1$.*

4   Connecting Habitats at Small Diameter

Lastly, we consider requiring short pairwise reachability in 1-REACH GBP.

Problem 4. d-DIAMATER GREEN BRIDGES PLACEMENT (d-DIAM GBP)

Input: An undirected graph $G = (V, E)$, a set $\mathcal{H} = \{V_1, \ldots, V_r\}$ of habitats where $V_i \subseteq V$ for all $i \in \{1, \ldots, r\}$, and an integer $k \in \mathbb{N}_0$.

Question: Is there a subset $F \subseteq E$ with $|F| \leq k$ such that for every $i \in \{1, \ldots, r\}$ it holds that $G[F][V_i]$ has diameter d?

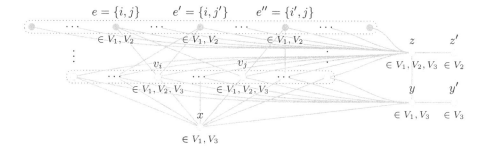

Fig. 5. Illustration for 2-DIAM GBP with $r = 3$. ($k' = 2m + 2n + k + 4$)

In particular, $G[F][V_i]$ is connected. Note that 1-REACH GBP reduces to DIAM GBP (where d is part of the input and then set to the number of vertices in the input instance's graph). We have the following.

Theorem 3. d-DIAM GBP *is (i) if $d = 1$, solvable in linear time; (ii) if $d = 2$, NP-hard even if $r = 3$; (iii) if $d = 3$, NP-hard even if $r = 2$. Moreover, d-DIAM GBP admits a problem kernel with at most $2k$ vertices and at most 2^{2k} habitats.*

1-DIAM GBP is equivalent to 1-CLOSED GBP, which is linear-time solvable by Observation 1. Thus, Theorem 3(i) follows. Applying Reduction Rules 2 & 5 and deleting all non-habitat vertices yields the problem kernel.

4.1 Via at Most One ($d = 2$) or Two ($d \geq 3$) Patches to Every Other

2-DIAM GBP and 3-DIAM GBP turn out to be NP-hard even for three habitats and two habitats, respectively.

Proposition 10. 2-DIAM GBP *is NP-hard even if $r = 3$.*

Construction 4. Let $\mathcal{I} = (G, k)$ with $G = (V, E)$ be an instance of VERTEX COVER and assume without loss of generality that $V = \{1, \ldots, n\}$ and $E = \{e_1, \ldots, e_m\}$. Construct an instance $\mathcal{I}' := (G', \{V_1, V_2, V_3\}, k')$ with $k' = 2m + 2n + k + 4$ as follows (see Fig. 5 for an illustration). Add the vertex sets $V_E := \{v_e \mid e \in E\}$ and $V_G = \{v_i \mid i \in V\}$, as well as the vertices x, y, y', z, z'. Add all vertices except for y' and z' to V_1. Let $V_2 := V_E \cup V_G \cup \{z, z'\}$ and $V_3 := V_G \cup \{x, z, y, y'\}$. Next, for each $e = \{i, j\} \in E$, connect v_e with v_i, v_j, and z. For each $i \in V$, connect v_i with x, y, and z. Lastly, add the edge set $E^* := \{\{x, y\}, \{y, y'\}, \{z, z'\}, \{z, y\}\}$ to E'. Let $E_y := \{\{y, v_i\} \mid i \in V\}$, $E_E := \{\{v_e, z\} \mid e \in E\}$, and $E_V := \{\{v_i, z\} \mid i \in V\}$. ◇

Proposition 11. 3-DIAM GBP *is NP-hard even if $r = 2$.*

5   Conclusion, Discussion, and Outlook

We modeled the problem of placing wildlife crossings with three different problem (families): d-REACH GBP, d-CLOSED GBP, and d-DIAM GBP. We studied the practically desired cases $d = 1$ and $d = 2$, as well as the cases $d \geq 3$. For all three problems, we settled the classical as well as the parameterized complexity (regarding the number k of wildlife crossings and the number r of habitats), except for the parameterized complexity of d-REACH GBP regarding r.

Discussion. We derived an intriguing interrelation of connection requirements, data quality, and computational and parameterized complexity. While each problem admits its individual complexity fingerprint, each of them depends highly on the value of d, the level of the respective connectivity constraint. This value can reflect the quality of the given data, since naturally we assume that habitats are connected. The worse the data, the stronger are the relaxations according to the connectivity of habitats, and thus the larger is the value of d. Our results show that having very small ($d = 2$) data gaps already leads to the problems becoming NP-hard, and that even larger gaps ($d \geq 3$) yield W[1]-hardness (when parameterized by k). Hence, knowledge about habitats, connections, and data quality decide which problem models can be applied, thus influencing the computation power required to determine an optimal placement of wildlife crossings. For instance, for larger networks, we recommend to ensure data quality such that one of our proposed problems for $d \leq 2$ becomes applicable. This in turn emphasizes the importance of careful habitat recognition.

   In our models, we neglected that different positions possibly lead to *different* costs of building bridges (i.e., edge costs). This neglect is justified when differentiating between types of bridges (and thus their costs) is not necessary (e.g., if the habitat's species share preferred types of green bridges, and the underlying human-made transportation lines are homogeneous). In other scenarios, additionally considering these costs may be beneficial for decision-making.

Outlook and Open Problems. For a final version, we plan to continue our study with approximation and (refined) data reduction for our three problems, as well as planar input graphs, and to settle 1-REACH GBP's complexity regarding r. Note that we obtained an $\mathcal{O}(rd)$-approximation for d-REACH GBP, which does not directly transfer to the other two problems. FPT approximations may be lucrative. For small $d \geq 2$, all problems allow for problems kernels where the number of vertices only depends on k. If more effective preprocessing is possible, then data reduction on the habitats is required. If the underlying street network is planar, then the input graphs to our problems can be seen as their planar dual. Thus, input graphs may be planar in the applications.

   Moreover, interesting directions for future work are, for instance, distinguishing types of green bridges to place, taking into account possible movement directions within habitats (connectivity in directed graphs), identifying real-world driven problem parameters leading to tractability, or the problem of maintaining and servicing green bridges over time under a possible seasonal change of wildlife habitats (temporal graph modeling could fit well).

References

1. Bateni, M., Hajiaghayi, M.T., Marx, D.: Approximation schemes for Steiner forest on planar graphs and graphs of bounded treewidth. J. ACM **58**(5), 21:1–21:37 (2011). https://doi.org/10.1145/2027216.2027219
2. Cygan, M., et al.: Parameterized Algorithms. Springer, Cham (2015). https://doi.org/10.1007/978-3-319-21275-3
3. Dahl, G., Johannessen, B.: The 2-path network problem. Networks **43**(3), 190–199 (2004). https://doi.org/10.1002/net.20003
4. Ding, W., Qiu, K.: A 2-approximation algorithm and beyond for the minimum diameter k-Steiner forest problem. Theor. Comput. Sci. **840**, 1–15 (2020). https://doi.org/10.1016/j.tcs.2019.12.012
5. Downs, J.A., Horner, M.W., Loraamm, R.W., Anderson, J., Kim, H., Onorato, D.: Strategically locating wildlife crossing structures for Florida panthers using maximal covering approaches. Trans. GIS **18**(1), 46–65 (2014). https://doi.org/10.1111/tgis.12005
6. Garey, M.R., Johnson, D.S., Stockmeyer, L.J.: Some simplified NP-complete graph problems. Theor. Comput. Sci. **1**(3), 237–267 (1976). https://doi.org/10.1016/0304-3975(76)90059-1
7. Gassner, E.: The Steiner forest problem revisited. J. Discrete Algorithms **8**(2), 154–163 (2010). https://doi.org/10.1016/j.jda.2009.05.002
8. Gomes, C.P.: Computational sustainability: computational methods for a sustainable environment, economy, and society. Bridge **39**(4), 5–13 (2009)
9. Huijser, M.P., Duffield, J.W., Clevenger, A.P., Ament, R.J., McGowen, P.T.: Cost–benefit analyses of mitigation measures aimed at reducing collisions with large ungulates in the United States and Canada: a decision support tool. Ecol. Soc. **14**(2) (2009). http://www.jstor.org/stable/26268301
10. Huijser, M.P., et al.: Wildlife-vehicle collision reduction study: report to congress. (2008). https://www.fhwa.dot.gov/publications/research/safety/08034/08034.pdf
11. Kerivin, H., Mahjoub, A.R.: Design of survivable networks: a survey. Networks **46**(1), 1–21 (2005). https://doi.org/10.1002/net.20072
12. Lai, K.J., Gomes, C.P., Schwartz, M.K., McKelvey, K.S., Calkin, D.E., Montgomery, C.A.: The Steiner multigraph problem: wildlife corridor design for multiple species. In: Proceedings of 25th AAAI. AAAI Press (2011). http://www.aaai.org/ocs/index.php/AAAI/AAAI11/paper/view/3768
13. Loraamm, R.W., Downs, J.A.: A wildlife movement approach to optimally locate wildlife crossing structures. Int. J. Geogr. Inf. Sci. **30**(1), 74–88 (2016). https://doi.org/10.1080/13658816.2015.1083995
14. Plesník, J.: The complexity of designing a network with minimum diameter. Networks **11**(1), 77–85 (1981). https://doi.org/10.1002/net.3230110110
15. Ravi, R., Sundaram, R., Marathe, M.V., Rosenkrantz, D.J., Ravi, S.S.: Spanning trees - short or small. SIAM J. Discret. Math. **9**(2), 178–200 (1996). https://doi.org/10.1137/S0895480194266331
16. Richey, M.B., Parker, R.G.: On multiple Steiner subgraph problems. Networks **16**(4), 423–438 (1986). https://doi.org/10.1002/net.3230160408

17. Sawaya, M.A., Kalinowski, S.T., Clevenger, A.P.: Genetic connectivity for two bear species at wildlife crossing structures in Banff national park. Proc. Royal Soc. B: Biol. Sci. **281**(1780), 20131705 (2014). https://www.ncbi.nlm.nih.gov/pmc/articles/PMC4027379/
18. Zheng, R., Luo, Z., Yan, B.: Exploiting time-series image-to-image translation to expand the range of wildlife habitat analysis. In: Proceedings of 33rd AAAI, pp. 825–832. AAAI Press (2019). https://doi.org/10.1609/aaai.v33i01.3301825

A Church-Turing Thesis for Randomness?

Johanna N.Y. Franklin$^{(\boxtimes)}$ (iD)

Hofstra University, Hempstead, NY 11590, USA
johanna.n.franklin@hofstra.edu
http://www.johannafranklin.net

Abstract. We discuss the difficulties in stating an analogue of the Church-Turing thesis for algorithmic randomness. We present one possibility and argue that it cannot occupy the same position in the study of algorithmic randomness that the Church-Turing thesis does in computability theory. We begin by observing that some evidence comparable to that for the Church-Turing thesis does exist for this statement: in particular, there are other reasonable formalizations of the intuitive concept of randomness that lead to the same class of random sequences (the Martin-Löf random sequences). However, we consider three properties that we would like a random sequence to satisfy and find that the Martin-Löf random sequences do not necessarily possess these properties to a greater degree than other types of random sequences, and we further argue that there is no more appropriate version of the Church-Turing thesis for algorithmic randomness. This suggests that consensus around a version of the Church-Turing thesis in this context is unlikely.

Keywords: Church-Turing thesis · Algorithmic randomness · Computability theory

1 A Potential Parallel

In 1948, Turing wrote that "[I]t is found in practice that [Turing machines] can do anything that could be described as 'rule of thumb' or 'purely mechanical.' This is sufficiently well established that it is now agreed amongst logicians that 'calculable by means of a [Turing machine]' is the correct accurate rendering of such phrases" ([30], p. 4). We take this as our formulation of the Church-Turing thesis and discuss the prospects for identifying an analogous statement in the context of algorithmic randomness.

Algorithmic randomness is the study of the formalization of the intuitive concept of randomness using concepts from computability theory. We begin by considering elements of the Cantor space (the space of infinite binary sequences, or 2^ω). Therefore, a statement of the following form would be a direct parallel to Church's thesis:

Supported in part by Simons Foundation Collaboration Grant #420806.

L. De Mol et al. (Eds.): CiE 2021, LNCS 12813, pp. 217–226, 2021.
https://doi.org/10.1007/978-3-030-80049-9_20

S: If an infinite binary sequence can be described as random, then it
_____.

The question, of course, is which mathematical property should be used to complete this statement. Is there one that occupies the same space in the universe of randomness as calculability by a Turing machine does in the universe of functions? We argue that, while there seems to be such a characterization at first glance, it appears inappropriate on further consideration, as do all other reasonable possibilities.

1.1   An Initial Characterization of Randomness

The first formal definitions of randomness were provided in the mid-1960s and early 1970s. We review these definitions briefly below. First, though, we explain our notation. We will typically denote finite binary strings by lowercase Greek letters and infinite binary sequences by uppercase Roman letters. The length of a finite binary string τ is denoted by $|\tau|$, and the measure of a class C in Cantor space, $\mu(C)$, is given by the Lebesgue measure in which the basic open set $[\sigma]$ consisting of all infinite binary sequences extending the finite string σ has measure $2^{-|\sigma|}$ (in other words, the "coin-flip" measure).

The first approach to be fully defined is due to Martin-Löf and is based on effectivized statistical tests [19]. We recall that a Σ_1^0 class in Cantor space is one that is definable as $\{A \mid (\exists n) R(A{\restriction} n)\}$ for a computable relation R.

Definition 1. *A* Martin-Löf test *is a sequence* $\langle V_i \rangle$ *of uniformly* Σ_1^0 *classes of Cantor space such that* $\mu(V_i) \leq 2^{-i}$. *An infinite sequence A is said to be* Martin-Löf random *if for any Martin-Löf test* $\langle V_i \rangle$, $A \notin \cap V_i$. *We say that such a sequence is not captured by any Martin-Löf test and thus* passes *all of them.*

This is the candidate I propose to complete S: "passes a Martin-Löf test."

The second approach is based on Kolmogorov complexity, initially defined by Kolmogorov in 1965. The prefix-free variant is due to Levin and Chaitin.

Definition 2. *[4, 16, 17] The* prefix-free Kolmogorov complexity *of a finite binary string σ is defined as* $K(\sigma) = \min\{|\tau| \mid U(\tau) = \sigma\}$, *where U is a universal prefix-free machine.*

The third approach is probabilistic and is based on Lévy's definition of a martingale [18]:

Definition 3. *A c.e. function* $d : 2^{<\omega} \to \mathbb{R}^{\geq 0}$ *is a c.e. martingale if it obeys the fairness condition*

$$d(\sigma) = \frac{d(\sigma 0) + d(\sigma 1)}{2}$$

for all σ.

Martin-Löf randomness can be defined using all of these approaches.

Theorem 1 ([4,27,28]). *The three following properties are equivalent for an infinite binary sequence A:*

1. *A is Martin-Löf random.*
2. *There is a constant c such that for all $n \in \omega$, $K(A{\upharpoonright}n) \geq n - c$: the Kolmogorov complexity of each initial segment of A is never much smaller than its length.*
3. *For any c.e. martingale d, $\limsup_n d(A{\upharpoonright}n)$ is finite: it is not possible to win arbitrarily large amounts of capital by betting on A using a c.e. martingale, and thus no such d succeeds on A.*

We note that there is not only a universal prefix-free machine but also a universal Martin-Löf test and a universal c.e. martingale.

This suggests that we can justify completing S with "passes a Martin-Löf test" with evidence similar to that given for Church's thesis: these very different approaches to formalizing the intuitive notion of randomness result in the same class of sequences being considered random just as Turing machines, register machines, general recursive functions, and the λ-calculus do for partial computable functions; Porter refers to this as an *equivalence-as-evidence-of-capturing (EEC) claim* [23]. However, we find, as Porter did, that the existence of other "loci of definitional equivalence" greatly weakens the value of this fact.

2 Other Randomness Notions

It was recognized in the early 1970s that the three characterizations of Martin-Löf randomness given above could be modified slightly to obtain other classes of sequences that could also reasonably be called random. For instance, we can define Schnorr randomness in each of these three ways by making each component of the definition computable rather than merely approximable from below:

– Rather than consider all Martin-Löf tests, we consider only *Schnorr tests*: those whose components have measure exactly 2^{-i} for the appropriate i rather than no larger than 2^{-i} [28].
– Rather than consider all prefix-free machines, we consider only prefix-free machines whose domains have computable measure [5].
– Rather than consider all c.e. martingales, we consider only computable martingales, and our success condition changes slightly: a martingale d *h-succeeds* on A if $\limsup_n \frac{d(A{\upharpoonright}n)}{h(n)} = \infty$. Here, h is taken to be an order function: a computable, nondecreasing, unbounded function on ω [27,28].

The resulting characterizations of randomness are, at worst, only slightly more complicated than the characterizations of Martin-Löf randomness we have already seen. Any increased complexity of the definitions results from the following facts: (1) there is no universal prefix-free computable measure machine, Schnorr test, or computable martingale, and (2) we have modified the definition of a martingale's success to reflect the idea that it may be possible for a martingale's values to increase unboundedly but so slowly that we cannot recognize this increase computably.

Schnorr randomness is certainly no stronger than Martin-Löf randomness: if we consider the test definitions, we can see that a sequence has to pass fewer tests in order to be Schnorr random than to be Martin-Löf random. In fact, it is a strictly weaker notion [28].

There is also a well-studied randomness notion strictly between Martin-Löf and Schnorr randomness: computable randomness, first described by Schnorr in [27,28]. Its characterizations in terms of Kolmogorov complexity and tests are far more complicated than those of either Martin-Löf or Schnorr randomness (see Sects. 7.1.4 and 7.1.5 of [6]), but its martingale characterization is as simple as that of Martin-Löf randomness: one simply substitutes "computable martingale" for "c.e. martingale."

The existence of these other notions of randomness would not necessarily preclude the analogy to Church's thesis given above. After all, computability theorists routinely investigate weak truth table reducibility and truth table reducibility in addition to Turing reducibility and don't consider this to contraindicate Church's thesis. We may ask whether the same sort of relationship holds between Turing, weak truth table, and truth table functionals as between Martin-Löf, computable, and Schnorr randomness.

At first, this analogy seems reasonable. Turing reducibility places no requirements on the convergence of the functional; weak truth table reducibility requires convergence within a computable bound, if such exists; and truth table reducibility requires convergence for all inputs. We can see that the characterizations of Martin-Löf randomness involve c.e. martingales and tests with components and prefix-free machines with domains that need only have lower semicomputable measures. This degree of approximability is precisely that which can be obtained from a Turing functional. The characterizations of computable randomness and Schnorr randomness, on the other hand, require computable martingales, and the components of Schnorr tests have computable measures, which better corresponds to weak truth table or truth table functionals.

While this suggests that Martin-Löf randomness is analogous to Turing reducibility and thus that "passing a Martin-Löf test" seems analogous to "being calculable by a Turing machine," we should investigate further and determine how far this analogy extends. While there are rich structural results for the Turing, weak truth table, and truth table degrees, it is the Turing degrees that have found the widest applicability to branches of computability beyond degree theory. The author knows of no results concerning truth table degrees and computable structure theory or weak truth table degrees and computable analysis, for instance. This could be a further sort of evidence for Church's thesis: that the degree structure generated by Turing functionals is the most generally applicable to other aspects of computability theory. This leads us to ask whether a similar statement can be made about Martin-Löf randomness.

3   Three Desiderata

In this section, we will consider the question of applicability discussed above as well as the question of whether Martin-Löf randomness most aptly captures our intuitions about random sequences. It has certainly been argued that Martin-Löf randomness does not capture these intuitions, most notably (and earliest) by Schnorr in [28]. We discuss some of these considerations here.

3.1 Decompositions and Combinations of Random Sequences

We begin by considering what happens if we computably decompose a random sequence: must this result in random sequences? Or, if we interleave two random sequences, under what circumstances will the resulting sequence be random? These questions were answered for Martin-Löf randomness by van Lambalgen [15] and are closely related to the role of relativization in computability.

Theorem 2 (van Lambalgen's Theorem). *The following are equivalent for any two Martin-Löf random sequences A and B:*

1. *$A \oplus B$ is Martin-Löf random.*
2. *A is Martin-Löf random relative to B and B is Martin-Löf random relative to A.*

In short, a sequence is Martin-Löf random if, when you decompose it into its "even" and "odd" bits, each half is not only Martin-Löf random but Martin-Löf random relative to the other. This result is frequently mentioned as a desideratum for a randomness notion (see, for instance, Sect. 7.1.2 in [6]) and is thus a reasonable place for us to begin. We should now ask if computable and Schnorr randomness have this property as well.

The answer is complicated. It is straightforward to see that the forward direction of the theorem does not hold for computable or Schnorr randomness (see, for instance, Kjos-Hanssen's argument in [22]). However, it becomes more complicated when we consider the backward direction. This direction was long claimed to hold for both computable and Schnorr randomness with "essentially the same proof" as for Martin-Löf randomness ([6], p. 276). However, no proof was provided until Franklin and Stephan gave one for Schnorr randomness [10], and a few years later, Bauwens proved that this direction does not actually hold for computable randomness [1].

However, in keeping with the analogy between stronger reductions and weaker forms of randomness described in Sect. 2, it turns out that this theorem holds for all three of these randomness notions if we apply a different relativization [20, 21]. This suggests that while Martin-Löf randomness initially seems to satisfy one of our intuitions about randomness in a way that computable and Schnorr randomness simply don't, it seems that our intuition is satisfied for the latter two as well when we use a more appropriate framework. Whether this more appropriate framework is as natural, though, is not apparent.

3.2 Computational Strength

Now we consider the computational strength of random sequences. It is fair to say that no random sequence should be computable: if it were, then we could predict it perfectly. Thus, random sequences should possess some noncomputable information. However, we can also argue that no random sequence should be very powerful computationally: if a sequence is random, then we should not be able to make any practical use of the information it possesses, and therefore it should not be contained in a powerful Turing degree.

However, Kučera proved that every Turing degree computing $0'$ contains a Martin-Löf random sequence [14]. Furthermore, Stephan proved that the Martin-Löf random sequences that cannot compute $0'$ are computationally weaker in another way: they are precisely the Martin-Löf random sequences that cannot compute a complete extension of Peano Arithmetic [29]. These results led to Hirschfeldt's argument that there are two types of Martin-Löf random sequences: those that are computationally weak and thus truly random, and those that are computationally strong and therefore "know enough" to pretend to be random (see [6], pp. 228–229).

If we take a measure of computational uselessness as a desideratum for a randomness notion, it is clear that not only does Martin-Löf randomness not meet this criterion, but neither do Schnorr and computable randomness since every Martin-Löf random sequence is Schnorr random and computably random. However, there are other randomness notions, and one of these satisfies this intuition perfectly.

Difference randomness was introduced by Franklin and Ng in [9]. This notion is most naturally defined using the test approach: while each component of a Martin-Löf test is a Σ_1^0 class, each component of a difference test is a difference of two Σ_1^0 classes. This means that, rather than creating a component by simply adding open neighborhoods $[\sigma]$ to the class, we create a component by adding such neighborhoods and then perhaps removing them or their subneighborhoods. Franklin and Ng further proved that the difference random sequences are precisely the Martin-Löf random sequences that cannot compute $0'$ and thus that difference randomness satisfies this intuition in a way no other notion does [9].[1] However, difference randomness does not satisfy many of the other criteria we have discussed so far: while its test definition is straightforward to state, its martingale definition is rather complicated, and no Kolmogorov complexity-based definition of it is known at this point.

3.3   Applications

Finally, we turn our attention to the last desideratum: we would like our notion of randomness to appear naturally in other branches of computability theory. We consider the case of computable analysis, the subfield that is most closely connected to algorithmic randomness as of this writing. Since many theorems in analysis hold on a conull set and all but measure 0 many points in a space

---

[1] We note that there are other randomness notions that also exhibit computational weakness, e.g., weak 2-randomness. However, we discuss difference randomness here because the difference random sequences can be identified as the Martin-Löf random sequences that are computationally weak in these two standard senses.

are random by any reasonable definition of randomness, it is natural to try to characterize the points in a computable probability space for which a certain theorem holds as the points in that space that are random under a certain definition.[2]

We consider several theorems as case studies, beginning with Birkhoff's ergodic theorem; this theorem states that for any measurable subset of a probability space, an ergodic transformation will map almost every point into that subset with a frequency proportional to the measure of the subset.

Theorem 3 (Birkhoff's ergodic theorem). *Let (X, μ) be a probability space, let $T : X \to X$ be ergodic, and let E be a measurable subset of X. Then for almost all $x \in X$,*

$$\lim_{n \to \infty} \frac{|\{i \mid i < n \ and \ T^i(x) \in E\}|}{n} = \mu(E).$$

To connect this theorem to algorithmic randomness, we must first frame it as a statement about individual points in the space. While the definition of a Birkhoff point arises naturally from Birkhoff's ergodic theorem, weak Birkhoff points are appropriate for a generalization of Birkhoff's ergodic theorem for measure-preserving functions.

Definition 4. *A point $x \in X$ is a* Birkhoff *point for T with respect to a class of sets $\mathcal{C}$ if for all $E \in \mathcal{C}$,*

$$\lim_{n \to \infty} \frac{|\{i \mid i < n \ and \ T^i(x) \in E\}|}{n} = \mu(E).$$

A point $x \in X$ is a weak Birkhoff *point for T with respect to a class of sets $\mathcal{C}$ if for all $E \in \mathcal{C}$, the above limit simply exists.*

We can now consider a theorem template; note that in order to state such a theorem precisely, we must include the type of transformation under consideration (ergodic or measure preserving) and the class of sets under consideration (computable or lower semicomputable).

Theorem template 1. *A point is a (weak) Birkhoff point for computable _____ T with respect to _____ sets if and only if it is _____ random.*

We synthesize the known results in Table 1.

We now turn our attention to differentiability and convergence of Fourier series; differentiability is considered in more depth in this context by Porter in [24]. We again have a theorem template, and in these cases, we only need to know what sort of functions we are considering the differentiability of or the Fourier series of.

---

[2] The reader may have noted that we are working in a general computable probability space rather than the Cantor space. This is possible because any computable probability space is isomorphic to the Cantor space in every relevant way and our notions of randomness transfer naturally [13].

Table 1. Birkhoff's ergodic theorem and randomness

	Transformation	
	Ergodic	Measure-preserving
Computable	Schnorr [12]	Martin-Löf [11,31]
Lower semicomputable	Martin-Löf [2,7]	?

Theorem template 2. *Every computable _____ function f is differentiable at z ∈ [0, 1] if and only if z is _____ random.*

Theorem template 3. *Every computable _____ function f's Fourier series converges at t_0 if and only if t_0 is _____ random.*

Brattka, J. Miller, and Nies proved that each computable nondecreasing function $f : [0, 1] \to \mathbb{R}$ is differentiable at a point z if and only if z is computably random and that each computable function $f : [0, 1] \to \mathbb{R}$ of bounded variation is differentiable at a point z if and only if z is Martin-Löf random [3]; Rute has a similar result for Schnorr randomness that is more complicated to state [25]. Later, Franklin, McNicholl, and Rute proved that the convergence of a Fourier series for a computable function f in $L^p[-\pi, \pi]$ at a point t_0 is essentially equivalent to the Schnorr randomness of t_0 [8].[3]

Both Schnorr randomness and Martin-Löf randomness make repeated appearances in this area; computable randomness has appeared less often. It does not appear that Martin-Löf randomness is primary in this context, and in fact Rute has argued that Schnorr randomness "stands out" as having "very strong connections to constructive and computable measure theory" ([26], p. 60).

4   Conclusion

It seems clear that Martin-Löf randomness does not hold the primacy of place in the context of algorithmic randomness that Turing functionals do in the context of basic computability. While Turing functionals are by far the most useful kind in classical computability theory, it seems that the same is not true for Martin-Löf random sequences in algorithmic randomness. While Martin-Löf randomness is straightforwardly defined in all the frameworks we consider and Martin-Löf random sequences can be decomposed or combined into other Martin-Löf random sequences as expected, it lacks the desired computational weakness and certainly does not stand out in the context of applications to computable analysis.

This suggests that giving a formal definition of Martin-Löf randomness is not the correct way to complete our statement S. It does not seem, though, that a formal definition of any other randomness notion would be correct, either: there is

---

[3] There is a subtlety in this result in that an incomputable function may be computable as a vector, hence the "essentially."

no more consistent evidence for any of the other notions we've discussed. There-fore, one of the most important forms of evidence for the Church-Turing thesis is missing in the context of algorithmic randomness, and we cannot reasonably provide an equivalent version for this context.

I suggest that this failure is due to the fact that randomness is a higher-order property than computability. To define a randomness notion formally, we need to state the level of computability of the measures of the test components, the measures of the prefix-free machines, or the martingales and we may need to consider the martingale's rate of growth. Porter presents an excellent analysis of the ingredients of a formal definition of a randomness notion in [23]: each defi-nition must have a *hallmark of randomness*, a *collection of underlying resources*, and an *implementation of these resources*. With so many factors in play, it seems unlikely that we will ever have consensus around a version of the Church-Turing thesis for algorithmic randomness.

References

1. Bauwens, B.: Uniform van Lambalgen's theorem fails for computable randomness. ArXiv e-prints (2015)
2. Bienvenu, L., Day, A., Mezhirov, I., Shen, A.: Ergodic-type characterizations of algorithmic randomness. In: Ferreira, F., Löwe, B., Mayordomo, E., Mendes Gomes, L. (eds.) CiE 2010. LNCS, vol. 6158, pp. 49–58. Springer, Heidelberg (2010). https://doi.org/10.1007/978-3-642-13962-8_6
3. Brattka, V., Miller, J.S., Nies, A.: Randomness and differentiability. Trans. Amer. Math. Soc. **368**(1), 581–605 (2016)
4. Chaitin, G.J.: A theory of program size formally identical to information theory. J. Assoc. Comput. Mach. **22**, 329–340 (1975)
5. Downey, R.G., Griffiths, E.J.: Schnorr randomness. J. Symbolic Logic **69**(2), 533–554 (2004)
6. Downey, R.G., Hirschfeldt, D.R.: Algorithmic Randomness and Complexity. Springer, New York (2010)
7. Franklin, J.N., Greenberg, N., Miller, J.S., Ng, K.M.: Martin-Löf random points satisfy Birkhoff's ergodic theorem for effectively closed sets. In: Proceedings of the American Mathematical Society, vol. 140, no. 10, pp. 3623–3628 (2012)
8. Franklin, J.N., McNicholl, T.H., Rute, J.: Algorithmic randomness and Fourier analysis. Theory Comput. Syst. **63**(3), 567–586 (2019)
9. Franklin, J.N., Ng, K.M.: Difference randomness. In: Proceedings of the American Mathematical Society, vol. 139, no. 1, pp. 345–360 (2011)
10. Franklin, J.N., Stephan, F.: Van Lambalgen's theorem and high degrees. Notre Dame J. Formal Logic **52**(2), 173–185 (2011)
11. Franklin, J.N., Towsner, H.: Randomness and non-ergodic systems. Mosc. Math. J. **14**(4), 711–744 (2014)
12. Gács, P., Hoyrup, M., Rojas, C.: Randomness on computable probability spaces–a dynamical point of view. Theory Comput. Syst. **48**(3), 465–485 (2011). https://doi.org/10.1007/s00224-010-9263-x. http://dx.doi.org/10.1007/s00224-010-9263-x
13. Hoyrup, M., Rojas, C.: Computability of probability measures and Martin-Löf randomness over metric spaces. Inform. Comput. **207**(7), 830–847 (2009). https://doi.org/10.1016/j.ic.2008.12.009. http://dx.doi.org/10.1016/j.ic.2008.12.009

14. Kučera, A.: Measure, π_1^0, -classes and complete extensions of PA. In: Ebbinghaus, H.-D., Müller, G.H., Sacks, G.E. (eds.) Recursion Theory Week. LNM, vol. 1141, pp. 245–259. Springer, Heidelberg (1985). https://doi.org/10.1007/BFb0076224

15. van Lambalgen, M.: The axiomatization of randomness. J. Symbolic Logic **55**(3), 1143–1167 (1990)

16. Levin, L.A.: Some Theorems on the Algorithmic Approach to Probability Theory and Information Theory. Ph.D. thesis, Moscow University (1971)

17. Levin, L.A.: Laws on the conservation (zero increase) of information, and questions on the foundations of probability theory. Problemy Peredači Informacii **10**(3), 30–35 (1974)

18. Lévy, P.: Téorie de l'Addition des Variables Aléatoires. Gauthier-Villars, Paris (1937)

19. Martin-Löf, P.: The definition of random sequences. Inf. Control **9**, 602–619 (1966)

20. Miyabe, K.: Truth-table Schnorr randomness and truth-table reducible randomness. MLQ Math. Log. Q. **57**(3), 323–338 (2011). https://doi.org/10.1002/malq.200910128. http://dx.doi.org/10.1002/malq.200910128

21. Miyabe, K., Rute, J.: van Lambalgen's theorem for uniformly relative Schnorr and computable randomness. In: Downey, R., et al. (eds.) Proceedings of the Twelfth Asian Logic Conference, pp. 251–270. World Scientific (2013)

22. Nies, A.: Computability and Randomness. Clarendon Press, Oxford (2009)

23. Porter, C.P.: The equivalence of definitions of algorithmic randomness. Philosophia Mathematica To appear

24. Porter, C.P.: On analogues of the Church-Turing thesis in algorithmic randomness. Rev. Symb. Log. **9**(3), 456–479 (2016). https://doi.org/10.1017/S1755020316000113. https://doi.org.ezproxy.hofstra.edu/10.1017/S17550203160000113

25. Rute, J.: Topics in Algorithmic Randomness and Computable Analysis. Ph.D. thesis, Carnegie Mellon University (2013)

26. Rute, J.: Algorithmic randomness and computable measure theory. In: Franklin, J.N., Porter, C.P. (eds.) Algorithmic Randomness: Progress and Prospects, pp. 58–114. Cambridge University Press, New York, NY (2020)

27. Schnorr, C.: A unified approach to the definition of random sequences. Math. Syst. Theory **5**, 246–258 (1971)

28. Schnorr, C.P.: Invarianzeigenschaften von Zufallsfolgen. Zufälligkeit und Wahrscheinlichkeit. LNM, vol. 218, pp. 83–88. Springer, Heidelberg (1971). https://doi.org/10.1007/BFb0112470

29. Stephan, F.: Martin-Löf Random and PA-complete Sets. Tech. Rep. 58, Matematisches Institut, Universität Heidelberg, Heidelberg (2002)

30. Turing, A.M.: Intelligent machinery. Tech. rep, National Physical Laboratory (1948)

31. V'yugin, V.V.: Effective convergence in probability, and an ergodic theorem for individual random sequences. Teor. Veroyatnost. i Primenen. **42**(1), 35–50 (1997). https://doi.org/10.1137/S0040585X97975915. http://dx.doi.org/10.1137/S0040585X97975915

Probabilistic Models of k-mer Frequencies (Extended Abstract)

Askar Gafurov, Tomáš Vinař, and Broňa Brejová[✉]

Faculty of Mathematics, Physics, and Informatics, Comenius University,
Mlynská dolina, 842 48 Bratislava, Slovakia
{askar.gafurov,tomas.vinar,bronislava.brejova}@fmph.uniba.sk

Abstract. In this article, we review existing probabilistic models for modeling abundance of fixed-length strings (k-mers) in DNA sequencing data. These models capture dependence of the abundance on various phenomena, such as the size and repeat content of the genome, heterozygosity levels, and sequencing error rate. This in turn allows to estimate these properties from k-mer abundance histograms observed in real data. We also briefly discuss the issue of comparing k-mer abundance between related sequencing samples and meaningfully summarizing the results.

Keywords: k-mer abundance · DNA sequencing · Genome size

1 Introduction

Rapid growth of the volume and complexity of available DNA sequencing data encourages research into efficient algorithms and data structures. A very fruitful approach is to represent individual sequences (usually sequencing reads) as sets of their subwords of a fixed length k called k-mers.

Many efficient methods, both exact and approximate, were developed for counting the occurrences of all k-mers in large sequencing datasets [9]. However, the focus has lately shifted to representing the sets of constituent k-mers without the abundance counts [3]. This leads to reduced memory requirements, allowing representation of large collections of data sets [11]. This capability is important in the field of pangemomics, with its focus on replacing a single reference genome with a collection of individual genomes, often represented in the form of a graph built from k-mers occurring in these genomes [25].

In this paper, however, we focus on k-mer abundance. Abundances are clearly essential for studying transcript abundance in RNA-seq data or large-scale copy number variation [12,18]. However, usefulness of k-mer abundance information is not limited to these applications, but can also be used to assess fundamental properties of newly sequenced genomes.

To demonstrate this point, we review existing probabilistic models of k-mer abundance, which can be used to estimate genome size and other properties based on a very succinct summary of the data set—the histogram of k-mer abundance. In Sect. 2, we define k-mer abundance and its spectrum. In Sect. 3, we

© Springer Nature Switzerland AG 2021
L. De Mol et al. (Eds.): CiE 2021, LNCS 12813, pp. 227–236, 2021.
https://doi.org/10.1007/978-3-030-80049-9_21

outline probabilistic models for capturing various genome and read set properties, including genome size, repeat content, heterozygosity, and sequencing error rate. Finally, in Sect. 4, we concentrate on comparing k-mer abundances in two different datasets and summarizing the results in a meaningful way.

2   Preliminaries

A k-mer is a string of a length k over a given finite alphabet Σ; in this paper we consider the DNA alphabet $\Sigma = \{A, C, G, T\}$. We say that a k-mer w matches a sequence S at position ℓ, if w is equal to the substring of S of length k starting at position ℓ. The number of matching positions of the k-mer w in the sequence S is called the abundance of w in S.

Given a sequence S, we define an absolute k-mer spectrum of S as the function $h_{S,k} : \mathbb{N} \to \mathbb{N}$, where $h_{S,k}(j)$ is the number of k-mers that have absolute abundance in S equal to j. If we normalize the absolute k-mer spectrum so that the sum of all values is one, we obtain the relative k-mer spectrum of S, which we denote $hr_{S,k}$. These definitions can be easily extended from a single string S to (multi)sets of strings, such as the set of chromosomes in a known genome or a set of sequencing reads.

For example, string $S = ACTACGCG$ contains dimers CT, TA, and GC once, and dimers AC and CG twice. Therefore, the absolute dimer spectrum has values $h_{S,2}(1) = 3, h_{S,2}(2) = 2$; the relative dimer spectrum has values $hr_{S,2}(1) = 3/5, hr_{S,2}(2) = 2/5$. For $j > 2$, we have $h_{S,2}(j) = hr_{S,2}(j) = 0$.

Several variations of k-mers and their abundances were considered in the literature. For example, the quality-adjusted version of k-mer abundance takes into account the probabilities of sequencing errors occurring at individual positions in the sequencing reads, which are typically available in the form of base quality scores. Each occurrence of a k-mer thus can be weighted by the probability that this occurrence is indeed correct [5, 7].

The notion of canonical k-mers helps to handle the double-stranded structure of DNA molecules. Both strands are usually sequenced with roughly equal probability, and therefore, it is not necessary to distinguish between a k-mer and its reverse complement. The canonical representation of a k-mer w is the lexicographically smaller string among w and its reverse complement. The absolute abundance of a canonical k-mer is defined as the sum of the absolute abundances of the k-mers it represents. While most canonical k-mers represent two k-mers, for even k there are $4^{k/2}$ palindromic canonical k-mers that represent only one k-mer. To avoid this unevenness, it is common to use only odd values of k for canonical representations.

Finally, spaced k-mers are motifs over an extended DNA alphabet $\Sigma = \{A, C, G, T, N\}$, where N stands for a "blank" or "don't care" nucleotide [1, 15, 22]. We say that a spaced k-mer w matches sequence S at position ℓ, if the substring of S of length k starting at position ℓ agrees with w at all non-blank positions. We can consider abundances of all spaced k-mers that have blank symbols at predefined locations. The main advantage of spaced k-mers is a smaller dependence between k-mer occurrences at adjacent positions of a sequence, resulting in a smaller variance of statistics used in phylogeny [15].

3 Models of k-mer Spectra

Spectra of k-mer abundance represent a very compact summary of large sequencing data sets. Assuming that genome sequencing can be modeled as a stochastic process, the corresponding k-mer spectrum will reflect important properties of the genome, such as its size, repeat content, the level of heterozygosity in diploid genomes, and will also depend on the parameters of the sequencing process, such as the length of sequencing reads, error rate, or sequencing biases.

All of these factors make modeling k-mer spectra an interesting problem. In particular, given basic parameters of the genome and the sequencing process, collectively denoted as θ, we would like to predict the corresponding k-mer spectrum hr_θ.

Such a model can then be used to interpret observed spectrum hr. In particular, our goal is to find parameters θ, for which the predicted k-mer spectrum hr_θ will be as close as possible to the observed k-mer spectrum hr. This is typically done by searching for θ^* minimizing a loss function, such as cross-entropy $-\sum_i hr(i) \log hr_{\theta^*}(i)$ or L_2-norm $\sum_i (hr(i) - hr_{\theta^*}(i))^2$. Both criteria can be optimized by general-purpose optimization algorithms supporting box constraints on parameter values, such as L-BFGS-B [29], and this process is typically very efficient due to the compact data representation.

In this way, just based on the observed k-mer spectrum, one can estimate key parameters of an unknown genome, such as the genome size, without attempting a complex process of genome assembly. Figure 1 shows 21-mer spectra for Illumina reads produced from *E. coli* genome at $10\times$ coverage and $2\times$ coverage. The model used to analyze this data set contained parameters for genome size, sequencing errors, and a simple model of genome repeat content [6]. For high-coverage data sets, low-abundance k-mers originating from sequencing errors are clearly separable from correct k-mers, and thus an estimate of read coverage can be obtained from the mode of the error-free k-mers (Fig. 1 left). For low-coverage data sets such a task is no longer easy (Fig. 1 right).

In the rest of this section, we discuss models of k-mer spectra, incorporating a variety of parameters representing properties of genome sequences or the sequencing process itself.

A simple model. In the simplest model, we assume that the target genome is a single circular chromosome of length L with no repeating k-mers, starting positions of N reads are sampled uniformly independently, the reads contain no errors, and have the same length r. Under these assumptions, the probability that a given read will cover a given k-mer from the genome is $p = (r - k + 1)/L$, and thus the absolute abundance of each of the L k-mers from the genome is a random variable from the binomial distribution with parameters N and p. A genome with linear chromosomes behaves similarly, only k-mers near chromosome ends will

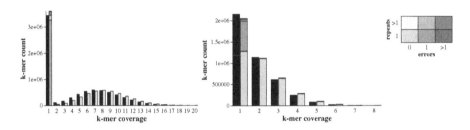

Fig. 1. Absolute 21-mer spectrum of Illumina reads for the *E. coli* genome at coverage 10 (left) and 2 (right) shown in black, and the fit of a model including sequence repeats and sequencing errors in colors. Data and model are taken from Hozza et al. [6]. (Color figure online)

have a smaller probability of being covered by a read. If the read length is much smaller than the chromosome lengths, this effect is negligible.

Given an observed k-mer spectrum (and assuming values N, k, and r are known), we may therefore seek parameter p of the binomial distribution that would well match the spectrum and then estimate the genome size L using the value that corresponds to this value of p. Note that the binomial distribution gives a non-zero probability to the event that a k-mer in a genome will be covered by zero reads, but such k-mers are not included in our observed spectra. For low-coverage data sets, we may need to account for this observation bias by using a truncated binomial distribution. More precisely, let X be a variable from the binomial distribution, and Y from the truncated distribution, where 0 cannot be observed, then $P(Y = k) = P(X = k)/P(X > 0)$.

In practice, the binomial distribution can be approximated by the Poisson distribution [6,24] or replaced by more complex distributions to compensate for unmodelled biases. These include the Gaussian distribution [4,7] and the negative binomial distribution [26,27]. In the rest of this section, we discuss extensions of this basic model that take into account important phenomena influencing the observed spectra.

Modeling Genomic Repeats. A significant fraction of k-mers in real genomes occurs in the genome more than once, due to the presence of transposons, simple tandem repeats, and segmental duplications. Thus we have to consider the k-mer spectrum of the genome itself, which is usually unknown. The absolute abundance of a k-mer in a genome is usually referred to as its *copy number*. Under the assumption of uniform sequencing, k-mers with higher copy numbers should appear proportionally more frequently in sequencing data. The k-mer spectrum of a read set can be thus represented as a mixture of simple distributions. Each component of the mixture corresponds to a certain copy number and its weight is defined by the relative genome spectrum. Therefore, a relative read set spectrum model can be written as

$$hr_R(j) = \sum_{i=1}^{\infty} hr_S(i) \cdot \phi(j; i, \theta),$$

(a)

(b)

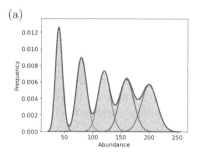
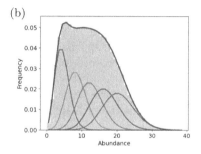

Fig. 2. A theoretical spectrum of a genome with repeats. A genome with equal proportions of 21-mers with copy numbers $1, \ldots, 5$ is modeled as a mixture of binomial distributions. Individual distributions scaled by their weights and the mixture are shown as lines. The relative spectrum sampled from the mixture is shown as a light blue histogram. Single-copy regions of the genome have expected coverage 50 (a) and 5 (b). (Color figure online)

where hr_R is the relative read set spectrum, hr_S is the relative genome spectrum, and $\phi(j; i, \theta)$ is the probability of a k-mer with copy-number i and some shared parameters θ having exactly j occurrences in the read set. The distribution $\phi(j; i, \theta)$ can be modeled by one of the distributions discussed for modeling genomes without repeats. For example, when using the (truncated) Poisson distribution, parameter λ_i for copy number i will have form $\lambda_i = ic$, where c is a free parameter representing coverage of single-copy k-mers. At high coverage, individual components of the mixture create clearly visible peaks in the relative spectrum, but at lower coverage levels, these peaks get closer together and are more difficult to identify (Fig. 2).

It remains to model the k-mer spectrum of an unknown genome. The simplest option is to let the whole genome spectrum up to some maximum value be free parameters estimated from the data [24, 26, 27]. However, this approach has a high number of parameters, including some configurations that are not plausible. For example, a fit of similar quality can be obtained by using copy numbers 1, 2, 3, ..., or by lowering the baseline coverage and using copy numbers 2, 4, 6, ... (assigning very low weights to odd copy numbers).

Another approach is to model copy numbers by a distribution, reducing the number of free parameters. A popular choice is the Zeta (ζ) distribution, which has only one free parameter, governing the shape [4, 7]. It is also possible to employ a hybrid technique, where the lowest copy-numbers are left as free parameters and the rest is modeled as a simple parametric distribution [6].

Modeling Sequencing Errors. Sequencing errors heavily influence k-mer spectra of a read set by lowering the coverage of the true genomic k-mers and creating spurious k-mers, which typically have a low abundance (Fig. 1). One way to handle this problem is to discard the lowest abundances from the spectrum and to assume that higher abundances correspond to true k-mers [21, 26, 27]. Another option is to include spurious k-mers explicitly in the model of the spectrum by a mixture model with two components corresponding to true and spurious k-mers, respectively.

The distribution of spurious k-mers can be modeled as a parametric distribution without any interpretation and its weight can be also left as a free parameter [4,7,24]. Another option is to use a simple substitution error model [6,13], which assumes that every base has a probability ε to be sequenced incorrectly. Under these assumptions, the probability of observing k-mer y when reading k-mer x is equal to $\varepsilon^s(1-\varepsilon)^{k-s}3^{-s}$, where s is the Hamming distance of x and y. If k-mer x is read c times, the expected number of times we observe k-mer y is equal to $\lambda_s := c\varepsilon^s(1-\varepsilon)^{k-s}3^{-s}$. The resulting distribution is then a mixture of $k+1$ distributions with means λ_s, where s goes from 0 to k.

Modeling Polyploid Genomes. Many organisms, including humans, are diploid, meaning that they have two sets of homologous chromosomes. The heterozygous sites in a diploid organism have two different alleles, producing two different k-mers instead of one. Therefore, homozygous k-mers should have on average twice the coverage of the heterozygous ones. This can be again represented as a mixture model, in which the homozygous component has the coverage parameter fixed as twice the coverage parameter of the heterozygous [4,26]. The weights of these components are governed by a free parameter related to the heterozygosity of the genome. The situation is more complex in organisms with higher ploidy, where a single position occurs in more than two homologous chromosomes and may contain more than two alleles [21].

4   Comparison of k-mer Frequencies Between Samples

The problem of differential analysis of sequencing samples arises in many areas: identifying differences between two individuals [23], comparing cancer samples with healthy tissues from the same individual [16], identification of differentially expressed transcripts in RNA-seq samples [2,18], or comparing a control sample with the sample biochemically treated to enrich or deplete particular functional elements (such as chromatin immunoprecipitation [14,28] or a treatment by enzymes depleting telomeric sequences [19]).

Typically, one first aligns the reads from two or more samples to an assembled genome or transcriptome reference and then identifies regions (genes, transcripts, or sequence windows) with significant differences in read coverage between samples. Such *alignment-first* approaches work well assuming that we have a reliable reference sequence and that we are able to map the reads to the reference uniquely. However, in cases where these conditions are violated, such approach may fail. This may be because there is a structural difference between the reference and both sequenced samples or the reference may be improperly assembled in some regions (such as highly repetitive regions).

In this section, we concentrate on a different approach that can at least partially overcome these limitations by comparing abundances of individual k-mers between two sequencing experiments instead of mapping the sequencing reads to the reference. Such approaches avoid potentially time-consuming alignment step, but more importantly, they can handle repetitive regions where reads cannot be reliably mapped and where even the assembly quality can be lower.

For example, if we sequence samples from two individuals, which share the same expansion of a particular locus, but the number of repeats is different from the reference sequence, the traditional reference-based methods would still identify this region and report a false positive, since both samples differ from the reference. On the other hand, the alignment-free approach would correctly identify that there is no significant difference in the k-mer abundances corresponding to that locus, and thus there is no reason to highlight this repeat.

In methods based on k-mer abundances, we first compare the abundances of k-mers in the two sets and identify k-mers which are significantly underrepresented in one of the read sets [2]. Only then the results are interpreted in the context of the reference sequence, for example by mapping k-mers back to the reference genome and identifying windows of the sequence that are significantly enriched for the underrepresented k-mers.

The window-based interpretation of the results helps us to relate found windows to annotated genome features, such as genes, and thus to assign them a biological meaning, which would be difficult for individual unmapped k-mers. Second, by requiring multiple underrepresented k-mers near each other, we filter out many false positives, that is, individual k-mers that appear underrepresented purely by chance. Conversely, considering whole windows allows us to avoid potential problems with local assembly errors which introduce incorrect k-mers to a window. If a window is sufficiently long, these will be compensated by the remaining correct k-mers. Note that this approach is not completely alignment-free, but the alignment to the reference is only used to interpret the results. Therefore we call this an *alignment-last* approach. Note that this window-based method is unsuitable when searching for very short differences, such as single nucleotide polymorphisms.

To illustrate the advantages of alignment-last methods, we demonstrate their simple application to a simulated dataset using chromosome IV of the yeast *Saccharomyces cerevisiae* as a starting point. We simulated two sequencing data sets by selecting random substrings of length 100 and adding substitution errors with probability 0.1% at each position. The *control read set* used the reference chromosome as an underlying string, while the second, *depleted set*, was generated from an underlying string featuring several large-scale deletions.

To simulate inaccuracies typical for draft genomes of newly sequenced species, we used a *draft assembly* produced from simulated nanopore sequencing reads by standard methods. This draft assembly has a single contig covering 99.8% of the original chromosome IV and has 0.5% error rate.

In the baseline alignment-first method, we have mapped both read sets to the draft assembly using Bowtie [8] and assigned the ratio between depleted and control coverages to each base pair. Base pairs having this ratio lower than a user-selected threshold are marked as depleted regions. In the k-mer based alignment-last method, k-mer abundances were counted using Jellyfish [10].

For each k-mer in the genome, ratio of abundance between the depleted and control set was computed. The draft assembly was split into non-overlapping 100bp windows, and the score of each window was computed as the median ratio of k-mers starting in this window. Windows having the score below a user-selected threshold are then marked as depleted regions. The accuracy of each method is summarized by using the area under curve (AUC) statistics.

Figure 3 (left) shows the results for depletion of a 6 kbp long unique sequence. For such long unique sequences both methods can reliably identify the depleted region even at small coverages. In Fig. 3 (right) we show the results for a more complex case, where two retrotransposons and one duplicated region of lengths between 6–7 Kbp were depleted. Here, alignment-last k-mer method shows clear advantage over the baseline alignment-first method.

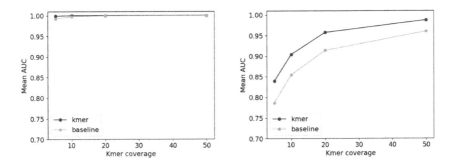

Fig. 3. Comparison of searching for depleted regions either by the alignment-first baseline method or by k-mer abundance in windows of size 100 bp on simulated data sets. On the left, the depleted region is a 6 kb long single-copy sequence, on the right, two retrotransposons and one duplicated region of lengths 6–7 were depleted. We report the AUC measure for different k-mer coverage levels averaged over five data sets.

5   Conclusion

In this paper, we summarized techniques used in various published models of k-mer spectra. Although the models cover many phenomena influencing k-mer abundance, some issues still remain to be explored. One example is the influence of GC content on read coverage, which is taken into account in RNA-seq studies [17]. More complex errors models, taking into account indels and context biases would also be appropriate, particularly for third-generation sequencing data.

An important practical issue involves DNA molecules that are present in cells in high copy numbers, leading to increased read coverage in sequencing. Examples include mitochondrial genomes in eukaryotes and plasmids in prokaryotes. Repeat-aware models consider such molecules as repeats present in many copies and thus inflate the estimated genome size. Pflug et al. filter out mitochondrial reads before applying k-mer models for genome size estimation [20], but perhaps a simple model of these short chromosomes could be incorporated instead.

Finally, it would be worthwhile to apply abundance models developed for k-mer spectra to the task of read set comparison.

Acknowledgments. Our research was supported by grants from the Slovak Research and Development Agency APVV-18-0239, the Scientific Grant Agency VEGA 1/0463/20 to BB and VEGA 1/0458/18 to TV, the European Union's Horizon 2020 research and innovation program (PANGAIA project #872539 and ALPACA project #956229) and Comenius University Grant UK/278/2020 to AG.

References

1. Břinda, K., Sykulski, M., Kucherov, G.: Spaced seeds improve k-mer-based metagenomic classification. Bioinformatics **31**(22), 3584–3592 (2015)
2. Chan, C.K.K., et al.: A differential k-mer analysis pipeline for comparing RNA-seq transcriptome and meta-transcriptome datasets without a reference. Funct. Integr. Genomics **19**(2), 363–371 (2019)
3. Chikhi, R., Holub, J., Medvedev, P.: Data structures to represent sets of k-long DNA sequences. arXiv preprint arXiv:1903.12312 (2019)
4. Chikhi, R., Medvedev, P.: Informed and automated k-mer size selection for genome assembly. Bioinformatics **30**(1), 31–37 (2014)
5. Comin, M., Leoni, A., Schimd, M.: Clustering of reads with alignment-free measures and quality values. Algorithms Mol. Biol. **10**(1), 4 (2015)
6. Hozza, M., Vinař, T., Brejová, B.: How big is that genome? Estimating genome size and coverage from k-mer abundance spectra. In: Iliopoulos, C., Puglisi, S., Yilmaz, E. (eds.) SPIRE 2015. LNCS, vol. 9309, pp. 199–209. Springer, Cham (2015). https://doi.org/10.1007/978-3-319-23826-5_20
7. Kelley, D.R., Schatz, M.C., Salzberg, S.L.: Quake: quality-aware detection and correction of sequencing errors. Genome Biol. **11**(11), R116 (2010)
8. Langmead, B., Trapnell, C., Pop, M., Salzberg, S.L.: Ultrafast and memory-efficient alignment of short DNA sequences to the human genome. Genome Biol. **10**(3), 1–10 (2009)
9. Manekar, S.C., Sathe, S.R.: Estimating the k-mer coverage frequencies in genomic datasets: a comparative assessment of the state-of-the-art. Curr. Genomics **20**(1), 2–15 (2019)
10. Marçais, G., Kingsford, C.: A fast, lock-free approach for efficient parallel counting of occurrences of k-mers. Bioinformatics **27**(6), 764–770 (2011)
11. Marchet, C., Boucher, C., Puglisi, S.J., Medvedev, P., Salson, M., Chikhi, R.: Data structures based on k-mers for querying large collections of sequencing data sets. Genome Res. **31**(1), 1–12 (2021)
12. Marchet, C., Iqbal, Z., Gautheret, D., Salson, M., Chikhi, R.: REINDEER: efficient indexing of k-mer presence and abundance in sequencing datasets. Bioinformatics **36**(Supplement_1), i177–i185 (2020)
13. Melsted, P., Halldórsson, B.V.: KmerStream: streaming algorithms for k-mer abundance estimation. Bioinformatics **30**(24), 3541–3547 (2014)
14. Menzel, M., Hurka, S., Glasenhardt, S., Gogol-Döring, A.: NoPeak: k-mer-based motif discovery in ChIP-Seq data without peak calling. Bioinformatics **37**(5), 596–602 (2020)
15. Morgenstern, B., Zhu, B., Horwege, S., Leimeister, A.A.: Estimating evolutionary distances between genomic sequences from spaced-word matches. Algorithms Mol. Biol. **10**(1), 5 (2015)

16. Narzisi, G., et al.: Genome-wide somatic variant calling using localized colored de Bruijn graphs. Commun. Biol. **1**(1), 1–9 (2018)
17. Patro, R., Duggal, G., Love, M.I., Irizarry, R.A., Kingsford, C.: Salmon provides fast and bias-aware quantification of transcript expression. Nature Methods **14**(4), 417–419 (2017)
18. Patro, R., Mount, S.M., Kingsford, C.: Sailfish enables alignment-free isoform quantification from RNA-seq reads using lightweight algorithms. Nature Biotechnol. **32**(5), 462–464 (2014)
19. Peška, V., Fajkus, P., Fojtová, M., et al.: Characterisation of an unusual telomere motif (TTTTTTAGGG)n in the plant Cestrum elegans (Solanaceae), a species with a large genome. Plant J. **82**(4), 644–654 (2015)
20. Pflug, J.M., Holmes, V.R., Burrus, C., Johnston, J.S., Maddison, D.R.: Measuring genome sizes using read-depth, k-mers, and flow cytometry: methodological comparisons in beetles (Coleoptera). G3: Genes Genomes Genet. **10**(9), 3047–3060 (2020)
21. Ranallo-Benavidez, T.R., Jaron, K.S., Schatz, M.C.: GenomeScope 2.0 and Smudgeplot for reference-free profiling of polyploid genomes. Nature Commun. **11**(1), 1–10 (2020)
22. Röhling, S., Linne, A., Schellhorn, J., Hosseini, M., Dencker, T., Morgenstern, B.: The number of k-mer matches between two DNA sequences as a function of k and applications to estimate phylogenetic distances. Plos One **15**(2), e0228070 (2020)
23. Shajii, A., Yorukoglu, D., William, Yu, Y., Berger, B.: Fast genotyping of known SNPs through approximate k-mer matching. Bioinformatics **32**(17), i538–i544 (2016)
24. Simpson, J.T.: Exploring genome characteristics and sequence quality without a reference. Bioinformatics **30**(9), 1228–1235 (2014)
25. The Computational Pan-Genomics Consortium: Computational pan-genomics: status, promises and challenges. Briefings Bioinform. **19**(1), 118–135 (2018)
26. Vurture, G.W., et al.: GenomeScope: fast reference-free genome profiling from short reads. Bioinformatics **33**(14), 2202–2204 (2017)
27. Williams, D., Trimble, W.L., Shilts, M., Meyer, F., Ochman, H.: Rapid quantification of sequence repeats to resolve the size, structure and contents of bacterial genomes. BMC Genomics **14**(1), 537 (2013)
28. Zhang, Y., et al.: Model-based analysis of ChIP-Seq (MACS). Genome Biol. **9**(9), 1–9 (2008)
29. Zhu, C., Byrd, R.H., Lu, P., Nocedal, J.: Algorithm 778: L-BFGS-B: fortran subroutines for large-scale bound-constrained optimization. ACM Trans. Math. Softw. **23**(4), 550–560 (1997)

Defining Formal Explanation in Classical Logic by Substructural Derivability

Francesco A. Genco and Francesca Poggiolesi[✉]

IHPST Université Paris 1 Panthéon-Sorbonne and CNRS,
13 Rue du Four, 75006 Paris, France

Abstract. Precisely framing a formal notion of explanation is a hard problem of great relevance for several areas of scientific investigation such as computer science, philosophy and mathematics. We study a notion of formal explanation according to which an explanation of a formula F must contain all and only the true formulae that concur in determining the truth of F. Even though this notion of formal explanation is defined by reference to derivability in classical logic, the relation that holds between the explained formula and the formulae explaining it has a distinct substructural flavour, due to the fact that no redundancy is admitted among the explaining formulae. We formalise this intuition and prove that this notion of formal explanation is essentially connected, in a very specific sense, to derivability in a substructural calculus.

Keywords: Formal explanation · Substructural logics · Proof theory

1 Introduction

Precisely framing a notion of explanation in formal contexts is a hard problem of great relevance for several areas of scientific investigation, as, for instance, computer science [10,12], philosophy [1,3], mathematics [9,19], and the natural sciences [8,17]. A research line essentially concerned with the formal definition and study of explanation relations is the one investigating the relation of *logical grounding*. According to this approach, a multiset of formulae Δ formally explains a complex formula F if Δ is a logical ground of F, that is, if F is true in virtue of the truth of the elements of Δ. Several endeavours along these lines have appeared in the contemporary literature, see, for instance, [4,5,13,14,16,18].

According to the strictest, and hence most informative, of these technical notions of formal explanation [13,14], a formal explanation of a true statement F must mention all and only the true statements that concur in determining the truth of F. This characterisation is captured by two formal conditions on the multiset Δ of explaining formulae. First, truth of the formulae in Δ must determine the truth of F. Second, the multiset Δ of explaining formulae must be a maximal multiset of, possibly negated, disjoint occurrences of subformulae

Supported by project IBS (ANR-18-CE27-0012-01).

L. De Mol et al. (Eds.): CiE 2021, LNCS 12813, pp. 237–247, 2021.
https://doi.org/10.1007/978-3-030-80049-9_22

of the explained formula F. Even though this notion of formal explanation is defined by reference to derivability in classical logic [13,15], the relation that holds between the explained formula and the multiset of formulae explaining it has a distinct substructural flavour, as discussed in [15]. In the present work we present a study of the formal connection between derivability in substructural logics and this notion of formal explanation.

Intuitively, the complexity constraints required to define this notion—which now on we will simply call *formal explanation*—impose a strict correspondence between the formulae occurring in the explanation and the syntactical parts of the formula explained. A formal explanation, indeed, is supposed to mention exactly those parts of the explained formula in virtue of which it is true, and no redundancy is admitted. Since the substructural logics in which the rules of weakening and contraction are not admissible can be seen as resource-aware reasoning systems in which the usage of redundant or duplicated hypotheses is limited, this feature of explanation raises the question whether there exists a substructural logic for which it is possible to spell out a rigorous connection with the relation of formal explanation. In the present work, we positively answer this question by defining a suitable substructural calculus and formally proving its connection with formal explanation. In order to do so, we define the substructural calculus SL which can be either seen as a fragment of FLe (Full Lambek Calculus with the Exchange rule) [6], or as a fragment of Linear Logic [7] extended by an axiom for the lattice bottom $\perp$.[1] We then show that the relation of formal explanation is essentially connected, in a very specific and formal sense, to derivability in SL.

The article is structured as follows. In Sect. 2, we define the notion of formal explanation in the traditional way, that is, by employing derivability classical logic. In Sect. 3, we present the substructural calculus SL that we will employ to show the relationship between formal explanation and substructural derivability. In Sect. 4, we show that formal explanation can be characterised by employing derivability conditions formulated in the calculus SL. In Sect. 5, we conclude with some remarks and a discussion of future work directions.

2   Formal Explanation

We present now the relation of formal explanation for classical logic formulae as introduced in [13]. The language that we will employ for classical logic is the following:

$$\varphi ::= \psi \mid \varphi \wedge \varphi \mid \varphi \vee \varphi \mid \varphi \rightarrow \varphi \mid \neg\varphi$$

$$\psi ::= p_1 \mid p_2 \mid p_3 \ldots$$

where $p_1, p_2, p_3 \ldots$ is a list of all the propositional variables of the language. In the following, we will employ capital Latin letters as metavariables for formulae and capital Greek letters as metavariables for multisets of formulae.

---

[1] The symbol $\perp$ is used in [7] for a weaker falsity constant which is often denoted by 0 in other works, see [11, p. 42] for a comparison table of different notations used for linear logic. In SL, no constant corresponds to Girard's constant 0.

In order to introduce the relation of formal explanation, we have to define some technical notions. We start from the converse of a formula.

Definition 1. *For any natural number n, we indicate by $\neg^n A$ the formula $\neg \ldots \neg A$ where A does not have $\neg$ as outermost connective and is preceded by a sequence of n negations $\neg$.*

For any formula A, the notation A^* represents the converse of A:

- *if $A = \neg^{2n} B$ then $A^* = \neg A$*
- *if $A = \neg^{2n+1} B$ then $A^* = \neg^{2n} B$*

For any multiset of formulae Γ, we denote by Γ^* the multiset $\{A^* \mid A \in \Gamma\}$.

Notice in particular that, for any formula A, $(A^*)^* = A$ since $((\neg^{2n} B)^*)^* = (\neg^{2n+1} B)^* = \neg^{2n} B$ and $((\neg^{2n+1} B)^*)^* = (\neg^{2n} B)^* = \neg^{2n+1} B$.

We can now define the notion of explanatory subformulae of a formula (exp-subformulae, for short). Intuitively, a set of exp-subformulae of a formula A contains either a positive occurrence or a negative occurrence (formalised by the notion of converse) of each maximal proper subformula of A the truth of which determines the truth of A.[2]

Definition 2. *The set of sets of exp-subformulae $es(A)$ of a classical formula A is defined as follows:*

- $es(B \wedge C) = es(B \vee C) = es(B \rightarrow C) = es(\neg(B \wedge C)) = es(\neg(B \vee C)) = es(\neg(B \rightarrow C)) = \{\{B, C\}, \{B, C^*\}, \{B^*, C\}, \{B^*, C^*\}\}$
- $es(p) = es(\neg p) = \emptyset$
- $es(\neg\neg B) = \{\{B\}, \{B^*\}\}$

We can finally define our notion of formal explanation for classical logic formulae. In order to do so, we first introduce the relation of *immediate formal explanation*, an instance of which intuitively corresponds to a basic step of explanation which displays the immediate reasons why a formula is true. Then, we introduce the relation of *mediate formal explanation*. This relation generalises that of immediate formal explanation and captures the idea that the process of explaining a statement can be iterated in order to find simpler and simpler reasons for our original statement.

Definition 3. *For any consistent multiset $\{C\} \cup \Gamma$ where C is a formula of classical logic and Γ is a multiset of formulae of classical logic, we say that, under the robust condition C (which might not occur), Γ is an immediate formal explanation of A, in symbols $[C]\Gamma \Vdash A$, if and only if:*

---

[2] For the philosophical motivation behind the definition of exp-subformulae, see the notion of *g-subformulae* in [13]. Notice also that the notion of g-subformula is closed under commutativity and associativity of conjunction and disjunction. In the present work, for the sake of simplicity, we omit this requirement. This, nevertheless, does not imply a loss of generality since also the corresponding substructural connectives $\otimes$ and $\sqcup$ of SL are commutative and associative.

- $\Gamma \vdash A$ (positive derivability),
- $C, \Gamma^* \vdash A^*$ (negative derivability),
- $\{C\} \cup \Gamma$ is a set of exp-subformulae of A, see Definition 2.

where $\vdash$ indicates derivability in any system for classical logic.

The notion of mediate formal explanation can be defined as follows via the notion of immediate formal explanation.

Definition 4. *For any consistent multiset of classical formulae $\Gamma \cup \Delta$ and classical formula A, $[\Gamma]\Delta \Vdash^m A$ if, and only if, one of the following conditions is satisfied:*

- $[\Gamma]\Delta \Vdash A$
- $[G]\Delta' \Vdash B$ and $[\Gamma']\Delta'', B \Vdash^m A$ where
  - $\Delta = \Delta' \cup \Delta''$
  - $\Gamma = \Gamma' \cup \{G\}$.

As shown in [15, pp. 14–16], positive and negative derivability characterize mediate formal explanation as well.

Proposition 1. *For any classical formula A and pair of multisets of classical formulae Γ and Δ, we have that if $[\Gamma]\Delta \Vdash^m A$ holds, then $\Delta \vdash A$ and $\Gamma, \Delta^* \vdash A^*$.*

Given Definition 3, it is possible to list all potential immediate formal explanations for the complex formulae of classical logic.[3] A complete enumeration of all valid schemata of explanations of this kind is the following:

$A, B \Vdash A \wedge B$		
$A, B \Vdash A \vee B$	$[A^*]B \Vdash A \vee B$	$[B^*]A \Vdash A \vee B$
$A^*, B \Vdash A \rightarrow B$	$[A]B \Vdash A \rightarrow B$	$[B^*]A^* \Vdash A \rightarrow B$
$A \Vdash \neg\neg A$		
$A^*, B^* \Vdash \neg(A \wedge B)$	$[A]B^* \Vdash \neg(A \wedge B)$	$[B]A^* \Vdash \neg(A \wedge B)$
$A^*, B^* \Vdash \neg(A \vee B)$		
$A, B^* \Vdash \neg(A \rightarrow B)$		

3   The Substructural Calculus

We introduce now the substructural calculus SL which will be used to show that the derivability conditions defining formal explanations in classical logic can be expressed in a substructural calculus.

---

[3] We talk here of potential explanations because, in order to have an actual explanation, we must also guarantee that all formulae occurring in the explanation are true.

The language that we will employ for the substructural calculus SL is the following:

$$\varphi ::= \psi \mid \bot \mid \varphi \otimes \varphi \mid \varphi \sqcup \varphi \mid \varphi \multimap \varphi$$

$$\psi ::= p_1 \mid p_2 \mid p_3 \ldots$$

where $p_1, p_2, p_3 \ldots$ is a list of all the propositional variables of the language.

The constant $\bot$ for falsity corresponds to the lattice bottom, see [6, p. 83] and [11, p. 42]; the connective $\otimes$ is the multiplicative conjunction;[4] $\sqcup$ is the additive disjunction;[5] and $\multimap$ is the multiplicative implication.

Rules and axioms of the substructural calculus SL are shown in Table 1.

Table 1. The substructural calculus SL

$$A \Rightarrow A \qquad \Gamma, \bot \Rightarrow C \qquad \frac{\Gamma \Rightarrow A\Delta, \ A \Rightarrow B}{\Gamma, \Delta \Rightarrow B} \ cut$$

$$\frac{\Gamma \Rightarrow A\Delta \Rightarrow B}{\Gamma, \Delta \Rightarrow A \otimes B} \ \otimes r \qquad \frac{\Gamma, A, B \Rightarrow C}{\Gamma, A \otimes B \Rightarrow C} \ \otimes l$$

$$\frac{\Gamma \Rightarrow A}{\Gamma \Rightarrow A \sqcup B} \ \sqcup r \qquad \frac{\Gamma \Rightarrow B}{\Gamma \Rightarrow A \sqcup B} \ \sqcup r \qquad \frac{\Gamma, A \Rightarrow C\Gamma, B \Rightarrow C}{\Gamma, A \sqcup B \Rightarrow C} \ \sqcup l$$

$$\frac{\Gamma, A \Rightarrow B}{\Gamma \Rightarrow A \multimap B} \ \multimap r \qquad \frac{\Gamma \Rightarrow A\Delta, \ B \Rightarrow C}{\Gamma, \Delta, A \multimap B \Rightarrow C} \ \multimap l$$

The calculus SL can be directly seen as a fragment of FLe (Full Lambek Calculus with the Exchange rule)—see, for instance, [6]. Otherwise, it is possible to consider SL as a fragment of Linear Logic—see, for instance, [11]—to which we add the axiom for the lattice bottom $\bot$.

Definition 5. *We define* $\sim A$ *as* $A \multimap \bot$, *and we introduce the following two rules for* $\sim$ *by simplifying the rules for* $\multimap$:

$$\frac{\Gamma, A \Rightarrow \bot}{\Gamma \Rightarrow \sim A} \ \sim r \qquad \frac{\Gamma \Rightarrow A}{\Gamma, \Delta, \sim A \Rightarrow C} \ \sim l$$

Notice that we omitted a premise of the form $\Delta, \bot \Rightarrow C$ from the $\sim r$ rule since this premise is always an axiom of the calculus.

We extend the definition of converse formula also for the language of SL.

Definition 6. *For any formula A in the language of* SL, *the converse* A^* *of* A *is defined according to Definition 1 where instead of the classical negation* $\neg$ *we use the substructural negation* $\sim$.

---

[4] The multiplicative conjunction is sometimes also called group-theoretical conjunction [11].

[5] The additive disjunction is sometimes also called lattice-theoretical disjunction [11]. This connective is often denoted by $\vee$, but here we reserve the symbol $\vee$ for the disjunction of classical logic.

4   Formal Explanation by Substructural Proofs

We show that the positive and negative derivability conditions used to define the formal explanation relation for classical logic can be expressed as derivability conditions in SL. In order to do this, we interpret classical formulae as substructural formulae according to the following recursive translation t:

$$
\begin{aligned}
t(A \wedge B) &= t(A) \otimes t(B) \\
t(A \vee B) &= t(A) \sqcup t(B) \sqcup (t(A) \otimes t(B)) \\
t(A \to B) &= t(\neg A \vee B) = \; \sim t(A) \sqcup t(B) \sqcup (\sim t(A) \otimes t(B)) \\
t(\neg A) &= \; \sim t(A)
\end{aligned}
$$

For any multiset Γ of classical formulae, we denote by $t(\Gamma)$ the multiset $\{t(A) \mid A \in \Gamma\}$ of substructural formulae.

The translations of conjunction and negation are self-explanatory. The case of disjunction requires a clarification. Indeed, it is not enough to translate classical disjunction $\vee$ by the substructural $\sqcup$ because, while the multiset of formulae $\{A, B\}$ is a legitimate immediate formal explanation of $A \vee B$, the substructural sequent $A, B \Rightarrow A \sqcup B$ is not derivable. This is due to the fact that the calculus SL enables us to prove a complex formula only if we have the exact amount of hypotheses required to introduce the outermost connective of the formula, and not one hypothesis more. For $A \sqcup B$, in particular, we have that $A \Rightarrow A \sqcup B$ and $B \Rightarrow A \sqcup B$ are derivable, but $A, B \Rightarrow A \sqcup B$ is not, because, in the latter sequent, one formula between A and B is redundant. In more conceptual terms, if A and B are true, both formulae have to be mentioned in order to provide the complete reason why $A \vee B$ is true, and hence both of them have to be in the formal explanation of $A \vee B$; from a substructural perspective, on the other hand, using both A and B to obtain $A \sqcup B$ is a waste of resources and hence is not allowed. Therefore, in order to faithfully encode the behavior of classical disjunction with respect to formal explanation in our substructural calculus SL, we need to translate classical disjunction by a weakened version of the substructural disjunction $\sqcup$. For similar reasons, we translate a classical implication $A \to B$ via its traditional encoding as the classical formula $\neg A \vee B$.

Before showing that the translation t enables us to formulate positive and negative derivability as derivability in SL, we need to prove two propositions concerning the converse relation.

Proposition 2. *For any formula A and multiset Γ, the sequent $\Gamma, A, A^* \Rightarrow \bot$ is derivable.*

Proof. By Definition 6, we have two cases: (i) $A =\sim^{2n} B$ where B does not have $\sim$ as outermost connective, and (ii) $A =\sim^{2n+1} B$, where B does not have $\sim$ as outermost connective.

If (i), $A^* =\sim A$ and $\Gamma, A, A^* \Rightarrow \bot$ can be derived as shown below on the left. If (ii), $A^* = (\sim^{2n+1} B)^* =\sim^{2n} B$ and $\Gamma, A, A^* \Rightarrow \bot$ can be derived as shown below on the right.

$$\frac{A \Rightarrow A}{\Gamma, A, \sim A \Rightarrow \perp} \sim l \qquad \frac{\sim^{2n} B \Rightarrow \sim^{2n} B}{\Gamma, \sim^{2n+1} B, \sim^{2n} B \Rightarrow \perp} \sim l$$

Proposition 3. *For any formula A, the sequent $A^* \Rightarrow \sim A$ is derivable.*

Proof. By Definition 6, we have two cases: (i) $A =\sim^{2n} B$ where B does not have $\sim$ as outermost connective, and (ii) $A =\sim^{2n+1} B$, where B does not have $\sim$ as outermost connective.

If (i), $A^* =\sim A$ and $A^* \Rightarrow \sim A$ is an axiom and hence derivable.

If (ii), $A^* = (\sim^{2n+1} B)^* =\sim^{2n} B$ and $A^* \Rightarrow \sim A$ can be derived as follows:

$$\frac{\dfrac{\sim^{2n} B \Rightarrow \sim^{2n} B}{\sim^{2n} B, \sim^{2n+1} B \Rightarrow \perp} \sim l}{\sim^{2n} B \Rightarrow \sim^{2n+2} B} \sim r$$

Theorem 1. *The positive and negative derivability conditions for formal explanation in classical logic can be expressed as derivability conditions in* SL.

Proof. Since the rules of SL are clearly sound with respect to classical logic, we have that, if the conditions of positive and negative derivability are met with respect to SL through the translation t, then they are also met with respect to classical logic. Which means that if a suitably partitioned multiset of formulae C, Γ, D verifies the complexity conditions of Definition 3 and its translation enjoys positive and negative derivability in SL, then $[C]\Gamma \Vdash D$ holds. Therefore, in order to prove the statement it is enough to show that, if $[C]\Gamma \Vdash D$ holds, then the conditions of positive and negative derivability are met in SL for the translation of the formulae C, D and of the multiset Γ. Hence, we consider all valid instances of the formal explanation relation $[C]\Gamma \Vdash D$, we reason by cases on the logical structure of the classical formula D, and we show in each case that $t(\Gamma) \Rightarrow t(D)$ and $t(C), t(\Gamma)^* \Rightarrow t(D)^*$ are derivable sequents. We only show some interesting cases.

– $A, B \Vdash A \vee B$

Positive derivability:
$$\frac{\dfrac{A \Rightarrow A \quad B \Rightarrow B}{A, B \Rightarrow A \otimes B}}{A, B \Rightarrow (A \sqcup B) \sqcup (A \otimes B)} \sqcup r$$

Negative derivability:

$$\frac{\dfrac{\dfrac{A^*, B^*, A \Rightarrow \perp \quad A^*, B^*, B \Rightarrow \perp}{A^*, B^*, A \sqcup B \Rightarrow \perp} \sqcup l \quad \dfrac{A^*, B^*, A, B \Rightarrow \perp}{A^*, B^*, A \otimes B \Rightarrow \perp} \otimes l}{A^*, B^*, (A \sqcup B) \sqcup (A \otimes B) \Rightarrow \perp} \sqcup l}{A^*, B^* \Rightarrow \sim ((A \sqcup B) \sqcup (A \otimes B))} \sim r$$

where, by Proposition 2, the sequents $A^*, B^*, A \Rightarrow \perp$ and $A^*, B^*, B \Rightarrow \perp$, and the sequent $A^*, B^*, A, B \Rightarrow \perp$ are derivable.

– $[B^*]A \Vdash A \vee B$

Positive derivability:
$$\frac{\dfrac{\dfrac{A \Rightarrow A}{A \Rightarrow A \sqcup B} \sqcup r}{A \Rightarrow (A \sqcup B) \sqcup (A \otimes B)}}{} \sqcup r$$

Negative derivability:

$$\frac{\dfrac{B^*,A^*,A \Rightarrow \bot \quad B^*,A^*,B \Rightarrow \bot}{B^*,A^*,A \sqcup B \Rightarrow \bot} \sqcup l \quad \dfrac{B^*,A^*,A,B \Rightarrow \bot}{B^*,A^*,A \otimes B \Rightarrow \bot} \otimes l}{\dfrac{B^*,A^*,(A \sqcup B) \sqcup (A \otimes B) \Rightarrow \bot}{B^*,A^* \Rightarrow \sim ((A \sqcup B) \sqcup (A \otimes B))} \sim r} \sqcup l$$

where, by Proposition 2, the sequents $B^*, A^*, A \Rightarrow \bot$ and $B^*, A^*, B \Rightarrow \bot$, and the sequent $B^*, A^*, A, B \Rightarrow \bot$ are all derivable.

- $A \Vdash \neg\neg A$

Positive derivability:
$$\dfrac{\dfrac{A \Rightarrow A}{A, \sim A \Rightarrow \bot} \sim l}{A \Rightarrow \sim\sim A} \sim r$$

Negative derivability: we have to derive the sequent $A^* \Rightarrow (\sim\sim A)^*$. By Definition 6, we have two cases: either $A^* = (\sim^{2n} B)^* =\sim A$ and hence $(\sim\sim A)^* =\sim\sim\sim A$; or $A^* = (\sim^{2n+1} B)^* =\sim^{2n} B$ and hence $(\sim\sim A)^* =\sim A$. If $A^* =\sim A$ and $(\sim\sim A)^* =\sim\sim\sim A$, we have the derivation below on the left. If $A^* = (\sim^{2n+1} B)^* =\sim^{2n} B$ and hence $(\sim\sim A)^* =\sim A =\sim^{2n+2} B$, we have the derivation below on the right.

$$\dfrac{\dfrac{\sim A \Rightarrow \sim A}{\sim A, \sim\sim A \Rightarrow \bot} \sim l}{\sim A \Rightarrow \sim\sim\sim A} \sim r \qquad\qquad \dfrac{\dfrac{\sim^{2n} B \Rightarrow \sim^{2n} B}{\sim^{2n} B, \sim^{2n+1} B \Rightarrow \bot} \sim l}{\sim^{2n} B \Rightarrow \sim^{2n+2} B} \sim r$$

- $A^*, B^* \Vdash \neg(A \wedge B)$

Positive derivability: derivation below on the left. Negative derivability: derivation below on the right.

$$\dfrac{\dfrac{A^*,B^*,A,B \Rightarrow \bot}{A^*,B^*,A \otimes B \Rightarrow \bot} \otimes l}{A^*,B^* \Rightarrow \sim (A \otimes B)} \sim r \qquad\qquad \dfrac{(A^*)^* \Rightarrow A \quad (B^*)^* \Rightarrow B}{(A^*)^*,(B^*)^* \Rightarrow A \otimes B} \otimes r$$

where, by Proposition 2, $A^*, B^*, A, B \Rightarrow \bot$ is a derivable sequent; and, since by Definition 6 we have that $(A^*)^* = A$ and $(B^*)^* = B$, both sequents $(A^*)^* \Rightarrow A$ and $(B^*)^* \Rightarrow B$ are derivable.

- $A^*, B^* \Vdash \neg(A \vee B)$

Positive derivability:

$$\dfrac{\dfrac{A^*,B^*,A \Rightarrow \bot \quad A^*,B^*,B \Rightarrow \bot}{A^*,B^*,A \sqcup B \Rightarrow \bot} \sqcup l \quad \dfrac{A^*,B^*,A,B \Rightarrow \bot}{A^*,B^*,A \otimes B \Rightarrow \bot} \otimes l}{\dfrac{A^*,B^*,(A \sqcup B) \sqcup (A \otimes B) \Rightarrow \bot}{A^*,B^* \Rightarrow \sim ((A \sqcup B) \sqcup (A \otimes B))} \sim r} \sqcup l$$

where, by Proposition 2, the sequents $A^*, B^*, A \Rightarrow \bot$ and $A^*, B^*, B \Rightarrow \bot$, and the sequent $A^*, B^*, A, B \Rightarrow \bot$ are derivable.

Negative derivability:
$$\dfrac{\dfrac{(A^*)^* \Rightarrow A \quad (B^*)^* \Rightarrow B}{(A^*)^*,(B^*)^* \Rightarrow A \otimes B} \otimes r}{(A^*)^*,(B^*)^* \Rightarrow (A \sqcup B) \sqcup (A \otimes B)} \sqcup r$$

where, since by Definition 6 we have that $(A^*)^* = A$ and $(B^*)^* = B$, the sequents $(A^*)^* \Rightarrow A$ and $(B^*)^* \Rightarrow B$ are derivable.

The previous proof also indicates that the other obvious choices of substructural connectives would not be suitable to characterise immediate formal explanation. Indeed, for additive conjunction $\sqcap$ and multiplicative disjunction $\oplus$, positive derivability does not hold since the leaves of the following derivations are not derivable:

$$\frac{A, B \Rightarrow A \quad A, B \Rightarrow B}{A, B \Rightarrow A \sqcap B} \qquad \frac{A \Rightarrow A, B}{A \Rightarrow A \oplus B}$$

Moreover, if we used the weaker falsity constant 0 instead of the lattice bottom $\perp$, Proposition 2 would fail, because $\Gamma, 0 \Rightarrow 0$ with $\Gamma \neq \emptyset$ is not derivable, and thus the rule $\sim l$ would not be strong enough to prove what is required.

We show that the generalisation of immediate formal explanation into mediate formal explanation preserves the embeddability of formal explanations in SL.

Theorem 2. *The construction of mediate formal explanations in classical logic preserves positive and negative derivability in the calculus* SL.

Proof. Consider any valid instance $[\Gamma]\Delta \Vdash^m A$ of the mediate formal explanation relation. We show that the substructural sequents $t(\Delta) \Rightarrow t(A)$ and $t(\Gamma), t(\Delta)^* \Rightarrow t(A)^*$ are derivable in SL.

According to Definition 4, the instance $[\Gamma]\Delta \Vdash^m A$ is valid if, and only if, it can be justified by a finite number of instances of the immediate formal explanation relation. The proof is by induction on the number of instances of the immediate formal explanation relation used to justify $[\Gamma]\Delta \Vdash^m A$.

In the base case, $[\Gamma]\Delta \Vdash^m A$ itself is a valid instance $[\Gamma]\Delta \Vdash A$ of immediate formal explanation and, by Theorem 1, we have that $t(\Delta) \Rightarrow t(A)$ and $t(\Gamma), t(\Delta)^* \Rightarrow t(A)^*$ are derivable in SL.

Suppose now that $[\Gamma]\Delta \Vdash^m A$ is justified by $n > 1$ instances of the immediate formal explanation relation. Suppose, moreover, that all instances of mediate formal explanation that are justified by less than n instances of immediate formal explanation correspond to sequents derivable in SL. We show that the SL sequents $t(\Delta) \Rightarrow t(A)$ and $t(\Gamma), t(\Delta)^* \Rightarrow t(A)^*$ are derivable as well.

If $[\Gamma]\Delta \Vdash^m A$ is justified by n instances of the immediate formal explanation relation, then, by Definition 4, we have that there is a valid immediate formal explanation $[G]\Delta' \Vdash B$ and a valid mediate formal explanation $[\Gamma']\Delta'', B \Vdash^m A$ which is justified by less than n instances of the immediate formal explanation relation. Moreover, it holds that $\Delta = \Delta' \cup \Delta''$ and $\Gamma = \Gamma' \cup \{G\}$. By induction hypothesis, we have that the pair of sequents $t(\Delta') \Rightarrow t(B)$ and $t(\Delta''), t(B) \Rightarrow t(A)$, and the pair of sequents $t(G), t(\Delta')^* \Rightarrow t(B)^*$ and $t(\Gamma'), t(\Delta'')^*, t(B)^* \Rightarrow t(A)^*$ are derivable in SL. By applying the cut rule, we can obtain the desired derivation of $t(\Gamma), t(\Delta) \Rightarrow t(A)$:

$$\frac{t(\Delta') \Rightarrow t(B) \quad t(\Delta''), t(B) \Rightarrow t(A)}{t(\Delta'), t(\Delta'') \Rightarrow t(A)} \; cut$$

since $t(\Delta) = t(\Delta') \cup t(\Delta'')$; and the desired derivation of $t(\Gamma), t(\Delta)^* \Rightarrow t(A)^*$:

$$\frac{t(G), t(\Delta')^* \Rightarrow t(B)^*   t(\Gamma'), t(\Delta'')^*, t(B)^* \Rightarrow t(A)^*}{t(G), t(\Gamma'), t(\Delta')^*, t(\Delta'')^* \Rightarrow t(A)^*} \ cut$$

since $t(\Delta)^* = t(\Delta')^* \cup t(\Delta'')^*$ and $t(\Gamma) = t(\Gamma') \cup t(G)$.

Since formal explanations are often associated with analytic proofs [16], we also prove that formal explanations in classical logic can always be represented as analytic SL derivations.

Corollary 1. *Formal explanations in classical logic can be represented as cut-free SL derivations.*

Proof. Since cut-elimination holds for SL, see, for instance, [2, p. 213], we have that for each SL derivation used in Theorem 2, there is a cut-free SL derivation with the same end-sequent.

5   Conclusions

We have fully characterised the notion of immediate formal explanation for classical logic by the substructural calculus SL, which is both a fragment of FLe (Full Lambek Calculus with the Exchange rule) and a fragment of Linear Logic extended by an axiom for the lattice bottom. Moreover, we have proved that the notion of mediate formal explanation—that is, the transitive closure of immediate formal explanation—preserves all relevant derivability conditions in SL without the *cut* rule. Thus, we have showed that formal explanation, in general, can be fully characterised by complexity constraints and by analytic derivability in SL, without any reference to derivability in classical logic. This notion of explanation, as a consequence, does not only have a substructural flavour, but is actually rigorously connected in an essential way to substructural derivability.

A further question remains open, though. Indeed, we conjecture that instead of defining mediate formal explanation as the transitive closure of immediate formal explanation, it should be possible to define it in a direct way by derivability conditions formulated in SL and suitably adapted complexity conditions. In order to prove this, since we already proved that mediate formal explanation preserves positive and negative derivability in *cut*-free SL, it is enough to prove that, if positive and negative derivability in SL hold for the translation of a suitable multiset of formulae Γ, Δ, A, then $[\Gamma]\Delta \Vdash^m A$ holds as well. Even though we conjecture that this is true, a formal proof of the result seems to require a rather lengthy argument. We leave it, therefore, for future work.

References

1. Barnes, J. (ed.): The Complete Works of Aristotle. Princeton University Press, Princeton (1984)

2. Belardinelli, F., Jipsen, P., Ono, H.: Algebraic aspects of cut elimination. Studia Logica Int. J. Symbolic Logic **77**(2), 209–240 (2004)
3. Bolzano, B.: Theory of Science. Oxford University Press, Oxford(2014), translated by Rolf George and Paul Rusnok
4. Correia, F.: Logical grounds. The review of symbolic logic **7**(1), 31–59 (2014)
5. Fine, K.: Guide to Ground, pp. 37–80. Cambridge University Press, Cambridge (2012)
6. Galatos, N., Jipsen, P., Kowalski, T., Ono, H.: Residuated lattices: an algebraic glimpse at substructural logics. Elsevier (2007)
7. Girard, J.Y.: Linear logic. Theor. Comput. Sci. **50**(1), 1–101 (1987)
8. Hempel, C.G., Oppenheim, P.: Studies in the logic of explanation. Philos. Sci. **15**(2), 135–175 (1948)
9. Mancosu, P.: Mathematical explanation: problems and prospects. Topoi **20**(1), 97–117 (2001)
10. Miller, T.: Explanation in artificial intelligence: insights from the social sciences. Artif. Intell. **267**, 1–38 (2019)
11. Paoli, F.: Substructural Logics: A Primer, vol. 13. Springer Science & Business Media (2013). https://doi.org/10.1007/978-94-017-3179-9
12. Pearl, J., Mackenzie, D.: The book of why: the new science of cause and effect. Basic Books, New York (2018)
13. Poggiolesi, F.: On defining the notion of complete and immediate formal grounding. Synthese **193**(10), 3147–3167 (2015). https://doi.org/10.1007/s11229-015-0923-x
14. Poggiolesi, F.: On constructing a logic for the notion of complete and immediate formal grounding. Synthese **195**(3), 1231–1254 (2016). https://doi.org/10.1007/s11229-016-1265-z
15. Poggiolesi, F.: A proof-based framework for several types of grounding. Logique et Analyse (2020)
16. Rumberg, A.: Bolzano's concept of grounding (Abfolge) against the background of normal proofs. Rev. Symbolic Logic **6**(3), 424–459 (2013)
17. Salmon, W.C.: Four Decades of Scientific Explanation. University of Pittsburgh Press, Pittsburgh (2006)
18. Schnieder, B.: A logic for because. Rev. Symbolic Logic **4**(3), 445–465 (2011)
19. Steiner, M.: Mathematical explanation. Philos. Stud. Int. J. Philos. Analytic Tradit. **34**(2), 135–151 (1978)

Dedekind Cuts and Long Strings of Zeros in Base Expansions

Ivan Georgiev$^{(\boxtimes)}$ (ID)

Department of Mathematics and Physics, Faculty of Natural Sciences,
University "Prof. D-r Asen Zlatarov", 8010 Burgas, Bulgaria
ivandg@btu.bg

Abstract. In this paper, we study the complexity of irrational numbers under different representations. It is well-known that they can be computably transformed from one into another, but in general not subrecursively (with respect to many natural subrecursive classes). Our present focus is mainly on Dedekind cuts and base-b expansions in some base b. There exists a simple algorithm, which converts the Dedekind cut of an irrational number into its base-b expansion, but the opposite conversion is not subrecursively possible. This is why we want to enforce some natural conditions on the distribution of digits in the base-b expansion, under which we can obtain low complexity of the Dedekind cut. Our first theorem states that, for a subrecursive class $\mathcal{S}$ with natural closure properties, the Dedekind cut of an irrational number will belong to $\mathcal{S}$, whenever its base-b expansion has only very small chunks of non-zero digits, which can be generated by means of the class $\mathcal{S}$. But much more interesting is the case, when long strings of zero digits alternate with long strings of non-zero digits. We give such an example, in which the Dedekind cut does not belong to $\mathcal{S}$, but after properly inserting zeros, its complexity lowers to belong to $\mathcal{S}$. We also give a construction of a real number, which has similar long stretches of zeros, but whose Dedekind cut can be made arbitrarily complex.

Keywords: Computable analysis · Irrational number representations · Dedekind cuts · Base expansions · Subrecursive classes

1 Introduction

One of the first directions in the development of computable analysis is the study of different representations of the real numbers. Among them are the more popular ones, such as Cauchy sequences, Dedekind cuts or base-b expansions, but there are also many others, not so well-known, such as Hurwitz characteristics or general sum approximations. From the point of view of computability, each of these representations defines one and the same class of computable real numbers. Moreover, restricted to irrational numbers, any two of them can be uniformly transformed into one another. Our research is concerned with the complexity

© Springer Nature Switzerland AG 2021
L. De Mol et al. (Eds.): CiE 2021, LNCS 12813, pp. 248–259, 2021.
https://doi.org/10.1007/978-3-030-80049-9_23

of these conversions. Our main question is: *which conversions remain possible if we disallow unbounded search?* More formally, we study the computational complexity of the representations of irrational numbers with respect to natural subrecursive classes.

Among the pioneering papers in the area are Specker [13], Mostowski [10], Lehman [9], Lachlan [8], which consider the class of primitive recursive functions. Later the focus shifted to the class of functions computable in polynomial time, for example, in Ko [2,3] and Labhalla and Lombardi [7]. More recently, questions about subclasses of the second Grzegorczyk class are studied by Skordev, Weiermann and Georgiev [12] and Georgiev [1].

The present paper is a continuation of an extensive research, initiated by Kristiansen [4,5], which aims to investigate all known representations of irrational numbers and introduce many new ones in the complexity framework of honest functions. More concretely, we focus on the interplay between Dedekind cuts and base-b expansions in a fixed base b. Our first aim is to generalize some positive results from [4,5] on Dedekind cuts of irrational numbers with sparse non-zero base-b digits. Our next more ambitious aim is to better understand the connections between the complexity of the Dedekind cut and the presence of zeros in the base-b expansion. To our knowledge, these connections have not been studied profoundly from the perspective of complexity theory. Our work may be seen as providing tools for a more comprehensive treatment of the matter. We pose some concrete questions whose answers are far from obvious and require technically involved arguments.

In more detail, for every irrational number α, we define the number α' by inserting infinitely many regular long strings of zeros in the base-b expansion of α. For a subrecursive class $\mathcal{S}$, we denote by $\mathcal{S}_D$ the set of irrational numbers with Dedekind cut in $\mathcal{S}$. We prove that the implication $\alpha' \in \mathcal{S}_D \implies \alpha \in \mathcal{S}_D$ does not hold by using a number α, constructed in [5]. Our conjecture is that the converse implication $\alpha \in \mathcal{S}_D \implies \alpha' \in \mathcal{S}_D$ is also false, but here we only construct a real number in support of the conjecture.

2   Preliminaries

Without loss of generality all irrational numbers we consider will be in $(0, 1)$.

Any natural number $b \geq 2$ will be called *a base* and the numbers in the set $\{0, 1, \ldots, b - 1\}$ will be called *base-b digits*. For any base-b digit D we define the complement digit $\overline{\mathsf{D}} = b - 1 - \mathsf{D}$. For any finite sequence $\mathsf{D}_1, \mathsf{D}_2, \ldots, \mathsf{D}_n$ of base-b digits, we denote by $(0.\mathsf{D}_1\mathsf{D}_2 \ldots \mathsf{D}_n)_b$ the rational number $\sum_{i=1}^{n} \mathsf{D}_i b^{-i}$.

Definition 1. *For an infinite sequence $\mathsf{D}_1, \mathsf{D}_2, \ldots$ of base-b digits, we will say that $(0.\mathsf{D}_1\mathsf{D}_2 \ldots)_b$ is the* base-b expansion *of the real number α if for all $n \geq 1$*

$$(0.\mathsf{D}_1\mathsf{D}_2 \ldots \mathsf{D}_n)_b \leq \alpha < (0.\mathsf{D}_1\mathsf{D}_2 \ldots \mathsf{D}_n)_b + b^{-n}.$$

We identify the base-b expansion with the function $E_b^\alpha : \mathbb{N} \to \{0, 1, \ldots, b - 1\}$, such that

$$E_b^\alpha(0) = 0, \quad E_b^\alpha(n) = \mathsf{D}_n \quad (for \ n \geq 1).$$

Since the second inequality is strict, every real number $\alpha \in [0,1)$ has a unique base-b expansion and E_b^α is well defined.

If for some n we have $\alpha = (0.\mathrm{D}_1\mathrm{D}_2 \ldots \mathrm{D}_n)_b$ with $\mathrm{D}_n \neq 0$, then α has *a finite base-b expansion* and the number n is called the *base-b length* of α. If such an n does not exist, then α has *an infinite base-b expansion*. It is well known that if α is rational, then its base-b expansion is either finite or periodic (where the period contains at least one non-zero digit).

Definition 2. *For a real number α, the function $D^\alpha : \mathbb{Q} \to \{0,1\}$ will be called the* Dedekind cut *of α if $D^\alpha(q) = 0$ for $q < \alpha$ and $D^\alpha(q) = 1$ for $q \geq \alpha$.*

For two functions ϕ, ψ we denote $\phi \leq_E \psi$ (ϕ *is elementary in* ψ), if ϕ can be generated from ψ and the initial elementary functions (constant 0, successor and projections) by composition and bounded primitive recursion. We say that ϕ is *elementary*, if ϕ is elementary in the constant 0. Usually ϕ and ψ are total functions of several arguments in $\mathbb{N}$, but we also allow the integers $\mathbb{Z}$ and the rationals $\mathbb{Q}$ under some fixed natural codings into $\mathbb{N}$. As an example, we can define an elementary ternary function $\mathbf{digit} : \mathbb{N} \times \mathbb{Q} \times \mathbb{N} \to \mathbb{Z}$, such that $\mathbf{digit}_0(q,b)$ is the integer part of q and for $n \geq 1$, $\mathbf{digit}_n(q,b) = E_b^{q'}(n)$, where q' is the fractional part of q.

For the purposes of this paper, a *subrecursive class* $\mathcal{S}$ is an efficiently enumerable class of total computable functions, which contains the elementary functions and is closed under composition and bounded primitive recursion. A function f is *honest*, if $f(x) \geq 2^x$, $f(x+1) \geq f(x)$ for all $x \in \mathbb{N}$ and the graph of f is elementary (more precisely, the characteristic function of the binary relation $f(x) = y$ is elementary). For any honest function f we define the *jump* f' of f by iteration: $f'(x) = f(f(\ldots f(x)\ldots))$ ($x+1$ times) for all $x \in \mathbb{N}$. It is not hard to show that the jump f' is also honest and that if $\phi \leq_E f$, then $\phi(x) < f'(x)$ for all sufficiently large x.

For any subrecursive class $\mathcal{S}$ there exists an honest function f, such that $\phi \leq_E f$ for all $\phi \in \mathcal{S}$. Therefore, the jump $f' \notin \mathcal{S}$.

For full proofs of the above results see Sects. 2, 8 in [4]. See also [6] for more details on honest functions and [11] on subrecursive classes. The reader interested in computable real numbers and real functions should consult the introductory textbook [14] on computable analysis.

For a subrecursive class $\mathcal{S}$, we denote

$$\mathcal{S}_{bE} = \{ \alpha \in (0,1) \setminus \mathbb{Q} \mid E_b^\alpha \in \mathcal{S} \}, \quad \mathcal{S}_D = \{ \alpha \in (0,1) \setminus \mathbb{Q} \mid D^\alpha \in \mathcal{S} \}.$$

A simple algorithm, which computes the base-b digits of α successively using less than b calls to D^α for each digit, shows that $E_b^\alpha \leq_E D^\alpha$ (even uniformly in b) and therefore $\mathcal{S}_D \subseteq \mathcal{S}_{bE}$. But as proven in Sect. 6 in [5], the inclusion is strict.

Lemma 1. *Let $q = \frac{m}{n}$ be a rational number, such that $q \in (0,1)$. Let b be a base and $(0.\mathrm{D}_1\mathrm{D}_2 \ldots)_b$ be the base-b expansion of q. Assume also that $s \in \mathbb{N}$ satisfies $b^s \geq n$. If for some x the consecutive digits $\mathrm{D}_{x+1}, \mathrm{D}_{x+2}, \ldots, \mathrm{D}_{x+s}$ are all equal to 0, then q has a finite base-b expansion of length $\leq x$. If they are all equal to $\overline{0}$, then $x = 0$, $m = b^s - 1$, $n = b^s$.*

The lemma gives an upper bound on the number of consecutive 0-s or $\bar{0}$-s in case we know q has an infinite base-b expansion. Note that $x = 0$ is the only possibility in the last case, because $x > 0$ implies $n > b^s$.

3   Dedekind Cuts of Numbers with Sparse Non-zero Digits

Our first theorem generalizes parts of Theorem 7.6 in [5] (and also Theorem 5.2 in [4] as a special case), where an irrational number with very sparse non-zero base-b digits is shown to have an elementary Dedekind cut.

Let us fix a subrecursive class S and as in the previous section, let f be an honest function, such that $\phi \in S$ implies $\phi \leq_E f$. Then the jump f' of f grows faster than any function in S.

Theorem 1. *Let b be any base and let the functions $a, e, h : \mathbb{N} \to \mathbb{N}$ satisfy the following conditions: the graph of e is an S-relation, $h \in S$ and for all $i \in \mathbb{N}$:*

$$0 < a(i+1) = h(e(i)), \quad e(i+1) \geq f'(e(i)).$$

Let $\alpha = \sum_{k=0}^{\infty} a(k) b^{-e(k)}$ and assume $\alpha \in (0,1)$. Then $\alpha \in S_D$.

Proof. Observe that e grows very fast so that $e \notin S$, but the graph of e is an S-relation. Note also that the assumption $\alpha \in (0,1)$ is not essential, it is made to be in agreement with the definition of S_D. Let us take $g(x) = \sum_{y=0}^{x} h(y)$. Obviously, $h(x) \leq g(x)$ for all $x \in \mathbb{N}$ and g is non-decreasing. We also have $g \in S$, because $h \in S$, thus we may choose i_0, such that

$$e(i+1) \geq f'(e(i)) > g(e(i)) + 2e(i) \tag{1}$$

holds for all $i \geq i_0$. Using (1) we obtain

$$a(k+1)b^{-e(k+1)} = h(e(k))b^{-e(k+1)} < b^{g(e(k))}b^{-e(k+1)} \leq b^{-2e(k)-1}$$

for all $k \geq i_0$ and this easily implies $\sum_{k=i+1}^{\infty} a(k)b^{-e(k)} < b^{-e(i)}$ for all $i \geq i_0$. It follows that the series is indeed convergent. Moreover, by the same token

$$\sum_{k=i+1}^{\infty} a(k)b^{-e(k)} < a(i+1)b^{-e(i+1)} + b^{-e(i+1)}$$

$$= (h(e(i)) + 1) b^{-e(i+1)} \leq b^{-e(i+1)+g(e(i))}$$

for all $i \geq i_0$. Note that the rational number $q_i = \sum_{k=0}^{i} a(k)b^{-e(k)}$ has a finite base-b expansion of length $\leq e(i)$ and we have

$$q_i < \alpha < q_i + b^{-e(i+1)+g(e(i))}.$$

Therefore the first $e(i)$ base-b digits of α and q_i coincide and then a long stretch of zeros follows, more precisely, for all $i \geq i_0$

$$e(i) + 1 \leq x \leq e(i+1) - g(e(i)) \implies E_b^{\alpha}(x) = 0. \tag{2}$$

Clearly its length grows very fast (indeed, for any function $s \in S$ we have $e(i+1) - g(e(i)) - e(i) \geq s(e(i))$ for all sufficiently large i). Since $a(i+1) > 0$, the stretch of zeros is followed by at least one non-zero digit in the base-b expansion of α. In particular, α is irrational and now we will describe an algorithm for the Dedekind cut of α, which shows that $\alpha \in S_D$.

The input is a rational number q.

Step 1. Compute m, n such that $q = mn^{-1}$ and $n \geq e(i_0)$.
Step 2. Search for the unique $i \geq i_0$, such that $e(i) \leq g(n) + 2n < e(i+1)$.
Step 3. Compute $q_i = \sum_{k=0}^{i} a(k) b^{-e(k)}$.
Step 4. If $q \leq q_i$ give output 0, otherwise give output 1. *End of algorithm.*

All the steps are formalizable in S. In Step 1 $e(i_0)$ is a fixed number and can be used as a constant. In Step 2 the search for i is bounded as $i < e(i)$. Furthermore, $g \in S$ and the graph of e is an S-relation, which allows to check the two inequalities. In Step 3 we have $e(k) \leq g(n) + 2n$ for all $k \leq i$, thus we can compute $e(0), e(1), \ldots, e(i)$ using the graph of e and also $a(0), a(1), \ldots, a(i)$, since $h \in S$ and $a(k) = h(e(k-1))$ for all non-zero k (think of $a(0)$ as constant). Therefore, we can also compute q_i. Step 4 is obviously elementary in q and q_i.

It remains to prove that the algorithm correctly computes the Dedekind cut D^α of α. The easy case is when $q \leq q_i$, because obviously $q \leq q_i < \alpha$ and $D^\alpha(q) = 0$, which coincides with the output 0. Suppose now that $q_i < q$. Then the algorithm outputs 1 and we must prove that $D^\alpha(q) = 1$. Assume by way of a contradiction that $D^\alpha(q) = 0$, so we have $q_i < q < \alpha$. We will prove the inequality

$$e(i) + n \leq e(i+1) - g(e(i)).$$

Indeed, if $n \leq e(i)$ it follows easily from (1) and if $n > e(i)$ we have

$$e(i) + g(e(i)) + n < g(n) + 2n < e(i+1),$$

due to the choice of i and the fact that g is non-decreasing. Now using (2) we obtain that the base-b digits of α in positions $e(i)+1, e(i)+2, \ldots, e(i)+n$ are all equal to 0 and thus the same is true for q. But by Lemma 1 this is only possible if q has a finite base-b expansion of length $\leq e(i)$, because q has denominator n. This clearly contradicts the strict inequalities $q_i < q < \alpha$ and the fact that the first $e(i)$ base-b digits of q_i and α coincide (thus $q_i \leq q \leq \alpha$ implies $q = q_i$).   □

Observe that intuitively (the base-b length of) $a(k+1) = h(e(k))$ is very small compared to $e(k+1) \geq f'(e(k))$, because f' grows faster than any function in S. For all sufficiently large k the base-b expansion of α is a concatenation of strings of the following kind:

$$\overbrace{}^{a(k+1)}$$

$$000000\ldots\ldots\ldots\ldots 000000\underbrace{** \ldots *}$$

$$\uparrow D_{e(k)+1} \qquad\qquad\qquad D_{e(k+1)} \uparrow$$

and this is the reason why long stretches of zeros are followed by (relatively) small chunks of non-zero digits. The examples from [5] are with h a constant, but we can obtain many others, one for each choice of the function $h \in \mathcal{S}$.

Now what happens if we allow longer parts of non-zero digits? The starting point for our motivation will be the number α, defined in the last example on page 22 in [5]. Besides long stretches of 0-s in positions $[d_{2i}, d_{2i+1})$ for all i, α also has long stretches of $\overline{0}$-s in the remaining positions $[d_{2i+1}, d_{2i+2})$. Moreover, the base-b expansion of α is elementary and Kristiansen gives a simple algorithm to prove that the Dedekind cut of α is also elementary.

Our question is: is it possible to modify Theorem 1 in a proper way, so that we could prove $\alpha \in \mathcal{S}_D$ by only assuming that $E_b^\alpha \in \mathcal{S}$ and $E_b^\alpha(x) = 0$ for all $x \in [d_{2i}, d_{2i+1})$ and all i? In other words, can we prove $\alpha \in \mathcal{S}_D$ by only assuming we have access to the base-b expansion of α and that α has consecutive long strings of zeros? It turns out that the answer is negative (see Theorem 3 below).

In the example from [5], all digits of α in the remaining positions $[d_{2i+1}, d_{2i+2})$ are equal to $\overline{0}$ and this is crucial to deduce that $\alpha \in \mathcal{S}_D$. On the other hand, we shall see in the next section that the remaining digits may constitute a number of high complexity, but still we may have elementary Dedekind cut.

4   Dedekind Cuts of Numbers with Alternating Long Strings of Zero and Non-zero Digits

Let again $\mathcal{S}$ be a subrecursive class and f be an honest functions, such that f' grows faster than any function in $\mathcal{S}$. Below we will use a sequence d, such that $d_{i+1} = f'(d_i)$ for all i. We consider the intervals $[d_i, d_{i+1})$ as *long*, because one cannot reach d_{i+1} when having d_i by means of a function from $\mathcal{S}$ for sufficiently large i. However, an easy modification of Lemma 4.7 in [5] shows that the graph of d is elementary (since f' is honest).

Definition 3. *For an arbitrary irrational number α we define α' in the following way:*

$$E_b^{\alpha'}(x) = \begin{cases} E_b^\alpha(x), & \text{if } x < d_0, \\ 0, & \text{if } d_{2i} \leq x < d_{2i+1}, \\ E_b^\alpha(x'), & \text{if } d_{2i+1} \leq x < d_{2i+2}, \end{cases}$$

where $x' = x - \sum_{j \leq i} (d_{2j+1} - d_{2j})$.

In other words, a string of 0-s is inserted in the base-b expansion of α in positions $[d_{2i}, d_{2i+1})$ for all i and the other base-b digits of α are properly shifted to the right (the sum gives the number of added 0-s preceding the position x). Observe that α' is irrational, because its base-b expansion is neither finite, nor periodic. Note also that $E_b^{\alpha'} \leq_E E_b^\alpha$, which follows easily from the fact that the graph of d is elementary. We can compute x' elementarily in the last case, since $x \geq d_{2i+1}$.

Our aim is to explore the connection between the Dedekind cuts of α and α'. In the next theorem we will show that the implication $\alpha' \in \mathcal{S}_D \implies \alpha \in \mathcal{S}_D$ is not true.

In order to do that we turn to the construction from Theorem 5.3 in [5]. The reader is invited to study the proof in [5] for better understanding of the rest of the paper. We have two bases a, b and a prime factor p of a, such that p does not divide b. An irrational number $\alpha = \lim_{i \to \infty} q_i$ is constructed, where:

1. $q_{2i} = q_{2i+1} = p^{-1} + m_1 p^{-bk_1} + m_2 p^{-bk_2} + \ldots + m_i p^{-bk_i}$
   with $m_i \in \{-1, 1\}$ and k_i a very fast increasing sequence;
2. q_{2i} can be computed elementarily in d_{2i};
3. the base-b expansions of q_{2i} and α coincide in all positions less than d_{2i+2};
4. E_b^α is elementary and $\alpha \notin \mathcal{S}_{aE}$, so that $\alpha \notin \mathcal{S}_D$.

Theorem 2. *The Dedekind cut of α' is elementary, where α is the irrational number, constructed in Theorem 5.3 in [5].*

Proof. We will describe an algorithm to compute the Dedekind cut of α', prove its correctness and leave the reader to check that it is elementary (note that $E_b^{\alpha'}$ is elementary, because so is E_b^α).

The input is a rational number q. We may assume $q \in (0, 1)$.

Step 0. If q has an infinite base-b expansion proceed with Step 1. Otherwise, compute the base-b length s of q and $\tilde{q} = \sum_{j \leq s} E_b^{\alpha'}(j) b^{-j}$. If $q \leq \tilde{q}$ give output 0 and if $q > \tilde{q}$ give output 1.

Step 1. Compute m, n, such that $q = mn^{-1}$ and $n \geq d_0$. Search for the unique i, such that $d_{2i} \leq n < d_{2i+2}$. Go to Step 2.

Step 2. Search for the least $k < d_{2i}$, such that $\mathbf{digit}_k(q, b) \neq E_b^{\alpha'}(k)$. If the search is successful, return output 0 if $\mathbf{digit}_k(q, b) < E_b^{\alpha'}(k)$ and output 1 if $\mathbf{digit}_k(q, b) > E_b^{\alpha'}(k)$. If the search is not successful proceed with Step 3.

Step 3. Check whether the following relation holds: $d_{2i} + n < d_{2i+1}$. If it is true give output 1, otherwise go to Step 4.

Step 4. Compute d_{2i+1} and q_{2i} (the sequence from [5]). Insert 0-s in the base-b expansion of q_{2i} in positions $[d_{2j}, d_{2j+1})$ for all $j \leq i$ to obtain q'_{2i}. Let $s = |q - q'_{2i}|$. If $s = 0$ give output 1, otherwise: Search for the least $y < s^{-1}$, such that $y = d_{2i+2}$ & $b^{y-1} < s^{-1}$. If the search is successful go to Step 5, if it is not proceed with Step 6.

Step 5. Search for the least $k < d_{2i+2}$, such that $\mathbf{digit}_k(q, b) \neq E_b^{\alpha'}(k)$. If the search is successful, give output 0 if $\mathbf{digit}_k(q, b) < E_b^{\alpha'}(k)$ and output 1 if $\mathbf{digit}_k(q, b) > E_b^{\alpha'}(k)$. If the search is not successful give output 1.

Step 6. If $q < q'_{2i}$ output 0, otherwise output 1. *End of algorithm.*

In Step 0 if q has a finite base-b expansion of length s, then clearly we need to check only the first s base-b digits of α'. The output in Step 2 is obviously correct when the search is successful.

Let us assume that $d_{2i} + n < d_{2i+1}$ in Step 3. On one hand by Lemma 1 at least one base-b digit of q in positions $d_{2i}, \ldots, d_{2i} + n$ is different from 0. On the other hand, α' has base-b digit 0 in all positions in $[d_{2i}, d_{2i} + n]$. Thus $q > \alpha'$ and the output 1 is correct (after Step 2 the base-b expansions of q and α' coincide on all positions less than d_{2i}).

In Step 4 we know that the first d_{2i+2} base-b digits of α and q_{2i} coincide. The same is true for α' and q'_{2i}, because they are obtained by inserting 0-s in the same positions (and of course in α' more 0-s are inserted, but from position d_{2i+2} onwards). Now $s = 0$ means $q = q'_{2i}$ and therefore q and α' coincide in base b in all positions less than d_{2i+2}. But $d_{2i+2} + n < 2d_{2i+2} < d_{2i+3}$, thus by Lemma 1 at least one base-b digit of q in positions $[d_{2i+2}, d_{2i+3})$ is not 0, whereas α' has only 0-s in the same positions. We conclude that $q > \alpha'$ and the output 1 is correct.

In Step 5, the search for y from Step 4 is successful, so we have computed d_{2i+2}. If the search for k is successful, the output is obviously correct. If it is not, we can reason as in the case $s = 0$ to see that the output 1 is correct.

In Step 6, the search for y from Step 4 is unsuccessful and $s \geq b^{-d_{2i+2}+1}$. Indeed, if $s < b^{-d_{2i+2}+1}$ then $d_{2i+2} \leq b^{d_{2i+2}-1} < s^{-1}$ and the search for y would be successful. So we have $s = |q - q'_{2i}| \geq b^{-d_{2i+2}+1}$, but this can only happen when there is a position $k < d_{2i+2}$ with $\mathbf{digit}_k(q, b) \neq \mathbf{digit}_k(q'_{2i}, b)$. Since q'_{2i} and α' coincide in all positions less than d_{2i+2} the output is correct. □

Now we turn to the converse implication $\alpha \in \mathcal{S}_D \implies \alpha' \in \mathcal{S}_D$, which seems somewhat easier at first sight. Our conjecture is that it is also false, but we do not have a proof yet. In the last theorem we will construct a number α'_Q, such that $\alpha'_Q \in \mathcal{S}_{bE}$, $\alpha'_Q \notin \mathcal{S}_D$ and the base-b expansion of α'_Q has 0-s in positions $[d_{2i}, d_{2i+1})$ for all i. Observe that if such a number α'_Q did not exist, the implication would hold. Indeed, if $\alpha \in \mathcal{S}_D$, then $\alpha \in \mathcal{S}_{bE}$ and therefore $\alpha' \in \mathcal{S}_{bE}$ and α' by definition has 0-s in positions $[d_{2i}, d_{2i+1})$ for all i. So if the real number in question did not exist, we would conclude $\alpha' \in \mathcal{S}_D$. We regard this observation as evidence in support of the conjecture. We will again exploit the sequence q_i and its limit α from Theorem 5.3 in [5].

Theorem 3. *There exists a number* α'_Q*, such that* $\alpha'_Q \in \mathcal{S}_{bE}$*,* $\alpha'_Q \notin \mathcal{S}_D$ *and* $E_b^{\alpha'_Q}(s) = 0$ *for all* $s \in [d_{2i}, d_{2i+1})$ *and all* i*.*

Proof. The idea is the following: construct α'_N from the base-b expansion of α by simultaneously inserting 0-s in all positions $[d_{2i}, d_{2i+1})$ with $q_{2i} < q_{2i-2}$ and $\overline{0}$-s in all positions $[d_{2i}, d_{2i+1})$ with $q_{2i} \geq q_{2i-2}$. A diagonalization argument will show that $\alpha'_N \notin \mathcal{S}_D$. Now to obtain α'_Q we add 1 to the last digit in each string of $\overline{0}$-s in positions $[d_{2i}, d_{2i+1})$ with $q_{2i} \geq q_{2i-2}$. Clearly, each of these strings is transformed into a string of 0-s and only a small portion of digits before the string is altered. Moreover, we will see that the complexity of the Dedekind cut does not change, because the added 1-s are sparse enough.

We begin by defining α'_N:

$$
E_b^{\alpha'_N}(x) = \begin{cases}
E_b^{\alpha}(x), & \text{if } x < d_0, \\
0, & \text{if } d_{2i} \leq x < d_{2i+1} \ \& \ q_{2i} < q_{2i \dot- 2}, \\
\overline{0}, & \text{if } d_{2i} \leq x < d_{2i+1} \ \& \ q_{2i} \geq q_{2i \dot- 2}, \\
E_b^{\alpha}(x'), & \text{if } d_{2i+1} \leq x < d_{2i+2},
\end{cases}
$$

where $x' = x - \sum_{j \leq i}(d_{2j+1} - d_{2j})$. First of all $E_b^{\alpha'_N}$ is elementary: the graph of d and E_b^{α} are elementary, q_{2i}, $q_{2i \dot- 2}$ are elementary in x in the case $d_{2i} \leq x$. Suppose now that $D^{\alpha'_N} \leq_E f$. Then it is rather straightforward to define a unary function γ, such that $\gamma \leq_E f$ and for all i

$$
\gamma(d_{2i+1}) \geq q_{2i} \iff q'_{2i} < \alpha'_N, \tag{3}
$$

where q'_{2i} is obtained from q_{2i} by inserting 0-s if $q_{2j} < q_{2j \dot- 2}$ and $\overline{0}$-s if $q_{2j} \geq q_{2j \dot- 2}$ in positions $[d_{2j}, d_{2j+1})$ for all $j \leq i$. Now the argument proceeds as in the proof of clause (ii) in Theorem 5.3 in [5]. Since $\gamma \leq_E f$, we can choose e, k, such that

$$
\gamma(n) = \mathcal{U}\left(\mu t \leq f^k(n)[\mathcal{T}_1(e, n, t)]\right)
$$

for all n. We also pick i, j, such that $i = \langle e, j \rangle$ and $d_{2i+1} \geq k$, therefore by the definition of q_{2i} we have

$$
q_{2i+2} < q_{2i} \iff \gamma(d_{2i+1}) \geq q_{2i+1} = q_{2i}. \tag{4}
$$

Combining (3) and (4) gives:

$$
q_{2i+2} < q_{2i} \iff q'_{2i} < \alpha'_N.
$$

Now if $q_{2i+2} < q_{2i}$, then α'_N has 0-s in positions $[d_{2i+2}, d_{2i+3})$, but q'_{2i} duplicates digits of q_{2i} in the same positions and thus one of them is non-zero by an application of Lemma 1. Since α'_N and q'_{2i} coincide in all positions less than d_{2i+2} we obtain $\alpha'_N < q'_{2i}$, which is a contradiction. The other case $q_{2i+2} > q_{2i}$ is handled similarly. We conclude that the Dedekind cut of α'_N is not elementary in f and thus $\alpha'_N \notin \mathcal{S}_D$. There are missing details here which are not hard to fill in.

We are ready to define α'_Q by the equality

$$
\alpha'_Q = \alpha'_N + \sum_{j=0}^{\infty} \chi(j) b^{-d_{2j+1}+1},
$$

where $\chi(j) = 0$ if $q_{2j} < q_{2j \dot- 2}$ and $\chi(j) = 1$ if $q_{2j} \geq q_{2j \dot- 2}$.

Firstly we give a sketch for an elementary algorithm to compute $E_b^{\alpha'_Q}(n)$: given input n, compute i, such that $d_{2i} \leq n < d_{2i+2}$. If $n < d_{2i+1}$, output $E_b^{\alpha'_Q}(n) = 0$. Now let $n \geq d_{2i+1}$. If n is not "close" to d_{2i+2}, $E_b^{\alpha'_Q}(n) = E_b^{\alpha'_N}(n)$ and if n is "close" to d_{2i+2}, we can compute d_{2i+2} and $E_b^{\alpha'_Q}(n)$ is obtained from $E_b^{\alpha'_N}$ (after adding $b^{-d_{2i+2}+1}$ if $\chi(i+1) = 1$). The formalization of this is relatively easy. Lemma 1 gives a precise bound how "close" n must be.

Secondly we prove that the Dedekind cut of α'_Q does not belong to $\mathcal{S}$. In order to do this we will describe an algorithm, which shows that $D^{\alpha'_N} \leq_E D^{\alpha'_Q}$. Since we know that $\alpha'_N \notin \mathcal{S}_D$, this will imply $\alpha'_Q \notin \mathcal{S}_D$, which is what we need. We only prove the correctness of the algorithm and again leave the reader to check it is elementary.

Let us denote $q''_i = \sum_{j<i} \chi(j) b^{-d_{2j+1}+1}$ for $i \in \mathbb{N}$.

The input of the algorithm is a rational number q. We assume $q \in (0, 1)$.

Step 0. If q has an infinite base-b expansion proceed with Step 1. Otherwise, compute $q' = \sum_{j \leq s} E_b^{\alpha'_N}(j) b^{-j}$, where s is the base-b length of q. If $q \leq q'$ give output 0 and if $q > q'$ give output 1.

Step 1. Find m, n, such that $q = mn^{-1}$ and $n \geq d_0$. Go to Step 2.

Step 2. Compute the unique i, such that $d_{2i} \leq n < d_{2i+2}$. Then compute $\widetilde{q}_i = q + q''_i$. Go to Step 3.

Step 3. Use the Dedekind cut of α'_Q to check if $\widetilde{q}_i < \alpha'_Q$. If it is false, give output 1. If it is true, go to Step 4.

Step 4. Check the inequality $d_{2i} + 2nb^n < d_{2i+1}$. If it holds, give output 0. Otherwise, go to Step 5.

Step 5. Compute d_{2i+1} and $\widetilde{q_{i+1}} = q + q''_{i+1}$. Use the Dedekind cut of α'_Q to check $\widetilde{q_{i+1}} < \alpha'_Q$. If it is false, output 1. If it is true, output 0. *End of algorithm.*

The output in Step 0 is clearly correct in case q has a finite base-b expansion of length s, because α'_N is irrational.

Note that both $\widetilde{q}_i$ (in Step 2) and $\widetilde{q_{i+1}}$ (in Step 5) have infinite base-b expansion, because so does q after Step 0.

In Step 3, case $\alpha'_Q \leq \widetilde{q}_i$, we have $\alpha'_Q \leq q + q''_i \leq q + \sum_{j=0}^{\infty} \chi(j) b^{-d_{2j+1}+1}$, therefore $\alpha'_N \leq q$ and the output 1 is correct.

Suppose the inequality in Step 4 holds and that the output 0 is wrong, that is $\alpha'_N \leq q$. We also have $\widetilde{q}_i < \alpha'_Q$ after Step 3, hence

$$\widetilde{q}_i < \alpha'_Q \leq q + \sum_{j=0}^{\infty} \chi(j) b^{-d_{2j+1}+1} = \widetilde{q}_i + \sum_{j=i}^{\infty} \chi(j) b^{-d_{2j+1}+1} < \widetilde{q}_i + b^{-d_{2i+1}+2}.$$

Let us denote by n' the denominator of $\widetilde{q}_i$. Clearly we have $n' \leq nb^{d_{2i}} \leq nb^n$. By Lemma 1 we can choose j, such that $j \in [d_{2i+1} - n' - 1, d_{2i+1} - 2]$ and $\mathbf{digit}_j(\widetilde{q}_i, b) \neq \overline{0}$. It follows that for all s

$$s < d_{2i+1} - n' - 1 \implies \mathbf{digit}_s(\widetilde{q}_i, b) = \mathbf{digit}_s(\widetilde{q}_i + b^{-d_{2i+1}+2}, b) = E_b^{\alpha'_Q}(s).$$

On one hand, we know that the base-b expansion of α'_Q has only 0-s in positions $[d_{2i}, d_{2i+1})$. Therefore, the same is true for $\widetilde{q}_i$ in positions $[d_{2i}, d_{2i+1} - n' - 2]$. On the other hand, we have $[d_{2i}, d_{2i} + n' - 1] \subseteq [d_{2i}, d_{2i+1} - n' - 2]$, according to the inequalities $d_{2i} + 2n' \leq d_{2i} + 2nb^n < d_{2i+1}$ assumed in Step 4. Applying again Lemma 1 we infer that $\widetilde{q}_i$ has a finite base-b expansion, which contradicts the assumption on q in Step 0. We conclude that the output 0 is correct.

If $\widetilde{q_{i+1}} \geq \alpha'_Q$ in Step 5, we have $\alpha'_Q \leq q + q''_{i+1} \leq q + \sum_{j=0}^{\infty} \chi(j) b^{-d_{2j+1}+1}$, therefore $\alpha'_N \leq q$ and the output 1 is correct. Now let $\widetilde{q_{i+1}} < \alpha'_Q$ and assume the output 0 is wrong, that is $\alpha'_N \leq q$. Similarly to Step 4 we have

$$\widetilde{q_{i+1}} < \alpha'_Q \leq q + \sum_{j=0}^{\infty} \chi(j) b^{-d_{2j+1}+1} < \widetilde{q_{i+1}} + b^{-d_{2i+3}+2}.$$

Let us denote by n'' the denóminator of $\widetilde{q_{i+1}}$. By applying Lemma 1 in the same way as in Step 4, we obtain that for all s

$$s < d_{2i+3} - n'' - 1 \implies \mathbf{digit}_s(\widetilde{q_{i+1}}, b) = \mathbf{digit}_s(\widetilde{q_{i+1}} + b^{-d_{2i+3}+2}, b) = E_b^{\alpha'_Q}(s).$$

Since α'_Q has 0-s in positions $[d_{2i+2}, d_{2i+3})$ in its base-b expansion, $\widetilde{q_{i+1}}$ also has 0-s in positions $[d_{2i+2}, d_{2i+3} - n'' - 2]$. Now n'' is the product of the denominators of q and q''_{i+1} and $n < d_{2i+2}$, therefore:

$$\begin{aligned}
d_{2i+2} + 2n'' + 1 \quad &< \quad d_{2i+2} + 2n \cdot b^{d_{2i+1}} + 1 \\
&< \quad d_{2i+2} + 2d_{2i+2} \cdot f(f(d_{2i+1})) + 1 \\
&< \quad d_{2i+2} + 2d_{2i+2} \cdot f'(d_{2i+1}) + 1 \;=\; 2d_{2i+2}^2 + d_{2i+2} + 1 \\
&< \quad 3d_{2i+2}^2 \;\leq\; f(f(d_{2i+2})) \;<\; f'(d_{2i+2}) \;=\; d_{2i+3}.
\end{aligned}$$

So we have $[d_{2i+2}, d_{2i+2} + n'' - 1] \subseteq [d_{2i+2}, d_{2i+3} - n'' - 2]$ and $\widetilde{q_{i+1}}$ has 0-s in these positions. By Lemma 1, $\widetilde{q_{i+1}}$ has a finite base-b expansion, which contradicts the assumption on q in Step 0. Therefore, the output 0 is correct. $\square$

Acknowledgements. This work was partially supported by Sofia University Science Fund, contract 80-10-136/26.03.2021.

References

1. Georgiev, I.: Continued fractions of primitive recursive real numbers. Math. Log. Q. **61**, 288–306 (2015)
2. Ko, K.: On the definitions of some complexity classes of real numbers. Math. Syst. Theory **16**, 95–109 (1983)
3. Ko, K.: On the continued fraction representation of computable real numbers. Theor. Comput. Sci. **47**, 299–313 (1986)
4. Kristiansen, L.: On subrecursive representability of irrational numbers. Computability **6**, 249–276 (2017)
5. Kristiansen, L.: On subrecursive representability of irrational numbers, part II. Computability **8**, 43–65 (2019)
6. Kristiansen, L., Schlage-Puchta, J.-C., Weiermann, A.: Streamlined subrecursive degree theory. Ann. Pure Appl. Log. **163**, 698–716 (2012)
7. Labhalla, S., Lombardi, H.: Real numbers, continued fractions and complexity classes. Ann. Pure Appl. Log. **50**, 1–28 (1990)

8. Lachlan, A.H.: Recursive real numbers. J. Symb. Log. **28**(1), 1–16 (1963)
9. Lehman, R.S.: On primitive recursive real numbers. Fundam. Math. **49**(2), 105–118 (1961)
10. Mostowski, A.: On computable sequences. Fundam. Math. **44**, 37–51 (1957)
11. Rose, H.E.: Subrecursion. Functions and hierarchies. Clarendon Press, Oxford (1984)
12. Skordev, D., Weiermann, A., Georgiev, I.: $\mathcal{M}^2$-computable real numbers. J. Log. Comput. **22**(4), 899–925 (2008)
13. Specker, E.: Nicht konstruktiv beweisbare Satze der analysis. J. Symbol. Log. **14**, 145–158 (1949)
14. Weihrauch, K.: Computable Analysis. Springer, Berlin (2000)

On the Impact of Treewidth in the Computational Complexity of Freezing Dynamics

Eric Goles[1], Pedro Montealegre[1], Martín Ríos Wilson[2,3(✉)],
and Guillaume Theyssier[4]

[1] Facultad de Ingeniería y Ciencias, Universidad Adolfo Ibáñez, Santiago, Chile
{eric.chacc,p.montealegre}@uai.cl
[2] Departamento de Ingeniería Matemática, Universidad de Chile, Santiago, Chile
mrios@dim.uchile.cl
[3] Aix Marseille Univ, Université de Toulon, CNRS, LIS, Marseille, France
[4] Aix-Marseille Université, CNRS, I2M (UMR 7373), Marseille, France
guillaume.theyssier@cnrs.fr

Abstract. An automata network is a network of entities, each holding a state from a finite set and evolving according to a local update rule which depends only on its neighbors in the network's graph. It is freezing if there is an order on states such that the state evolution of any node is non-decreasing in any orbit. They are commonly used to model epidemic propagation, diffusion phenomena like bootstrap percolation or cristal growth. In this paper we establish how alphabet size, treewidth and maximum degree of the underlying graph are key parameters which influence the overall computational complexity of finite freezing automata networks. First, we define a general specification checking problem that captures many classical decision problems such as prediction, nilpotency, predecessor, asynchronous reachability. Then, we present a fast-parallel algorithm that solves the general problem when the three parameters are bounded, hence showing that the problem is in **NC**. Finally, we show that these problems are hard from two different perspectives. First, the general problem is **W**[2]-hard when taking either treewidth or alphabet as single parameter and fixing the others. Second, the classical problems are hard in their respective classes when restricted to families of graphs with sufficiently large treewidth.

Keywords: Freezing automata networks · Treewidth · Fast parallel algorithm · Prediction · Nilpotency · Asynchronous reachability · Predecessor problem

This reasearch was partially supported by French ANR project FANs ANR-18-CE40-0002 (G.T., M.R.W.) and ECOS project C19E02 (G.T., M.R.W.), ANID via PAI + Convocatoria Nacional Subvención a la Incorporación en la Academia Año 2017 + PAI77170068 (P.M.), FONDECYT 11190482 (P.M.), FONDECYT 1200006 (E.G., P.M.), STIC- AmSud CoDANet project 88881.197456/2018-01 (E.G., P.M.), ANID via PFCHA/DOCTORADO NACIONAL/2018 – 21180910 + PIA AFB 170001 (M.R.W).

L. De Mol et al. (Eds.): CiE 2021, LNCS 12813, pp. 260–272, 2021.
https://doi.org/10.1007/978-3-030-80049-9_24

1 Introduction

An automata network is a network of n entities, each holding a state from a finite set Q and evolving according to a local update rule which depends only on its neighbors in the network's graph. More concisely, it can be seen as a dynamical system (deterministic or not) acting on the set Q^n. The model can be seen as a non-uniform generalization of (finite) cellular automata. Automata networks have been used as modelization tools in many areas [17] and they can also be considered as a distributed computational model with various specialized definitions like in [33]. An automata network is *freezing* if there is an order on states such that the state evolution of any node is non-decreasing in any orbit. Several models that received a lot of attention in the literature are actually freezing automata networks, for instance: bootstrap percolation which has been studied on various graphs [1], epidemic [12] or forest fire [3] propagation models, cristal growth models [31] and more recently self-assembly tilings [32]. On the other hand, their complexity as computational models has been studied from various point of view: as language recognizers where they correspond to bounded change or bounded communication models [25], for their computational universality [4,15], as well as for various associated decision problems [19].

A major topic of interest in automata networks theory is to determine how the network graph affects dynamical or computational properties [13,19]. In the freezing case, it was for instance established that one-dimensional freezing cellular automata, while being Turing universal via Minsky machine simulation, have striking computational limitations when compared to bi-dimensional ones: they are **NL**-predictable (instead of **P**-complete) [21,27], can only produce computable limit fixed points starting from computable initial configurations (instead of non-computable ones starting from finite configurations) [27], and have a polynomial time decidable nilpotency problem (instead of uncomputable) [27].

The present paper aims at understanding what are the key parameters which influence the overall computational complexity of finite freezing automata networks. A natural first parameter is the alphabet size, as automata networks are usually considered as simple machines having a number of states that is independent of the size of the network. For the same reasons, a second parameter that we consider is the maximum degree of the network, as a simple machine might not be able to handle the information incoming from a large number of neighbors. Finally the results mentioned earlier show a gap between bi-dimensional grids and one-dimensional grids (i.e. paths or rings). Since Courcelle's theorem on MSO properties [7], graph parameters like treewidth [29] are used to measure a sort of distance to a grid. Indeed, it is known that paths or rings have constant treewidth, and the treewidth of a graph is polynomially related to the size of its largest grid minor [6]. Therefore treewidth is a natural parameter for our study.

2 Preliminaries

Given a graph $G = (V, E)$ and a vertex v we will call $N(v)$ to the neighborhood of v and δ_v to the degree of v. In addition, we define the closed neighborhood of

v as the set $N[v] = N(v) \cup \{v\}$ and we use the following notation $\Delta(G) = \max_{v \in V} \delta_v$ for the maximum degree of G. We will use the letter n to denote the order of G, i.e. $n = |V|$. Also, if G is a graph and the set of vertices and edges is not specified we use the notation $V(G)$ and $E(G)$ for the set of vertices and the set of edges of G respectively. In addition, we will assume that if $G = (V, E)$ is a graph then, there exist an ordering of the vertices in V from 1 to n. During the rest of the text, every graph G will be assumed to be connected and undirected. We define a *class* or a *family* of graphs as a set $\mathcal{G} = \{G_n\}_{n \geq 1}$ such that $G_n = (V_n, E_n)$ is a graph and $|V_n| = n$.

Non-deterministic Freezing Automata Networks. Let Q be a finite set that we will call an *alphabet*. We define a non-deterministic automata network in the alphabet Q as a tuple $(G = (V, E), \mathcal{F} = \{F_v : Q^{N(v)} \to \mathcal{P}(Q) | v \in V\}))$ where $\mathcal{P}(Q)$ is the power set of Q. To every non-deterministic automata network we can associate a non-deterministic dynamics given by the global function $F : Q^n \to \mathcal{P}(Q^n)$ defined by $F(x) = \{x \in Q^n | x_v \in F_v(x), \forall v\}$.

Definition 1. *Given a a non-deterministic automata network $(G, \mathcal{F})$ we define an orbit of a configuration $x \in Q^n$ at time t as a sequence $(x^s)^{0 \leq s \leq t}$ such that $x_0 = x$ and $x^s \in F(x_{s-1})$. In addition, we call the set of all possible orbits at time t for a configuration x as $\mathcal{O}(x, t)$. Finally, we also define the set of all possible orbits at time t as $\mathcal{O}(\mathcal{A}, t) = \bigcup_{x \in Q^n} \mathcal{O}(x, t)$.*

A non-deterministic automata network $(G, \mathcal{F})$ defined in the alphabet Q is *freezing* if there exists a partial order $\leq$ in Q such that for every $t \in \mathbb{N}$ and for every orbit $y = (x^s)_{0 \leq s \leq t} \in \mathcal{O}(\mathcal{A}, t)$ we have that $x_v^s \leq x_v^{s+1}$ for every $0 \leq s \leq t$ and for every $0 \leq v \leq n$. Let $y = (x^s)_{0 \leq s \leq t}$ be an orbit of a non-deterministic automata network $(G, \mathcal{F})$ and $S \subseteq V$ we define the restriction of y to S as the sequence $y|_S \in (Q^t)^{|S|}$ such that $(y|_S)_v = x_v^s$ for every $v \in V$. In the case in which $S = \{v\}$ we write y_v in order to denote the restriction of y to the singleton $\{v\}$. Finally, if $\mathcal{A} = (G, \mathcal{F})$ is a non-deterministic freezing automata network such that for every $v \in V(G)$, $F_v \in \mathcal{F}$ is such that $|F_v(x)| = 1$, for all $x \in Q^{N(v)}$ then, we say that $\mathcal{A}$ is deterministic and we consider local rules as maps $F_v : Q^{N(v)} \to Q$.

Tree Decompositions and Treewidth. Let $G = (V, E)$ be a connected graph. We say that G is a tree-graph or simply a tree if it does not have cycles as subgraphs. Usually, we will distinguish certain node in $r \in V(G)$ that we will call the root of G. Whenever G is a tree and there is a fixed vertex $r \in V(G)$ we will call G a rooted tree-graph. In addition, we will say that $v \in V(G)$ is a leaf if $\delta_v = 1$. The choice of r induces a partial order in the vertices of G given by the distance (length of the unique path) between a node $v \in V(G)$ and the root r. We define the height of G (and we write it as $h(G)$) as the longest path between a leaf and r. We say that a node v is in the $(h(G) - k)$-th level of a tree-graph G if the distance between v and r is k and we write $v \in \mathcal{L}_{h(G)-k}$. We will call the *children* of a node $v \in \mathcal{L}_k$ to all $w \in N(v)$ such that w is in level $k - 1$.

Definition 2. *Given a graph $G = (V, E)$ a tree decomposition is pair $\mathcal{D} = (T, \Lambda)$ such that T is a tree graph and Λ is a family of subsets of nodes $\Lambda = \{X_t \subseteq V \mid t \in V(T)\}$, called bags, such that:*

- *Every node in G is in some X_t, i.e.:* $\bigcup\limits_{t \in V(T)} X_t = V$
- *For every $e = uv \in E$ there exists $t \in V(T)$ such that $u, v \in X_t$*
- *For every $u, v \in V(T)$ if $w \in V(T)$ is in the v-y path in T, then $X_u \cap X_v \subseteq X_w$*

We define the width of a tree decompostion $\mathcal{D}$ as the amount width$(\mathcal{D}) = \max\limits_{t \in V(T)} |X_t| - 1$. Given a graph $G = (V, E)$, we define its treewidth as the parameter $\text{tr}(G) = \min\limits_{\mathcal{D}} \text{width}(\mathcal{D})$. In other words, the treewidth is the minimum width of a tree decomposition of G. Note that, if G is a connected graph such that $|E(G)| \geq 2$ then, G is a tree if and only if $\text{tw}(G) = 1$.

It is well known that, given an arbitrary graph G, and $k \in \mathbb{N}$, the problem of deciding if $\text{tw}(G) \leq k$ is **NP**-complete [2]. Nevertheless, if k is fixed, that is to say, it is not part of the input of the problem then, there exist efficient algorithms that allow us to compute a tree-decomposition of G. More precisely, it is shown that for every constant $k \in \mathbb{N}$ and a graph G such that $\text{tw}(G) \leq k$, there exist a log-space algorithm that computes a tree-decomposition of G [11]. In addition, in Lemma 2.2 of [5] it is shown that given any tree decomposition of a graph G, there exist a fast parallel algorithm that computes a slightly bigger width binary tree decomposition of G. More precisely, given a tree decomposition of width k, the latter algorithm computes a binary tree decomposition of width at most $3k + 2$. We outline these results in the following proposition:

Proposition 1. *Let $n \geq 2, k \geq 1$ and let $G = (V, E)$ with $|V| = n$ be a graph such that $\text{tw}(G) \leq k$. There exists a CREW PRAM algorithm using $\mathcal{O}(\log^2 n)$ time, $n^{\mathcal{O}(1)}$ processors and $\mathcal{O}(n)$ space that computes a binary treewidth decomposition of width at most $3k + 2$ for G.*

Parametrized Complexity. A parameterized language is defined by $L \subseteq \{0, 1\}^* \times \mathbb{N}$. Whenever we take an instance (x, k) of a parameterized problem we will call k a *parameter*. The objective behind parameterized complexity is to identify which are the key parameters in an intractable problem that make it hard. We say that a parameterized language L is *slice-wise polynomial* if $(x, k) \in L$ is decidable in polynomial time for every fixed $k \in \mathbb{N}$. More precisely, when $(x, k) \in L$ can be decided in time $|x|^{f(k)}$ for some arbitrary function f depending only on k. The class of slice-wise polynomial parameterized languages is called **XP**. An important subclass of **XP** is the set of parameterized languages L that are *fixed-parameter tractable*, denoted **FPT**. A parameterized language L is in **FPT** if there exist an algorithm deciding if $(x, k) \in L$ in time $f(k)|x|^{\mathcal{O}(1)}$ where f an arbitrary function depending only in k. It is known that **XP** is not equal to **FPT**, however showing that some problem in **XP** is not in **FPT** seems currently out of reach for many natural examples (see [10] for more details and context).

3   Localized Trace Properties

In this section we formalize a general decision problem on our dynamical systems. Finally, we define the set of all possible orbits at time t as $\mathcal{O}(\mathcal{A}, t)$. Freezing automata network have temporally sparse orbits, however the set of possible configurations is still exponential. Our formalism takes this into account by considering properties that are spatially localized but without restriction in their temporal expressive power. It is based on the following concept of specification.

Definition 3. *Consider $t \in \mathbb{N}$ and $\mathcal{A} = (G = (V, E), \mathcal{F})$ a non-deterministic freezing automata network in some partially-ordered alphabet Q. A $(Q, t, \mathcal{A})$-specification (or simply a t-specification when the context is clear) is a function $\mathcal{E}_t : V \to \mathcal{P}(Q^t)$ such that, for every $v \in V$, the sequences in $\mathcal{E}_t(v)$ are non-decreasing.*

The following lemma shows that for all freezing automata networks the set of orbits of any length restricted to a set of nodes is determined by the set of orbits of fixed (polynomial) length restricted to these nodes. Moreover, if the set of considered nodes is finite, then the fixed length can be chosen linear.

Lemma 1. *Let Q be an alphabet, V a set of nodes with $|V| = n$ and $U \subseteq V$. Let $L = |U||Q|(|Q|n + 1)$. Then if two non-deterministic freezing automata have the same set of orbits restricted to U of length L then they have the same set of orbits restricted to U of any length.*

Note that, as a consequence of the above lemma, for any freezing non-deterministic automata network it suffices to consider t-specifications with t being linear in the size of the interaction graph defining the network.

Specification Checking Problem. We observe also that the number of possible t-specifications can be represented in polynomial space (as a Boolean vector indicating the allowed t-specifications). Also, in the absence of explicit mention, all the considered graphs will have bounded degree Δ by default, so a freezing automata network rule can be represented as the list of local update rules for each node which are maps of the form $Q^\Delta \to \mathcal{P}(Q)$ whose representation as transition table is of size $O(|Q|^{\Delta+1})$. The specification checking problem we consider asks whether a given freezing automata network verifies a given localized trace property on the set of orbits whose restriction on each node adheres to a given t-specification. In order to do that, we introduce the concept of a satisfiable t-specification

Definition 4. *Let $\mathcal{A} = (G, \mathcal{F})$ be a non-deterministic automata network and let $\mathcal{E}_t$ a t-specification. We say that $\mathcal{E}_t$ is satisfiable by $\mathcal{A}$ if there exists an orbit $O \in \mathcal{O}(\mathcal{A}, t)$ such that $O_v \in \mathcal{E}_t(v)$ for every $v \in V$.*

If $\mathcal{E}_t$ is a satisfiable t-specification for some automata network $\mathcal{A}$ we write $\mathcal{A} \models \mathcal{E}_t$. We present now the *Specification checking problem* as the problem of verifying whether a given t-specification is satisfiable by some automata network $\mathcal{A}$.

Problem 1 (Specification checking problem (SPEC))

Parameters: Alphabet Q, family of graphs $\mathcal{G}$ of max degree Δ.
Input:
    1. a non-deterministic freezing automata network $\mathcal{A} = (G, F)$ on alphabet Q, with set of nodes V and $G \in \mathcal{G}$;
    2. a time $t \in \mathbb{N}$.
    3. a t-specification $\mathcal{E}_t$
Question: $\mathcal{A} \models \mathcal{E}_t$

Four Canonical Problems. When studying a dynamical system, one is often interested in determining properties of the future state of the system given its initial state. In the context of automata networks, various decision problems have been studied where a question about the evolution of the dynamics at a given node is asked. Usually, the computational complexity of such problems is compared to the complexity of simulating the automata network. Roughly, one can observe that some systems are complex in some way if the complexity of latter problems are "as much as hard" as simply simulating the system.

Note that prediction problem (see full version [18]) is clearly a subproblem of SPEC. Also, observe that a specification allows us to ask various questions considered in the literature: what will be the state of the node at a given time [15,21], will the node change its state during the evolution [16], or, thanks to Lemma 1, what will be state of the node once a fixed point is reached [27, section 5]. Note that the classical circuit value problem for Boolean circuits easily reduces to the prediction problem above when we take G to be the DAG of the Boolean circuit and choose local rules at each node that implement circuit gates. Theorem 35 in [18] gives a much stronger result using such a reduction where the graph and the rule are independent of the circuit.

We now turn to the classical problem of finding predecessors back in time to a given configuration [20,23]. Detailed definition is available in the full version [18]. Note that, analogously to the previous case, the final configuration in the input can be given through a particular t-specification $\mathcal{E}_t$, such that for all $y \in \mathcal{E}_t(v) : y_t = c$ for any $v \in V$. Thus, by considering $\mathcal{E}_t$ we can see predecessor problem as a subproblem of SPEC.

Deterministic automata networks have ultimately periodic orbits. When they are freezing, any configuration reaches a fixed point. Nilpotency asks whether their is a unique fixed point whose basin of attraction is the set of all configurations. It is a fundamental problem in finite automata networks theory [14,28] as well as in cellular automata theory where the problem is undecidable for any space dimension [22], but whose decidability depends on the space dimension in the freezing case [27]. Detailed definition can be found in the full version [18].

In this case, it is not clear that Nilpotency is actually a subproblem of SPEC. However, we will show that we can solve Nilpotency by solving a polynomial amount of instances (linear on the size of the interaction graph of the network $|G|$) for SPEC in parallel. More precisely, we show that there exist a **NC** Turing reduction. In order to do that note first that we can use Lemma 1 to fix $t = \lambda(n)$, where $\lambda(n)$ is an appropriate polynomial. Then, we express that $F^t(Q^V)$ is a singleton as the following formula, which intuitively says that for each node there is a state such that all orbits terminate in that state at this node: $\bigwedge_{s \in V} \bigvee_{q_0 \in Q} \mathcal{A} \models \mathcal{E}_t^{q_0,s}$, where $\mathcal{E}_t^{q_0,s}$ are t-specifications satisfying $\mathcal{E}_t^{q_0,s}(v) = Q^t$, for every $v \neq s$ and $\mathcal{E}_t^{q_0,s}(s)$ is the set of orbits y such that $y_t = q_0$. The reduction holds.

It is straightforward to reduce coloring problems (does the graph admit a proper coloring with colors in Q) and more generally tilings problems to nilpotency using an error state that spread across the network when a local condition is not satisfied (note that tiling problem are known to be tightly related to nilpotency in cellular automata [22]). Using the same idea one can reduce SAT to nilpotency by choosing G to be the DAG of a circuit computing the given SAT formula (see Theorem 2 below for a stronger reduction that works on any family of graphs with polynomial treewidth).

Given a deterministic freezing automata network of global rule $F : Q^V \to Q^V$, we define the associated non-deterministic global rule F^* where each node can at each step apply F or stay unchanged, formally: $F_v^*(c) = \{F_v(c), c_v\}$. It represents the system F under totally asynchronous update mode.

In this context, asynchronous reachability problem is defined (see full version [18]). Latter problem consists in decide if a particular configuration can be reached by the dynamics starting from a fixed initial condition. Note that no bound is given in the problem for the time needed to reach the target configuration. However, Lemma 1 ensures that c_1 can be reached from c_0 if and only if it can be reach in a polynomial number of steps (in n). Thus this problem can again be seen as a sub-problem of our SPEC by defining a $\lambda(n)$-specification $\mathcal{E}_{\lambda(n)}$ such that for any $y \in \mathcal{E}_{\lambda(n)} : y_0 = c_0 \wedge y_{\lambda(n)} = c_1$. This bound on the maximum time needed to reach the target ensures that the problem is **NP** (a witness of reachability is an orbit of polynomial length). Note that the problem is **PSPACE**-complete for general automata networks: in fact it is **PSPACE**-complete even when the networks considered are one-dimensional (network is a ring) cellular automata (same local rule everywhere) [8].

4   A Fast-Parallel Algorithm for the Specification Checking Problem

In this section we present a fast-parallel algorithm for solving the Specification Checking Problem when the input graph is restricted to the family of graphs with bounded degree and treewidth. More precisely, we show that the problem can be solved by a CREW PRAM that runs in the time $\mathcal{O}(\log^2(n))$ where n is the amount of nodes of the network. Thus, restricted to graphs of bounded degree and bounded treewidth, Specification Checking Problem belongs to the class **NC**.

We define a locally-valid trace of a vertex v as a sequence of state-transitions of all the vertices in $N[v]$ which are consistent with local rule of v, but not necessarily consistent with the local-rules of the vertices in $N(v)$. We also ask that the state-transitions of v satisfy the t-specification $\mathcal{E}_t$.

In addition, we define a partially-valid trace for a set U as a sequence of state-transition of all the vertices in $N[U]$, which are consistent with the local rules of all vertices in U, but not necessarily consistent with the local-rules of the vertices in $N(U)$. We call the set of all partially-valid traces of U as $PVT(U)$

Let $(W, F, \{X_w : w \in W\})$ be a rooted binary-tree-decomposition of graph G with root r, that we assume that has width at most $(3\mathrm{tw}(G) + 2)$. For $w \in W$, we call T_w the set of all the descendants of w, including w.

Our algorithm consists in a dynamic programming scheme over the bags of the tree. For each bag $w \in T$ and $\beta^w \in PVT(X_w)$ we call $\mathrm{Sol}_w(\beta^w)$ the partial answer of the problem on the vertices contained bags in T_w, when the locally-valid traces of the vertices in X_w are induced by β^w. We say that $\mathrm{Sol}_w(\beta^w) = $ **accept** when it is possible to extend β^w into a partially-valid trace of all the vertices in bags of T_w, and **reject** otherwise. More precisely, if w is a leaf of T, we define $\mathrm{Sol}_w(\beta^w) = $ **accept** for all $\beta^w \in PVT(X_w)$. For the other bags, $\mathrm{Sol}_w(\beta^w) = $ **accept** if and only if exists a $\beta \in PVT(\bigcup_{z \in T_w} X_z)$ such that $\beta(u) = \beta^w(u)$, for all $u \in X_w$. Observe that the instance of the Specification Checking problem is accepted when there exists a $\beta^r \in PVT(X_r)$ such that $\mathrm{Sol}_r(\beta^r) = $ **accept**.

In order to solve our problem efficiently in parallel, we define a data structure that allows us efficiently encode locally-valid traces and partially-valid traces. More precisely, in $N[v]$ there are at most $|Q|^\Delta$ possible state transitions. Therefore, when t is comparable to n, most of the time the vertices in $N[v]$ remain in the same state. Then, in order to efficiently encode a trace, it is enough to keep track only of the time-steps on which some state-transition occurs. We are now ready to present the main result of present section. Full algorithm is available in the full version [18].

Theorem 1. *Specification Checking problem can be solved by an CREW PRAM algorithm running in time $\mathcal{O}(\log^2 n)$ and using $n^{\mathcal{O}(1)}$ processors on graphs of bounded treewidth.*

The proof of previous Theorem 1 shows that SPEC can be solved in time $f(|Q| + \Delta(G) + \mathrm{tw}(G)) \log n$ using $n^{f(|Q|+\Delta(G)+\mathrm{tw}(G))}$ processors in a PRAM machine, hence in time $n^{g(|Q|+\Delta(G)+\mathrm{tw}(G))}$ on a sequential machine, for some computable functions f and g. In other words, when the alphabet, the maximum degree and the tree-width of the input automata network are parameters, our result shows that SPEC is in **XP**. In the next section, we show that SPEC is not in **FPT**, unless **FPT = W**[2].

Constraint Satisfaction Problem. We remark that problem SPEC can be interpreted as a specific instance of the Constraint Satisfaction Problem (CSP). The problem CSP is a sort of generalization of SAT into a set of more versatile variable constraint. It is formally defined as a triple (X, D, C), where $X = \{X_1, \ldots, X_n\}$ is a set of *variables*, $D = \{D_1, \ldots, D_n\}$ is a set of *domains* where are picked each variable, and a set $C = \{C_1, \ldots, C_m\}$ of *constraints*, which are k-ary relations of some set of k variables. The question is whether exists a set of values of each variables in their corresponding domains, in order to satisfy each one of the constraints. As we mentioned, SPEC can be seen as a particular instance of CSP, where we choose one variable for each node of the input graph. The domain of each variable is the set of all locally valid traces of the corresponding node. Finally, we define one constraint for each node, where the variable involved are all the vertices in the close-neighborhood of the corresponding node, and the relation corresponds to the consistency in the information of the locally-valid traces involved.

Now consider an instance of SPEC with constant tree-width, maximum degree and size of the alphabet, and construct the instance of CSP with the reduction described in the previous paragraph. Then, the obtained instance of CSP has polynomially-bounded domains and constant tree-width, where the tree-width of a CSP instance is defined as the tree-width of the graph where each variable is a node, and two nodes are adjacent if the corresponding variables appear in some restriction. Interestingly, it is already known that in these conditions CSP can be solved in polynomial time [26,30] . This implies that, subject to the given restrictions, SPEC is solvable in polynomial time using the given algorithm for CSP as a blackbox.

The algorithm given in the proof of Theorem 1 is better than the use of the CSP blackbox in two senses. First, we obtain explicit dependencies on the size of the alphabet, maximum degree and tree-width. Second, the *Prediction Problem* is trivially solvable in polynomial time, and then the use of the CSP blackbox gives no new information for this problem. Moreover our algorithm does not decides SPEC, but also can be used to obtain a coding of the orbit satisfying the given specification, and moreover, the possibility to test any **NC**-property on deterministic freezing automata networks.

5   W[2]-Hardness Results

The goal of this section is to show that, even when alphabet and degree are fixed and treewidth is considered as the only parameter, then the SPEC problem is **W**[2]-hard (see [10] for an introduction to the **W** hierarchy) and thus not believed to be fixed parameter tractable. This is in contrast with classical results of Courcelle establishing that model-checking of MSO formulas parametrized by treewidth is fixed-parameter tractable [7].

Lemma 2. *There is a fixed alphabet Q and an algorithm which, given $k \in \mathbb{N}$ and a graph G of size n, produces in time $O(k \cdot n^{O(1)})$:*

- *a deterministic freezing automata network $\mathcal{A} = (G', \mathcal{F})$ with alphabet Q and where G' has treewidth $O(k)$ and degree 4*
- *a $O(n^2)$-specification $\mathcal{E}$*

such that G admits a dominating set of size k if and only if $\mathcal{A} \models \mathcal{E}$.

The construction of the lemma works by producing a freezing automata network on a $O(k) \times n^2$-grid together with a specification which intuitively work as follows. A row of the grid is forced (by the specification) to contain the adjacency matrix of the graph, k rows serve as selection of a subset of k nodes of G, and another row is used to check domination of the candidate subset. The key of the construction is to use the dynamics of the network to test that the information in each row is encoded coherently as intended, and raise an error if not. The specification serves both as a partial initialization (graph adjacency matrix and tests launching are forced, but the choice in selection rows is free) and a check that no error are raised by the tests.

From Lemma 2 and $\mathbf{W}[2]$-hardness of the k-Dominating-Set problem [9], we immediately get the following corollary.

Corollary 1. *The* SPEC *problem with fixed degree and fixed alphabet and with treewidth as unique parameter is* $\boldsymbol{W[2]}$-*hard.*

A freezing automata network on a $O(k) \times n^2$-grid with alphabet Q can be seen as a freezing automata network on a line of length n^2 with alphabet $Q^{O(k)}$. One might therefore want to adapt the above result to show $\mathbf{W}[2]$-hardness in the case where treewidth and degree are fixed while alphabet is the parameter. However, the specification which is part of the input, has an exponential dependence on the alphabet (a t-specification is of size $O(n \cdot t^{|Q|})$). Therefore **FPT** reductions are not possible when the alphabet is the parameter. We can circumvent this problem by considering a new variant of the SPEC problem where specification are given in a more succinct way through regular expressions. A regular (Q, V)-specification is a map from V to regular expressions over alphabet Q. We therefore consider the problem REGSPEC which is the same as SPEC except that the specification must be a regular specification. With this modified settings, the construction of Lemma 2 can be adapted to deal with the alphabet as parameter.

Corollary 2. *The* REGSPEC *problem with fixed degree and fixed treewidth and with alphabet as unique parameter is* $\boldsymbol{W[2]}$-*hard.*

6 Hardness Results for Polynomial Treewidth Networks

We say a family of graphs $\mathcal{G}$ has polynomial treewidth if the graphs of the family are of size at most polynomial in their treewidth, precisely: if there is a non-constant polynomial map $p_{\mathcal{G}}$ (with rational exponents in $(0, 1)$) such that for any $G = (V, E) \in \mathcal{G}$ it holds $\mathrm{tw}(G) \geq p_{\mathcal{G}}(|V|)$. Moreover, we say the family is *constructible* if there is a polynomial time algorithm that given n produces a connex graph $G_n \in \mathcal{G}$ with n nodes. The following results are based on a polynomial time algorithm to find large perfect brambles in graphs [24]. This structure allows to embed any digraph in an input graph with sufficiently large treewidth via path routing while controlling the maximum number of intersections per node of the set of paths.

Theorem 2. *For any family $\mathcal{G}$ of constructible graphs of polynomial treewidth, the problem* nilpotency *is* **coNP***-complete.*

When giving an automata network as input, the description of the local functions depends on the underlying graph (and in particular the degree of each node). However, some local functions are completely isotropic and blind to the number of neighbors and therefore can be described once for all graphs. This is the case of local functions that only depends on the set of states present in the neighborhood. Indeed, given a map $\rho : Q \times 2^Q \to Q$ and any graph $G = (V, E)$, we define the automata network on G with local functions $F_v : Q^{N(v)} \to Q$ such that $F_v(c) = \rho\big(c(v), \{c(v_1), \ldots, c(v_k)\}\big)$ where $N(v) = \{v_1, \ldots, v_k\}$ is the neighborhood of v which includes v. We then say that the automata network is *set defined* by ρ.

Theorem 3. *There exists a map $\rho : Q \times 2^Q \to Q$ such that for any family $\mathcal{G}$ of constructible graphs of polynomial treewidth and bounded degree, the problems* predecessor *and* asynchronous reachability *are both* **NP***-complete when restricted to $\mathcal{G}$ and automata networks set-defined by ρ.*

References

1. Amini, H., Fountoulakis, N.: Bootstrap percolation in power-law random graphs. J. Stat. Phys. **155**(1), 72–92 (2014)
2. Arnborg, S., Corneil, D.G., Proskurowski, A.: Complexity of finding embeddings in a k-tree. SIAM J. Algebraic Disc. Methods **8**(2), 277–284 (1987)
3. Bak, P., Chen, K., Tang, C.: A forest-fire model and some thoughts on turbulence. Phys. Lett. A **147**(5), 297–300 (1990)
4. Becker, F., Maldonado, D., Ollinger, N., Theyssier, G.: Universality in freezing cellular automata. In: Manea, F., Miller, R.G., Nowotka, D. (eds.) CiE 2018. LNCS, vol. 10936, pp. 50–59. Springer, Cham (2018). https://doi.org/10.1007/978-3-319-94418-0_5
5. Bodlaender, H.L., Hagerup, T.: Parallel algorithms with optimal speedup for bounded treewidth. SIAM J. Comput. **27**(6), 1725–1746 (1998)
6. Chekuri, C., Chuzhoy, J.: Polynomial bounds for the grid-minor theorem. J. ACM **63**(5), 1–65 (2016)
7. Courcelle, B.: The monadic second-order logic of graphs. i. recognizable sets of finite graphs. Inf. Comput. **85**(1), 12–75 (1990)
8. Dennunzio, A., Formenti, E., Manzoni, L., Mauri, G., Porreca, A.E.: Computational complexity of finite asynchronous cellular automata. Theor. Comput. Sci. **664**, 131–143 (2017)
9. Downey, R.G., Fellows, M.R.: Fixed-parameter tractability and completeness i: basic results. SIAM J. Comput. **24**(4), 873–921 (1995)
10. Downey, R.G., Fellows, M.R.: Appendix 2: Menger's theorems. Fundamentals of Parameterized Complexity. TCS, pp. 705–707. Springer, London (2013). https://doi.org/10.1007/978-1-4471-5559-1_35

11. Elberfeld, M., Jakoby, A., Tantau, T.: Logspace versions of the theorems of bodlaender and courcelle. In: 2010 IEEE 51st Annual Symposium on Foundations of Computer Science. IEEE (2010)
12. Fuentes, M., Kuperman, M.: Cellular automata and epidemiological models with spatial dependence. Physica A Stat. Mech. Appl. **267**(3–4), 471–486 (1999)
13. Gadouleau, M.: On the influence of the interaction graph on a finite dynamical system. Natural Computing (to appear)
14. Gadouleau, M., Richard, A.: Simple dynamics on graphs. Theor. Comput. Sci. **628**, 62–77 (2016)
15. Goles, E., Ollinger, N., Theyssier, G.: Introducing freezing cellular automata. In: Exploratory Papers of Cellular Automata and Discrete Complex Systems (AUTOMATA 2015), pp. 65–73 (2015)
16. Goles, E., Maldonado, D., Montealegre-Barba, P., Ollinger, N.: Fast-parallel algorithms for freezing totalistic asynchronous cellular automata. In: Mauri, G., El Yacoubi, S., Dennunzio, A., Nishinari, K., Manzoni, L. (eds.) ACRI 2018. LNCS, vol. 11115, pp. 406–415. Springer, Cham (2018). https://doi.org/10.1007/978-3-319-99813-8_37
17. Goles, E., Martínez, S.: Neural and Automata Networks: Dynamical Behavior and Applications. Kluwer Academic Publishers, Norwell (1990)
18. Goles, E., Montealegre, P., Ríos-Wilson, M., Theyssier, G.: On the impact of treewidth in the computational complexity of freezing dynamics. arXiv preprint arXiv:2005.11758 (2020)
19. Goles, E., Montealegre-Barba, P., Todinca, I.: The complexity of the bootstraping percolation and other problems. Theor. Comput. Sci. **504**, 73–82 (2013)
20. Green, F.: NP-complete problems in cellular automata. Complex Syst. 1 (1987)
21. Griffeath, D., Moore, C.: Life without death is P-complete. Complex Syst. **10** (1996)
22. Kari, J.: The nilpotency problem of one-dimensional cellular automata. SIAM J. Comput. **21**, 571–586 (1992)
23. Kawachi, A., Ogihara, M., Uchizawa, K.: Generalized predecessor existence problems for boolean finite dynamical systems on directed graphs. Theor. Comput. Sci. **762**, 25–40 (2019)
24. Kreutzer, S., Tazari, S.: On brambles, grid-like minors, and parameterized intractability of monadic second-order logic. In: SODA 2010, pp. 354–364. SIAM (2010)
25. Kutrib, M., Malcher, A.: Cellular automata with sparse communication. Theor. Comput. Sci. **411**(38–39), 3516–3526 (2010)
26. Marx, D.: Can you beat treewidth? In: 48th Annual IEEE Symposium on Foundations of Computer Science (FOCS 2007), pp. 169–179. IEEE (2007)
27. Ollinger, N., Theyssier, G.: Freezing, bounded-change and convergent cellular automata. CoRR abs/1908.06751 (2019)
28. Richard, A.: Nilpotent dynamics on signed interaction graphs and weak converses of thomas' rules. Disc. Appl. Math. **267**, 160–175 (2019)
29. Robertson, N., Seymour, P.: Graph minors. v. excluding a planar graph. J. Comb. Theory Series B **41**(1), 92–114 (1986)
30. Samer, M., Szeider, S.: Constraint satisfaction with bounded treewidth revisited. J. Comput. Syst. Sci. **76**(2), 103–114 (2010)
31. Ulam, S.M.: On some mathematical problems connected with patterns of growth of figures. In: Bukrs, A.W. (ed.) Essays on Cellular Automata, pp. 219–231. U. of Illinois Press (1970)

32. Winslow, A.: A brief tour of theoretical tile self-assembly. In: Cook, M., Neary, T. (eds.) AUTOMATA 2016. LNCS, vol. 9664, pp. 26–31. Springer, Cham (2016). https://doi.org/10.1007/978-3-319-39300-1_3
33. Wu, A., Rosenfeld, A.: Cellular graph automata. i. basic concepts, graph property measurement, closure properties. Inf. Control **42**(3), 305–329 (1979)

Towards a Map for Incremental Learning in the Limit from Positive and Negative Information

Ardalan Khazraei[1], Timo Kötzing[2], and Karen Seidel[2]([✉])

[1] University of Potsdam, Potsdam, Germany
[2] Hasso-Plattner-Institute, Potsdam, Germany
karen.seidel@hpi.de

Abstract. In order to model an efficient learning paradigm, iterative learning algorithms access data one by one, updating the current hypothesis without regress to past data. Prior research investigating the impact of additional requirements on iterative learners left many questions open, especially in learning from informant, where the input is binary labeled.

We first compare learning from positive information (text) with learning from informant. We provide different concept classes learnable from text but not by an iterative learner from informant. Further, we show that totality restricts iterative learning from informant.

Towards a map of iterative learning from informant, we prove that strongly non-U-shaped learning is restrictive and that iterative learners from informant can be assumed canny for a wide range of learning criteria.

Finally, we compare two syntactic learning requirements.

Keywords: Learning in the limit · Map for iterative learners from informant · (Strongly) Non-U-shaped learning

1 Introduction

We are interested in the problem of algorithmically learning a description for a formal language (a computably enumerable subset of the set of natural numbers) when presented successively all information about that language; this is sometimes called *inductive inference*, a branch of (algorithmic) learning theory.

Many criteria for deciding whether a learner M is *successful* on a language L have been proposed in the literature. Gold, in his seminal paper [Gol67], gave a first, simple learning criterion, **Ex**-*learning*[1], where a learner is *successful* iff, on every complete information about L it eventually stops changing its conjectures, and its final conjecture is a correct description for the input sequence. Trivially, each single, describable language L has a suitable constant function as a **Ex**-learner (this learner constantly outputs a description for L). As we want

---

[1] **Ex** stands for *explanatory*.

© Springer Nature Switzerland AG 2021
L. De Mol et al. (Eds.): CiE 2021, LNCS 12813, pp. 273–284, 2021.
https://doi.org/10.1007/978-3-030-80049-9_25

algorithms for more than a single learning task, we are interested in analyzing for which *classes of languages* $\mathcal{L}$ there is a *single learner* M learning *each* member of $\mathcal{L}$. This framework is also sometimes known as *language learning in the limit* and has been studied using a wide range of learning criteria in the flavor of **Ex**-learning (see, for example, the textbook [JORS99]).

One major criticism of the model suggested by Gold, see for example [CM08], is its excessive use of memory: for each new hypothesis the entire history of past data is available. Iterative learning [Wie76], is the most common variant of learning in the limit which addresses memory constraints: the memory of the learner on past data is just its current hypothesis. Due to the padding lemma, this memory is still not void, but finitely many data can be memorized in the hypothesis.

Prior work on iterative learning [CK10, CM08, JKMS16, JMZ13, JORS99] focused on learning from text, that is, from positive data only. Hence, in **TxtEx**-learning the complete information is a listing of all and only the elements of L. In this paper we are mainly interested in the paradigm of learning from both positive and negative information. For example, when learning half-spaces, one could see data declaring that $\langle 1, 1 \rangle$ is in the target half-space, further is $\langle 3, 2 \rangle$, but $\langle 1, 7 \rangle$ is not, and so on. This setting is called *learning from informant* (in contrast to learning from *text*) (Fig. 1).

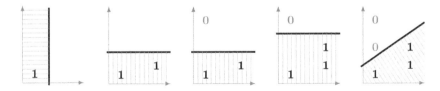

Fig. 1. Example Learning Process with binary labeled data and half-spaces as hypotheses.

Iterative learning from informant was analyzed by [JLZ07], where various natural restrictions have been considered and the authors focused on the case of learning indexable families (classes of languages which are uniformly decidable). In this paper we are looking at other established restrictions and also consider learning of arbitrary classes of computably enumerable languages.

In Sect. 3 we consider the two aforementioned restrictions on learning from informant: learning from text and learning iteratively. Both restrictions render fewer classes of languages learnable; in fact, the two restrictions yield two incomparable sets of language classes being learnable, which also shows that learning iteratively from text is weaker than supposing just one of the two restrictions.

Towards a better understanding of iterative learners we analyze which normal forms can be assumed in Sect. 4. First we show that, analogously to the case of learning from text (as analyzed in [CM09]), we cannot assume learners to be total (i.e. always giving an output).

However, from [CM08] we know that we can assume iterative text learners to be *canny* (also defined in Sect. 4); we adapt this normal form for the case of iterative learning from informant and show that it can be assumed to hold for iterative learners generally.

Many works in inductive inference, see for example [JKMS16, KP16, KS16, KSS17], focus on relating different additional learning requirements for a fixed learning model. In particular, [JKMS16] mapped out all pairwise relations for an established choice of learning restrictions for iterative learning from text. The complete map of all pairwise relations between for full-information learners from informant can be found in [AKS18]. A similar map for the case of iterative learning from informant is not known. Canniness is central in investigating the learning power of iterative learning from texts. Hence, it is an important stepping stone to understand iterative learners better and determine such pairwise relations. We argue in Lemma 3 that the normal form of canniness still can be assumed in case we pose additional semantic learning requirements.

In Sect. 5 we collect all previously known results for such a map, see [LZ92, JLZ07]. We observe that it decreases learning power to require the learner to never change its hypothesis, once it is correct. The proof for separating this notion, called strong non-U-shapedness, relies on the ORT recursion theorem [Cas74]. We close this section by comparing two syntactic learning requirements for iterative learners from informant that proved important to derive the equivalence of all syntactic requirements for iterative learners from text.

We continue this paper with some mathematical preliminaries in Sect. 2 before discussing our results in more detail.

2   Iterative Learning from Informant

Notation and terminology on the learning theoretic side follow [OSW86, JORS99] and [LZZ08], whereas on the computability theoretic side we refer to [Odi99] and [Rog67]. For both we also recommend [Köt09].

A *language* L is a recursively enumerable subset of $\mathbb{N}$. We denote the characteristic function for $L \subseteq \mathbb{N}$ by $f_L : \mathbb{N} \to \{0, 1\}$.

Gold in his seminal paper [Gol67], distinguished two major different kinds of information presentation. A function $I : \mathbb{N} \to \mathbb{N} \times \{0, 1\}$ is an *informant for language* L, if there is a surjection $n : \mathbb{N} \to \mathbb{N}$ such that $I(t) = (n(t), f_L(n(t)))$ holds for every $t \in \mathbb{N}$. Moreover, for an informant I let

$$\mathrm{pos}(I) := \{y \in \mathbb{N} \mid \exists x \in \mathbb{N} \colon \mathrm{pr}_1(I(x)) = y \wedge \mathrm{pr}_2(I(x)) = 1\} \text{ and}$$
$$\mathrm{neg}(I) := \{y \in \mathbb{N} \mid \exists x \in \mathbb{N} \colon \mathrm{pr}_1(I(x)) = y \wedge \mathrm{pr}_2(I(x)) = 0\}$$

denote the sets of all natural numbers, about which I gives some positive or negative information, respectively. A *text for language* L is a function $T : \mathbb{N} \to \mathbb{N} \cup \{\#\}$ with range L after removing $\#$. The symbol $\#$ is interpreted as pause symbol.

Therefore, when learning from informant, the *set of admissible inputs to the learning algorithm* $\mathbb{S}$ is the set of all finite sequences

$$\sigma = ((n_0, y_0), \ldots, (n_{|\sigma|-1}, y_{|\sigma|-1}))$$

of *consistently* binary labeled natural numbers. When learning from text (positive data only), we encounter inputs to the learning algorithm from the set $\mathbb{T}$ of finite sequences $\tau = (n_0, \ldots, n_{|\tau|-1})$ of natural numbers and the pause symbol #. The initial subsequence relation is denoted by $\sqsubseteq$.

A set $\mathcal{L} = \{L_i \mid i \in \mathbb{N}\}$ of languages is called *indexable family* if there is a computer program that on input $(i, n) \in \mathbb{N}^2$ returns 1 if $n \in L_i$ and 0 otherwise. Examples are **Fin** and **CoFin**, the set of all finite subsets of $\mathbb{N}$ and the set of all complements of finite subsets of $\mathbb{N}$, respectively.

Let $\mathcal{L}$ be a collection of languages we seek a provably correct learning algorithm for. We will refer to $\mathcal{L}$ as the *concept class* which will often be an indexable family. Further, let $\mathcal{H} = \{L_i \mid i \in \mathbb{N}\}$ with $\mathcal{L} \subseteq \mathcal{H}$ be a collection of languages called the *hypothesis space*. In general we do *not* assume that for every $L \in \mathcal{L}$ there is a unique index $i \in \mathbb{N}$ with $L_i = L$. Indeed, ambiguity in the hypothesis space helps memory-resticted learners to remember data.

A *learner M from informant (text)* is a computable function

$$M : \mathbb{S} \to \mathbb{N} \cup \{?\} \qquad (M : \mathbb{T} \to \mathbb{N} \cup \{?\})$$

with the output i interpreted with respect to $\mathcal{H} = \{L_i \mid i \in \mathbb{N}\}$, a prefixed hypothesis space. The output ? often serves as initial hypothesis or is interpreted as no new hypothesis. Often $\mathcal{H}$ is an indexable class or the established W-hypothesis space defined in Subsect. 4.

Let I be an informant (T be a text) for L and $\mathcal{H} = \{L_i \mid i \in \mathbb{N}\}$ a hypothesis space. A learner $M : \mathbb{S} \to \mathbb{N} \cup \{?\}$ ($M : \mathbb{T} \to \mathbb{N} \cup \{?\}$) is *successful on I (on T)* if it eventually settles on $i \in \mathbb{N}$ with $L_i = L$. This means that when receiving increasingly long finite initial segments of I (of T) as inputs, it will from some time on be correct and not change the output on longer initial segments of I (of T). *M learns L wrt $\mathcal{H}$* if it is successful on every informant I (on every text T) for L. *M learns $\mathcal{L}$* if there is a hypothesis space $\mathcal{H}$ such that M learns every $L \in \mathcal{L}$ wrt $\mathcal{H}$. We denote the collection of all $\mathcal{L}$ learnable from informant (text) by [**InfEx**] ([**TxtEx**]). If we fix the hypothesis space, we denote this by a subscript for **Ex**.

According to [Wie76, LZ96, CJLZ99] a learner M is *iterative* if its output on $\sigma \in \mathbb{S}$ ($\tau \in \mathbb{T}$) only depends on the last input $\mathrm{last}(\sigma)$ and the hypothesis $M(\sigma^-)$ after observing σ without its last element $\mathrm{last}(\sigma)$. The collection of all concept classes $\mathcal{L}$ learnable by an iterative learner from informant (text) is denoted by [**ItInfEx**] ([**ItTxtEx**]).

The s-m-n theorem gives finite and infinite recursion theorems, see [Cas94, Odi99]. We will refer to Case's Operator Recursion Theorem ORT in its 1-1-form, see [Cas74, JORS99, Köt09].

3   Comparison with Learning from Text

By ignoring negative information every informant incorporates a text for the language presented and we gain $[\mathbf{ItTxtEx}] \subseteq [\mathbf{ItInfEx}]$.

It has been observed in [OSW86] that the superfinite language class $\mathbf{Fin} \cup \{\mathbb{N}\}$ is in $[\mathbf{InfEx}] \setminus [\mathbf{ItInfEx}]$. With $L_k = 2\mathbb{N} \cup \{2k+1\}$ and $L'_k = L_k \setminus \{2k\}$ the indexable family $\mathcal{L} = \{2\mathbb{N}\} \cup \{L_k, L'_k \mid k \in \mathbb{N}\}$ lies in $[\mathbf{TxtEx}] \cap [\mathbf{ItInfEx}]$ but not in $[\mathbf{ItTxtEx}]$. In [JORS99] the separations are witnessed by the indexable family $\{\mathbb{N} \setminus \{0\}\} \cup \{D \cup \{0\} : D \in \mathbf{Fin}\}$.

It can easily be verified that $\mathbf{CoFin} \in [\mathbf{ItInfEx}] \setminus [\mathbf{TxtEx}]$ and with the next result $[\mathbf{ItInfEx}]$ and $[\mathbf{TxtEx}]$ are incomparable by inclusion.

Lemma 1. *There is an indexable family in* $[\mathbf{TxtEx}] \setminus [\mathbf{ItInfEx}]$.

Summing up, we know $[\mathbf{ItTxtEx}] \subsetneq [\mathbf{TxtEx}] \perp [\mathbf{ItInfEx}] \subsetneq [\mathbf{InfEx}]$, where $\perp$ stands for incomparability with respect to set inclusion, meaning (1) there is a concept class learnable from text but not by an iterative learner from informant and (2) there is a concept class learnable by an iterative learner from informant but not from text.

Moreover, with a Boolean function we can show that every concept class separating $[\mathbf{ItInfEx}]$ and $[\mathbf{InfEx}]$ yields a separating class for $[\mathbf{ItInfEx}]$ and $[\mathbf{TxtEx}]$. We generalize this further in the full version.

4   Total and Canny Learners

For the rest of the paper, without further notation, all results are understood with respect to the W-hypothesis space defined in the following. We fix a programming system φ as introduced in [RC94]. Briefly, in the φ-system, for a natural number p, we denote by φ_p the partial computable function with program code p. We also call p an *index* for W_p defined as $\mathrm{dom}(\varphi_p)$.

We show that totality, denoted by $\mathcal{R}$, restricts iterative learning from informant. The proof uses an easy ORT argument.

Theorem 1. $[\mathbf{ItInfEx}] \setminus [\mathcal{R}\mathbf{ItInfEx}] \neq \varnothing$.

Proof. Let o be an index for $\varnothing$ and define the iterative learner M for all $\xi \in \mathbb{N} \times \{0, 1\}$ by

$$M(\varnothing) = o;$$

$$h_M(h, \xi) = \begin{cases} \varphi_{\mathrm{pr}_1(\xi)}(0), & \text{else if } \mathrm{pr}_2(\xi) = 1 \text{ and } h \notin \mathrm{ran}(\mathrm{ind}); \\ h, & \text{otherwise.} \end{cases}$$

We argue that $\mathcal{L} := \{L \subseteq \mathbb{N} \mid L \in \mathbf{ItInfEx}(M)\}$ is not learnable by a total learner from informants. Assume towards a contradiction M' is such a learner. For a finite informant sequence σ we denote by $\overline{\sigma}$ the corresponding canonical finite informant sequence, ending with σ's datum with highest first coordinate.

Then by padded ORT there are $e \in \mathbb{N}$ and a strictly increasing computable function $a : \mathbb{N}^{<\omega} \to \mathbb{N}$, such that for all $\sigma \in \mathbb{N}^{<\omega}$ and all $i \in \mathbb{N}$

$$\sigma_0 = \varnothing;$$

$$\sigma_{i+1} = \sigma_i \,{}^\frown \begin{cases} (a(\sigma_i), 1), & \text{if } M'(\overline{\sigma_i \,{}^\frown (a(\sigma_i), 1)}) \neq M'(\overline{\sigma_i}); \\ \varnothing, & \text{otherwise;} \end{cases} \tag{1}$$

$$W_e = \bigcup_{i \in \mathbb{N}} \text{pos}(\overline{\sigma_i});$$

$$\varphi_{a(\sigma)}(x) = \begin{cases} e, & \text{if } M'(\overline{\sigma \,{}^\frown (a(\sigma), 1)}) \neq M'(\overline{\sigma}); \\ \text{ind}_{\text{pos}(\sigma) \cup \{a(\sigma)\}}, & \text{otherwise;} \end{cases}$$

Clearly, we have $W_e \in \mathcal{L}$ and thus M' also **InfEx**-learns W_e. By the **Ex**-convergence there are $e', t_0 \in \mathbb{N}$, where t_0 is minimal, such that $W_{e'} = W_e$ and for all $t \geq t_0$ we have $M'(\bigcup_{i \in \mathbb{N}} \overline{\sigma_i}[t]) = e'$ and hence by (1) for all i with $|\overline{\sigma_i}| \geq t_0$

$$M'(\overline{\sigma_i \,{}^\frown (a(\sigma_i), 1)}) = M'(\overline{\sigma_i}) = M'(\overline{\sigma_i \,{}^\frown (a(\sigma_i), 0)}).$$

It is easy to see, that $W_e = \text{pos}(\sigma_i)$ and $W_e \cup \{a(\sigma_i)\} \in \mathcal{L}$. Moreover, M' is iterative and hence does not learn W_e and $W_e \cup \{a(\sigma_i)\}$. □

We transfer the notion of canny learners to learning from informant.

Definition 1. *A learner M from informant is called* canny *in case for every finite informant sequence σ holds*

1. *if $M(\sigma)$ is defined then $M(\sigma) \in \mathbb{N}$;*
2. *for every $x \in \mathbb{N} \setminus (\text{pos}(\sigma) \cup \text{neg}(\sigma))$ and $i \in \{0, 1\}$ a mind change $M(\sigma \,{}^\frown (x, i)) \neq M(\sigma)$ implies for all finite informant sequences τ with $\sigma \,{}^\frown (x, i) \sqsubseteq \tau$ that $M(\tau \,{}^\frown (x, i)) = M(\tau)$.*

Hence, the learner is canny in case it always outputs a hypotheses and no datum twice causes a mind change of the learner. Also for learning from informant the learner can be assumed canny by a simulation argument.

Lemma 2. *For every iterative learner M, there exists a canny iterative learner N such that $\mathbf{InfEx}(M) \subseteq \mathbf{InfEx}(N)$.*

Proof. Let f be a computable 1-1 function mapping every finite informant sequence σ to a natural number encoding a program with $W_{f(\sigma)} = W_{M(\sigma)}$ if $M(\sigma) \in \mathbb{N}$ and $W_{f(\sigma)} = \varnothing$ otherwise. Clearly, σ can be reconstructed from $f(\sigma)$. We define the canny learner M' by letting

$$M'(\varnothing) = f(\varnothing)$$

$$h_{M'}(f(\sigma), (x, i)) = \begin{cases} f(\sigma \,{}^\frown (x, i)), & \text{if } x \notin \text{pos}(\sigma) \cup \text{neg}(\sigma) \wedge \\ & \quad M(\sigma \,{}^\frown (x, i)) \!\downarrow\, \neq M(\sigma) \!\downarrow; \\ f(\sigma), & \text{if } M(\sigma \,{}^\frown (x, i)) \!\downarrow\, = M(\sigma) \!\downarrow\, \vee \\ & \quad x \in \text{content}(\sigma); \\ \uparrow, & \text{otherwise.} \end{cases}$$

M' mimics M via f on a possibly finite informant subsequence of the originally presented informant with ignoring data not causing mind changes of M or that has already caused a mind change.

Let $L \in \mathbf{InfEx}(M)$ and $I' \in \mathbf{Inf}(L)$. As M has to learn L from every informant for it, M' will always be defined. Further, let $\sigma_0 = \varnothing$ and

$$\sigma_{t+1} = \begin{cases} \sigma_t {}^\frown I'(t), & \text{if } I'(t) \notin \operatorname{ran}(\sigma_t) \wedge M(\sigma_t {}^\frown I'(t))\downarrow \neq M(\sigma_t)\downarrow; \\ \sigma_t, & \text{otherwise.} \end{cases}$$

Then by induction for all $t \in \mathbb{N}$ holds $M'(I'[t]) = f(\sigma_t)$.

The following function translates between the two settings

$$\mathfrak{r}(0) = 0;$$
$$\mathfrak{r}(t+1) = \min\{r > \mathfrak{r}(t) \mid I'(r-1) \notin \operatorname{ran}(\sigma_{\mathfrak{r}(t)})\}.$$

Intuitively, the infinite range of $\mathfrak{r}$ captures all points in time r at which a datum that has not caused a mind change so far, is seen and a mind-change of M' is possible. Thus the mind change condition is of interest in order to decide whether $\sigma_{\mathfrak{r}(t+1)} \neq \sigma_{\mathfrak{r}(t)}$. Note that $\sigma_r = \sigma_{\mathfrak{r}(t)}$ for all r with $\mathfrak{r}(t) \leqslant r < \mathfrak{r}(t+1)$.

Let $I(t) = I'(\mathfrak{r}(t+1) - 1)$ for all $t \in \mathbb{N}$. Since only already observed data is ommited, I is an informant for L.

We next argue that $M(I[t]) = M(\sigma_{\mathfrak{r}(t)})$ for all $t \in \mathbb{N}$. As $I[0] = \varnothing = \sigma_0$, the claim holds for $t = 0$. Now we assume $M(I[t]) = M(\sigma_{\mathfrak{r}(t)})$ and obtain $M(I[t+1]) = M(I[t]{}^\frown I(t)) = M(\sigma_{\mathfrak{r}(t)}{}^\frown I(t)) = M(\sigma_{\mathfrak{r}(t+1)})$.

As by the definitions of I and $\mathfrak{r}$ we have $I(t) = I'(\mathfrak{r}(t+1) - 1) \notin \operatorname{ran}(\sigma_{\mathfrak{r}(t)})$ there are two cases:

1. If $M(\sigma_{\mathfrak{r}(t)}{}^\frown I(t)) = M(\sigma_{\mathfrak{r}(t)})$, then from $\sigma_{\mathfrak{r}(t+1)-1} = \sigma_{\mathfrak{r}(t)}$ and the definition of M' we obtain $\sigma_{\mathfrak{r}(t+1)} = \sigma_{\mathfrak{r}(t)}$. Putting both together the claimed equality $M(\sigma_{\mathfrak{r}(t)}{}^\frown I(t)) = M(\sigma_{\mathfrak{r}(t+1)})$ follows.
2. If $M(\sigma_{\mathfrak{r}(t)}{}^\frown I(t)) \neq M(\sigma_{\mathfrak{r}(t)})$, the definition of M' yields $\sigma_{\mathfrak{r}(t+1)} = \sigma_{\mathfrak{r}(t)}{}^\frown I(t)$. Hence the claimed equality also holds in this case.

We now argue that M' explanatory learns L from I'. In order to see this, first observe $\sigma_{\mathfrak{r}(t+1)} = \sigma_{\mathfrak{r}(t)}$ if and only if $M(I'[t+1]) = M(I'[t])$ for every $t \in \mathbb{N}$. This is because

$$\sigma_{\mathfrak{r}(t+1)} = \sigma_{\mathfrak{r}(t)} \Leftrightarrow M(\sigma_{\mathfrak{r}(t)}{}^\frown I(t)) = M(\sigma_{\mathfrak{r}(t)})$$
$$\Leftrightarrow M(I[t]{}^\frown I(t)) = M(I[t])$$
$$\Leftrightarrow M(I[t+1]) = M(I[t]).$$

As I is an informant for L, the learner M explanatory learns L from I. Hence there exists some t_0 such that $W_{M(I[t_0])} = L$ and for all $t \geqslant t_0$ holds $M(I[t]) = M(I[t_0])$. With this follows $\sigma_{\mathfrak{r}(t)} = \sigma_{\mathfrak{r}(t_0)}$ for all $t \geqslant t_0$. As for every r there exists some t with $\mathfrak{r}(t) \leqslant r$ and $\sigma_r = \sigma_{\mathfrak{r}(t)}$, we obtain $\sigma_r = \sigma_{\mathfrak{r}(t_0)}$ for all $r \geqslant \mathfrak{r}(t_0)$. We conclude $M'(I'[t]) = f(\sigma_t) = f(\sigma_{\mathfrak{r}(t_0)})$ for all $t \geqslant \mathfrak{r}(t_0)$ and by the definition of f finally $W_{f(\sigma_{\mathfrak{r}(t_0)})} = W_{M(\sigma_{\mathfrak{r}(t_0)})} = W_{M(I[t_0])} = L$. $\qquad\square$

5 Additional Requirements

In the following we review additional properties one might require the learning process to have in order to consider it successful. For this, we employ the following notion of consistency when learning from informant.

As in [LZZ08] according to [BB75] and [Bär77] for $A \subseteq \mathbb{N}$ we define

$$\mathbf{Cons}(f, A) \quad :\Leftrightarrow \quad \mathrm{pos}(f) \subseteq A \ \wedge \ \mathrm{neg}(f) \subseteq \mathbb{N} \setminus A$$

and say f *is consistent with* A or f *is compatible with* A.

Learning restrictions incorporate certain desired properties of the learners' behavior relative to the information being presented. We state the definitions for learning from informant here.

Definition 2. *Let M be a learner and I an informant. We denote by $h_t = M(I[t])$ the hypothesis of M after observing $I[t]$ and write*

1. $\mathbf{Conv}(M, I)$ *([Ang80]), if M is* conservative *on I, i.e., for all s, t with $s \leqslant t$ holds $\mathbf{Cons}(I[t], W_{h_s}) \Rightarrow h_s = h_t$.*
2. $\mathbf{Dec}(M, I)$ *([OSW82]), if M is* decisive *on I, i.e., for all r, s, t with $r \leqslant s \leqslant t$ holds $W_{h_r} = W_{h_t} \Rightarrow W_{h_r} = W_{h_s}$.*
3. $\mathbf{Caut}(M, I)$ *([OSW86]), if M is* cautious *on I, i.e., for all s, t with $s \leqslant t$ holds $\neg W_{h_t} \subsetneq W_{h_s}$.*
4. $\mathbf{WMon}(M, I)$ *([Jan91, Wie91]), if M is* weakly monotonic *on I, i.e., for all s, t with $s \leqslant t$ holds $\mathbf{Cons}(I[t], W_{h_s}) \Rightarrow W_{h_s} \subseteq W_{h_t}$.*
5. $\mathbf{Mon}(M, I)$ *([Jan91, Wie91]), if M is* monotonic *on I, i.e., for all s, t with $s \leqslant t$ holds $W_{h_s} \cap \mathrm{pos}(I) \subseteq W_{h_t} \cap \mathrm{pos}(I)$.*
6. $\mathbf{SMon}(M, I)$ *([Jan91, Wie91]), if M is* strongly monotonic *on I, i.e., for all s, t with $s \leqslant t$ holds $W_{h_s} \subseteq W_{h_t}$.*
7. $\mathbf{NU}(M, I)$ *([BCM+08]), if M is* non-U-shaped *on I, i.e., for all r, s, t with $r \leqslant s \leqslant t$ holds $W_{h_r} = W_{h_t} = \mathrm{pos}(I) \Rightarrow W_{h_r} = W_{h_s}$.*
8. $\mathbf{SNU}(M, I)$ *([CM11]), if M is* strongly non-U-shaped *on I, i.e., for all r, s, t with $r \leqslant s \leqslant t$ holds $W_{h_r} = W_{h_t} = \mathrm{pos}(I) \Rightarrow h_r = h_s$.*
9. $\mathbf{SDec}(M, I)$ *([KP16]), if M is* strongly decisive *on I, i.e., for all r, s, t with $r \leqslant s \leqslant t$ holds $W_{h_r} = W_{h_t} \Rightarrow h_r = h_s$.*

When additional requirements apply to the definition of learning success, we write them between $\mathbf{Inf}$ and $\mathbf{Ex}$. For example, Theorem 1 proves $[\mathbf{ItInfConvSDecSMonEx}] \setminus [\mathcal{R}\mathbf{ItInfEx}] \neq \varnothing$ because the non-total learner acts conservatively, strongly decisively and strongly monotonically when learning $\mathcal{L}$.

The text variants can be found in [JKMS16] where all pairwise relations $=$, $\subsetneq$ or $\perp$ between the sets $[\mathbf{ItTxt}\delta\mathbf{Ex}]$ (iterative learners from text) for $\delta \in \Delta$, where $\Delta = \{\mathbf{Conv}, \mathbf{Dec}, \mathbf{Caut}, \mathbf{WMon}, \mathbf{Mon}, \mathbf{SMon}, \mathbf{NU}, \mathbf{SNU}, \mathbf{SDec}\}$, are depicted. We sum up the current status regarding the map for iterative learning from informant in the following.

For all $\delta \in \Delta \setminus \{\mathbf{SMon}\}$ with a locking sequence argument we can observe $[\mathbf{ItInfSMonEx}] \subsetneq [\mathbf{ItInf}\delta\mathbf{Ex}]$. If we denote by $\mathbf{Inf}_{can}$ the set of all informants labelling the natural numbers according to their canonical order, which corresponds to the characterisic function of the respective language, we obtain **Fin** $\cup$ $\{\mathbb{N}\}$ $\in$ [$\mathcal{R}$**ItInf**$_{\text{can}}$**ConsConvSDecMonEx**] and thus it holds in contrast to full-information learning from informant [**ItInf**$_{\text{can}}$**Ex**] $\neq$ [**ItInfEx**], see [AKS18]. [LZ92] observed that requiring a monotonic behavior of the learner is restrictive, i.e. there exists an indexable family in [**ItInfMonEx**] $\subsetneq$ [**ItInfEx**]. The indexable family $\{\mathbb{N}\}$ $\cup$ $\{\mathbb{N} \setminus \{x\} \mid x \in \mathbb{N}\}$ is clearly not cautiously learnable but conservatively, strongly decisively and monotonically learnable by a total iterative learner from informant. Hence, [**ItInfCautEx**] $\perp$ [**ItInfMonEx**]. Moreover, [JLZ07] observed that requiring a conservative learning behavior is also restrictive. Indeed, they provide an indexable family in [**ItInfCautWMonNUDecEx**] \ [**ItInfConvEx**] and another indexable family in [$\mathcal{R}$**ItTxtCautConvSDecEx**] \ [**ItInfMonEx**].

Hence, the map on iterative learning from informant differs from the map on iterative learning from text in [JKMS16] as **Caut** is restrictive and also from the map of full-information learning in [AKS18] from informant as **Conv** is restrictive too. It has been open how **WMon**, **Dec**, **NU**, **SDec** and **SNU** relate to each other and the other requirements. We show that also **SNU** restricts **ItInfEx** with an intricate **ORT**-argument.

Theorem 2. [**ItInfSNUEx**] $\subsetneq$ [**ItInfEx**]

In the following we provide a lemma that might help to investigate **WMon**, **Dec** and **NU**.

Definition 3. *Denote the set of all unbounded and non-decreasing functions by* $\mathfrak{S}$, *i.e.,*

$$\mathfrak{S} := \{\, \mathfrak{s} : \mathbb{N} \to \mathbb{N} \mid \forall x \in \mathbb{N} \exists t \in \mathbb{N} \colon \mathfrak{s}(t) \geqslant x \text{ and } \forall t \in \mathbb{N} \colon \mathfrak{s}(t+1) \geqslant \mathfrak{s}(t) \,\}.$$

Then every $\mathfrak{s} \in \mathfrak{S}$ *is a so called* admissible simulating function.

A predicate $\beta \subseteq \mathfrak{P} \times \mathcal{I}$, *where* $\mathfrak{P}$ *stands for the set of all learners, is* semantically delayable, *if for all* $\mathfrak{s} \in \mathfrak{S}$, *all* $I, I' \in \mathcal{I}$ *and all learners* $M, M' \in \mathfrak{P}$ *holds: Whenever we have* $\text{pos}(I'[t]) \supseteq \text{pos}(I[\mathfrak{s}(t)])$, $\text{neg}(I'[t]) \supseteq \text{neg}(I[\mathfrak{s}(t)])$ *and* $W_{M'(I'[t])} = W_{M(I[\mathfrak{s}(t)])}$ *for all* $t \in \mathbb{N}$, *from* $\beta(M, I)$ *we can conclude* $\beta(M', I')$.

It is easy to see that every $\delta \in \{\text{\bf Caut}, \text{\bf Dec}, \text{\bf WMon}, \text{\bf Mon}, \text{\bf SMon}, \text{\bf NU}\}$ is semantically delayable and Lemma 2 can be restated as follows.

Lemma 3. *For every iterative learner* M *and every semantically delayable learning restriction* δ, *there exists a canny iterative learner* N *such that* $\text{Inf}\delta\text{Ex}(M) \subseteq \text{Inf}\delta\text{Ex}(N)$.

Proof. Add any semantically delayable δ in front of **Ex** in the proof of Lemma 2. We define a simulating function (Definition 3) by $\mathfrak{s}(t) = \max\{s \in \mathbb{N} \mid \mathfrak{r}(s) \leqslant t\}$. It is easy to check that $\mathfrak{s}$ is unbounded and clearly it is non-decreasing. Then by the definitions of I and $\mathfrak{s}$ we have $\text{pos}(I[\mathfrak{s}(t)]) \subseteq \text{pos}(I'[\mathfrak{r}(\mathfrak{s}(t))]) \subseteq \text{pos}(I'[t])$ and similarly $\text{neg}(I[\mathfrak{s}(t)]) \subseteq \text{neg}(I'[t])$ for all $t \in \mathbb{N}$. As $M'(I'[t]) = f(\sigma_t)$ and

$M(\sigma_{\mathfrak{r}(\mathfrak{s}(t))}) = M(I[\mathfrak{s}(t)])$ for all $t \in \mathbb{N}$, in order to obtain $W_{M'(I'[t])} = W_{M(I[\mathfrak{s}(t)])}$ it suffices to show $W_{f(\sigma_t)} = W_{M(\sigma_{\mathfrak{r}(\mathfrak{s}(t))})}$. Since $W_{f(\sigma_t)} = W_{M(\sigma_t)}$ for all $t \in \mathbb{N}$, this can be concluded from $\sigma_t = \sigma_{\mathfrak{r}(\mathfrak{s}(t))}$. But this obviously holds because $\mathfrak{r}(\mathfrak{s}(t)) \leqslant t < \mathfrak{r}(\mathfrak{s}(t)+1)$ follows from the definition of $\mathfrak{s}$.

Finally, from $\delta(M, I)$ we conclude $\delta(M', I')$. $\square$

Two other learning restrictions that might be helpful to understand the syntactic learning criteria **SNU**, **SDec** and **Conv** better are the following.

Definition 4. *Let M be a learner and I an informant. We denote by $h_t = M(I[t])$ the hypothesis of M after observing $I[t]$ and write*

1. **LocConv**(M, I) *([JLZ07]), if M is locally conservative on I, i.e., for all t holds $h_t \neq h_{t+1} \Rightarrow \neg$**Cons**$(I(t), W_{h_t})$.*
2. **Wb**(M, I) *([KS16]), if M is witness-based on I, i.e., for all r, s, t with $r < s \leqslant t$ the mind-change $h_r \neq h_s$ implies $\mathrm{pos}(I[s]) \cap W_{h_t} \setminus W_{h_r} \neq \varnothing \vee \mathrm{neg}(I[s]) \cap W_{h_r} \setminus W_{h_t} \neq \varnothing$.*

Hence, in a locally conservative learning process every mind-change is justified by the datum just seen. Moreover, a in witness-based learning process each mind-change is witnessed by some false negative or false positive datum. Obviously, **LocConv** $\Rightarrow$ **Conv** and **Wb** $\Rightarrow$ **Conv**.

As for learning from text, see [JKMS16], we gain that every concept class locally conservatively learnable by an iterative learner from informant is also learnable in a witness-based fashion by an iterative learner.

Theorem 3. **[ItInfLocConvEx]** $\subseteq$ **[ItInfWbEx]**

Proof. Let $\mathcal{L}$ be a concept class learned by the iterative learner M in a locally conservative manner. As we are interested in a witness-based learner N, we always enlarge the guess of M by all data witnessing a mind-change in the past. As we want N to be iterative, this is done via padding the set of witnesses to the hypothesis and a total computable function g adding this information to the hypothesis of M as follows:

$$W_{g(\mathrm{pad}(h, \langle MC \rangle))} = (W_h \cup \mathrm{pos}[MC]) \setminus \mathrm{neg}[MC];$$
$$N(\varnothing) = g(\mathrm{pad}(M(\varnothing), \langle \varnothing \rangle));$$

$$h_N(g(\mathrm{pad}(h, \langle MC \rangle)), \xi) = \begin{cases} g(\mathrm{pad}(h, \langle MC \rangle)), & \text{if } h_M(h, \xi) = h \vee \\ & \qquad \xi \in MC; \\ g(\mathrm{pad}(h_M(h, \xi), \\ \langle MC \cup \{\xi\} \rangle)), & \text{otherwise.} \end{cases}$$

Clearly, N is iterative. Further, whenever M is locked on h and $W_h = L$, since MC is consistent with L, we also have $W_{g(\mathrm{pad}(f(h), \langle MC \rangle))} = L$. As N simulates M on an informant omitting all data that already caused a mind-change beforehand, N does explanatory learn $\mathcal{L}$. As M learns locally conservatively and by employing g, the learner N acts witness-based. $\square$

6 Suggestions for Future Research

Future work should address the complete map for iterative learners from informant. In particular, **WMon**, **Dec** and **NU** seem to be challenging as the proofs in related settings fail without an obvious fix. We hope that Lemma 3 is a helping hand in this endeavour. Also the equivalence of the syntactic criteria **SNU**, **SDec** and **Conv** does not trivially hold. Theorem 3 might be helpful regarding the latter.

Maps for other models of memory-limited learning, such as **BMS**, see [CCJS07], or **Bem**, see [FJO94, LZ96] and [CJLZ99], would help to rate models.

Last but not least we encourage to investigate the learnability of indexable classes motivated by grammatical inference or machine learning research. The pattern languages often serve as a helpful example to refer to and we hope for even more examples of this kind. As a starting point, in the full version, we prove the learnability of half-spaces.

Acknowledgements. This work was supported by DFG Grant Number KO 4635/1-1. We are grateful to the people supporting us.

References

[AKS18] Aschenbach, M., Kötzing, T., Seidel, K.: Learning from informants: Relations between learning success criteria (2018). arXiv preprint arXiv:1801.10502

[Ang80] Angluin, D.: Inductive inference of formal languages from positive data. Inf. Control **45**(2), 117–135 (1980)

[Bār77] Bānrzdiņš, J.: Inductive inference of automata, functions and programs. Amer. Math. Soc. Transl., 107–122 (1977)

[BB75] Blum, L., Blum, M.: Toward a mathematical theory of inductive inference. Inf. Control **28**, 125–155 (1975)

[BCM+08] Baliga, G., Case, J., Merkle, W., Stephan, F., Wiehagen, R.: When unlearning helps. Inf. Comput. **206**, 694–709 (2008)

[Cas74] Case, J.: Periodicity in generations of automata. Math. Syst. Theory. **8**(1), 15–32 (1974)

[Cas94] Case, J.: Infinitary self-reference in learning theory. J. Exp. Theor. Artif. Intell. **6**, 3–16 (1994)

[CCJS07] Carlucci, L., Case, J., Jain, S., Stephan, F.: Results on memory-limited U-shaped learning. Inf. Comput. **205**, 1551–1573 (2007)

[CJLZ99] Case, J., Jain, S., Lange, S., Zeugmann, T.: Incremental concept learning for bounded data mining. Inf. Comput. **152**, 74–110 (1999)

[CK10] Case, J., Kötzing, T.: Strongly non-U-shaped learning results by general techniques. COLT **2010**, 181–193 (2010)

[CM08] Case, J., Moelius, S.E.: U-shaped, iterative, and iterative-with-counter learning. Mach. Learn. **72**, 63–88 (2008)

[CM09] Case, J., Moelius, S.: Parallelism increases iterative learning power. Theor. Comput. Sci. **410**(19), 1863–1875 (2009)

[CM11] Case, J., Moelius, S.: Optimal language learning from positive data. Inf. Comput. **209**, 1293–1311 (2011)

[FJO94] Fulk, M., Jain, S., Osherson, D.: Open problems in systems that learn. J. Comput. Syst. Sci. **49**(3), 589–604 (1994)

[Gol67] Gold, E.: Language identification in the limit. Inf. Control **10**, 447–474 (1967)

[Jan91] Jantke, K.P.: Monotonic and nonmonotonic inductive inference of functions and patterns. In: 1st International Workshop on Nonmonotonic and Inductive Logic, Proceedings, pp. 161–177 (1991)

[JKMS16] Jain, S., Kötzing, T., Ma, J., Stephan, F.: On the role of update constraints and text-types in iterative learning. Inf. Comput. **247**, 152–168 (2016)

[JLZ07] Jain, S., Lange, S., Zilles, S.: Some natural conditions on incremental learning. Inf. Comput. **205**, 1671–1684 (2007)

[JMZ13] Jain, S., Moelius, S., Zilles, S.: Learning without coding. Theor. Comput. Sci. **473**, 124–148 (2013)

[JORS99] Jain, S., Osherson, D., Royer, J., Sharma, A.: Systems that Learn: An Introduction to Learning Theory, 2nd edn. MIT Press, Cambridge (1999)

[Köt09] Kötzing, T.: Abstraction and Complexity in Computational Learning in the Limit. PhD thesis, University of Delaware (2009)

[KP16] Kötzing, T., Palenta, R.: A map of update constraints in inductive inference. Theor. Comput. Sci. **650**, 4–24 (2016)

[KS16] Kötzing, T., Schirneck, M.: Towards an atlas of computational learning theory. In: 33rd Symposium on Theoretical Aspects of Computer Science (2016)

[KSS17] Kötzing, T., Schirneck, M., Seidel, K.: Normal forms in semantic language identification. In: Proceedings of Algorithmic Learning Theory, pp. 493–516. PMLR (2017)

[LZ92] Lange, S., Zeugmann, T.: Types of monotonic language learning and their characterization. In: Proceedings 5th Annual ACM Workshop on Computing Learning Theory, New York, NY, pp. 377–390 (1992)

[LZ96] Lange, S., Zeugmann, T.: Incremental learning from positive data. J. Comput. Syst. Sci. **53**, 88–103 (1996)

[LZZ08] Lange, S., Zeugmann, T., Zilles, S.: Learning indexed families of recursive languages from positive data: a survey. Theor. Comput. Sci. **397**(1), 194–232 (2008)

[Odi99] Odifreddi, P.: Classical Recursion Theory, vol. II. Elesivier, Amsterdam (1999)

[OSW82] Osherson, D., Stob, M., Weinstein, S.: Learning strategies. Inf. Control **53**, 32–51 (1982)

[OSW86] Osherson, D., Stob, M., Weinstein, S.: Systems that Learn: An Introduction to Learning Theory for Cognitive and Computer Scientists. MIT Press, Cambridge (1986)

[RC94] Royer, J., Case, J.: Subrecursive Programming Systems: Complexity and Succinctness. Research monograph in Progress in Theoretical Computer Science, Birkhäuser Boston (1994)

[Rog67] Rogers, H.: Theory of Recursive Functions and Effective Computability. McGraw Hill, New York (1967). Reprinted, MIT Press (1987)

[Wie76] Wiehagen, R.: Limes-erkennung rekursiver funktionen durch spezielle strategien. J. Inf. Process. Cybern. **12**(1–2), 93–99 (1976)

[Wie91] Wiehagen, R.: A thesis in inductive inference. In: 1st International Workshop on Nonmonotonic and Inductive Logic, Proceedings, pp. 184–207 (1991)

On Preserving the Computational Content of Mathematical Proofs: Toy Examples for a Formalising Strategy

Angeliki Koutsoukou-Argyraki[✉]

Department of Computer Science and Technology (Computer Laboratory),
University of Cambridge, Cambridge, UK
ak2110@cam.ac.uk

Abstract. Instead of using program extraction mechanisms in various theorem provers, I suggest that users opt to create a database of formal proofs whose computational content is made explicit; this would be an alternative approach which, as libraries of formal mathematical proofs are constantly growing, would rely on future advances in automation and machine learning tools, so that as blocks of (sub)proofs get generated automatically, the preserved computational content would get recycled, recombined and would eventually manifest itself in different contexts. To this end, I do not suggest restricting to only constructive proofs, but I suggest that proof mined, possibly also non-constructive proofs with some explicit computational content should be preferable, if possible. To illustrate what kind of computational content in mathematical proofs may be of interest I give several very elementary examples (to be regarded as building blocks of proofs) and some samples of formalisations in Isabelle/HOL. Given the state of the art in automation and machine learning tools currently available for proof assistants, my suggestion is rather speculative, yet starting to build a database of formal proofs with explicit computational content would be a potentially useful first step.

Keywords: Proof assistants · Computational content · Proof mining · Isabelle/HOL · Proof theory · Interactive theorem provers · Formalisation · Machine learning

1 Motivation

A very active and significant area of modern proof theory is devoted to the specification of the computational content of mathematical methods, both from a foundational and a structural point of view, and the extraction of computational content from mathematical proofs. Avigad's survey [1] summarises various such methods in classical first order arithmetic. In applied proof theory, within the past couple of decades, Kohlenbach's proof mining school, originating from Kreisel's program of "unwinding of proofs" from the 1950's [22–24], has succeeded in obtaining computational content from mathematical proofs in a vast

© Springer Nature Switzerland AG 2021
L. De Mol et al. (Eds.): CiE 2021, LNCS 12813, pp. 285–296, 2021.
https://doi.org/10.1007/978-3-030-80049-9_26

number of applications, mainly within nonlinear analysis (fixed point theory, ergodic theory, topological dynamics, convex optimisation, Cauchy problems, nonlinear semigroup theory in Banach spaces and more). The main reference is Kohlenbach's monograph from 2008 [12] while the recent reviews [13,14] give a general overview of the main recent results since 2008 and discuss how proof-theoretic methods can be applied for extracting explicit bounds in each family of applications. The proof-theoretic tools applied for the pen-and-paper extraction of computational content are variations and/or combinations of certain so-called proof interpretations, such as Gödel's functional "Dialectica" (monotone) interpretation, the Gödel–Gentzen double-negation translation and Friedman's A-translation. Within proof mining, certain logical metatheorems guarantee the extractability of effective information (computational content) in specific contexts even for classical (i.e. non-constructive) proofs, as long as the statement proved is of a certain logical form.

So far we have referred to pen-and-paper extraction of computational content. On a parallel note, in the world of interactive theorem provers (proof assistants), computational content, in the sense of algorithms contained in constructive proofs, can be obtained by various program extraction mechanisms that implement the aforementioned proof interpretations. These are available for instance in Nuprl and Coq, which are based on dependent type theories. Constable and Murthy have studied the effectiveness of proof transformations which reveal the computational content of even classical proofs using the A-translation and discuss their implementation in Nuprl [9]. A program extraction mechanism based on A-translation and modified realizability is also available in Schwichtenberg's system MINLOG[1] which is based on minimal first order logic [27]. Berghofer developed the foundations for a program extraction mechanism in Isabelle[2] in his PhD thesis [5] given that Isabelle, as a generic theorem prover, is based on simply-typed minimal higher order logic, which is purely constructive.

Meanwhile, the popularity of proof assistants among working mathematicians in recent years has been rapidly increasing. An important milestone reflecting this is the inclusion of the new class 68VXX referring to formalisation of mathematics and proof assistants in mathematical practice in the 2020 Mathematics Subject Classification. The field has come a long way since the significant first attempt of the formal language AUTOMATH by de Bruijn in the late 1960's [8][3] but we are still very far from achieving the goals outlined in the QED Manifesto from the 1990's [7] or the vision of an interactive assistant that would "converse" with human mathematicians and assist them in the discovery of new theorems by giving ideas and providing counterexamples, as described e.g. by Timothy Gowers [10]. However, the body of formalised mathematical material in various proof assistants is growing very rapidly thanks to contributions by

---

[1] http://www.mathematik.uni-muenchen.de/~logik/minlog/.

[2] http://www.cl.cam.ac.uk/research/hvg/Isabelle/index.html.

[3] L. S. van Benthem Jutting, as part of his PhD thesis in the late 1970s translated Edmund Landau's Foundations of Analysis into AUTOMATH and checked its correctness [4].

an international community of users and developers. For instance, the Archive of Formal Proofs (AFP)[4] that contains material formalised in Isabelle based on the Isabelle Libraries[5] as of 25 April 2021 includes 594 articles by 379 authors.

On a different note, machine learning technology is considered promising when it comes to providing tools complementary to these of the axiomatic approach in many areas in computer science, thanks to work towards the integration of deep neural networks with logic and symbolic computation. The communities of machine learning and formal verification in particular have been growing increasingly close during the past few years, e.g. note the very successful "Artificial Intelligence and Theorem Proving" (AITP)[6] and "Intelligent Computer Mathematics"(CICM)[7] conference series, as well as the upcoming MATH-AI workshop[8]. Pattern recognition tools from machine learning could find applications not only in searching the libraries of formal proofs, but also in recognising proof patterns and providing proof recommendation methods. Recent efforts in machine learning (e.g. by Li *et al.* [25], Bansal *et al.* [3], Polu and Sutskever [26]) are very promising first steps towards the long-term goal of having proof assistants generate human-readable proofs automatically.

2   My Suggestion

Given the promising prospects of applications of machine learning in theorem proving, I suggest[9] that extracting computational content from mathematical proofs could be another area where machine learning could help.

To this end, a database of proofs whose computational content is made explicit, whenever possible, would be necessary to have. It would be therefore meaningful for users to opt to formalise proofs with explicit computational content[10]. Enriching the (extensive and fast-growing) libraries of formal proofs with such proofs would not only help preserve their computational content so that the mathematician user could "manually" make use of it, but also it would provide a body of data to be used for machine learning: detecting computational content thanks to pattern matching and as automation improves and blocks of (sub)proofs get generated automatically, the preserved computational content would get recycled, recombined and would eventually manifest itself in different contexts. To create this database of formal proofs I do not suggest restricting to only strictly constructive proofs; proofs with some explicit computational content

---

[4] https://www.isa-afp.org/index.html.

[5] http://www.cl.cam.ac.uk/research/hvg/Isabelle/dist/library/HOL/index.html
http://www.cl.cam.ac.uk/research/hvg/Isabelle/dist/library/ZF/index.html.

[6] http://aitp-conference.org.

[7] https://easychair.org/smart-program/CICM-13/.

[8] https://mathai-iclr.github.io.

[9] I have very briefly mentioned this suggestion as a comment in a number of talks since [21] including in CICM 2018, AITP 2019, Big Proof 2019 as well as in [17].

[10] A recent work in this direction is the formalisation of several results from computable analysis in Coq by Steinberg, Théry and Thies [28].

(in the sense of quantitative information expressed as computable functionals) would be appropriate too even if they are not done within some fully constructive system (see Remark 1). This would constitute a way of "automating" proof mining without using program extraction mechanisms as in MINLOG or Nuprl.

The manually constructed dataset of formal proofs would include the general definitions of various aspects of computational content in mathematical applications, plus as many as possible examples. It could be potentially helpful for the AI tools if the users make the qualitative information explicit as well; e.g. in the case of Theorem 1 in Sect. 3, the user could prove the explicit qualitative statement "$\sqrt{2}$ is irrational" in addition to proving the bound implying the irrationality of $\sqrt{2}$. While the future human mathematician users would be in the process of writing a new proof, an AI interactive assistant would notify them of related proofs carrying explicit computational content (recommended via pattern matching on mathematical formulas) so that these could be manually used. Another useful feature to implement would be an AI "diagnostic" tool to automatically check whether the statement proved fits the requirements for the application of a general proof mining metatheorem (thus guaranteeing the extractability of computational content from a proof, regardless of whether it is constructive or not, as long as the mathematical statement at hand can be reduced to a certain logical form and the input data fulfill certain conditions [12]). At a later, much more advanced stage, where proof blocks would be generated automatically thanks to sophisticated code generation mechanisms, (some of) the quantitative information would be computed automatically too, using the available material from the library that would be detected via machine learning.

A prerequisite for the realisation of this idea is having achieved the stage of an interactive proof assistant that could automatically generate intermediate small lemmas and results "at the level of a capable graduate student", assisting working mathematicians in their daily research work. To this end, the main two areas requiring a great deal of work are (1) automation and (2) search features, both for proof patterns and algorithms and names of required facts. The state of the art is still far from this goal; Avigad in [2] writes on this: *"we are not there yet, but such technology seems to be within reach. There are no apparent conceptual hurdles that need to be overcome, though getting to that point will require a good deal of careful thought, clever engineering, experimentation, and hard work"*. My plan sketched here is thus purely speculative, and I do not have more specific technical details to offer about how this highly non-trivial goal could be achieved, except from suggesting that two different communities come together: (a) experts who formalise mathematics who could opt for formalising proofs with explicit computational content contributing to the creation of a large body of data and (b) machine learning experts who could build the tools for the automated "identification" of this explicit computational content and the proof recommendation tools that would make use of it in an appropriate way.

Clearly, my suggestion could not be applicable for all proofs as not every mathematical proof carries meaningful computational content. At the same time, even pen-and-paper proof mining works only for statements of the logical form ∀∃ so one cannot expect that *any* mathematical proof could by "automatically" proof-mined either. But I suggest that it would be meaningful to at least capture the computational content of *some* proofs wherever possible. At this point it is important to stress that different proofs of the same statement give different computational content. For example, regarding the infinitude of primes, in [12] in addition to the proof presented later in this paper, two other, different proofs, by Euclid and Euler respectively, are given; these proofs are all proof mined in [12] and each of them gives a different bound. Another instance of this phenomenon I have encountered in my work on approximate common fixed points of one-parameter nonexpansive semigroups on subsets of a Banach space [16,18,20]. Usually it is unclear how to evaluate which proof gives "better" computational content. "Better" computational content might mean a numerically more precise bound, or a bound of lower complexity (e.g. polynomial instead of exponential), or a bound that is more "elegant", or a bound with fewer parameters– and there is no reason why the aforementioned properties would coincide, nor is there any a priori relationship between them clear.

3   Toy Examples

In proof mining, the quantitative versions of properties like convergence/metastability/asymptotic regularity [12,15,19,20] convey the related computational information through the respective rates. Similarly, the quantitative versions of properties such as uniform continuity, uniqueness, uniform convexity [12,20] convey the computational information through the respective moduli, and irrationality conveys the computational information through the irrationality measure (see e.g. next section or [16,18,20]). In a similar spirit, other interesting notions (referring to certain operators in Banach spaces) are the modulus of accretivity/modulus of accretivity at zero introduced by Kohlenbach and the author [15,20], and the modulus of ϕ-accretivity/modulus of ϕ-accretivity at zero introduced by the author [19,20]. Other such notions are the moduli of total boundedness (for compactness), uniform closedness, uniform Fejér monotonicity, approximate fixed points bound [13]. All these various moduli/rates/measures/bounds can be seen as "black box information" entering the assumptions (and can be computed by choosing values for their parameters as they are number-theoretic functionals). The final bound referring to the conclusion of the theorem which may be obtained via proof mining on the original proof will take the functionals of the moduli/rates/measures/bounds originating from the assumptions as inputs.

I now proceed to give a few very elementary examples to illustrate aspects of computational content in mathematical proofs that we may be interested in. The proofs are trivial and they are omitted for the sake of brevity (except from the proof of Theorem 1). For Theorem 1, Lemmas 1 and 2 we give formalisations in Isabelle/HOL. Such elementary proofs could be included as ingredients of building blocks of more elaborate proofs. The examples presented here are as simple as possible; for the general underlying logical form of the statements in each family of mathematical applications as well as the statements in full generality (e.g. in general metric spaces etc.) we refer to [12–14]. In the following, $\mathbb{N}$ denotes the set of natural numbers $\{1, 2, 3...\}$, $\mathbb{Z}$, $\mathbb{Z}^+$, $\mathbb{Z}^*$, $\mathbb{Q}$, $\mathbb{R}$ denote the sets of integers, positive integers without zero, integers without zero, rationals and reals respectively.

3.1 $\sqrt{2}$ Is Irrational

The following proof of the irrationality of $\sqrt{2}$ due to Bishop [6] does not make any use of the law of the excluded middle and moreover provides quantitative information showing that $\sqrt{2}$ is "constructively" irrational.

Theorem 1. *(Bishop [6])* $\sqrt{2} \in \mathbb{R} \setminus \mathbb{Q}$. *In particular:*

$$\forall a, b \in \mathbb{Z}^+ \ (a/b \in (0, 2] \rightarrow |\sqrt{2} - \frac{a}{b}| \geq \frac{1}{4b^2}).$$

Proof (Bishop [6]). The first step is to show that, for all $a, b \in \mathbb{Z}^+$, $a^2 \neq 2b^2$ (e.g. by using the argument that as the highest power of 2 dividing $2b^2$ is odd, while the highest power of 2 dividing a^2 is even, they must be distinct integers). Therefore, for all $a, b \in \mathbb{Z}^+$, $|a^2 - 2b^2| \geq 1$. Then have that $|\frac{a}{b} - \sqrt{2}||\frac{a}{b} + \sqrt{2}| = |\frac{a^2}{b^2} - 2| = \frac{1}{b^2}|a^2 - 2b^2| \geq \frac{1}{b^2}$. Finally notice that, assuming $a/b \in (0, 2]$ we have $|\frac{a}{b} - \sqrt{2}| \geq \frac{1}{|\frac{a}{b} + \sqrt{2}|} \frac{1}{b^2} \geq \frac{1}{4} \frac{1}{b^2}$.

A formal proof in Isabelle/HOL (due to Wenda Li and the author) is given below:

```
lemma identity_square_dif:
  fixes a b:: "'a:: comm_ring"
  shows " a*a - b*b = (a-b)*(a+b)   "
  by (auto simp add:algebra simps)

lemma valuation_dif:
  fixes a b::int
  assumes "a>0" and "b>0"
  shows "a^2 ≠2* (b^2)"
proof-
  have is_even: "multiplicity 2 (a^2) = 2* multiplicity 2 a"
    apply (subst prime_elem_multiplicity_power_distrib)
    using ‹a>0› by (auto simp add:prime_imp_prime_elem)
  have val1: "even( multiplicity 2 (a^2))" using is_even by auto
  have *: "multiplicity 2 (2*b^2) = (multiplicity 2 (b^2)) +1"
    apply (subst multiplicity_times_same)
    using ‹b>0› by auto
  have **: "(multiplicity 2 (b^2))+1 = (2* (multiplicity 2 b))+1"
    apply (subst prime_elem_multiplicity_power_distrib)
    using ‹b>0› by (auto simp add:prime_imp_prime_elem)
  have is_odd: "multiplicity 2 (2*b^2) = (2* (multiplicity 2 b))+1"
    using * ** by auto
  have val2: "odd(multiplicity 2 (2*b^2))" using is_odd by auto
  have dif: "multiplicity 2 (a^2) ≠ multiplicity 2 (2*b^2) "
    using val1 val2 by auto
  show ?thesis using dif by auto
qed

theorem sqrt2isirrational_Bishop:
  fixes a b ::int assumes "a >0" and "b >0" and "a/b ≤2"
  shows "¦ a/b - sqrt(2)¦ ≥  1/(4*b^2)"
proof-
  have *:"¦ a/b - sqrt(2)¦ *¦ a/b + sqrt(2)¦ ≥ 1/b^2"
  proof -
    have "¦ a/b - sqrt(2)¦ *¦ a/b + sqrt(2)¦ =¦ (a/b)^2 - 2¦ "
      using identity_square_dif
      by (metis abs_mult abs_numeral real_sqrt_mult_self semiring_normalization_rules(29))
    also have "... = (1/b^2)*¦ a^2 -(b^2)* 2 ¦"
      using ‹b>0› by (simp add: power_divide divide simps)
    also have "... ≥ (1/b^2)"
    proof-
      have **: "¦ a^2 -2* (b^2) ¦ ≥ 1" using valuation_dif[OF ‹a>0› ‹b>0›]
        by fastforce
      then show ?thesis using **
        by (metis (no_types, hide_lams) divide_right_mono mult.commute
mult.left_neutral of_int_1 le_iff of_int_mult semiring_normalization_rules(29)
times_divide_eq_left zero_le_power2)
    qed
    finally show ?thesis.
  qed
  show "¦ a/b - sqrt(2)¦ ≥  1/(4*b^2)"
  proof -
    have "1/(4*b^2) ≤ (1/¦2 + sqrt(2)¦) *( 1/b^2)"
    proof -
      have "2 + sqrt(2) < 4"
        by (simp add: sqrt2_less_2)
      then show ?thesis using ‹b>0›
        apply (simp add:divide simps)
        by (smt mult_less_0_iff not_sum_power2_lt_zero real_sqrt_gt_0_iff)
    qed
    also have "... ≤ (1/¦ a/b + sqrt(2)¦) *( 1/b^2)"
    proof -
      define t1 t2 where "t1=¦ a/b + sqrt(2)¦" and "t2=¦2 + sqrt 2¦"
      have "t1 ≤ t2" unfolding t1_def t2_def
        using assms by force
      moreover have "t1 >0"
        unfolding t1_def using "*" by auto
      ultimately show ?thesis
      apply (fold t1_def t2_def)
        using ‹b>0 ›by (simp add:divide simps)
    qed
    also have "... ≤ ¦ a/b - sqrt(2)¦"
    proof -
      have "¦ a/b + sqrt(2)¦ > 0" using * by auto
      then show ?thesis using * ‹b>0› by (auto simp add:divide simps)
  qed
    finally show ?thesis.
  qed
qed
```

Remark 1. Here we are not concerned with the use of the label "constructive" but with the computational information that emerges together with the property of irrationality. Taking a case distinction and assuming that $a/b > 2$, it is easy to estimate that $\forall a, b \in \mathbb{Z}^+$ $(a/b > 2 \rightarrow |\sqrt{2} - \frac{a}{b}| \geq 1/4 \geq 1/4b^2)$ so in any case $\forall a, b \in \mathbb{Z}^+$ $(|\sqrt{2} - \frac{a}{b}| \geq 1/4b^2)$. To characterise a proof as "constructive" according to Bishop it should not make use of the trichotomy $(a/b > 2) \vee (a/b = 2) \vee (a/b < 2)$, while the proof given here where it is a priori assumed $a/b \in (0, 2]$ is constructive as it is. But we could have alternatively assumed $a, b \in \mathbb{Z}^+$ (or $a \in \mathbb{Z}, b \in \mathbb{Z}^*$) and written the proof using non-constructive trichotomy case distinctions while still obtaining computational content. Thus, for our database we do not need to restrict to proofs that are constructive (performed within a constructive calculus); proofs giving some explicit computational content (i.e. quantitative information expressed as computable functionals) even if not fully constructive, could be appropriate too. In fact, in actual mathematical practice most proofs make heavy use of non-constructive principles, like the law of the excluded middle (proof by contradiction, case distinctions) and the axiom of choice, so excluding such proofs would be very impractical. Isabelle/HOL in particular is a proof assistant that allows for such non-constructive proofs.

3.2 There Exist Infinitely Many Prime Numbers

The following statement not only attests the infinitude of primes [12] but moreover gives quantitative information on the value of a prime P as a function of the number of primes that are smaller than P. For the proof see [12].

Theorem 2 *(Kohlenbach [12]).* *There exist infinitely many prime numbers. In particular, given the first r many prime numbers, there exists a prime number* $P > p_r$ *and moreover* $P \leq 4^r + 1$.

3.3 Uniform Continuity

Lemma 1. *The function* $f : \mathbb{R} \rightarrow \mathbb{R}$ *defined as* $f(x) := x$ *is uniformly continuous. In particular,*

$$\forall k \in \mathbb{N} \; \forall x, y \in \mathbb{R} \; (|x - y| < 2^{-\omega(k)} \rightarrow |f(x) - f(y)| < 2^{-k})$$

where a bound (depending on k) on $\omega : \mathbb{N} \rightarrow \mathbb{N}$ *is a modulus of continuity for* f *and here we may trivially take* $\omega(k) := k$.

3.4 Uniqueness

Lemma 2. *The function* $f : \mathbb{R} \rightarrow \mathbb{R}$ *defined as* $f(x) := x$ *has a unique zero. In particular, given* $z_1, z_2 \in \mathbb{R}$

$$\forall k \in \mathbb{N} \; (|f(z_1)| < 2^{-\omega(k)} \wedge |f(z_2)| < 2^{-\omega(k)} \rightarrow |z_1 - z_2| < 2^{-k})$$

where a bound (depending on k) on $\omega : \mathbb{N} \rightarrow \mathbb{N}$ *is a modulus of uniqueness*[11] *for the zero of* f *and here we may trivially take* $\omega(k) := k + 1$.

---

[11] see [11,12,14] for the general logical form.

Formal proofs of Lemma 1 and Lemma 2 in Isabelle/HOL are given below:

```
lemma uniformlycontinuous:
  fixes f ::"real⇒real" assumes "∀ x. f x = x"
  shows "∀ k ∈ N. ∀ x y.(⦃x-y⦄ < 2 powr (- real_of_nat (k)) ⟶⦃f x- f y⦄<2 powr(-real_of_nat(k)))"
  using assms by auto


lemma uniquezero:
  fixes f ::"real⇒real" and z1::real and z2::real  assumes "∀ x. f x = x"
  shows "∀ k ∈ N.(⦃f z1⦄ <  2 powr (-real_of_nat (k+1))∧⦃ f z2⦄ <  2 powr(-real_of_nat (k+1))
  ⟶⦃z1- z2⦄ <2 powr(-real_of_nat(k)))"
proof-
  have 1:"∀ k ∈ N. (⦃ z1⦄ <  2 powr (- real_of_nat (k+1))∧ ⦃z2⦄ < 2 powr (- real_of_nat (k+1))
  ⟶⦃z1- z2⦄ <(2 powr 1) *2 powr(-real_of_nat(k+1)))"
  using abs_triangle_ineq  by auto
  have 2: "∀ k ∈ N. (⦃ z1⦄ < 2 powr (- real_of_nat (k+1))∧ ⦃z2⦄ < 2 powr (-real_of_nat (k+1))
  ⟶⦃z1- z2⦄ <2 powr(1 - real_of_nat(k+1)))"
  using 1 assms by (metis minus_real_def powr_add)
show ?thesis using 2  using assms by fastforce
qed
```

3.5   Inclusions Between Sets of Solutions

See [14] for a discussion on the general form. The following is a trivial example:

Lemma 3. *Let* $f, g : \mathbb{R} \to \mathbb{R}$ *defined as* $f(x) := x$ *and* $g(x) := 4x$. *Then* $z \in \mathbb{R}$ *is a zero of* f *if and only if it is a zero of* g. *In particular:*[12]

$$\forall k \in \mathbb{N} \ (|f(z)| \leq 2^{-\omega(k)} \to |g(z)| < 2^{-k})$$

$$\forall k \in \mathbb{N}, k > 2 \ (|g(z)| \leq 2^{-\tilde{\omega}(k)} \to |f(z)| < 2^{-k})$$

where we may take $\omega(k) := k + 2$ *and* $\tilde{\omega}(k) := k - 2$.

4   Propagation of Computational Information: More Toy Examples

Thanks to the modularity of the bounds, the computational content "propagates". We illustrate how this may happen, and what kind of computational information we may be looking for, in the following elementary examples.

4.1   The Product and Ratio of an Irrational and a Rational Are Irrationals

Given $\gamma \in \mathbb{R} \setminus \mathbb{Q}$, as definition of irrationality we consider the following:

$$\forall a \in \mathbb{Z}, b \in \mathbb{Z}^* \ \exists z \in \mathbb{N} \ |\gamma - \frac{a}{b}| \geq \frac{1}{z}.$$

Skolemizing, $\exists z \in \mathbb{Z} \times \mathbb{Z}^* \to \mathbb{N} \ \forall a \in \mathbb{Z}, b \in \mathbb{Z}^* \ |\gamma - \frac{a}{b}| \geq \frac{1}{z(a,b)}$
and the Skolem function z plays the role of the irrationality measure.

---

[12] in this totally trivial example the set of zeros for both f and g is of course the one-element set $\{0\}$.

Lemma 4. *Let $\delta \in \mathbb{Q}$ with $\delta > 0$. Then $\sqrt{2}\delta \in \mathbb{R} \setminus \mathbb{Q}$. In particular, assuming*
$\delta = \frac{c}{d}$ *for $c, d \in \mathbb{Z}^+$*

$$\forall a \in \mathbb{Z}, b \in \mathbb{Z}^* \ (|\sqrt{2}\delta - \frac{a}{b}| \geq \frac{1}{4dcb^2}).$$

More generally:

Lemma 5. *Let $\gamma \in \mathbb{R} \setminus \mathbb{Q}$ and $\delta \in \mathbb{Q}$ with $\delta \neq 0$. Then $\gamma\delta \in \mathbb{R} \setminus \mathbb{Q}$ and $\frac{\gamma}{\delta} \in \mathbb{R} \setminus \mathbb{Q}$.*
In particular, assuming

$$\exists z \in \mathbb{Z} \times \mathbb{Z}^* \rightarrow \mathbb{N} \ \forall a \in \mathbb{Z}, b \in \mathbb{Z}^* \ |\gamma - \frac{a}{b}| \geq \frac{1}{z(a,b)}$$

and $\delta = \frac{c}{d}$ for some $c, d \in \mathbb{Z}^*$, we have

$$\exists \tilde{z} \in \mathbb{Z} \times \mathbb{Z}^* \rightarrow \mathbb{N} \ \forall a \in \mathbb{Z}, b \in \mathbb{Z}^* \ |\gamma\delta - \frac{a}{b}| \geq \frac{1}{\tilde{z}(a,b)},$$

$$\exists \overline{z} \in \mathbb{Z} \times \mathbb{Z}^* \rightarrow \mathbb{N} \ \forall a \in \mathbb{Z}, b \in \mathbb{Z}^* \ |\frac{\gamma}{\delta} - \frac{a}{b}| \geq \frac{1}{\overline{z}(a,b)}$$

and we may take $\tilde{z}(a,b) := |d|z(ad, bc)$ and $\overline{z}(a,b) := |c|z(ac, bd)$.

4.2   Example on Uniformly Continuous Bounded Functions

Lemma 6. *Let $f : \mathbb{R} \rightarrow \mathbb{R}$ be a uniformly continuous function with a modulus of continuity $\omega : \mathbb{N} \rightarrow \mathbb{N}$ i.e.*

$$\forall k \in \mathbb{N} \ \forall x, y \in \mathbb{R} \ (|x - y| < 2^{-\omega(k)} \rightarrow |f(x) - f(y)| < 2^{-k})$$

and let $\forall x \in \mathbb{R} \ (0 < f(x) < 2^B)$ for some $B \in \mathbb{N}$. Let $g : \mathbb{R} \rightarrow \mathbb{R}$ defined as $g(x) := (f(x))^2$. Then g is uniformly continuous and in particular as a modulus of continuity for g we can take $\tilde{\omega}(k) := \omega(k + 1 + B)$ i.e. :

$$\forall k \in \mathbb{N} \ \forall x, y \in \mathbb{R} \ (|x - y| < 2^{-\omega(k+1+B)} \rightarrow |g(x) - g(y)| < 2^{-k}).$$

4.3   Rates of Convergence

Lemma 7. *Let $f_1, f_2 : \mathbb{R} \rightarrow \mathbb{R}$ converging to zero with rates of convergence Φ, Ψ respectively, i.e.:*

$$\forall k \in \mathbb{N} \ \exists n \leq \Phi(k) \ \forall m \geq n \ |f_1(m)| \leq 2^{-k}$$

and

$$\forall k \in \mathbb{N} \ \exists n \leq \Psi(k) \ \forall m \geq n \ |f_2(m)| \leq 2^{-k}.$$

Then, the function $g : \mathbb{R} \rightarrow \mathbb{R}$ defined as $g := f_1 + f_2$, converges to zero and in particular as a rate of convergence for g we can take Θ such as

$$\forall k \in \mathbb{N} \ \exists n \leq \Theta(k + 1) \ \forall m \geq n \ |g(m)| \leq 2^{-k}$$

where, for given $k \in \mathbb{N}$, $\Theta(k) := \max\{\Phi(k), \Psi(k)\}$.

4.4   Asymptotic Regularity

See [14] for a discussion on the general form.

Lemma 8. *Let $T : \mathbb{R} \to \mathbb{R}$, $S : \mathbb{R} \to \mathbb{R}$ with $S^n x := \frac{1}{4} T^n x$ and $t_n := T^n x$, $s_n := S^n x$. Then if T is asymptotically regular (i.e. if $|t_n - Tt_n| \to 0$) and in a nonincreasing way, S is asymptotically regular. In particular, by*

$$\forall k \in \mathbb{N} \; \exists n \le \Phi(k) \; (|t_n - Tt_n| < 2^{-k})$$

it follows that

$$\forall k \in \mathbb{N} \; \exists n \le \tilde{\Phi}(k) \; (|s_n - Ss_n| < 2^{-k})$$

where for the rate of asymptotic regularity we may take $\tilde{\Phi}(k) := \Phi(k - 2)$.

Ackowledgements. The author was supported by the ERC Advanced Grant ALEXANDRIA (Project 742178) led by Professor Lawrence C. Paulson FRS. I thank Wenda Li, Yiannos Stathopoulos and Lawrence Paulson for their very useful comments on a previous draft of this paper and Tobias Nipkow for informing me of reference [5].

References

1. Avigad, J.: The computational content of classical arithmetic. In: Feferman, S., Sieg, W. (eds.) Proofs, Categories, and Computations: Essays in Honor of Grigori Mints, pp. 15–30. College Publications (2010)
2. Avigad, J.: The mechanization of Mathematics **65**(6) (2018). https://www.ams.org/journals/notices/201806/rnoti-p681.pdf
3. Bansal, K., Loos, S., Rabe, M., Szegedy, C., Wilcox, S.: HOList: an environment for machine learning of higher order logic theorem proving. In: Proceedings of the 36th International Conference on Machine Learning, vol. 97, pp. 454–463. PMLR (2019)
4. van Benthem Jutting, L.S.: Checking Landau's "Grundlagen" in the Automath system. PhD thesis, Eindhoven University of Technology, 1977. Published as Mathematical Centre Tracts nr. 83 (1979)
5. Berghofer, S.: Proofs, Programs and Executable Specifications in Higher Order Logic. PhD thesis, Technische Universität München, Institut für Informatik (2003)
6. Bishop, E.: Schizophrenia in contemporary mathematics. Am. Math. Soc. (1973)
7. Boyer, R., et al.: The QED manifesto. In: Bundy, A. (ed.) Automated Deduction - CADE 12, LNAI, vol. 814, pp. 238–251. Springer-Verlag (1994)
8. de Bruijn, N.G.: AUTOMATH, a language for mathematics. Technical Report 68-WSK-05, T.H.-Reports, Eindhoven University of Technology (1968)
9. Constable, R., Murthy, C.: Finding computational content in classical proofs. In: Huet, G., Plotkin, G. (eds.) Logical Frameworks, pp. 341–362. Cambridge University Press, Cambridge (1991)
10. Gowers, W.T.: Rough structure and classification. In: Alon, N., Bourgain, J., Connes, A., Gromov, M., Milman, V. (eds.) Visions in Mathematics. Modern Birkhäuser Classics. Birkhäuser Basel (2010)
11. Kohlenbach, U.: Effective moduli from ineffective uniqueness proofs. An unwinding of de La Vallée Poussin's proof for Chebycheff approximation. Ann. Pure Appl. Logic **64**, 27–94 (1993). https://doi.org/10.1016/0168-0072(93)90213-W

12. Kohlenbach, U.: Applied Proof Theory: Proof Interpretations and their use in Mathematics. Springer Monographs in Mathematics. Springer-Verlag, Berlin Heidelberg (2008)

13. Kohlenbach, U.: Recent progress in proof mining in nonlinear analysis. IFCoLog J. Logics Appl. **10**(4), 3357–3406 (2017)

14. Kohlenbach, U.: Proof-theoretic methods in nonlinear analysis. In: Sirakov, B., Ney de Souza, P., Viana, M. (eds.) Proceedings of the International Congress of Mathematicians -2018, Rio de Janeiro, vol. 2, pp. 61–82. World Scientific (2019)

15. Kohlenbach, U., Koutsoukou-Argyraki, A.: Rates of convergence and metastability for abstract Cauchy problems generated by accretive operators. J. Math. Anal. Appl. **423**(2), 1089–1112 (2015)

16. Kohlenbach, U., Koutsoukou-Argyraki, A.: Effective asymptotic regularity for one-parameter nonexpansive semigroups. J. Math. Anal. Appl. **433**(2), 1883–1903 (2016)

17. Koutsoukou-Argyraki, A.: Formalising mathematics-in praxis: a mathematician's first experiences with Isabelle/HOL and the why and how of getting started. Jahresbericht der Deutschen Mathematiker-Vereinigung **123**, 3–26 (2021)

18. Koutsoukou-Argyraki, A.: New effective bounds for the approximate common fixed points and asymptotic regularity of nonexpansive semigroups. J. Log. Anal. **10**(7), 1–30 (2018)

19. Koutsoukou-Argyraki, A.: Effective rates of convergence for the resolvents of accretive operators. Numer. Funct. Anal. Optim. **38**(12), 1601–1613 (2017)

20. Koutsoukou-Argyraki, A.: Proof Mining for Nonlinear Operator Theory: Four Case Studies on Accretive Operators, the Cauchy Problem and Nonexpansive Semigroups. PhD thesis, TU Darmstadt (2017). URN:urn:nbn:de:tuda-tuprints-61015

21. Koutsoukou-Argyraki, A.: Proof mining mathematics, formalizing mathematics. In: Proceedings of the North American Annual Meeting of the Association for Symbolic Logic, University of Western Illinois, Macomb, Illinois, USA, 16–19 May 2018, the Bulletin of Symbolic Logic, vol. 24, no. 4 (2018)

22. Kreisel, G.: On the interpretation of non-finitist proofs. I. J. Symb. Log. **16**, 241–267 (1951)

23. Kreisel, G.: On the interpretation of non-finitist proofs. II. Interpretation of number theory. Applications. J. Symb. Log. **17**, 43–58 (1952)

24. Kreisel, G.: Interpretation of analysis by means of constructive functionals of finite types. In: Heyting, A. (ed.) Constructivity in Mathematics: Proceedings of the Colloquium held at Amsterdam, 1957. Studies in Logic and the Foundations of Mathematics, pp. 101–128. North-Holland Publishing Co., Amsterdam (1959)

25. Li, W., Yu, L., Wu, Y., Paulson, L.C.: IsarStep: a benchmark for high-level mathematical reasoning. In: International Conference on Learning Representations (2021). https://openreview.net/forum?id=Pzj6fzU6wkj

26. Polu, S., Sutskever, I.: Generative Language Modeling for Automated Theorem Proving. arXiv:2009.03393v1 (2020)

27. Schwichtenberg, H., Wainer, S.S.: Proofs and Computations, Perspectives in Logic, Association for Symbolic Logic and Cambridge University Press (2012)

28. Steinberg, F., Théry, L., Thies, H.: Computable analysis and notions of continuity in Coq. Log. Methods Comput. Sci. **17**(2), 16:1–16:43 (2021)

In Search of the First-Order Part of Ramsey's Theorem for Pairs

Leszek Aleksander Kołodziejczyk[1] and Keita Yokoyama[2(⊠)]

[1] Institute of Mathematics, University of Warsaw, Warsaw, Poland
lak@mimuw.edu.pl
[2] School of Information Science, Japan Advanced Institute of Science and Technology, Nomi, Japan
y-keita@jaist.ac.jp

Abstract. In reverse mathematics, determining the first-order consequences of Ramsey's theorem for pairs and two colors is a long-standing open problem. In this paper, we give an overview of some recent developments related to this problem.

Keywords: Reverse mathematics · Ramey's theorem · First-order strength

1 Introduction

Given $n, k \geq 1$, Ramsey's theorem for n-tuples and k colours can be formalized in the language of second-order arithmetic as the following Π_2^1 statement:

RT_k^n: for any $c : [\mathbb{N}]^n \to k$, there exists an infinite set $H \subseteq \mathbb{N}$ such that c is constant on $[H]^n$,

where $[X]^n = \{F \subseteq X : |F| = n\}$. We also let RT^n be $\forall k\, \mathrm{RT}_k^n$. Ramsey's theorem and its variants have been a major topic of interest in the research programme known as reverse mathematics, which tries to determine the strength of mathematical theorems by comparing them with some prominent fragments of second-order arithmetic axiomatized by set existence principles. (For background on second-order arithmetic and reverse mathematics, refer to [25] or [10].)

It is not difficult to check that the implications $\mathrm{RT}_k^n \to \mathrm{RT}_{k+1}^n$ and $\mathrm{RT}_2^{n+1} \to \mathrm{RT}^n$ are provable in RCA_0 (be aware that the former does not imply $\mathrm{RT}_2^n \to \mathrm{RT}^n$

This paper surveys some material that forms a common background for the first author's special session talk 'Reverse mathematics of combinatorial principles over a weak base theory' and the second author's plenary talk 'Reverse mathematics and proof and model theory of arithmetic'. The authors thank Marta Fiori-Carones, Katarzyna Kowalik, Ludovic Patey, and Tin Lok Wong for helpful discussions and comments related to various results described here. The first author was partially supported by grant no. 2017/27/B/ST1/01951 of the National Science Centre, Poland. The second author was partially supported by JSPS KAKENHI grant no. 19K03601.

© Springer Nature Switzerland AG 2021
L. De Mol et al. (Eds.): CiE 2021, LNCS 12813, pp. 297–307, 2021.
https://doi.org/10.1007/978-3-030-80049-9_27

in the absence of sufficiently strong induction). So, we obtain a hierarchy of principles of increasing logical strength: $\mathrm{RT}^1_2 \leq \mathrm{RT}^1 \leq \mathrm{RT}^2_2 \leq \mathrm{RT}^2 \leq \mathrm{RT}^3_2 \leq \dots$. By formalizing work of Jockusch [13], Simpson [25] showed that RT^n_2 is in fact equivalent to ACA_0 over RCA_0 for each $n \geq 3$. On the other hand, Hirst [12] showed that RT^1 is equivalent to $\mathrm{B}\Sigma^0_2$ over RCA_0, hence it is essentially a first-order statement. Determining the strength of Ramsey's theorem for pairs, that is, of RT^2_2 and RT^2, is a much more complicated matter.

Various computability-theoretic arguments have clarified the relationship of Ramsey's theorem for pairs to the usual set existence principles appearing in reverse mathematics. By the counterexample to Ramsey's theorem in computable mathematics provided by Specker [27], RT^2_2 is not provable in RCA_0, and by Jockusch [13] it is also unprovable in WKL_0, even in the presence of the full mathematical induction scheme. In [24], Seetapun showed by means of a cone avoidance theorem that $\mathsf{RCA}_0 + \mathrm{RT}^2$ does not imply ACA_0. Finally, the unprovability of WKL_0 in $\mathsf{RCA}_0 + \mathrm{RT}^2$ was shown by Liu [21].

What about consequences of Ramsey's theorem for pairs expressible in the language of first-order arithmetic? Recall that Peano Arithmetic is the union of a chain of weaker theories $\mathrm{B}\Sigma_1 < \mathrm{I}\Sigma_1 < \mathrm{B}\Sigma_2 < \mathrm{I}\Sigma_2 < \dots$, where $\mathrm{I}\Sigma_n$ is axiomatized by induction for Σ_n formulas, and $\mathrm{B}\Sigma_n$ is axiomatized by the collection (or bounding) principle for Σ_n formulas. RCA_0 is a conservative extension of $\mathrm{I}\Sigma_1$, and ACA_0 is a conservative extension of full PA, so the first-order consequences of RT^2_2 and RT^2 must lie somewhere in between. Where exactly? A very closely related problem concerns Π^1_1 consequences, which are intuitively "first-order consequences with a set parameter". If we write Σ^0_n for the class of formulas defined like Σ_n but allowing set parameters, and we let $\mathrm{I}\Sigma^0_n$ stand for the universal closure of the induction scheme for Σ^0_n formulas, where in the interval between $\mathrm{I}\Sigma^0_1$ and $\bigcup_n \mathrm{I}\Sigma^0_n$ do the Π^1_1 consequences of $\mathrm{RT}^2_{(2)}$ lie?

The first significant result about the first-order consequences of Ramsey's theorem for pairs was obtained by Hirst [12], who showed the following.

Theorem 1 ([12]). $\mathsf{RCA}_0 + \mathrm{RT}^2_2$ *implies* $\mathrm{B}\Sigma^0_2$, *and* $\mathsf{RCA}_0 + \mathrm{RT}^2$ *implies* $\mathrm{B}\Sigma^0_3$.

This paper provides an overview of recent work aimed at characterizing the first-order (and Π^1_1-) strength of Ramsey's theorem for pairs. Theorem 1 will be our starting point. In other words, the main question is: *"Does* $\mathrm{RT}^2_{(2)}$ *imply sentences stronger than the* Σ^0_2 *(resp.* Σ^0_3*) collection principle?"*

Notational convention. The symbol ω stands for the set of standard natural numbers, while $\mathbb{N}$ stands for the set of natural numbers as formalized in arithmetic theories (which will have nonstandard models). To refer to the smallest infinite ordinal as formalized in arithmetic, we use the symbol $\upomega$.

2 Π^1_1-Conservation via $\upomega$-Extensions

In the study of first-order consequences of fragments of second-order arithmetic, results stating that one theory is Π^1_1-conservative over another play a major role. A standard approach to showing Π_1^1-conservation is based on the following simple model-theoretic criterion.

Proposition 1. *Let T, T' be Π_2^1-axiomatized theories such that $T \subseteq T'$. Then the following are equivalent.*

(1) T' is a Π_1^1-conservative extension of T.
(2) For every countable recursively saturated model $(M, \mathcal{S}) \models T$ and every set $A \in \mathcal{S}$, there exists $\mathcal{S}' \subseteq \mathcal{P}(M)$ such that $A \in \mathcal{S}'$ and $(M, \mathcal{S}') \models T'$.

In particular, (2) is satisfied if each countable $(M, \mathcal{S}) \models T$ can be extended to $(M, \mathcal{S}') \models T'$ with $\mathcal{S} \subseteq \mathcal{S}'$. In such a case, we say that $(M, \mathcal{S}')$ is an ω-*extension* of $(M, \mathcal{S})$, to emphasize that the first-order universes of the two structures coincide. (Usually one would write simply "ω-extension", but this conflicts with our convention that ω stands for the standard natural numbers.)

2.1   Extending Models via Second-Jump Control

In [3], Cholak, Jockusch, and Slaman proved a low$_2$-basis theorem for Ramsey's theorem for pairs: for any computable coloring $c : [\omega]^2 \to k$, there exists an infinite c-homogenous set $H \subseteq \omega$ such that $H'' \equiv_T \mathbf{0}''$. They gave two proofs of this result, and converted the one based on "second-jump control" (directly controlling the Σ_2 theory of H) to a forcing construction, thus obtaining the following ω-extension theorem for RT_2^2 and RT^2.

Theorem 2 (Cholak/Jockusch/Slaman [3]).

(i) For any countable model $(M, \mathcal{S}) \models \mathsf{RCA}_0 + \mathrm{I}\Sigma_2^0$, there exists $\widetilde{\mathcal{S}} \supseteq \mathcal{S}$ such that $(M, \widetilde{\mathcal{S}}) \models \mathsf{RCA}_0 + \mathrm{I}\Sigma_2^0 + \mathrm{RT}_2^2$.
(ii) For any countable model $(M, \mathcal{S}) \models \mathsf{RCA}_0 + \mathrm{I}\Sigma_3^0$, there exists $\widetilde{\mathcal{S}} \supseteq \mathcal{S}$ such that $(M, \widetilde{\mathcal{S}}) \models \mathsf{RCA}_0 + \mathrm{I}\Sigma_3^0 + \mathrm{RT}^2$.

Corollary 1 ([3]). $\mathsf{RCA}_0 + \mathrm{I}\Sigma_2^0 + \mathrm{RT}_2^2$ *is a Π_1^1-conservative extension of $\mathrm{I}\Sigma_2^0$, and $\mathsf{RCA}_0 + \mathrm{I}\Sigma_3^0 + \mathrm{RT}^2$ is a Π_1^1-conservative extension of $\mathrm{I}\Sigma_3^0$.*

Combining this with Theorem 1, we can conclude that the Π_1^1-part of RT_2^2 lies between $\mathrm{B}\Sigma_2^0$ and $\mathrm{I}\Sigma_2^0$, and that of RT^2 lies between $\mathrm{B}\Sigma_3^0$ and $\mathrm{I}\Sigma_3^0$. Corresponding results for the purely first-order parts follow.

In [26], it was shown that the second-jump control argument for the low$_2$-basis theorem for RT^2 can be formalized within $\mathrm{B}\Sigma_3^0$ by adapting a technique known as Shore's blocking argument. Note that given sets X, Y, the relation "X is Σ_n relative to Y" can be expressed using a Σ_n^0-universal formula, and "X is low$_n$ relative to Y" (i.e., $(X \oplus Y)^{(n)} \equiv_T Y^{(n)}$) can be formalized as "any $\Sigma_{n+1}^{X \oplus Y}$-set is Σ_{n+1}^Y". (For a set Z and a number m, a set is Σ_m^Z if it is Σ_m^0-definable with Z as the only set parameter.)

Theorem 3 (Slaman/Yokoyama [26]). $\mathsf{RCA}_0 + \mathrm{B}\Sigma_3^0$ *proves the following statement.*

For every $k \in \mathbb{N}$, every set X, and every X-computable coloring $c : [\mathbb{N}]^2 \to k$, there exists (an index of) a Σ_3^X-set H such that H is an infinite homogeneous set for c and $(X \oplus H)'' \equiv_T X''$.

Given a countable model $(M, \mathcal{S}) \models \mathsf{RCA}_0 + \mathsf{B}\Sigma_3^0$ and $X \in \mathcal{S}$, one may apply this theorem repeatedly, since adding a low_2 set to a model of $\mathsf{B}\Sigma_3^0$ preserves $\mathsf{B}\Sigma_3^0$. This produces a model of RT^2 in which the second–order part consists of low_2 sets relative to X. Thus, by Proposition 1, we have the following.

Corollary 2 ([26]). $\mathsf{RCA}_0 + \mathrm{RT}^2$ *is a Π_1^1-conservative extension of* $\mathsf{B}\Sigma_3^0$.

In other words, the Π_1^1-part of $\mathsf{RCA}_0 + \mathrm{RT}^2$ is exactly the same as $\mathsf{B}\Sigma_3^0$.

2.2   Extending Models via First-Jump Control

It seems likely that the ideas developed in order to prove the low_2-basis theorem of [3] will also be useful in the search for the Π_1^1-part of $\mathsf{RCA}_0 + \mathrm{RT}_2^2$. In that context, however, we have to deal with structures that might not satisfy $\mathsf{I}\Sigma_2^0$. For this reason, it makes more sense to focus on the "first-jump control" proof, which leads to the following result: for any computable coloring $c : [\omega]^2 \to 2$ and any set Z such that $0' \ll Z'$, there exists an infinite c-homogenous set $H \subseteq \omega$ such that $H' \ll Z'$. Here $X \ll Y$ stands for "Y has PA-degree relative to X", that is, Y computes a path in any X-computable infinite 0–1 tree.

Using a jump inversion argument due to Belanger [1], and an appropriate formalization of the $\ll$ relation in $\mathsf{RCA}_0 + \mathsf{B}\Sigma_2^0$, we get the following.

Proposition 2. *Let $\theta \equiv \forall X \exists Y \, \theta_0(X, Y)$ be a Π_2^1 sentence. Let $n \geq 1$. Assume that $\mathsf{RCA}_0 + \mathsf{B}\Sigma_{n+1}^0$ (resp. $\mathsf{RCA}_0 + \mathsf{I}\Sigma_{n+1}^0$) proves*

for any sets X, Z such that $X^{(n)} \ll Z^{(n)}$, there exists (an index of) a Σ_{n+1}^Z-set Y such that $\theta_0(X, Y)$ and $(X \oplus Y)^{(n)} \ll Z^{(n)}$.

then θ is Π_1^1-conservative over $\mathsf{RCA}_0 + \mathsf{B}\Sigma_{n+1}^0$ (resp. $\mathsf{RCA}_0 + \mathsf{I}\Sigma_{n+1}^0$).

Thus, the following question is highly relevant to the problem of characterizing the Π_1^1-part of $\mathsf{RCA}_0 + \mathrm{RT}_2^2$.

Question 1. Does $\mathsf{RCA}_0 + \mathsf{B}\Sigma_2^0$ prove the following?

(†)   For any sets X, Z such that $X' \ll Z'$ and any coloring $c : [\mathbb{N}]^2 \to 2$ satisfying $c \leq_T X$, there exists (an index of) a Σ_2^Z-set H such that H is an infinite homogeneous set for c and $(X \oplus H)' \ll Z'$.

Unfortunately, it is not clear whether $\mathsf{RCA}_0 + \mathsf{B}\Sigma_2^0$ is strong enough to formalize the first-jump control argument of [3] for RT_2^2. However, it is sufficient to formalize the argument in restricted cases, such as that of the ascending-descending sequence principle ADS, which states that every infinite linear order contains either an infinite ascending sequence or an infinite descending sequence. ADS is equivalent to the restriction of RT_2^2 to so-called transitive colourings [11].

Theorem 4 (Chong/Slaman/Yang [5]). $\mathsf{RCA}_0 + \mathsf{B}\Sigma_2^0$ *proves the following statement.*

For any sets X, Z such that $X' \ll Z'$ and any transitive coloring $c : [\mathbb{N}]^2 \to 2$ satisfying $X' \ll Z'$, there exists (an index of) a Σ_2^Z-set H such that H is an infinite homogeneous set for c and $(X \oplus H)' \ll Z'$.

Corollary 3 ([5]). $\mathsf{RCA}_0 + \mathsf{ADS}$ *is a Π_1^1-conservative extension of $\mathsf{B}\Sigma_2^0$.*

A similar conservation result was also proved in [5] for the somewhat stronger chain-antichain principle CAC, which says that any infinite partial order contains an infinite chain or an infinite antichain.

On the other hand, $\mathsf{I}\Sigma_2^0$ is enough to prove (†) (this is possibly a folklore result). To show this, one may combine Theorem 4 with a natural adaptation of the first-jump control construction for the so-called stable Erdös-Moser principle, which formalizes within $\mathsf{I}\Sigma_2^0$ without any extra tricks. This argument provides a low-like solution for RT_2^2 witnessing (†).

Theorem 5. $\mathsf{RCA}_0 + \mathsf{I}\Sigma_2^0$ *proves* (†).

We say more about (†) in Sect. 5.

An interesting variation of single-jump control was introduced by Chong, Slaman and Yang [6], who showed that in a very special model of $\mathsf{B}\Sigma_2 + \neg\mathsf{I}\Sigma_2$ it is possible to construct a low infinite homogeneous set for any computable $c : [\mathbb{N}]^2 \to 2$ that is *stable*, which means that for any given x the colour $c(x, y)$ is the same for all but finitely many y. This result has no counterpart over the standard model, because there exists a computable stable $c : [\omega]^2 \to 2$ with no low homogeneous set. In [7], the argument of [6] was combined with an additional technical construction to prove:

Theorem 6 ([7]). $\mathsf{RCA}_0 + \mathsf{RT}_2^2$ *does not prove* $\mathsf{I}\Sigma_2$.

It is possible to carry out the constructions behind Theorems 3, 4, 5 effectively enough to obtain theorems on lack of proof speedup corresponding to the conservation results. For theories T, T' and a class of formulas Γ, let us say that T *polynomially simulates* T' with respect to consequences from Γ if there is a polytime procedure which, given a proof of a formula from Γ in T', outputs a proof of that formula in T. In particular this means that any formula from Γ with a proof in T' also has a proof in T that is at most polynomially longer.

Theorem 7. *With respect to Π_1^1 consequences:*

(i) $\mathsf{RCA}_0 + \mathsf{B}\Sigma_3^0$ *polynomially simulates* $\mathsf{RCA}_0 + \mathsf{RT}^2$,
(ii) $\mathsf{RCA}_0 + \mathsf{I}\Sigma_2^0$ *polynomially simulates* $\mathsf{RCA}_0 + \mathsf{RT}_2^2$,
(iii) $\mathsf{RCA}_0 + \mathsf{B}\Sigma_2^0$ *polynomially simulates* $\mathsf{RCA}_0 + \mathsf{ADS}$.

The details will appear in a forthcoming paper.

3   Partial-Conservation and Indicator Arguments

In the study of first-order arithmetic, indicator arguments, developed by Paris, Kirby and many others (see e.g. [14, Chapter 14]), are a well-understood tool for constructing interesting initial segments of models. Among other things, indicators were used to prove the unprovability in Peano Arithmetic of combinatorial statements such as the Paris-Harrington principle and its variants, including one of the earliest examples by Paris [22]. In [2], Bovykin and Weiermann proved some results witnessing the usefulness of an indicator-based approach in the study of the first-order part of Ramsey's theorem for pairs.

Here, we adopt a definition of indicators in the spirit of [23]. We formulate the definition over the base theory RCA_0^*, which is weaker than RCA_0 in that the Σ_1^0-induction axiom of RCA_0 is replaced by Σ_0^0-induction and the axiom exp which states that the exponential function is total. The Π_1^1-part of RCA_0^* is $\mathsf{B}\Sigma_1^0 + \exp$.

Definition 1. *Let $(M, \mathcal{S}) \models \mathsf{RCA}_0^*$, and let T be a set of $\mathcal{L}_2$-sentences possibly with parameters from $(M, \mathcal{S})$. A Σ_0^0-definable function $Y : [\mathbb{N}]^{<\mathbb{N}} \to \mathbb{N}$ is said to be an* indicator *for T in $(M, \mathcal{S})$ if*

- *$Y(F') \leq Y(F) \leq \max F$ if $F \subseteq F'$,*
- *$Y(F)$ is nonstandard if and only if there exists a cut $I \subseteq_e M$ with $\min F \in I < \max F$ such that $F \cap I$ is unbounded in I and $(I, \mathrm{Cod}(M/I)) \models T$.*

Proposition 3. *Let T be a set of $\mathcal{L}_2$-sentences, and let a Σ_0-definable function $Y : [\mathbb{N}]^{<\mathbb{N}} \to \mathbb{N}$ be an indicator for T in each $(M, \mathcal{S}) \models \mathsf{RCA}_0^*$. Then the theory $\mathsf{WKL}_0^* + T$ is a Π_3^0-conservative extension of*

$$\mathsf{RCA}_0^* + \{\forall X \subseteq_{\inf} \mathbb{N} \, \exists F \subseteq_{\text{fin}} X \, (Y(F) \geq n) : n \in \omega\}.$$

Indicators are often provided by finite combinatorial statements related to the Paris-Harrington principle, as in [22], and they are closely connected to ordinal analysis. In an important early paper of Ketonen and Solovay [16], the unprovability of the Paris-Harrington principle in PA was tied to a quantitative analysis of finite Ramsey's theorem in which the "largeness" of finite subsets of $\mathbb{N}$ is measured in terms of ordinals. For instance, a set X is ω-large if $|X| > \min X$, and it is ω^{n+1}-large if $X \setminus \min X$ can be written as $X_1 \sqcup \ldots \sqcup X_k$, where $k \geq \min X$ and for each i, the set X_i is ω^n-large and $\max X_i < \min X_{i+1}$. It turns out that such amounts of largeness are enough to define an indicator for RT_2^2.

Theorem 8 (Patey/Yokoyama [23]). *The function Y defined by $Y(F) = \max(\{m : F \text{ is } \omega^m\text{-large}\})$ is an indicator for $\mathsf{RCA}_0 + \mathrm{RT}_2^2$.*

For each $n \in \omega$, RCA_0 proves $\forall X \subseteq_{\inf} \mathbb{N} \, \exists F \subseteq_{\text{fin}} X \, (F \text{ is } \omega^n\text{-large})$. This gives:

Corollary 4 ([23]). *$\mathsf{RCA}_0 + \mathrm{RT}_2^2$ is a Π_3^0-conservative extension of RCA_0.*

Corollary 4 gives a new proof that $\mathsf{RCA}_0 + \mathsf{RT}_2^2$ does not imply $\mathsf{I}\Sigma_2^0$ (cf. Theorem 6). Additionally, a nontrivial refinement of Theorem 8 provided in [19], together with a reformulation of the general conservation criterion from Proposition 3 as a formalized forcing construction, shows that in fact RCA_0 polynomially simulates $\mathsf{RCA}_0 + \mathsf{RT}_2^2$ with respect to Π_3^0 consequences [18].

Indicator arguments can be used to describe consequences of $\mathsf{RCA}_0 + \mathsf{RT}_2^2$ of higher first-order quantifier complexity. For example, the Π_5^0-part of $\mathsf{RCA}_0 + \mathsf{RT}_2^2$ can be described as follows. Within RCA_0, for a given double sequence of sets $\mathcal{X} = \langle X_{ij} \rangle_{i,j \in \mathbb{N}}$ such that $\forall i \exists j \, (X_{ij}$ is infinite$)$, a finite set F is said to be $(n, \mathcal{X})$-dense if

- $n = 0$ and $|F| > \min F$, or
- $n = m + 1$ and
    - for any $i < \min F$, there exists $H \subseteq F$ and $j < \min H$ such that $H \subseteq X_{ij}$ and H is $(m, \mathcal{X})$-dense,
    - for any $c : [F]^2 \to 2$, there exists $H \subseteq F$ such that H is c-homogeneous and H is $(m, \mathcal{X})$-dense.

Then, in any model $(M, \mathcal{S}) \models \mathsf{RCA}_0^*$ and any $\mathcal{X} = \langle X_{ij} \rangle_{i,j \in \mathbb{N}} \in \mathcal{S}$, the function $Y_{\mathcal{X}}(F) = \max(\{m : F$ is $(m, \mathcal{X})$-dense$\})$ is an indicator for the theory $\mathsf{RCA}_0 + \mathsf{RT}_2^2 + \forall i \exists j \, (X_{ij}$ is infinite$)$. Since any Π_4^0 statement is equivalent over RCA_0 to "$\forall i \exists j \, (X_{ij}$ is infinite$)$" for an appropriate family $\langle X_{ij} \rangle_{i,j \in \mathbb{N}}$, this indicator can preserve any Π_4^0 statements. From this, we obtain the following characterization.

Theorem 9. *For each $n \in \omega$, let γ_n be the sentence*

$$\forall \mathcal{X} = \langle X_{ij} \rangle_{i,j \in \mathbb{N}} \, (\forall i \exists j \, (X_{ij} \text{ is infinite}) \to \exists F \, (Y_{\mathcal{X}}(F) > n)).$$

Then $\mathsf{RCA}_0 + \mathsf{RT}_2^2$ is Π_5^0-conservative over $\mathsf{RCA}_0 + \{\gamma_n : n \in \omega\}$.

The Π_5^0-part of $\mathsf{RCA}_0 + \mathsf{RT}_2^2$ seems particularly important, for reasons that will be explained in Sect. 5.

One other application of an indicator-style analysis is to show that the model construction of [7] also corresponds to a partial conservation result. Let BME stand for the *bounded enumeration scheme* of [7], a set of sentences that is now known to be equivalent to the well-orderedness of $\omega, \omega^\omega, \dots$ [20]. Building an appropriate initial segment by means of an indicator-style argument, one can show that any Σ_4^0 statement consistent with $\mathsf{RCA}_0 + \mathsf{B}\Sigma_2^0 + \mathsf{BME}$ can be satisfied in a model satisfying the special conditions described in [7, Proposition 2.5]. This leads to the following conservation theorem.

Theorem 10. $\mathsf{RCA}_0 + \mathsf{RT}_2^2 + \mathsf{BME}$ *is a Π_4^0-conservative extension of* $\mathsf{RCA}_0 + \mathsf{B}\Sigma_2^0 + \mathsf{BME}$.

4   The Strength of Ramsey's Theorem over RCA_0^*

We now turn to a discussion of the first-order strength of Ramsey's theorem for pairs over the weaker base theory RCA_0^*. A novel phenomenon in this setting is that within RCA_0^*, it can happen that an infinite subset of $\mathbb{N}$ has cardinality strictly smaller than $\mathbb{N}$. More precisely, it follows from $\mathsf{RCA}_0^* + \neg \mathsf{I}\Sigma_1^0$ that there exists a set $X \subseteq \mathbb{N}$ such that X is unbounded in $\mathbb{N}$, but for some number $k \in \mathbb{N}$ the set X does not contain a k-element finite subset.

It remains provable in RCA_0^* that RT^1 is equivalent to $\mathsf{B}\Sigma_2^0$, and RT^2 trivially implies RT^1. Thus, $\mathsf{RCA}_0^* + \mathsf{RT}^2$ is simply the same as $\mathsf{RCA}_0 + \mathsf{RT}^2$. On the other hand, RT_2^2 no longer implies RT^1 within RCA_0^*, and in fact it does not imply $\mathsf{I}\Sigma_1^0$ either. Actually, by slightly reformulating the indicator argument used in [28] to prove Π_2-conservation of $\mathsf{RCA}_0^* + \mathsf{RT}_2^2$ over $\mathsf{I}\Delta_0 + \exp$, we get the following.

Theorem 11. *Let* $Y(F) = \max(\{|F'| : F' \subseteq F \wedge \forall x, y \in F'(x < y \to 2^x < y)\})$. *Then the function* Y *is an indicator for* $\mathsf{RCA}_0^* + \mathsf{RT}_2^2$.

Corollary 5 ([17,28]). $\mathsf{RCA}_0^* + \mathsf{RT}_2^2$ *is a* Π_3^0-*conservative extension of* RCA_0^*.

The next question is whether RT_2^2 is Π_1^1-conservative over RCA_0^*. The answer is "no", and there are many interesting Π_1^1 and first-order consequences of RT_2^2 over RCA_0^*. We give two examples.

Theorem 12 ([17]). $\mathsf{RCA}_0^* + \mathsf{RT}_2^2$ *proves the following:*

(i) rec-RT_2^2: *for any* X *and* $c : [\mathbb{N}]^2 \to 2$, *if* $\mathsf{I}\Sigma_1^X$ *fails, then there exists (an index of) a* Δ_1^X-*set which is infinite and homogeneous for* c.
(ii) $\mathsf{C}\Sigma_2^0$: *for every set* X *and every* $k \in \mathbb{N}$, *there is no* Σ_2^X-*definable one-to-one function from* $\mathbb{N}$ *into* $\{0, \ldots, k\}$.

Let us discuss these two consequences of $\mathsf{RCA}_0^* + \mathsf{RT}_2^2$, neither of which is provable in RCA_0^* alone.

The first statement, rec-RT_2^2, can be regarded as a first-order version of RT_2^2. In the absence of $\mathsf{I}\Sigma_1^0$, computability-theoretic notions can behave in a strange way. For instance, one can use a coding lemma due to Chong and Mourad [4] to prove the following within RCA_0^*:

($\ddagger$) if $\mathsf{I}\Sigma_1^X$ fails, then there exists an infinite set $Z \subseteq \mathbb{N}$ such that every set $Y \subseteq Z$ is X-computable.

Given ($\ddagger$), it is not hard to show that RT_2^2 implies rec-RT_2^2: it suffices to check within $\mathsf{RCA}_0^* + \mathsf{RT}_2^2$ that any 2-colouring of pairs from an infinite subset of $\mathbb{N}$ has an infinite homogeneous set. On the other hand, it is also the case that rec-RT_2^2 implies RT_2^2 assuming the negation of $\mathsf{I}\Sigma_1^0$. In other words, RT_2^2 is equivalent to a Π_1^1 statement over $\mathsf{RCA}_0^* + \neg \mathsf{I}\Sigma_1^0$.

The second consequence, $\mathsf{C}\Sigma_2^0$, was also one of the earliest known Π_1^1 consequences of $\mathsf{RCA}_0 + \mathsf{RT}_2^2$: the fact that $\mathsf{C}\Sigma_2^0$ follows from $\mathsf{RCA}_0 + \mathsf{RT}_2^2$ was pointed out in [24], independently of the result of [12] on $\mathsf{B}\Sigma_2^0$. However, the proof of $\mathsf{C}\Sigma_2^0$ from RT_2^2 over RCA_0^* is very different from the one in [24]. One first observes that, by [12], RT_2^2 proves the implication $\mathsf{I}\Sigma_1^0 \to \mathsf{B}\Sigma_2^0$ (note that this statement is not Π_1^1, it is an implication between two Π_1^1 sentences). Then, one can use a model-theoretic argument to show that this implication actually implies $\mathsf{C}\Sigma_2^0$.

Interestingly, the consequences of the purely first-order implication $I\Sigma_n \to B\Sigma_{n+1}$ are studied by Kaye [15] with the aim of understanding the theory of cardinal-like models of arithmetic. The following question is still open.

Question 2. Does $RCA_0^* + RT_2^2$ prove $I\Sigma_1 \to B\Sigma_2$?

Note that both of rec-RT_2^2 and $C\Sigma_2^0$ are Π_4^0 statements, which means that Corollary 5 is tight in the following sense.

Corollary 6. $RCA_0^* + RT_2^2$ *is not* Π_4^0-*conservative over* RCA_0^*.

It may be worth pointing out that the versions of rec-RT_2^2 and $C\Sigma_2^0$ obtained by replacing universal quantification over sets by quantification over (indices of) computable sets remain unprovable in RCA_0^*. Thus, Corollary 6 also holds for purely first-order ("lightface") Π_4 sentences.

Corollary 6 should not be viewed as a clear indication that RT_2^2 is unlikely to be Π_1^1-conservative over $B\Sigma_2^0$. Indeed, it was shown in [8] that various weakenings of RT_2^2 known to be Π_1^1-conservative over $B\Sigma_2^0$, such as ADS and CAC, or even over $I\Sigma_1^0$, as the cohesive Ramsey's theorem CRT_2^2, also fail to be Π_1^1-conservative over RCA_0^*.

5   First-Jump Control and an Isomorphism Argument

To conclude our discussion, we return to Question 1 in light of some new model-theoretic results inspired by the work on RCA_0^*. A recent argument [9] leads to an isomorphism theorem for countable models of WKL_0^* sharing the same first-order universe and a common witness to $\neg I\Sigma_1^0$. That theorem has some consequences concerning models of $B\Sigma_n^0 + \neg I\Sigma_n^0$ for higher n, including the following.

Theorem 13. *Let* $(M, \mathcal{S}) \models RCA_0 + B\Sigma_2^0 + \neg I\Sigma_2$ *be countable, and let* $A \in \mathcal{S}$ *be such that* $(M, \mathcal{S}) \models \mathbf{0}' \ll A'$. *Then there exists* $\widetilde{\mathcal{S}} \subseteq \mathcal{P}(M)$ *such that*

(i) for every $X \in \widetilde{\mathcal{S}}$, *the set* X *is* Σ_2^A-*definable and* $X' \ll A'$ *in* $(M, \mathcal{S})$,
(ii) $(M, \widetilde{\mathcal{S}})$ *is isomorphic to* $(M, \mathcal{S})$.

Note that the isomorphism in (ii) does not fix M pointwise. We can infer from Theorem 13 that (†) in Question 1 holds in every model of $RCA_0 + RT_2^2 + \neg I\Sigma_2$. Relativizing the argument and combining it with Theorem 5, we can obtain:

Theorem 14. $RCA_0 + RT_2^2$ *proves* (†).

Combining this with Proposition 2, we get the following.

Corollary 7. $RCA_0 + RT_2^2$ *is a* Π_1^1-*conservative extension of* $RCA_0 + B\Sigma_2^0$ *if and only if* $RCA_0 + B\Sigma_2^0$ *proves* (†).

Loosely speaking, Corollary 7 says that if the Π_1^1-part of $RCA_0 + RT_2^2$ is in fact $B\Sigma_2^0$, then over $\neg I\Sigma_2^0$ this *has to* be proved by a single-jump control argument.

A careful analysis shows that (†) can be formulated as a Π_5^0 statement. Hence $RCA_0 + RT_2^2$ is Π_1^1-conservative over of $RCA_0 + B\Sigma_2^0$ if it is Π_5^0-conservative over that theory. In other words, to answer the original question on Π_1^1-conservation, we only need to check conservation up to the level of Π_5^0.

References

1. Belanger, D.A.: Conservation theorems for the cohesiveness principle (2015). Preprint
2. Bovykin, A., Weiermann, A.: The strength of infinitary Ramseyan principles can be accessed by their densities. Ann. Pure Appl. Logic **168**(9), 1700–1709 (2017)
3. Cholak, P.A., Jockusch, C.G., Slaman, T.A.: On the strength of Ramsey's theorem for pairs. J. Symb. Log **66**(1), 1–15 (2001)
4. Chong, C.T., Mourad, K.J.: The degree of a Σ_n cut. Ann. Pure Appl. Logic **48**(3), 227–235 (1990)
5. Chong, C.T., Slaman, T.A., Yang, Y.: Π_1^1-conservation of combinatorial principles weaker than Ramsey's theorem for pairs. Adv. Math. **230**(3), 1060–1077 (2012)
6. Chong, C.T., Slaman, T.A., Yang, Y.: The metamathematics of stable Ramsey's theorem for pairs. J. Amer. Math. Soc. **27**(3), 863–892 (2014)
7. Chong, C.T., Slaman, T.A., Yang, Y.: The inductive strength of Ramsey's theorem for pairs. Adv. Math. **308**, 121–141 (2017)
8. Fiori-Carones, M., Kołodziejczyk, L.A., Kowalik, K.W.: Weaker cousins of Ramsey's theorem over a weak base theory (2021). Preprint arxiv.org/abs/2105.11190
9. Fiori-Carones, M., Kołodziejczyk, L.A., Wong, T.L., Yokoyama, K.: An isomorphism theorem for models of Weak König's Lemma without primitive recursion. (in preparation)
10. Hirschfeldt, D.R.: Slicing the Truth. World Scientific Publishing Co. (2015)
11. Hirschfeldt, D.R., Shore, R.A.: Combinatorial principles weaker than Ramsey's theorem for pairs. J. Symb. Log. **72**(1), 171–206 (2007)
12. Hirst, J.L.: Combinatorics in subsystems of second order arithmetic. Ph.D. thesis. The Pennsylvania State University, August 1987
13. Jockusch, C.G.: Ramsey's theorem and recursion theory. J. Symb. Log. **37**(2), 268–280 (1972)
14. Kaye, R.: Models of Peano Arithmetic. Oxford University Press, Oxford (1991)
15. Kaye, R.: Constructing κ-like models of arithmetic. J. London Math. Soc. (2) **55**(1), 1–10 (1997)
16. Ketonen, J., Solovay, R.: Rapidly growing Ramsey functions. Ann. of Math. **113**(2), 267–314 (1981)
17. Kołodziejczyk, L.A., Kowalik, K.W., Yokoyama, K.: How strong is Ramsey's theorem if infinity can be weak? Preprint arxiv.org/abs/2011.02550
18. Kołodziejczyk, L.A., Wong, T.L., Yokoyama, K.: Ramsey's theorem for pairs, collection, and proof size. Preprint arxiv.org/abs/2005.06854
19. Kołodziejczyk, L.A., Yokoyama, K.: Some upper bounds on ordinal-valued Ramsey numbers for colourings of pairs. Selecta Mathematica **26**(4), 1–18 (2020). https://doi.org/10.1007/s00029-020-00577-3
20. Kreuzer, A.P., Yokoyama, K.: On principles between Σ_1- and Σ_2-induction, and monotone enumerations. J. Math. Log. **16**(1), 1650004, 21 pages (2016)
21. Liu, J.: RT_2^2 does not imply WKL_0. J. Symbolic Logic **77**(2), 609–620 (2012)
22. Paris, J.B.: Some independence results for Peano Arithmetic. J. Symb. Log. **43**(4), 725–731 (1978)
23. Patey, L., Yokoyama, K.: The proof-theoretic strength of Ramsey's theorem for pairs and two colors. Adv. Math. **330**, 1034–1070 (2018)
24. Seetapun, D., Slaman, T.A.: On the strength of Ramsey's theorem. Notre Dame J. Form. Log. **36**(4), 570–582 (1995)
25. Simpson, S.G.: Subsystems of Second Order Arithmetic. Springer-Verlag (1999)

26. Slaman, T.A., Yokoyama, K.: The strength of Ramsey's theorem for pairs and arbitrarily many colors. J. Symb. Log. **83**(4), 1610–1617 (2018)
27. Specker, E.: Ramsey's theorem does not hold in recursive set theory. In: Logic Colloquium 1969, pp. 439–442. North-Holland, Amsterdam (1971)
28. Yokoyama, K.: On the strength of Ramsey's theorem without Σ_1-induction. Math. Log. Q. **59**(1–2), 108–111 (2013)

On Subrecursive Representation of Irrational Numbers: Contractors and Baire Sequences

Lars Kristiansen[1,2]($\boxtimes$)

[1] Department of Mathematics, University of Oslo, Oslo, Norway
`larsk@math.uio.no`
[2] Department of Informatics, University of Oslo, Oslo, Norway

Abstract. We study the computational complexity of three representations of irrational numbers: standard Baire sequences, dual Baire sequences and contractors. Our main results: Irrationals whose standard Baire sequences are of low computational complexity might have dual Baire sequences of arbitrarily high computational complexity, and vice versa, irrationals whose dual Baire sequences are of low complexity might have standard Baire sequences of arbitrarily high complexity. Furthermore, for any subrecursive class $\mathcal{S}$ closed under primitive recursive operations, the class of irrationals that have a contractor in $\mathcal{S}$ is exactly the class of irrationals that have both a standard and a dual Baire sequence in $\mathcal{S}$. Our results implies that a subrecursive class closed under primitive recursive operations contains the continued fraction of an irrational number α if and only if there is a contractor for α in the class.

Keywords: Computable analysis · Representation of irrationals · Subrecursion · Computational complexity · Baire sequences · Contraction maps

1 Introduction

The theorems proved below complement the picture drawn in Kristiansen [4,5] and, particularly, Georgiev et al. [1]. Our investigations are motivated by the question: *Do we need, or do we not need, unbounded search in order to convert one representation of an irrational number into another representation?* A computation that does not apply unbounded search is called a *subrecursive* computation. Primitive recursive computations and (Kalmar) elementary computations are typical examples of subrecursive computations. A representation R_1 (of irrational numbers) is *subrecursive in* a representation R_2 if the R_1-representation of α can be subrecursively computed in the R_2-representation of α.

The reader that wants to know more about our motivations, or want further explanations, should consult the first few sections of Georgiev et al. [1]. This is a

© Springer Nature Switzerland AG 2021
L. De Mol et al. (Eds.): CiE 2021, LNCS 12813, pp. 308–317, 2021.
https://doi.org/10.1007/978-3-030-80049-9_28

technical paper where our main concern is to give reasonably full proofs of some new theorems.[1]

What we will call a *Baire sequence* is an infinite sequence of natural numbers. Such a sequence $a_0, a_1, a_2, \ldots$ represents an irrational number α in the interval $(0, 1)$. We split the interval $(0, 1)$ into infinitely many open subintervals with rational endpoints. We may, e.g., use the splitting

$$(\; 0/1 \; , \; 1/2 \;) \; (\; 1/2 \; , \; 2/3 \;) \; (\; 2/3 \; , \; 3/4 \;) \; \ldots \; (\; n/(n+1) \; , \; (n+1)/(n+2) \;) \ldots \; .$$

The first number of the sequence a_0 tells us in which of these intervals we find α. Thus if $a_0 = 17$, we find α in the interval $(17/18, 18/19)$. Then we split the interval $(17/18, 18/19)$ in a similar way. The second number of the sequence a_1 tells us in which of these intervals we find α, and thus we proceed.

In general, in order to split the interval (q, r), we need a strictly increasing sequence of rationals $s_0, s_1, s_2 \ldots$ such that $s_0 = q$ and $\lim_i s_i = r$. We will use the splitting $s_i = (a + ic)/(b + id)$ where a, b are (the unique) relatively prime natural numbers such that $q = a/b$ and c, d are (the unique) relatively prime natural numbers such that $r = c/d$ (let $0 = 0/1$ and $1 = 1/1$). This particular splitting makes our proof smooth and transparent, but our main results are invariant over all natural splittings.

We will say that the Baire sequences explained above are *standard*. The standard Baire sequence of the irrational number α will lexicographically precede standard Baire sequence of the irrational number β iff $\alpha < \beta$. We will also work with what we will call *dual* Baire sequences. The dual sequence of α will lexicographically precede the dual sequence of β iff $\alpha > \beta$. We get the dual sequences by using decreasing sequences of rationals to split intervals, e.g., the interval $(0, 1)$ may be split into the intervals

$$(\; 1/1 \; , \; 1/2 \;) \; (\; 1/2 \; , \; 1/3 \;) \; (\; 1/3 \; , \; 1/4 \;) \; \ldots \; (\; 1/n \; , \; 1/(n+1) \;) \; \ldots \; .$$

Definition 1. *Let $f : \mathbb{N} \to \mathbb{N}$ be any function, and let $n \in \mathbb{N}$. We define the interval I_f^n by $I_f^0 = (0/1, 1/1)$ and*

$$I_f^{n+1} = \left(\frac{a + f(n)c}{b + f(n)d} \; , \; \frac{a + f(n)c + c}{b + f(n)d + d} \right)$$

if $I_f^n = (a/b, c/d)$. We define the interval J_f^n by $J_f^0 = (0/1, 1/1)$ and

$$J_f^{n+1} = \left(\frac{a + f(n)a + c}{b + f(n)b + d} \; , \; \frac{f(n)a + c}{f(n)b + d} \right)$$

if $J_f^n = (a/b, c/d)$. The function $B : \mathbb{N} \to \mathbb{N}$ is the standard Baire representation *of the irrational number $\alpha \in (0, 1)$ if we have $\alpha \in I_B^n$ for every n. The function $A : \mathbb{N} \to \mathbb{N}$ is the* dual Baire representation *of the irrational number $\alpha \in (0, 1)$ if we have $\alpha \in J_A^n$ for every n.*

---

[1] The author wants to thank Dag Normann for enlightening discussions which lead up to this paper. The author wants to thank Eyvind Briseid for helpful advice and for pinpointing weaknesses in an early version of this paper.

Before we discuss contractors, we will recall the *trace functions* introduced in Kristiansen [4]. A trace function for α is a function that move any rational number closer to α. The formal definition, which follows, is straightforward.

Definition 2. *A function* $T : [0,1] \cap \mathbb{Q} \to (0,1) \cap \mathbb{Q}$ *is a* trace function *for the irrational number* α *if we have* $|\alpha - q| > |\alpha - T(q)|$ *for any rational* q.

We will say that a trace function T moves q to the right (left) if $q < T(q)$ $(T(q) < q)$. The easiest way to realize that a trace function indeed defines a unique real number, is probably to observe that a trace function T for α yields the Dedekind cut of α: if T moves q the right, then we know that q lies below α; if T moves q the left, then we know that q lies above α. Obviously, T cannot yield the Dedekind cut for any other number than α. It is proved in [4] that trace functions are subrecursively equivalent to continued fractions.

Intuitively, a *contractor* is a function that moves two (rational) numbers closer to each other. It turns out that also contractors can be used to represent irrational numbers.

Definition 3. *A function* $F : [0,1] \cap \mathbb{Q} \to (0,1) \cap \mathbb{Q}$ *is a* contractor *if we have*

$$F(q) \neq q \quad and \quad |F(q_1) - F(q_2)| < |q_1 - q_2|$$

for any rationals q, q_1, q_2 *where* $q_1 \neq q_2$.

Theorem 4. *Any contractor is a trace function for some irrational number.*

Proof. Let F be a contractor. If F moves q to the right (left), then F also move any rational less (greater) than q to the right (left); otherwise F would not be a contractor. We define two sequences $q_0, q_1, q_2 \ldots$ and $p_0, p_1, p_2 \ldots$ of rationals. Let $q_0 = 0$ and $p_0 = 1$. Let $q_{i+1} = (q_i + p_i)/2$ if F moves $(q_i + p_i)/2$ to the right; otherwise, let $q_{i+1} = q_i$. Let $p_{i+1} = (q_i + p_i)/2$ if F moves $(q_i + p_i)/2$ to the left; otherwise, let $p_{i+1} = p_i$ (Definition 3 requires that a contractor moves any rational number). Obviously, we have $\lim_i q_i = \lim_i p_i$, and obviously, this limit is an irrational number α. It is easy to see that F is a trace function for α. $\square$

Definition 5. *A contractor* F *is a* contractor for *the irrational number* α *if* F *is a trace function for* α *(Theorem 4 shows that this definition makes sense).*

Contractors, also known as contraction maps, come in a number of variants. The variant given by Definition 3 is tailored for our purposes. Computational aspects of contractors have also been studied in proof mining, see Kohlenbach and Olivia [3] and Gerhardy and Kohlebach [2].

2   Technical Preliminaries

Definition 6. *For any string* $\tau \in \{L, R\}^*$, *we define the* interval addressed by τ *inductively over the structure of* τ: *The empty sequence addresses the interval* $(0/1, 1/1)$. *Furthermore*

$$\tau L \text{ addresses } \left(\frac{a}{b}, \frac{a+c}{b+d} \right) \quad and \quad \tau R \text{ addresses } \left(\frac{a+c}{b+d}, \frac{c}{d} \right)$$

if τ *addresses* $(a/b, c/d)$. *We will use* $I[\tau]$ *to denote the interval addressed by* τ.

Definition 7. *Let α be an irrational number in the interval $(0, 1)$. Let a and b be relatively prime natural numbers with $b > 0$. The fraction a/b is a left best approximant of α if we have $c/d \leq a/b < \alpha$ or $\alpha < c/d$ for any natural numbers c, d with $0 < d \leq b$. The fraction a/b is a right best approximant of α if we have $\alpha < a/b \leq c/d$ or $c/d < \alpha$ for any natural numbers c, d with $0 < d \leq b$.*

Lemma 8. *Assume that the interval $(a/b, c/d)$ is addressed by some $\tau \in \{L, R\}^*$. Then, (i) a/b and c/d are, respectively, left and right best approximants of any irrational number in the interval $(a/b, c/d)$, and (ii) we have $c/d - a/b = 1/(db)$.*

Proof. If an interval $(a/b, c/d)$ is addressed by some $\tau \in \{L, R\}^*$, then a/b and c/d will be a Farey pair, that is, neighbors in the Farey series of order $\max(b, d)$. It is well know that the *mediant* of the pair, that is, $(a + c)/(b + d)$ will be in its lowest terms and lie in the interval, moreover, for any other vulgar fraction m/n that lie in the interval, we have $n > b + d$, see Richards [8]. It follows that (i) holds. Moreover, it well know that we have $cb - ad = 1$, or equivalently $c/d - a/b = 1/(db)$, for any Farey pair $(a/b, c/d)$, and thus (ii) also holds. $\square$

The next lemma is the key to the proof of one of our main theorems.

Lemma 9. *(i) The string $R^{f(0)}LR^{f(1)}L \ldots R^{f(n)}L$ addresses the interval I_f^{n+1}. (ii) The string $L^{f(0)}RL^{f(1)}R \ldots L^{f(n)}R$ addresses the interval J_f^{n+1}.*

Proof. We prove (i). The proof of (ii) is symmetric.

Let $\tau = R^{f(0)}LR^{f(1)}L \ldots R^{f(n-1)}L$. Observe that we have $I[\tau] = (0/1, 1/1) = I_f^0$ when τ is the empty sequence.

Assume that $I[\tau] = I_f^n = (a/b, c/d)$. We need to prove that

$$I[\tau R^{f(n)}L] = I_f^{n+1} . \tag{1}$$

Let $k = f(n)$. We prove (1) by a secondary induction on k.

Assume $k = 0$. By Definition 6, we have

$$I[\tau R^{f(n)}L] = I[\tau R^0 L] = I[\tau L] = (a/b, (a + c)/(b + d)) .$$

By Definition 1, we have

$$I_f^{n+1} = ((a+kc)/(b+kd), (a+kc+c)/(b+kd+d)) = (a/b, (a+c)/(b+d)) .$$

Thus (1) holds when $f(n) = 0$. Now, assume by induction hypothesis that

$$I[\tau R^k L] = \left(\frac{a + kc}{b + kd}, \frac{a + kc + c}{b + kd + d} \right) . \tag{2}$$

Observe that the right hand side of (2) is the definition of I_f^{n+1} with k for $f(n)$. Now, by (2) and Definition 6, we have

$$I[\tau R^k] = \left(\frac{a + kc}{b + kd}, \frac{c}{d} \right) . \tag{3}$$

Furthermore, by (3) and Definition 6, we have

$$I[\tau R^{k+1}] = \left(\frac{a+kc+c}{b+kd+d}, \frac{c}{d} \right) = \left(\frac{a+(k+1)c}{b+(k+1)d}, \frac{c}{d} \right) \tag{4}$$

and by (4) and Definition 6, we have

$$I[\tau R^{k+1}L] = \left(\frac{a+(k+1)c}{b+(k+1)d}, \frac{a+(k+1)c+c}{b+(k+1)d+d} \right). \tag{5}$$

Observe that the right hand side of (5) is the definition of I_f^{n+1} with $k+1$ for $f(n)$. This proves that (1) holds. □

Note that it follows from the two lemmas above that the endpoints of the interval I_f^n (for any n and any f) will be best approximants of every irrational in the interval. The same goes for and J_f^n.

Lemma 10. *For any n and any f, let r_n denote the right endpoint of the interval I_f^n, and let ℓ_n denote the left endpoint of the J_f^n. Then, we have (i) $r_n - r_{n+1} > r_{n+1} - r_{n+2}$ and (ii) $\ell_{n+1} - \ell_n > \ell_{n+2} - \ell_{n+1}$.*

Proof. We prove (i). Assume $I_f^n = (a/b, c/d) = I[\tau]$. By Definition 1 and Lemma 9, we have

$$I_f^{n+1} = \left(\frac{a+f(n)c}{b+f(n)d}, \frac{a+f(n)c+c}{b+f(n)d+d} \right) = I[\tau R^{f(n)}L]. \tag{6}$$

Let $\mathbf{a} = a + f(n)c$, let $\mathbf{b} = b + f(n)d$ and let $k = f(n)$. We can now rewrite (6) as

$$I_f^{n+1} = \left(\frac{\mathbf{a}}{\mathbf{b}}, \frac{\mathbf{a}+c}{\mathbf{b}+d} \right) = I[\tau R^k L]. \tag{7}$$

By (7) and Definition 6, we have

$$I[\tau R^k] = (\mathbf{a}/\mathbf{b}, c/d) \quad \text{and} \quad I[\tau R^k R] = ((\mathbf{a}+c)/(\mathbf{b}+d), c/d).$$

This shows that $((\mathbf{a}+c)/(\mathbf{b}+d), c/d)$ is addressed by some string in $\{L, R\}^*$. Thus, by Lemma 8 (ii), we have

$$\frac{c}{d} - \frac{\mathbf{a}+c}{\mathbf{b}+d} = \frac{1}{d(\mathbf{b}+d)}. \tag{8}$$

By Lemma 9, we have $I_f^{n+2} = I[\tau R^k L R^m L]$ where $m = f(n+1)$. We can assume that $m = 0$ since $m = 0$ yields the maximal distance between r_{n+1} and r_{n+2}. Thus, by Definition 6, $I_f^{n+2} = I[\tau R^k L L] = (\mathbf{a}/\mathbf{b}, (2\mathbf{a}+c)/(2\mathbf{b}+d))$. Moreover, again by Definition 6, we have

$$I[\tau R^k L] = \left(\frac{\mathbf{a}}{\mathbf{b}}, \frac{\mathbf{a}+c}{\mathbf{b}+d} \right) \quad \text{and} \quad I[\tau R^k L R] = \left(\frac{2\mathbf{a}+c}{2\mathbf{b}+d}, \frac{\mathbf{a}+c}{\mathbf{b}+d} \right).$$

This shows that $((2\mathbf{a} + c)/(2\mathbf{b} + d), (\mathbf{a} + c)/(\mathbf{b} + d)$ is addressed by a string in $\{L, R\}^*$, and thus, by Lemma 8 (ii), we have

$$\frac{\mathbf{a} + c}{\mathbf{b} + d} - \frac{2\mathbf{a} + c}{2\mathbf{b} + d} = \frac{1}{(\mathbf{b} + d)(2\mathbf{b} + d)} . \tag{9}$$

Now we can conclude our proof of (i) with

$$r_n - r_{n+1} = \frac{c}{d} - \frac{\mathbf{a} + c}{\mathbf{b} + d} \overset{(8)}{=} \frac{1}{d(\mathbf{b} + d)} > \frac{1}{(\mathbf{b} + d)(2\mathbf{b} + d)} \overset{(9)}{=}$$

$$\frac{\mathbf{a} + c}{\mathbf{b} + d} - \frac{2\mathbf{a} + c}{2\mathbf{b} + d} = r_{n+1} - r_{n+2} .$$

The proof of (ii) is symmetric. □

The *Hurwitz characteristic* of an irrational $\alpha \in (0, 1)$ is the (unique) infinite sequence Σ over the alphabet $\{L, R\}$ such that we have $\alpha \in I[\sigma]$ for any finite prefix σ of Σ. Hurwitz characteristics, which are subrecursively equivalent to Dedekind cuts, have been studied by Lehman [7] and, more recently, by Kristiansen and Simonsen [6].

3   Main Results

Theorem 11. *Let B and A be, respectively, the standard and the dual Baire sequence of α, and let F be any contractor for α. (i) We can compute B primitive recursively in F. (ii) We can compute A primitive recursively in F.*

Proof. We will show that the interval I_B^{n+1} and the value of $B(n)$ can be computed primitive recursively in F. It is trivial to compute the interval I_B^0. Assume that we have computed the interval $I_B^n = (a/b, c/d)$. First, we compute $c'/d' = F(c/d)$. Since F is a contractor for α, we have $a/b < \alpha < c'/d' < c/d$. Next, we find j such that

$$\frac{a + jc}{b + jd} < \frac{c'}{d'} \leq \frac{a + jc + c}{b + jd + d} .$$

Observe that $(\frac{a+jc}{b+jd}, \frac{a+jc+c}{b+jd+d})$ is an addressable interval and that c'/d' either lies inside, or is the right endpoint of, the interval. Thus, by Lemma 8, we have $d' \geq b + jd + d$. No unbounded search is needed to determine j. Indeed, j has to be less than d'. Thus we can primitive recursively compute j such that

$$\frac{a}{b} < \alpha < \frac{a + (j + 1)c}{b + (j + 1)d} .$$

Finally, we search for the least i less than or equal to $j + 1$ such that F moves $(a + ic + c)/(b + id + d)$ to the left, and then we let $B(n)$ equal that i. This shows that we can compute $B(n)$ primitive recursively in F, and thus (i) holds. The proof of (ii) is symmetric. Use the contractor at the left endpoint of intervals in place of the right endpoint. □

Theorem 12. *Let B and A be, respectively, the standard and the dual Baire representation of α. We can compute a contractor for α primitive recursively in B and A (and we will need both oracles).*

Proof. Let r_i denote the right endpoint of the interval I_B^i, and let ℓ_i denote the left endpoint of the interval J_A^i. For every rational number $x \in [0,1]$, we define

$$
F(x) = \begin{cases}
r_{i+1} - (r_i - x)\dfrac{r_{i+1} - r_{i+2}}{r_i - r_{i+1}} & \text{if } r_{i+1} < x \le r_i \\[3mm]
\ell_{i+1} + (x - \ell_i)\dfrac{\ell_{i+2} - \ell_{i+1}}{\ell_{i+1} - \ell_i} & \text{if } \ell_i \le x < \ell_{i+1}.
\end{cases}
$$

First we will prove that F is contractor, that is, we will prove that we have

$$
|F(x) - F(y)| < |x - y| \tag{10}
$$

for any rationals x, y where $x \ne y$. Once we have established that F is a contractor, it will be clear that F is a contractor for α.

Assume that one of the rationals x and y lies below α and that the other lies above. It is easy to see that F will move one of the numbers to the right and the other one to the left, and thus, (10) holds. Assume that both x and y lie at same side of α. We can w.l.o.g. assume that both lie below and that we have $x < y < \alpha$. The proof splits into two cases: (i) $\ell_i \le x < y < \ell_{i+1}$ for some i, and (ii) $\ell_i \le x < \ell_{i+1} \le \ell_j \le y < \ell_{j+1}$ for some i, j where $j \ge i + 1$.

Case (i). Let $k = (\ell_{i+2} - \ell_{i+1})/(\ell_{i+1} - \ell_i)$. By Lemma 10, we have $k < 1$, and then by the definition of F, we have

$$
F(y) - F(x) = \ell_{i+1} + (y - \ell_i)k - (\ell_{i+1} + (x - \ell_i)k) = (y - x)k < y - x
$$

and thus (10) holds.

Case (ii). This case is slightly more involved, but in the end everything is straightforward. We omit the details.

This proves that F is a contractor for α. It remains to argue that F can be computed primitive recursively in B and A. Let q be an arbitrary rational in the interval $[0,1]$, and let m/n be q written in lowest terms.

(Claim) There exists $j < n$ such that $\ell_j \le q < \ell_{j+1}$ or $r_{j+1} < q \le r_j$.

In order to see that the claim holds, assume that $\alpha < q = m/n$. It follows from the lemmas in Sect. 2 that each $r_j = c_j/d_j$ is a right best approximant to α. Thus we have $n \ge d_j$ whenever $m/n \le c_j/d_j$. Moreover, as $d_j > j$, we have $j < n$ such that $r_{j+1} < q = m/n \le r_j$ if $\alpha < q$. If $q = m/n < \alpha$, a symmetric argument yields $j < n$ such that $\ell_j \le q < \ell_{j+1}$. This proves the claim.

The sequence $r_0, r_1, r_2, \ldots$ can be computed primitive recursively in B, and the sequence $\ell_0, \ell_1, \ell_2, \ldots$ can be computed primitive recursively in A. Thus, it follows from the claim that F can be computed primitive recursively in B and A. □

It follows from the next theorem that we cannot compute the standard Baire sequence of an irrational α subrecursively in the dual Baire sequence of α. That requires unbounded search.

Theorem 13. *Let S be any subrecursive class. There exists an irrational number α such that (i) the standard Baire sequence of α is not in S, and (ii) the dual Baire sequence of α is (Kalmar) elementary.*

Proof. A function f is *honest*, by definition, if $2^x \leq f(x)$, $f(x) \leq f(x+1)$ and the relation $f(x) = y$ is elementary. Let B be the an honest function which is not in S. Such a B exists (a proof can be found in Georgiev et al. [1]). Now, B is the standard Baire sequence of some irrational number α, and since an irrational number only has one standard Baire sequence, the standard Baire sequence of α is not in S. It remains to prove that the dual Baire sequence of α is elementary.

Let $a_n = B(0) + (\sum_{i=1}^{n} B(i) + 1)$. Let $A(x) = 1$ if $x = a_n$ for some n; otherwise, let $A(x) = 0$. Since B is an honest function, we can check by elementary means if there exists n such that $x = a_n$. Hence A is an elementary function. We will prove that A is the dual Baire sequence of α.

For any natural number n, we define the strings σ_n and τ_n by

$$\sigma_n = L^{A(0)} R L^{A(1)} R \ldots L^{A(a_n - 1)} R L^{A(a_n)} \text{ and } \tau_n = R^{B(0)} L R^{B(1)} L \ldots R^{B(n)} L \, .$$

We will prove the following claim by induction on n: $\sigma_n = \tau_n$ (claim).

Let $n = 0$. We have $a_0 = B(0)$ and thus, by the definition of A, we have

$$\sigma_0 = L^{A(0)} R L^{A(1)} R \ldots L^{A(a_0 - 1)} R L^{A(a_0)} = R^{a_0} L = R^{B(0)} L = \tau_0 \, .$$

Let $n > 0$. By the definition of a_n, we have $a_n = a_{n-1} + B(n) + 1$, and thus $B(n) = a_n - (a_{n-1} + 1)$. Furthermore, we have

$$\sigma_n \overset{(1)}{=} \sigma_{n-1} R L^{A(a_{n-1}+1)} R L^{A(a_{n-1}+2)} \ldots R L^{A(a_n - 1)} R L^{A(a_n)} \overset{(2)}{=}$$

$$\sigma_{n-1} R^{a_n - (a_{n-1}+1)} L \overset{(3)}{=} \sigma_{n-1} R^{B(n)} L \overset{(4)}{=} \tau_{n-1} R^{B(n)} L \overset{(5)}{=} \tau_n$$

where (1) holds by the definition of σ_n; (2) holds by the definition of A; (3) holds by the definition of a_n; (4) holds by the induction hypothesis; and (5) holds by the definition of τ_n. This proves (claim).

It follows from (claim) and Lemma 9 that the inclusion $J_A^{a_n} \subseteq I_B^n$ holds for all n. This proves that A is the dual Baire sequence of α.   □

Just for the record, the proof of the next theorem is symmetric to the proof of the preceding theorem.

Theorem 14. *Let S be any subrecursive class. There exists an irrational number α such that (i) the dual Baire sequence of α is not in S, and (ii) the standard Baire sequence of α is (Kalmar) elementary.*

4   The Big Picture

Definition 15. *For any subrecursive class S, let S_F denote the class of irrational numbers that have a* contractor *in S, let S_{sB} denote the class of irrational numbers that have a* standard Baire sequence *in S, and S_{dB} denote the class of irrational numbers that have a* dual Baire sequence *in S.*

Corollary 16. *Let S be any subrecursive class closed under primitive recursive operations. Then, (i) $S_{sB} \not\subseteq S_{dB}$, (ii) $S_{dB} \not\subseteq S_{sB}$ and (iii) $S_F = S_{dB} \cap S_{sB}$.*

Proof. Theorem 13 entails (i). Theorem 14 entails (ii). Theorem 11 entails $S_F \subseteq S_{dB} \cap S_{sB}$. Theorem 12 entails $S_{dB} \cap S_{sB} \subseteq S_F$. Thus, (iii) holds.   □

Definition 17. *Let α be an irrational number in the interval $(0,1)$. A* left best approximation *of α is a sequence of fractions $\{a_i/b_i\}_{i\in\mathbb{N}}$ such that $(0/1) = (a_0/b_0) < (a_1/b_1) < (a_2/b_2) < \ldots$ and each a_i/b_i is a left best approximant of α (see Definition 7). A* right best approximation *of α is a sequence of fractions $\{a_i/b_i\}_{i\in\mathbb{N}}$ such that $(1/1) = (a_0/b_0) > (a_1/b_1) > (a_2/b_2) > \ldots$ and each a_i/b_i is a right best approximant of α. Clearly, both sequences converge to α.*

Let $S_<$ denote the class of irrational numbers that have a left best approximation in the subrecursive class S, and let $S_>$ denote the class of irrational numbers that have a right best approximation in S.

Theorem 18. *For any subrecursive class S closed under primitive recursion, we have $S_< = S_{dB}$ and $S_> = S_{sB}$.*

Proof. We say that a right best approximation of α is *complete* if every right best approximant occurs in the approximation. Note that the complete best approximation of an irrational α in the interval $(0,1)$ is unique.

Let B be the standard Baire sequence of α. Consider the interval I addressed by $R^{f(0)}LR^{f(1)}L\ldots R^{f(n)}L$. By Lemma 9, we have $I = I_B^{n+1}$. The right endpoint of I will be the n'th best approximant in the complete right best approximation of α. These considerations make it easy to see that the inclusion $S_{sB} \subseteq S_>$ holds.

Let $\{a_i/b_i\}_{i\in\mathbb{N}}$ be a right best approximation of α. We can w.l.o.g. assume that $\{a_i/b_i\}_{i\in\mathbb{N}}$ is complete since a complete right best approximation can be computed primitive recursively in an arbitrary right best approximation. We can primitive recursively in $\{a_i/b_i\}_{i\in\mathbb{N}}$ compute a (unique) string of the form $R^{k_0}LR^{k_1}L\ldots R^{k_n}L$ such that the right endpoint of the interval addressed by $R^{k_0}L\ldots R^{k_i}L$ equals a_{i+1}/b_{i+1} (for all $i \leq n$). Let B be the standard Baire sequence of α. By Lemma 9, we have $B(i) = k_i$ (for all $i \leq n$). These considerations make it easy to see that the inclusion $S_> \subseteq S_{sB}$ holds. This proves $S_> = S_{sB}$. The proof of $S_< = S_{dB}$ is of course symmetric.   □

For any subrecursive class S, let $S_{g\uparrow}$ denote the class of irrational numbers that have a *general sum approximation from below* in S, let $S_{g\downarrow}$ denote the class of irrational numbers that have a *general sum approximation from above* in S, furthermore, let $S_{[]}$ denote the class of irrational numbers that have a *continued fraction* in $\mathcal{S}$. Definitions of general sum approximations from below and above can be found in [4] and [1]. It is proved in [4] that we have $\mathcal{S}_{[]} = \mathcal{S}_{g\uparrow} \cap \mathcal{S}_{g\downarrow}$ for any $\mathcal{S}$ closed under primitive recursive operations. It is proved in [1] that we have $\mathcal{S}_< = \mathcal{S}_{g\uparrow}$ and $\mathcal{S}_> = \mathcal{S}_{g\downarrow}$ for any $\mathcal{S}$ closed under primitive recursive operations. Thus, we have the following corollary.

Corollary 19. *For any subrecursive class $\mathcal{S}$ closed under primitive recursive operations, we have $\mathcal{S}_{[]} = \mathcal{S}_F$ and $\mathcal{S}_< = \mathcal{S}_{g\uparrow} = \mathcal{S}_{dB}$ and $\mathcal{S}_> = \mathcal{S}_{g\downarrow} = \mathcal{S}_{sB}$.*

References

1. Georgiev, I., Kristiansen, L., Stephan, F.:Computable Irrational Numbers with Representations of Surprising Complexity. Ann. Pure Appl. Logic **172**, 102893 (2021). https://doi.org/10.1016/j.apal.2020.102893
2. Gerhardy, P., Kohlebach, U.: Strongly uniform bounds from semi-constructive proofs. Ann. Pure Appl. Logic **141**, 89–107 (2006). https://doi.org/10.1016/j.apal.2005.10.003
3. Kohlebach, U., Olivia, P.: Proof mining: a systematic way of analysing proofs in mathematics. Proc. Steklov Inst. Math. **242**, 136–164 (2003)
4. Kristiansen, L.: On subrecursive representability of irrational numbers. Computability **6**, 249–276 (2017). https://doi.org/10.3233/COM-160063
5. Kristiansen, L.: On subrecursive representability of irrational numbers, part II. Computability **8**, 43–65 (2019). https://doi.org/10.3233/COM-170081
6. Kristiansen, L., Simonsen, J.G.: on the complexity of conversion between classic real number representations. In: Anselmo, M., Della Vedova, G., Manea, F., Pauly, A. (eds.) CiE 2020. LNCS, vol. 12098, pp. 75–86. Springer, Cham (2020). https://doi.org/10.1007/978-3-030-51466-2_7
7. Lehman, R.S.: On primitive recursive real numbers. Fundamenta Mathematica **49**(2), 105–118 (1961)
8. Richards, I.: Continued fractions without tears. Math. Mag. **54**, 163–171 (1981)

Learning Languages in the Limit from Positive Information with Finitely Many Memory Changes

Timo Kötzing and Karen Seidel[(✉)]

Hasso-Plattner-Institute, University of Potsdam, Potsdam, Germany
`karen.seidel@hpi.de`

Abstract. We investigate learning collections of languages from texts by an inductive inference machine with access to the current datum and a bounded memory in form of states. Such a bounded memory states (**BMS**) learner is considered successful in case it eventually settles on a correct hypothesis while exploiting only finitely many different states.

We give the complete map of all pairwise relations for an established collection of criteria of successful learning. Most prominently, we show that non-U-shapedness is not restrictive, while conservativeness and (strong) monotonicity are. Some results carry over from iterative learning by a general lemma showing that, for a wealth of restrictions (the *semantic* restrictions), iterative and bounded memory states learning are equivalent. We also give an example of a non-semantic restriction (strongly non-U-shapedness) where the two settings differ.

Keywords: Memory restricted learning algorithms ·
Map for bounded memory states learners ·
(Strongly) non-U-shaped learning

1 Introduction

We are interested in the problem of algorithmically learning a description for a formal language (a computably enumerable subset of the set of natural numbers) when presented successively all and only the elements of that language; this is sometimes called *inductive inference*, a branch of (algorithmic) learning theory. For example, a learner M might be presented more and more even numbers. After each new number, M outputs a description for a language as its conjecture. The learner M might decide to output a program for the set of all multiples of 4, as long as all numbers presented are divisible by 4. Later, when M sees an even number not divisible by 4, it might change this guess to a program for the set of all multiples of 2.

Many criteria for deciding whether a learner M is *successful* on a language L have been proposed in the literature. Gold, in his seminal paper [Gol67], gave a first, simple learning criterion, **TxtEx**-*learning*[1], where a learner is *successful*

---

[1] **Txt** stands for learning from a *text* of positive examples; **Ex** stands for *explanatory*.

© Springer Nature Switzerland AG 2021
L. De Mol et al. (Eds.): CiE 2021, LNCS 12813, pp. 318–329, 2021.
https://doi.org/10.1007/978-3-030-80049-9_29

iff, on every *text* for L (listing of all and only the elements of L) it eventually stops changing its conjectures, and its final conjecture is a correct description for the input sequence. Trivially, each single, describable language L has a suitable constant function as an **TxtEx**-learner (this learner constantly outputs a description for L). Thus, we are interested in analyzing for which *classes of languages* $\mathcal{L}$ there is a *single learner* M learning *each* member of $\mathcal{L}$. Sometimes, this framework is called *language learning in the limit* and has been studied extensively. For an overview see for example, the textbook [JORS99].

One major criticism of the model suggested by Gold is its excessive use of memory, see for example [CM08]: for each new hypothesis the entire history of past data is available. Iterative learning is the most common variant of learning in the limit which addresses memory constraints: the memory of the learner on past data is just its current hypothesis. Due to the padding lemma [JORS99], this memory is not necessarily void, but only finitely many data can be memorized in the hypothesis. There is a comprehensive body of work on iterative learning, see, e.g., [CK10, CM08, JKMS16, JMZ13, JORS99].

Another way of modelling restricted memory learning is to grant the learner access to not their current hypothesis, but a *state* which can be used in the computation of the next hypothesis (and next state). This was introduced in [CCJS07] and called *bounded memory states (BMS)* learning. It is a reasonable assumption to have a countable reservoir of states. Assuming a computable enumeration of these states, we use natural numbers to refer to them. Note that allowing arbitrary use of all natural numbers as states would effectively allow a learner to store all seen data in the state, thus giving the same mode as Gold's original setting. Probably the minimal way to restrict the use of states is to demand for successful learning that a learner must stop using new states eventually (but may still traverse among the finitely many states produced so far, and may use infinitely many states on data for a non-target language). It was claimed that this setting is equivalent to iterative learning [CCJS07, Remark 38] (this restriction is called *ClassBMS* there, we refer to it by **TxtBMS$_*$Ex**). However, this was only remarked for the plain setting of explanatory learning; for further restrictions, the setting is completely unknown, only for explicit constant state bounds a few scattered results are known, see [CCJS07, CK13].

In this paper, we consider a wealth of restrictions, described in detail in Sect. 2 (after an introduction to the general notation of this paper). Following the approach of giving *maps* of pairwise relations suggested in [KS16], we give a complete map. We note that this map is the same as the map for iterative learning given in [JKMS16], but partially for different reasons.

In Lemma 31 we show that, for many restrictions (the so-called *semantic* restrictions, where only the semantics of hypotheses are restricted) the learning setting with bounded memory states is equivalent to learning iteratively. This proves and generalizes the aforementioned remark in [CCJS07] to a wide class of restrictions.

However, if restrictions are not semantic, then iterative and bounded memory states learning can differ. We show this concretely for *strongly non-U-shaped* learning in Theorem 45. Inspired by cognitive science research [SS82], [MPU+92] a semantic version of this requirement was defined in [BCM+08] and later the syntactic variant was introduced in [CM11]. Both requirements have been extensively studied, see [CC13] for a survey and moreover [CK13], [CK16], [KSS17]. The proof combines the techniques for showing that strong non-U-shapedness restricts iterative learning, as proved in [CK13, Theorem 5.7], and that not every class strongly monotonically learnable by an iterative learner is strongly non-U-shapedly learnable by an iterative learner, see [JKMS16, Theorem 5]. Moreover, it relies on showing that state decisiveness can be assumed in Lemma 41.

The remainder of Sect. 4 completes the map for the case of syntactic restrictions (since these do not carry over from the setting of iterative learning). All syntactic learning requirements are closely related to strongly locking learners. The fundamental concept of a locking sequence was introduced by [BB75]. For a similar purpose than ours [JKMS16] introduced strongly locking learners. We generalize their construction for certain syntactically restricted iterative learners from a strongly locking iterative learner. Finally, we obtain that all non-semantic learning restrictions also coincide for $\mathbf{BMS}_*$-learning.

2 Learners, Success Criteria and Other Terminology

As far as possible, we follow [JORS99] on the learning theoretic side and [Odi99] for computability theory. We recall the most essential notation and definitions.

We let $\mathbb{N}$ denote the *natural numbers* including 0. For a function f we write $\mathrm{dom}(f)$ for its *domain* and $\mathrm{ran}(f)$ for its *range*.

Further, $X^{<\omega}$ denotes the *finite sequences* over the set X and X^ω stands for the *countably infinite sequences* over X. For every $\sigma \in X^{<\omega}$ and $t \leq |\sigma|$, $t \in \mathbb{N}$, we let $\sigma[t] := \{(s, \sigma(s)) \mid s < t\}$ denote the *restriction of σ to t*. Moreover, for sequences $\sigma, \tau \in X^{<\omega}$ their concatenation is denoted by $\sigma^\frown \tau$. Finally, we write $\mathrm{last}(\sigma)$ for the last element of σ, $\sigma(|\sigma| - 1)$, and σ^- for the initial segment of σ without $\mathrm{last}(\sigma)$, i.e. $\sigma[|\sigma| - 1]$. Clearly, $\sigma = \sigma^{-\frown}\mathrm{last}(\sigma)$.

For a finite set $D \subseteq \mathbb{N}$ and a finite sequence $\sigma \in X^{<\omega}$, we denote by $\langle D \rangle$ and $\langle \sigma \rangle$ a canonical index for D or σ, respectively. Further, we fix a Gödel pairing function $\langle ., . \rangle$ with two arguments.

If we deal with (a subset of) a cartesian product or Gödel pairs, we are going to refer to the *projection functions* to the first or second coordinate by pr_1 and pr_2, respectively.

Let $L \subseteq \mathbb{N}$. We interpret every $n \in \mathbb{N}$ as a code for a word. If L is recursively enumerable, we call L a *language*.

We fix a programming system φ as introduced in [RC94]. Briefly, in the φ-system, for a natural number p, we denote by φ_p the partial computable function with program code p. We call p an *index* for W_p defined as $\mathrm{dom}(\varphi_p)$.

In reference to a Blum complexity measure Φ_p, for all $p, t \in \mathbb{N}$, we denote by $W_p^t \subseteq W_p$ the recursive set of all natural numbers less or equal to t, on which the machine executing p halts in at most t steps, i.e. $W_p^t = \{x \mid x \leq t \wedge \Phi_p(x) \leq t\}$.

Moreover, the well-known s-m-n theorem gives finite and infinite recursion theorems, see [Cas94], [Odi99]. We will refer to Case's Operator Recursion Theorem ORT in its 1-1-form, [Cas74].

Throughout the paper, we let $\Sigma = \mathbb{N} \cup \{\#\}$ be the input alphabet with $n \in \mathbb{N}$ interpreted as code for a word in the language and $\#$ interpreted as pause symbol, i.e. no new information. Further, let $\Omega = \mathbb{N} \cup \{?\}$ be the output alphabet with $p \in \mathbb{N}$ interpreted as φ-index and ? as no hypothesis or repetition of the last hypothesis, if existent. A function with range Ω is called a hypothesis generating function.

A *learner* is a (partial) computable function $M : \mathrm{dom}(M) \subseteq \Sigma^{<\omega} \to \Omega$. The set of all total computable functions $M : \Sigma^{<\omega} \to \Omega$ is denoted by $\mathcal{R}$.

Let $f \in \Sigma^{<\omega} \cup \Sigma^{\omega}$, then the *content of* f, defined as $\mathrm{content}(f) := \mathrm{ran}(f) \setminus \{\#\}$, is the set of all natural numbers, about which f gives some positive information. $\mathbf{Txt}(L) := \{T \in \Sigma^{\omega} \mid \mathrm{content}(T) = L\}$ denotes *set of all texts for* L.

Definition 21. *Let M be a learner. M is an* iterative learner *or* **It***-learner, for short $M \in \mathbf{It}$, if there is a computable (partial) hypothesis generating function $h_M : \Omega \times \Sigma \to \Omega$ such that $M = h_M^{\ddagger}$ where $h_M^{\ddagger}$ is defined on finite sequences by*

$$h_M^{\ddagger}(\epsilon) = ?; \qquad h_M^{\ddagger}(\sigma^\frown x) = h_M(h_M^{\ddagger}(\sigma), x).$$

Definition 22. *Let M be a learner. M is a* bounded memory states learner *or* **BMS***-learner, for short $M \in \mathbf{BMS}$, if there are a computable (partial) hypothesis generating function $h_M : \mathbb{N} \times \Sigma \to \Omega$ and a computable (partial) state transition function $s_M : \mathbb{N} \times \Sigma \to \mathbb{N}$ such that $\mathrm{dom}(h_M) = \mathrm{dom}(s_M)$ and $M = h_M^*$ where h_M^* and s_M^* are defined on finite sequences by*

$$s_M^*(\epsilon) = 0; \qquad h_M^*(\sigma^\frown x) = h_M(s_M^*(\sigma), x); \qquad s_M^*(\sigma^\frown x) = s_M(s_M^*(\sigma), x).$$

We now clarify what we mean by successful learning.

Definition 23. *Let M be a learner and $\mathcal{L}$ a collection of languages.*

1. *Let $L \in \mathcal{L}$ be a language and $T \in \mathbf{Txt}(L)$ a text for L presented to M.*
   (a) *We call $h = (h_t)_{t \in \mathbb{N}} \in \Omega^{\omega}$, where $h_t := M(T[t])$ for all $t \in \mathbb{N}$, the learning sequence of M on T.*
   (b) *M learns L from T in the limit, for short M **Ex**-learns L from T or **Ex**(M, T), if there exists $t_0 \in \mathbb{N}$ such that $W_{h_{t_0}} = \mathrm{content}(T)$ and $\forall t \geq t_0$ ($h_t \neq ? \Rightarrow h_t = h_{t_0}$).*
2. *M learns $\mathcal{L}$ in the limit, for short M **Ex**-learns $\mathcal{L}$, if **Ex**(M, T) for every $L \in \mathcal{L}$ and every $T \in \mathbf{Txt}(L)$.*

Definition 24. *Let $\mathcal{L}$ be a collection of languages. $\mathcal{L}$ is* learnable in the limit *or* **Ex***-learnable, if there exists a learner M that **Ex***-learns $\mathcal{L}$.*

In our investigations, the most important additional requirement on a successful learning process for a **BMS**-learner is to use finitely many states only, as stated in the following definition.

Definition 25. *Let M be a **BMS**-learner and $T \in$ **Txt**. We say that M uses finitely many memory states on T, for short $\mathbf{BMS}_*(M, T)$, if $\{ s_M^*(T[t]) \mid t \in \mathbb{N} \}$ is finite.*

Let L be a language. M is said to $\mathbf{BMS}_*\mathbf{Ex}$-learn L, if $\mathbf{BMS}_*\mathbf{Ex}(M, T)$ for every text $T \in \mathbf{Txt}(L)$.

In [CCJS07, Rem. 38] it is claimed that $\mathbf{BMS}_*$-learners and iterative learners are equally powerful on texts. This also follows from our more general Lemma 31.

We list the most common additional requirements regarding the learning sequence, which may tag a learning process just like $\mathbf{BMS}_*$ above. For this we first recall the notion of consistency of a sequence with a set. For $f \in \Sigma^{<\omega} \cup \Sigma^\omega$ and $A \subseteq \Sigma$ we say f *is consistent with A* if and only if content$(f) \subseteq A$.

The listed properties of the learning sequence have been at the center of different investigations. Studying how they relate to one another did begin in [KP16, KS16, JKMS16] and [AKS18].

Definition 26. *Let M be a learner, $T \in$ **Txt** and $h = (h_t)_{t \in \mathbb{N}} \in \Omega^\omega$ the learning sequence of M on T, i.e. $h_t = M(T[t])$ for all $t \in \mathbb{N}$. We write*

1. $\mathbf{Cons}(T[t], W_{h_t})$, if $\{T(s) \mid s < t\} \setminus \{\#\} \subseteq W_{h_t}$.
2. $\mathbf{Cons}(M, T)$ *[Ang80], if M is consistent on T, i.e., for all t holds $\mathbf{Cons}(T[t], W_{h_t})$.*
3. $\mathbf{Conv}(M, T)$ *[Ang80], if M is conservative on T, i.e., for all s, t with $s \leq t$ holds $\mathbf{Cons}(T[t], W_{h_s}) \Rightarrow h_s = h_t$.*
4. $\mathbf{Dec}(M, T)$ *[OSW82], if M is decisive on T, i.e., for all r, s, t with $r \leq s \leq t$ holds $W_{h_r} = W_{h_t} \Rightarrow W_{h_r} = W_{h_s}$.*
5. $\mathbf{Caut}(M, T)$ *[OSW86], if M is cautious on T, i.e., for all s, t with $s \leq t$ holds $\neg W_{h_t} \subsetneq W_{h_s}$.*
6. $\mathbf{WMon}(M, T)$ *[Jan91, Wie91], if M is weakly monotonic on T, i.e., for all s, t with $s \leq t$ holds $\mathbf{Cons}(T[t], W_{h_s}) \Rightarrow W_{h_s} \subseteq W_{h_t}$.*
7. $\mathbf{Mon}(M, T)$ *[Jan91, Wie91], if M is monotonic on T, i.e., for all s, t with $s \leq t$ holds $W_{h_s} \cap$ content$(T) \subseteq W_{h_t} \cap$ content(T).*
8. $\mathbf{SMon}(M, T)$ *[Jan91, Wie91], if M is strongly monotonic on T, i.e., for all s, t with $s \leq t$ holds $W_{h_s} \subseteq W_{h_t}$.*
9. $\mathbf{NU}(M, T)$ *[BCM+08], if M is non-U-shaped on T, i.e., for all r, s, t with $r \leq s \leq t$ holds $W_{h_r} = W_{h_t} = $ content$(T) \Rightarrow W_{h_r} = W_{h_s}$.*
10. $\mathbf{SNU}(M, T)$ *[CM11], if M is strongly non-U-shaped on T, i.e., for all r, s, t with $r \leq s \leq t$ holds $W_{h_r} = W_{h_t} = $ content$(T) \Rightarrow h_r = h_s$.*
11. $\mathbf{SDec}(M, T)$ *[KP16], if M is strongly decisive on T, i.e., for all r, s, t with $r \leq s \leq t$ holds $W_{h_r} = W_{h_t} \Rightarrow h_r = h_s$.*
12. $\mathbf{Wb}(M, T)$ *[KS16], if M is witness-based on T, i.e., for all r, t such that for some s with $r < s \leq t$ holds $h_r \neq h_s$ holds content$(T[s]) \cap (W_{h_t} \setminus W_{h_r}) \neq \varnothing$.*

It is easy to see that $\mathbf{Conv}(M, T)$ implies $\mathbf{SNU}(M, T)$ and $\mathbf{WMon}(M, T)$; $\mathbf{SDec}(M, T)$ implies $\mathbf{Dec}(M, T)$ and $\mathbf{SNU}(M, T)$; $\mathbf{SMon}(M, T)$ implies all of $\mathbf{Caut}(M, T)$, $\mathbf{Dec}(M, T)$, $\mathbf{Mon}(M, T)$, $\mathbf{WMon}(M, T)$ and finally $\mathbf{Dec}(M, T)$, $\mathbf{WMon}(M, T)$ and $\mathbf{SNU}(M, T)$ imply $\mathbf{NU}(M, T)$. Further, $\mathbf{Wb}(M, T)$ implies $\mathbf{Conv}(M, T)$, $\mathbf{SDec}(M, T)$ and $\mathbf{Caut}(M, T)$.

In order to characterize what successful learning means, these predicates may be combined with the explanatory convergence criterion. For this, we let $\Delta := \{\mathbf{Caut}, \mathbf{Conv}, \mathbf{Dec}, \mathbf{SDec}, \mathbf{WMon}, \mathbf{Mon}, \mathbf{SMon}, \mathbf{NU}, \mathbf{SNU}, \mathbf{T}\}$ denote the set of *admissible learning restrictions*, with $\mathbf{T}$ standing for no restriction. Further, a *learning success criterion* is a predicate being the intersection of the convergence criterion $\mathbf{Ex}$ with arbitrarily many admissible learning restrictions. This means that the sequence of hypotheses has to converge and in addition has the desired properties. Therefore, the collection of all learning success criteria is $\{\bigcap_{i=0}^{n} \delta_i \cap \mathbf{Ex} \mid n \in \mathbb{N}, \forall i \leq n (\delta_i \in \Delta)\}$. Note that plain explanatory convergence is a learning success criterion by letting $n = 0$ and $\delta_0 = \mathbf{T}$.

We refer to all $\delta \in \{\mathbf{Caut}, \mathbf{Cons}, \mathbf{Dec}, \mathbf{Mon}, \mathbf{SMon}, \mathbf{WMon}, \mathbf{NU}, \mathbf{T}\}$ also as *semantic* learning restrictions, as they do not require the learner to settle on exactly one hypothesis. More formally, if texts T_1, T_2 are such that for all $t \in \mathbb{N}$ holds $W_{M(T_1[t])} = W_{M(T_2[t])}$, then $\delta(M, T_1)$ and $\delta(M, T_2)$ are equivalent.

In order to state observations about how two ways of defining learning success relate to each other, the learning power of the different settings is encapsulated in notions $[\alpha\mathbf{Txt}\beta]$. A collection of languages $\mathcal{L}$ is in $[\alpha\mathbf{Txt}\beta]$, if there is a learner with property α that β-*learns* $\mathcal{L}$. We do not use separators in the notation to stay consistent with established notation in the field that was inspired by [JORS99]. Whenever β includes $\mathbf{BMS}_*$ it is understood that we are only considering $\mathbf{BMS}$-learners.

The proofs of Lemmata 31 and 41 employ the following property of learning requirements and learning success criteria, that applies to all such considered in this paper.

Definition 27. *Denote the set of all unbounded and non-decreasing functions by* $\mathfrak{S}$, *i.e.*, $\mathfrak{S} := \{\mathfrak{s} : \mathbb{N} \to \mathbb{N} \mid \forall x \in \mathbb{N} \exists t \in \mathbb{N} : \mathfrak{s}(t) \geq x \text{ and } \forall t \in \mathbb{N} : \mathfrak{s}(t+1) \geq \mathfrak{s}(t)\}$. *Then every* $\mathfrak{s} \in \mathfrak{S}$ *is a so called* simulating function.

A predicate β *on pairs of learners and texts allows for simulation on equivalent text, if for all simulating functions* $\mathfrak{s} \in \mathfrak{S}$, *all texts* $T, T' \in \mathbf{Txt}$ *and all learners* M, M' *holds: Whenever we have* $\mathrm{content}(T'[t]) = \mathrm{content}(T[\mathfrak{s}(t)])$ *and* $M'(T'[t]) = M(T[\mathfrak{s}(t)])$ *for all* $t \in \mathbb{N}$, *from* $\beta(M, T)$ *we can conclude* $\beta(M', T')$.

Intuitively, as long as the learner M' conjectures $h'_t = h_{\mathfrak{s}(t)} = M(T[\mathfrak{s}(t)])$ at time t and has, in form of $T'[t]$, the same data available as was used by M for this hypothesis, M' on T' is considered to be a simulation of M on T.

It is easy to see that all learning success criteria considered in this paper allow for simulation on equivalent text.

3   Relations Between Semantic Learning Requirements

The following lemma formally establishes the equal learning power of iterative and $\mathbf{BMS}_*$-learning for all learning success criteria but $\mathbf{Conv}$, $\mathbf{SDec}$ and $\mathbf{SNU}$. We are going to prove in Sect. 4 that this is not true for these three non-semantic additional requirements.

Lemma 31. *Let δ allow for simulation on equivalent text.*

1. We have $[\mathbf{TxtBMS}_*\delta\mathbf{Ex}] \supseteq [\mathbf{ItTxt}\delta\mathbf{Ex}]$.
2. If δ is semantic then $[\mathbf{TxtBMS}_*\delta\mathbf{Ex}] = [\mathbf{ItTxt}\delta\mathbf{Ex}]$.

While 1. and "$\supseteq$" in 2. are easy to verify by using the hypotheses as states, the other inclusion in 2. is more challenging. The iterative learner constructed from the **BMS**-learner M uses the hypotheses of M on an equivalent text and additionally pads a subgraph of the translation diagram of M to it.

With Lemma 31 the following results transfer from learning with iterative learners and it remains to investigate the relations to and between the non-semantic requirements **Conv, SDec** and **SNU**.

Theorem 32. *1.* $[\mathbf{TxtBMS}_*\mathbf{NUEx}] = [\mathbf{TxtBMS}_*\mathbf{Ex}]$
2. $[\mathbf{TxtBMS}_*\mathbf{DecEx}] = [\mathbf{TxtBMS}_*\mathbf{WMonEx}] = [\mathbf{TxtBMS}_*\mathbf{CautEx}] = [\mathbf{TxtBMS}_*\mathbf{Ex}]$
3. $[\mathbf{TxtBMS}_*\mathbf{MonEx}] \subsetneqq [\mathbf{TxtBMS}_*\mathbf{Ex}]$
4. $[\mathbf{TxtBMS}_*\mathbf{SMonEx}] \subsetneqq [\mathbf{TxtBMS}_*\mathbf{MonEx}]$

Proof. The respective results for iterative learners can be found in [CM08, Theorem 2], [JKMS16, Theorem 10], [JKMS16, Theorem 3] and [JKMS16, Theorem 2]. □

4 Relations to and Between Syntactic Learning Requirements

The following lemma establishes that we may assume **BMS**$_*$-learners to never go back to withdrawn states. This is essential in almost all of the following proofs. It can also be used to simplify the proof of Lemma 31.

Lemma 41. *Let β be a learning success criterion allowing for simulation on equivalent text and $\mathcal{L} \in [\mathbf{TxtBMS}_*\beta]$. Then there is a **BMS**-learner N such that N never returns to a withdrawn state and **BMS**$_*\beta$-learns $\mathcal{L}$ from texts.*

With the latter result we can show that strongly monotonically **BMS**$_*$-learnability does not imply strongly non-U-shapedly **BMS**$_*$-learnability.

Theorem 42. $[\mathbf{TxtBMS}_*\mathbf{SMonEx}] \not\subseteq [\mathbf{TxtBMS}_*\mathbf{SNUEx}]$

In the proof a self-learning **BMS**-learner M is defined and with a tailored ORT-argument there can not be a **BMS**-learner strongly non-U-shapedly learning all languages that M learns strongly monotonically.

For inferring the relations between the syntactic learning requirements **SNU**, **SDec** and **Conv**, we refer to **Wb**. All these criteria are closely related to strongly locking learners. The learnability of every language L by a learner M is witnessed by a sequence σ, consistent with L, such that $M(\sigma)$ is an index for L and no extension of σ consistent with L will lead to a mind-change of M. Such a sequence

σ is called *(sink-)locking sequence for M on L*. A learner M acts strongly locking on a language L, if for every text T for L there is an initial segment σ of T that is a locking sequence for M on L.

The proof of the following theorem generalizes the construction of a conservative and strongly decisive iterative learner from a strongly locking iterative learner in [JKMS16, Theorem 8]. With it we obtain in the Corollary thereafter, that all non-semantic learning restrictions coincide.

Theorem 43. *Let $\mathcal{L}$ be a set of languages **BMS**$_*$**Ex**-learned by a strongly locking **BMS**-learner. Then $\mathcal{L} \in$ [**TxtBMS**$_*$**WbEx**].*

The construction of the witness-based learner proceeds in two steps. First, we construct a learner **BMS**$_*$-learning $\mathcal{L}$ locally conservatively, as defined in [JLZ07], requiring the last datum to violate consistency with the former hypothesis. Second, from the aforementioned locally conservative learner, we obtain a new learner that **BMS**$_*$**Ex**-learns $\mathcal{L}$ in a witness-based fashion. We will do this by keeping track of all data having caused a mind-change so far. More concretely, we alter the text by excluding mind-change data causing another mind-change and make sure that the witness for the mind-change is contained in all future hypotheses.

With the latter theorem it is straightforward to observe that in the **BMS**$_*$**Ex**-setting conservative, strongly decisive and strongly non-U-shaped **Ex**-learning are equivalent.

Corollary 44. *For all $\gamma, \delta \in \{$**Conv**, **SDec**, **SNU**$\}$ holds [**TxtBMS**$_*$$\gamma$**Ex**] = [**TxtBMS**$_*$$\delta$**Ex**].*

By [JKMS16, Theorem 2] and Lemma 31 we obtain [**TxtBMS**$_*$**ConvEx**] $\not\subseteq$ [**TxtBMS**$_*$**SMonEx**]. From this we conclude with Theorem 42 and Corollary 44 that [**TxtBMS**$_*$**ConvEx**] $\perp$ [**TxtBMS**$_*$**SMonEx**].

Similarly, with [JKMS16, Theorem 3] and Lemma 31 [**TxtBMS**$_*$**ConvEx**] $\not\subseteq$ [**TxtBMS**$_*$**MonEx**]. As [**TxtBMS**$_*$**MonEx**] $\not\subseteq$ [**TxtBMS**$_*$**SNUEx**] by Theorem 42, with Corollary 44 [**TxtBMS**$_*$**ConvEx**] $\perp$ [**TxtBMS**$_*$**MonEx**].

Because Theorem 42 also reproves [**TxtBMS**$_*$**SNUEx**] $\subsetneq$ [**TxtBMS**$_*$**Ex**], first observed in [CK13, Th. 3.10], we completed the map for **BMS**$_*$**Ex**-learning from texts.

As the relations equal the ones for **It**-learning, naturally the question arises, whether a result similar to Lemma 31 can be observed for the syntactic learning criteria. In the following we show that this is not the case.

Theorem 45. [**ItTxtSNUEx**] $\subsetneq$ [**TxtBMS**$_*$**SNUEx**]

Proof. By Lemma 31 we have [**ItTxtSNUEx**] $\subseteq$ [**TxtBMS**$_*$**SNUEx**].

We consider the **BMS**-learner M initialized with state $\langle\langle ?,0\rangle,\langle\varnothing\rangle\rangle$ and h_M and s_M for every $\langle e,\xi\rangle \in \Omega$, $D \subseteq \mathbb{N}$ finite and $x \in \Sigma$ defined by:

$$s_M(\langle\langle e,\xi\rangle,\langle D\rangle\rangle,x) = \begin{cases} \langle\langle e,\xi\rangle,\langle D\rangle\rangle, & \text{if } x \in D \cup \{\#\} \vee \\ & \quad \mathrm{pr}_1(\varphi_x(\langle e,\xi\rangle)\!\downarrow) = e; \\ \langle\varphi_x(\langle e,\xi\rangle),\langle D \cup \{x\}\rangle\rangle, & \text{else if } \mathrm{pr}_1(\varphi_x(\langle e,\xi\rangle)\!\downarrow) \neq e; \\ \uparrow, & \text{otherwise.} \end{cases}$$

$$h_M(\langle\langle e,\xi\rangle,\langle D\rangle\rangle,x) = \begin{cases} e, & \text{if } x \in D \cup \{\#\} \vee \\ & \quad \mathrm{pr}_1(\varphi_x(\langle e,\xi\rangle)\!\downarrow) = e; \\ \mathrm{pr}_1(\varphi_x(\langle e,\xi\rangle)), & \text{else if } \mathrm{pr}_1(\varphi_x(\langle e,\xi\rangle)\!\downarrow) \neq e; \\ \uparrow, & \text{otherwise.} \end{cases}$$

Additionally to the last hypothesis as well as exactly the data that already lead to a mind-change of M, some parameter ξ is stored, indicating whether a further mind-change may cause a syntactic U-shape.

Let $\mathcal{L} = \mathbf{TxtBMS}_*\mathbf{SNUEx}(M)$. We will show that there is no iterative learner $\mathbf{ItTxtSNUEx}$-learning $\mathcal{L}$. Assume N is an iterative learner with hypothesis generating function h_N and $\mathcal{L} \subseteq \mathbf{ItTxtEx}(N)$.

We obtain $L \in \mathcal{L} \setminus \mathbf{ItTxtSNUEx}(N)$ by applying 1-1 ORT [Cas74] referring to the Σ_1-predicates MC and NoMC, expressing that N does (not) perform a mind-change on a text built from parameters $a,b \in \mathcal{R}$. More specifically, the predicates state that N does converge and (not) make a mind-change when observing $\sigma \in \Sigma^{<\omega}$ after having observed $a[i]^\frown b(i)^\frown \#^{\ell_i}$, with $i \in \mathbb{N}$.

$$\psi_i(\ell) \Leftrightarrow N(a[i]^\frown b(i)^\frown \#^\ell) = N(a[i]^\frown b(i)^\frown \#^{\ell+1});$$

$$\mathrm{NoMC}(i,\sigma) \Leftrightarrow \exists \ell_i \in \mathbb{N}\,(\,\psi_i(\ell_i) \wedge \forall \ell < \ell_i\, \neg\psi_i(\ell) \wedge$$
$$N(a[i]^\frown b(i)^\frown \#^{\ell_i}{}^\frown \sigma)\!\downarrow = N(a[i]^\frown b(i)^\frown \#^{\ell_i})\,);$$

$$\mathrm{MC}(i,\sigma) \Leftrightarrow \exists \ell_i \in \mathbb{N}\,(\,\psi_i(\ell_i) \wedge \forall \ell < \ell_i\, \neg\psi_i(\ell) \wedge$$
$$N(a[i]^\frown b(i)^\frown \#^{\ell_i}{}^\frown \sigma)\!\downarrow \neq N(a[i]^\frown b(i)^\frown \#^{\ell_i})\,).$$

By 1-1 ORT [Cas74], applied to the recursive operator implicit in the following case distinction, there are recursive total functions a,b,e_1,e_2 with pairwise disjoint ranges and $e_0 \in \mathbb{N}$, such that for all $i,\xi \in \mathbb{N}$, $e \in \Omega$

$$\varphi_{a(i)}(\langle e,\xi\rangle) = \begin{cases} \langle e_0,\xi\rangle, & \text{if } e \in \{?,e_0\}; \\ \langle e_1(k),1\rangle, & \text{else if } \xi = 0, i \text{ even and } \exists k \leq i\,(\,e = e_1(k)\,); \\ \langle e_1(k),2\rangle, & \text{else if } \xi = 0, i \text{ odd and } \exists k \leq i\,(\,e = e_1(k)\,); \\ \langle e_2(k),0\rangle, & \text{else if } \xi = 1, i \text{ odd and } \exists k \leq i\,(\,e = e_1(k)\,); \\ \langle e_2(k),0\rangle, & \text{else if } \xi = 2, i \text{ even and } \exists k \leq i\,(\,e = e_1(k)\,); \\ \langle e,\xi\rangle, & \text{otherwise}; \end{cases}$$

$$\varphi_{b(i)}(\langle e,\xi\rangle) = \begin{cases} \langle e_1(i),\xi\rangle, & \text{if } e \in \{?,e_0\}; \\ \langle e,\xi\rangle, & \text{otherwise}; \end{cases}$$

$$W_{e_0} = \begin{cases} \mathrm{ran}(a[t_0]), & \text{if } t_0 \text{ is minimal with } \forall t \geq t_0 \, N(a[t]) = N(a[t_0]); \\ \mathrm{ran}(a), & \text{no such } t_0 \text{ exists}; \end{cases}$$

$$W_{e_1(i)} = \mathrm{ran}(a[i]) \cup \{b(i)\} \cup \begin{cases} \{a(j)\} & \text{for first } j \geq i \text{ found} \\ & \text{with } \mathrm{MC}(i, a(j)); \\ \varnothing, & \text{no such } j \text{ exists}; \end{cases}$$

$$W_{e_2(i)} = \mathrm{ran}(a) \cup \{b(i)\}.$$

As the learner constantly puts out e_0 on every text for W_{e_0}, we have $W_{e_0} \in \mathcal{L}$. Thus, also N learns the finite language W_{e_0} and t_0 exists. Note that by the iterativeness of N we obtain $N(a[t_0]) = N(a[t_0]^\frown a(i))$ for all $i \geq t_0$ and with this $N(a[t_0]^\frown b(t_0)^\frown \#^{\ell_{t_0}}) = N(a[t_0]^\frown a(i)^\frown b(t_0)^\frown \#^{\ell_{t_0}})$ for all $i \geq t_0$.

$W_{e_1(t_0)}$ and $W_{e_2(t_0)}$ also lie in $\mathcal{L}$. To see that M explanatory learns both of them, note that, after having observed $b(t_0)$, M only changes its mind from $e_1(t_0)$ to $e_2(t_0)$ after having seen $a(i)$ and $a(j)$ with $i, j \geq t_0$ and $i \in 2\mathbb{N}$ as well as $j \in 2\mathbb{N}+1$. This clearly happens for every text for the infinite language $W_{e_2(t_0)}$. As $|W_{e_1(t_0)} \setminus (\mathrm{content}(a[t_0]) \cup \{b(t_0)\})| \leq 1$, this mind change never occurs for any text for $W_{e_1(t_0)}$.

The syntactic non-U-shapedness of M's learning processes can be easily seen as for all $k, l \in \mathbb{N}$ the languages W_{e_0}, $W_{e_1(k)}$ and $W_{e_2(l)}$ are pairwise distinct, the learner never returns to an abandoned hypothesis and M only leaves hypothesis $\langle e_1(k), 0\rangle$ for $\langle e_1(k), \xi\rangle$, $\xi \neq 0$, if $W_{e_1(k)}$ is not correct.

Next, we show the existence of $j \geq t_0$ with $\mathrm{MC}(t_0, a(j))$. Assume towards a contradiction that j does not exist. Then $W_{e_1(t_0)} = \mathrm{content}(a[t_0]) \cup \{b(t_0)\}$. As M learns this language from the text $a[t_0]^\frown b(t_0)^\frown \#^\infty$, so does N. The convergence of N implies the existence of ℓ_{t_0}. Thus, for every $j \in \mathbb{N}$ we either have $N(a[t_0]^\frown b(t_0)^\frown \#^{\ell_{t_0}}^\frown a(j)) = N(a[t_0]^\frown b(t_0)^\frown \#^{\ell_{t_0}})$ or the computation of $N(a[t_0]^\frown b(t_0)^\frown \#^{\ell_{t_0}}^\frown a(j))$ does not terminate. Because N is iterative and learns $W_{e_2(t_0)}$, it may not be undefined and therefore always the latter is the case. But then N will not learn $W_{e_1(t_0)}$ and $W_{e_2(t_0)}$ as they are different but N does not make a mind-change on the text $a[t_0]^\frown b(t_0)^\frown \#^{\ell_{t_0}}^\frown a$ after having observed the initial segment $a[t_0]^\frown b(t_0)^\frown \#^{\ell_{t_0}}$, due to its iterativeness. Hence, j exists and $W_{e_1(t_0)} = \mathrm{ran}(a[t_0]) \cup \{b(t_0), a(j)\}$.

Finally, by the choice of j, the learner N does perform a syntactic U-shape on the text $a[t_0]^\frown a(j)^\frown b(t_0)^\frown \#^{\ell_{t_0}}^\frown a(j)^\frown \#^\infty$ for $W_{e_1(t_0)}$. More precisely, t_0 and ℓ_{t_0} were chosen such that $N(a[t_0]^\frown a(j)^\frown b(t_0)^\frown \#^{\ell_{t_0}})$ has to be correct and the characterizing property of j assures

$$N(a[t_0]^\frown a(j)^\frown b(t_0)^\frown \#^{\ell_{t_0}}) \neq N(a[t_0]^\frown a(j)^\frown b(t_0)^\frown \#^{\ell_{t_0}}^\frown a(j)).$$

Thus, no iterative learner can explanatory syntactically non-U-shapedly learn the language $\mathcal{L}$. □

By Corollary 44 we also obtain $[\mathbf{ItTxtSDecEx}] \subsetneq [\mathbf{TxtBMS_*SDecEx}]$ and $[\mathbf{ItTxtConvEx}] \subsetneq [\mathbf{TxtBMS_*ConvEx}]$.

5 Related Open Problems

We have given a complete map for learning with bounded memory states, where, on the way to success, the learner must use only finitely many states. Future work can address the complete maps for learning with an a priori bounded number of memory states, which needs very different combinatorial arguments. Results in this regard can be found in [CCJS07] and [CK13]. We expect to see trade-offs, for example allowing for more states may make it possible to add various learning restrictions (just as non-deterministic finite automata can be made deterministic at the cost of an exponential state explosion).

Also memory-restricted learning from positive and negative data (so-called informant) has only partially been investigated for iterative learners and not at all for other models of memory-restricted learning. Very interesting also in regard of 1-1 hypothesis spaces that prevent coding tricks is the **Bem**-hierarchy, see [FJO94], [LZ96] and [CJLZ99].

Acknowledgements. This work was supported by DFG Grant Number KO 4635/1-1. We are grateful to the people supporting us.

References

AKS18. Aschenbach, M., Kötzing, T., Seidel, K.: Learning from informants: relations between learning success criteria. arXiv preprint arXiv:1801.10502 (2018)

Ang80. Angluin, D.: Inductive inference of formal languages from positive data. Inf. Control **45**(2), 117–135 (1980)

BB75. Blum, L., Blum, M.: Toward a mathematical theory of inductive inference. Inf. Control **28**, 125–155 (1975)

BCM+08. Baliga, G., Case, J., Merkle, W., Stephan, F., Wiehagen, R.: When unlearning helps. Inf. Comput. **206**, 694–709 (2008)

Cas74. Case, J.: Periodicity in generations of automata. Math. Syst. Theory **8**(1), 15–32 (1974). https://doi.org/10.1007/BF01761704

Cas94. Case, J.: Infinitary self-reference in learning theory. J. Exp. Theor. Artif. Intell. **6**, 3–16 (1994)

CC13. Carlucci, L., Case, J.: On the necessity of U-shaped learning. Top. Cogn. Sci. **5**, 56–88 (2013)

CCJS07. Carlucci, L., Case, J., Jain, S., Stephan, F.: Results on memory-limited U-shaped learning. Inf. Comput. **205**, 1551–1573 (2007)

CJLZ99. Case, J., Jain, S., Lange, S., Zeugmann, T.: Incremental concept learning for bounded data mining. Inf. Comput. **152**, 74–110 (1999)

CK10. Case, J., Kötzing, T.: Strongly non-U-shaped learning results by general techniques. In: Kalai, A.T., Mohri, M. (eds.) COLT 2010, pp. 181–193 (2010)

CK13. Case, J., Kötzing, T.: Memory-limited non-U-shaped learning with solved open problems. Theoret. Comput. Sci. **473**, 100–123 (2013)

CK16. Case, J., Kötzing, T.: Strongly non-U-shaped language learning results by general techniques. Inf. Comput. **251**, 1–15 (2016)

CM08. Case, J., Moelius, S.: U-shaped, iterative, and iterative-with-counter learning. Mach. Learn. **72**, 63–88 (2008). https://doi.org/10.1007/s10994-008-5047-9

CM11. Case, J., Moelius, S.: Optimal language learning from positive data. Inf. Comput. **209**, 1293–1311 (2011)

FJO94. Fulk, M., Jain, S., Osherson, D.: Open problems in Systems That Learn. J. Comput. Syst. Sci. **49**(3), 589–604 (1994)

Gol67. Gold, E.: Language identification in the limit. Inf. Control **10**, 447–474 (1967)

Jan91. Jantke, K.P.: Monotonic and non-monotonic inductive inference of functions and patterns. In: Dix, J., Jantke, K.P., Schmitt, P.H. (eds.) NIL 1990. LNCS, vol. 543, pp. 161–177. Springer, Heidelberg (1991). https://doi.org/10.1007/BFb0023322

JKMS16. Jain, S., Kötzing, T., Ma, J., Stephan, F.: On the role of update constraints and text-types in iterative learning. Inf. Comput. **247**, 152–168 (2016)

JLZ07. Jain, S., Lange, S., Zilles, S.: Some natural conditions on incremental learning. Inf. Comput. **205**, 1671–1684 (2007)

JMZ13. Jain, S., Moelius, S., Zilles, S.: Learning without coding. Theoret. Comput. Sci. **473**, 124–148 (2013)

JORS99. Jain, S., Osherson, D., Royer, J., Sharma, A.: Systems that Learn: An Introduction to Learning Theory, 2nd edn. MIT Press, Cambridge (1999)

KP16. Kötzing, T., Palenta, R.: A map of update constraints in inductive inference. Theoret. Comput. Sci. **650**, 4–24 (2016)

KS16. Kötzing, T., Schirneck, M.: Towards an atlas of computational learning theory. In 33rd Symposium on Theoretical Aspects of Computer Science (2016)

KSS17. Kötzing, T., Schirneck, M., Seidel, K.: Normal forms in semantic language identification. In: Proceedings of Algorithmic Learning Theory, pp. 493–516. PMLR (2017)

LZ96. Lange, S., Zeugmann, T.: Incremental learning from positive data. J. Comput. Syst. Sci. **53**, 88–103 (1996)

MPU+92. Marcus, G., Pinker, S., Ullman, M., Hollander, M., Rosen, T.J., Xu, F.: Overregularization in language acquisition. monographs of the society for research in child development, vol. 57, no. 4. University of Chicago Press (1992). Includes commentary by H. Clahsen

Odi99. Odifreddi, P.: Classical Recursion Theory, vol. II. Elsivier, Amsterdam (1999)

OSW82. Osherson, D., Stob, M., Weinstein, S.: Learning strategies. Inf. Control **53**, 32–51 (1982)

OSW86. Osherson, D., Stob, M., Weinstein, S.: Systems that Learn: An Introduction to Learning Theory for Cognitive and Computer Scientists. MIT Press, Cambridge (1986)

RC94. Royer, J., Case, J.: Subrecursive Programming Systems: Complexity and Succinctness. Research monograph in Progress in Theoretical Computer Science. Birkhäuser, Boston (1994). https://doi.org/10.1007/978-1-4612-0249-3

SS82. Strauss, S., Stavy, R. (eds.): U-Shaped Behavioral Growth. Developmental Psychology Series. Academic Press (1982)

Wie91. Wiehagen, R.: A thesis in inductive inference. In: Dix, J., Jantke, K.P., Schmitt, P.H. (eds.) NIL 1990. LNCS, vol. 543, pp. 184–207. Springer, Heidelberg (1991). https://doi.org/10.1007/BFb0023324

Compression Techniques in Group Theory

Markus Lohrey[✉]

University of Siegen, Siegen, Germany
lohrey@eti.uni-siegen.de

Abstract. This paper gives an informal overview over applications of compression techniques in algorithmic group theory.

1 Algorithmic Problems in Group Theory

The study of computational problems in group theory goes back more than 100 years. In a seminal paper from 1911, Dehn posed three decision problems [15]: The *word problem*, the *conjugacy problem*, and the *isomorphism problem*. The word and conjugacy problem are defined for a finitely generated group G. This means that there exists a finite subset $\Sigma \subseteq G$ such that every element of G can be written as a finite product of elements from Σ. This allows to represent elements of G by finite words over the alphabet Σ. For the word problem, the input consists of such a finite word $w \in \Sigma^*$ and the goal is to check whether w represents the identity element of G. For the conjugacy problem, the input consists of two finite words $u, v \in \Sigma^*$ and the question is whether the group elements represented by u and v are conjugated. For the isomorphism problem the input consists of two finite group presentations (roughly speaking, two finite descriptions of groups in terms of generators and defining relations) and the question is whether these presentations describe isomorphic groups. Dehn's motivation for studying these abstract group theoretical problems came from topology. In his paper from 1912 [16], Dehn gave an algorithm that solves the word problem for fundamental groups of orientable closed 2-dimensional manifolds, but also realized that his three problems seem to be very hard in general. In [15], he wrote *"Die drei Fundamentalprobleme für alle Gruppen mit zwei Erzeugenden ... zu lösen, scheint einstweilen noch sehr schwierig zu sein."* (*Solving the three fundamental problems for all groups with two generators seems to be very difficult at the moment.*) When Dehn wrote this sentence, a formal definition of computability was still missing. So, it is not surprising that it took more than 40 years until Novikov [56] and independently Boone [11] proved that the word problem and hence also the conjugacy problem are in general undecidable for finitely presented groups. The isomorphism problems was shown to be undecidable by Adjan [1].

In this paper we are mainly interested in the word problem. Despite the undecidability results from [11,56], for many groups the word problem is decidable. Dehn's result for fundamental groups of orientable closed 2-dimensional manifolds was extended to one-relator groups (finitely generated groups that can be

© Springer Nature Switzerland AG 2021
L. De Mol et al. (Eds.): CiE 2021, LNCS 12813, pp. 330–341, 2021.
https://doi.org/10.1007/978-3-030-80049-9_30

defined by single defining relation) by his student Magnus [49]. Other important classes of groups with a decidable word problem are:

- automatic groups [21] (including important classes like braid groups [4], Coxeter groups, right-angled Artin groups, hyperbolic groups [24]),
- finitely generated linear groups, i.e., finitely generated groups that can be faithfully represented by matrices over a field [58] (including polycyclic groups and nilpotent groups), and
- finitely generated metabelian groups (they can be embedded in direct products of linear groups [65]).

With the rise of computational complexity theory in the 1960's, also the computational complexity of group theoretic problems moved into the focus of research. From the very beginning, this field attracted researchers from mathematics as well as computer science. One of the early results in this context was that for every given $n \geq 0$ there exist groups for which the word problem is decidable but does not belong to the n-th level of the Grzegorczyk hierarchy (a hierarchy of decidable problems) [13]. On the other hand, for many prominent classes of groups the complexity of the word problem is quite low. For instance, for automatic groups, the word problem can be solved in quadratic time [21], and for the subclass of hyperbolic groups the word problem can be solved in linear time (even real time) [31].

For finitely generated linear groups Lipton and Zalcstein [39] (for fields of characteristic zero) and Simon [62] (for prime characteristic) proved in 1977 (resp., 1979) that deterministic logarithmic space (L for short) suffices to solve the word problem. This was the first result putting the word problem for an important class of groups into a complexity class below polynomial time. The class L is located between the classes NC^1 and NC^2 (NC stands for Nick's class after Nicolas Pippenger). The circuit complexity class NC^k consists of all problems that can be solved by uniform polynomial size boolean circuits of bounded fan-in and depth $(\log n)^k$. The class $NC = \bigcup_{k \geq 1} NC^k$ is usually identified with the class of problems that have an efficient parallel algorithm. It is a subclass of Ptime and it is a famous open problem whether $NC = Ptime$. In his thesis [59] from 1993, Robinson investigated the parallel complexity of word problems in more detail. He proved that for several important classes of groups (nilpotent groups, polycyclic groups, solvable linear groups) the word problem belongs to (subclasses of) NC^1. For the free group of rank two, he proved that the word problem is hard for NC^1 (and since it is linear, the word problem belongs to L). Other groups with low complexity word problems are hyperbolic groups (NC^2 due to Cai [12], which was improved to $LOGCFL \subseteq NC^2$ in [40]), Thompson's group V (NC^2 due to Birget [10]), Baumslag-Solitar groups[1] (L due to Weiß [66]) and of course finite groups. A famous result of Barrington [7] says that for every finite non-solvable group the word problem is NC^1-complete. In recent years, also the class $TC^0 \subseteq NC^1$ came into the focus of group theorists. Roughly speaking,

---

[1] These are the one-relator groups $BS(p,q) = \langle a, t \mid t^{-1}a^p t = a^q \rangle$.

uniform TC^0 captures the complexity of multiplying two binary encoded integers. It turned out that for many interesting groups the word problem belongs to uniform TC^0. This includes finitely generated solvable linear groups [37] and all subgroups of groups that can be obtained from finitely generated solvable linear groups using direct products and wreath products [53]. This includes for instance all metabelian groups and free solvable groups.

2   Compression with Straight-Line Programs

In recent years, compression techniques have led to important breakthroughs concerning the complexity of word problems. The general strategy (which is not restricted to word problems) is to use data compression to avoid the storage of huge intermediate data structures. For solving the word problem in automorphism groups and certain group extensions (in particular, semi-direct products), so called *straight-line programs* turned out to be the right compressed representation. A straight-line program is a context-free grammar that produces only a single word. A typical example is the context-free grammar $S \to ABA$, $A \to CBC$, $B \to CcC$, $C \to DaD$, $D \to bb$. The nonterminal C produces the word $bbabb$, hence B produces $bbabb\,c\,bbabb$. Then, A produces $bbabb\,bbabbcbbabb\,bbabb$. Finally, the start nonterminal S produces

$$bbabbbbabbcbbabbbbabb\ bbabbcbbabb\ bbabbbbabbcbbabbbbabb$$

The length of the word produced by a straight-line program $\mathcal{G}$ can be exponential in the length of $\mathcal{G}$, where the latter is usually defined as the sum of the lengths of all right-hand sides of the grammar (14 in the above example). In other words, straight-line programs allow for exponential compression rates in the best case. Let us just mention that straight-line programs are a very active area in string algorithms and data compression, see for instance [14,42].

Here, we are interested in group theoretical applications of straight-line programs. One of the first such applications is the so-called reachability theorem of Babai and Szemerédi for finite groups [6]. It says that if G is a finite group of order n and $S \subseteq G$ is any generating set of G such that $S = S^{-1}$, then every element $g \in G$ can be defined by a straight-line program with terminal alphabet S and size $\mathcal{O}((\log n)^2)$. Babai and Szemerédi used this result for the solution of subgroup membership problems in finite black-box groups.

2.1   Compressed Word Problems

Here, we are mainly interested in finitely generated infinite groups. Straight-line programs entered this area with the so-called *compressed word problem*. The compressed word problem for a finitely generated group G is the variant of the word problem for G where the input word is represented by a straight-line program. The compressed word problem can be also explained in terms of circuits. Define a circuit over the group G as a directed acyclic graph, where the nodes of indegree 0 are labelled with group generators and all other nodes have exactly two incoming edges (they have to be ordered in the sense that there is a left and a right incoming edge). Moreover, there is a distinguished output node. Such a circuit computes an element of G in the natural way (every inner node computes the product of the two incoming group elements). Then, the compressed word problem for G is equivalent to the problem whether a given circuit over the group G evaluates to the group identity.

Schleimer [61] observed that the (standard) word problem for every finitely generated subgroup of the automorphism group of a group G is polynomial time reducible to the compressed word problem for G. Similar transfer results hold for semi-direct products and other group extensions. For instance, the word problem for a semi-direct product $K \rtimes Q$ is logspace reducible to (i) the word problem for Q and (ii) the compressed word problem for K [43]. These results make the compressed word problem interesting for the efficient solution of standard word problems. It has been shown before Schleimer's work that the compressed word problem for a free group can be solved in polynomial time (the problem is in fact Ptime-complete) [41]. As a consequence, the word problem for the automorphism group of a free group (which is finitely generated) can be solved in polynomial time [61]. This solved an open problem from [36]. Schleimer's result has drawn interest on the compressed word problem in the combinatorial group theory community. In general, the complexity of the compressed word problem is higher than the complexity of the standard word problem, since the input is given in a more succinct way (we will see concrete examples later). Nevertheless, there are, in addition to free groups, many groups with a polynomial time compressed word problem:

(i) finite groups. It is easy to see that the compressed word problem for a finite group can be solved in polynomial time. Less trivial is the fact that for every finite non-solvable group the compressed word problem is Ptime-complete [9].
(ii) hyperbolic groups [32] and, more generally, groups that are hyperbolic relative to a collection of free abelian subgroups [33]
(iii) fully residually free groups [48],
(iv) right-angled Artin groups [28,45], and, more generally, virtually special groups (finite extensions of subgroups of graph groups) [43]. By the work of Agol, Haglund and Wise [2,26,67], virtually special groups are tightly connected to low dimensional topology and contain many other important classes of groups (Coxeter groups, one-relator groups with torsion, fully residually free groups, and fundamental groups of hyperbolic 3-manifolds).

The polynomial time algorithms from (ii), (iii) and (iv) are all based on the following important result: for two straight-line programs one can check in polynomial time whether they produce the same word. This result has been shown independently in [30,52,57].

For finitely generated virtually nilpotent groups, the compressed word problem belongs to the parallel complexity class NC^2 [37]. Finitely generated virtually nilpotent groups are in fact the larges class of infinite groups, for which the compressed word problem is known to be in NC.

If we allow randomization, we find further examples of groups where the compressed word problem can be parallelized efficiently: for finitely generated free metabelian groups and wreath products of the form $\left(\prod_{i=1}^{k} A_i \right) \wr \mathbb{Z}^n$, where every A_i is either $\mathbb{Z}$ or a cyclic group of prime order, the compressed word problem belongs to the class coRNC^2 (the complement of the randomized version of NC^2) [38]. To show this result, the compressed word problem for $\left(\prod_{i=1}^{k} A_i \right) \wr \mathbb{Z}^n$ is reduced to a special case of *polynomial identity testing* (PIT for short). This is the question, whether a given algebraic circuit over a polynomial ring evaluates to the zero polynomial [60]. It is known that for polynomials over the rings $\mathbb{Z}$ and $\mathbb{Z}_n$, PIT belongs to coRP (the complement of randomized polynomial time) [3,34]. In [38] it was shown that a special case of PIT, where the input circuit is a so-called powerful skew circuit over a polynomial ring $\mathbb{Z}[x]$ or $\mathbb{F}_p[x]$ (p a prime), belongs to coRNC^2. The compressed word problem for $\left(\prod_{i=1}^{k} A_i \right) \wr \mathbb{Z}^n$ is logspace reducible to this special case of PIT.

Using a reduction to the general PIT problem, the compressed word problems for the following groups were shown to be in coRP:

- finitely generated linear groups (which contain the above mentioned virtually special groups), [43,45]
- wreath products of the form $G \wr H$, where G is finitely generated abelian and H is finitely generated virtually abelian [38].

PIT is a famous problem in complexity theory. Proving $\mathsf{PIT} \in \mathsf{Ptime}$ would imply spectacular progress on circuit complexity lower bounds [35]. Therefore, complexity theorists believe that proving $\mathsf{PIT} \in \mathsf{Ptime}$ will be extremely difficult. In [43] it was shown that PIT can be reduced in logspace to the compressed word problem for the linear group $\mathsf{SL}(3, \mathbb{Z})$ (all (3×3)-matrices over the integers with determinant 1), showing that the two problems are equivalent with respect to logspace reductions. Hence, proving that the compressed word problem for $\mathsf{SL}(3, \mathbb{Z})$ belongs to Ptime seems to be very difficult.

Besides specific classes of groups, also constructions that allow to build new groups from existing groups are important in group theory. For the following important group theoretical constructions the compressed word problem for the constructed group is polynomial time Turing-reducible to the compressed word problems for the constitutent groups: finite group extensions [43,45], HNN extensions with finite associated subgroups [27], amalgamated free products with finite amalgamated subgroups [27], graph products [28].

Another important construction in group theory is the wreath product. We have already seen some positive results for wreath products of abelian groups (at least if we allow randomization). It turns out that the wreath product does not preserve the complexity of the compressed word problem in general. Based on a characterization of the class PSPACE in terms of so-called leaf languages [29], it was shown in [8] that for many groups G the compressed word problem for the wreath product $G \wr \mathbb{Z}$ is PSPACE-complete. Concrete examples of such groups G are finite non-solvable groups and free groups of rank at least two.[2] Since the compressed word problem for these groups as well as for $\mathbb{Z}$ belongs to L, one obtains two important consequences: (i) wreath products may strictly increase the complexity of the compressed word problem (L is a proper subclass of PSPACE) and (ii) there exist groups for which the compressed word problems is strictly more difficult than the standard word problem (for this one needs the fact that the word problem for a wreath product $G \wr H$ is logspace reducible to the word problems for G and H [63]).

Using the same technique as for wreath products, it was also shown in [8] that the compressed word problem is PSPACE-complete for the Grigorchuk group and Thompson's group F. These groups are famous for their quite unusual properties. Let us just mention that the Grigorchuk group was the first example of a group of intermediate growth. The Grigorchuk group belongs to the rich class of automaton groups (which should not be confused with the class of automatic groups). Recently, examples of automaton groups with an EXPSPACE-complete compressed word problem (and PSPACE-complete word problem) were constructed in [64].

2.2 Power Words

In some group theoretical applications, the straight-line programs that appear have a very restricted form: a *power word* has the form $w_1^{n_1} w_2^{n_2} \cdots w_k^{n_k}$, where the exponents $n_1, \ldots, n_k$ are integers that are given in binary encoding and the words $w_1, \ldots, w_k$ are given explicitly (uncompressed). Using the iterated squaring trick, one can translate a power word into an equivalent straight-line program in logspace. Power words were used in order to solve algorithmic problems for (2×2)-matrix groups. Consider the group $\mathsf{GL}(2, \mathbb{Z})$ of all (2×2)-matrices over the integers with determinant ± 1. The natural representation of elements in this group consists of 4-tuples of binary encoded integers. In [44] it was shown that for this input representation the subgroup membership problem (does a given element of $\mathsf{GL}(2, \mathbb{Z})$ belong to a given finitely generated subgroup of $\mathsf{GL}(2, \mathbb{Z})$?) can be solved in polynomial time. An analogous result was shown in [25] for the modular group $\mathsf{PSL}(2, \mathbb{Z})$. Let us briefly sketch the proof for $\mathsf{GL}(2, \mathbb{Z})$. It is a well-known fact that $\mathsf{GL}(2, \mathbb{Z})$ is virtually-free, i.e., it has a free subgroup of finite index. The connection to power words is made by the observation that a matrix $A \in \mathsf{GL}(2, \mathbb{Z})$ can be translated into a power word $w_1^{n_1} w_2^{n_2} \cdots w_k^{n_k}$ over a fixed (but arbitrarily chosen) finite generating set of $\mathsf{GL}(2, \mathbb{Z})$. Thus, evaluating $w_1^{n_1} w_2^{n_2} \cdots w_k^{n_k}$ in the group $\mathsf{GL}(2, \mathbb{Z})$ yields the matrix A. Therefore, it suffices to show that for every virtually-free group G, the so called power subgroup membership problem for G belongs to Ptime. The power subgroup membership problem for G is the subgroup membership problem for G, where all input elements of G

---

[2] In fact, PSPACE-hardness of the compressed word problem for $G \wr \mathbb{Z}$ holds for a quite large class of non-solvable groups, namely all so-called uniformly SENS groups G [8], whereas for every non-abelian group G, the compressed word problem for $G \wr \mathbb{Z}$ is already coNP-hard [43].

are represented by power words. One can easily get rid off the finite extension, which leaves the power subgroup membership problem for a free group. This problem is finally solved in polynomial using an adaptation of Stallings folding procedure. The ordinary subgroup membership problem for a free group, where all group elements are given by finite words, is known to the Ptime-complete [5].

The proof for $\mathsf{PSL}(2, \mathbb{Z})$ [25] follows the same strategy as for $\mathsf{GL}(2, \mathbb{Z})$. Due to the simpler algebraic structure of $\mathsf{PSL}(2, \mathbb{Z})$ (it is isomorphic to the free product $\mathbb{Z}_2 * \mathbb{Z}_3$), it suffices to solve the power subgroup membership problem for a finitely generated free group, where the input power words have the form $a_1^{n_1} a_2^{n_2} \cdots a_k^{n_k}$ for free generators $a_1, \ldots, a_k$, in polynomial time.

Power words have been also studied in the context of the word problem. The power word problem for a finitely generated group G is the word problem for G, where the input word is given as a power word. In [46] it was shown that the power word problem for a finitely generated free group F_k is logspace reducible to the standard word problem for F_k. Since F_k is a finitely generated linear group, the result of Lipton and Zalcstein [39] implies that the word problem, and hence also the power word problem, for every finitely generated free group can be solved in logspace.

For the following groups, the power word problem even belongs to TC^0:

- wreath products of the form $G \wr \mathbb{Z}$ with G finitely generated nilpotent [22],
- right iterated wreath products of the form $\mathbb{Z}^{n_1} \wr (\mathbb{Z}^{n_2} \wr (\mathbb{Z}^{n_3} \wr \cdots \wr \mathbb{Z}^{n_k}))$ and, as a consequence of the Magnus embedding [50], free solvable groups [22],
- solvable Baumslag-Solitar groups $\mathsf{BS}(1, q)$ [47].

Interestingly, it was shown in [46] that the power word problems for Thompson's group F and all wreath products $G \wr \mathbb{Z}$ with G free of rank at least two or finite non-solvable are coNP-complete.[3] Recall that the compressed word problems for these groups are PSPACE-complete [8]. For the Grigorchuk group the power word problem belongs to L [46], whereas the compressed word problem is again PSPACE-complete [8]. This yields an example of a group, where the compressed word problem is strictly more difficult than the power word problem.

In the commutative setting, power words can be traced back to work from the 1990's. Ge [23] showed that one can verify in polynomial time an identity $\alpha_1^{n_1} \alpha_2^{n_2} \cdots \alpha_n^{n_n} = 1$, where the α_i are elements of an algebraic number field and the n_i are binary encoded integers.

3   Compression Beyond Straight-Line Programs

Recall that straight-line programs were applied to word problems for automorphism groups (and certain group extensions) and yield in some cases polynomial time algorithms. This is achieved by representing long words that appear as intermediate results in computations succinctly by straight-line programs. In the best case, a straight-line program allows to represent a word of length n in space

---

[3] coNP-hardness holds for every uniformly SENS group G.

$\log n$. For some word problems, this exponential compression is not enough. This holds in particular for groups with extremely fast growing Dehn functions like the Baumslag group or Higman's group. The Dehn functions for these groups have recursive but non-elementary growth. If one tries to solve the word problem naively, one obtains intermediate words of non-elementary length. Therefore, it was conjectured that these groups may have very hard word problems. But this turned out to be wrong. For both the Baumslag group [54] as well as Higman's group [17], the word problem can be solved in polynomial time. To prove these results, *power circuits* were introduced in [55]. Power circuits allow to represent huge integers, which arise as exponent towers, succinctly. Moreover, comparison and the arithmetic operations $x + y$ and $x \cdot 2^y$ on numbers that are represented by power circuits can be carried out in polynomial time. Recently, the power circuit technique has been further developed in [51], where it was shown that the word problem for the Baumslag group belongs to NC. Further work on power circuits in the context of group theory can be found in [18].

An even more extreme integer compression is used in [19]. Using the so-called hydra groups, a family of groups G_k $(k \geq 1)$ was constructed in [20] such that the Dehn functions of the groups G_k are arbitrarily high in the Ackermann hierarchy. Nevertheless, the word problem for every group G_k can be solved in polynomial time [19].

4   Open Problems

Let us conclude with some open problems related to compression in algorithmic group theory:

Linear Groups. Recall that the compressed word problem for a finitely generated linear group belongs to coRP. Showing that the compressed word problem for finitely generated linear groups belongs to Ptime seems to be very difficult (it would imply that polynomial identity testing belongs to Ptime). But what about restricted classes of linear groups? Braid groups and solvable linear groups might be good candidates to look at. Within the class of solvable linear groups one might first investigate polycyclic groups or solvable Baumslag-Solitar groups $BS(1, q)$. Also the power word problem for linear groups might be interesting to look at. The author is not aware of any better upper bound than coRP (the same upper bound as for the compressed word problem for linear groups). Recall that for the solvable and linear Baumslag-Solitar groups $BS(1, q)$ the power word problem belongs to TC^0 [47]. Is it possible to extend this result to all solvable linear groups?

Baumslag-Solitar Groups. Weiß [66] showed that the word problem for every Baumslag-Solitar group $BS(p, q)$ can be solved in logspace by reducing it in logspace to the word problem for a free group. The same reduction does not work in logspace for the compressed word problem. Currently, the best upper bound for the compressed word problem of a non-solvable Baumslag-Solitar group is PSPACE.

Right Iterated Wreath Products of Free Abelian Groups. Recall that for right iterated wreath products of free abelian groups the power word problem belongs to TC^0 [22]. This gives hope that the compressed word problem for these groups should be not too difficult. Since the word problem belongs to TC^0, a standard argument shows that the compressed word problem for every right iterated wreath product of free abelian groups lies in the counting hierarchy. This makes PSPACE-hardness quite unlikely. The compressed word problem for a wreath product of two free abelian groups belongs to coRP [38]. It would be interesting to see whether this result can be extended to all right iterated wreath products of free abelian groups.

Subgroup Membership Problems. In [44] it is shown that the subgroup membership problem for a free group can be solved in polynomial time, when all group element are specified by power words. Is it possible to extend this result to the case where all group element are specified by straight-line programs. Straight-line programs are strictly more succinct than power words. On could try to come up with an extension of Stallings' folding procedure to the case where edges are labelled with straight-line programs (the same strategy with power words instead of straight-line programs was successful in [44]).

References

1. Adjan, S.I.: The unsolvability of certain algorithmic problems in the theory of groups. Trudy Moskov. Mat. Obsc. **6**, 231–298 (1957). in Russian
2. Agol, I.: The virtual Haken conjecture. Documenta Mathematica **18**, 1045–1087 (2013). With an appendix by Ian Agol, Daniel Groves, and Jason Manning
3. Agrawal, M., Biswas, S.: Primality and identity testing via Chinese remaindering. J. Assoc. Comput. Mach. **50**(4), 429–443 (2003)
4. Artin, E.: Theorie der Zöpfe. Abh. Math. Semin. Univ. Hambg. **4**(1), 47–72 (1925)
5. Avenhaus, J., Madlener, K.: The Nielsen reduction and P-complete problems in free groups. Theoret. Comput. Sci. **32**(1–2), 61–76 (1984)
6. Babai, L., Szemerédi, E.: On the complexity of matrix group problems I. In: Proceedings of the 25th Annual Symposium on Foundations of Computer Science, FOCS 1984, pp. 229–240 (1984)
7. Barrington, D.A.M.: Bounded-width polynomial-size branching programs recognize exactly those languages in NC^1. J. Comput. Syst. Sci. **38**, 150–164 (1989)
8. Bartholdi, L., Figelius, M., Lohrey, M., Weiß, A.: Groups with ALOGTIME-hard word problems and PSPACE-complete circuit value problems. In: Proceedings of the 35th Computational Complexity Conference, CCC 2020. LIPIcs, vol. 169, pp. 29:1–29:29. Schloss Dagstuhl - Leibniz-Zentrum für Informatik (2020)
9. Beaudry, M., McKenzie, P., Péladeau, P., Thérien, D.: Finite monoids: from word to circuit evaluation. SIAM J. Comput. **26**(1), 138–152 (1997)

10. Birget, J.-C.: The groups of Richard Thompson and complexity. Int. J. Algebra Comput. **14**(5–6), 569–626 (2004)
11. Boone, W.W.: The word problem. Ann. Math. Second Ser. **70**, 207–265 (1959)
12. Cai, J.-Y.: Parallel computation over hyperbolic groups. In: Proceedings of the 24th Annual Symposium on Theory of Computing, STOC 1992, pp. 106–115. ACM Press (1992)
13. Cannonito, F.B.: Hierarchies of computable groups and the word problem. J. Symb. Log. **31**, 376–392 (1966)
14. Charikar, M., et al.: The smallest grammar problem. IEEE Trans. Inf. Theory **51**(7), 2554–2576 (2005)
15. Dehn, M.: Über unendliche diskontinuierliche Gruppen. Math. Ann. **71**, 116–144 (1911). In German
16. Dehn, M.: Transformation der Kurven auf zweiseitigen Flächen. Math. Ann. **72**, 413–421 (1912). In German
17. Diekert, V., Laun, J., Ushakov, A.: Efficient algorithms for highly compressed data: the word problem in Higman's group is in P. Int. J. Algebra Comput. **22**(8) (2012)
18. Diekert, V., Myasnikov, A., Weiß, A.: Conjugacy in Baumslag's group, generic case complexity, and division in power circuits. Algorithmica **76**(4), 961–988 (2016)
19. Dison, W., Einstein, E., Riley, T.R.: Taming the hydra: the word problem and extreme integer compression. Int. J. Algebra Comput. **28**(7), 1299–1381 (2018)
20. Dison, W., Riley, T.R.: Hydra groups. Commentarii Mathematici Helvetici **88**(3), 507–540 (2013)
21. Epstein, D.B.A., Cannon, J.W., Holt, D.F., Levy, S.V.F., Paterson, M.S., Thurston, W.P.: Word Processing in Groups. Jones and Bartlett, Boston (1992)
22. Figelius, M., Ganardi, M., Lohrey, M., Zetzsche, G.: The complexity of knapsack problems in wreath products. In: Proceedings of the 47th International Colloquium on Automata, Languages, and Programming, ICALP 2020. LIPIcs, vol. 168, pp. 126:1–126:18. Schloss Dagstuhl - Leibniz-Zentrum für Informatik (2020)
23. Ge, G.: Testing equalities of multiplicative representations in polynomial time (extended abstract). In: Proceedings of the 34th Annual Symposium on Foundations of Computer Science, FOCS 1993, pp. 422–426 (1993)
24. Gromov, M.: Hyperbolic groups. In: Gersten, S.M. (ed.) Essays in Group Theory. MSRI, vol. 8, pp. 75–263. Springer, New York (1987). https://doi.org/10.1007/978-1-4613-9586-7_3
25. Gurevich, Y., Schupp, P.E.: Membership problem for the modular group. SIAM J. Comput. **37**(2), 425–459 (2007)
26. Haglund, F., Wise, D.T.: Coxeter groups are virtually special. Adv. Math. **224**(5), 1890–1903 (2010)
27. Haubold, N., Lohrey, M.: Compressed word problems in HNN-extensions and amalgamated products. Theory Comput. Syst. **49**(2), 283–305 (2011). https://doi.org/10.1007/s00224-010-9295-2
28. Haubold, N., Lohrey, M., Mathissen, C.: Compressed decision problems for graph products of groups and applications to (outer) automorphism groups. Int. J. Algebra Comput. **22**(8) (2013)
29. Hertrampf, U., Lautemann, C., Schwentick, T., Vollmer, H., Wagner, K.W.: On the power of polynomial time bit-reductions. In: Proceedings of the 8th Annual Structure in Complexity Theory Conference, pp. 200–207. IEEE Computer Society Press (1993)
30. Hirshfeld, Y., Jerrum, M., Moller, F.: A polynomial algorithm for deciding bisimilarity of normed context-free processes. Theoret. Comput. Sci. **158**(1 & 2), 143–159 (1996)

31. Holt, D.: Word-hyperbolic groups have real-time word problem. Int. J. Algebra Comput. **10**, 221–228 (2000)
32. Holt, D., Lohrey, M., Schleimer, S.: Compressed decision problems in hyperbolic groups. In: Proceedings of the 36th International Symposium on Theoretical Aspects of Computer Science, STACS 2019. LIPIcs, vol. 126, pp. 37:1–37:16. Schloss Dagstuhl - Leibniz-Zentrum für Informatik (2019)
33. Holt, D., Rees, S.: The compressed word problem in relatively hyperbolic groups. Technical report (2020). arxiv:2005.13917
34. Ibarra, O.H., Moran, S.: Probabilistic algorithms for deciding equivalence of straight-line programs. J. Assoc. Comput. Mach. **30**(1), 217–228 (1983)
35. Kabanets, V., Impagliazzo, R.: Derandomizing polynomial identity tests means proving circuit lower bounds. Comput. Complex. **13**(1–2), 1–46 (2004)
36. Kapovich, I., Myasnikov, A., Schupp, P., Shpilrain, V.: Generic-case complexity, decision problems in group theory, and random walks. J. Algebra **264**(2), 665–694 (2003)
37. König, D., Lohrey, M.: Evaluation of circuits over nilpotent and polycyclic groups. Algorithmica **80**(5), 1459–1492 (2018)
38. König, D., Lohrey, M.: Parallel identity testing for skew circuits with big powers and applications. Int. J. Algebra Comput. **28**(6), 979–1004 (2018)
39. Lipton, R.J., Zalcstein, Y.: Word problems solvable in logspace. J. Assoc. Comput. Mach. **24**(3), 522–526 (1977)
40. Lohrey, M.: Decidability and complexity in automatic monoids. Int. J. Found. Comput. Sci. **16**(4), 707–722 (2005)
41. Lohrey, M.: Word problems and membership problems on compressed words. SIAM J. Comput. **35**(5), 1210–1240 (2006)
42. Lohrey, M.: Algorithmics on SLP-compressed strings: a survey. Groups Complex. Cryptol. **4**(2), 241–299 (2012)
43. Lohrey, M.: The Compressed Word Problem for Groups. SM, Springer, New York (2014). https://doi.org/10.1007/978-1-4939-0748-9
44. Lohrey, M.: Subgroup membership in GL(2, Z). In: Proceedings of the 38th International Symposium on Theoretical Aspects of Computer Science, STACS 2021. LIPIcs, vol. 187, pp. 51:1–51:17. Schloss Dagstuhl - Leibniz-Zentrum für Informatik (2021)
45. Lohrey, M., Schleimer, S.: Efficient computation in groups via compression. In: Diekert, V., Volkov, M.V., Voronkov, A. (eds.) CSR 2007. LNCS, vol. 4649, pp. 249–258. Springer, Heidelberg (2007). https://doi.org/10.1007/978-3-540-74510-5_26
46. Lohrey, M., Weiß, A.: The power word problem. In: Proceedings of the 44th International Symposium on Mathematical Foundations of Computer Science, MFCS 2019. LIPIcs, vol. 138, pp. 43:1–43:15. Schloss Dagstuhl - Leibniz-Zentrum für Informatik (2019)
47. Lohrey, M., Zetzsche, G.: Knapsack and the power word problem in solvable Baumslag-Solitar groups. In: Proceedings of the 45th International Symposium on Mathematical Foundations of Computer Science, MFCS 2020. LIPIcs, vol. 170, pp. 67:1–67:15. Schloss Dagstuhl - Leibniz-Zentrum für Informatik (2020)
48. Macdonald, J.: Compressed words and automorphisms in fully residually free groups. Int. J. Algebra Comput. **20**(3), 343–355 (2010)
49. Magnus, W.: Das Identitätsproblem für Gruppen mit einer definierenden Relation. Math. Ann. **106**(1), 295–307 (1932)
50. Magnus, W.: On a theorem of Marshall Hall. Ann. Math. Second Ser. **40**, 764–768 (1939)

51. Mattes, C., Weiß, A.: Parallel algorithms for power circuits and the word problem of the Baumslag group. CoRR, abs/2102.09921 (2021)
52. Mehlhorn, K., Sundar, R., Uhrig, C.: Maintaining dynamic sequences under equality-tests in polylogarithmic time. In: Proceedings of the 5th Annual ACM-SIAM Symposium on Discrete Algorithms, SODA 1994, pp. 213–222. ACM/SIAM (1994)
53. Miasnikov, A., Vassileva, S., Weiß, A.: The conjugacy problem in free solvable groups and wreath products of Abelian groups is in TC^0. Theory Comput. Syst. **63**(4), 809–832 (2019)
54. Myasnikov, A., Ushakov, A., Won, D.W.: The word problem in the Baumslag group with a non-elementary Dehn function is polynomial time decidable. J. Algebra **345**(1), 324–342 (2011)
55. Myasnikov, A.G., Ushakov, A., Won, D.W.: Power circuits, exponential algebra, and time complexity. Int. J. Algebra Comput. **22**(6) (2012)
56. Novikov, P.S.: On the algorithmic unsolvability of the word problem in group theory. Am. Math. Soc. Transl. II Ser. **9**, 1–122 (1958)
57. Plandowski, W.: Testing equivalence of morphisms on context-free languages. In: van Leeuwen, J. (ed.) ESA 1994. LNCS, vol. 855, pp. 460–470. Springer, Heidelberg (1994). https://doi.org/10.1007/BFb0049431
58. Rabin, M.O.: Computable algebra, general theory and theory of computable fields. Trans. Am. Math. Soc. **95**, 341–360 (1960)
59. Robinson, D.: Parallel algorithms for group word problems. Ph.D. thesis, University of California, San Diego (1993)
60. Saxena, N.: Progress on polynomial identity testing-II. In: Agrawal, M., Arvind, V. (eds.) Perspectives in Computational Complexity. PCSAL, vol. 26, pp. 131–146. Springer, Cham (2014). https://doi.org/10.1007/978-3-319-05446-9_7
61. Schleimer, S.: Polynomial-time word problems. Commentarii Mathematici Helvetici **83**(4), 741–765 (2008)
62. Simon, H.-U.: Word problems for groups and contextfree recognition. In: Proceedings of Fundamentals of Computation Theory, FCT 1979, pp. 417–422. Akademie-Verlag (1979)
63. Waack, S.: The parallel complexity of some constructions in combinatorial group theory. J. Inf. Process. Cybern. EIK **26**, 265–281 (1990)
64. Wächter, J.P., Weiß, A.: An automaton group with PSPACE-complete word problem. In: Proceedings of the 37th International Symposium on Theoretical Aspects of Computer Science, STACS 2020. LIPIcs, vol. 154, pp. 6:1–6:17. Schloss Dagstuhl - Leibniz-Zentrum für Informatik (2020)
65. Wehrfritz, B.A.F.: On finitely generated soluble linear groups. Mathematische Zeitschrift **170**, 155–167 (1980)
66. Weiß, A.: A logspace solution to the word and conjugacy problem of generalized Baumslag-Solitar groups. In: Algebra and Computer Science, volume 677 of Contemporary Mathematics. American Mathematical Society (2016)
67. Wise, D.T.: The Structure of Groups with a Quasiconvex Hierarchy. Princeton University Press, Princeton (2021)

Computable Procedures for Fields

Russell Miller[1,2]([⊠])

[1] Queens College, 65-30 Kissena Blvd, Queens, NY 11367, USA
`Russell.Miller@qc.cuny.edu`
[2] C.U.N.Y. Graduate Center, 365 Fifth Avenue, New York, NY 10016, USA

Abstract. This tutorial will introduce listeners to many questions that can be asked about computable processes on fields, and will present the answers that are known, sometimes with proofs. This is not original work. The questions in greatest focus here include decision procedures for the existence of roots of polynomials in specific fields, for the irreducibility of polynomials over those fields, and for transcendence of specific elements over the prime subfield. Several of these questions are related to the construction of algebraic closures, making Rabin's Theorem prominent.

Keywords: Computability · Computable structure theory ·
Factorization · Field · Hilbert's Tenth Problem · Irreducibility ·
Polynomials · Rabin's Theorem · Root set · Splitting set

1 Introduction

The classic problems of factoring polynomials and finding their roots arose long before any formal notion of decidability existed. In Greece, geometric problems led to it, such as finding side lengths in right triangles: the Greeks knew that $\sqrt{2}$ was irrational, which is to say, that $X^2 - 2$ does not factor over $\mathbb{Q}$. In India, the sixth-century mathematician Brahmagupta investigated integer solutions to what came to be known in the West as the Pell equations $X^2 - dY^2 = 1$. (Not only did Pell trail Brahmagupta by a millennium in this study, but he was also not even the first on his own continent to consider these equations, being preceded in this by Fermat.) In 1900, the tenth of the problems posed by Hilbert for the new century was to find a method of determining whether an arbitrary diophantine equation $f = 0$, with $f \in \mathbb{Z}[X_1, X_2, \ldots]$, has a solution in integers.

Algorithms for answering various of these questions had been discovered over the centuries, of course, often independently in different cultures. With Alan Turing's 1936 definition of an algorithm – using what came to be known as a *Turing machine* – questions of decidability could be studied more rigorously. Not only could one ask whether classical algorithms really could be implemented on a Turing machine – for the most part, the answers were affirmative, although some of these algorithms require prohibitive time and memory resources – but

The author was partially supported by Grant #581896 from the Simons Foundation and by the City University of New York PSC-CUNY Research Award Program.

L. De Mol et al. (Eds.): CiE 2021, LNCS 12813, pp. 342–352, 2021.
https://doi.org/10.1007/978-3-030-80049-9_31

in certain cases, one could now prove that no algorithm at all could succeed. Hilbert appears not to have anticipated the possibility that his Tenth Problem would be resolved in this way, but indeed, in 1970, Matiyasevich [15] completed work by Davis, Putnam, and Robinson [2], proving that no Turing machine at all can accomplish the task demanded by Hilbert.

In this tutorial we will examine both decidability and undecidability results within these topics. We will focus on fields of characteristic 0, and on results in pure computability, rather than considering results from theoretical computer science and other disciplines. In doing so, we omit a significant and intriguing body of knowledge, but paradoxically, the time and space constraints on this abstract and tutorial preclude consideration of time and space constraints on the algorithms.

Apart from Sect. 6, we will restrict ourselves to countable (infinite) fields, the most natural subjects for our questions. A *presentation* of such a field is a first-order structure F with domain $\{x_0, x_1, \ldots\}$, in the signature of rings, that is isomorphic to F. (One can use $\mathbb{N}$ itself as the domain, but when studying fields this would create much confusion.) The *atomic diagram* $\Delta(F)$ of F essentially consists of the addition and multiplication tables for F, coded using a Gödel coding so that we may regard $\Delta(F)$ as a subset of $\mathbb{N}$. If $\Delta(F)$ is decidable, then F is a *computable presentation* of the field. A single field will have many presentations; some may be computable, but not all, and many fields have no computable presentation at all. In order to consider all fields, we often give ourselves the set $\Delta(F)$ as an oracle. The most basic countable field is the field $\mathbb{Q}$ of rational numbers, and we fix a single computable presentation of it to be used hereafter. (Often we will conflate the isomorphism type, the presentation, and the atomic diagram of a field, when it seems safe to do so.)

The work described here is not original in this article, although certain standard facts may go uncited. Historically important sources on computable fields includes work by van der Waerden [27], Fröhlich and Shepherdson [8], Rabin [23], Ershov [6], Metakides and Nerode [16], Fried and Jarden [7], and Stoltenberg-Hansen and Tucker [26], while [17] gives a helpful basic introduction to these topics.

2   Rabin's Theorem

The natural starting point is the simplest case: polynomials in a single variable X, over $\mathbb{Q}$. Systematic algorithms factoring polynomials in the polynomial ring $\mathbb{Q}[X]$ and finding their roots date back at least as far as Kronecker [14], who began with $\mathbb{Z}[X]$ and moved on to $\mathbb{Q}[X]$ and then to field extensions of $\mathbb{Q}$. It is worthwhile to examine his work: a good modern presentation appears in Edwards's *Galois Theory* [5, §§55–57]. In fact, determining whether an $f \in \mathbb{Z}[X]$ has a root x in $\mathbb{Z}$ is fairly trivial: x itself, if it exists, must divide the constant coefficient c_0, as it divides every other term in $0 = c_0 + c_1 x + \cdots + c_d x^d = f(x)$. If $c_0 = 0$, then 0 is a root; otherwise this leaves only finitely many possible values $x_1, \ldots, x_n$ for x, each of which can be tested by computing $f(x_i)$.

A similar procedure applies when $f \in \mathbb{Q}[X]$: after one clears the denominators in the coefficients, each prime power that divides the denominator of the possible root must also divide the leading coefficient. The more challenging problem is to determine reducibility: which $f \in \mathbb{Z}[X]$ can be factored there? Kronecker's algorithm for answering this question also appears in [5], and extends readily to $\mathbb{Q}[X]$.

We choose here to present a more general result: the theorem of Michael Rabin from 1960 [23, Thms. 7 & 8] that relates these questions to constructions of the algebraic closure of a given field. The version given here includes a fairly trivial extension to fields that have no computable presentation, as such fields are omitted from Rabin's own statement of the theorem.

Theorem 1 (Rabin's Theorem [23]). *There exist Turing functionals Φ and Ψ, such that, for every presentation F of any countable field,*

- *$\Phi^{\Delta(F)}$ computes $\Delta(K)$ for a presentation K of some algebraically closed field;*
- *and $\Psi^{\Delta(F)}$ computes an embedding $i : F \to K$ such that K is algebraic over the image $i(F)$.*

Thus K may be regarded as the algebraic closure of (the isomorphic image of) F, being both algebraically closed and algebraic over that image. Moreover, the following sets, each computably enumerable relative to $\Delta(F)$, are all Turing-equivalent (relative to $\Delta(F)$):

- *The image $i(F)$ of F, as a subset of the domain of K;*
- *the image $j(F)$ of F within any computable presentation K_0 of K, for an arbitrary F-computable embedding $j : F \to K_0$ with K_0 algebraic over $j(F)$;*
- *the root set $R_F = \{f \in F[X] : (\exists x \in F)\, f(x) = 0\}$ of F;*
- *the splitting set $S_F = \{f \in F[X] : (\exists \text{ nonconstant } g, h \in F[X])\, f = g \cdot h\}$.*

The Turing-equivalence of R_F and S_F may be surprising. It is quickly seen that $R_F \leq_T S_F$ (relative to $\Delta(F)$: this really means $R_F \leq_T S_F \oplus \Delta(F)$). Indeed, with an S_F-oracle, we can determine whether a given f factors over F, and if so, we can find a factorization (using $\Delta(F)$) and repeat the question for each factor until we have found the irreducible factors of f in $F[X]$. Then $f \in R_F$ just if one of these factors is linear. The reverse reduction is not so clear. However, it is soon seen that $S_F \leq_T i(F)$ (relative to $\Delta(F)$, again), since for $f \in F[X]$, we can factor the image of f in $K[X]$ into linear factors in $K[X]$, and the products of these linear factors yield all possible factorizations of f in $K[X]$. Then $f \in S_F$ just if one of those (finitely many) factorizations in $K[X]$ has all its coefficients in $i(F)$. To complete the equivalence, one reduces $i(F)$ to R_F: for any $x \in K$, we can find some $g \in F[X]$ with $(i \circ g)(x) = 0$, as K is algebraic over $i(F)$. Using R_F, we can determine whether g has roots in F – and if so, how many roots, by finding a root $a \in F$ and repeating the process for $\frac{g(X)}{X-a}$. Then compute $i(a)$ for each root a of g in F: x lies in $i(F)$ just if it is equal to one of these $i(a)$.

Rabin's Theorem is a classic example of computable structure theory. It reveals exactly how much one needs to know about F in order to construct the algebraic closure of F "around" F, with F as a decidable subfield. The pleasing

Turing-equivalence of R_F and S_F is a byproduct. Those readers who still feel that S_F is somehow more difficult to compute than R_F will find an affirmation of their intuition in [19,25], where it is shown that (for computable fields F algebraic over $\mathbb{Q}$) R_F is always 1-reducible to S_F, uniformly in $\Delta(F)$, whereas S_F can fail to be 1-reducible to R_F, or even bounded-Turing-reducible to R_F.

Useful corollaries of Rabin's Theorem include several theorems first proven by Kronecker. For example, Kronecker gave an algorithm for deciding $S_{\mathbb{Q}}$, but this now follows directly from Rabin's Theorem and Kronecker's algorithm for $R_{\mathbb{Q}}$. (Irreducibility in $\mathbb{Z}[X]$ is also now quickly seen to be decidable.) Edwards [5] gives Kronecker's actual algorithm, which enables him to construct algebraic closures from an intuitionistic point of view. Rabin's proof of his theorem may be seen as nonconstructive in certain respects: it essentially assumes the existence of the algebraic closure and builds a computable presentation of that closure, rather than constructing the algebraic closure directly.

Furthermore, if $E = F(a)$ is an algebraic field extension of F, then by Rabin's Theorem $S_E \leq_T S_F$, uniformly relative to $\Delta(E)$ and the embedding of F into E. This follows by giving an algorithm for deciding membership (of each $x \in K$) in the image of $F(a)$ from membership in $i(F)$, once the embedding $i : F \to K$ is extended to $F(a)$. Thus all number fields have decidable splitting sets and root sets. In turn, this allows one to compute the Galois group G of a finite algebraic extension E/F, viewed as a set of automorphisms of E: one can determine the order n of G and name its elements $g_1, \ldots, g_n$ so that $g_m(x)$ is computable uniformly from $m \leq n$ and $x \in E$. All that is needed is an S_F-oracle and the minimal polynomial of a primitive generator of E over F.

3   Polynomials in Several Variables

The reader will notice that, while we sketched a proof of the Turing-equivalence claims of Rabin's Theorem above, we never addressed the initial claim of the theorem: the uniform method of producing an algebraic closure K of the given field F and of situating F inside K, via an embedding $i : F \to K$, so as to view K accurately as the algebraic closure of F. This claim is not that difficult to prove when F is an *algebraic field*, by which we mean an algebraic extension of its prime subfield (which here is always $\mathbb{Q}$; if we considered characteristics $p > 0$, it would be the p-element subfield $\mathbb{F}_p$). However, the proof is significantly more difficult for non-algebraic fields. For those with no computable transcendence basis over $\mathbb{Q}$, even a non-uniform construction of K and i requires real work. We content ourselves here with referring the reader to the original paper [23].

However, another algorithm of Kronecker is worthy of notice here. We mentioned above that his method of deciding the splitting set of an algebraic extension $F(a)$, given the splitting set for F, was superseded by Rabin's Theorem (apart from intuitionistic considerations). Kronecker showed the same for a purely transcendental extension $F(t)$ of F, and this does not follow from Rabin's Theorem. Here $F(t)$ can be presented (given $\Delta(F)$) as the set of all rational functions in one variable t over F, i.e., quotients of polynomials in $F[t]$. Of course,

$F(t)$ does not sit inside the algebraic closure of F, so we appeal instead to Kronecker [14].

Kronecker found a trick for deciding whether a polynomial $f \in F(t)[X]$ factors there. (Once again, the modern source [5, §59] expounds his method well.) First he argued that we can clear out the denominators of the rational functions in $F(t)$ serving as coefficients of f, reducing the problem to the situation where f can be viewed as an element of $F[t][X]$, or equivalently, a polynomial in two variables in $F[T, X]$. If n is the degree of T in f, then any factorization of f in $F[T, X]$ produces a factorization of $f(T, T^{n+1})$ in $F[T]$. Using the splitting set of F as an oracle, we find all (finitely many) factorizations of $f(T, T^{n+1})$ in $F[T]$, and check whether any of them arises from a factorization of $f(T, X)$, thus deciding $S_{F(t)}$. Moreover, Rabin's Theorem, with $F(t)$ as the given field, shows that $R_{F(t)} \equiv_T S_{F(t)}$, so $R_{F(t)} \equiv_T S_F \equiv_T R_F$ as well.

This result allows us to move beyond single-variable polynomials when considering irreducibility.

Proposition 1. *Irreducibility of polynomials in $F[X_0, X_1, X_2, \ldots]$ is decidable by a uniform procedure using the splitting set S_F of F (and the atomic diagram $\Delta(F)$, if F is not computable) as an oracle.*

Proof. Applying Kronecker's trick recursively, we derive procedures for deciding irreducibility in $R_n = F[X_0, \ldots, X_n]$ for each n, uniformly in n. So, given $f \in F[X_0, X_1, \ldots]$, we simply find an n with $f \in R_n$ and apply the algorithm for that R_n. Of course, if f factors in $F[X_0, X_1, \ldots]$ at all, both factors must lie in this R_n, so the algorithm for R_n gives the correct answer.   □

Proposition 1 reveals a significant distinction between the single-variable and multi-variable situations. In $F[X]$, the questions of irreducibility and having a root are Turing-equivalent (relative to the atomic diagram $\Delta(F)$), by Rabin's Theorem: for example, with $F = \mathbb{Q}$, both $R_{\mathbb{Q}}$ and $S_{\mathbb{Q}}$ are decidable. However, in the multivariable situation, this fails. Irreducibility of polynomials in $\mathbb{Q}[X_0, X_1, \ldots]$ is decidable (and for F in general, it remains Turing-equivalent to S_F, so we do not even bother to give a separate name to the multivariable problem). However, the question of whether an $f \in \mathbb{Q}[X_0, X_1, \ldots]$ has a solution in $\mathbb{Q}$ poses a huge open problem. We refer to it as *Hilbert's Tenth Problem*, generalizing the original question posed by Hilbert.

Definition 1. *For a field F (or more generally a ring), Hilbert's Tenth Problem for F is the set*

$$\mathrm{HTP}(F) = \bigcup_n \{f \in F[X_0, \ldots, X_n] : (\exists (x_0, \ldots, x_n) \in F^{n+1}) \ f(x_0, \ldots, x_n) = 0\}.$$

So R_F is just the single-variable case of $\mathrm{HTP}(F)$. Of course $R_F \leq_T \mathrm{HTP}(F)$, indeed via a 1-reduction, but the converse in general is false. Indeed, the decidability of $\mathrm{HTP}(\mathbb{Q})$ itself is unknown: this set is computably enumerable, but there is no proof yet whether its Turing degree is the computable degree $\mathbf{0}$, or the degree $\mathbf{0}'$ of the Halting Problem – or conceivably even a different c.e. degree in between these two! Of course, Hilbert's original Tenth Problem was to give an algorithm deciding HTP($\mathbb{Z}$), which is now known from [15] to be undecidable, having degree $\mathbf{0}'$. It also remains unknown whether there is any existential formula defining the set $\mathbb{Z}$ within the field $\mathbb{Q}$: if such a definition exists, then HTP($\mathbb{Q}$) would have degree $\mathbf{0}'$ too, as membership questions about HTP($\mathbb{Z}$) could then be reduced to membership questions about HTP($\mathbb{Q}$) using that definition. Julia Robinson [24] created the first definition of $\mathbb{Z}$ in $\mathbb{Q}$, in 1949, by a $\forall\exists\forall\exists$-formula, thus showing that the theory of the field $\mathbb{Q}$ is undecidable. Within the past decade, Koenigsmann [13] gave a definition of $\mathbb{Z}$ in $\mathbb{Q}$ by a purely universal formula, but there are reasons to doubt whether an existential definition exists.

Although the situation of HTP($\mathbb{Q}$) remains unresolved, one certainly can build fields F (with $\Delta(F)$ computable) for which HTP(F) $\not\leq_T R_F$. Thus R_F can be strictly easier than HTP(F) under Turing reducibility. In Sect. 5 we will mention some further results concerning this question.

4   Transcendence Bases

For fields in general, the most basic question about an element is whether it is algebraic or transcendental. These questions can be asked relative to any subfield, but for us they will always refer to transcendence over the prime subfield: either $\mathbb{Q}$ or $\mathbb{F}_p$, depending on the characteristic. In every presentation F of any countable field, the prime subfield is always computably enumerable relative to $\Delta(F)$.

The prime subfield may be undecidable relative to $\Delta(F)$, but this can only occur if F has characteristic 0 and contains transcendental elements. For an algebraic x, Kronecker's decision procedure for $S_{\mathbb{Q}}$ allows us to find the minimal polynomial of x over $\mathbb{Q}$ and thus decide whether $x \in \mathbb{Q}$. Furthermore, it almost defines x within F, as only finitely many other elements of f can have the same minimal polynomial. (These other elements are the $\mathbb{Q}$-*conjugates* of x in F.)

A *transcendence basis* B for F (over $\mathbb{Q}$) is a way of extending this situation to all of F. By definition, B is a maximal subset of F algebraically independent over $\mathbb{Q}$, and so every $x \in F$ has a minimal polynomial over the subfield $\mathbb{Q}(B)$, which identifies x (relative to B) up to finitely many conjugates, in the same way that the minimal polynomial of an element of an algebraic field identifies that element. If we can enumerate a particular transcendence basis B, therefore, we are largely back in the comfortable situation of an algebraic field.

In [16], Metakides and Nerode provided the first example of a computable field with no computable transcendence basis. This result is sharp, as there is always a transcendence basis B that is co-c.e. relative to $\Delta(F)$: it consists of each element x_i in the domain of F that is independent over $\mathbb{Q}(x_0, \ldots, x_{i-1})$. The proof involves ensuring that F has infinite transcendence degree, but using a priority construction to guarantee that every infinite c.e. subset W_e of the domain contains an algebraic element. This is not difficult: the key is that the type of a transcendental x in a field is a nonprincipal type, i.e., not generated by any single formula, and therefore certainly not generated by any single existential formula. It follows that, no matter what finite amount of $\Delta(F)$ has been defined so far, it will always be consistent with that amount for x to be algebraic. So, when it comes time to make some element of W_e algebraic, it is always possible to do so.

The further complication in transcendental fields is that there is no canonical transcendence basis. The prime subfield of F is always c.e. relative to $\Delta(F)$, by a uniform enumeration procedure, and is rigid, so elements in each copy of an algebraic field F can be identified, up to conjugacy, by their minimal polynomials. However, even in fields such as the purely transcendental extension $K = \mathbb{Q}(t_0, t_1, \ldots)$ of $\mathbb{Q}$ (which is just the field of all rational functions over $\mathbb{Q}$ in the variables t_i), this property no longer holds. It is quickly seen that there are presentations of this K in which no generating set of transcendentals is c.e., and therefore, even the natural candidate $\{t_0, t_1, \ldots\}$ for a canonical transcendence basis does not succeed in the way one requires. Indeed, this is a significant open question.

Question 1 (Melnikov & Miller). For the field $K = \mathbb{Q}(t_0, t_1, \ldots)$, find the lowest complexity level $\mathcal{S}$ such that every presentation F of K is generated by some transcendence basis of complexity $\mathcal{S}$.

It is known that Π_1^1 is such a complexity level, and that Π_1^0 is the least candidate for $\mathcal{S}$, but this leaves a wide spread of possibilities. This question is closely related to the categoricity spectrum of the field $\mathbb{Q}(t_0, t_1, \ldots)$; see [9] for basic definitions and [18] for some results involving fields.

One might hope that, for each computable field K, there might at least exist a computable copy F of K with a computable transcendence basis. However, this hope was dashed by Kalimullin, Schoutens, and the author in [11].

Theorem 2 (Corollary 3 from [11]). *For every Turing degree $\mathbf{c} \leq \mathbf{0}'$, there exists a computable field K such that, in every computable copy $F \cong K$, every transcendence basis for F has degree $\geq \mathbf{c}$. If $\mathbf{c}$ is itself a c.e. degree, then we can also ensure that every computable copy actually has a basis of degree $\mathbf{c}$.*

Oddly, the results here were based largely on work in [22] that produced a computable field K, of infinite transcendence degree, such that every copy F of K has a transcendence basis computable from $\Delta(F)$. (In particular, every computable copy has a computable transcendence basis.) The argument there used existential formulas to define the elements of one particular transcendence basis: the basis elements were those x such that, for some y in the field, (x, y) formed a nontrivial solution to a Fermat polynomial $X^p + Y^p = 1$. This technique of "tagging" basis elements by adjoining roots of polynomials over those elements was subsequently extended by Poonen, Schoutens, Shlapentokh, and the author in [21]. Discussion of that work is beyond the scope of this tutorial, but we provide a short version of the relevant theorem here. Roughly, it states that fields in general are just as complex as any other class of structures. Graphs, groups, and partial orders are also maximally complex, whereas linear orders and trees (for example) are not.

Theorem 3 (Theorem 1.8 from [21]). *For every countable first-order struc-*
ture $\mathcal{M}$ in a finite signature, there exists a countable field K with the same
computable-structure-theoretic properties as $\mathcal{M}$.

To give a more precise, though incomplete, list of the properties preserved:
K has the same Turing degree spectrum as $\mathcal{M}$, the same categoricity spectrum
as $\mathcal{M}$, the same computable dimension as $\mathcal{M}$, and the same automorphism
spectrum as $\mathcal{M}$. Moreover, for every relation R on $\mathcal{M}$, there is a relation on K
with the same degree spectrum. (All of these properties are described in [21], and
most in [10]. Some of them require $\mathcal{M}$ to be a computable structure; if it is, then
K can also be taken to be computable.) Indeed, even properties unknown when
Theorem 3 was proven have turned out to carry over from $\mathcal{M}$ to K, such as the
degree of categoricity on a cone, defined in [1]. The theorem in general holds for
countable structures in computable signatures, not just finite signatures, with the
exception of certain simple but pathological structures known as *automorphically*
trivial structures; see [12] for those details.

5   Algebraic Fields

Algebraic fields are fields in which every element is algebraic, i.e., is the root of
some polynomial over the prime subfield, which in this section will always be
$\mathbb{Q}$. The class $\mathfrak{A}$ of such fields is very far from satisfying Theorem 3: procedures
involving fields in $\mathfrak{A}$ are in general much closer to computable, because each
element of such a field can be effectively identified up to conjugacy over $\mathbb{Q}$,
thanks to the decidability of $S_{\mathbb{Q}}$. The means that, for two presentations of fields
in $\mathfrak{A}$, the property of being isomorphic is far simpler than for fields in general.
(Theorem 3 shows the isomorphism relation to be $\mathit{\Pi}_1^1$-complete for fields in
general.)

Theorem 4. *Two fields $E, F \in \mathfrak{A}$ are isomorphic just if*

$$\{f \in \mathbb{Q}[X] : f \text{ has a root in } E\} = \{f \in \mathbb{Q}[X] : f \text{ has a root in } F\}.$$

Similarly, the elements $x_0, x_1 \in F$ lie in the same orbit under automorphisms of
F just if they have the same minimal polynomial over $\mathbb{Q}$ and

$$\{f \in \mathbb{Q}[X,Y] : f(x_0, Y) \in R_F\} = \{f \in \mathbb{Q}[X,Y] : f(x_1, Y) \in R_F\}.$$

Theorem 4 suggests that $\{f \in \mathbb{Q}[X] : f \text{ has a root in } F\}$ can serve as an index
for the isomorphism type of F, for each $F \in \mathfrak{A}$. This is the foundation of work in
[20] and [3,4], which uses these indices to place a topology on the space $\mathfrak{A}$ of all
algebraic field extensions of $\mathbb{Q}$. The same topology has been discovered by various
field theorists independently over the years: it is sometimes known as the *étale*
topology, or seen as the Vietoris topology on the space of all closed subgroups
of $\mathrm{Aut}(\overline{\mathbb{Q}})$. Each of these is the same topology on the space of all subfields
of $\overline{\mathbb{Q}}$; one mods out by the relation of isomorphism and imposes the quotient
topology in order to topologize the space of isomorphism types in $\mathfrak{A}$. Both the

étale topology and the quotient modulo isomorphism have the pleasing property of being homeomorphic to the set $2^{\mathbb{N}}$ under the usual Cantor topology, and this allows one to use elements of Cantor space as indices for the isomorphism types. Each index essentially specifies $\{f \in \mathbb{Q}[X] : f \text{ has a root in } F\}$, as described above, although a certain amount of coding is necessary. This creates an effective classification of $\mathfrak{A}$ by the elements of $2^{\mathbb{N}}$.

Theorem 5 (see [20]). *There exist Turing functionals Φ and Ψ such that, for all presentations E and F of fields in $\mathfrak{A}$ and all $S \in 2^{\mathbb{N}}$:*

- $\Phi^{E \oplus R_E} \in 2^{\mathbb{N}}$, *with* $\Phi^{E \oplus R_E} = \Phi^{F \oplus R_F}$ *if and only if* $E \cong F$; *and*
- Ψ^S *computes* $\Delta(F) \oplus R_F$ *for some presentation F of a field in $\mathfrak{A}$; and*
- $\Phi^{(\Psi^S)} = S$.

Eisenträger, Springer, Westrick, and the author have recently exploited this homeomorphism, using the Baire-category property of co-meagerness in $2^{\mathbb{N}}$ to prove the following.

Theorem 6 (see [4]). *In $\mathfrak{A}$ under the topology described above, the (isomorphism types of) fields satisfying all of the following properties form a comeager set.*

- *in some presentation F of the field, $R_F \not\leq_T \Delta(F)$; but*
- *in every presentation F of the field, $(R_F)' \leq_T (\Delta(F))'$; and*
- *in every presentation F of the field, $R_F \oplus \Delta(F) \equiv_T \mathrm{HTP}(F) \oplus \Delta(F)$.*

Thus the "generic" situation for algebraic extensions of $\mathbb{Q}$ is that the root set is noncomputable but always low relative to the atomic diagram. Moreover, the question of solvability of polynomial equations in several variables is "generically" only as hard as the same question for polynomials in a single variable (i.e., the root set), hence also low but generally noncomputable relative to $\Delta(F)$.

6   The Field $\mathbb{R}$

Abstractly, it is natural to consider the problem of whether a polynomial $f \in \mathbb{R}[X_0, \ldots, X_n]$ has a real solution $\boldsymbol{x} \in \mathbb{R}^n$ with $f(\boldsymbol{x}) = 0$. In practice, since $\mathbb{R}$ is uncountable, the techniques used here are entirely different from those for countable fields, and we content ourselves with a brief summary.

It has been known since the work of Tarski that the theory of the field $\mathbb{R}$ is decidable. From this one directly infers a decision procedure for the question of whether an $f \in \mathbb{Q}[X_0, \ldots, X_n]$ has a solution in $\mathbb{R}^n$. Indeed, we can describe it succinctly: if we find $\boldsymbol{x}$ and $\boldsymbol{y}$ in the dense subset $\mathbb{Q}^n$ with $f(\boldsymbol{x}) < 0 < f(\boldsymbol{y})$, then the Intermediate Value Theorem guarantees a solution of f in $\mathbb{R}^n$; whereas, if no such pair $(\boldsymbol{x}, \boldsymbol{y})$ exists, then by a theorem of Artin f is either a sum of squares of polynomials in $\mathbb{Q}[X_0, \ldots, X_n]$ or else the negation of a sum of such squares. For a sum of squares, f will have a solution only if the absolute minimum value of f is 0, and so basic calculus yields the endgame.

When the polynomial f is allowed to have arbitrary real coefficients, one must first explain how those coefficients are to be presented. The usual procedure, in computable analysis, is to use *fast-converging Cauchy sequences* $\langle q_n \rangle_{n \in \mathbb{N}}$ of rational numbers, with the limit $c \in \mathbb{R}$ of the sequence satisfying $|c - q_n| < 2^{-n}$ for all n. This is best viewed as an approximation of c by open intervals $(q_n - 2^{-n}, q_n + 2^{-n})$, all containing c, whose lengths decrease effectively to 0. Of course, a single c will have many such representations, including noncomputable ones. The book [28] is a standard source for computable analysis.

Over countable fields, the only important aspect of the root set is determining whether a root exists: if it does, one can simply search through the field until a root is found. Over $\mathbb{R}^n$, this is no longer applicable, so this problem bifurcates: the first problem is to decide the existence of a solution, and if one exists, the second problem is to produce a solution. The first of these is undecidable, even for $n = 1$, and the proof is fairly quick. Suppose Φ were a Turing functional that, when given an oracle containing $(d+1)$ Cauchy sequences converging fast to real numbers $c_0, \ldots, c_d$, outputs either "yes" if $\sum c_i X^i = 0$ has a solution in $\mathbb{R}$, or "no" if it has no solution. Run Φ on the monomial X^2, given by constant Cauchy sequences $(1, 1, 1, \ldots)$, $(0, 0, 0, \ldots)$ and $(0, 0, 0, \ldots)$ to represent $1X^2 + 0X^1 + 0X^0$. Φ must output "yes" after examining the first u terms of each sequence, for some finite "use" $u \in \mathbb{N}$. But then, if we run it again and replace the coefficient in the X^0 term by $(0, 0, \ldots, 0, 2^{-(u+1)}, 2^{-(u+1)}, \ldots)$ with u initial 0's, it will give the same output "yes," which will be incorrect: the polynomial is now $X^2 + \frac{1}{2^{u+1}}$, which has no root in $\mathbb{R}$.

The second problem is also undecidable, and again tangency is the culprit. For example, the polynomial $f(X) = X^4 - 2X^2 + 1$ has real roots ± 1, but an arbitrarily small nonzero linear coefficient c can make either of them disappear: for $c > 0$, $X^4 - 2X^2 + cX + 1$ has only negative roots, while when $c < 0$ it has only positive roots. This allows us to use a strategy similar to the above: wait for a functional Φ to compute its first approximation q_0 to a root of $f(X)$, and then perturb the linear coefficient just slightly, making it either positive (if $q_0 \geq 0$) or negative (otherwise).

References

1. Csima, B.F., Harrison-Trainor, M.: Degrees of categoricity on a cone via eta-systems. J. Symbolic Logic **82**(1), 325–346 (2017)
2. Davis, M., Putnam, H., Robinson, J.: The decision problem for exponential diophantine equations. Ann. Math. **74**(3), 425–436 (1961)
3. Eisenträger, K., Miller, R., Springer, C., Westrick, L.: A topological approach to undefinability in algebraic fields, submitted for publication
4. Eisenträger, K., Miller, R., Springer, C., Westrick, L.: Genericity and forcing for algebraic fields, in preparation
5. Edwards, H.M.: Galois Theory. Springer, New York (1984)
6. Ershov, Y.L.: Theorie der Numerierungen. Zeits. Math. Logik Grund. Math. **23**, 289–371 (1977)
7. Fried, M.D., Jarden, M.: Field Arithmetic. Springer, Berlin (1986)

8. Frohlich, A., Shepherdson, J.C.: Effective procedures in field theory. Phil. Trans. R. Soc. Lond. Ser. A **248**(950), 407–432 (1956)
9. Fokina, E., Kalimullin, I., Miller, R.: Degrees of categoricity of computable structures. Arch. Math. Logic **49**, 51–67 (2010)
10. Hirschfeldt, D.R., Khoussainov, B., Shore, R.A., Slinko, A.M.: Degree spectra and computable dimensions in algebraic structures. Ann. Pure Appl. Logic **115**, 71–113 (2002)
11. Kalimullin, I., Miller, R., Schoutens, H.: Degree spectra for transcendence in fields. In: Manea, F., Martin, B., Paulusma, D., Primiero, G. (eds.) CiE 2019. LNCS, vol. 11558, pp. 205–216. Springer, Cham (2019). https://doi.org/10.1007/978-3-030-22996-2_18
12. Knight, J.F.: Degrees coded in jumps of orderings. J. Symbolic Logic **51**, 1034–1042 (1986)
13. Koenigsmann, J.: Defining $\mathbb{Z}$ in $\mathbb{Q}$. Ann. Math. **183**(1), 73–93 (2016)
14. Kronecker, L.: Grundzüge einer arithmetischen Theorie der algebraischen Größen. Journal für die reine und angewandte Mathematik **92**, 1–122 (1882)
15. Matiyasevich, Y.V.: The Diophantineness of enumerable sets. Dokl. Akad. Nauk SSSR **191**, 279–282 (1970)
16. Metakides, G., Nerode, A.: Effective content of field theory. Ann. Math. Logic **17**, 289–320 (1979)
17. Miller, R.: Computable fields and Galois theory. Not. Am. Math. Soc. **55**(7), 798–807 (2008)
18. Miller, R.: d-Computable categoricity for algebraic fields. J. Symbolic Logic **74**(4), 1325–1351 (2009)
19. Miller, R.: Is it easier to factor a polynomial or to find a root? Trans. Am. Math. Soc. **362**(10), 5261–5281 (2010)
20. Miller, R.: Isomorphism and classification for countable structures. Computability **8**(2), 99–117 (2019)
21. Miller, R., Poonen, B., Schoutens, H., Shlapentokh, A.: A computable functor from graphs to fields. J. Symbolic Logic **83**(1), 326–348 (2018)
22. Miller, R., Schoutens, H.: Computably categorical fields via Fermat's last theorem. Computability **2**, 51–65 (2013)
23. Rabin, M.: Computable algebra, general theory, and theory of computable fields. Trans. Am. Math. Soc. **95**, 341–360 (1960)
24. Robinson, J.: Definability and decision problems in arithmetic. J. Symbolic Logic **14**, 98–114 (1949)
25. Steiner, R.M.: Computable fields and the bounded Turing reduction. Ann. Pure Appl. Logic **163**, 730–742 (2012)
26. Stoltenberg-Hansen, V., Tucker, J.V.: Computable rings and fields. In: Griffor, E.R. (ed.) Handbook of Computability Theory, pp. 363–447. Elsevier, Amsterdam (1999)
27. van der Waerden, B.L.: Algebra, volume I, trans. F Blum and J.R. Schulenberger. Springer, New York (1991) (1970 hardcover, 2003 softcover)
28. Weihrauch, K.: Computable Analysis: An Introduction. Springer, Berlin (2000)

Minimum Classical Extensions of Constructive Theories

Joan Rand Moschovakis[1,2]($\boxtimes$) and Garyfallia Vafeiadou[2]

[1] Department of Mathematics, Occidental College, Los Angeles, CA, USA
joan@math.ucla.edu
[2] Graduate Program in Logic, Algorithms and Computation,
Department of Mathematics, University of Athens, Athens, Greece
gvf@math.uoa.gr

Abstract. Reverse constructive mathematics, based on the pioneering work of Kleene, Vesley, Kreisel, Troelstra, Bishop, Bridges and Ishihara, is currently under development. Bishop constructivists tend to emulate the classical reverse mathematics of Friedman and Simpson. Veldman's reverse intuitionistic analysis and descriptive set theory split notions in the style of Brouwer. Kohlenbach's proof mining uses interpretations and translations to extract computational information from classical proofs. We identify the *classical content* of a constructive mathematical theory with the Gentzen negative interpretation of its classically correct part. In this sense **HA** and **PA** have the same classical content but intuitionistic and classical two-sorted recursive arithmetic with quantifier-free countable choice do not; Σ_1^0 numerical double negation shift expresses the precise difference. Other double negation shift and weak comprehension principles clarify the classical content of stronger constructive theories. Any consistent axiomatic theory **S** based on intuitionistic logic has a *minimum classical extension* $\mathbf{S}^{+g}$, obtained by adding to **S** the negative interpretations of its classically correct consequences. Subsystems of Kleene's intuitionistic analysis and supersystems of Bishop's constructive analysis provide interesting examples, with the help of constructive decomposition theorems.

1 There Is Virtue in Simplicity

The negative translations proposed by Gödel [7] and Gentzen [6] are straightforward syntactic methods for converting formulas of a full logical language into classically equivalent formulas not involving $\vee$ or $\exists$. The Gentzen negative translation E^g of a formula E replaces $\vee$ and $\exists$ by their classical equivalents in terms of $\neg$, & and $\forall$, but does not change $\rightarrow$. The Gödel translation also replaces $\rightarrow$ by its classical equivalent in terms of $\neg$ and &, but the simpler Gentzen version is more transparent and will be used in what follows. When necessary to guarantee the intuitionistic equivalence of $\neg\neg E^g$ and E^g, prime formulas are replaced by their double negations; this step is omitted in applications where prime formulas are stable under double negation.

© Springer Nature Switzerland AG 2021
L. De Mol et al. (Eds.): CiE 2021, LNCS 12813, pp. 353–362, 2021.
https://doi.org/10.1007/978-3-030-80049-9_33

The negative translations of classical logical axioms and rules are correct by intuitionistic logic, so if E follows from Γ by classical logic then E^g follows from Γ^g by intuitionistic logic. With classical logic E and E^g are equivalent. Even with intuitionistic logic, $\neg\neg E^g$ and E^g are equivalent.

1.1 Classical Content in Arithmetic and Analysis

In what follows, the "language of arithmetic" may be any first-order language with equality, constants $0, ', +, \cdot$ and possibly other constants for primitive recursive functions. Prime formulas are equations $s = t$ between terms. Classical arithmetic **PA** and intuitionistic arithmetic **HA** are expressed in this language.

The "language of analysis" is any two-sorted language extending the language of arithmetic, with variables $m, n, \ldots, x, y, z$ over natural numbers and variables $\alpha, \beta, \gamma, \ldots$ over infinite sequences of natural numbers (i.e. one-place number-theoretic functions), with constants for additional primitive recursive functions and functionals. Function application is denoted by $\alpha(x)$, and equality at type 1 is defined extensionally. Troelstra's **EL** and Kleene and Vesley's **I** are expressed in the language of analysis; cf. [12,19,21].

Some applications involve intuitionistic arithmetic of arbitrary finite types **HA**$^\omega$, an extension of **HA** in a language with variables over, and terms for, primitive recursive functions of all finite types. Prime formulas are equations between terms of the same type. There are intensional and extensional versions of **HA**$^\omega$; for details see [19].

Definition 1. *The* **classical content** *of a formula* E *in the language of arithmetic, analysis, or the arithmetic of finite types is its Gentzen negative translation* E^g. *The* **classical content** Γ^g *of a collection* Γ *of formulas consistent with classical logic is the closure under intuitionistic logic of the set* $\{E^g \colon E \in \Gamma\}$.

Remark 1. If **S** is a formal system, based on intuitionistic logic but consistent with classical logic, and if **T** comes from **S** by adding one or more classically correct *logical* axiom schemas, then $\mathbf{S}^g = \mathbf{T}^g$ so the classical content of **S** is determined by the negative translations of its *mathematical* axioms. Following Kleene [11] we denote $\mathbf{S} + (\neg\neg A \to A)$ by $\mathbf{S}^\circ$.

1.2 Minimum Classical Extension of a Constructive Theory

Definition 2. *If* **S** *is a formal system based on intuitionistic logic, in a language including* $\&, \vee, \to, \neg$, *and quantifiers* $\forall$ *and* $\exists$ *of one or more sorts, then the* **classical subtheory** cls(**S**) *of* **S** *is the set of all classically correct theorems of* **S***; the* **classical content** *of* **S** *is* $(\mathrm{cls}(\mathbf{S}))^g$*; and the* **minimum classical extension** $\mathbf{S}^{+g}$ *of* **S** *is the closure under intuitionistic logic of* $\mathbf{S} \cup (\mathrm{cls}(\mathbf{S}))^g$.

Remark 2. If **S** is an intuitionistic subsystem of a (consistent) classical theory then cls(**S**) = **S**, so $\mathbf{S}^{+g}$ is the closure under intuitionistic logic of $\mathbf{S} \cup \mathbf{S}^g$. Intuitionistic arithmetic is its own minimum classical extension because the negative interpretations of all the mathematical axioms of **HA** (and **PA**) are provable in **HA**. The prime formulas of arithmetic are equations between terms of type 0, which do not change under the translation because **HA** proves that they are decidable, hence stable under double negation.

However, the neutral (classically correct) "basic" subsystem **B** of Kleene and Vesley's formal system **I** for intuitionistic analysis does not contain its classical content, nor does Troelstra's recursive analysis **EL**. In each of these cases the classical content is nevertheless completely determined by the negative translations of the mathematical axioms of the system.

In the case that **S** is consistent but **S**° is not, cls(**S**) may include more than the consequences of the classically correct axioms of **S**. Intuitionistic analysis **I** differs from **B** by just one axiom schema (which conflicts with **B**°), but **I** proves monotone bar induction (which is consistent with **B**°) while **B** does not. The question then is how to determine the classical subtheory of a subsystem **S** of **I** with a classically false continuity axiom or schema, in order to identify $\mathbf{S}^{+g}$.

2 Double Negation Shift and Weak Comprehension Axioms

Double negation shift principles have long been studied as weaker alternatives to constructively questionable axioms like Markov's Principle (cf. [1,5,16,18]). In [2] Brouwer himself used double negation shift to prove that the intuitionistic real numbers form a closed species, though he later rejected this argument.

The most general double negation shift schema, whose addition to intuitionistic predicate logic would suffice to prove Glivenko's Theorem, is

$$\text{DNS}: \quad \forall x \neg\neg A(x) \rightarrow \neg\neg \forall x A(x).$$

The converse holds by intuitionistic predicate logic. The strength of an instance of double negation shift depends on the logical complexity of the formula $A(x)$ and the domain of the variable x.

2.1 "Double Negation Shift for Numbers" DNS_0

DNS_0 denotes the restriction of DNS to cases where x is a number variable. AC_0 denotes the axiom schema of countable choice, which was accepted by Bishop and Brouwer. Danko Ilik argues in [8] that $\mathbf{HA}^\omega + \text{AC}_0 + \text{DNS}_0$ "is a distinct variety of Constructive Mathematics" because it satisfies existential instantiation, proves the double negation of Bishop's Limited Principle of Omniscience for numbers, refutes the recursive choice principle CT_0, and contains its classical content. He also observes that DNS_0 can replace Markov's Principle in consistency proofs for classical analysis (cf. [15]).

Proposition 1. $\mathbf{HA}^\omega + \text{AC}_0 + \text{DNS}_0$ *is its own minimum classical extension.*

Proof. $(\mathbf{HA}^\omega + \text{AC}_0 + \text{DNS}_0)^g = (\mathbf{HA}^\omega + \text{AC}_0)^g$ by Remark 1 since DNS_0 is a classical logical schema. $(\mathbf{HA}^\omega)^g \subseteq \mathbf{HA}^\omega$ and $(\text{AC}_0)^g \subseteq \mathbf{HA}^\omega + \text{AC}_0 + \text{DNS}_0$.

2.2 Σ_1^0 Double Negation Shift for Numbers

Definition 3. *In the language of arithmetic or analysis, Σ_1^0-double negation shift for numbers is the schema*

$$\Sigma_1^0\text{-DNS}_0 : \quad \forall x \neg\neg \exists y A(x, y) \to \neg\neg \forall x \exists y A(x, y)$$

where $A(x, y)$ is a formula with only bounded number quantifiers and no sequence quantifiers, but perhaps containing additional free variables.

Proposition 2. **EL** $+\ \Sigma_1^0$-DNS$_0$ *is its own minimum classical extension.*

Proof. The only axiom or schema of **EL** $+\ \Sigma_1^0$-DNS$_0$ whose negative interpretation is not provable in **EL** is the quantifier-free countable choice schema

$$\text{QF-AC}_{00} : \quad \forall x \exists y A(x, y) \to \exists \alpha \forall x A(x, \alpha(x))$$

where $A(x, y)$ may have additional free variables of both sorts but only bounded numerical quantifiers. Its negative interpretation is intuitionistically equivalent to $\forall x \neg\neg \exists y A^g(x, y) \to \neg\neg \exists \alpha \forall x A^g(x, \alpha(x))$, which follows easily from Σ_1^0-DNS$_0$ and QF-AC$_{00}$ by intuitionisitic logic.

Definition 4. Two-sorted intuitionistic arithmetic IA$_1$ *is the subsystem of Kleene's* **B** *obtained by omitting the axiom schemas of countable choice and bar induction.* **Intuitionistic recursive analysis IRA** *comes from* **IA$_1$** *by adding the recursive comprehension axiom*[1]

$$\forall \rho [\forall x \exists y \rho(\langle x, y \rangle) = 0 \to \exists \alpha \forall x \rho(\langle x, \alpha(x) \rangle) = 0]$$

Remark 3. This special case is equivalent, over **IA$_1$**, to QF-AC$_{00}$. Vafeiadou proved in [23] that **IRA** and **EL** are mathematically equivalent, in the sense of having a common definitional extension. See [17] for a precise description of **IA$_1$** and [22] for her comparison of **EL** with **IRA**.

Proposition 3. *Over* **EL** *and* **IRA**, Σ_1^0-DNS$_0$ *is interderivable with the case*

$$\forall \rho [\forall x \neg\neg \exists y \rho(\langle x, y \rangle) = 0 \to \neg\neg \forall x \exists y \rho(\langle x, y \rangle) = 0].$$

Proof. Each formula $A(x, y)$ of the language of analysis with only bounded number quantifiers and no sequence quantifiers expresses a primitive recursive relation of its free variables, such that **EL** and **IRA** both prove

$$\exists \rho \forall x \forall y [A(x, y) \leftrightarrow \rho(\langle x, y \rangle) = 0].$$

Corollary 1. *(to Propositions 2, 3):*
(a) $(\mathbf{EL}_0)^{+g} = \mathbf{EL}_0$, *where* $\mathbf{EL}_0$ *comes from* **EL** *by omitting* QF-AC$_{00}$.
(b) $\mathbf{EL}^{+g} = \mathbf{EL} + \Sigma_1^0$-DNS$_0$.
(c) $(\mathbf{IA}_1)^{+g} = \mathbf{IA}_1$.
(d) $\mathbf{IRA}^{+g} = \mathbf{IRA} + \Sigma_1^0$-DNS$_0$

Proof. $\mathbf{EL}_0$ and $\mathbf{IA}_1$ prove the converse of QF-AC$_{00}$, and so Σ_1^0-DNS$_0$ follows from $(\text{QF-AC}_{00})^g$ in **EL** and **IRA**.

---

[1] $\langle x, y \rangle = 2^x \cdot 3^y$ is Kleene's code for the ordered pair of x and y; similarly for n-tuples.

2.3   Stronger Restricted Versions of DNS_0

The full strength of DNS_0 is not needed to negatively interpret AC_0. In constructive or intuitionistic arithmetic and analysis, if $A(x)$ is a *negative* formula (i.e. has no occurrences of $\exists$ or $\vee$) then $\neg\neg A(x) \leftrightarrow A(x)$ is provable, so double negation shift holds trivially for all negative formulas $A(x)$.

Definition 5. *In the language of arithmetic or analysis, DNS_{00}^- denotes the restriction of DNS to the case where x is a number variable and $A(x)$ is of the form $\exists y A(x,y)$ where $A(x,y)$ is negative. In the language of analysis, DNS_{01}^- denotes the restriction of DNS to the case where x is a number variable and $A(x)$ is of the form $\exists \alpha A(x,\alpha)$ where $A(x,\alpha)$ is negative. In the language of $\mathbf{HA}^\omega$, $DNS_{0\infty}^-$ includes all $DNS_{0\sigma}^-$ for finite types σ.*

These restricted double negation shift schemas characterize the minimum classical extensions of theories with AC_{00}, AC_{01}, or the collection $AC_{0\infty}$ of all $AC_{0\sigma}$ for finite types σ. In particular, AC_{01} is the strong countable choice axiom schema of Kleene's **B**:

$$AC_{01}: \quad \forall x \exists \alpha A(x,\alpha) \rightarrow \exists \alpha \forall x A(x, \lambda y.\alpha(\langle x, y\rangle))$$

where x is free for α in $A(x,\alpha)$, and AC_{00} is like $QF\text{-}AC_{00}$ but with no restriction on $A(x,y)$ except that the substitution of $\alpha(x)$ for y must be free.

Proposition 4. *Each of $\mathbf{HA}^\omega + AC_{0\infty} + DNS_{0\infty}^-$, $\mathbf{EL}_0 + AC_{00} + DNS_{00}^-$, $\mathbf{IA}_1 + AC_{00} + DNS_{00}^-$, $\mathbf{EL}_0 + AC_{01} + DNS_{01}^-$ and $\mathbf{IA}_1 + AC_{01} + DNS_{01}^-$ is its own minimum classical extension.*

Proof. As for Propositions 1 and 2.

Corollary 2. *(to Proposition 4):*

(a) $(\mathbf{HA}^\omega + AC_{0\infty})^{+g} = \mathbf{HA}^\omega + AC_{0\infty} + DNS_{0\infty}^-$.
(b) $(\mathbf{EL}_0 + AC_{0i})^{+g} = (\mathbf{EL} + AC_{0i})^{+g} = \mathbf{EL}_0 + AC_{0i} + DNS_{0i}^-$ *for $i = 0,1$.*
(c) $(\mathbf{IA}_1 + AC_{0i})^{+g} = (\mathbf{IRA} + AC_{0i})^{+g} = \mathbf{IA}_1 + AC_{0i} + DNS_{0i}^-$ *for $i = 0,1$.*

Proof. As for Corollary 1.

Remark 4. Many syntactic refinements of these results are possible. For example, Proposition 15 in [4] shows that (b) holds for $\Pi_1^0\text{-}AC_{00}$ (with the hypothesis $\forall x \exists y \forall z \rho(\langle x,y,z\rangle) = 0$) and $\Sigma_2^0\text{-}DNS_0$ (with hypothesis $\forall x \neg\neg \exists y \forall z \rho(\langle x,y,z\rangle) = 0$) in place of AC_{00} and DNS_{00}^-, respectively. Observe that $\Sigma_1^0\text{-}DNS_0$ and $\Sigma_2^0\text{-}DNS_0$ are instances of DNS_{00}^-, but e.g. $\Sigma_3^0\text{-}DNS_0$ is not.

2.4   Weak Comprehension Principles

Over $\mathbf{EL}$ or $\mathbf{IRA}$, a number-theoretic relation $A(x)$ (perhaps with number and sequence parameters) has a characteristic function for x only if it satisfies $\forall x(A(x) \vee \neg A(x))$. The weak comprehension schema

$$\neg\neg\, CF_0: \quad \neg\neg \exists \zeta \forall x(\zeta(x) = 0 \leftrightarrow A(x))$$

asserts only that it is *consistent* to assume that $A(x)$ has a characteristic function for x. Here we consider two restricted versions of $\neg\neg CF_0$. The first one is[2]

$$\neg\neg \Pi_1^0\text{-}CF_0 : \quad \forall\alpha\neg\neg\exists\zeta\forall x(\zeta(x) = 0 \leftrightarrow \forall y\alpha(\langle x, y\rangle) = 0).$$

By formula induction, **IRA** $+ \neg\neg \Pi_1^0\text{-}CF_0$ proves $\neg\neg\exists\zeta\forall x(\zeta(x) = 0 \leftrightarrow A(x))$ for all negative arithmetical formulas $A(x)$, with universal number quantifiers and free variables of both types allowed. The same holds with **EL** in place of **IRA**.

The second is $\neg\neg CF_0^-$, the restriction of $\neg\neg CF_0$ to negative formulas $A(x)$. Over **IA₁** or **EL₀**, $\neg\neg CF_0^-$ is equivalent to the negative translation of the schema

$$CF_d : \quad \forall x(A(x) \vee \neg A(x)) \rightarrow \exists\alpha\forall x[\alpha(x) \leq 1 \ \& \ (\alpha(x) = 0 \leftrightarrow A(x))].$$

Over **IA₁** or **EL₀** the conjunction of CF_d and $QF\text{-}AC_{00}$ is equivalent (cf. [23]) to the countable comprehension ("unique choice") schema $AC_{00}!$ which is like AC_{00} but with hypothesis $\forall x\exists! y A(x, y)$, where in general $\exists! y B(y)$ abbreviates $\exists y B(y) \ \& \ \forall y\forall z(B(y) \ \& \ B(z) \rightarrow y = z)$. Over **IA₁** or **EL₀**, AC_{00} is stronger than $AC_{00}!$ (which is stronger than $QF\text{-}AC_{00}$), but the negative translations of AC_{00} and $AC_{00}!$ are equivalent. Putting these facts together gives refinements of Corollary 2(b), (c) (for $i = 0$), and additional characterizations.

Theorem 1. *Let AC_{00}^{Ar} be the restriction of AC_{00} to arithmetical predicates $A(x, y)$, with number quantifiers and free variables of both types allowed. Then*

(a) $(\mathbf{IA_1} + AC_{00}^{Ar})^{+g} = \mathbf{IA_1} + AC_{00}^{Ar} + \Sigma_1^0\text{-}DNS_0 + \neg\neg \Pi_1^0\text{-}CF_0.$
(b) $(\mathbf{IA_1} + AC_{00})^{+g} = \mathbf{IA_1} + AC_{00} + \Sigma_1^0\text{-}DNS_0 + \neg\neg CF_0^-.$
(c) $(\mathbf{IA_1} + AC_{00}!)^{+g} = (\mathbf{IRA} + CF_d)^{+g} = \mathbf{IRA} + CF_d + \Sigma_1^0\text{-}DNS_0 + \neg\neg CF_0^-.$

Each of these results remains true with **EL₀** *in place of* **IA₁**, *and* **EL** *in place of* **IRA**.

3   Bar Induction, a Weak Continuity Principle, and BD-N

As Iris Loeb [13] observes, constructive reverse mathematics currently lacks a unifying methodology. According to Ishihara [9] its aim is "to classify various theorems in intuitionistic, constructive recursive and classical mathematics by logical principles, function existence axioms and their combinations" over a weak constructive base built on intuitionistic logic. Resulting decomposition theorems can help to extract and compare the classical content of constructive and semi-constructive theories. Two examples, one involving bar induction and the other involving a weak continuity principle, illustrate the method.

Brouwer's bar theorem, although not accepted by Bishop, is of interest to constructive mathematicians. The fan theorem FT, which follows from the bar theorem but is conservative over Heyting arithmetic by [20], has the property that the minimum classical extension of **IRA** + FT proves that intuitionistic predicate logic is complete for its intended interpretation ([3,14]).

---

[2] Over **EL** or **IRA**, $\neg\neg \Pi_1^0\text{-}CF_0$ entails the principle $\neg\neg\Pi_1^0\text{-}LEM$ in [5], and similarly for $\neg\neg \Sigma_1^0\text{-}CF_0$ and $\neg\neg\Sigma_1^0\text{-}LEM$.

3.1   Three Versions of Bar Induction

Kleene chose to axiomatize his neutral basic system $\mathbf{B}$ by $\mathbf{IA_1} + AC_{01} + BI_d$, where BI_d is "*decidable* bar induction:"

$$BI_d : \quad \forall\alpha\exists xR(\overline{\alpha}(x)) \ \& \ \forall w(R(w) \vee \neg R(w)) \ \& \ \forall w(R(w) \rightarrow A(w))$$
$$\& \ \forall w(\forall xA(w * \langle x+1 \rangle) \rightarrow A(w)) \rightarrow A(1).$$

Classical bar induction BI° simply drops the premise $\forall w(R(w) \vee \neg R(w))$, and *monotone* bar induction (which is provable in $\mathbf{I}$ but not in $\mathbf{B}$) is

$$BI_{mon} : \quad \forall\alpha\exists xR(\overline{\alpha}(x)) \ \& \ \forall w(R(w) \rightarrow \forall uR(w * u)) \ \& \ \forall w(R(w) \rightarrow A(w))$$
$$\& \ \forall w(\forall xA(w * \langle x+1 \rangle) \rightarrow A(w)) \rightarrow A(1).$$

Here $\overline{\alpha}(0) = 1$ and $\overline{\alpha}(x+1) = \langle \alpha(0) + 1, \ldots, \alpha(x) + 1 \rangle$. We let w, u vary over Kleene's "sequence numbers" (so w determines the length $lh(w)$ of the sequence w codes); $w * v$ codes the concatenation of the sequences coded by w and v, $\langle x+1 \rangle$ codes the sequence whose only term is x, and 1 codes the empty sequence.

Kleene proved ([12] p. 79) that $\mathbf{IA_1} + AC_{00} + BI_{mon} \vdash BI_d$, so BI_{mon} lies between BI_d and BI° in strength over $\mathbf{IA_1} + AC_{00}$.

Proposition 5. BI_d *has the same classical content as* BI° *over* $\mathbf{IA_1}$ *or* $\mathbf{EL_0}$.

Proof. The only difference between BI_d and BI° is a classically provable premise $\forall w(R(w) \vee \neg R(w))$ whose negative interpretation is provable intuitionistically.

3.2   A Double Negation Shift Principle for Functions

In the absence of countable choice, the double negation shift principle

$$DNS_1^- : \quad \forall\alpha\neg\neg\exists xR(\overline{\alpha}(x)) \rightarrow \neg\neg\forall\alpha\exists xR(\overline{\alpha}(x)),$$

where $R(w)$ is a negative formula of the language of analysis, is a sufficient addition to prove the double negation translation of BI_d and BI_{mon}.[3]

Theorem 2. *The minimum classical extensions of* $\mathbf{B}$ *and its subsystems with* AC_{01} *replaced by* AC_{00} *or by* $QF\text{-}AC_{00}$ *or omitted altogether are computed as follows. Similar results hold with* $\mathbf{EL_0}$ *in place of* $\mathbf{IA_1}$.

(a) $\mathbf{B}^{+g} \equiv (\mathbf{IA_1} + AC_{01} + BI_d)^{+g} = \mathbf{B} + (AC_{01})^g = \mathbf{B} + DNS_{01}^-$.
(b) $(\mathbf{IA_1} + AC_{00} + BI_d)^{+g} = \mathbf{IA_1} + AC_{00} + BI_d + DNS_{00}^-$.
(c) $(IRA + BI_d)^{+g} = IRA + BI_d + (BI^\circ)^g + \Sigma_1^0\text{-}DNS_0 \subseteq IRA + BI_d + DNS_1^-$.
(d) $(\mathbf{IA_1} + BI_d)^{+g} = \mathbf{IA_1} + BI_d + (BI^\circ)^g \subseteq \mathbf{IA_1} + BI_d + DNS_1^-$.

Proof. $\mathbf{IA_1} + AC_{00} + (\neg\neg A \rightarrow A) \vdash BI^\circ$ (*26.1° on p. 53 of [12]) and therefore $(\mathbf{IA_1} + AC_{00})^g \vdash (BI^\circ)^g$. Proposition 5, Corollary 2(b), (c), Corollary 1(c), (d) and the (easy) fact that $\mathbf{IA_1} + DNS_1^- \vdash \Sigma_1^0\text{-}DNS_0$ complete the argument.

---

[3] Only the special case $\Sigma_1^0\text{-}DNS_1$ is needed for the version of bar induction labeled ^{x}26.3b in [12]; cf. [14,15].

3.3   Applying a Typical Constructive Decomposition Theorem

Kleene proved (*27.23 on p. 87 of [12]) that $\mathbf{IRA} + \mathrm{BI}^\circ$ entails the "weak limited principle of omniscience" WLPO, which is inconsistent with $\mathbf{I}$. In [4] Fujiwara proved that BI° is equivalent over $\mathbf{EL}_0$ to $\mathrm{BI}_{\mathrm{mon}} + \mathrm{CD}$, where CD is the constant domain axiom schema $\forall x(A(x) \vee B) \rightarrow (\forall x A(x) \vee B)$ (with x not free in B).

Proposition 6. $\mathrm{BI}_{\mathrm{mon}}$ *has the same classical content as* BI° *over* $\mathbf{IA}_1$ *or* $\mathbf{EL}_0$, *so* $(\mathbf{IA}_1 + \mathrm{BI}_{\mathrm{d}})^g = (\mathbf{IA}_1 + \mathrm{BI}_{\mathrm{mon}})^g = \mathbf{IA}_1 + (\mathrm{BI}^\circ)^g$ *and similarly with* $\mathbf{EL}_0$ *in place of* $\mathbf{IA}_1$.

Proof. CD is a classical logical schema whose negative interpretation is provable by intuitionistic logic. Use Fujiwara's decomposition theorem and Proposition 5.

Remark 5. It follows that the neutral subsystem $\mathbf{B}$ of Kleene and Vesley's intuitionistic analysis I has the same classical content as the variant $\mathbf{B}'$ with $\mathrm{BI}_{\mathrm{mon}}$ replacing BI_{d}, and so $(\mathbf{B}')^{+g} \equiv (\mathbf{IA}_1 + \mathrm{AC}_{01} + \mathrm{BI}_{\mathrm{mon}})^{+g} = \mathbf{B}' + \mathrm{DNS}_{01}^-$.

3.4   Applying an Atypical Constructive Decomposition Theorem

In [10], over a constructive base theory $\mathbf{EL}' \equiv \mathbf{EL} + \Pi_1^0\text{-}\mathrm{AC}_{00}$, Ishihara and Schuster decompose a restricted version

WC-N′ : $\forall \alpha \exists n \forall k\, \sigma(\langle \overline{\alpha}(k), n \rangle) = 0$
$$\& \;\forall w \forall m \forall n (\sigma(\langle w, m \rangle) = 0 \;\&\; m \leq n \rightarrow \sigma(\langle w, n \rangle) = 0)$$
$$\rightarrow \forall \alpha \exists n \exists m \forall \beta \in \overline{\alpha}(m) \forall k\, \sigma(\langle \overline{\beta}(k), n \rangle) = 0$$

of weak continuity into a classically correct mathematical principle

BD-N : $\forall \alpha \exists m \forall n \geq m\, \beta(\alpha(n)) < n \rightarrow \exists m \forall n\, \beta(n) \leq m$

and a classically false logical principle $\neg \forall \alpha (\exists x\, \alpha(x) \neq 0 \vee \forall x\, \alpha(x) = 0)$ negating the limited principle of omniscience LPO.

Proposition 7. *The minimum classical extensions of* $\mathbf{EL}'$ *and* $\mathbf{EL}' + \mathrm{BD\text{-}N}$ *are computed as follows, and similarly for* $\mathbf{IRA} + \Pi_1^0\text{-}\mathrm{AC}_{00}$ ($\equiv \mathbf{IA}_1 + \Pi_1^0\text{-}\mathrm{AC}_{00}$) *in place of* $\mathbf{EL}'$.

(a) $\mathbf{EL}'^{+g} \equiv (\mathbf{EL}_0 + \Pi_1^0\text{-}\mathrm{AC}_{00})^{+g} = \mathbf{EL}' + \Sigma_2^0\text{-}\mathrm{DNS}_0$.
(b) $(\mathbf{EL}' + \mathrm{BD\text{-}N})^{+g} = \mathbf{EL}' + \mathrm{BD\text{-}N} + \Sigma_2^0\text{-}\mathrm{DNS}_0$.

Proof. (a) holds by Corollary 1(a) and Remark 4. $\mathbf{EL}' + \mathrm{BD\text{-}N}$ is consistent with classical logic and satisfies (b) because $\mathbf{EL}^{+g}$ proves the contrapositive of $(\mathrm{BD\text{-}N})^g$, which is equivalent to $(\mathrm{BD\text{-}N})^g$ over $\mathbf{EL}$.

Remark 6. $\mathbf{IA}_1 + \Pi_1^0\text{-}\mathrm{AC}_{00} + \mathrm{WC\text{-}N}'$ is a subsystem of Kleene's $\mathbf{I}$ which is consistent (by [12]) but is not consistent with classical logic. The next theorems, discovered by the second author, essentially trivialize the notion of minimum classical extension for intuitionistic systems inconsistent with classical logic.

Theorem 3. $(\mathbf{EL}' + \mathrm{WC\text{-}N}')^{+g} = \mathbf{EL}' + \mathrm{WC\text{-}N}' + (\Gamma^\circ)^g$ *where* Γ° *is the set of all classically true sentences in the language of* $\mathbf{EL}'$. *A corresponding result holds with* $\mathbf{IA}_1 + \Pi_1^0\text{-}\mathrm{AC}_{00}$ *in place of* $\mathbf{EL}'$.

Proof. $\mathbf{EL}' + \text{WC-N}' \vdash \neg\text{LPO}$ by Ishihara and Schuster's decomposition theorem, therefore $\mathbf{EL}' + \text{WC-N}' \vdash (\neg\text{LPO} \vee \text{E})$ for every formula E. If $\text{E} \in \Gamma^0$ then $(\neg\text{LPO} \vee \text{E}) \in \Gamma^\circ$, so $(\neg\text{LPO} \vee \text{E}) \in \text{cls}(\mathbf{EL}' + \text{WC-N}')$. But $(\neg\text{LPO} \vee \text{E})^g$ is just $\neg(\neg\neg\text{LPO}^g \& \neg\text{E}^g)$, which is equivalent by intuitionistic logic to $\neg\neg\text{E}^g$ and hence to E^g. So $(\Gamma^\circ)^g \subseteq (\text{cls}(\mathbf{EL}' + \text{WC-N}'))^g$, and the reverse inclusion is immediate from the definitions.

Theorem 4. $\mathbf{I}^{+g} = \mathbf{I} + (\Gamma^\circ)^g$ *where now* Γ° *is the set of all classically true sentences in the language of* $\mathbf{I}$.

Proof. $(\text{cls}(\mathbf{I}))^g = (\Gamma^\circ)^g$ by an argument similar to the proof of Theorem 3, but with WLPO $(\equiv \forall\alpha(\forall x\alpha(x) = 0 \vee \neg\forall x\alpha(x) = 0))$ in place of LPO, using the fact that $\mathbf{I} \vdash \neg\text{WLPO}$ by *27.17 on p. 84 of [12]. Thus $\mathbf{I}^{+g} = \mathbf{I} + (\Gamma^\circ)^g$.

Remark 7. By Lemma 8.4a in [12] every classically true negative sentence of the language of analysis is realizable by a primitive recursive function, so (by Theorem 9.3 of [12]) Kleene's function-realizability guarantees the consistency of $(\mathbf{EL}' + \text{WC-N}')^{+g}$ and of $\mathbf{S}^{+g}$ for every subsystem $\mathbf{S}$ of $\mathbf{I}$.

4   Conclusion

We have suggested a way to define the minimum classical extension $\mathbf{S}^{+g}$ of a mathematical theory $\mathbf{S}$ based on intuitionistic logic, with examples from arithmetic, analysis and the arithmetic of finite types. If $\mathbf{S} + (\neg\neg A \to A)$ is consistent, then classical and constructive mathematics coexist in $\mathbf{S}^{+g}$ exactly as far as the mathematical axioms of $\mathbf{S}$ permit.

For example, if $\mathbf{S}$ is a classically correct subsystem of Kleene's intuitionistic analysis $\mathbf{I} \equiv \mathbf{B} + \text{CC}_{11}$, then by viewing the choice sequence variables $\alpha, \beta, \ldots$ alternatively as variables over classical one-place number-theoretic functions, restricting the language and logic by omitting the symbols $\vee$ and $\exists$ with their axioms and rules, and replacing each mathematical axiom of $\mathbf{S}$ by its negative translation, one obtains a faithful copy of $\mathbf{S}^\circ \equiv \mathbf{S} + (\neg\neg A \to A)$ within the extended intuitionistic system $\mathbf{S}^{+g}$. In particular, $\mathbf{B}^{+g}$ includes the negative translation of a system $\mathbf{C} \equiv \mathbf{B}^\circ$ of classical analysis with countable choice.

On the other hand, if $\mathbf{S}$ refutes a classical logical principle, then $\mathbf{S}^{+g}$ includes the negative translations of *all classically true sentences in the language of* $\mathbf{S}$. In particular, $\mathbf{I}^{+g}$ contains a negative version of true classical analysis.

We conclude that only constructive and semi-constructive systems consistent with classical logic have interesting minimum classical extensions, and typical constructive decomposition theorems assist in comparing their classical content.

References

1. Berardi, S., Bezem, M., Coquand, T.: On the computational content of the axiom of choice. J. Symb. Logic **63**(2), 600–622 (1998)

2. Brouwer, L.E.J.: Begründung der Mengenlehre unabhängig vom logischen Satz vom ausgeschlossenen Dritten. In: Heyting, A. (ed.), L. E. J. Brouwer: Collected Works, vol. I, pp. 150–190. North-Holland/American Elsevier (1975)
3. Dyson, V., Kreisel, G.: Analysis of Beth's semantic construction of intuitionistic logic. Technical report 3, Applied mathematics and statistics laboratory, Stanford University (1961)
4. Fujiwara, M.: Bar induction and restricted classical logic. In: Iemhoff, R., Moortgat, M., de Queiroz, R., (eds.) Logic, Language, Information and Computation: WoLLIC Proceedings, pp. 236–247 (2019)
5. Fujiwara, M., Kohlenbach, U.: Interrelation between weak fragments of double negation shift and related principles. J. Symb. Logic **81**(3), 991–1012 (2018)
6. Gentzen, G.: Über das Verhältnis zwischen intuitionistischer und klassischer Logik. Arch. Math. Logik Grund. **16**, 119–132 (1974). Accepted by Math. Annalen in 1933, but withdrawn. English trans. In: Szabo (ed.), Gentzen: Collected Papers
7. Gödel, K.: Zur intuitionistischen Arithmetik und Zahlentheorie. Ergebnisse eines math. Koll. **4**, 34–38 (1933)
8. Ilik, D. Double-negation shift as a constructive principle. arXiv:1301.5089v1
9. Ishihara, H.: Constructive reverse mathematics: compactness properties. In: Crosilla, L., Schuster, P. (eds.) From Sets and Types to Topology and Analysis, pp. 245–267. Clarendon Press, Oxford (2005)
10. Ishihara, H., Schuster, P.: A continuity principle, a version of Baire's theorem and a boundedness principle. J. Symb. Logic **73**, 1354–1360 (2008)
11. Kleene, S.C.: Introduction to Metamathematics. D. van Nostrand Company Inc., Princeton (1952)
12. Kleene, S.C., Vesley, R.E.: The Foundations of Intuitionistic Mathematics, Especially in Relation to Recursive Functions, North Holland (1965)
13. Loeb, I.: Questioning constructive reverse mathematics. Constructivist Found. **7**, 131–140 (2012)
14. Moschovakis, J.R.: Calibrating the negative interpretation. arXiv:2101.10313. Expanded extended abstract for 12th Panhellenic Logic Symposium, June 2019
15. Moschovakis, J.R.: Solovay's relative consistency proof for FIM and BI. arXiv:2101.05878v1. Historical note, to appear in Notre Dame J. Formal Logic
16. Moschovakis, J.R.: Classical and constructive hierarchies in extended intuitionistic analysis. J. Symb. Logic **68**, 1015–1043 (2003)
17. Moschovakis, J.R., Vafeiadou, G.: Some axioms for constructive analysis. Arch. Math. Logik **51**, 443–459 (2012). https://doi.org/10.1007/s00153-012-0273-z
18. Scedrov, A., Vesley, R.: On a weakening of Markov's principle. Arch. Math. Logik **23**, 153–160 (1983). https://doi.org/10.1007/BF02023022
19. Troelstra, A. S.: Intuitionistic formal systems. In: Troelstra, A.S. (ed.) Metamathematical Investigation of Intuitionistic Arithmetic and Analysis. LNM, vol. 344. Springer, Heidelberg (1973). https://doi.org/10.1007/BFb0066740
20. Troelstra, A.S.: Note on the fan theorem. J. Symb. Logic **39**, 584–596 (1974)
21. Troelstra, A.S.: Corrections to some publications. University of Amsterdam, 10 December 2018. https://eprints.illc.uva.nl/1650/1/CombiCorr101218.pdf
22. Vafeiadou, G.: A comparison of minimal systems for constructive analysis. arXiv:1808.000383
23. Vafeiadou, G.: Formalizing Constructive Analysis: a comparison of minimal systems and a study of uniqueness principles. Ph.D. thesis, University of Athens (2012)

Subrecursive Equivalence Relations and (non-)Closure Under Lattice Operations

Jean-Yves Moyen[1] and Jakob Grue Simonsen[2(✉)]

[1] LIPN, UMR CNRS 7030 – Université Paris 13, 99, Avenue Jean-Baptiste Clément, 93430 Villetaneuse, France
[2] Department of Computer Science, University of Copenhagen (DIKU), Universitetsparken 5, 2100 Copenhagen Ø, Denmark
simonsen@diku.dk

Abstract. The set of equivalence relations on any non-empty set is equipped with a natural order that makes it a complete lattice. The lattice structure only depends on the cardinality of the set, and thus the study of the lattice structure on any countably infinite set is (up to order-isomorphism) the same as studying the lattice of equivalence relations on the set of natural numbers.

We investigate closure under the meet and join operations in the lattice of equivalence relations on the set of natural numbers. Among other results, we show that no set of co-r.e. equivalence relations that contains all logspace-decidable equivalence relations is a lattice.

Keywords: Lattice theory · Equivalence relations · Subrecursive sets

1 Introduction

The set of equivalence relations $\mathrm{Equ}(S)$ on any set S is ordered by the relation $\leq$ defined by $\mathcal{E} \leq \mathcal{E}'$ iff $m\mathcal{E}n \Rightarrow m\mathcal{E}'n$. It is known that for any non-empty set S, $(\mathrm{Equ}(S), \leq)$ is a complete, algebraic, simple, semimodular, and relatively complemented lattice [6,8,12,24], and the behaviour of complements and (anti-)chains in S has been studied for both finite and infinite-cardinality sets S [5]. The lattice structure of $\mathrm{Equ}(S)$ is order-isomorphic to $\mathrm{Equ}(T)$ for any sets S and T of the same cardinality, whence we may wlog. consider $\mathrm{Equ}(\mathbb{N})$, the set of equivalence relations on the set of natural numbers[1].

It is known that the set of r.e. equivalence relations form a sublattice of $\mathrm{Equ}(\mathbb{N})$ [9], and sublattices of $\mathbb{N}$ appear in different guises in multiple places in computability theory; we give three examples: (I) every countable partial order

---

[1] Computable *reducibility* between elements of $\mathrm{Equ}(\mathbb{N})$ gives rise to an order structure distinct from the classical ordering we consider, and has been investigated in various settings, in particular for Σ_1^0 equivalence relations (called *ceers*) [1–3,10,11,13].

© Springer Nature Switzerland AG 2021
L. De Mol et al. (Eds.): CiE 2021, LNCS 12813, pp. 363–372, 2021.
https://doi.org/10.1007/978-3-030-80049-9_34

embeds in the lattice of r.e. lambda theories[2] [25]; (II) let $\mathfrak{R}$ be the equivalence relation on (Gödel numbers of) programs such that $m\mathfrak{R}n$ iff $\phi_m = \phi_n$; then, Rice's Theorem [22] can be seen as the statement that any non-trivial equivalence relation $\mathcal{E}$ on programs such that $\mathfrak{R} \leq \mathcal{E}$ is undecidable [18]; (III) let Φ be any Blum complexity measure [7], and let $\mathfrak{A}$ be the equivalence relation on programs defined by $m\mathfrak{A}n$ iff $(\phi_m = \phi_n \wedge \Phi_m \in \Theta(\Phi(n)))$ (i.e., "two programs are equal if they compute the same function and have the same complexity"), then a *complexity clique* is an equivalence relation $\mathcal{C}$ on programs that respects $\mathfrak{A}$, and the set of complexity cliques is a complete complemented distributive lattice under $\subseteq$ and $\vee$ and $\wedge$ [4], and is thus a sublattice of Equ($\mathbb{N}$).

While the sublattice of r.e. equivalence relations is well-understood, and reasonably well-behaved [9], Examples (II) above suggests that interesting lattice-theoretic phenomena may appear *below* $\mathfrak{R}$ and $\mathfrak{A}$ in Equ($\mathbb{N}$), in particular for sets of *decidable* equivalence relations—such as the ones induced by common subrecursive classes in complexity theory. We hope that this paper can provide impetus for future investigations of computability-related properties of (subsets of) Equ($\mathbb{N}$) in addition to the well-known case of the sublattice Equ($\mathbb{N}$)$_{\text{r.e.}}$ of r.e. equivalence relations.

Scope and Contributions: The purpose of this paper is to study sets of decidable equivalence relations on countable sets and whether they are closed under the lattice operations, i.e. meet ($\wedge$) and join ($\vee$). In particular, show that almost none of the classes of equivalence relations corresponding to the usual set of subrecursive classes of interest in computability and (coarse-grained) complexity theory–e.g., classes induced by complexity classes, primitive recursive relations, Kalmár elementary relations–are closed under the meet operation. This thwarts any naïve hope of using Equ($\mathbb{N}$) as an "easy" means of studying subrecursive analogues of Rice's theorem in the vein of [18], but will hopefully pave the way for further study of computability in Equ($\mathbb{N}$).

2   Preliminaries

Throughout the paper we let $\mathbb{N} = \mathbb{N}_0 = \{0, 1, 2, \ldots\}$. For $n \in \mathbb{N}$, we write $\uparrow\{n\} = \{n, n+1, n+2, \ldots\}$. Symbols $k, l, m, n, \ldots, x, y, \ldots$ range over $\mathbb{N}$ unless otherwise noted. If $m \in \mathbb{N}$, we denote by $\mathtt{m}$ its standard representation as an element of $\{0, 1\}^*$.

Equivalence Relations. We let cursive letters $\mathcal{E}, \mathcal{F}, \mathcal{G}$ range over equivalence relations, and let bold capital roman letters $\mathbf{C}, \mathbf{D}, \mathbf{E}$ designate sets of equivalence relations. If $\mathcal{E}$ is an equivalence relation, we write $m\mathcal{E}n$ to indicate that m and n are related by $\mathcal{E}$. Equ($\mathbb{N}$) is ordered by $\leq$, defined by $\mathcal{E} \leq \mathcal{E}'$ iff $m\mathcal{E}n \Rightarrow m\mathcal{E}'n$. Note that if an equivalence relation is seen as a subset of $\mathbb{N} \times \mathbb{N}$, this order is exactly the subset ordering $\subseteq$ on $\mathbb{N}\times\mathbb{N}$. By the standard correspondence between

---

[2] A lambda theory is an equivalence relation on the set of closed terms in lambda calculus containing β-equivalence and satisfying a few other natural constraints; see [16] for basic properties.

equivalence relation and partitions, this is isomorphic with the usual refinement ordering on partitions. It is easily seen that $\mathcal{E} \leq \mathcal{E}'$ iff each class of $\mathcal{E}'$ is the union of one or more classes of $\mathcal{E}$.

An *atom* is an equivalence relations such that exactly one class contains two elements, and all other classes are singletons. An equivalence relation $\mathcal{E}$ where exactly one class is not a singleton is called *singular*[3].

Lattices and Lattice Operations. Standard results for Equ($\mathbb{N}$) were laid out in the seminal paper by Ore [19]; we briefly recapitulate basic definitions here. (Equ($\mathbb{N}$), $\leq$) is a lattice with the following operations:

- The meet (greatest lower bound) of $\mathcal{E}$ and $\mathcal{F}$ is $\mathcal{G} = \mathcal{E} \wedge \mathcal{F}$ such that $m\mathcal{G}n$ iff $m\mathcal{E}n$ and $m\mathcal{F}n$. In this case, the classes of $\mathcal{G}$ are exactly the (non-empty) intersections of one class of $\mathcal{E}$ and one class of $\mathcal{F}$.
- The join (least upper bound) of $\mathcal{E}$ and $\mathcal{F}$ is $\mathcal{G} = \mathcal{E} \vee \mathcal{F}$ such that $m\mathcal{G}n$ iff there exists a finite sequence $a_1, \ldots, a_k$ such that $m\mathcal{E}a_1\mathcal{F}a_2\mathcal{E} \cdots \mathcal{F}n$.

As equivalence relations are subsets of $\mathbb{N} \times \mathbb{N}$ we have $\mathcal{E} \wedge \mathcal{F} = \mathcal{E} \cap \mathcal{F}$ while $\mathcal{E} \vee \mathcal{F}$ is the least (wrt. $\subseteq$) equivalence relation that contains $\mathcal{E} \cup \mathcal{F}$. For the lattice of *relations* with the same order, $\mathcal{E} \vee \mathcal{F} = \mathcal{E} \cup \mathcal{F}$. However, not all the relations are equivalence relations and transitivity is closed under intersection but not under union.

Observe that if $\mathbf{E} = \{\mathcal{E}_k\} \subseteq \text{Equ}(\mathbb{N})$, then $\mathcal{E} = \bigwedge \mathbf{E}$ is the equivalence relation where each class is obtained as $\bigcap_k C_k$ where each C_k is a class of $\mathbf{E}_k$. That is, $m\mathcal{E}n$ iff $m\mathcal{E}_kn$ for all k. Likewise, $\mathcal{E} = \bigvee \mathbf{E}$ is the equivalence relation defined by $m\mathcal{E}n$ iff $m(\bigcup_k \mathbf{E}_k)^*n$, that is, $m\mathcal{E}n$ iff there are finite sequences $a_1, \cdots, a_{p-1} \in \mathbb{N}$ and $\mathcal{E}_1, \cdots, \mathcal{E}_p \in \mathbf{E}$ such that $m\mathcal{E}_1a_1, a_1\mathcal{E}_2a_2, \cdots, a_{p-1}\mathcal{E}_pn$. If $\mathbf{A} \subseteq \text{Equ}(\mathbb{N})$, we say that $\mathbf{A}$ is *closed under finite* join (resp. finite meet) if, for any $\mathcal{E}, \mathcal{F} \in \mathbf{A}$ we have $\mathcal{E} \vee \mathcal{F} \in \mathbf{A}$ (resp. $\mathcal{E} \wedge \mathcal{F} \in \mathbf{A}$), and *closed under arbitrary* join (resp. arbitrary meet) if, for any $\mathbf{B} \subseteq \mathbf{A}$, we have $\bigvee \mathbf{B} \in \mathbf{A}$ (resp. $\bigwedge \mathbf{B} \in \mathbf{A}$). Closure of other classes under join and meet of are defined *mutatis mutandis*.

Computability and Complexity. We refer to standard textbooks covering computability and complexity theory (e.g., [14,20,23]) and presuppose knowledge of the basic notions; we recapitulate the most necessary concepts and notation below. Unless otherwise stated, all Turing Machines are multi-tape machines with one read-only input tape, one write-only output tape, and any number of work tapes. Machines are assumed to be deterministic unless otherwise stated. Using standard efficient computable pairing functions, for every $k \in \mathbb{N}$, every Turing machine may be assumed to compute a partial function on $(\{0,1\}^*)^k \longrightarrow \{0,1\}$ or $\mathbb{N}^k \longrightarrow \{0,1\}$. We assume a fixed Gödel numbering of the Turing machines and designate by M_i the machine with Gödel number i and by ϕ_i the partial function computed by M_i. For $i, j \in \mathbb{N}$, we write $\phi_i = \phi_j$ if $\text{dom}(\phi_i) = \text{dom}(\phi_j)$, and

---

[3] "Singular" is the term originally used in the seminal paper [19] and in the literature on lattice theory; in work on reducibility between equivalence relations, the term "1-dimensional" is sometimes used [10].

$\phi_i(n) = \phi_j(n)$ for all $n \in \text{dom}(\phi_i)$. If i is the Gödel number of a Turing machine taking no input, we write $\phi_i\uparrow$ if i does not halt and $\phi_i\downarrow$ if it halts; the notation is extended in the obvious way to Turing machines with input.

An equivalence relation $\mathcal{E} \in \text{Equ}(\mathbb{N})$ is *decidable* if there is a Turing Machine M such that for every $(m, n) \in \mathbb{N}$, M halts with output 1 on input $(\mathtt{m}, \mathtt{n})$ if $m\mathcal{E}n$ and halts with output 0 otherwise, when M is started with the (suitably encoded) pair of binary representations of m and n on the input tape. Using highly efficient pairing functions, both pairing and unpairing can be performed in linear time and constant space [21].

Similarly, $\mathcal{E}$ is *recursively enumerable* (abbreviated r.e.) if there is a TM that halts on the corresponding input (or equivalently if there is a single-tape Turing machine that halts on input $(\mathtt{m}, \mathtt{n})$ iff $m\mathcal{E}n$). We denote by $\text{Equ}(\mathbb{N})_{\text{r.e.}}$ (resp. $\text{Equ}(\mathbb{N})_{\text{co-r.e.}}$, resp. $\text{Equ}(\mathbb{N})_{\text{dec}}$) the set of all r.e. equivalence relations (resp. co-r.e., resp. decidable equivalence relations). As usual, we use the term Σ_1^0-(hard,complete) instead of r.e.-(hard, complete), and the same terms *mutatis mutandis* for Π_1^0. We say that $\mathcal{E}$ is Σ_1^0-hard if there is a computable total function $f : \mathbb{N} \longrightarrow \mathbb{N} \times \mathbb{N}$ and a Σ_1^0-hard subset $H \subseteq \mathbb{N}$ such that, for all $i \in \mathbb{N}$, we have $i \in H$ iff $f(i) = (m, n)$ such that $m\mathcal{E}n$.

If $s, t : \mathbb{N} \longrightarrow \mathbb{N}$ are non-decreasing functions and M decides $\mathcal{E}$ in space s and time t in the size of the input, we say that $\mathcal{E}$ is s-space and t-time decidable. Observe that if $(m, n) \in \mathbb{N} \times \mathbb{N}$, the input given to M is $(\mathtt{m}, \mathtt{n})$ where $(\cdot, \cdot)$ is an efficient pairing function. Thus, for example, as $|\mathtt{m}| = O(\log m)$, an equivalence relation decidable in logarithmic space uses space at most $O(\log \log \max\{m, n\})$.

The sets of equivalence relations on $\mathbb{N}$ consisting of primitive recursive, Kalmár elementary, exponential time, polynomial space, polynomial time, and logarithmic space-decidable equivalence relations are denoted by $\text{Equ}(\mathbb{N})_{\text{p.r.}}$, $\text{Equ}(\mathbb{N})_{\text{KE}}$, $\text{Equ}(\mathbb{N})_{\text{EXPTIME}}$, $\text{Equ}(\mathbb{N})_{\text{PSPACE}}$, $\text{Equ}(\mathbb{N})_{\text{P}}$, and $\text{Equ}(\mathbb{N})_{\text{L}}$, respectively. Observe that each of these classes contains $\text{Equ}(\mathbb{N})_{\text{L}}$.

If $\mathbf{A} \subseteq \text{Equ}(\mathbb{N})$ is a set of decidable equivalence relations, we say that $\mathbf{A}$ is r.e. if there is a Turing machine that enumerates a sequence of Turing machines M_i such that (i) for each $\mathcal{E} \in \mathbf{A}$ there is a (not necessarily unique) i such that M_i decides $\mathcal{E}$, and (ii) each M_i decides some $\mathcal{E} \in \mathbf{A}$.

3   (Absence of) Closure Under the Lattice Operations

We wish to treat classes of equivalence relations where each relation is *subrecursive*, i.e. is decidable by machinery less extensionally powerful than Turing machines and typically avoiding unbounded search. We consider any class of decidable equivalence relations that contains $\text{Equ}(\mathbb{N})_{\text{L}}$, and this section is devoted to prove the (lack of) closure properties for any such class.

There are simple counterexamples to closure under finite $\wedge$ for sets of decidable equivalence relations (even those containing $\text{Equ}(\mathbb{N})_{\text{L}}$): By the Space Hierarchy Theorem, pick a decidable equivalence relation $\mathcal{E} \notin \text{Equ}(\mathbb{N})_{\text{L}}$ having at least three equivalence classes. If A, B, and C are distinct equivalence classes of $\mathcal{E}$, let $\mathcal{F}$ be the relation with the same equivalence classes as $\mathcal{E}$ except for the class $A \cup B$, and $\mathcal{F}$ the same relation as $\mathcal{E}$ except for the class $B \cup C$. Then, $\mathbf{C} = \mathrm{Equ}(\mathbb{N})_{\mathrm{L}} \cup \{\mathcal{F}, \mathcal{G}\}$ is a set of decidable equivalence relations containing $\mathrm{Equ}(\mathbb{N})_{\mathrm{L}}$, but $\mathcal{F} \wedge \mathcal{G} = \mathcal{E} \notin \mathbf{C}$.

However, with mild closure conditions, we can easily prove a a positive result:

Proposition 1. *Let $\mathcal{E}$ and $\mathcal{F}$ be equivalence relations and $M_{\mathcal{E}}$ and $M_{\mathcal{F}}$ be Turing machines such that $M_{\mathcal{E}}$ decides $\mathcal{E}$ in time $T(\mathcal{E})$ and space $S(\mathcal{E})$, and $M_{\mathcal{F}}$ decides $\mathcal{F}$ in time $T(\mathcal{F})$ and space $S(\mathcal{F})$ Then $\mathcal{E} \wedge \mathcal{F}$ is decidable in time $O(n(T(\mathcal{E}) + T(\mathcal{F})))$ and space $O(\max(S(\mathcal{E}), S(\mathcal{F})) + \log n)$.*

Proof. Observe that $x(\mathcal{E} \wedge \mathcal{F})y$ iff $x\mathcal{E}y$ and $x\mathcal{F}y$. Hence to check whether $x(\mathcal{E} \wedge \mathcal{F})y$, it is sufficient to first check whether $x\mathcal{E}y$ and then check whether $x\mathcal{F}y$. Using built-in copies of $M_{\mathcal{E}}$ and $M_{\mathcal{F}}$ as subroutines, a Turing machine M can perform the simulations with exactly the same time and space resources as $M_{\mathcal{E}}$ and $M_{\mathcal{F}}$, using linear time overhead and constant space overhead to unpair the input [21]. To avoid storing the binary representations of m and n on an auxiliary tape (which would require linear space), M maintains two counters to the current bits being read in each binary representation (taking logarithmic space in the input size); note that this requires recomputing the unpairing function each time any bit of the input is queried by $M_{\mathcal{E}}$ or $M_{\mathcal{F}}$, and hence each step of the subroutines for $M_{\mathcal{E}}$ and $M_{\mathcal{F}}$ potentially uses linear time and constant space. Finally, M compares the results of running $M_{\mathcal{E}}$ and $M_{\mathcal{F}}$.   □

Corollary 1. *Each of the sets $Equ(\mathbb{N})_{\mathrm{r.e.}}$, $Equ(\mathbb{N})_{\mathrm{dec}}$, $Equ(\mathbb{N})_{\mathrm{p.r.}}$, $Equ(\mathbb{N})_{\mathrm{KE}}$, $Equ(\mathbb{N})_{\mathrm{EXPTIME}}$, $Equ(\mathbb{N})_{\mathrm{PSPACE}}$, $Equ(\mathbb{N})_{\mathrm{P}}$, and $Equ(\mathbb{N})_{\mathrm{L}}$ are closed under finite meet.*

The $\vee$ operation is more intricate than $\wedge$; in stark contrast to Proposition 1, we have the following negative result:

Proposition 2. *There are equivalence relations $\mathcal{E}, \mathcal{F} \in Equ(\mathbb{N})_{\mathrm{L}}$ such that $\mathcal{E} \vee \mathcal{F}$ is Σ_1^0-complete.*

Proof. If $\mathcal{E}, \mathcal{F} \in \mathrm{Equ}(\mathbb{N})_{\mathrm{L}}$, a fortiori, $\mathcal{E}, \mathcal{F} \in \mathrm{Equ}(\mathbb{N})_{\mathrm{r.e.}}$, and as $\mathrm{Equ}(\mathbb{N})_{\mathrm{r.e.}}$ is a lattice, we have $\mathcal{E} \vee \mathcal{F} \in \mathrm{Equ}(\mathbb{N})_{\mathrm{r.e.}}$, thus proving containment in Σ_1^0.

Proving Σ_1^0-hardness is significantly more involved; the rest of the proof is devoted to that task.

Let M be any deterministic Turing machine. We may assume wlog. that the machine is not stuck on any of its configurations unless the state in the configuration is either "accept" or "reject".

Let c, c' be any two of the configurations of M, and let c, c' be the binary representations of these under some standard one-to-one encoding. We assume wlog. that each representation starts with 1. Note in particular that no configuration is represented by the string "0" and that the representation of any configuration is also a representation of some positive integer. Thus, we may wlog. define equivalence relations on pairs of binary strings on this form instead of equivalence relations on $\mathbb{N}$.

Define a partial binary relation $\to \,\subseteq \{0,1\}^+ \times \{0,1\}^+$ as follows: Write $\mathsf{c} \to \mathsf{c}'$ if c and c' are configurations of M and c' is the result of executing one step of M from configuration c. Moreover, we *define* $\mathsf{c} \to 0$ whenever c is a final configuration of M (i.e., c is in an accept or reject state). The relation $\to$ can be decided in logarithmic space in the size of c and c' by standard techniques: If M is a Turing machine and configurations d are of the form $\mathsf{d} = (\text{state}, \text{tape content}, \text{tape head position})$, there is a Turing machine R_M that on input $(\mathsf{c}, \mathsf{c}')$ decides whether $\mathsf{c} \to \mathsf{c}'$ using a fixed number of counters to (a) keep track of where in each of c and c' the tape heads of R_M are currently reading[4] and (b) to check whether the tape contents of c and c' are identical except for, possibly, at the tape head positions in c and c'. R_M carries a copy of M in its internal logic as a lookup table and simply performs a lookup to see whether there is a transition in M that would allow the change observed at the tape head positions in c and c' (and checks whether the tape positions differ by at most one). We denote by $\mathsf{c} \approx \mathsf{c}'$ the reflexive transitive symmetric closure of $\to$. By construction, $\approx$ is an equivalence relation, and clearly $\mathsf{c} \approx 0$ iff the execution starting in configuration c terminates.

Let U be a universal Turing machine, and let $\approx_U$ be the equivalence relation induced by U as defined above. For each Turing machine N and input x, let $\mathsf{c}_{N,x}$ be the configuration of U where U is in the start state, all work tapes are empty, and the input tape contains a binary representation of the pair (N, x). Then, $\mathsf{c}_{N,x} \approx_U 0$ iff N halts on input x. As $\{(N, x) : \phi_N(x)\!\downarrow\}$ is Σ_1^0-complete, and $\mathsf{c}_{N,x}$ can clearly be obtained from (N, x) by computable many-one reduction, the set $\{c : c \approx 0\}$ is Σ_1^0-hard.

Now, given a non-negative integer n with binary representation n, and a configuration c, we build the pair (n, c) as the base-3 integer $(\mathsf{n}, \mathsf{c}) = \mathsf{n}2\mathsf{c}$. We define the *clocked one-step relation* $\to'$ on $\{0,1\}^+2\{0,1\}^+$ by $(\mathsf{n}, \mathsf{c}) \to' (\mathsf{m}, \mathsf{c}')$ iff either $m = n + 1$ and $\mathsf{c} \to \mathsf{c}'$ or if c is final and $(\mathsf{m}, \mathsf{c}') = 020$ ($= 6$, in base-3). Note that as $\to$ is decidable in logarithmic space in the size of c and c', and as it can be checked in logarithmic space whether $m = n + 1$ and in constant space whether $(\mathsf{m}, \mathsf{c}') = 020$, then $\to'$ is decidable in logarithmic space in (n, c) and $(\mathsf{m}, \mathsf{c}')$. We define $\approx'$ as the reflexive transitive symmetric closure of $\to'$. Observe that, for each n, $(\mathsf{n}, \mathsf{c}) \approx' 020$ iff the execution of U starting at c terminates. Hence, $\approx'$ is a Σ_1^0-hard equivalence relation.

We now define two further binary relations $\to_{\text{even}}$ and $\to_{\text{odd}}$ as follows: $(\mathsf{n}, \mathsf{c}) \to_{\text{even}} (\mathsf{m}, \mathsf{c}')$ (resp. $(\mathsf{n}, \mathsf{c}) \to_{\text{odd}} (\mathsf{m}, \mathsf{c}')$) if either (i) $(\mathsf{n}, \mathsf{c}) \to' (\mathsf{m}, \mathsf{c}')$ and n is even (resp. odd) or (ii) c is final, n is even (resp. odd) and $(\mathsf{m}, \mathsf{c}') = 020$. It is obvious that $\to_{\text{even}}$ and $\to_{\text{odd}}$ are still decidable in logarithmic space.

---

[4] Note that c does contain the whole tape, of size N and the head position in it. Because the head must point into the tape, the head position is at most N and thus can be stored in binary using only $\log N$ space. Therefore, counting up to N to actually find the position in the tape requires space $\log N$ (which is indeed logarithmic in the size of c).

Note that by construction, if $(\mathbf{n}, \mathbf{c}) \rightarrow_{\text{even}} (\mathbf{m}, \mathbf{c'})$ then n must be even, hence $m = n+1$ must be odd (except for the special case of final configuration). Hence, there is no $(\mathbf{p}, \mathbf{c''})$ such that $(\mathbf{m}, \mathbf{c'}) \rightarrow_{\text{even}} (\mathbf{p}, \mathbf{c''})$.

Now, consider $\approx_{\text{even}}$, the reflexive transitive symmetric closure of $\rightarrow_{\text{even}}$. We claim that $\approx_{\text{even}}$ is decidable in logarithmic space. Indeed, the only cases where we can have $x \approx_{\text{even}} y$ are: (i) $x = y$, (ii) $x \rightarrow_{\text{even}} y$ (this is decidable in logarithmic space), (iii) $y \rightarrow_{\text{even}} x$ (this is also decidable in logarithmic space), and (iv) there exists z such that $x \rightarrow_{\text{even}} z$ and $y \rightarrow_{\text{even}} z$. It is decidable in logarithmic space whether case (iv) holds by simulating one step of the execution on the configurations in x and y and checking whether the results are the same. As the Turing machine M is deterministic, at most one step per configuration must be simulated. Observe that in case (iv) we cannot directly simulate one step of the machine from each of the configurations x and y and compare the results in logarithmic space (because we cannot store the results). However, we can check that x and y only differ by the state of the machine and the symbol read by the tape head; then, the step can be performed by changing only local information, i.e., compute the new state, tape symbol, and move. All of these can clearly be done in logarithmic space.

As it is not possible to have $x \rightarrow_{\text{even}} z \rightarrow_{\text{even}} y$, it is never necessary to simulate several steps in a row in order to check whether $x \approx_{\text{even}} y$. Thus, *at most one* step from each of x and y must be simulated. Hence, $\approx_{\text{even}}$ is decidable in logarithmic space. Similarly, $\approx_{\text{odd}}$ is decidable in logarithmic space. However, $\approx_{\text{even}} \vee \approx_{\text{odd}} = \approx'$. □

Corollary 2. *Let* $\mathbf{B}$ *be a set of equivalence relations such that* $Equ(\mathbb{N})_{\text{L}} \subseteq \mathbf{B} \subseteq Equ(\mathbb{N})_{\text{co-r.e.}}$. *Then there are* $\mathcal{E}, \mathcal{F} \in \mathbf{B}$ *such that* $\mathcal{E} \vee \mathcal{F} \notin \mathbf{B}$ *(in fact,* $\mathcal{E} \vee \mathcal{F} \notin Equ(\mathbb{N})_{\text{co-r.e.}}$*).*

Hence, none of $Equ(\mathbb{N})_{\text{co-r.e.}}$, $Equ(\mathbb{N})_{\text{dec}}$, $Equ(\mathbb{N})_{\text{p.r.}}$, $Equ(\mathbb{N})_{\text{KE}}$, $Equ(\mathbb{N})_{\text{EXPTIME}}$, $Equ(\mathbb{N})_{\text{PSPACE}}$, $Equ(\mathbb{N})_{\text{P}}$, *or* $Equ(\mathbb{N})_{\text{L}}$ *are closed under finite join, and hence none of these sets is a sublattice of* $Equ(\mathbb{N})$.

Remark 1. Corollary 2 contains the folklore result that $Equ(\mathbb{N})_{\text{dec}}$ is not a lattice. For completeness, observe that this also implies that the set of co-r.e. equivalence relations is not a lattice either: Assume that the join of every pair of co-r.e. equivalence relations were co-r.e., consider any two decidable equivalence relations. *A fortiori*, both of these relations are both r.e. and co-r.e., and thus the join of the two relations is r.e., and by the assumption above also co-r.e. and hence decidable. But this is impossible by Corollary 2.

Observe also that the Σ_1^0-completeness of Proposition 2 is, in a sense, an upper bound on the complexity of $\mathcal{E} \vee \mathcal{F}$ for decidable equivalence relations $\mathcal{E}, \mathcal{F}$, because $\mathcal{E} \vee \mathcal{F}$ is necessarily r.e. as $Equ(\mathbb{N})_{\text{r.e.}}$ is a lattice.

3.1 On Meet and Join of r.e Subsets of $Equ(\mathbb{N})_{\text{L}}$

Let $\mathbf{B}$ be any set of decidable equivalence relations such that $Equ(\mathbb{N})_{\text{L}} \subseteq \mathbf{B}$. Then $\mathbf{B}$ is not closed under taking the join of a finite number of elements, and $\mathbf{B}$

is not closed under taking the meet of an arbitrary set of elements (as any subset of $\mathbb{N}$ can be realized as the meet of a subset of $Equ(\mathbb{N})_L$). However, there remains the possibility that a "middle ground" could be found where **B** could be closed under taking the meet of r.e. subsets. This would be *unexpected* though, as it is well-known that the set of decidable sets is not closed under r.e. intersection. Indeed, the following proposition settles the matter, and is straightforwardly proved:

Proposition 3. *There is an r.e. set **A** of elements of $Equ(\mathbb{N})_L$ such that $\wedge\mathbf{A}$ is Π_1^0-complete.*

Proof. First observe that if **A** is an r.e. set of decidable equivalence relations (so, for instance, an r.e. subset of $Equ(\mathbb{N})_L$), $\wedge\mathbf{A}$ is co-r.e.: Let M be a Turing machine that, on input (m, n), uses the enumeration of **A** to generate a sequence of (encodings of) Turing machines that decide each of the elements A_j of **A**, and then uses a universal Turing machine as a subroutine that asks each of the generated machines whether mA_jn. If a j is encountered such that $\neg(mA_jn)$, M outputs 'no'. As $m(\wedge\mathbf{A})n$ iff mA_jn for all $A_j \in \mathbf{A}$, we conclude that $\wedge\mathbf{A}$ is co-r.e.

It remains to prove existence of an r.e. set $\mathbf{A} \subseteq Equ(\mathbb{N})_L$ such that $\wedge\mathbf{A}$ is Π_1^0-hard. Consider a standard representation of Turing machines as elements of $\{0,1\}^*$ such that it can be checked in logarithmic space whether $x \in \{0,1\}^*$ is a valid representation of a Turing machine (see e.g., [20, Ch. 3]). Consider, for every $j \in \mathbb{N}$, the singular equivalence relation $\mathcal{E}_j$ whose unique non-singleton class A_j consists of the Gödel numbers of Turing machines that *do not* halt in at most j steps. Observe that $A_1 \supseteq A_2 \supseteq \cdots$. Set $\mathbf{A} = \{\mathcal{E}_j : j \in \mathbb{N}\}$.

Define $\mathcal{E}_\infty$ as the singular equivalence relation whose unique non-singleton class, A_∞, consists of those non-negative integers whose binary expansion is the encoding of an inputless Turing machine that does not halt. Then, $\mathcal{E}_\infty = \wedge\mathcal{E}_j$. Furthermore, $\mathcal{E}_\infty$ is Π_1^0-hard: $\{i : \phi_i\uparrow\}$ is a well-known Π_1^0-complete set and $m\mathcal{E}_\infty n$ iff $m = n$ or ($\phi_m\uparrow$ and $\phi_n\uparrow$). Let l be (the Gödel number of) an inputless Turing machine that does not halt. Then, $i \in \{i : \phi_i\uparrow\}$ iff $i\mathcal{E}_\infty l$.

Now, for each $j \in \mathbb{N}$, we have $\mathcal{E}_j \in Equ(\mathbb{N})_L$, by the following reasoning: Fix a non-negative integer j; for every pair $(m, n) \in \mathbb{N}$, a Turing machine may check in logarithmic space whether the binary representation of m and n represents a Turing machine and then use a universal Turing machine to simulate running of both m and n for j steps. By proper construction of the universal machine, the space overhead required can be made constant in the space used by the machines that m and n represents, which is bounded above by j cells as both machines are run for at most j steps; the universal machine needs only a fixed number of counters beyond this overhead (see, e.g. [20, Ch. 3]; in essence, the universal machine works by simulating one step of the simulated machine at a time and keeping a representation of its configuration in memory). Hence, each $\mathcal{E}_j \in Equ(\mathbb{N})_L$.

Observe that there is a Turing machine that, on input j produces a(n encoding of a) Turing machine that decides $\mathcal{E}_j$ in logarithmic space: it simply specializes a universal Turing machine to j and outputs the specialization along with some fixed (i.e., independent of j) operations for comparing binary representations and checking whether the inputs are correct encodings of Turing machines. Hence, the set $\{\mathcal{E}_j : j \in \mathbb{N}\}$ is r.e.

However, $\mathcal{E}_\infty$ is not decidable (indeed, is not even r.e.: if it were, we could enumerate the set of all non-halting Turing machines by fixing a single such machine M_i and recursively enumerate all pairs $(i, n) \in \mathcal{E}_\infty$, hence recursively enumerate the set of all non-halting machines, a contradiction). $\qquad \square$

Proposition 3 immediately implies the below corollary.

Corollary 3. *Let* **B** *be a set of equivalence relations such that* $Equ(\mathbb{N})_{\mathrm{L}} \subseteq \mathbf{B} \subseteq Equ(\mathbb{N})_{\mathrm{r.e.}}$. *Then, there is an r.e. set* **A** *of elements of* **B** *such that* $\wedge \mathbf{A} \notin \mathbf{B}$ *(in fact,* $\wedge \mathbf{A} \notin Equ(\mathbb{N})_{\mathrm{r.e.}}$*).*

4   Future Work

While we laid the groundwork for study of closure under the lattice operations for some of the most obvious subsets of $Equ(\mathbb{N})$ related to computability, there are many possible extensions. For example, performing a systematic study of closure properties of sublattices decided by automata (e.g., multi-tape or synchronous automata), and lattices corresponding to subrecursive classes not necessarily containing $Equ(\mathbb{N})_{\mathrm{L}}$, possibly using an axiomatization of the notion of subrecursive class in the style of [15]. In addition to the basic lattice operations of join and meet, $Equ(\mathbb{N})$ is known to be closed under other operations, notably taking lattice complements, and closure under these operations should be considered for the classes treated in this paper. Finally, the connection between sets of equivalence relations in the lattice and existing complexity theory should be investigated, e.g. whether $Equ(\mathbb{N})_{\mathrm{P}} = Equ(\mathbb{N})_{\mathrm{PSPACE}}$ implies $\mathrm{P} = \mathrm{PSPACE}$.

References

1. Andrews, U., Badaev, S., Sorbi, A.: A survey on universal computably enumerable equivalence relations. In: Day, A., Fellows, M., Greenberg, N., Khoussainov, B., Melnikov, A., Rosamond, F. (eds.) Computability and Complexity. LNCS, vol. 10010, pp. 418–451. Springer, Cham (2017). https://doi.org/10.1007/978-3-319-50062-1_25
2. Andrews, U., Sorbi, A.: The complexity of index sets of classes of computably enumerable equivalence relations. J. Symbolic Logic **81**(4), 1375–1395 (2016)
3. Andrews, U., Sorbi, A.: Joins and meets in the structure of ceers. Computability **8**(3–4), 193–241 (2019)
4. Asperti, A.: The intensional content of rice's theorem. In: Proceedings of the 35th Annual ACM SIGPLAN - SIGACT Symposium on Principles of Programming Languages (POPL 2008) (2008)
5. Avery, J.E., Moyen, J.Y., Růžička, P., Simonsen, J.G.: Chains, antichains, and complements in infinite partition lattices. Algebra Univers. **79**(2), 37 (2018)
6. Birkhoff, G.: Lattice Theory, Colloquium Publications, vol. 25. American Mathematical Society (1940)

7. Blum, M.: A machine-independent theory of the complexity of recursive functions. J. ACM **14**(2), 322–336 (1967)
8. Burris, S., Sankappanavar, H.P.: A Course in Universal Algebra, Graduate Texts in Mathematics, vol. 78. Springer, New York (1981)
9. Carroll, J.S.: Some undecidability results for lattices in recursion theory. Pacific J. Math. **122**(2), 319–331 (1986)
10. Gao, S., Gerdes, P.: Computably enumerable equivalence relations. Studia Logica: Int. J. Symbolic Logic **67**(1), 27–59 (2001)
11. Gavryushkin, A., Khoussainov, B., Stephan, F.: Reducibilities among equivalence relations induced by recursively enumerable structures. Theoretical Computer Science **612**, 137–152 (2016)
12. Grätzer, G.: General Lattice Theory. Birkhäuser, second edn. (2003)
13. Ianovski, E., Miller, R., Ng, K.M., Nies, A.: Complexity of equivalence relations and preorders from computability theory. J. Symbolic Logic **79**(3), 859–881 (2014)
14. Jones, N.D.: Computability and Complexity, from a Programming Perspective. MIT press (1997)
15. Kozen, D.: Indexings of subrecursive classes. Theor. Comput. Sci. **11**, 277–301 (1980)
16. Lusin, S., Salibra, A.: The lattice of lambda theories. J. Log. Comput. **14**(3), 373–394 (2004)
17. Moyen, J., Simonsen, J.G.: Computability in the lattice of equivalence relations. In: Proceedings of DICE-FOPARA@ETAPS 2017, pp. 38–46 (2017)
18. Moyen, J., Simonsen, J.G.: More intensional versions of Rice's theorem. In: Proceedings of the 15th Conference on Computability in Europe (CiE 2019), pp. 217–229 (2019)
19. Ore, Ø.: Theory of equivalence relations. Duke Math. J. **9**(3), 573–627 (1942)
20. Papadimitriou, C.H.: Computational Complexity. Addison-Wesley (1994)
21. Regan, K.W.: Minimum-complexity pairing functions. J. Comput. Syst. Sci. **45**(3), 285–295 (1992)
22. Rice, H.G.: Classes of Recursively Enumerable Sets and Their Decision Problems. Trans. Am. Math. Soc. **74**, 358–366 (1953)
23. Rogers, H.: Theory of Recursive Functions and Effective Computability. McGraw-Hill (1967). (reprint, MIT press 1987)
24. Stern, M.: Semimodular Lattices. Cambridge University Press (1999)
25. Visser, A.: Numerations, λ-calculus and arithmetic. In: To H.B. Curry: Essays on Combinatory Logic, Lambda-Calculus and Formalism, pp. 259–284. Academic Press (1980)

Interactive Physical ZKP for Connectivity: Applications to Nurikabe and Hitori

Léo Robert[1]([⊠]) [ID], Daiki Miyahara[2,4] [ID], Pascal Lafourcade[1] [ID], and Takaaki Mizuki[3,4] [ID]

[1] LIMOS, University Clermont Auvergne, CNRS UMR 6158, Aubière, France
leo.robert@uca.fr
[2] Graduate School of Information Sciences, Tohoku University, Sendai, Japan
[3] Cyberscience Center, Tohoku University, Sendai, Japan
[4] National Institute of Advanced Industrial Science and Technology, Tokyo, Japan

Abstract. During the last years, many Physical Zero-knowledge Proof (ZKP) protocols for Nikoli's puzzles have been designed. In this paper, we propose two ZKP protocols for the two Nikoli's puzzles called Nurikabe and Hitori. These two puzzles have some similarities, since in their rules at least one condition requires that some cells are connected to each other, horizontally or vertically. The novelty in this paper is to propose two techniques that allow us to prove such connectivity without leaking any information about a solution.

Keywords: Zero-knowledge proofs · Card-based secure two-party protocols · Puzzle · Nurikabe · Hitori

1 Introduction

Zero-Knowledge Proofs (ZKP) were introduced by Goldwasser et al. [7]. Such a protocol has two parties: a prover P and a verifier V. The prover P wants to convince the verifier V that P knows the solution s of a problem without revealing any information about s. A ZKP must satisfy the following properties:

Completeness. If P knows s, then P can convince V.
Soundness. If P does not know s, then P cannot convince V.
Zero-Knowledge. V learns nothing about s. Formally, outputs of a simulator and outputs of the real protocol follow the same probability distribution.

In [5], the authors proved that for any NP-complete problem there exists an interactive ZKP protocol. A physical ZKP uses only physical algorithms with day-to-day objects such as cards, envelopes or bags while prohibiting large computations (i.e., no computer allowed). In 2007, the first physical ZKP was introduced for Sudoku [8], which is the most famous Nikoli's[1] puzzle. In this paper we focus on two other Nikoli's puzzles, *Nurikabe* and *Hitori*.

---

[1] Nikoli is a game publisher famously known for its Sudoku puzzle.

© Springer Nature Switzerland AG 2021
L. De Mol et al. (Eds.): CiE 2021, LNCS 12813, pp. 373–384, 2021.
https://doi.org/10.1007/978-3-030-80049-9_37

In [10] solving simple versions of Nurikabe was proven to be NP-complete. In [9] the authors proved that Hitori is also NP-complete. One might think that physical ZKP protocols for Nurikabe and Hitori could be constructed by transforming a known physical ZKP protocol for an NP-complete problem, such as a lockable-box-based ZKP protocol for 3-Colorability [6]; however, such a transformation is not practical because the overhead must be included in the transformation. Besides, the transformed ZKP protocol does not capture the property of a puzzle.

Contributions: In this paper, we present physical ZKP protocols for Nurikabe and Hitori using a deck of cards. Our protocols achieve no soundness error. That is, no malicious P who does not have a solution can convince V that it has a solution. Our work is inspired by [12], where P has to convince V of a single loop property. For Nurikabe and Hitori, we take a similar strategy to [12]. That is, P first increases the number of black (or white) cells one by one so that the resulting cells are guaranteed to satisfy the constraint of connectivity; then V verifies all the remaining constraints. We note that our protocols in this paper could not be constructed by simply adapting the existing technique [12].

We emphasize that our proposed protocols can be applied to a situation where Bob cannot solve by hand a Nurikabe or Hitori puzzle Alice created. In addition to such really practical applications, we believe that one can add our protocols (with others such as 3-Colorability one for instance) to introduce the notion of a ZKP system to non-experts such as high school students.

Related Work: Efficient physical ZKP protocols for Nikoli puzzles have been proposed: Sudoku [8,20], Akari [2], Takuzu [2,13], Kakuro [2,14], Kenken [2], Makaro [3], Norinori [4], Slitherlink [12], Juosan [13], Suguru [16], Ripple Effect [19], and Numberlink [18]. An important step in this line of research is to achieve no soundness error.

Nurikabe's Rule: This puzzle is formed by a rectangular grid where some cells contain numbers (Fig. 1). The goal is to color some cells in black as follows:

1. Each numbered cell tells the number of continuous white cells surrounded by black cells. Such a region is called an *island*.
2. An island must contain only one numbered cell.
3. The black cells form a connected figure (called a *sea*).
4. The *sea* cannot form a 2×2 area.

Fig. 1. Initial Nurikabe grid on the left and its solution on the right.

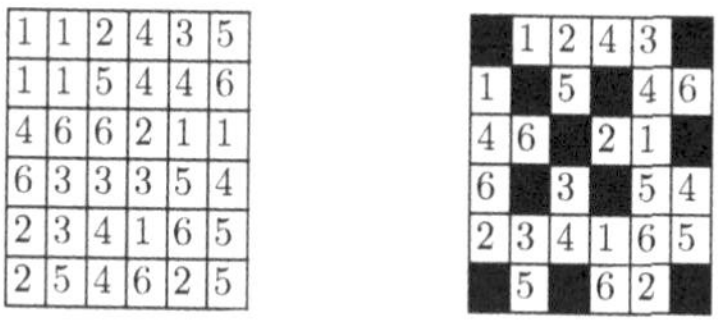

1	1	2	4	3	5
1	1	5	4	4	6
4	6	6	2	1	1
6	3	3	3	5	4
2	3	4	1	6	5
2	5	4	6	2	5

■	1	2	4	3	■
1	■	5	■	4	6
4	6	■	2	1	■
6	■	3	■	5	4
2	3	4	1	6	5
■	5	■	6	2	■

Fig. 2. Initial Hitori grid on the left and its solution on the right.

Hitori's Rule: This puzzle is a grid where each cell contains a number as in the example of Fig. 2. The goal is to color in black some cells with the following constraints:

1. Each row and each column must contain only one occurrence of a number.
2. The black cells cannot touch side to side although they can be diagonal.
3. The numbered cells must be connected to each other, horizontally or vertically.

2 Preliminaries

We introduce some notations of cards and shuffles and explain simple physical sub-protocols used in our constructions.

Card: A deck of cards used in our protocols consists of clubs ♣ ♣ $\cdots$, hearts ♡ ♡ $\cdots$, and number cards 1 2 $\cdots$, whose backs are identical ?. We encode three colors with the order of two cards as follows:

$$\text{black} \leftarrow \boxed{♣}\boxed{♡}, \quad \text{white} \leftarrow \boxed{♡}\boxed{♣}, \quad \text{red} \leftarrow \boxed{♡}\boxed{♡}. \tag{1}$$

We call such a face-down two cards ? ? corresponding to a color according to the above encoding rule a *commitment* to the respective color. We also use the terms, a *black commitment*, a *white commitment*, and a *red commitment*.

Pile-shifting Shuffle [15,21]: This shuffling action means to *cyclically* shuffle piles of cards. More formally, given m piles, each of which consists of the same number of face-down cards, denoted by $(p_1, p_2, \ldots, p_m)$, applying a *pile-shifting shuffle* (denoted by $< \cdot | \cdots | \cdot >$) results in $(p_{s+1}, p_{s+2}, \ldots, p_{s+m})$:

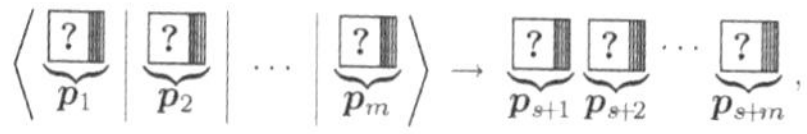

$$\left\langle \underset{p_1}{\boxed{?}}\ \underset{p_2}{\boxed{?}}\ \cdots\ \underset{p_m}{\boxed{?}} \right\rangle \rightarrow \underset{p_{s+1}}{\boxed{?}}\ \underset{p_{s+2}}{\boxed{?}}\ \cdots\ \underset{p_{s+n}}{\boxed{?}},$$

where s is uniformly and randomly chosen from $\mathbb{Z}/m\mathbb{Z}$. Implementing a pile-shifting shuffle is simple: We use physical cases that can store a pile of cards, such as boxes and envelopes; a player (or players) cyclically shuffle them by hand until nobody traces the offset.

Chosen Pile Protocol [4]: This is an extended version of the "chosen pile cut" proposed in [11]. Given m piles $(\boldsymbol{p}_1, \boldsymbol{p}_2, \ldots, \boldsymbol{p}_m)$ with $2m$ additional cards, the *chosen pile protocol* enables a prover P to choose the i-th pile $\boldsymbol{p}_i$ and replace back the sequence of m piles to their original order.

1. Using $m-1$ ♣s and one ♡, P places m face-down cards (denoted *row 2*) below the given piles such that only the i-th card is ♡. We further put m cards (denoted *row 3*) below the cards such that only the first card is ♡:

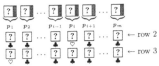

2. Considering the cards in the same column as a pile, apply a pile-shifting shuffle to the sequence of piles.
3. Reveal all the cards in the *row 2*. Then, one ♡ appears, and the pile above the revealed ♡ is the i-th pile (and hence, P can obtain $\boldsymbol{p}_i$). When this protocol is invoked, certain operations are applied to the chosen pile. Then, the chosen pile is placed back to the i-th position in the sequence.
4. Remove the revealed cards, i.e., the cards in the *row 2*. (Note, therefore, that we do not use the card ♡ revealed in Step 3.) Then, apply a pile-shifting shuffle.
5. Reveal all the cards in the *row 3*. Then, one ♡ appears, and the pile above the revealed ♡ is $\boldsymbol{p}_1$. Therefore, by shifting the sequence of piles (such that $\boldsymbol{p}_1$ becomes the first pile in the sequence), we can obtain a sequence of piles whose order is the same as the original one without revealing any information about the order of input sequence.

Input-Preserving Five-Card Trick [13]: Given two commitments to $a, b \in \{0, 1\}$ based on the encoding: ♣♡ $= 0$ and ♡♣ $= 1$, this sub-protocol [1,13] starts by adding extra cards and rearranging the commitment to a so that we have the negation $\bar{a}$, as follows: $\boxed{?}\boxed{?}\ \boxed{?}\boxed{?} \rightarrow \boxed{?}\boxed{?}\ \boxed{♡}\ \boxed{?}\boxed{?}\ \boxed{1}\boxed{2}\boxed{3}\boxed{4}\boxed{5}$.

The sub-protocol proceeds as follows to reveal only the value of $a \wedge b$ as well as restore commitments to a and b:

1. Rearrange the sequence of cards and then turn over the face-up cards as:

$$\boxed{?}\boxed{?}\boxed{♡}\boxed{?}\boxed{?}\boxed{1}\boxed{2}\boxed{3}\boxed{4}\boxed{5} \rightarrow \begin{matrix} \boxed{?}\boxed{?}\boxed{♡}\boxed{?}\boxed{?} \\ \boxed{1}\boxed{2}\boxed{3}\boxed{4}\boxed{5} \end{matrix} \rightarrow \begin{matrix} \boxed{?}\boxed{?}\boxed{?}\boxed{?}\boxed{?} \\ \boxed{?}\boxed{?}\boxed{?}\boxed{?}\boxed{?} \end{matrix}.$$

2. Regarding cards in the same column as a pile, apply a pile-shifting shuffle to the sequence: $\left\langle \begin{matrix} \boxed{?} \\ \boxed{?} \end{matrix} \begin{matrix} \boxed{?} \\ \boxed{?} \end{matrix} \begin{matrix} \boxed{?} \\ \boxed{?} \end{matrix} \begin{matrix} \boxed{?} \\ \boxed{?} \end{matrix} \begin{matrix} \boxed{?} \\ \boxed{?} \end{matrix} \right\rangle \rightarrow \begin{matrix} \boxed{?}\boxed{?}\boxed{?}\boxed{?}\boxed{?} \\ \boxed{?}\boxed{?}\boxed{?}\boxed{?}\boxed{?} \end{matrix}.$
3. Reveal all the cards in the first row, if the resulting sequence is:
   (a) ♣♣♡♡♡ (up to cyclic shifts), then we have $a \wedge b = 1$.
   (b) ♡♣♡♣♡ (up to cyclic shifts), then we have $a \wedge b = 0$.
4. After turning over all the face-up cards, apply a pile-shifting shuffle.
5. Reveal all the cards in the second row, i.e., all the number cards. Then, rearrange the sequence of piles so that the revealed number cards are in ascending order again to restore commitments to a and b.

3   ZKP Protocol for Nurikabe

We propose a ZKP protocol for Nurikabe, which is composed of three phases: the setup phase, the sea formation phase, and the verification phase. The full security proof is provided in [17], we only give here a sketch in Sect. 3.4.

Consider a puzzle instance of a $p \times q$ grid containing m numbered cells such that the ith numbered cell (in any order) has a number x_i for every i, $1 \leq i \leq m$. Remember that an island of a Nurikabe puzzle must contain exactly one numbered cell, and the number of white cells inside the island is indicated by the number written on the numbered cell. Thus, the number of (filled) black cells in the solution, denoted by N_b, is the difference between the number of total cells and the white cells (including the numbered cells), so $N_b = pq - \sum_{i=1}^{m} x_i$.

Thus, this number N_b can be regarded as public information, and indeed, we use the number N_b explicitly in our protocol.

Before going into the details of our protocol, let us define a *neighbour* cell and show a sub-protocol called the *4-neighbour protocol* that is important for constructing our ZKP protocols.

Neighbour Cell: Consider a target cell c_t on a grid. A cell is a *neighbour* of c_t if it is next to c_t, on the left, the right, the top, or the bottom but not in diagonal.

4-Neighbour Protocol: Given pq commitments placed on a $p \times q$ grid, a prover P wants to reveal a target commitment and another one that lies next to the target commitment. Here, a verifier V is convinced that the second commitment is a neighbour of the first one (without knowing which one) as well as V confirms the colors of both the commitments. To handle the case where the target commitment is at the edge of the grid, we add red commitments (as "dummy" commitments) around the grid to prevent P from choosing a commitment that is not a neighbour. Thus, the size of the new grid is $(p + 2) \times (q + 2)$.

This protocol uses the chosen pile protocol (Sect. 2) twice. P first uses the chosen pile protocol to reveal a target commitment. Since a pile-shifting shuffle is a cyclic reordering, the distance between commitments are kept (up to a given modulo). That is, for a target commitment (not at the edge), the possible four neighbours are at distance 1 for the left or right one, and $p + 2$ for the bottom or top one. Therefore, V and P can determine the positions of all the four neighbours. Among these, P chooses one commitment by using the chosen pile protocol again, and reveals it. This convinces V that the second commitment is indeed a neighbour. The rest of the protocol is to end the second and first chosen pile protocols.

3.1   Setup Phase

The verifier V and the prover P place a white commitment on each cell of a given $p \times q$ grid and place red commitments (as "dummy" commitments) around the grid so that we have $(p + 2)(q + 2)$ commitments on the board.

3.2   Sea Formation Phase

In this phase, P forms a sea on the board, i.e., P replaces a white commitment with a black commitment one by one according to the solution which only P knows, while hiding any information about the solution to V.

Let N_b be the number of black cells in the solution. This phase proceeds as follows.

1. P uses the chosen pile protocol to choose one white commitment which P wants to replace.
   (a) V reveals the chosen commitment; if it corresponds to white, V swaps the two cards constituting it so that the two cards become a black commitment. Otherwise, V aborts.
   (b) P and V end the chosen pile protocol to return the commitments to their original positions.
2. Repeat the following steps exactly $N_b - 1$ times:
   (a) P chooses one black commitment as a target and one white commitment among its neighbours using the 4-neighbour protocol; the neighbour is chosen such that P wants to make it black.

   (b) V reveals the target commitment. If it corresponds to black, V continues; otherwise V aborts.
   (c) V reveals the neighbour commitment (chosen by P). If it corresponds to white, V swaps the two cards constituting it to make it be a black commitment; otherwise V aborts.
   (d) P and V end the 4-neighbour protocol.
3. P and V replace every red commitment (i.e., dummy commitment) with a black commitment.

After this process, V is convinced that all the black commitments form a connected sea (rule 3).

3.3   Verification Phase

V first verifies that the current commitments placed on the grid (after the sea formation phase) satisfy the rule 4 (forbidden 2×2 area). Then, V verifies the rules 1 and 2, relating to the white commitments (island constraints).

Sea Rule: Forbidden Area. The prover P wants to convince V that any 2×2 area contains at least one white cell. Note that all 2×2 areas are determined given an initial grid. Indeed, for a given $p \times q$ grid, there are $(p - 1)(q - 1)$ possible squares.

Thus, P and V consider each 2×2 area of commitments one by one (in any order) and will repeat the following for each possible square:

1. P chooses a white commitment on this square via the chosen-pile protocol applied to the four commitments.
2. V reveals the commitment marked by P. If the revealed commitment corresponds to white, then V is convinced that the square is not formed by only black commitments. Otherwise, V aborts.

$$\boxed{?}\boxed{?}\;\boxed{?}\boxed{?}\atop \boxed{?}\boxed{?}\;\boxed{?}\boxed{?} \quad \rightarrow \quad \text{Chosen pile protocol} \quad \rightarrow \quad \boxed{?}\boxed{?}\;\boxed{\heartsuit}\boxed{\clubsuit}\atop \boxed{?}\boxed{?}\;\boxed{?}\boxed{?} .$$

Island Rules. P wants to convince V that the white cells respect the constraints. There are two verifications to make. Only one numbered cell for a given region and all white commitments are connected inside the region. Those two constraints are verified in the following protocol:

Let $n \geq 2$ be the number written on a given numbered cell.[2]

1. V reveals the commitment on the numbered cell. If it corresponds to white, V replaces it with a red commitment; otherwise V aborts.
2. Repeat the following steps exactly $n-1$ times.
   (a) P uses the 4-neighbour protocol to choose a red commitment as a target and one white commitment among its neighbours.
   (b) V reveals the target commitment. If it corresponds to red, V continues; otherwise V aborts.
   (c) V reveals the neighbour commitment (chosen by P). If it corresponds to white, V replaces it with a red commitment; otherwise V aborts.
   (d) P and V end the 4-neighbour protocol to return the commitments to their original positions.

Now, V is convinced that the size of the island consisting of white cells is greater than or equal to n. To show that it is equal to n, it suffices to prove that there exists no white cell around them, as follows.

3. V replaces the commitment on the numbered cell with a black commitment.
4. Repeat the following steps exactly $n-1$ times.
   (a) P uses the chosen pile protocol to choose a red commitment.
   (b) V reveals the chosen commitment. If it corresponds to red, V continues; otherwise V aborts.
   (c) Remember that P wants to show that any of four neighbour commitments is not white. Recall also the encoding (1), i.e., note that the right card of a black or red commitment is a heart $\boxed{\heartsuit}$. V reveals the right card of each of the four neighbours. If all of them are hearts (which means that all the commitments do not correspond to white), V replaces the chosen commitment with a black commitment; otherwise V aborts.

$$\begin{array}{ccc}\boxed{?}\boxed{?}\\ \boxed{?}\boxed{?}\boxed{\heartsuit}\boxed{\heartsuit}\boxed{?}\boxed{?}\\ \boxed{?}\boxed{?}\end{array} \rightarrow \begin{array}{c}\boxed{?}\boxed{\heartsuit}\\ \boxed{?}\boxed{\heartsuit}\boxed{\heartsuit}\boxed{\heartsuit}\boxed{?}\boxed{\heartsuit}\\ \boxed{?}\boxed{\heartsuit}\end{array} \rightarrow \begin{array}{c}\boxed{?}\boxed{?}\\ \boxed{?}\boxed{?}\boxed{?}\boxed{?}\boxed{?}\boxed{?}\\ \boxed{?}\boxed{?}\end{array}.$$

---

[2] For a numbered cell where 1 is written, V simply reveals the commitment on it and its four neighbours to confirm that the island is surrounded by the sea.

(d) P and V end the chosen pile protocol to return the commitments to their original positions.

By applying the above steps to all the numbered cells, V is convinced that the placement of the commitments satisfies all the constraints, i.e., P has the solution.

3.4   Security Proofs

We give the following theorems to show that our protocol respects the security properties. All the proofs of our theorems are given in [17], we only give here a proof sketch.

Theorem 1 (Completeness). *If P knows a solution of a Nurikabe grid, then it can convince V.*

Proof (sketch). We suppose that P knows the solution s of the grid and runs the setup phase. P is able to perform the proofs for the sea formation since all the black cells are connected. P is also able to end the verification phase. Basically, since s is a solution, all the rules are verified.

Theorem 2 (Soundness). *If P does not provide a solution of the $p \times q$ Nurikabe grid G, it is not able to convince V.*

Proof (sketch). We suppose that P does not know the solution s and the proof is about showing that V will always detect it. Notice that the commitments of P form a connected figure (otherwise the protocol is ended without any verification). There are two cases to consider for the verification; (1) the forbidden area, if all the commitment are black on a 2×2 square then V will detect it since P cannot choose a white commitment, and (2) the island rules, where two invalid shapes can occur, when a region is completely covered with another region and, a part of a region is covered with another one. In both cases, we show that V will detect it using the protocol.

Theorem 3 (Zero-knowledge). *V learns nothing about P's solution of the given grid G.*

Proof (sketch). We use the same technique as in [8]; zero-knowledge is induced by a description of an efficient *simulator* which simulates interaction between a cheating verifier and a real prover. However, the simulator does not have a solution but it can swap cards for different ones during shuffles. The aim of the proof is to describe the behaviour of this simulator. Basically, the simulator creates a random connected figure of size N_b and during the verification, it swaps the cards to verify the rules.

4   ZKP Protocol for Hitori

We present a ZKP protocol for Hitori. The full security proof is provided in [17]; we only give here a sketch presented in Sect. 4.4. Similar to our protocol for Nurikabe presented in Sect. 3, we let P choose a commitment which P wants to make white so that V is convinced that the resulting numbered cells are connected each other. However, we note that for Hitori the size of numbered cells could be information about the solution. That is, we cannot simply use the sea formation phase shown in Sect. 3.2. Therefore, we construct a sub-protocol called the *still-black protocol* as follows.

Still-black Protocol: Given a black commitment, P can choose either changing it (i.e., swapping the two cards constituting the commitment) or not without V noticing it, as follows.

1. V reveals the given commitment to confirm that it is surely a black commitment.
2. If P wants to change the commitment, P places face-down club-to-heart below it; otherwise heart-to-club: $\boxed{?}\boxed{?}$ → $\boxed{?}\boxed{?}$ or $\boxed{?}\boxed{?}$.
3. Regarding cards in the same column as a pile, V applies a pile-shifting shuffle to the sequence of piles.
4. V reveals all the cards in the second row. If the revealed card on the right is a heart $\boxed{\heartsuit}$, V swaps the two cards in the first row; otherwise V does nothing.

4.1   Setup Phase

Put a black commitment on each cell of the $p \times q$ grid and red commitments around the grid.

4.2   Connectivity Phase

This phase follows the same steps as the ones in the sea formation phase shown in Sect. 3.2 (where a white commitment is regarded as a black one and vice versa) except for Step 2c; instead of swapping the two cards, V and P use the still-black protocol so that P can choose either swapping the two cards or not. (Remember that P cannot change a white commitment into black.) Note that the steps are repeated exactly $pq - 1$ times.

After the above process, V is convinced that the resulting commitments represent a connected (white) figure (rule 3) and information about the number of the white commitments is hidden from V.

4.3   Verification Phase

One Occurrence for Each Row/Column. Here, V checks if each row and column contains only one occurrence of a number. The idea is that for a given row or column it suffices to look at only numbered cells that appear $k > 1$ times and confirm that the k commitments on the numbered cells correspond to either k blacks or $k - 1$ blacks. For a given row or column, this verification proceeds as follows.

1. V looks for numbered cells that appear more than once; take such a number which appears exactly $k > 1$ times. Then, V picks the corresponding k commitments.
2. P uses the chosen pile protocol to choose a white commitment among the k commitments if it exists; otherwise P uses the one to choose any commitment.
3. V reveals the $k - 1$ commitments that are not chosen by P. If all of them correspond to black (this means that the k commitments correspond to k or $k - 1$ blacks), V continues; otherwise V aborts.
4. V and P end the chosen pile protocol to return the k commitments to their original places.
5. V and P repeat the above steps for all numbers that appear twice or more.

Lonely Black. V checks that black cells are isolated from each other. Let a white commitment correspond to bit 0 and a black to 1. For each pair of adjacent commitments, V applies the input-preserving five-card trick (Sect. 2) to the two commitments. If the output is 0, V continues; otherwise V aborts.

4.4   Security Proofs

Theorem 4 (Completeness). *If P knows a solution of a Hitori grid, then it can convince V.*

Proof (sketch). Suppose that P knows the solution; thus P can perform the connectivity and verification phases without aborting. There are two cases to consider, when P wants to change the black commitment and when P wants a black commitments to be still black.

Theorem 5 (Soundness). *If P does not provide a solution of the $p \times q$ Hitori grid G, then it is not able to convince V.*

Proof (sketch). Suppose that P does not know the solution. The proof consists in showing that V will detect it using the protocol. Without loss of generality, suppose that P gives correct commitments (i.e., white cells are connected) without corresponding to the solution, we show that V detects that the uniqueness and the lonely black constraints are not respected.

Theorem 6 (Zero-knowledge). *V learns nothing about P's solution of the given grid G.*

Proof (sketch). The same technique as for Nurikabe is used, namely the presence of a simulator that does not know the solution but can swap cards randomly.

5   Conclusion

We proposed two ZKP protocols for Nurikabe and Hitori. These two Nikoli's puzzles require that some cells of the solution are continuous without any precision on the number of cells in Hitori and without an exact number of cells in Nurikabe. We designed two methods and encoding for solving this continuity challenge and also respecting the other rules of the puzzles.

In the future, we aim at solving more challenging puzzles with other rules that also involve a kind of continuity property. For instance, in the puzzles Shikaku and Shakashaka, the goal is to draw rectangles of a certain size, which does not seem easy.

Acknowledgements. This work was supported in part by JSPS KAKENHI Grant Numbers JP19J21153 and JP21K11881. This study was partially supported by the French ANR project ANR-18-CE39-0019 (MobiS5), by the research program "Investissements d'Avenir" through the IDEX-ISITE initiative 16-IDEX-0001 (CAP 20-25), by the IMobS3 Laboratory of Excellence (ANR-10-LABX-16-01), by the French ANR project DECRYPT (ANR-18-CE39-0007) and SEVERITAS (ANR-20-CE39-0009).

References

1. Boer, B.: More efficient match-making and satisfiability *the five card trick*. In: Quisquater, J.-J., Vandewalle, J. (eds.) EUROCRYPT 1989. LNCS, vol. 434, pp. 208–217. Springer, Heidelberg (1990). https://doi.org/10.1007/3-540-46885-4_23
2. Bultel, X., Dreier, J., Dumas, J., Lafourcade, P.: Physical zero-knowledge proofs for Akari, Takuzu, Kakuro and KenKen. In: Demaine, E.D., Grandoni, F. (eds.) Fun with Algorithms. LIPIcs, vol. 49, pp. 8:1–8:20. Schloss Dagstuhl, Dagstuhl (2016). https://doi.org/10.4230/LIPIcs.FUN.2016.8
3. Bultel, X., et al.: Physical zero-knowledge proof for Makaro. In: Izumi, T., Kuznetsov, P. (eds.) SSS 2018. LNCS, vol. 11201, pp. 111–125. Springer, Cham (2018). https://doi.org/10.1007/978-3-030-03232-6_8
4. Dumas, J.-G., Lafourcade, P., Miyahara, D., Mizuki, T., Sasaki, T., Sone, H.: Interactive physical zero-knowledge proof for Norinori. In: Du, D.-Z., Duan, Z., Tian, C. (eds.) COCOON 2019. LNCS, vol. 11653, pp. 166–177. Springer, Cham (2019). https://doi.org/10.1007/978-3-030-26176-4_14
5. Goldreich, O., Kahan, A.: How to construct constant-round zero-knowledge proof systems for NP. J. Cryptology **9**(3), 167–189 (1996). https://doi.org/10.1007/BF00208001
6. Goldreich, O., Micali, S., Wigderson, A.: Proofs that yield nothing but their validity for all languages in NP have zero-knowledge proof systems. J. ACM **38**(3), 691–729 (1991). https://doi.org/10.1145/116825.116852
7. Goldwasser, S., Micali, S., Rackoff, C.: Knowledge complexity of interactive proof-systems. In: Annual ACM Symposium on Theory of Computing, pp. 291–304 (1985). https://doi.org/10.1145/3335741.3335750
8. Gradwohl, R., Naor, M., Pinkas, B., Rothblum, G.N.: Cryptographic and physical zero-knowledge proof systems for solutions of Sudoku puzzles. Theory Comput. Syst. **44**(2), 245–268 (2009). https://doi.org/10.1007/s00224-008-9119-9

9. Hearn, R.A., Demaine, E.D.: Games, Puzzles, and Computation. A. K. Peters Ltd., USA (2009)

10. Holzer, M., Klein, A., Kutrib, M., Ruepp, O.: Fundamenta. Informaticae. Comput. Complex. NURIKABE. **110**(1–4), 159–174 (2011). https://doi.org/10.3233/FI-2011-534

11. Koch, A., Walzer, S.: Foundations for actively secure card-based cryptography. In: Farach-Colton, M., Prencipe, G., Uehara, R. (eds.) Fun with Algorithms. LIPIcs, vol. 157, pp. 17:1–17:23. Schloss Dagstuhl, Dagstuhl (2021). https://doi.org/10.4230/LIPIcs.FUN.2021.17

12. Lafourcade, P., Miyahara, D., Mizuki, T., Sasaki, T., Sone, H.: A physical ZKP for Slitherlink: how to perform physical topology-preserving computation. In: Heng, S.-H., Lopez, J. (eds.) ISPEC 2019. LNCS, vol. 11879, pp. 135–151. Springer, Cham (2019). https://doi.org/10.1007/978-3-030-34339-2_8

13. Miyahara, D., et al.: Card-based ZKP protocols for Takuzu and Juosan. In: Farach-Colton, M., Prencipe, G., Uehara, R. (eds.) Fun with Algorithms. LIPIcs, vol. 157, pp. 20:1–20:21. Schloss Dagstuhl, Dagstuhl (2021). https://doi.org/10.4230/LIPIcs.FUN.2021.20

14. Miyahara, D., Sasaki, T., Mizuki, T., Sone, H.: Card-based physical zero-knowledge proof for Kakuro. IEICE Trans. Fundam. Electron. Commun. Comput. Sci. **102-A**(9), 1072–1078 (2019). https://doi.org/10.1587/transfun.E102.A.1072

15. Nishimura, A., Hayashi, Y., Mizuki, T., Sone, H.: Pile-shifting scramble for card-based protocols. IEICE Trans. Fundam. Electron. Commun. Comput. Sci. **101-A**(9), 1494–1502 (2018). https://doi.org/10.1587/transfun.E101.A.1494

16. Robert, L., Miyahara, D., Lafourcade, P., Mizuki, T.: Physical zero-knowledge proof for Suguru puzzle. In: Devismes, S., Mittal, N. (eds.) SSS 2020. LNCS, vol. 12514, pp. 235–247. Springer, Cham (2020). https://doi.org/10.1007/978-3-030-64348-5_19

17. Robert, L., Miyahara, D., Lafourcade, P., Mizuki, T.: Interactive physical ZKP for connectivity: applications to Nurikabe and Hitori (long version). In: CiE. à distance, Belgium (2021). https://hal.uca.fr/hal-03209911

18. Ruangwises, S., Itoh, T.: Physical zero-knowledge proof for Numberlink puzzle and k vertex-disjoint paths problem. New Gener. Comput. **39**, 3–17 (2021). https://doi.org/10.1007/s00354-020-00114-y

19. Ruangwises, S., Itoh, T.: Physical zero-knowledge proof for ripple effect. In: Uehara, R., Hong, S.-H., Nandy, S.C. (eds.) WALCOM 2021. LNCS, vol. 12635, pp. 296–307. Springer, Cham (2021). https://doi.org/10.1007/978-3-030-68211-8_24

20. Sasaki, T., Miyahara, D., Mizuki, T., Sone, H.: Efficient card-based zero-knowledge proof for Sudoku. Theor. Comput. Sci. **839**, 135–142 (2020). https://doi.org/10.1016/j.tcs.2020.05.036

21. Shinagawa, K., et al.: Card-based protocols using regular polygon cards. IEICE Trans. Fundam. Electron. Commun. Comput. Sci. **100-A**(9), 1900–1909 (2017). https://doi.org/10.1587/transfun.E100.A.1900

Positive Enumerable Functors

Barbara F. Csima[1], Dino Rossegger[1(✉)] [iD], and Daniel Yu[2]

[1] Department of Pure Mathematics, University of Waterloo, Waterloo, Canada
{csima,dino.rossegger}@uwaterloo.ca
[2] University of Waterloo, Waterloo, Canada
zy3yu@uwaterloo.ca

Abstract. We study reductions well suited to compare structures and classes of structures with respect to properties based on enumeration reducibility. We introduce the notion of a positive enumerable functor and study the relationship with established reductions based on functors and alternative definitions.

1 Introduction

In this article we study notions of reductions that let us compare classes of structures with respect to their computability theoretic properties. Computability theoretic reductions between classes of structures can be formalized using effective versions of the category theoretic notion of a functor. While computable functors have already been used in the 80's by Goncharov [1], the formal investigation of this notion was only started recently after R. Miller, Poonen, Schoutens, and Shlapentokh [3] explicitly used a computable functor to obtain a reduction from the class of graphs to the class of fields. Their result shows that fields are universal with respect to many properties studied in computable structure theory.

In [4] the third author studied effective versions of functors based on enumeration reducibility and their relation to notions of interpretability. There, it was shown that the existence of an enumerable functor implies the existence of a computable functor effectively isomorphic to it. In that article there also appeared an unfortunately incorrect claim that enumerable functors are equivalent to a variation of effective interpretability. Indeed, it was later shown in Rossegger's thesis [5], that the existence of a computable functor implies the existence of an enumerable functor and thus enumerable functors are equivalent to the original notion of effective interpretability, which was shown to be equivalent to computable functors in [2]. Hence, enumerable functors are equivalent to this original version. We provide a simple proof that computable functors imply enumerable functors in Sect. 2. The equivalence of these two types of functors is not very surprising, as the enumeration operators witnessing the effectiveness of an enumerable functor are given access to the atomic diagrams of structures, which are total sets.

The main objective of this article is the study of positive enumerable functors, an effectivization of functors that grants the involved enumeration operators

© Springer Nature Switzerland AG 2021
L. De Mol et al. (Eds.): CiE 2021, LNCS 12813, pp. 385–394, 2021.
https://doi.org/10.1007/978-3-030-80049-9_38

access to the positive diagrams of structures instead of their atomic diagrams. While computable functors are well suited to compare structures with respect to properties related to relative computability and the Turing degrees, positive enumerable functors provide the right framework to compare structures with respect to their enumerations and properties related to the enumeration degrees.

The paper is organized as follows. In Sect. 2 we introduce the necessary notions and show that computable functors and enumerable functors are equivalent. Section 3 is dedicated to the study of positive enumerable functors and reductions based on them. We show that reductions by positive enumerable bitransformations preserve enumeration degree spectra, a generalization of degree spectra considering all enumerations of a structure introduced by Soskov [6]. We then exhibit an example consisting of two structures which are computably bitransformable but whose enumeration degree spectra are different. This implies that positive enumerable functors and computable functors are independent notions. Towards the end of the section we compare different possible definitions of positive enumerable functors. At last, in Sect. 4 we extend our results to reductions between arbitrary classes of structures based on effectivizations of functors.

2   Computable and Enumerable Functors

In this article we assume that our structures are in a relational language $(R_i)_{i\in\omega}$ where each R_i has arity a_i and the map $i \mapsto a_i$ is computable. We furthermore only consider countable structures with universe ω. We view classes of structures as categories where the objects are structures in a given language $\mathcal{L}$ and the morphisms are isomorphisms between them. Recall that a functor $F : \mathfrak{C} \to \mathfrak{D}$ maps structures from $\mathfrak{C}$ to structures in $\mathfrak{D}$ and maps isomorphisms $f : \mathcal{A} \to \mathcal{B}$ to $F(f) : F(\mathcal{A}) \to F(\mathcal{B})$ preserving composition and identity.

The smallest classes we consider are isomorphism classes of a single structure $\mathcal{A}$,

$$Iso(\mathcal{A}) = \{\mathcal{B} : \mathcal{B} \cong \mathcal{A}\}.$$

We will often talk about a functor from $\mathcal{A}$ to $\mathcal{B}$, $F : \mathcal{A} \to \mathcal{B}$ when we mean a functor $F : Iso(\mathcal{A}) \to Iso(\mathcal{B})$. Depending on the properties that we want our functor to preserve we may use different effectivizations, but they will all be of the following form. Generally, an effectivization of a functor $F : \mathfrak{C} \to \mathfrak{D}$ will consist of a pair of operators (Φ, Φ_*) and a suitable coding C such that

1. for all $\mathcal{A} \in \mathfrak{C}$, $\Phi(C(\mathcal{A})) = C(F(\mathcal{A}))$,
2. for all $\mathcal{A}, \mathcal{B} \in \mathfrak{C}$ and $f \in Hom(\mathcal{A}, \mathcal{B})$, $\Phi_*(C(\mathcal{A}), C(f), C(\mathcal{B})) = C(F(f))$.

In this article the operators will either be enumeration or Turing operators. If the coding is clear from context we will omit the coding function, i.e., we write $\Phi(\mathcal{A})$ instead of $\Phi(C(\mathcal{A}))$. The most common coding in computable structure theory is the following.

Definition 1. *Let $\mathcal{A}$ be a structure in relational language $(R_i)_{i \in \omega}$. Then the atomic diagram $D(\mathcal{A})$ of $\mathcal{A}$ is the set*

$$\bigoplus_{i \in \omega} R_i^{\mathcal{A}} \oplus \bigoplus_{i \in \omega} \neg R_i^{\mathcal{A}}.$$

In the literature one can often find different definitions of the atomic diagram. It is easy to show that all of these notions are Turing and enumeration equivalent. The reason why we chose this definition is that it is conceptually easier to define the positive diagram and deal with enumerations of structures like this. We are now ready to define various effectivizations of functors.

Definition 2 ([3],[2]). *A functor $F : \mathfrak{C} \to \mathfrak{D}$ is computable if there is a pair of Turing operators (Φ, Φ_*) such that for all $\mathcal{A}, \mathcal{B} \in \mathfrak{C}$*

1. *$\Phi^{D(\mathcal{A})} = D(F(\mathcal{A}))$,*
2. *for all $f \in Hom(\mathcal{A}, \mathcal{B})$, $\Phi_*^{D(\mathcal{A}) \oplus Graph(f) \oplus D(\mathcal{B})} = F(f)$.*

Definition 3. *A functor $F : \mathfrak{C} \to \mathfrak{D}$ is enumerable if there is a pair (Ψ, Ψ_*) where Ψ and Ψ_* are enumeration operators such that for all $\mathcal{A}, \mathcal{B} \in \mathfrak{C}$*

1. *$\Psi^{D(\mathcal{A})} = D(F(\mathcal{A}))$,*
2. *for all $f \in Hom(\mathcal{A}, \mathcal{B})$, $\Psi_*^{D(\mathcal{A}) \oplus Graph(f) \oplus D(\mathcal{B})} = Graph(F(f))$.*

In [4] enumerable functors were defined differently, using a Turing operator instead of an enumeration operator for the homomorphisms. The definition was as follows.

Definition 4. ([4]). *A functor $F : \mathfrak{C} \to \mathfrak{D}$ is $\star$-enumerable if there is a pair (Ψ, Φ_*) where Ψ is an enumeration operator and Φ_* is a Turing operator such that for all $\mathcal{A}, \mathcal{B} \in \mathfrak{C}$*

1. *$\Psi^{D(\mathcal{A})} = D(F(\mathcal{A}))$,*
2. *for all $f \in Hom(\mathcal{A}, \mathcal{B})$, $\Phi_*^{D(\mathcal{A}) \oplus Graph(f) \oplus D(\mathcal{B})} = Graph(F(f))$.*

It turns out that the two definitions are equivalent and we will thus stick with Definition 3 which seems to be more natural.

Proposition 5. *A functor $F : \mathcal{A} \to \mathcal{B}$ is enumerable if and only if it is $\star$-enumerable.*

Proof. Say we have an enumerable functor given by (Ψ, Ψ_*) and an isomorphism $f : \tilde{\mathcal{A}} \to \hat{\mathcal{A}}$ for $\tilde{\mathcal{A}} \cong \hat{\mathcal{A}} \cong \mathcal{A}$. We can compute the isomorphism $F(f)$ by enumerating $Graph(F(f))$ using $\Psi_*^{\tilde{\mathcal{A}} \oplus f \oplus \hat{\mathcal{A}}}$. For every x we are guaranteed to enumerate $(x, y) \in Graph(F(f))$ for some y as the domain of $\mathcal{A}$ is ω. This is uniform in $\tilde{\mathcal{A}}$, f and $\hat{\mathcal{A}}$. Thus there is a Turing operator Φ_* such that (Ψ, Φ_*) witnesses that F is $\star$-enumerable.

Now, say F is $\star$-enumerable as witnessed by (Ψ, Φ_*). For every σ, x, y with $\Phi_*^\sigma(x) \downarrow = y$ such that σ can be split into $\sigma_0 \oplus \sigma_1 \oplus \sigma_2$ where σ_0, σ_2 are partial characteristic functions of finite structures in a finite sublanguage L of the language of $\mathcal{A}$ and σ_1 is the partial graph of a function, consider the set

$$X_\sigma^{x,y} = \{(B \oplus Graph(\tau) \oplus C, \langle x, y \rangle) : B, C \text{ are atomic diagrams of finite}$$
$$L\text{-structures}, B \text{ compatible with } \sigma_0, C \text{ compatible with } \sigma_2,$$
$$\sigma_1(u, v) = 1 \to \tau(u) = v, \text{ and } \sigma_1(u, v) = 0 \to \tau(u) = z$$
$$\text{where } z \notin range(\sigma_1)\}.$$

We can now define our enumeration operator as $\Psi_\star = \bigcup_{x,y,\sigma : \Phi_*^\sigma(x)\downarrow = y} X_\sigma^{x,y}$. Given an enumeration of Φ_* we can produce an enumeration of $\Psi_\star$, so $\Psi_\star$ is c.e. It remains to show that $\Psi_\star^{\hat{\mathcal{A}} \oplus f \oplus \tilde{\mathcal{A}}} = \Phi_*^{\hat{\mathcal{A}} \oplus f \oplus \tilde{\mathcal{A}}}$.

Say $\Phi_*^{\tilde{\mathcal{A}} \oplus f \oplus \hat{\mathcal{A}}}(x) = y$. Then there is $\sigma \preceq \tilde{\mathcal{A}} \oplus f \oplus \hat{\mathcal{A}}$ such that $(\sigma, x, y) \in \Phi_*$ and thus by the construction of X_σ there is $B \subseteq D(\tilde{\mathcal{A}})$, $C \subseteq D\hat{\mathcal{A}})$ and $Graph(\tau) \subseteq Graph(f)$ such that $(B \oplus Graph(\tau) \oplus C, \langle x, y \rangle) \in X_\sigma$. Thus $\langle x, y \rangle \in \Psi_\star^{\tilde{\mathcal{A}} \oplus f \oplus \hat{\mathcal{A}}}$.

On the other hand say $\langle x, y \rangle \in \Psi_\star^{\tilde{\mathcal{A}} \oplus f \oplus \hat{\mathcal{A}}}$. Then, there is $(B \oplus Graph(\tau) \oplus C, \langle x, y \rangle) \in \Psi_\star$ with $B \subseteq \tilde{\mathcal{A}}$, $Graph(\tau) \subseteq Graph(f)$ and $C \subseteq \hat{\mathcal{A}}$. Furthermore, there is $\sigma \preceq \chi_{B \oplus Graph(\tau) \oplus C}$ such that $(\sigma, x, y) \in \Phi_*$. Thus $\Psi_\star^{\tilde{\mathcal{A}} \oplus f \oplus \hat{\mathcal{A}}} = Graph(F(f))$ for any $\hat{\mathcal{A}} \cong \tilde{\mathcal{A}} \cong \mathcal{A}$ and $f : \tilde{\mathcal{A}} \cong \hat{\mathcal{A}}$ and hence F is enumerable. $\square$

In [4] it was shown that the existence of an enumerable functor implies the existence of a computable functor and in [5] the converse was shown. We give a simple proof of the latter.

Theorem 6. *If $F : \mathcal{A} \to \mathcal{B}$ is a computable functor, then it is enumerable.*

Proof. Given a computable functor F we will show that F is $\star$-enumerable. That F is then also enumerable follows from Proposition 5.

Let $D(L_\mathcal{A})$ be the collection of finite atomic diagrams in the language of $\mathcal{A}$. To every $p \in D(L_\mathcal{A})$ we associate a finite string α_p in the alphabet $\{0, 1, \uparrow\}$ so that if p specifies that R_i holds on elements coded by u, then we set that $\neg R_i$ does not hold on these elements. More formally, $\alpha_p(x) = 1$ if $x \in p$, $\alpha_p(x) = 0$ if $x = 2\langle i, u \rangle$ and $2\langle i, u \rangle + 1 \in p$ or $x = 2\langle i, u \rangle + 1$ and $2\langle i, u \rangle \in p$, and $\alpha_p(x) = \uparrow$ if x is less than the largest element of p and none of the other cases fits. We also associate a string $\tilde{\alpha}_p \in 2^{|\alpha_p|}$ with p where $\tilde{\alpha}_p(x) = 1$ if and only if $\alpha_p(x) = 1$ and $\tilde{\alpha}_p(x) = 0$ if and only if $\alpha_p(x) = 0$ or $\alpha_p(x) \uparrow$.

Let the computability of F be witnessed by (Φ, Φ_*). We build the enumeration operator Ψ as follows. For every $p \in D(L_\mathcal{A})$ and every x if $\Phi^{\tilde{\alpha}_p}(x) \downarrow = 1$ and every call to the oracle during the computation is on an element z such that $\alpha_p(z) \neq \uparrow$, then enumerate (p, x) into Ψ. This finishes the construction of Ψ.

Now, let $\hat{\mathcal{A}} \cong \mathcal{A}$. We have that $x \in \Psi^{\hat{\mathcal{A}}}$ if and only if there exists $p \in D(L_\mathcal{A})$ such that $p \subseteq D(\hat{\mathcal{A}})$ and $(p, x) \in \Psi$. We further have that $(p, x) \in \Psi$ if and only if $\Phi^{\tilde{\alpha}_p}(x) \downarrow = 1$ if and only if $\Phi^{\hat{\mathcal{A}}}(x) = 1$. Thus F is enumerable using (Ψ, Φ_*). $\square$

Combining Theorem 6 with the results from [4] we obtain that enumerable functors and computable functors defined using the atomic diagram of a structure as input are equivalent notions. This is not surprising. After all, the atomic diagram of a structure always has total enumeration degree and there is a canonical isomorphism between the total enumeration degrees and the Turing degrees. In order to make this equivalence precise we need another definition.

Definition 7 ([2]). *A functor $F : \mathfrak{C} \to \mathfrak{D}$ is effectively isomorphic to a functor $G : \mathfrak{C} \to \mathfrak{D}$ if there is a Turing functional Λ such that for any $\mathcal{A} \in \mathfrak{C}$, $\Lambda^{\mathcal{A}} : F(\mathcal{A}) \to G(\mathcal{A})$ is an isomorphism. Moreover, for any morphism $h \in Hom(\mathcal{A}, \mathcal{B})$ in $\mathfrak{C}$, $\Lambda^{\mathcal{B}} \circ F(h) = G(h) \circ \Lambda^{\mathcal{A}}$. That is, the diagram below commutes.*

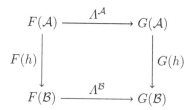

The following is an immediate corollary of Theorem 6 and [4, Theorem 2].

Theorem 8. *Let $F : \mathcal{A} \to \mathcal{B}$ be a functor. Then F is computable if and only if there is an enumerable functor $G : \mathcal{A} \to \mathcal{B}$ effectively isomorphic to F.*

Definition 9 ([2]). *Suppose $F : \mathfrak{C} \to \mathfrak{D}$, $G : \mathfrak{D} \to \mathfrak{C}$ are functors such that $G \circ F$ is effectively isomorphic to $Id_{\mathfrak{C}}$ via the Turing functional $\Lambda_{\mathfrak{C}}$ and $F \circ G$ is effectively isomorphic to $Id_{\mathfrak{D}}$ via the Turing functional $\Lambda_{\mathfrak{D}}$. If furthermore, for any $\mathcal{A} \in \mathfrak{C}$ and $\mathcal{B} \in \mathfrak{D}$, $\Lambda_{\mathfrak{D}}^{F(\mathcal{A})} = F(\Lambda_{\mathfrak{C}}^{\mathcal{A}}) : F(\mathcal{A}) \to F(G(F(\mathcal{A})))$ and $\Lambda_{\mathfrak{C}}^{G(\mathcal{B})} = G(\Lambda_{\mathfrak{D}}^{\mathcal{B}}) : G(\mathcal{B}) \to G(F(G(\mathcal{B})))$, then F and G are said to be* pseudo inverses.

Definition 10 ([2]). *Two structures $\mathcal{A}$ and $\mathcal{B}$ are* computably bi-transformable *if there are computable pseudo-inverse functors $F : \mathcal{A} \to \mathcal{B}$ and $G : \mathcal{B} \to \mathcal{A}$.*

If the functors in Definition 10 are enumerable instead of computable then we say that $\mathcal{A}$ and $\mathcal{B}$ are *enumerably bi-transformable*. As an immediate corollary of Theorem 8 we obtain the following.

Corollary 11. *Two structures $\mathcal{A}$ and $\mathcal{B}$ are enumerably bi-transformable if and only if they are computably bi-transformable.*

3   Effectivizations Using Positive Diagrams

We now turn our attention to the setting where we only have positive information about the structures. We follow Soskov [6] in our definitions. See also the survey paper by Soskova and Soskova [7] on computable structure theory and enumeration degrees.

Definition 12. *Let $\mathcal{A}$ be a structure in relational language $(R_i)_{i \in \omega}$. The positive diagram of $\mathcal{A}$, denoted by $P(\mathcal{A})$, is the set*

$$= \oplus \neq \oplus \bigoplus_{i \in \omega} R_i^{\mathcal{A}}.$$

We are interested in the degrees of enumerations of $P(\mathcal{A})$. To be more precise let f be an enumeration of ω and for $X \subseteq \omega^n$ let

$$f^{-1}(X) = \{\langle x_1, \ldots, x_n \rangle : (f(x_1), \ldots, f(x_n)) \in X\}.$$

Given $\mathcal{A}$ let $f^{-1}(\mathcal{A}) = f^{-1}(=) \oplus f^{-1}(\neq) \oplus f^{-1}(R_0^{\mathcal{A}}) \oplus \ldots$. Notice that if $f = id$, then $f^{-1}(\mathcal{A})$ is just the positive diagram of $\mathcal{A}$.

Definition 13. *The* enumeration degree spectrum *of $\mathcal{A}$ is the set*

$$eSp(\mathcal{A}) = \{d_e(f^{-1}(\mathcal{A})) : f \text{ is an enumeration of } \omega\}.$$

If $\mathbf{a}$ is the least element of $eSp(\mathcal{A})$, then $\mathbf{a}$ is called the enumeration degree *of $\mathcal{A}$.*

In order to obtain a notion of reduction that preserves enumeration spectra we need an effectivization of functors where we use positive diagrams of structures as coding. It is clear that for computable functors this makes no difference as $P(\mathcal{A}) \equiv_T D(\mathcal{A})$. For enumerable functors it does make a difference. We also need to replace the Turing operators in the definition of pseudo inverses with enumeration operators. The new notions are as follows.

Definition 14. *A functor $F : \mathfrak{C} \to \mathfrak{D}$ is* positive enumerable *if there is a pair (Ψ, Ψ_*) where Ψ and Ψ_* are enumeration operators such that for all $\mathcal{A}, \mathcal{B} \in \mathfrak{C}$*

1. $\Psi^{P(\mathcal{A})} = P(F(\mathcal{A}))$,
2. *for all $f \in Hom(\mathcal{A}, \mathcal{B})$, $\Psi_*^{P(\mathcal{A}) \oplus Graph(f) \oplus P(\mathcal{B})} = Graph(F(f))$.*

Definition 15. *A functor $F : \mathfrak{C} \to \mathfrak{D}$ is* enumeration isomorphic *to a functor $G : \mathfrak{C} \to \mathfrak{D}$ if there is an enumeration operator Λ such that for any $\mathcal{A} \in \mathfrak{C}$, $\Lambda^{P(\mathcal{A})} : F(\mathcal{A}) \to G(\mathcal{A})$ is an isomorphism. Moreover, for any morphism $h \in Hom(\mathcal{A}, \mathcal{B})$ in $\mathfrak{C}$, $\Lambda^{P(\mathcal{B})} \circ F(h) = G(h) \circ \Lambda^{P(\mathcal{A})}$.*

Definition 16. *Suppose $F : \mathfrak{C} \to \mathfrak{D}$, $G : \mathfrak{D} \to \mathfrak{C}$ are functors such that $G \circ F$ is enumeration isomorphic to $Id_{\mathfrak{C}}$ via the enumeration operator $\Lambda_{\mathfrak{C}}$ and $F \circ G$ is enumeration isomorphic to $Id_{\mathfrak{D}}$ via the enumeration operator $\Lambda_{\mathfrak{D}}$. If, further- more, for any $\mathcal{A} \in \mathfrak{C}$ and $\mathcal{B} \in \mathfrak{D}$, $\Lambda_{\mathfrak{D}}^{P(F(\mathcal{A}))} = F(\Lambda_{\mathfrak{C}}^{P(\mathcal{A})}) : F(\mathcal{A}) \to F(G(F(\mathcal{A})))$ and $\Lambda_{\mathfrak{C}}^{P(G(\mathcal{B}))} = G(\Lambda_{\mathfrak{D}}^{P(\mathcal{B})}) : G(\mathcal{B}) \to G(F(G(\mathcal{B})))$, then F and G are said to be* enumeration pseudo inverses.

Definition 17. *Two structures $\mathcal{A}$ and $\mathcal{B}$ are* positive enumerably bi-transform- able *if there are positive enumerable enumeration pseudo-inverse functors $F : \mathcal{A} \to \mathcal{B}$ and $G : \mathcal{B} \to \mathcal{A}$.*

Theorem 18. *Let $\mathcal{A}$ and $\mathcal{B}$ be positive enumerably bi-transformable. Then $eSp(\mathcal{A}) = eSp(\mathcal{B})$.*

Proof. Say $\mathcal{A}$ and $\mathcal{B}$ are positive enumerably bi-transformable by $F : \mathcal{A} \to \mathcal{B}$ and $G : \mathcal{B} \to \mathcal{A}$. Let f be an arbitrary enumeration of ω, then, viewing $f^{-1}(\mathcal{A})/f^{-1}(=)$ as a structure on ω by pulling back a canonical enumeration of the least elements in its $=$-equivalence classes, we have that there is $\hat{\mathcal{A}} \cong \mathcal{A}$ such that $P(\hat{\mathcal{A}}) = f^{-1}(\mathcal{A})/f^{-1}(=)$ and $P(\hat{\mathcal{A}}) \leq_e f^{-1}(\mathcal{A})$. As F is positive enumerable we have that $f^{-1}(\mathcal{A}) \geq_e P(F(\hat{\mathcal{A}}))$. Furthermore, we shall see that $f^{-1}(F(\hat{\mathcal{A}})) \leq_e f^{-1}(\mathcal{A})$ and that $f^{-1}(\mathcal{A})/f^{-1}(=) = P(F(\hat{\mathcal{A}}))$. Given an enumeration of $f^{-1}(\mathcal{A})$ and an enumeration of $P(F(\hat{\mathcal{A}}))$, we may first order the equivalence classes of $f^{-1}(=)$ by their least elements and then, if $R_i(a_1, \ldots, a_n) \in P(F(\hat{\mathcal{A}}))$ we enumerate $R_i(b_1, \ldots, b_n)$ for all $b_1, \ldots, b_n \in \omega$ such that b_j is in the $a_j{}^{\text{th}}$ equivalence class of $f^{-1}(=)$. It is not hard to see that this gives an enumeration of a set X such that $f^{-1}(=) \oplus f^{-1}(\neq) \oplus X = f^{-1}(F(\hat{\mathcal{A}}))$, that $f^{-1}(F(\hat{\mathcal{A}}))/f^{-1}(=) = P(F(\hat{\mathcal{A}}))$, and since by construction $f^{-1}(F(\hat{\mathcal{A}})) \leq_e P(F(\hat{\mathcal{A}})) \oplus f^{-1}(\mathcal{A})$ we have $f^{-1}(F(\hat{\mathcal{A}})) \leq_e f^{-1}(\mathcal{A})$.

We can apply the same argument with G in place of F and $F(\hat{\mathcal{A}})$ in place of $\mathcal{A}$ to get that $f^{-1}(G(F(\hat{\mathcal{A}})))/f^{-1}(=) = P(G(F(\hat{\mathcal{A}})))$ and

$$f^{-1}(G(F(\hat{\mathcal{A}}))) \leq_e f^{-1}(F(\hat{\mathcal{A}})) \leq_e f^{-1}(\mathcal{A}).$$

At last, recall that, as $\mathcal{A}$ and $\mathcal{B}$ are positive enumerably bi-transformable, there is an enumeration operator Ψ such that $\Psi^{P(G(F(\hat{\mathcal{A}})))}$ is the enumeration of the graph of an isomorphism $i : G(F(\hat{\mathcal{A}})) \cong \hat{\mathcal{A}}$. But then $(f \circ i)^{-1}(G(F(\hat{\mathcal{A}}))) = f^{-1}(\mathcal{A})$ and

$$f^{-1}(\mathcal{A}) \leq_e f^{-1}(G(F(\hat{\mathcal{A}}))) \leq_e f^{-1}(F(\hat{\mathcal{A}})) \leq_e f^{-1}(\mathcal{A}).$$

This shows that $eSp(\mathcal{A}) \subseteq eSp(\mathcal{B})$. The proof that $eSp(\mathcal{B}) \subseteq eSp(\mathcal{A})$ is analogous. $\qquad\square$

Proposition 19. *There are computably bi-transformable structures $\mathcal{A}$ and $\mathcal{B}$ such that $eSp(\mathcal{A}) \neq eSp(\mathcal{B})$. In particular, $\mathcal{A}$ and $\mathcal{B}$ are not positive enumerably bi-transformable.*

Proof. Let $\mathcal{A} = (\omega, \underline{0}, s, K)$ where s is the successor relation on ω, $\underline{0}$ the first element, and K the membership relation of the halting set. Assume $\mathcal{B} = (\omega, \underline{0}, s, \overline{K})$ is defined as $\mathcal{A}$ except that $\overline{K}(x)$ if and only if $\neg K(x)$. There is a computable functor $F : \mathcal{A} \to \mathcal{B}$ taking $\hat{\mathcal{A}} = (\omega, \underline{0}^{\hat{\mathcal{A}}}, s^{\hat{\mathcal{A}}}, K^{\hat{\mathcal{A}}}) \cong \mathcal{A}$ to $F(\hat{\mathcal{A}}) = (\omega, \underline{0}^{\hat{\mathcal{A}}}, s^{\hat{\mathcal{A}}}, \neg K^{\hat{\mathcal{A}}})$ and acting as the identity on isomorphisms. Furthermore, F has a computable inverse and thus $\mathcal{A}$ is computably bi-transformable to $\mathcal{B}$.

However, $\mathcal{A}$ has enumeration degree $\mathbf{0}_e$ and $\mathcal{B}$ has enumeration degree $\mathbf{0}'_e$. Thus there cannot be a positive enumerable functor from $\mathcal{B}$ to $\mathcal{A}$. $\qquad\square$

The following shows that computable functors and positive enumerable functors are independent.

Proposition 20. *There are structures $\mathcal{A}$ and $\mathcal{B}$ such that $\mathcal{A}$ is positive enumerably bi-transformable with $\mathcal{B}$ but $\mathcal{A}$ is not computably bi-transformable with $\mathcal{B}$.*

Proof. Let $\mathcal{A}$ be as in Proposition 19, i.e., $\mathcal{A} = (\omega, \underline{0}, s, K)$ and $\mathcal{B} = (\omega, \underline{0}, s)$. Then it is not hard to see that $\mathcal{A}$ is positive enumerably bi-transformable with $\mathcal{B}$. However, there can not be a computable functor from $\mathcal{B}$ to $\mathcal{A}$ as $\mathcal{B}$ has Turing degree $\mathbf{0}$ and $\mathcal{A}$ has Turing degree $\mathbf{0}'$.   □

We have seen in Proposition 5 that $\star$-enumerable functors and enumerable functors are equivalent. Positive enumerable functors also admit a different definition.

Definition 21. *A functor $F : \mathfrak{C} \to \mathfrak{D}$ is positive $\star$-enumerable if there is a pair (Ψ, Φ_*) where Ψ is an enumeration operator and Φ_* is a Turing operator such that for all $\mathcal{A}, \mathcal{B} \in \mathfrak{C}$*

1. $\Psi^{P(\mathcal{A})} = P(F(\mathcal{A}))$,
2. *for all $f \in Hom(\mathcal{A}, \mathcal{B})$, $\Phi_*^{P(\mathcal{A}) \oplus Graph(f) \oplus P(\mathcal{B})} = Graph(F(f))$.*

Proposition 22. *Every positive enumerable functor is positive $\star$-enumerable.*

Proof. Let $F : \mathcal{A} \to \mathcal{B}$ be given by (Ψ, Ψ_*) and let $f : \tilde{\mathcal{A}} \cong \hat{\mathcal{A}}$ for $\tilde{\mathcal{A}} \cong \hat{\mathcal{A}} \cong \mathcal{A}$. Now we can define a procedure computing $F(f)$ as follows. Given x, and $\tilde{\mathcal{A}} \oplus f \oplus \hat{\mathcal{A}}$ enumerate $\Psi_*^{\tilde{\mathcal{A}} \oplus f \oplus \hat{\mathcal{A}}}$ until $\langle x, y \rangle \searrow \Psi_*^{\tilde{\mathcal{A}} \oplus f \oplus \hat{\mathcal{A}}}$ for some y. This is uniform in $\tilde{\mathcal{A}} \oplus f \oplus \hat{\mathcal{A}}$ and thus there exists a Turing operator Φ_* with this behaviour. The pair (Ψ, Φ_*) then witnesses that F is $\star$-enumerable.   □

Theorem 23. *There is positive $\star$-enumerable functor that is not enumeration isomorphic to any positive enumerable functor.*

Proof. We will build two structures $\mathcal{A}$ and $\mathcal{B}$ such that there is a positive $\star$-enumerable functor $F : \mathcal{A} \to \mathcal{B}$ that is not positive enumerable. The structure $\mathcal{A}$ is a graph constructed as follows. It has a vertex a with a loop connected to a and a cycle of size n for every natural number n. If $n \in K$ then there is an edge between a and one element of the n cycle, otherwise there is no such edge. Clearly, $deg_T(P(\mathcal{A})) = \mathbf{0}'$ and $K \geq_e P(\mathcal{A})$. Thus, $deg_e(P(\mathcal{A})) = \mathbf{0}_e$ and, in particular, $P(\mathcal{A}) \not\geq_e \overline{K}$.

The structure $\mathcal{B}$ is a typical graph that witnesses that there is a structure with degree of categoricity $\mathbf{0}'$ (that is, $\mathbf{0}'$ is the least degree computing an isomorphism between any two computable copies of $\mathcal{B}$). Let us recap how we build two copies of $\mathcal{B}$, $\mathcal{B}_1$ and $\mathcal{B}_2$ such that $\mathbf{0}'$ is the least degree computing isomorphism between $\mathcal{B}_1$ and $\mathcal{B}_2$. Both graphs consist of an infinite ray with a loop at its first element. Let v_i be the ith element in the ray in $\mathcal{B}_1$ and $\hat{v}_i$ be the ith element in the ray in $\mathcal{B}_2$. Now for every v_i there are two elements a_i and b_i with $v_i E a_i$ and $v_i E b_i$. Likewise for every $\hat{v}_i$ there are two elements $\hat{a}_i$ and $\hat{b}_i$ with $\hat{v}_i E \hat{a}_i$ and $\hat{v}_i E \hat{b}_i$. Furthermore there are additional vertices $s_i, \hat{s}_i$ with $a_i E s_i$ and $\hat{a}_i E \hat{s}_i$.

Take an enumeration of K. If $i \searrow K$, then add vertices $b_i E \cdot E \cdot$ and $\hat{s}_i E \cdot$, $\hat{b}_i E \cdot$. This finishes the construction of $\mathcal{B}$. It is not hard to see that there is a unique isomorphism $f : \mathcal{B}_1 \to \mathcal{B}_2$ and that $deg(f) = \mathbf{0}'$ and $Graph(f) \geq_e \overline{K}$.

We now construct the functor F. Given an enumeration of $P(\hat{\mathcal{A}})$ for $\hat{\mathcal{A}} \cong \mathcal{A}$ we wait until we see the cycle containing 0 (any natural number would work). If it is of even length, or 0 is the special vertex a, we let $F(\hat{\mathcal{A}}) = \mathcal{B}_1$ and if it is of odd length we let $F(\hat{\mathcal{A}}) = \mathcal{B}_2$. Clearly given any enumeration of a copy of $\mathcal{A}$ this procedure produces an enumeration of a copy of $\mathcal{B}$.

As $\mathcal{B}$ is rigid we just let $F(f : \hat{\mathcal{A}} \to \tilde{\mathcal{A}}) = g : F(\hat{\mathcal{A}}) \to F(\tilde{\mathcal{A}})$ where g is the unique isomorphism between $F(\hat{\mathcal{A}})$ and $F(\tilde{\mathcal{A}})$. Note that there is a Turing operator Θ such that $\Theta^{P(\hat{\mathcal{A}})} = K$ for any $\hat{\mathcal{A}} \cong \mathcal{A}$ and that the isomorphism between $F(\hat{\mathcal{A}})$ and $F(\tilde{\mathcal{A}})$ can be computed uniformly from $P(F(\hat{\mathcal{A}})) \oplus P(F(\tilde{\mathcal{A}})) \oplus K$. Thus, there is an operator Φ_* witnessing that F is positive $\star$-enumerable.

To see that F is not positive enumerable consider two copies $\hat{\mathcal{A}}$ and $\tilde{\mathcal{A}}$ of $\mathcal{A}$ with $deg_e(P(\hat{\mathcal{A}})) = deg_e(P(\tilde{\mathcal{A}})) = \mathbf{0}_e$ such that 0 is part of an even cycle in $\hat{\mathcal{A}}$ and part of an odd cycle in $\tilde{\mathcal{A}}$. Notice that there is $f : \hat{\mathcal{A}} \to \tilde{\mathcal{A}}$ such that $P(\hat{\mathcal{A}}) \oplus P(\tilde{\mathcal{A}}) \geq_e P(\hat{\mathcal{A}}) \oplus Graph(f : \hat{\mathcal{A}} \to \tilde{\mathcal{A}}) \oplus P(\tilde{\mathcal{A}})$, and also that $P(\hat{\mathcal{A}}) \oplus P(\tilde{\mathcal{A}}) \not\geq_e \overline{K}$. But $Graph(g : F(\hat{\mathcal{A}}) \to F(\tilde{\mathcal{A}})) \geq_e \overline{K}$ as $F(\hat{\mathcal{A}}) = \mathcal{B}_1$ and $F(\tilde{\mathcal{A}}) = \mathcal{B}_2$. Thus there can not be an enumeration operator witnessing that F is positive enumerable.

Assume F was enumeration isomorphic to a positive enumerable functor G and that this isomorphism is witnessed by Λ. Then, taking $\hat{\mathcal{A}}, \tilde{\mathcal{A}}$ and $f : \hat{\mathcal{A}} \to \tilde{\mathcal{A}}$ as in the above paragraph we have that $P(\hat{\mathcal{A}}) \oplus P(\tilde{\mathcal{A}}) \geq_e Graph(G(f))$. But then

$$P(\hat{\mathcal{A}}) \oplus P(\tilde{\mathcal{A}}) \geq_e Graph(\Lambda^{P(\hat{\mathcal{A}})} \circ G(f) \circ \Lambda^{P(\tilde{\mathcal{A}})^{-1}}) = Graph(F(f)) \geq_e \overline{K}.$$

This is a contradiction since $deg_e(P(\hat{\mathcal{A}}) \oplus P(\tilde{\mathcal{A}})) = \mathbf{0}_e$. □

4 Reductions Between Arbitrary Classes

So far we have seen how we can compare structures with respect to computability theoretic properties. Our notions can be naturally extended to allow the comparison of arbitrary classes of structures.

Definition 24 ([2]). *Let $\mathfrak{C}$ and $\mathfrak{D}$ be classes of structures. The class $\mathfrak{C}$ is uniformly computably transformably reducible, short u.c.t. reducible, to $\mathfrak{D}$ if there are a subclass $\mathfrak{D}' \subseteq \mathfrak{D}$ and computable functors $F : \mathfrak{C} \to \mathfrak{D}' \subseteq \mathfrak{D}$ and $G : \mathfrak{D}' \to \mathfrak{C}$ such that F and G are pseudo-inverses.*

Definition 25. *Let $\mathfrak{C}$ and $\mathfrak{D}$ be classes of structures. The class $\mathfrak{C}$ is uniformly (positive) enumerably transformably reducible, short u.e.t., (u.p.e.t.) reducible, to $\mathfrak{D}$ if there is a subclass $\mathfrak{D}' \subseteq \mathfrak{D}$ and (positive) enumerable functors $F : \mathfrak{C} \to \mathfrak{D}' \subseteq \mathfrak{D}$ and $G : \mathfrak{D}' \to \mathfrak{C}$ such that F and G are pseudo-inverses.*

Propositions 19 and 20 show that u.p.e.t. and u.c.t reductions are independent notions.

Corollary 26. *There are classes of structures* $\mathfrak{C}_1, \mathfrak{C}_2$ *and* $\mathfrak{D}_1, \mathfrak{D}_2$ *such that*

1. $\mathfrak{C}_1$ *is u.c.t. reducible to* $\mathfrak{D}_1$ *but* $\mathfrak{C}_1$ *is not u.p.e.t. reducible to* $\mathfrak{D}_1$.
2. $\mathfrak{C}_2$ *is u.p.e.t. reducible to* $\mathfrak{D}_2$ *but* $\mathfrak{C}_2$ *is not u.c.t. reducible to* $\mathfrak{D}_2$.

Similar to Corollary 11 we obtain the equivalence of u.e.t. and u.c.t reductions.

Corollary 27. *Let* $\mathfrak{C}$ *and* $\mathfrak{D}$ *be arbitrary classes of countable structures. Then* $\mathfrak{C}$ *is u.e.t. reducible to* $\mathfrak{D}$ *if and only if it is u.c.t. reducible to* $\mathfrak{D}$.

References

1. Goncharov, S.S.: Problem of the number of non-self-equivalent constructivizations. Algebra Logic **19**(6), 401–414 (1980)
2. Harrison-Trainor, M., Melnikov, A., Miller, R., Montalbán, A.: Computable functors and effective interpretability. J. Symbolic Logic **82**(1), 77–97 (2017)
3. Miller, R., Poonen, B., Schoutens, H., Shlapentokh, A.: A computable functor from graphs to fields. J. Symbolic Logic **83**(1), 326–348 (2018)
4. Rossegger, D.: On functors enumerating structures. Siberian Electron. Math. Rep. **14**, 690–702 (2017)
5. Rossegger, D.: Computable structure theory with respect to equivalence relations (2019)
6. Soskov, I.N.: Degree spectra and co-spectra of structures. Ann. Univ. Sofia **96**, 45–68 (2004)
7. Soskova, A.A., Soskova, M.I.: Enumeration reducibility and computable structure theory. In: Day, A., Fellows, M., Greenberg, N., Khoussainov, B., Melnikov, A., Rosamond, F. (eds.) Computability and Complexity. LNCS, vol. 10010, pp. 271–301. Springer, Cham (2017). https://doi.org/10.1007/978-3-319-50062-1_19

Splittings and Robustness
for the Heine-Borel Theorem

Sam Sanders[✉]

Institute for Philosophy II, RUB, Bochum, Germany
sasander@me.com

Abstract. The Heine-Borel theorem for uncountable coverings has recently emerged as an interesting and central principle in higher-order Reverse Mathematics and computability theory, formulated as follows: HBU is the Heine-Borel theorem for uncountable coverings given as $\cup_{x \in [0,1]} (x - \Psi(x), x + \Psi(x))$ for arbitrary $\Psi : [0,1] \to \mathbb{R}^+$, i.e. the original formulation going back to Cousin (1895) and Lindelöf (1903). In this paper, we show that HBU is equivalent to its restriction to functions *continuous almost everywhere*, an elegant robustness result. We also obtain a nice splitting HBU $\leftrightarrow$ [WHBU$^+$ + HBC$_0$ + WKL] where WHBU$^+$ is a strengthening of Vitali's covering theorem and where HBC$_0$ is the Heine-Borel theorem for countable collections (and **not sequences**) of basic open intervals, as formulated by Borel himself in 1898.

Keywords: Higher-order Reverse Mathematics · Heine-Borel theorem · Vitali covering theorem · Splitting · Robustness

1 Introduction and Preliminaries

We sketch our aim and motivation within the *Reverse Mathematics* program (Sect. 1.1) and introduce some essential axioms and definitions (Sect. 1.2).

1.1 Aim and Motivation

Reverse Mathematics (RM hereafter) is a program in the foundations of mathematics initiated by Friedman [5,6] and developed extensively by Simpson and others [25,26]; an introduction to RM for the 'mathematician in the street' may be found in [27]. We assume basic familiarity with RM, including Kohlenbach's *higher-order* RM introduced in [10]. Recent developments in higher-order RM, including our own, are published in [13–19].

Now, a *splitting* $A \leftrightarrow [B+C]$ is a relatively rare phenomenon in second-order RM where a natural theorem A can be *split* into two *independent* natural parts B and C. Splittings are quite common in higher-order RM, as studied in some detail in [23]. An unanswered question here is whether the higher-order generalisations of the Big Five of RM (and related principles) have natural splittings.

Supported by the *Deutsche Forschungsgemeinschaft* via the DFG grant SA3418/1-1.

L. De Mol et al. (Eds.): CiE 2021, LNCS 12813, pp. 395–406, 2021.
https://doi.org/10.1007/978-3-030-80049-9_39

In this paper, we study the Vitali and Heine-Borel covering theorems for uncountable coverings with an eye on splittings. In particular, our starting point is HBU, defined in Sect. 1.2, which is the Heine-Borel theorem for uncountable coverings $\cup_{x \in [0,1]} I_x^\Psi$ for *arbitrary* third-order $\Psi : [0,1] \to \mathbb{R}^+$ and $I_x^\Psi \equiv (x - \Psi(x), x + \Psi(x))$. This kind of coverings was already studied by Cousin in 1895 [3] and Lindelöf in 1903 [12]. In Sect. 2.2, we obtain an elegant splitting involving HBU, namely as follows:

$$HBU \leftrightarrow [WHBU^+ + HBC_0 + WKL], \tag{1.1}$$

where $WHBU^+$ is a strengthening of the Vitali covering theorem and where HBC_0 is the Heine-Borel theorem for *countable collections* (and **not** sequences) of open intervals, as formulated by Borel himself in [1]. In Sect. 2.1, we prove $HBU \leftrightarrow HBU_{ae}$, where the latter is HBU restricted to functions $\Psi : [0,1] \to \mathbb{R}^+$ continuous *almost everywhere* on the unit interval. By contrast, the same restriction for the Vitali covering theorem results in a theorem equivalent to *weak weak König's lemma* WWKL. The results in Sect. 2.1 were obtained following the study of splittings involving 'continuity almost everywhere'. The proof of Theorem 2.2 (in a stronger system) was suggested to us by Dag Normann. In general, this paper constitutes a spin-off from our joint project with Dag Normann on the Reverse Mathematics and computability theory of the uncountable (see [13,17,19]).

Finally, the foundational and historical significance of our results is as follows.

Remark 1.1. First of all, as shown in [13,15,16], the third-order statements HBU and WHBU cannot be proved Z_2^ω, a conservative extension of Z_2 based on third-order comprehension functionals. A sceptic of third-order objects could 'downplay' this independence result by pointing to the outermost quantifier of HBU and WHBU and declare that the strength of these principles is simply due to the quantification over *all* third-order functions. This point is moot in light of $HBU \leftrightarrow HBU_{ae}$ proved in Theorem 2.2, and the central role of 'continuity almost everywhere' in e.g. the study of the Riemann integral and measure theory.

Secondly, our first attempt at obtaining a splitting for HBU was to decompose the latter as $HBU_{ae} + WHBU$, where WHBU allows one to reduce an arbitrary covering to a covering generated by a function that is continuous almost everywhere. Alas, this kind of splitting does not yield *independent* conjuncts, which is why we resort to stronger notions like *countability*, namely in Sect. 2.2.

Thirdly, the splitting in (1.1) has some historical interest as well: Borel himself formulates the Heine-Borel theorem in [1] using *countable collections* of intervals rather than *sequences* of intervals (as in second-order RM). In fact, Borel's proof of the Heine-Borel theorem in [1, p. 42] starts with: *Let us enumerate our intervals, one after the other, according to whatever law, but determined.* He then proceeds with the usual 'interval-halving' proof, similar to Cousin in [3].

1.2   Preliminaries

We introduce some axioms and definitions from (higher-order) RM needed below. We refer to [10, §2,] or [13, §2,] for the definition of Kohlebach's base theory

RCA_0^ω, and basic definitions like the real numbers $\mathbb{R}$ in RCA_0^ω. For completeness, some definitions are included in the technical appendix, namely Sect. A.

Some Axioms of Higher-Order Arithmetic. First of all, the functional φ in $(\exists^2)$ is clearly discontinuous at $f = 11\dots$; in fact, $(\exists^2)$ is equivalent to the existence of $F : \mathbb{R} \to \mathbb{R}$ such that $F(x) = 1$ if $x >_\mathbb{R} 0$, and 0 otherwise [10, §3,].

$$(\exists \varphi^2 \leq_2 1)(\forall f^1)\big[(\exists n)(f(n) = 0) \leftrightarrow \varphi(f) = 0\big]. \tag{$\exists^2$}$$

Related to $(\exists^2)$, the functional μ^2 in (μ^2) is also called *Feferman's μ* ([10]).

$$(\exists \mu^2)(\forall f^1)\big[(\exists n)(f(n) = 0) \to [f(\mu(f)) = 0 \wedge (\forall i < \mu(f))(f(i) \neq 0)] \tag{μ^2}$$
$$\wedge\, [(\forall n)(f(n) \neq 0) \to \mu(f) = 0]\big].$$

Intuitively, μ^2 is the least-number-operator, i.e. $\mu(f)$ provides the least $n \in \mathbb{N}$ such that $f(n) = 0$, if such there is. We have $(\exists^2) \leftrightarrow (\mu^2)$ over RCA_0^ω and $\mathsf{ACA}_0^\omega \equiv \mathsf{RCA}_0^\omega + (\exists^2)$ proves the same second-order sentences as ACA_0 by [9, Theorem 2.5,].

Secondly, the Heine-Borel theorem states the existence of a finite sub-covering for an open covering of certain spaces. Now, a functional $\Psi : \mathbb{R} \to \mathbb{R}^+$ gives rise to the *canonical cover* $\cup_{x \in I} I_x^\Psi$ for $I \equiv [0,1]$, where I_x^Ψ is the open interval $(x - \Psi(x), x + \Psi(x))$. Hence, the uncountable covering $\cup_{x \in I} I_x^\Psi$ has a finite sub-covering by the Heine-Borel theorem; in symbols:

Principle 1.2 (HBU). $(\forall \Psi : \mathbb{R} \to \mathbb{R}^+)(\exists y_0, \dots, y_k \in I)(\forall x \in I)(x \in \cup_{i \leq k} I_{y_i}^\Psi)$.

Cousin and Lindelöf formulate their covering theorems using canonical covers in [3,12]. This restriction does not make much of a difference, as studied in [24].

Thirdly, let WHBU be the following weakening of HBU:

Principle 1.3 (WHBU). For any $\Psi : \mathbb{R} \to \mathbb{R}^+$ and $\varepsilon >_\mathbb{R} 0$, there are pairwise distinct $y_0, \dots, y_k \in I$ with $1 - \varepsilon <_\mathbb{R} \sum_{i \leq k} |J_{y_i}^\Psi|$, where $J_{y_{i+1}}^\Psi := I_{y_{i+1}}^\Psi \setminus (\cup_{j \leq i} I_{y_i}^\Psi)$.

As discussed at length in [14], WHBU expresses the essence of the Vitali covering theorem for *uncountable* coverings; Vitali already considered the latter in [31]. Basic properties of the *gauge integral* [28] are equivalent to HBU while WHBU is equivalent to basic properties of the Lebesgue integral (without RM-codes; [14]). By [13,14,16], Z_2^Ω proves HBU and WHBU, but Z_2^ω cannot. The exact definition of Z_2^ω and Z_2^Ω is in the aforementioned references and Sect. A.2. What is relevant here is that Z_2^ω and Z_2^Ω are conservative extensions of Z_2 by [9, Cor. 2.6,], i.e. the former prove the same second-order sentences as the latter.

We note that HBU (resp. WHBU) is the higher-order counterpart of WKL (resp. WWKL), i.e. *weak König's lemma* (resp. *weak weak König's lemma*) from RM as the ECF-translation [10,29] maps HBU (resp. WHBU) to WKL (resp. WWKL), i.e. these are (intuitively) *weak* principles. We refer to [10, §2,] or Remark A.1 for a discussion of the relation between ECF and RCA_0^ω.

Finally, the aforementioned results suggest that (higher-order) comprehension as in Z_2^ω is not the right way of measuring the strength of HBU. As a better alternative, we have introduced the following axiom in [22].

Principle 1.4 (BOOT). $(\forall Y^2)(\exists X \subset \mathbb{N})(\forall n^0)\big[n \in X \leftrightarrow (\exists f^1)(Y(f,n) = 0)\big]$.

By [22, §3,], BOOT is equivalent to convergence theorems for *nets*, we have the implication BOOT $\rightarrow$ HBU, and $\mathsf{RCA}_0^\omega + \mathsf{BOOT}$ has the same first-order strength as ACA_0. Moreover, BOOT is a natural fragment of Feferman's *projection axiom* (Proj1) from [4]. Thus, BOOT is a natural axiom that provides a better 'scale' for measuring the strength of HBU and its ilk, as discussed in [17,22].

Some Basic Definitions. We introduce the higher-order definitions of 'open' and 'countable' set, as can be found in e.g. [15,17,19].

First of all, open sets are represented in second-order RM as countable unions of basic open sets [26, II.5.6,], and we refer to such sets as 'RM-open'. By [26, II.7.1,], one can effectively convert between RM-open sets and (RM-codes for) continuous characteristic functions. Thus, a natural extension of the notion of 'open set' is to allow *arbitrary* (possibly discontinuous) characteristic functions, as is done in e.g. [15,17,21], which motivates the following definition.

Definition 1.5 [Sets in RCA_0^ω]. We let $Y : \mathbb{R} \to \mathbb{R}$ represent subsets of $\mathbb{R}$ as follows: we write '$x \in Y$' for '$Y(x) >_\mathbb{R} 0$' and call a set $Y \subseteq \mathbb{R}$ 'open' if for every $x \in Y$, there is an open ball $B(x,r) \subset Y$ with $r^0 > 0$. A set Y is called 'closed' if the complement, denoted $Y^c = \{x \in \mathbb{R} : x \notin Y\}$, is open.

For open Y as in Definition 1.5, the formula '$x \in Y$' has the same complexity (modulo higher types) as for RM-open sets, while given $(\exists^2)$ it is equivalent to a 'proper' characteristic function, only taking values '0' and '1'. Hereafter, an '(open) set' refers to Definition 1.5; 'RM-open set' refers to the definition from second-order RM, as in e.g. [26, II.5.6,].

Secondly, the definition of 'countable set' (Kunen; [11]) is as follows in RCA_0^ω.

Definition 1.6 [Countable subset of $\mathbb{R}$]. A set $A \subseteq \mathbb{R}$ is *countable* if there exists $Y : \mathbb{R} \to \mathbb{N}$ such that $(\forall x, y \in A)(Y(x) =_0 Y(y) \to x =_\mathbb{R} y)$. If $Y : \mathbb{R} \to \mathbb{N}$ is also *surjective*, i.e. $(\forall n \in \mathbb{N})(\exists x \in A)(Y(x) = n)$, we call A *strongly countable*.

Hereafter, '(strongly) countable' refers to Definition 1.6, unless stated otherwise. We note that 'countable' is defined in second-order RM using *sequences* [26, V.4.2,], a notion we shall call 'enumerable'.

Thirdly, we have explored the connection between HBU, generalisations of HBU, and fragments of the *neighbourhood function principle* NFP from [30] in [22,24]. In each case, nice equivalences were obtained *assuming* A_0 *as follows*.

Principle 1.7 (A_0). For Y^2 and $A(\sigma) \equiv (\exists g \in 2^\mathbb{N})(Y(g,\sigma) = 0)$, we have

$$(\forall f \in \mathbb{N}^\mathbb{N})(\exists n \in \mathbb{N})A(\overline{f}n) \to (\exists G^2)(\forall f \in \mathbb{N}^\mathbb{N})A(\overline{f}G(f)),$$

where $\overline{f}n$ is the finite sequence $\langle f(0), f(1), \ldots, f(n-1)\rangle$.

As discussed in [22,24], the axiom A_0 is a fragment of NFP and can be viewed as a generalisation of $\mathsf{QF\text{-}AC}^{1,0}$, included in RCA_0^ω. As an alternative to A_0, one could add 'extra data' or moduli to the theorems to be studied.

2   Main Results

In Sect. 2.1, we show that HBU is equivalent to $HBU_{\text{æ}}$, i.e. the restriction to functions continuous almost everywhere, while the same restriction applied to WHBU results in a theorem equivalent to WWKL (see [26, X.1,] for the latter). In Sect. 2.2, we establish the splitting (1.1) involving HBU.

2.1   Ontological Parsimony and the Heine-Borel Theorem

We introduce $HBU_{\text{æ}}$, the restriction of HBU from Sect. 1.2 to functions *continuous almost everywhere*, and establish HBU $\leftrightarrow HBU_{\text{æ}}$ over RCA_0^ω. The same restriction for WHBU turns out to be equivalent to *weak weak König's lemma* (WWKL; see [26, X.1,]), well-known from second-order RM.

We first need the following definition, where we note that the usual[1] definition of 'measure zero' is used in RM.

Definition 2.1 [Continuity almost everywhere]. We say that $\Psi : [0,1] \to \mathbb{R}$ is *continuous almost everywhere* if it is continuous outside of an RM-closed set $E \subset [0,1]$ which has measure zero.

Let $HBU_{\text{æ}}$ be HBU restricted to functions continuous almost everywhere as in the previous definition. The proof of the following theorem (in a stronger system) was suggested by Dag Normann, for which we are grateful.

Theorem 2.2. *The system* RCA_0^ω *proves* HBU $\leftrightarrow HBU_{\text{æ}}$.

Proof. First of all, as noted in Sect. 1.2, $(\exists^2)$ is equivalent to the existence of a discontinuous $\mathbb{R} \to \mathbb{R}$-function, namely by [10, Prop. 3.12,]. Thus, in case $\neg(\exists^2)$, all functions on $\mathbb{R}$ are continuous. In this case, we trivially obtain HBU $\leftrightarrow HBU_{\text{æ}}$. Since RCA_0^ω is a classical system, we have the law of excluded middle as in $\neg(\exists^2) \vee (\exists^2)$. As we have provided a proof in the first case $\neg(\exists^2)$, it suffices to provide a proof assuming $(\exists^2)$, and the law of excluded middle finishes the proof. Hence, for the rest of the proof, we may assume $(\exists^2)$.

Secondly, the *Cantor middle third set* $\mathcal{C} \subset [0,1]$ is available in RCA_0 by (the proof of) [26, IV.1.2,] as an RM-closed set, as well as the well-known recursive homeomorphism from Cantor space $2^\mathbb{N}$ to $\mathcal{C}$ defined as $H : 2^\mathbb{N} \to [0,1]$ and $H(f) := \sum_{n=0}^{\infty} \frac{2f(n)}{3^{n+1}}$. Note that given $\exists^2$, we can decide whether $x \in \mathcal{C}$ or not.

Thirdly, we prove $HBU_{\text{æ}} \to HBU_c$, where the latter is HBU for $2^\mathbb{N}$ as follows:

$$(\forall G^2)(\exists f_0, \ldots, f_k \in 2^\mathbb{N})(\forall g \in 2^\mathbb{N})(\exists i \leq k)(g \in [\overline{f_i}G(f_i)]) \tag{HBU_c}$$

and where $[\sigma]$ is the open neighbourhood in $2^\mathbb{N}$ of sequences starting with the finite binary sequence σ. The equivalence HBU $\leftrightarrow HBU_c$ may be found in [13,16]. Now assume $HBC_{\text{æ}}$ and fix G^2 and define $\Psi : [0,1] \to \mathbb{R}^+$ using $(\exists^2)$ as:

$$\Psi(x) := \begin{cases} d(x,\mathcal{C}) & x \notin \mathcal{C} \\ \frac{1}{2^{G(\overline{1}(x))}} & \text{otherwise} \end{cases}, \tag{2.1}$$

---

[1] A set $A \subset \mathbb{R}$ is *measure zero* if for any $\varepsilon > 0$ there is a sequence of basic open intervals $(I_n)_{n \in \mathbb{N}}$ such that $\cup_{n \in \mathbb{N}} I_n$ covers A and has total length below ε.

where $I(x)$ is the unique $f \in 2^{\mathbb{N}}$ such that $H(f) = x$ in case $x \in \mathcal{C}$, and $00\ldots$ otherwise. Note that the distance function $d(x, \mathcal{C})$ exists given ACA_0 by [7, Theorem 1.2,]. Clearly, $\exists^2$ allows us to define this function as a third-order object that is continuous on $[0,1] \setminus \mathcal{C}$. Since $\mathcal{C}$ has measure zero (and is RM-closed), apply $\mathsf{HBU}_{\mathbf{æ}}$ to $\cup_{x \in [0,1]} I_x^{\Psi}$. Let $y_0, \ldots, y_k$ be such that $\cup_{i \leq k} I_{y_i}^{\Psi}$ covers $[0,1]$. By the definition of Ψ in (2.1), if $x \in [0,1] \setminus \mathcal{C}$, then $\mathcal{C} \cap I_x^{\Psi} = \emptyset$. Hence, let $z_0, \ldots, z_m$ be those $y_i \in \mathcal{C}$ for $i \leq k$ and note that $\cup_{j \leq m} I_{z_j}^{\Psi}$ covers $\mathcal{C}$. Clearly, $I(z_0), \ldots, I(z_m)$ yields a finite sub-cover of $\cup_{f \in 2^{\mathbb{N}}} [\overline{f}G(f)]$, and $\mathsf{HBU}_{\mathbf{c}}$ follows. $\square$

We could of course formulate $\mathsf{HBU}_{\mathbf{æ}}$ with the higher-order notion of 'closed set' from [15], and the equivalence from the theorem would still go through. The proof of the theorem also immediately yields the following.

Corollary 2.3 (ACA_0^{ω}). HBU *is equivalent to the Heine-Borel theorem for canonical coverings* $\cup_{x \in E} I_x^{\Psi}$, *where* $E \subset [0,1]$ *is RM-closed and has measure zero.*

As expected, Theorem 2.2 generalises to principles that imply HBU over RCA_0^{ω} (see [17, Fig. 1,] for an overview) and that boast a third-order functional to which the 'continuous almost everywhere' restriction can be naturally applied. An example is the following corollary involving BOOT.

Corollary 2.4. *The system* RCA_0^{ω} *proves* $\mathsf{BOOT} \leftrightarrow \mathsf{BOOT}_{\mathbf{æ}}$, *where the latter is*

$$(\exists X \subset \mathbb{N})(\forall n^0)\big[n \in X \leftrightarrow (\exists x \in [0,1])(Y(x,n) = 0)\big],$$

where $\lambda x.Y(x,n)$ *is continuous almost everywhere on* $[0,1]$ *for any fixed* $n \in \mathbb{N}$.

Proof. In case $\neg(\exists^2)$, all functions on $\mathbb{R}$ are continuous by [10, Prop. 3.12,]; in this case, the equivalence is trivial. In case $(\exists^2)$, the forward direction is immediate, modulo coding real numbers given $\exists^2$. For the reverse direction, fix Y^2 and note that we may restrict the quantifier $(\exists f^1)$ in BOOT to $2^{\mathbb{N}}$ without loss of generality. Indeed, μ^2 allows us to represent f^1 via its graph, a subset of $\mathbb{N}^2$, which can be coded as a binary sequence. Now define

$$Z(x,n) := \begin{cases} 0 & x \in \mathcal{C} \wedge Y(I(x),n) = 0 \\ 1 & \text{otherwise} \end{cases}, \tag{2.2}$$

where $\mathcal{C}$ and I are as in the theorem. Note that $\lambda x.Z(x,n)$ is continuous outside of $\mathcal{C}$. By $\mathsf{BOOT}_{\mathbf{æ}}$, there is $X \subset \mathbb{N}$ such that for all $n \in \mathbb{N}$, we have:

$$n \in X \leftrightarrow (\exists x \in [0,1])(Z(x,n) = 0) \leftrightarrow (\exists f \in 2^{\mathbb{N}})(Y(f,n) = 0),$$

where the last equivalence is by the definition of Z in (2.2). $\square$

Next, we show that the *Vitali covering theorem* as in WHBU behaves quite differently from the Heine-Borel theorem as in HBU. Recall that the Heine-Borel theorem applies to open coverings of compact sets, while the Vitali covering theorem applies to Vitali coverings[2] of any set E of finite (Lebesgue) measure. The former provides a finite sub-covering while the latter provides a sequence that covers E up to a set of measure zero. As argued in [14], WHBU is the combinatorial essence of Vitali's covering theorem.

Now, let WHBU$_{\text{æ}}$ be WHBU restricted to functions continuous almost everywhere, as in Definition 2.1; recall that Z_2^ω cannot prove WHBU.

Theorem 2.5. *The system* RCA$_0^\omega$ + WKL *proves* WHBU$_{\text{æ}}$.

Proof. Let $\Psi : [0,1] \to \mathbb{R}^+$ be continuous on $[0,1] \setminus E$ with $E \subset [0,1]$ of measure zero and RM-closed. Fix $\varepsilon > 0$ and let $\cup_{n \in \mathbb{N}} I_n$ be a union of basic open intervals covering E and with measure at most $\varepsilon/2$. Then $[0,1]$ is covered by:

$$\cup_{q \in \mathbb{Q} \setminus E} B(q, \Psi(q)) \bigcup \cup_{n \in \mathbb{N}} I_n. \tag{2.3}$$

Indeed, that the covering in (2.3) covers E is trivial, while $[0,1] \setminus E$ is (RM)-open. Hence, $x_0 \in [0,1] \setminus E$ implies that $B(x_0, r) \subset [0,1] \setminus E$ for $r > 0$ small enough and for $q \in \mathbb{Q} \cap [0,1]$ close enough to x_0, we have $x_0 \in B(q, \Psi(q))$. By [26, IV.1,], WKL is equivalent to the countable Heine-Borel theorem. Hence, there are $q_0, \ldots, q_k \in \mathbb{Q} \setminus E$ and $n_0 \in \mathbb{N}$ such that the finite union $\cup_{i=1}^k B(q_i, \Psi(q_i)) \bigcup \cup_{j=0}^{n_0} I_j$ covers $[0,1]$. Since the measure of $\cup_{j=0}^{n_0} I_j$ is at most $\varepsilon/2$, the measure of $\cup_{i=1}^k B(q_i, \Psi(q_i))$ is at least $1 - \varepsilon/2$, as required by WHBU$_{\text{æ}}$. $\square$

Corollary 2.6. *The system* RCA$_0^\omega$ *proves* WWKL $\leftrightarrow$ WHBU$_{\text{æ}}$.

Proof. The reverse implication is immediate in light of the RM of WWKL in [26, X.1,], which involves the Vitali covering theorem for countable coverings (given by a sequence). For the forward implication, convert the cover from (2.3) to a Vitali cover and use [26, X.1.13,]. $\square$

Finally, recall Remark 1.1 discussing the foundational significance of the above.

2.2 Splittings for the Heine-Borel Theorem

We establish a splitting for HBU as in Theorem 2.9 based on known principles formulated with *countable sets* as in Definition 1.6. As will become clear, there is also some historical interest in this study.

First of all, the following principle HBC$_0$ is studied in [17, §3,], while the (historical and foundational) significance of this principle is discussed in Remark 1.1. The aforementioned system Z_2^ω cannot prove HBC$_0$.

Principle 2.7 (HBC$_0$). For countable $A \subset \mathbb{R}^2$ with $(\forall x \in [0,1])(\exists (a,b) \in A)(x \in (a,b))$, there are $(a_0, b_0), \ldots, (a_k, b_k) \in A$ with $(\forall x \in [0,1])(\exists i \leq k)(x \in (a_i, b_i))$.

---

[2] An open covering V is a *Vitali covering* of E if any point of E can be covered by some open in V with arbitrary small (Lebesgue) measure.

Secondly, the second-order Vitali covering theorem has a number of equivalent formulations (see [26, X.1,]), including the statement *a countable covering of $[0,1]$ has a sub-collection with measure zero complement*. Intuitively speaking, the following principle WHBU$^+$ strengthens 'measure zero' to 'countable'. Alternatively, WHBU$^+$ can be viewed as a weakening of the Lindelöf lemma, introduced in [12] and studied in higher-order RM in [13,16].

Principle 2.8 (WHBU$^+$). For $\Psi : [0,1] \to \mathbb{R}^+$, there is a sequence $(y_n)_{n \in \mathbb{N}}$ in $[0,1]$ such that $[0,1] \setminus \cup_{n \in \mathbb{N}} I_{y_n}^{\Psi}$ is countable.

Note that WHBU$^+$ + HBC$_0$ yields a conservative[3] extension of RCA$_0^{\omega}$, i.e. the former cannot imply HBU without the presence of WKL. Other independence results are provided by Theorem 2.10.

We have the following theorem, where A$_0$ was introduced in Sect. 1.2. This axiom can be avoided by enriching[4] the antecedent of HBC$_0$.

Theorem 2.9. *The system* RCA$_0^{\omega}$ + A$_0$ *proves*

$$[\text{WHBU}^+ + \text{HBC}_0 + \text{WKL}] \leftrightarrow \text{HBU}, \tag{2.4}$$

where the axiom A$_0$ *is only needed for* HBU $\to$ HBC$_0$.

Proof. First of all, in case $\neg(\exists^2)$, all functions on $\mathbb{R}$ are continuous, rendering WHBU$^+$ + HBC$_0$ trivial while HBU reduces to WKL. Hence, for the rest of the proof, we may assume $(\exists^2)$, by the law of excluded middle as in $(\exists^2) \vee \neg(\exists^2)$.

For the reverse implication, assume A$_0$ + HBU and let A be as in HBC$_0$. The functional $\exists^2$ can uniformly convert real numbers to a binary representation. Hence (2.5) is equivalent to a formula as in the antecedent of A$_0$:

$$(\forall x \in [0,1])(\exists n \in \mathbb{N})\big[(\exists(a,b) \in A)(a < [x](n+1) - \tfrac{1}{2^n} \wedge [x](n+1) + \tfrac{1}{2^n} < b)\big], \tag{2.5}$$

where '$[x](n)$' is the n-th approximation of the real x, given as a fast-converging Cauchy sequence. Apply A$_0$ to (2.5) to obtain $G : [0,1] \to \mathbb{N}$ such that $G(x) = n$ as in (2.5). Apply HBU to $\cup_{x \in [0,1]} I_x^{\Psi}$ for $\Psi(x) := \frac{1}{2^{G(x)}}$. The finite sub-cover $y_0, \ldots, y_k \in [0,1]$ provided by HBU gives rise to $(a_i, b_i) \in A$ containing $I_{y_i}^{\Psi}$ for $i \leq k$ by the definition of G. Moreover, HBU implies WKL as the latter is equivalent to the 'countable' Heine-Borel theorem as in [26, IV.1,]. Clearly, the empty set is countable by Definition 1.6 and HBU $\to$ WHBU$^+$ is therefore trivial.

For the forward implication, fix $\Psi : [0,1] \to \mathbb{R}^+$ and let $(y_n)_{n \in \mathbb{N}}$ be as in WHBU$^+$. Define '$x \in B$' as $x \in [0,1] \setminus \cup_{n \in \mathbb{N}} I_{y_n}^{\Psi}$ and note that when B is empty, the theorem follows as WKL implies the second-order Heine-Borel theorem [26, IV.1,]. Now assume $B \neq \emptyset$ and define A as the set of (a,b) such that either

---

[3] The system RCA$_0^{\omega}$ + $\neg(\exists^2)$ is an L$_2$-conservative extension of RCA$_0^{\omega}$ and the former readily proves WHBU$^+$ + HBC$_0$. By constrast HBU $\to$ WKL over RCA$_0^{\omega}$.

[4] In particular, one would add a function $G : [0,1] \to \mathbb{R}^2$ to the antecedent of HBC$_0$ such that $G(x) \in A$ and $x \in \big(G(x)(1), G(x)(2)\big)$ for $x \in [0,1]$. In this way, the covering is given by $\cup_{x \in [0,1]}(G(x)(1), G(x)(2))$.

$(a, b) = I_x^\Psi$ for $x \in B$, or $(a, b) = I_{y_n}^\Psi$ for some $n \in \mathbb{N}$. Note that in the first case, $(a, b) \in A$ if and only if $\frac{a+b}{2} \in B$, i.e. defining A does not require quantifying over $\mathbb{R}$. Moreover, A is countable because B is: if Y is injective on B, then W defined as follows is injective on A:

$$W((a,b)) := \begin{cases} 2Y(\frac{a+b}{2}) & \frac{a+b}{2} \in B \\ H((a,b)) & \text{otherwise} \end{cases},$$

where $H((a, b))$ is the least $n \in \mathbb{N}$ such that $(a, b) = I_{y_n}^\Psi$, if such there is, and zero otherwise. The intervals in the set A cover $[0, 1]$ as in the antecedent of $\mathsf{HBC_0}$, and the latter now implies HBU. $\square$

The principles $\mathsf{WHBU^+}$ and $\mathsf{HBC_0}$ are 'quite' independent by the following theorem, assuming the systems therein are consistent.

Theorem 2.10. *The system* $\mathsf{Z_2^\omega} + \mathsf{QF\text{-}AC^{0,1}} + \mathsf{WHBU^+}$ *cannot prove* $\mathsf{HBC_0}$. *The system* $\mathsf{RCA_0^\omega} + \mathsf{HBC_0} + \mathsf{WHBU^+}$ *cannot prove* $\mathsf{WKL_0}$.

Proof. For the first part, suppose $\mathsf{Z_2^\omega} + \mathsf{QF\text{-}AC^{0,1}} + \mathsf{WHBU^+}$ does prove $\mathsf{HBC_0}$. The latter implies NIN as follows by [17, Cor. 3.2,]:

$$(\forall Y : [0, 1] \to \mathbb{N})(\exists x, y \in [0, 1])(Y(x) = Y(y) \wedge x \neq_{\mathbb{R}} y). \tag{NIN}$$

Clearly, $\neg\mathsf{NIN}$ implies $\mathsf{WHBU^+}$, and we obtain that $\mathsf{Z_2^\omega} + \mathsf{QF\text{-}AC^{0,1}} + \neg\mathsf{NIN}$ proves a contradiction, namely $\mathsf{WHBU^+}$ and its negation. Hence, $\mathsf{Z_2^\omega} + \mathsf{QF\text{-}AC^{0,1}}$ proves NIN, a contradiction by [17, Theorem 3.1,], and the first part follows.

For the second part, the ECF-translation (see Remark A.1) converts $\mathsf{HBC_0} + \mathsf{WHBU^+}$ into a triviality. $\square$

Finally, we discuss similar results as follows. Of course, the proof of Theorem 2.9 goes through *mutatis mutandis* for $\mathsf{WHBU^+} + \mathsf{HBC_0}$ formulated using *strongly* countable sets. Moreover, (2.6) can be proved in the same way as (2.4), assuming additional countable choice as in $\mathsf{QF\text{-}AC^{0,1}}$:

$$\mathsf{WHBU} \leftrightarrow [\mathsf{WHBU^+} + \mathsf{WHBC_0} + \mathsf{WWKL}], \tag{2.6}$$

where $\mathsf{WHBC_0}$ is $\mathsf{HBC_0}$ with the conclusion weakened to the existence of a sequence $(a_n, b_n)_{n \in \mathbb{N}}$ of intervals in A with measure at least one. Also, if we generalise HBU to coverings of any *separably closed* set in $[0, 1]$, the resulting version of (2.4) involves $\mathsf{ACA_0}$ rather than $\mathsf{WKL_0}$ in light of [8, Theorem 2].

A   Reverse Mathematics: Second- and Higher-Order

A.1   Reverse Mathematics

Reverse Mathematics (RM hereafter) is a program in the foundations of mathematics initiated around 1975 by Friedman [5,6] and developed extensively by

Simpson [26]. The aim of RM is to identify the minimal axioms needed to prove theorems of ordinary, i.e. non-set theoretical, mathematics. We refer to [27] for a basic introduction to RM and to [25, 26] for an overview of RM. The details of Kohlenbach's *higher-order* RM may be found in [10], including the base theory RCA_0^ω. The latter is connected to RCA_0 by the ECF-translation as follows.

Remark A.1 (The ECF-interpretation). The (rather) technical definition of ECF may be found in [29, p. 138, §2.6,]. Intuitively, the ECF-interpretation $[A]_{\mathsf{ECF}}$ of a formula $A \in L_\omega$ is just A with all variables of type two and higher replaced by type one variables ranging over so-called 'associates' or 'RM-codes'; the latter are (countable) representations of continuous functionals. The ECF-interpretation connects RCA_0^ω and RCA_0 (see [10, Prop. 3.1,]) in that if RCA_0^ω proves A, then RCA_0 proves $[A]_{\mathsf{ECF}}$, again 'up to language', as RCA_0 is formulated using sets, and $[A]_{\mathsf{ECF}}$ is formulated using types, i.e. using type zero and one objects.

In light of the widespread use of codes in RM and the common practise of identifying codes with the objects being coded, it is no exaggeration to refer to ECF as the *canonical* embedding of higher-order into second-order arithmetic.

We now introduce the usual notations for common mathematical notions.

Definition A.2 (Real numbers and related notions in RCA_0^ω)

a. Natural numbers correspond to type zero objects, and we use 'n^0' and '$n \in \mathbb{N}$' interchangeably. Rational numbers are defined as signed quotients of natural numbers, and '$q \in \mathbb{Q}$' and '$<_\mathbb{Q}$' have their usual meaning.

b. Real numbers are coded by fast-converging Cauchy sequences $q_{(.)} : \mathbb{N} \to \mathbb{Q}$, i.e. such that $(\forall n^0, i^0)(|q_n - q_{n+i}| <_\mathbb{Q} \frac{1}{2^n})$. We use Kohlenbach's 'hat function' from [10, p. 289,] to guarantee that every q^1 defines a real number.

c. We write '$x \in \mathbb{R}$' to express that $x^1 := (q_{(.)}^1)$ represents a real as in the previous item and write $[x](k) := q_k$ for the k-th approximation of x.

d. Two reals x, y represented by $q_{(.)}$ and $r_{(.)}$ are *equal*, denoted $x =_\mathbb{R} y$, if $(\forall n^0)(|q_n - r_n| \leq 2^{-n+1})$. Inequality '$<_\mathbb{R}$' is defined similarly. We sometimes omit the subscript '$\mathbb{R}$' if it is clear from context.

e. Functions $F : \mathbb{R} \to \mathbb{R}$ are represented by $\Pi^{1 \to 1}$ mapping equal reals to equal reals, i.e. extensionality as in $(\forall x, y \in \mathbb{R})(x =_\mathbb{R} y \to \Pi(x) =_\mathbb{R} \Pi(y))$.

f. Binary sequences are denoted '$f, g \in C$' or '$f, g \in 2^\mathbb{N}$'. Elements of Baire space are given by f^1, g^1, but also denoted '$f, g \in \mathbb{N}^\mathbb{N}$'.

Notation A.3 (Finite sequences). The type for 'finite sequences of objects of type ρ' is denoted ρ^*, which we shall only use for $\rho = 0, 1$. Since the usual coding of pairs of numbers goes through in RCA_0^ω, we shall not always distinguish between 0 and 0^*. Similarly, we assume a fixed coding for finite sequences of type 1 and shall make use of the type '1^*'. In general, we do not always distinguish between 's^ρ' and '$\langle s^\rho \rangle$', where the former is 'the object s of type ρ', and the latter is 'the sequence of type ρ^* with only element s^ρ'. The empty sequence for the type ρ^* is denoted '$\langle \rangle_\rho$', usually with the typing omitted. Furthermore, we denote by '$|s| = n$' the length of the finite sequence $s^{\rho^*} = \langle s_0^\rho, s_1^\rho, \ldots, s_{n-1}^\rho \rangle$, where $|\langle\rangle| = 0$, i.e. the empty sequence has length zero. For sequences s^{ρ^*}, t^{ρ^*}, we denote by '$s * t$' the concatenation of s and t, i.e. $(s * t)(i) = s(i)$ for $i < |s|$ and $(s * t)(j) = t(|s| - j)$ for $|s| \le j < |s| + |t|$. For a sequence s^{ρ^*}, we define $\overline{s}N := \langle s(0), s(1), \ldots, s(N-1)\rangle$ for $N^0 < |s|$. For a sequence $\alpha^{0 \to \rho}$, we also write $\overline{\alpha}N = \langle \alpha(0), \alpha(1), \ldots, \alpha(N-1)\rangle$ for *any* N^0. Finally, $(\forall q^\rho \in Q^{\rho^*})A(q)$ abbreviates $(\forall i^0 < |Q|)A(Q(i))$, which is (equivalent to) quantifier-free if A is.

A.2   Further Systems

We define some standard higher-order systems that constitute the counterpart of e.g. Π_1^1-CA_0 and Z_2. First of all, *the Suslin functional* S^2 is defined in [10] as:

$$(\exists \mathsf{S}^2 \le_2 1)(\forall f^1)\big[(\exists g^1)(\forall n^0)(f(\overline{g}n) = 0) \leftrightarrow \mathsf{S}(f) = 0\big]. \tag{S^2}$$

The system Π_1^1-$\mathsf{CA}_0^\omega \equiv \mathsf{RCA}_0^\omega + (\mathsf{S}^2)$ proves the same Π_3^1-sentences as Π_1^1-CA_0 by [20, Theorem 2.2,]. By definition, the Suslin functional S^2 can decide whether a Σ_1^1-formula as in the left-hand side of (S^2) is true or false. We similarly define the functional S_k^2 which decides the truth or falsity of Σ_k^1-formulas from L_2; we also define the system Π_k^1-CA_0^ω as $\mathsf{RCA}_0^\omega + (\mathsf{S}_k^2)$, where (S_k^2) expresses that S_k^2 exists. We note that the operators ν_n from [2, p. 129,] are essentially S_n^2 strengthened to return a witness (if existant) to the Σ_n^1-formula at hand.

Secondly, second-order arithmetic Z_2 readily follows from $\cup_k \Pi_k^1$-CA_0^ω, or from:

$$(\exists E^3 \le_3 1)(\forall Y^2)\big[(\exists f^1)(Y(f) = 0) \leftrightarrow E(Y) = 0\big], \tag{$\exists^3$}$$

and we therefore define $\mathsf{Z}_2^\Omega \equiv \mathsf{RCA}_0^\omega + (\exists^3)$ and $\mathsf{Z}_2^\omega \equiv \cup_k \Pi_k^1$-$\mathsf{CA}_0^\omega$, which are conservative over Z_2 by [9, Cor. 2.6,]. Despite this close connection, Z_2^ω and Z_2^Ω can behave quite differently, as discussed in e.g. [13, §2.2,]. The functional from $(\exists^3)$ is also called '$\exists^3$', and we use the same convention for other functionals.

References

1. Borel, E.: Leçons sur la théorie des fonctions. Gauthier-Villars, Paris (1898)
2. Buchholz, W., Feferman, S., Pohlers, W., Sieg, W.: Iterated Inductive Definitions and Subsystems of Analysis: Recent Proof-Theoretical Studies. LNM, vol. 897. Springer, Heidelberg (1981). https://doi.org/10.1007/BFb0091894
3. Cousin, P.: Sur les fonctions de n variables complexes. Acta Math. **19**, 1–61 (1895)
4. Feferman, S.: How a Little Bit goes a Long Way: Predicative Foundations of Analysis (2013). unpublished notes from 1977–1981 with updated introduction. https://math.stanford.edu/~feferman/papers/pfa.pdf
5. Friedman, H.: Some systems of second order arithmetic and their use. In: Proceedings of the ICM (Vancouver, B. C., 1974), vol. 1, pp. 235–242 (1975)
6. Friedman, H.: Systems of second order arithmetic with restricted induction, I & II (abstracts). J. Symbolic Logic **41**, 557–559 (1976)

7. Giusto, M., Simpson, S.G.: Located sets and reverse mathematics. J. Symbolic Logic **65**(3), 1451–1480 (2000)
8. Hirst, J.L.: A note on compactness of countable sets. In: Reverse Mathematics (2001). Lect. Notes Log., vol. 21. Assoc. Symbol. Logic 2005, pp. 219–221
9. Hunter, J.: Higher-order reverse topology, ProQuest LLC, Ann Arbor, MI (2008). Thesis (Ph.D.)-The University of Wisconsin - Madison
10. Kohlenbach, U.: Higher order reverse mathematics. In: Reverse Mathematics (2001). Lect. Notes Log., vol. 21. ASL 2005, pp. 281–295
11. Kunen, K.: Set theory, Studies in Logic, vol. 34. College Publications, London (2011)
12. Lindeöf, E.: Sur Quelques Points De La Théorie Des Ensembles. Comptes Rendus, pp. 697–700 (1903)
13. Normann, D., Sanders, S.: On the mathematical and foundational significance of the uncountable. J. Math. Logic (2019). https://doi.org/10.1142/S0219061319500016
14. Normann, D., Sanders, S.: Representations in measure theory. arxiv:1902.02756 (2019, Submitted)
15. Normann, D., Sanders, S.: Open sets in reverse mathematics and computability theory. J. Logic Computability **30**(8), 40 (2020)
16. Normann, D., Sanders, S.: Pincherle's theorem in reverse mathematics and computability theory. Ann. Pure Appl. Logic **171**(5), 102788, 41 (2020)
17. Normann, D., Sanders, S.: On the uncountability of $\mathbb{R}$, p. 37. arxiv:2007.07560 (2020, Submitted)
18. Normann, D., Sanders, S.: The axiom of choice in computability theory and reverse mathematics. J. Log. Comput. **31**(1), 297–325 (2021)
19. Normann, D., Sanders, S.: On robust theorems due to Bolzano, Weierstrass, and Cantor in Reverse Mathematics, p. 30. https://arxiv.org/abs/2102.04787 (2021)
20. Sakamoto, N., Yamazaki, T.: Uniform versions of some axioms of second order arithmetic. MLQ Math. Log. Q. **50**(6), 587–593 (2004)
21. Sanders, S.: Nets and reverse mathematics: a pilot study. Computability 34 (2019). https://doi.org/10.3233/COM-190265
22. Sanders, S.: Plato and the foundations of mathematics, p. 40. arxiv:1908.05676 (2019, Submitted)
23. Sanders, S.: Splittings and disjunctions in reverse mathematics. Notre Dame J. Form. Log. **61**(1), 51–74 (2020)
24. Sanders, S.: Reverse mathematics of topology: dimension, paracompactness, and splittings. Notre Dame J. Formal Logic **61**(4), 537–559 (2020)
25. Simpson, S.G. (ed.): Reverse Mathematics (2001). Lecture Notes in Logic, vol. 21, ASL, 2005
26. Simpson, S.G. (ed.): Subsystems of Second Order Arithmetic, 2nd edn. Perspectives in Logic. Cambridge University Press (2009)
27. Stillwell, J.: Reverse Mathematics, Proofs from the Inside Out. Princeton University Press, Princeton (2018)
28. Swartz, C.: Introduction to Gauge Integrals. World Scientific (2001)
29. Troelstra, A.S.: Metamathematical Investigation of Intuitionistic Arithmetic and Analysis. LNM, vol. 344. Springer, Heidelberg (1973). https://doi.org/10.1007/BFb0066739
30. Troelstra, A.S., van Dalen, D.: Constructivism in Mathematics. Vol. I. Studies in Logic and the Foundations of Mathematics, vol. 121. North-Holland (1988)
31. Vitali, G.: Sui gruppi di punti e sulle funzioni di variabili reali. Atti della Accademia delle Scienze di Torino, vol. XLIII **4**, pp. 229–247 (1907)

Non-collapse of the Effective Wadge Hierarchy

Victor Selivanov[✉]

A.P. Ershov Institute of Informatics Systems SB RAS and S.L. Sobolev Institute
of Mathematics SB RAS, Novosibirsk, Russia
vseliv@iis.nsk.su

Abstract. We study the recently suggested effective Wadge hierarchy
in effective spaces, concentrating on the non-collapse property. Along
with hierarchies of sets, we study hierarchies of k-partitions which are
interesting on their own. In particular, we establish sufficient conditions
for the non-collapse of the effective Wadge hierarchy and apply them to
some concrete spaces.

Keywords: Effective space · Computable quasi-Polish space · Effective
Wadge hierarchy · Fine hierarchy · k-partition · Non-collapse property

1 Introduction

Hierarchies are basic tools for calibrating objects according to their complexity,
hence the non-collapse of a natural hierarchy is fundamental for understanding
the corresponding notion of complexity. A lot of papers investigate the non-
collapse property in different contexts, see e.g. [16] for a survey of hierarchies
relevant to those studied in this paper.

The Wadge hierarchy (WH), which is fundamental for descriptive set theory
(DST), was developed for the Baire space $\mathcal{N}$, first for the case of sets [22], and
recently for the Q-valued Borel functions on $\mathcal{N}$, for any better quasiorder Q [10].
A convincing extension of this to arbitrary topological spaces was developed in
[20] (see also [13,19]). In [21] we introduced and studied the effective Wadge
hierarchy (EWH) in effective spaces as an instantiation of the fine hierarchy
(FH) [16]. Here we concentrate on the non-collapse property of this hierarchy.
As in [21], along with the EWH of sets we consider the EWH of k-partitions for
$k > 2$ (sets correspond to 2-partitions).

The non-collapse of EWH is highly non-trivial already for the discrete space
$\mathbb{N}$ of natural numbers. In fact, for the case of sets it follows from the results on the
non-collapse of a FH in [15]; m-degrees of complete sets in levels of this hierarchy
are among the "natural m-degrees" studied recently in [9]. For k-partitions with

---

V. Selivanov—The work is supported by Mathematical Center in Akademgorodok
under agreement No. 075-15-2019-1613 with the Ministry of Science and Higher Education of the Russian Federation.

L. De Mol et al. (Eds.): CiE 2021, LNCS 12813, pp. 407–416, 2021.
https://doi.org/10.1007/978-3-030-80049-9_40

$k > 2$, the non-collapse property was not proved in [15] because that time we did not have a convincing notion of a hierarchy of k-partitions (introduced only in [17,19]). Here we prove additional facts which, together with the results in [15], imply the non-collapse of EWH of k-partitions in $\mathbb{N}$. We also prove the non-collapse of EWH in $\mathcal{N}$, providing an effective version for the fundamental result in [10]; modulo this result, our proofs for $\mathcal{N}$ are easy.

Along with the spaces $\mathbb{N}$ and $\mathcal{N}$, which are central in computability theory, we discuss the non-collapse of EWH for other spaces which became popular in computable analysis and effective DST. The preservation property established in [20,21] implies that the non-collapse property is inherited by the (effective) continuous open surjections which suggests a method for proving non-collapse. Unfortunately, this method is less general than the dual inheritance method for the Hausdorff-Kuratowski property [20,21], that completely reduces this property in (computable) quasi-Polish spaces to that in the Baire space. Nevertheless, the method suggested here provides some insight which enables e.g. to show that the non-collapse property is hard to prove for the majority of spaces.

The FH (as most objects related to the WH) has inherent combinatorial complexity resulting in rather technical notions and involved proofs. For this reason, it was not possible to make this paper completely self-contained. But, with papers [10,15,17,21] at hand, the reader would have everything to understand the remaining technical details.

After preliminaries in the next section, we recall in Sect. 3 necessary information on the EWH. In Sect. 4 we define some versions of the non-collapse property and relate them to the preservation property. In Sect. 5 we state our main technical results on the EWH in $\mathbb{N}$ and the domain $\omega^{\leq \omega}$ of finite and infinite strings and apply them to some other spaces, illustrating the method of Sect. 4.

2   Preliminaries

We use standard set-theoretical notation, in particular, Y^X is the set of functions from X to Y, and $P(X)$ is the class of subsets of a set X. All (topological) spaces in this paper are countably based T_0 (cb$_0$-spaces, for short). An *effective* cb$_0$-*space* is a pair (X, β) where X is a cb$_0$-space, and $\beta : \omega \to P(X)$ is a numbering of a base in X such that there is a uniformly c.e. sequence $\{A_{ij}\}$ of c.e. sets with $\beta(i) \cap \beta(j) = \bigcup \beta(A_{ij})$ where $\beta(A_{ij})$ is the image of A_{ij} under β. We simplify (X, β) to X if β is clear from the context. The *effectively open sets* in X are the sets $\bigcup \beta(W)$, for some c.e. set $W \subseteq \mathbb{N}$. The standard numbering $\{W_n\}$ of c.e. sets [14] induces a numbering of the effectively open sets. The notion of effective cb$_0$-space allows to define e.g. computable and effectively open functions between such spaces [18,23].

Among effective cb$_0$-space are: the discrete space $\mathbb{N}$ of natural numbers, the Euclidean spaces $\mathbb{R}^n$, the Scott domain $P\omega$ (see [1] for information about domains), the Baire space $\mathcal{N} = \mathbb{N}^{\mathbb{N}}$, the Baire domain $\omega^{\leq \omega}$ of finite and infinite strings over ω with the Scott topology, the Cantor space 2^{ω} of binary infinite strings, the Cantor domains $n^{\leq \omega}$, $2 \leq n < \omega$, of finite and infinite strings over

$\{0,\ldots,n-1\}$ with the Scott topology; all these spaces come with natural numberings of bases. The space $\mathbb{N}$ is trivial topologically but very interesting for computability theory.

Quasi-Polish spaces (introduced in [4]) are important for DST and have several characterisations. Effectivizing one of them we obtain the following notion identified implicitly in [18] and explicitly in [5,7]: a *computable quasi-Polish space* is an effective cb$_0$-space (X,β) such that there exists a computable effectively open surjection from $\mathcal{N}$ onto (X,β). Most spaces of interest for computable analysis and effective DST, in particular the aforementioned ones, are computable quasi-Polish.

Effective hierarchies were studied by many authors, see e.g. [2,3,6,12,18]. Let $\{\Sigma^0_{1+n}(X)\}_{n<\omega}$ be the effective Borel hierarchy, and $\{\Sigma^{-1,m}_{1+n}(X)\}_n$ (with $\Sigma^{-1,1}$ usually simplified to Σ^{-1}) be the effective Hausdorff difference hierarchy over $\Sigma^0_m(X)$ in arbitrary effective cb$_0$-space X. We do not repeat standard definitions but mention that the effective hierarchies come with standard numberings of all levels, so we can speak e.g. about uniform sequences of sets in a given level. We use definitions based on set operations (see e.g. [18]); there is also an equivalent approach based on the Borel codes [8,11]. E.g., $\Sigma^0_1(X)$ is the class of effectively open sets in X, $\Sigma^{-1}_2(X)$ is the class of differences of $\Sigma^0_1(X)$-sets, and $\Sigma^0_2(X)$ is the class of effective countable unions of $\Sigma^{-1}_2(X)$-sets. A function $f : X \to Y$ is Σ^0_2-*measurable* if $f^{-1}(B) \in \Sigma^0_2$ for each $B \in \Sigma^0_1(Y)$, effectively on the indices.

Levels of effective hierarchies are denoted in the same manner as levels of the corresponding classical hierarchies, using the lightface letters Σ, Π instead of the boldface $\boldsymbol{\Sigma}, \boldsymbol{\Pi}$ used for the classical hierarchies [8,12]. Any lightface notion in this paper will have a classical boldface counterpart, as is standard in DST. In particular, $f : X \to Y$ is $\boldsymbol{\Sigma}^0_2$-*measurable* if $f^{-1}(B) \in \boldsymbol{\Sigma}^0_2$ for each $B \in \boldsymbol{\Sigma}^0_1(Y)$. Every Σ^0_2-measurable function is $\boldsymbol{\Sigma}^0_2$-measurable.

As preparation to the next section, let us recall a notation system for levels of the FH of k-partitions introduced in [17]. Let $\omega^{<\omega}$ be the set of finite strings of natural numbers including the empty string ε, and $|\sigma|$ be the length of a string σ. A *tree* is a nonempty initial segment of $(\omega^{<\omega}; \sqsubseteq)$ where $\sqsubseteq$ is the prefix relation; by default, all trees below are finite. A tree T is *normal* if $\tau(i+1) \in T$ implies that $\tau i \in T$. For any finite tree T and any $\tau \in T$, define the tree $T(\tau) = \{\sigma \mid \tau \cdot \sigma \in T\}$. Then any non-singleton tree T is determined by the singleton tree $\{\varepsilon\}$ and the trees $T(i)$, $i \in T$ of lesser tree ranks than T, then $T = \{\varepsilon\} \cup \bigcup_{i \in T} T(i)$. We will use this representation in the proofs by induction on ranks. By a *forest* we mean an initial segment of $(\omega^{<\omega} \setminus \{\varepsilon\}; \sqsubseteq)$. Note that there is a unique empty forest, and for any forest F there is a unique tree T with $F = T \setminus \{\varepsilon\}$. The non-empty forest F may be considered as the non-empty disjoint union of trees $F(i)$, $i \in \omega \cap F$.

Next we recall notation related to iterated labeled trees from [17]. Let $(Q; \leq)$ be a preorder; abusing notation we often denote it just by Q. A Q-*tree* is a pair (T,t) consisting of a tree $T \subseteq \omega^{<\omega}$ and a labeling $t : T \to Q$. Let $\mathcal{T}(Q)$ denote the set of all finite Q-trees. The h-*preorder* $\leq_h$ on $\mathcal{T}(Q)$ is defined as follows: $(T,t) \leq_h (V,v)$, if there is a monotone function $f : (T; \sqsubseteq) \to (S; \sqsubseteq)$

satisfying $\forall \tau \in T(t(\tau) \leq v(f(\tau)))$. For any $q \in Q$, let $s(q) = (\{\varepsilon\}, q)$ be the singleton tree labeled by q. The preorder Q is called WQO if it has neither infinite descending chains nor infinite antichains. An example of WQO is the antichain $\bar{k} = \{0, \ldots, k-1\}$ with k elements. A famous Kruskal's theorem implies that if Q is WQO then $(T_Q; \leq_h)$ is WQO.

Define the sequence $\{T_m(\bar{k})\}_{m<\omega}$ of preorders by induction on m as follows: $T_0(\bar{k}) = \bar{k}$ and $T_{m+1}(\bar{k}) = T_{T_m(\bar{k})}$. The sets $T_m(\bar{k})$, $m < \omega$, are pairwise disjoint but, identifying $i < k$ with $s(i)$, we may think that $T_0(\bar{k}) \sqsubseteq T_1(\bar{k})$, i.e. the quotient-poset of the first preorder is an initial segment of the quotient-poset of the second. This also induces an embedding of $T_m(\bar{k})$ into $T_{m+1}(\bar{k})$ as an initial segment, so (abusing notation) we may think that $T_0(\bar{k}) \sqsubseteq T_1(\bar{k}) \sqsubseteq \cdots$, hence $T_\omega(\bar{k}) = \bigcup_{m<\omega} T_m(\bar{k})$ is WQO w.r.t. the induced preorder which we also denote $\leq_h$. The embedding s is extended to $T_\omega(\bar{k})$ by defining $s(T)$ as the singleton tree labeled by T. For $k = 2$, the quotient-poset of $(T_\omega(\bar{2}); \leq_h)$ has order type $\bar{2} \cdot \varepsilon_0$ (see Proposition 8.28 in [17]).

The construction in the previous paragraph is made precise by using the known fact that the category of WQOs has arbitrary colimits and considering $T_\omega(\bar{k})$ as the colimit of the sequence $\{T_m(\bar{k})\}$.

3   Effective Wadge Hierarchy

Since the EWH is a special case of the FH, we first recall some information about the FH from [17,21]. We warn the reader that our definition of EWH uses set operations instead of the Wadge reducibility the reader could expect to see. The Wadge reducibility leads to complex degree structures in non-zero-dimensional spaces which hide the hierarchy (see [20,21] for detailed discussion).

By a *base in a set* X we mean a sequence $\mathcal{L}(X) = \{\mathcal{L}_n\}_{n<\omega}$ of subsets of $P(X)$ such that any $\mathcal{L}_n$ is closed under union and intersection, contains $\emptyset, X$, and $A \in \mathcal{L}_n$ implies that $A, X \setminus A \in \mathcal{L}_{n+1}$. With any base $\mathcal{L}(X)$ we associate some other bases as follows. For any $m < \omega$, let $\mathcal{L}^m(X) = \{\mathcal{L}_{m+n}(X)\}_n$; we call this base m-*shift of* $\mathcal{L}(X)$. For any $U \in \mathcal{L}_0$, let $\mathcal{L}(U) = \{\mathcal{L}_n(U)\}_{n<\omega}$ where $\mathcal{L}_n(U) = \{U \cap S \mid S \in \mathcal{L}_n(X)\}$; we call this base in U the U-*restriction of* $\mathcal{L}(X)$.

We define the FH not only of subsets of X but also of k-partitions $A : X \to \bar{k}$, $1 < k < \omega$. Note that 2-partitions of X are essentially subsets of X. For any finite tree $T \subseteq \omega^{<\omega}$ and any T-family $\{U_\tau\}$ of subsets of X, we define the T-family $\{\tilde{U}_\tau\}$ of subsets of X by $\tilde{U}_\tau = U_\tau \setminus \bigcup\{U_{\tau'} \mid \tau \sqsubset \tau' \in T\}$. The T-family $\{U_\tau\}$ is *monotone* if $U_\tau \supseteq U_{\tau'}$ for all $\tau \sqsubseteq \tau' \in T$. We associate with any T-family $\{U_\tau\}$ the monotone T-family $\{U'_\tau\}$ by $U'_\tau = \bigcup_{\tau' \sqsupseteq \tau} U_{\tau'}$. A T-family $\{V_\tau\}$ is *reduced* if it is monotone and satisfies $V_{\tau i} \cap V_{\tau j} = \emptyset$ for all $\tau i, \tau j \in T$. Obviously, for any reduced T-family $\{V_\tau\}$ the components $\tilde{V}_\tau$ are pairwise disjoint.

We will use the following technical notions. The first one is the notion "F is a T-family in $\mathcal{L}(X)$" defined by induction as follows: if $T \in T_0(\bar{k})$ then $F = \{X\}$; if $(T, t) \in T_{m+1}(\bar{k})$ then $F = (\{U_\tau\}, \{F_\tau\})$ where $\{U_\tau\}$ is a monotone T-family of $\mathcal{L}_0$-sets with $T_\varepsilon = X$ and, for each $\tau \in T$, F_τ is a $t(\tau)$-family in $\mathcal{L}^1(\tilde{U}_\tau)$. The version of this notion "F is a reduced T-family in $\mathcal{L}(X)$" is obtained by taking the reducible T-families in place of the monotone ones. The second is the notion "a T-family F in $\mathcal{L}(X)$ determines $A : X \to \bar{k}$" defined by induction as follows: if $T \in \mathcal{T}_0(\bar{k})$, $T = i < k$ (so $F = \{X\}$), then T determines the constant partition $A = \lambda x.i$; if $(T,t) \in \mathcal{T}_{m+1}(\bar{k})$ (so F is of the form $(\{U_\tau\}, \{F_\tau\})$) then T determines the k-partition A such that $A|_{\tilde{U}_\tau} = B_\tau$ for every $\tau \in T$, where $B_\tau : \tilde{U}_\tau \to \bar{k}$ is the k-partition of $\tilde{U}_\tau$ determined by F_τ.

As explained in [20], the T-family F that determines A provides a mind-change algorithm for computing $A(x)$ (see Section 3 of [21] for additional details). We are ready to give a precise definition of the FH of k-partitions over $\mathcal{L}(X)$.

Definition 1. *The FH of k-partitions over $\mathcal{L}(X)$ is the family $\{\mathcal{L}(X,T)\}_{T \in \mathcal{T}_\omega(\bar{k})}$ of subsets of k^X where $\mathcal{L}(X,T)$ is the set of $A : X \to \bar{k}$ determined by some T-family in $\mathcal{L}(X)$.*

As shown in [17], $T \leq_h S$ implies $\mathcal{L}(X,T) \subseteq \mathcal{L}(X,S)$, hence $(\{\mathcal{L}(X,T) \mid T \in \mathcal{T}_\omega(\bar{k})\}; \subseteq)$ is WQO. The FH of sets obtained from this construction for $k = 2$ is even semi-well-ordered since the quotient-poset of $(\mathcal{T}_2(\omega); \leq_h)$ has order type $\bar{2} \cdot \varepsilon_0$ (see Definition 8.27 and Proposition 8.28 in [17]).

The FH of k-partitions over the effective Borel base $\mathcal{L}(X) = \{\Sigma^0_{1+n}(X)\}$ in an effective cb$_0$-space X is written as $\{\Sigma(X,T)\}_{T \in \mathcal{T}_\omega(\bar{k})}$ and called the *effective Wadge hierarchy in X*. For $k = 2$ the structure of levels degenerate to the semi-well-ordered structure which enables the Σ, Π-notation for them. The EWH of sets subsumes many hierarchies including those mentioned in Sect. 2.

The corresponding boldface FH $\{\mathbf{\Sigma}(X,T)\}_{T \in \mathcal{T}_\omega(\bar{k})}$ over the finite Borel base $\mathcal{L}(X) = \{\mathbf{\Sigma}^0_{1+n}(X)\}$ is written as $\{\mathbf{\Sigma}(X,T)\}_{T \in \mathcal{T}_\omega(\bar{k})}$ and is called *finitary Wadge hierarchy in X*. It forms a small but important fragment of the whole (infinitary) Wadge hierarchy of k-partitions in X (which is constructed from the whole Borel hierarchy by taking countable well-founded trees T in place of the finite trees T). The latter hierarchy, which may be defined in arbitrary space, was introduced and studied in [20]. For the effective and boldface versions we have the obvious inclusions $\Sigma(X,T) \subseteq \mathbf{\Sigma}(X,T)$. In this paper we stick to levels of EWH corresponding to finite trees; the levels corresponding to computable well-founded trees (briefly discussed in [21]) are important on their on and we plan to investigate them in a separate publication. In [20,21] the following preservation property for levels of the introduced hierarchies was established.

Proposition 1. *Let $f : Y \to X$ be a computable effectively open surjection between effective cb$_0$-spaces. Then, for all $T \in \mathcal{T}_\omega(\bar{k})$ and $A \in k^X$, we have: $A \in \Sigma(X,T)$ iff $A \circ f \in \Sigma(Y,T)$. Similarly for the boldface versions and continuous open surjections between cb$_0$-spaces.*

4   Non-collapse Property

Here we establish some general facts about the non-collapse property. First we carefully define natural versions of this property.

We say that EWH $\{\Sigma(X,T)\}_{T\in\mathcal{T}_\omega(\bar{k})}$ *does not collapse at level* T if $\Sigma(X,T) \nsubseteq \Sigma(X,V)$ for each $V \in \mathcal{T}_\omega(\bar{k})$ with $T \nleq_h V$; it *strongly does not collapse at level* T if $\Sigma(X,T) \nsubseteq \bigcup\{\Sigma(X,V) \mid V \in \mathcal{T}_\omega(\bar{k}), T \nleq_h V\}$. We say that $\{\Sigma(X,T)\}_{T\in\mathcal{T}_\omega(\bar{k})}$ *(strongly) does not collapse* if it (strongly) does not collapse at any level $T \in \mathcal{T}_\omega(\bar{k})$. The latter non-strong version is equivalent to saying that the quotient-poset of $(\mathcal{T}_\omega(\bar{k}); \leq_h)$ is isomorphic to $(\{\Sigma(X,T) \mid T \in \mathcal{T}_\omega(\bar{k})\}; \subseteq)$.

Note that for the case of sets $k = 2$ these definitions are equivalent to the standard definition of non-collapse in DST (Σ-levels are distinct from the corresponding Π-levels), and the strong version is equivalent to the non-strong one.

The non-collapse for the boldface versions are defined in the same way. In the effective case, there are also the following uniform versions of non-collapse property which relate EWH to the corresponding WH. The EWH $\{\Sigma(X,T)\}$ *uniformly does not collapse at level* T if $\Sigma(X,T) \nsubseteq \boldsymbol{\Sigma}(X,V)$ for each $V \in \mathcal{T}_\omega(\bar{k})$ with $T \nleq_h V$. It *strongly uniformly does not collapse at level* T if $\Sigma(X,T) \nsubseteq \bigcup\{\boldsymbol{\Sigma}(X,V) \mid V \in \mathcal{T}_\omega(\bar{k}), T \nleq_h V\}$. It *strongly uniformly does not collapse* if $\Sigma(X,T) \nsubseteq \bigcup\{\boldsymbol{\Sigma}(X,V) \mid V \in \mathcal{T}_\omega(\bar{k}), T \nleq_h V\}$ for all $T \in \mathcal{T}_\omega(\bar{k})$.

The next assertion follows from the inclusions between levels.

Proposition 2. *For any effective cb$_0$-space X we have: if $\{\Sigma(X,T)\}$ (strongly) uniformly does not collapse (at level T) then both $\{\Sigma(X,T)\}$ and $\{\boldsymbol{\Sigma}(X,T)\}$ (strongly) do not collapse (at level T).*

For cb$_0$-spaces X and Y, let $X \leq_{co} Y$ mean that there is a continuous open surjection f from Y onto X. For effective cb$_0$-spaces X and Y, let $X \leq_{eco} Y$ mean that there is a computable effectively open surjection f from Y onto X. Clearly, both $\leq_{eco}$ and $\leq_{co}$ are preorders, and the first preorder is contained in the second. The non-collapse property is inherited w.r.t. these preorders:

Proposition 3. *1. If $X \leq_{co} Y$ and $\{\boldsymbol{\Sigma}(X,T)\}_{T\in\mathcal{T}_\omega(\bar{k})}$ (strongly) does not collapse (at level T) then $\{\boldsymbol{\Sigma}(Y,T)\}$ (strongly) does not collapse (at level T). The same holds for the infinitary version of WH in X.*

2. If $X \leq_{eco} Y$ and $\{\Sigma(X,T)\}_{T\in\mathcal{T}_\omega(\bar{k})}$ (strongly) does not collapse (at level T) then $\{\Sigma(Y,T)\}$ (strongly) does not collapse (at level T). The same holds for the uniform version of non-collapse property.

Proof. All assertions follow from the definitions and the preservation property, so consider only the finitary version in item (1). Let $X \leq_{co} Y$ via $f : Y \to X$, and $\{\boldsymbol{\Sigma}(X,T)\}$ does not collapse at level T. We have to show that $\boldsymbol{\Sigma}(Y,T) \nsubseteq \boldsymbol{\Sigma}(Y,V)$ for any fixed $V \in \mathcal{T}_\omega(\bar{k})$ with $T \nleq_h V$. Choose $A \in \boldsymbol{\Sigma}(X,T) \setminus \boldsymbol{\Sigma}(X,V)$. By Proposition 1 we get $A \circ f \in \boldsymbol{\Sigma}(Y,T) \setminus \boldsymbol{\Sigma}(Y,V)$. $\qquad\square$

Corollary 1. *1. If X is quasi-Polish and $\{\boldsymbol{\Sigma}(X,T)\}_{T\in\mathcal{T}_\omega(\bar{k})}$ (strongly) does not collapse (at level T) then $\{\boldsymbol{\Sigma}(\mathcal{N},T)\}$ (strongly) does not collapse (at level T). The same holds for the infinitary version of the WH.*

2. If X is computable quasi-Polish and $\{\Sigma(X,T)\}_{T\in\mathcal{T}_\omega(\bar{k})}$ (strongly) does not collapse (at level T) then $\{\Sigma(\mathcal{N},T)\}$ (strongly) does not collapse (at level T). The same holds for the uniform version.

3. If X is the product of a sequence $\{X_n\}$ of nonempty cb_0-spaces, and the finitary WH $\{\mathbf{\Sigma}(X_n, T)\}_{T \in \mathcal{T}_\omega(\bar{k})}$ (strongly) does not collapse (at level T) for some $n < \omega$, then $\{\mathbf{\Sigma}(X, T)\}$ (strongly) does not collapse (at level T). The same holds for the infinitary version of the WH.

4. If X is the product of a uniform sequence $\{X_n\}$ of nonempty effective cb_0-spaces, and $\{\Sigma(X_n, T)\}_{T \in \mathcal{T}_\omega(\bar{k})}$ (strongly) does not collapse (at level T) for some $n < \omega$, then $\{\Sigma(X, T)\}$ (strongly) does not collapse (at level T). The same holds for the uniform version.

Proof. (1) Follows from Proposition 3(1) since X is quasi-Polish iff $X \leq_{co} \mathcal{N}$.

(2) Follows from Proposition 3(2) since X is computable quasi-Polish iff $X \leq_{eco} \mathcal{N}$.

(3) Follows from Proposition 3(1) since $X_n \leq_{co} X$.

(4) Follows from Proposition 3(2) since $X_n \leq_{eco} X$.

$\square$

Although the assertion (1) is void (because the infinitary WH in $\mathcal{N}$ strongly does not collapse [10,22]), it is of some methodological interest because it shows that proving the non-collapse of WH in any quasi-Polish space is at least as complicated as proving it in $\mathcal{N}$, and the proof of the latter fact is highly non-trivial. The same applies to item (2) but this assertion is non-void because the non-collapse of EWH in $\mathcal{N}$ was open until this paper, to my knowledge. In the next section we give prominent examples of spaces with the non-collapse property. A good strategy to obtain broad classes of such spaces is to make them as low as possible w.r.t. $\leq_{co}, \leq_{eco}$, and use the preservation property.

5   Some Examples

In this section we illustrate the method of Proposition 3 by proving the non-collapse of EWH and WH in some concrete spaces.

First we consider the domain $\omega^{\leq \omega}$. For this space, the result is obtained by some observations and additions to the proofs in [10], so let us recall some information from that paper. Let $\hat{\omega} = \omega \cup \{\mathrm{p}\}$ be obtained from ω by adjoining a new element p; we endow $\hat{\omega}$ with the discrete topology. For $x \in \hat{\omega}^\omega$, let $\delta(x) \in \omega^{\leq \omega}$ be obtained from x by deleting all entries of p. A function $f : \hat{\omega}^\omega \to \hat{\omega}^\omega$ (resp. $A : \hat{\omega}^\omega \to \bar{k}$) is *conciliating* if $\delta \circ f = f^* \circ \delta$ (resp. $A = A^* \circ \delta$) for some (unique) $f^* : \omega^{\leq \omega} \to \omega^{\leq \omega}$ (resp. $A^* : \omega^{\leq \omega} \to \bar{k}$). The function f is *initializable* if, for every $\tau \in \hat{\omega}^{<\omega}$, there is a continuous function $h_\tau : \hat{\omega}^\omega \to \hat{\omega}^\omega$ such that $\delta(f(x)) = \delta(f(\tau h_\tau(x)))$ for all $x \in \hat{\omega}^\omega$. In Proposition 2.15 from [10], an initializable $\mathbf{\Sigma}_2^0$-measurable conciliating function $\mathcal{U} : \hat{\omega}^\omega \to \hat{\omega}^\omega$ was constructed which is universal in the sense that for every $\mathbf{\Sigma}_2^0$-measurable conciliating function $\mathcal{V} : \hat{\omega}^\omega \to \hat{\omega}^\omega$ there is a continuous function $h : \hat{\omega}^\omega \to \hat{\omega}^\omega$ such that $\delta \circ \mathcal{V} = \delta \circ \mathcal{U} \circ h$.

Theorem 1. *1. The WH $\{\mathbf{\Sigma}(\omega^{\leq\omega}, T)\}_{T \in \mathcal{T}_\omega(\bar{k})}$ strongly does not collapse. Similarly for the infinitary WH.*

2. The EWH $\{\Sigma(\omega^{\leq\omega}, T)\}_{T \in \mathcal{T}_\omega(\bar{k})}$ strongly uniformly does not collapse.

Proof. (1) For notation simplicity, we only consider the finitary case, in the infinitary case the argument is the same. The Definition 3.1.4 in [10], using the induction on trees and the universal function $\mathcal{U}$, associates with any tree T a conciliatory $\Omega_T : \hat{\omega}^\omega \to \bar{k}$. By the results in Sect. 3.3 of [10], Ω_T is in $\Sigma(\hat{\omega}^\omega, T) \setminus \bigcup \{\Sigma(\hat{\omega}^\omega, V) \mid V \in \mathcal{T}_\omega(\bar{k}), \; T \not\leq_h V\}$. As the function δ is a continuous open surjection and $\Omega_T = \Omega_T^* \circ \delta$, we get $\Omega_T^* \in \Sigma(\omega^{\leq\omega}, T) \setminus \bigcup \{\Sigma(\omega^{\leq\omega}, V) \mid V \in \mathcal{T}_\omega(\bar{k}), \; T \not\leq_h V\}$ by Proposition 1, completing the proof.

(2) Inspecting the proof of Proposition 2.15 (resp. Lemma 2.11) in [10] shows that $\mathcal{U}$ and $\mathcal{U}^*$ are in fact Σ_2^0-measurable. Inspecting the proof of Lemma 2.15 in [10] shows that Ω_T is in $\Sigma(\hat{\omega}^\omega, T)$. Clearly, δ is a computable effectively open surjection. Thus, Ω_T^* is in $\Sigma(\omega^{\leq\omega}, T) \setminus \bigcup \{\Sigma(\omega^{\leq\omega}, V) \mid V \in \mathcal{T}_\omega(\bar{k}), \; T \not\leq_h V\}$ by Proposition 1, completing the proof. □

This theorem and Proposition 1 imply some new information on the EWH in Baire and Cantor spaces:

Theorem 2. *The EWHs $\{\Sigma(\mathcal{N}, T)\}$ and $\{\Sigma(\mathcal{C}, T)\}_{T \in \mathcal{T}_\omega(\bar{k})}$ strongly uniformly do not collapse.*

Proof. As δ is a computable effectively open surjection, $\{\Sigma(\hat{\omega}^\omega, T)\}$ strong uniformly does not collapse by Theorem 1 and Proposition 3. As $\hat{\omega}^\omega$ is effectively homeomorpic to $\mathcal{N}$, the first assertion follows.

For the second assertion, consider the domain $n^{\leq\omega}$, $n \geq 2$, in place of $\omega^{\leq\omega}$. A slight modification of the notions from the beginning of this section apply to $n^{\leq\omega}$. Also, a slight modification of the proof of Theorem 1 shows that it remains true for $n^{\leq\omega}$. As in the previous paragraph, $\{\Sigma(\hat{n}^\omega, T)\}$ strong uniformly does not collapse. Since $\hat{n}^\omega$ is effectively homeomorpic to $(n+1)^\omega$, and thus to $\mathcal{C}$, this implies the second assertion. Note that the spaces $n^{\leq\omega}$ for distinct n are not homeomorphic. □

Next we discuss the EWH in $\mathbb{N}$. Since $\mathbb{N}$ is discrete, the WH $\{\Sigma(\mathbb{N}, T)\}$ collapses to very low levels (it has finitely many distinct levels), so the next result cannot be improved to the strong uniform version.

Theorem 3. *The EWH $\{\Sigma(\mathbb{N}, T)\}_{T \in \mathcal{T}_\omega(\bar{k})}$ strongly does not collapse.*

Proof sketch. Let G be the ternary operation on k^ω introduced in [15]. For $i < k$, let $\mathbf{i}$ be the constant function $\lambda n.i \in k^\omega$. Let $\mathbf{0}^{(0)} = \mathbf{0}$ and $\mathbf{0}^{(n+1)} = G(\mathbf{0}, \mathbf{1}, \mathbf{0}^{(n)})$, then $\{\mathbf{0}^{(n)}\}$ coincides (up to computable isomorphism) with the usual sequence of iterations of Turing jump starting with $\mathbf{0}$. For any $n < \omega$ we define the binary operation $\cdot^n$ on k^ω by $\nu \cdot^n \mu = G(\mu, \nu, \mathbf{0}^{(n)})$. We also define the family $\{\mathbf{f}_m^n\}_{n<\omega}$ of functions from $\mathcal{T}_m(\bar{k})$ to k^ω by induction on m as follows. Let $\mathbf{f}_0^n(i) = \mathbf{i}$ for all $i < k, n < \omega$. It remains to define $\mathbf{f}_{m+1}^n$ from $\mathbf{f}_m^{n+1}$. Let $(T, t) \in \mathcal{T}_k(m+1) = \mathcal{T}_{\mathcal{T}_k(m)}$. If T is singleton we set $\mathbf{f}_{m+1}^n(T) = \mathbf{f}_m^{n+1}(t(\varepsilon))$, otherwise we set $\mathbf{f}_{m+1}^n(T) = \mathbf{f}_m^{n+1}(t(\varepsilon)) \cdot^n (\bigoplus \{\mathbf{f}_{m+1}^n(T(i)) \mid i \in T\})$ (using induction on the rank of T) where $\bigoplus$ is the finitary join operation on k^ω.

Using algebraic properties of G established in [15] and some additional similar facts, it may be shown that for all $T, V \in \mathcal{T}_k(m)$ we have: $T \leq_h V$ iff $\mathbf{f}_m^n(T) \leq$

$\mathbf{f}_m^n(V)$, where $\leq$ is the reducibility of numberings in k^ω (i.e., $\mu \leq \nu$ iff $\mu = \nu \circ g$, for some computable function g on ω). Using the representation of $\mathcal{T}_\omega(\bar{k})$ as the colimit of the sequence of preorders $\{\mathcal{T}_m(\bar{k})\}$, the sequence $\{\mathbf{f}_m^0\}$ induces a function $\mathbf{f} : \mathcal{T}_\omega(\bar{k}) \to k^\omega$ such that $T \leq_h V$ iff $\mathbf{f}(T) \leq \mathbf{f}(V)$, for all $T, V \in \mathcal{T}_\omega(\bar{k})$.

Finally, from Definition 1 by induction on trees we deduce that, for any $T \in \mathcal{T}_\omega(\bar{k})$, we have: $\Sigma(\mathbb{N}, T) = \{B \in k^\omega \mid B \leq \mathbf{f}(T)\}$, i.e. $\mathbf{f}(T)$ is complete in $\Sigma(\mathbb{N}, T)$ w.r.t. the reducibility of numberings. Thus, $\mathbf{f}(T) \in \Sigma(\mathbb{N}, T) \setminus \bigcup\{\Sigma(\mathbb{N}, V) \mid V \in \mathcal{T}_\omega(\bar{k}), T \not\leq_h V\}$, completing the proof. $\qquad\square$

Theorem 3 implies non-collapse of EWH in many spaces:

Corollary 2. *Let $X = X_0 \sqcup X_1 \sqcup \cdots$ be the disjoint union of a uniform sequence $\{X_n\}$ of nonempty effective cb_0-spaces. Then the EWH $\{\Sigma(X, T)\}$ strongly does not collapse.*

Proof. For $x \in X$, let $g(x)$ be the unique number n with $x \in X_n$. Then $g : X \to \mathbb{N}$ is a computable effectively open surjection. By Theorem 3 and Proposition 1, $\{\Sigma(X, T)\}$ strongly does not collapse. $\qquad\square$

In particular, Corollary 2 applies to $\mathcal{N} \simeq \mathcal{N} \sqcup \mathcal{N} \sqcup \cdots$ which gives another proof of the fact that $\{\Sigma(\mathcal{N}, T)\}$ strongly does not collapse. But properties of witnesses for strong non-collapse given by Theorem 2 and Corollary 2 are quite different: the first ones have typically high topological complexity while the second ones are clopen (a k-partition A is clopen if $A^{-1}(i)$ is clopen for each $i < k$).

However, Corollary 2 still does not apply to many spaces, e.g. to $\mathcal{C}$ or $P\omega$ (because these spaces are compact), and to the intervals of $\mathbb{R}$ (because they are connected).

The EWH strongly does not collapse also in $P\omega$ and $\mathbb{R}$ but the proofs of this (which are not given in this paper) require additional modifications of our method. We hope that suitable modifications apply to many other natural spaces.

References

1. Abramsky S., Jung, A.: Domain theory. In: Handbook of Logic in Computer Science, vol. 3, pp. 1–168, Oxford (1994)
2. Becher, V., Grigorieff, S.: Borel and Hausdorff hierarchies in topological spaces of Choquet games and their effectivization. Math. Struct. Comput. Sci. **25**(7), 1490–1519 (2015)
3. Brattka, V.: Effective Borel measurability and reducibility of functions. Math. Logic Q. **51**(1), 19–44 (2005)
4. de Brecht, M.: Quasi-Polish spaces. Ann. Pure Appl. Logic **164**, 356–381 (2013)
5. de Brecht, M., Pauly, A., Schröder, M.: Overt choice. Computability **9**(3–4), 169–191 (2020)
6. Hemmerling, A.: The Hausdorff-Ershov hierarchy in Euclidean spaces. Arch. Math. Logic **45**, 323–350 (2006)
7. Hoyrup, M., Rojas, C., Selivanov, V., Stull, D.M.: Computability on Quasi-Polish spaces. In: Hospodár, M., Jirásková, G., Konstantinidis, S. (eds.) DCFS 2019. LNCS, vol. 11612, pp. 171–183. Springer, Cham (2019). https://doi.org/10.1007/978-3-030-23247-4_13

8. Kechris, A.S.: Classical Descriptive Set Theory. GTM, vol. 156. Springer, New York (1995). https://doi.org/10.1007/978-1-4612-4190-4

9. Kihara, T., Montalbán, A.: The uniform Martin's conjecture for many-one degrees. Trans. Am. Math. Soc. **370**(12), 9025–9044 (2018)

10. Kihara, T., Montalbán, A.: On the structure of the Wadge degrees of BQO-valued Borel functions. Trans. Am. Math. Soc. **371**(11), 7885–7923 (2019)

11. Louveau, A.: Recursivity and compactness. In: Müller, G.H., Scott, D.S. (eds.) Higher Set Theory. LNM, vol. 669, pp. 303–337. Springer, Heidelberg (1978). https://doi.org/10.1007/BFb0103106

12. Moschovakis, Y.N.: Descriptive Set Theory. North Holland, Amsterdam (2009)

13. Pequignot, Y.: A Wadge hierarchy for second countable spaces. Arch. Math. Logic **54**(5–6), 659–683 (2015). https://doi.org/10.1007/s00153-015-0434-y

14. Rogers Jr, H.: Theory of Recursive Functions and Effective Computability. McGraw-Hill, New York (1967)

15. Selivanov, V.L.: Hierarchies of hyperarithmetical sets and functions. Algebra Logic **22**, 473–491 (1983). https://doi.org/10.1007/BF01978879

16. Selivanov, V.L.: Fine hierarchies and m-reducibilities in theoretical computer science. Theor. Comput. Sci. **405**, 116–163 (2008)

17. Selivanov, V.L.: Fine hierarchies via Priestley duality. Ann. Pure Appl. Logic **163**, 1075–1107 (2012)

18. Selivanov, V.: Towards the effective descriptive set theory. In: Beckmann, A., Mitrana, V., Soskova, M. (eds.) CiE 2015. LNCS, vol. 9136, pp. 324–333. Springer, Cham (2015). https://doi.org/10.1007/978-3-319-20028-6_33

19. Selivanov, V.L.: Towards a descriptive theory of cb0-spaces. Mathematical Structures in Computer Science, vol. 28, no. 8, pp. 1553–1580 (2017). arXiv:1406.3942v1 [Math.GN], 16 June 2014

20. Selivanov, V.: A Q-Wadge hierarchy in quasi-Polish spaces. J. Symbolic Logic (2019). https://doi.org/10.1017/jsl.2020.52

21. Selivanov, V.L.: Effective Wadge hierarchy in computable quasi-Polish spaces. Siberian Electron. Math. Rep. **18**(1), 121–135 (2021). https://doi.org/10.33048/semi.2021.18.010. arxiv:1910.13220v2

22. Wadge, W.: Reducibility and determinateness in the Baire space. Ph.D. thesis, University of California, Berkely (1984)

23. Weihrauch, K.: Computable Analysis. TTCSAES. Springer, Heidelberg (2000). https://doi.org/10.1007/978-3-642-56999-9

Effective Inseparability
and Its Applications

Andrea Sorbi[(✉)] [iD]

Department of Information Engineering and Mathematics,
University of Siena, 53100 Siena, Italy
sorbi@unisi.it
http://www3.diism.unisi.it/sorbi/

Abstract. We survey some recent applications of the classical notion of effective inseparability to computably enumerable structures, formal systems and lattices of sentences.

Keywords: Effective inseparability · Lattices of sentences · Computably enumerable structures

The notion of effective inseparability for pairs of disjoint sets of natural numbers is due to Smullyan [14].

Definition 1. *A pair (A, B) of sets of natural numbers is said to be* effectively inseparable *(or, simply,* e.i.*) if the pair is* disjoint, *i.e.* $A \cap B = \emptyset$, *and there exists a partial computable function* $\psi(u, v)$ *(called a* productive function *for the pair) such that*

$$(\forall u, v)[A \subseteq W_u \,\&\, B \subseteq W_v \,\&\, W_u \cap W_v = \emptyset \Rightarrow \psi(u, v) \downarrow \,\&\, \psi(u, v) \notin W_u \cup W_v].$$

When applied to computably enumerable (or, simply, c.e.) sets, this notion provides a natural generalization to disjoint pairs of the notion of creativeness for a single set. In particular: each half of an e.i. pair of c.e. sets is creative; if (A, B) is an e.i. pair of c.e. sets then $(C, D) \leq_1 (A, B)$ for every disjoint pair of c.e. sets (C, D), i.e. there exists a 1-1 computable function f which simultaneously 1-reduces $C \leq_1 A$ and $D \leq_1 B$ (this shows that 1-completeness of creative sets generalizes to e.i. pairs of c.e. sets); every two pairs of e.i. pairs of c.e. sets are computably isomorphic (hence the Myhill Isomorphism Theorem for creative sets generalizes to e.i. pairs of c.e. sets). The proofs of these theorems rely on beautiful applications of the Recursion Theorem, in the form (due to Smullyan) of the Double Recursion Theorem.

Effective inseparability was exploited by Smullyan to shed new insights on the Gödel incompleteness phenomenon of formal systems. If T is even a very

Partially supported by PRIN 2017 Grant "Mathematical Logic: models, sets, computability". Sorbi is a member of INDAM.

L. De Mol et al. (Eds.): CiE 2021, LNCS 12813, pp. 417–423, 2021.
https://doi.org/10.1007/978-3-030-80049-9_41

weak c.e. consistent system of arithmetic such as Robinson's R or Q (see [15]), then one can show that the pair $(\mathsf{Thm_T}, \mathsf{Ref_T})$ is an e.i. pair of c.e. sets, where $\mathsf{Thm_T}$ consists of the (Gödel numbers of the) sentences which are theorems of T, and $\mathsf{Ref_T}$ consists of the (Gödel numbers of the) refutable sentences, i.e. the sentences which are negations of theorems of T. This implies that T is essentially undecidable, i.e. every consistent c.e. extension of T is undecidable, since disjoint pairs extending e.i. pairs are e.i., and thus consist of undecidable sets.

Effective inseparability and Smullyan's results for pairs were subsequently generalized by Cleave [4] from pairs to c.e. sequences of mutually disjoint sets computably listed without repetitions. More recently the notion of effective inseparability has been applied to uniformly c.e. sequences of sets which provide partitions of the set ω of natural numbers, or, equivalently, to computably enumerable equivalence relations (called also *ceers*) on ω. Namely, an equivalence relation R on ω is *uniformly effectively inseparable* (abbreviated as u.e.i.) if it is *nontrivial* (i.e. it has more than one equivalence class) and there is a computable function $f(x, y)$ such that if $x \not\!R y$ then $\varphi_{f(x,y)}$ is a productive function witnessing that the pair of equivalence classes $([x]_R, [y]_R)$ is effectively inseparable. Moreover, if R, S are equivalence relations on ω let us say that R is *computably reducible* to S (in symbols $R \leq_c S$) if there exists a computable function f such that $x \ R \ y$ if and only if $f(x) \ S \ f(y)$, for all x, y. In analogy with the 1-completeness results of Smullyan and Cleave, the following theorem holds.

Theorem 1 ([1]). *Every u.e.i. ceer R is universal, i.e. $S \leq_c R$ for every ceer S, with reduction provided by a 1-1 computable function.*

There are nice and popular properties of equivalence relations that imply u.e.i.-ness. The most useful one is given by the following definition (Shavrukov [12], after Montagna [7]).

Definition 2. *An equivalence relation S on ω is* uniformly finitely precomplete *(abbreviated as u.f.p.) if it is nontrivial and there exists a computable function of three variables $f(D, e, x)$ (where D is a finite set given by its canonical index) such that*

$$(\forall D, e, x)[\varphi_e(x) \downarrow \ \& \ (\exists y)[y \in D \ \& \ \varphi_e(x) \ S \ y] \Rightarrow \varphi_e(x) \ S \ f(D, e, x)]. \quad (1)$$

It is not difficult to see:

Lemma 1. *If S is u.f.p. then S is u.e.i..*

By Theorem 1 our generalization to ceers of the notion of effective inseparability preserves the classical 1-completeness result, but interestingly it does not preserve the Myhill Isomorphism Theorem, as there exist u.e.i. (even u.f.p.) ceers which are not computably isomorphic. To see this, an important example of a u.f.p. ceer is given by the relation $\leftrightarrow_T$ of provable equivalence (i.e. $x \leftrightarrow_T y$ if $\mathsf{T} \vdash x \leftrightarrow y$), where T is any c.e. consistent extension of R or Q ([7,8]). To provide examples of u.f.p. ceers which are not computably isomorphic with $\leftrightarrow_T$, let us first recall a notion from the theory of numberings [6]: an equivalence relation S on ω is *precomplete* if S is nontrivial and there is a computable function of two variables $f(e, x)$ such that if $\varphi_e(x) \downarrow$ then $\varphi_e(x) \, S \, f(e, x)$. Several examples of precomplete ceers are given in [16]: for instance, for every $n \geq 1$, the relation $\leftrightarrow_{\mathsf{PA},n}$ of provable equivalence in Peano Arithmetic PA restricted to the Σ_n sentences is a precomplete ceer. As clearly precompleteness implies u.f.p.-ness, we have that every precomplete ceer is u.f.p.. But $\leftrightarrow_\mathsf{T}$ cannot be computably isomorphic with any precomplete ceer. To show this, notice that the function induced by the connective $\neg$ provides a *computable diagonal* function for $\leftrightarrow_\mathsf{T}$ (i.e. a computable function d such that $x \not\leftrightarrow_\mathsf{T} d(x)$ for every x), whereas no precomplete equivalence relation R admits a computable diagonal function by the Ershov Fixed Point Theorem [6] stating that for every computable function f there exists a number e such that $e \, R \, f(e)$.

1   Does u.e.i. Imply u.f.p.?

We have observed that all u.f.p. equivalence relations are u.e.i.. Is the converse true, or is it at least true for ceers?

Question 1. *Does u.e.i.-ness coincide with u.f.p.-ness for ceers?*

It is known in this regard that u.e.i.-ness for ceers is equivalent ([1]) to a seemingly weaker version of u.f.p.-ness: namely, being u.e.i. is equivalent for ceers to being weakly u.f.p., where an equivalence relation S is said to be *weakly u.f.p.* if S is nontrivial and there exists a computable function $f(D, e, x)$ for which (1) is required to hold only if the elements of D are pairwise non-S-equivalent. This is perhaps evidence for a negative answer to Question 1.

However, if more structure is added to a ceer R, then there are cases in which u.e.i.-ness implies u.f.p-ness. In the following we agree that a *computably enumerable structure* (or, simply, a *c.e. structure*) A is a nontrivial algebraic-relational structure for which there exists a *c.e. presentation*, i.e. a structure A_ω of the same type as A but with universe ω and possessing uniformly computable operations, uniformly c.e. relations, and a ceer $=_A$ which is a congruence on A_ω such that A is isomorphic with A_ω divided by $=_A$. For more on c.e. structures see [11]. When talking about a c.e. structure A with some property, in the following we intend in fact a c.e. presentation of it with that property.

With the exception of Corollary 1, the results of this section (coming from [2]) are proved by the Recursion Theorem, or suitable generalized versions of it.

Lemma 2. *Let A be a c.e. algebra whose type contains two binary operations $+, \cdot$, and two constants (presented by the numbers) $0, 1$ such that $+$ is associative, the pair of sets $0_A = \{x : x =_A 0\}$, and $1_A = \{x : x =_A 1\}$ is e.i., and, for every a,*

$$a + 0 =_A a, \qquad a \cdot 0 =_A 0, \qquad a \cdot 1 =_A a.$$

Then $=_A$ is a u.f.p. ceer.

The usefulness of this lemma consists in the fact that from effective inseparability of just the pair $(0_A, 1_A)$ one can infer that $=_A$ is not only u.e.i., but
even u.f.p.. As a first application, we look at c.e. lattices. If L is a c.e. lattice
then its preordering relation $\leq_L$ is a c.e. preordering relation on ω, and $a =_L b$
if and only if $a \leq_L b$ and $b \leq_L a$.

Definition 3. *A c.e. lattice L is said to be* effectively inseparable *(or simply
e.i.) if L is bounded, with, say, the numbers 0 and 1 presenting the least element
and the greatest element, respectively, and the pair of sets $(0_L, 1_L)$ is e.i.. Let us
also say that a c.e. lattice L is u.e.i. or u.f.p. if so is the ceer $=_L$.*

By Lemma 2 we have:

Theorem 2. *If L is an e.i. c.e. lattice then L is u.f.p..*

Given two preordering relations R, S on ω, one says that R is *computably
reducible* to S (in symbols, $R \leq_c S$) if, as for equivalence relations, there exists a
computable function f such that $x \mathrel{R} y$ if and only if $f(x) \mathrel{S} f(y)$, for all x, y. A
c.e preorder R is *universal*, if $S \leq_c R$, for every c.e. preorder S. A c.e preorder R
is *locally universal*, if R is nontrivial and for every pair a, b of natural numbers
such that $a \mathrel{R} b$ but $b \mathrel{\bar{R}} a$, and every c.e. preorder S, we have that $S \leq_c R$ via
some reducing function whose range is contained in the interval $\{x : a \mathrel{R} x \mathrel{R} b\}$.

Theorem 3. *If L is a u.f.p. c.e. lattice then the associated c.e. pre-ordering
relation $\leq_L$ is locally universal.*

If L is a c.e. lattice such that its preordering relation is locally universal then
we say that L is *locally universal*. A close look at the proofs of the previous two
theorems, and their uniformity, enables us to show that for ceers of the form $=_L$,
where L is a c.e. lattice, Question 1 can be positively answered:

Corollary 1. *Il L is a u.e.i. c.e. lattice then L is u.f.p.. In other words, for any
c.e. lattice L, $=_L$ is u.e.i. if and only if $=_L$ is u.f.p..*

2 Applications to Lattices of Sentences

The following theorem from [2] is an interesting application of u.f.p.-ness. Its
proof is again by the Recursion Theorem.

Theorem 4. *If L is a u.f.p. c.e. lattice then the associated pre-ordering relation
$\leq_L$ is* uniformly dense, *i.e. there exists a computable function f such that for
every a, b if $a <_L b$ then $a <_L f(a, b) <_L b$, and if $a =_L a'$ and $b =_L b'$ then
$f(a, b) =_L f(a', b')$.*

A c.e. lattice L is called *uniformly dense* if so is $\leq_L$. Uniform density for lattices
of sentences of formal theories, where the preordering relation is presented by
provable implication, has been studied in [13]. Given a formal system T, suppose
that $\mathcal{C}$ is any c.e. set of sentences such that (via coding of sentences as numbers)

the set $L_{\mathcal{C},\mathsf{T}}$ (that is, $\mathcal{C}$ modulo provable equivalence in T) is a c.e. bounded lattice with its operations presented by the propositional connectives, and its preordering relation presented by provable implication. In view of Lemma 2, Theorem 2, Theorem 3 and Theorem 4, the following results can be proved by just showing that the pair $(0_{L_{\mathcal{C},\mathsf{T}}}, 1_{L_{\mathcal{C},\mathsf{T}}})$ is e.i.:

(1) If T is any c.e. consistent extension of Buss' weak arithmetical system S_2^1 (see [3]) then ([10]) the c.e. lattice $L_{\exists\Sigma_1^b/\mathsf{T}}$ of $\exists\Sigma_1^b$ sentences modulo provable equivalence in T, is uniformly dense (this answers a question in [13]), and locally universal. We recall that Σ_1^b is the smallest class of formulas containing the formulas in which all possibly existing quantifiers are sharply bounded, i.e. bounded by the length of a term, and is closed under sharply bounded quantification, the connectives $\vee, \wedge$ and bounded existential quantification. Then $\exists\Sigma_1^b$ is comprised of the formulas which arise from allowing a single unbounded existential quantifier over a Σ_1^b formula.

(2) If iT is a c.e. consistent intuitionistic extension of iR or iQ (the intuitionistic versions of Robinson's systems R or Q), and $\mathcal{C}$ is any c.e. set of sentences closed under $\wedge, \vee$ and containing the $\exists\Delta_0$ sentences (in fact one can isolate much smaller classes than $\exists\Delta_0$ which suffice for the claim) then the c.e. lattice $L_{\mathcal{C}/\mathsf{T}}$ is locally universal and uniformly dense, [10]. It also follows that if T is any c.e. (classical) consistent extension of R or Q, and $\mathcal{C}$ is as above, then the c.e. lattice $L_{\mathcal{C}/\mathsf{T}}$ is locally universal and uniformly dense (for this last observation on classical theories, see also [2]). Taking $\mathcal{C}$ to be the class of all sentences, this includes that if iT is a c.e. consistent intuitionistic extension of iR, or iQ, then the c.e. Lindenbaum Heyting algebra L_T is locally universal, and uniformly dense (for this observation see also [2]).

3   Diagonal Functions

A positive answer to Question 1 would be relevant also to the study of the complexity of word problems of c.e. structures. The *word problem* of a c.e. presentation of A is just the ceer $=_A$. For instance it is known [7] that all ceers R which are u.f.p. and possess a computable diagonal function are computably isomorphic with $\leftrightarrow_\mathsf{T}$, where T is any c.e. consistent extension of R or Q. We recall from the informal remarks following Lemma 1 that a computable diagonal function for an equivalence relation R on ω is a computable function d such that $x\cancel{R}d(x)$, for every $x \in \omega$. Therefore the word problem of every e.i. Boolean algebra (according to Definition 3) is computably isomorphic with $\leftrightarrow_\mathsf{T}$. This follows from the fact that an e.i. Boolean algebra is u.f.p. by Theorem 2, and obviously has a computable diagonal function, for instance the one that maps x to its complement; or it follows from the stronger result [8] that the e.i. Boolean algebras are computably isomorphic with each other, and thus with the c.e. Lindenbaum algebra of T. However the e.i. Boolean algebras do not cover all the cases of c.e. structures having word problem computably isomorphic with $\leftrightarrow_\mathsf{T}$. For instance, a non-commutative c.e ring A satisfying the hypotheses of Lemma 2 has been built in [5]. This implies that the word problem $=_A$ of A is u.f.p., which in turn implies that $=_A$ is computably isomorphic to $\leftrightarrow_T$, as clearly $=_A$ has a computable diagonal function, for instance the function $d(a) = a + v$ for any fixed $v \neq_A 0$. On the other hand it is known [9] that there is a finitely presented group G whose word problem $=_G$ is u.e.i.. Clearly $=_G$ has a computable diagonal function: again, take $d(a) = ab^{-1}$ for any fixed $b \neq_G 1$. So, should u.e.i.-ness coincide with u.f.p.-ness for ceers or at least for finitely presented groups, it would automatically follow that the group built in [9] has a word problem which is computably isomorphic to $\leftrightarrow_T$. The existence of such a finitely presented group is however still an open problem, since Lemma 2 does not apply to groups, and at the same time it is still open whether for ceers u.e.i.-ness plus the existence of a computable diagonal function implies u.f.p.-ness. A small approximation to this is given by the following observation (Andrews and Sorbi, unpublished), where a *computable strong diagonal function* for an equivalence relation R on ω is a computable function d such that for every (canonical index of a) finite set D, $d(D)$ outputs a number such that $d(D)\not\!R x$, for every $x \in D$:

Fact 1. *Every u.e.i. ceer with a computable strong diagonal function is u.f.p..*

References

1. Andrews, U., Lempp, S., Miller, J.S., Ng, K.M., San Mauro, L., Sorbi, A.: Universal computably enumerable equivalence relations. J. Symbolic Logic **79**(1), 60–88 (2014)
2. Andrews, U., Sorbi, A.: Effective inseparability, lattices, and pre-ordering relations. Rev. Symbolic Logic. **13**, 1–28 (2019). https://doi.org/10.1017/S1755020319000273
3. Buss, S.R.: Bounded Arithmetic, Studies in Proof Theory. Lecture Notes, vol. 3. Bibliopolis, Naples (1986)
4. Cleave, J.P.: Creative functions. Z. Math. Logik Grundlagen Math. **7**, 205–212 (1961)
5. Delle Rose, V., San Mauro, L.F., Sorbi, A.: Word problems and ceers. Math. Logic Q. **66**(3), 341–354 (2020)
6. Yu. L.: Ershov, Theory of Numberings, Nauka, Moscow (1977)
7. Montagna, F.: Relatively precomplete numerations and arithmetic. J. Philos. Logic **11**, 419–430 (1982)
8. Montagna, F., Sorbi, A.: Universal recursion theoretic properties of r.e. preordered structures. J. Symbolic Logic **50**(2), 397–406 (1985)
9. Nies, A., Sorbi, A.: Calibrating word problems of groups via the complexity of equivalence relations. Math. Struct. Comput. Sci. **28**(3), 1–15 (2018)
10. Pianigiani, D., Sorbi, A.: A note on uniform density in weak arithmetical theories. Arch. Math. Logic **60**(1), 211–225 (2020). https://doi.org/10.1007/s00153-020-00741-8
11. Selivanov, V.: Positive structures. In: Cooper, S.B., Goncharov, S.S. (eds.) Computability and Models. The University Series in Mathematics. Springer, Boston (2003). https://doi.org/10.1007/978-1-4615-0755-0_14
12. Shavrukov, V.Y.: Remarks on uniformly finitely precomplete positive equivalences. Math. Logic Q. **42**, 67–82 (1996)

13. Yu, V.: Shavrukov and Visser, A., Uniform density in Lindenbaum algebras. Notre Dame J. Formal Logic **55**(4), 569–582 (2014)

14. Smullyan, R.M.: Theory of Formal Systems, Revised Princeton University Press, Princeton (1961)

15. Tarski, A., Mostowsky, A., Robinson, T.M.: Undecidable Theories. North-Holland, Amsterdam (1953)

16. Visser, A.: Numerations, λ-calculus & arithmetic. In: Seldin, J.P., Hindley, J.R. (eds.) To H.B. Curry: Essays on Combinatory Logic, Lambda Calculus and Formalism, pp. 259–284. Academic Press, London (1980)

Simple Betting and Stochasticity

Tomasz Steifer[✉]

Institute of Fundamental Technological Research, Polish Academy of Sciences,
Warsaw, Poland
tsteifer@ippt.pan.pl

Abstract. A sequence of zeros and ones is called Church stochastic if all subsequences chosen in an effective manner satisfy the law of large numbers with respect to the uniform measure. This notion may be independently defined by means of simple martingales, i.e., martingales with restricted (constant) wagers (hence, simply random sequences). This paper is concerned with generalization of Church stochasticity for arbitrary (possibly non-stationary) measures. We compare two ways of doing this: (i) via a natural extension of the law of large numbers (for non-i.i.d. processes) and (ii) via restricted martingales, i.e., by redefining simple randomness for arbitrary measures. It is shown that in the general case of non-uniform measures the respective notions of stochasticity do not coincide but the first one is contained in the second.

Algorithmic randomness is a dynamically developing field of study at the crossroads between computability theory and foundations of probability. In a way, it builds up on the following question: what does it mean for an individual sequence of zeros and ones to be random? Approached from of a perspective of computability, this led to the discovery of a whole universe of notions of randomness and stochasticity (c.f. [2]). One of these is the notion of Church stochasticity. Its history dates back to von Mises' project to axiomatize probability theory in a frequentist spirit. His theory was based on the concept of *Kollektiv* [5]. But it was Church who gave it a formal mathematical interpretation grounded in recursion theory. Consider the class of sequences x of zeros and ones satisfying the following condition: every subsequence of x selected by a computable method (where selecting $n + 1$-th bit depends only on the previous bits) satisfies the law of large numbers, i.e. the ratio of zeros and ones in the subsequence converges to half. We call such sequences Church stochastic.

The existence of Church stochastic sequence may be proven in a constructive way (see e.g. [9]). Countability of computable functions guarantees that the set of all Church stochastic sequences is of λ-measure one, where λ denotes the uniform probability measure, i.e. the measure corresponding to unbiased coin tossing. In fact, Church stochasticity is one of the weakest known notions of randomness, containing other standard classes such as Martin-Löf randomness [4].

T. Steifer—This work was supported by the Polish National Foundation of Science grant 2018/31/B/HS1/04018. The author thanks Łukasz Debowski for his advice on how to simplify the proof of the main result.

L. De Mol et al. (Eds.): CiE 2021, LNCS 12813, pp. 424–433, 2021.
https://doi.org/10.1007/978-3-030-80049-9_42

The intuition behind the definition of Church stochasticity may be visualised by betting. As written by Church:

> If a fixed number of wagers of "heads" are to be made, at fixed odds and in fixed amount, on the tosses of a coin, no advantage is gained in the long run if the player, instead of betting at random, follows some system, such as betting on every seventh toss, or (more plausibly) betting on the next toss after the appearance of four tails in succession, or (still more plausibly) making his nth bet after the appearance of n+4c tails in succession.

As was shown by Ambos-Spies et al. [1], the idea of fixed wagers may be given a precise meaning which leads to an alternative definition of Church stochasticity. This is done by means of martingales (betting strategies). Consider a game in which the player starts with a finite amount of capital and the bits of a sequence are unraveled one by one. At each step, the player bets (or not) some amount of their money on one of the possible outcomes. After the next bit is revealed, the capital is modified in accordance, just as in a real casino. Our only requirement is that the expected (with respect to a given measure) change of the capital is zero, perhaps unlike in some casinos. In principle, the player should not be able to earn much but it may happen that the strategy (i.e. the martingale) performs so well that they can obtain unbounded amount of capital. In such case we say that the martingale succeeds on the sequence.

Now, we may imagine betting schemes which modify the wagers as they see fit, perhaps acting on a presupposition that the sequence exemplifies some sort of regularity. However, for now our attention will be limited to a very simple case of martingales which always bet a fixed fraction of capital (or pass the round). Consider a class of sequences on which no such martingale succeeds. We will call these simply random sequences. And what we already know is that this class coincides precisely with Church stochasticity. At least, this is how the story goes as long as we stay in the textbook case of the uniform measure.

Now, to get a better understanding of the present contribution, suppose we want to somehow generalize the notion of Church stochasticity for other probability measures. The law of the large numbers might be stated as follows: given a sequence of outcomes $x_1, x_2, \ldots$ from an independent and identically distributed (i.i.d) process[1], the average $(x_1 + \ldots + x_n)/n$ converges to the expected value of x_i. Consequently, defining Church stochasticity for i.i.d. measures is rather straightforward. To make this notion applicable in the general case, we treat the law of large numbers as a special case of the following general law: given a probability measure μ, for μ-almost every sequence $x = x_1, x_2, \ldots$ the average $(x_1 + \ldots + x_n)/n$ converges to the average of the expected values conditioned on the past, i.e., to $(\mu(1) + \mu(x_1 1)/\mu(x_1) + \ldots + \mu(x_1^{n-1} 1)/\mu(x_1^{n-1}))/n$. In fact, a very similar idea was proposed in a more general context of non-monotonic stochasticity in [6].

---

[1] Such process may be interpreted as performing the same experiment infinitely many times.

For a generalized definition of Church stochasticity with respect to an arbitrary computable measure we require that every effectively selected subsequence satisfies this extended law of large numbers. Indeed, as already stated, it might be shown using well-established probabilistic tools that the class of sequences Church stochastic with respect to a measure μ is of μ-measure one. Following this observation, we consider a natural generalization of the simple randomness for non-uniform probability measures. In general case of non-uniform measure μ, a betting strategy which uses constant wagers may not satisfy the fairness condition with respect to μ and hence, it may fail to define a martingale. Hence, to get a non-trivial class of sequences, we will have to weaken the condition of fixed wagers to a form of 'semi-fixation'. This will be given a precise meaning in the technical part of the paper along with the formal definition of simple randomness for an arbitrary measure. The main question tackled in the paper is whether Church stochasticity and simple randomness coincide in a general case as they do for uniform measure. As it turns out, this is not true. We prove the inclusion in one direction, i.e., that every simply random sequence satisfies the extended law of large numbers. The proof of this facts relies on an auxiliary theorem concerning Cesàro limitability. To our best knowledge, this result is novel. A simple example witnessing the failure of the other inclusion is also given.

1   Preliminaries

The set of finite words over $\{0,1\}$ is denoted by $2^{<\mathbb{N}}$. The set of all one-sided infinite sequences is denoted by $2^{\mathbb{N}}$. Following information-theoretic convention, bits of a one-sided infinite sequence are indexed from 1. It is assumed that $0 \in \mathbb{N}$. The set of all nonzero natural numbers is denoted as $\mathbb{N}^+$ (similarly, $\mathbb{R}^+$ denotes the set of positive real numbers). The empty word is denoted by $\square$.

Given a sequence (or a word) x we denote the i-th bit of x by x_i. To denote a string $x_j, x_{j+1}, \ldots, x_k$ (with $j < k$) we write x_j^k. In particular, a prefix of length n of x is denoted by x_1^n. By convention, for a sequence (or a word) x, we let x_1^0 denote the empty word as well.

Since every measure in this paper is a probability measure, we simply refer to these as *measures*. Every binary word $w \in 2^{<\mathbb{N}}$ corresponds to a cylinder set $[\![w]\!] = \{x : x_1^{|w|} = w\}$. We work with probability measures over the σ-algebra generated by all cylinder sets $[\![w]\!]$. By the Kolmogorov extension theorem, such measure is uniquely determined by the values $\mu([\![w]\!])$ for all $w \in 2^{<\mathbb{N}}$. We will abuse this notation by writing $\mu(w)$ instead of $\mu([\![w]\!])$. The Kolmogorov extension theorem guarantees the existence of the canonical stochastic process $X = X_1, X_2, \ldots$ with $\mu(X_1^{|w|} = w) = \mu(w)$ for all $w \in 2^{<\mathbb{N}}$.

We will give a special attention to conditional probabilities $\mu(X_n = b | X_1^{n-1})$. These random variables obey the elementary definition, i.e.,

$$\mu(X_n = b | X_1^{n-1})(x) = \mu(X_n = b | X_1^{n-1} = x_1^{n-1}) = \frac{\mu(x_1^{n-1}b)}{\mu(x_1^{n-1})}.$$

Similarly, we have conditional expectations $\mathbb{E}(X_n|X_1^{n-1})$ reducing to

$$\mathbb{E}(X_n|X_1^{n-1}) = \mu(X_n = 1|X_1^{n-1}).$$

In order to compress some equations we often write $\mu(1|w)$ instead of $\mu(X_n = 1|X_1^{n-1} = w)$.

If an event A has μ-probability one, i.e., $\mu(A) = 1$, we say that A happens μ-almost surely. If $x \in 2^{\mathbb{N}}$ and $\mu(\{x\}) > 0$ we say that x is a μ-atom and μ is an atomic measure. A measure with no atoms is called continuous. A stochastic process Y is independent and identically distributed (i.i.d) if $Y_1, Y_2, \ldots$ are mutually independent and $\mu(Y_i = b)$ is constant for all i. If the canonical process X satisfies this condition, then we say that μ is an i.i.d. measure. In particular, we single out the uniform measure λ with $\lambda(w) = 2^{-|w|}$ for all $w \in 2^{<\mathbb{N}}$.

We adapt the following convention: a partial computable function is called computable if it is total. A function $g : 2^{<\mathbb{N}} \to \mathbb{R}^{0\leq}$ is called computable if there exists a computable $f : 2^{<\mathbb{N}} \times \mathbb{N} :\to \mathbb{Q}$ such that for every $\sigma \in 2^{<\mathbb{N}}$ and every $n \in \mathbb{N}$ we have

$$|g(\sigma) - f(\sigma, n)| < 2^{-n}.$$

In particular, this allows us to talk about computable measures.

1.1   Martingales

The notion of a martingale (or a betting strategy) (as defined in algorithmic randomness theory) formalizes the idea of prediction through betting. A gambler starts with some amount of capital. Again, at each step, information about the past observations is available. The gambler bets some amount of capital on the outcome. This notion may be also seen as representing the degree of confidence that the gambler has in his prediction. In fact, the gambler may consider both outcomes to be equiprobable and effectively abstain from prediction by making a zero bet.

As the new bit is unraveled, the gambler either gets richer or loses part of his wealth. The evolution of the capital is governed by a simple fairness condition.

Definition 1. *Let μ be a computable probability measure. A function $d : 2^{<\mathbb{N}} \to \mathbb{R}^{\geq 0}$ is called a μ-martingale if for all $\sigma \in 2^{<\mathbb{N}}$:*

$$d(\sigma)\mu(\sigma) = d(\sigma 0)\mu(\sigma 0) + d(\sigma 1)\mu(\sigma 1)$$

A μ-martingale d succeeds on a sequence x if

$$\limsup_n d(x_1^n) = \infty.$$

In probabilistic terms, the fairness condition says that the conditional expectation of the capital after the bet is equal to the capital before the bet. Given a class of martingales we get a corresponding class of random sequences by considering these sequences for which no relevant martingale succeeds. As long as we limit ourselves to a countable set of martingales, the obtained class of sequences is of measure one, as was shown by Ville [9].

Hence, sequences corresponding to some countable class of martingales form natural candidates for notions of algorithmic randomness. In particular, one of such classes, namely, computably enumerable martingales, gives a characterization of the familiar notion of Martin Löf randomness as witnessed by Schnorr's theorem [7]. Other classes may be also of interest, as we will discuss soon.

1.2   Church Stochasticity for the Uniform Measure

We have already signalled that every sufficiently effective subsequence of a Church stochastic sequence must satisfy the law of large numbers. To make it more precise, we need to explain how the subsequence selection works.

Definition 2. *A selection rule is a partial function* $f : 2^{<\mathbb{N}} \to \{yes, no\}$. *Given a sequence* x *and a selection rule* f, *we denote the n-th number k such that* $f(x_1^k)=yes$ *as* $s_f(x,n)$ *and say that* $s_f(x,n)$ *is the n-th bit selected by* f.

Definition 3. *We will say that a sequence x is Church stochastic if for every total computable selection rule f either f selects only a finite number of bits or we have:*

$$\lim_{n\to\infty} \frac{1}{n} \sum_{i=1}^{n} x_{s_f(x,n)} = \frac{1}{2}.$$

1.3   Simple λ-martingales

It is now time to turn our attention back to martingales. Recall the idea that a notion of randomness may be defined via a restricted class of martingales. A particular case of such a restriction is given by the following condition: an admissible betting strategy always bets a fixed proportion of capital. This is exactly the concept of a simple martingale as defined by Ambos-Spies et al. [1].

Definition 4. *A computable λ-martingale d is simple if there exists some $q \in \mathbb{Q}$ with $0 < q < 1$ such that for every $\sigma \in 2^{<\mathbb{N}}$ and $j \in \{1,0\}$ we have $d(\sigma j) \in \{d(\sigma), (1-q)d(\sigma), (1+q)d(\sigma)\}$*

In other words, a simple λ-martingale always bets a fraction q of the accumulated capital or does nothing. The next definition comes with no surprise:

Definition 5. *A sequence x is called λ-simply random if no simple λ-martingale succeeds on x.*

We might now state the theorem of interest:

Theorem 1 ([1]). *A sequence x is Church λ-stochastic if and only if it is λ-simply random.*

In what follows we will provide a sound candidate for generalization of the notion of Church stochasticity for non-uniform measures.

2   Technical Developments

2.1   Beyond LLN

In the elementary probability theory, the law of large numbers (LLN) is defined for i.i.d. processes. This corresponds to an ideal scenario in which we are performing the same experiment infinitely many times with the assumption that no trial influences another one. Then, LLN simply tells us that almost surely the average of the outcomes is asymptotically equal to the expected outcome. In particular, the empirical average is a consistent estimator of the expectation. Some might consider this to be quite intuitive. In fact, von Mises wanted to treat the law of large numbers as an axiom (c.f. [8]).

In the general case, the limit of empirical averages may not exist. For an example, consider a measure μ of the following form. Let f be some fast growing function, e.g., $f(n) = 2^{2^n}$. Set $\mu(X_i = 1) = \frac{3}{4}$ if for some odd n we have $f(n) \leq i < f(n+1)$. Otherwise, let $\mu(X_i = 1) = \frac{1}{4}$. Assume that $X_1, X_2, \ldots$ are pairwise independent. It is almost routine to show that the average of $X_1, \ldots, X_n$ does not converge μ-almost surely. Indeed, this follows from the fact that μ-almost surely

$$\lim_{n \to \infty} \frac{1}{n} \sum_{i=1}^{n} \left(X_i - \mathbb{E}(X_i | X_1^{i-1}) \right) = 0. \tag{1}$$

If $X_1, X_2, \ldots$ are pairwise independent binary variables, this reduces to

$$\lim_{n \to \infty} \frac{1}{n} \sum_{i=1}^{n} (X_i - \mu(X_i = 1)) = 0.$$

In other words, the difference between the empirical average of first n outcomes and its expected value vanishes as n goes to infinity. Hence, as in the example, the empirical average may diverge if the corresponding expectations diverge.

As soon as we begin to realize that LLN is, in fact, a special case of (1), a way presents itself for a generalization of Church stochasticity for arbitrary measures. To be precise, we will require that (1) is satisfied by every subsequence selected by a computable method.

Definition 6. *Let μ be a computable measure. We will say that a sequence $x \in 2^{\mathbb{N}}$ is Church μ-stochastic if for every computable selection function f either the number of bits selected by f on x is finite or*

$$\lim_{n \to \infty} \frac{1}{n} \sum_{i=1}^{n} \left(x_{s_{f(x,i)}} - \mathbb{E}(X_{s_{f(x,i)}} | x_1^{s_{f(x,i)}-1}) \right) = 0.$$

We note that this definition is similar to what was proposed by Muchnik, Semenov and Uspensky [6] in the context of nonmonotonic stochasticity. However, those authors chose to condition the conditional expectation not on the whole prefix but only on the bits selected by the selection function in question— which makes sense in the case of nonmonotonic selection functions.

2.2   Semi-fixed Wagers

For many measures, fixation of wagers makes it impossible to satisfy fairness condition by a non-trivial martingale. Hence, we propose the following relaxed condition.

Definition 7. *Given a computable measure μ, let d be a μ-martingale. We will say that d is simple if it is computable and there is a rational $0 < q \leq 1$ such that for every $\sigma \in 2^{<\mathbb{N}}$ and $j \in \{0, 1\}$ we have*

$$\frac{d(\sigma j)}{d(\sigma)} \in \{1, 1 + q\mu(0|\sigma), 1 - q\mu(0|\sigma), 1 + q\mu(1|\sigma), 1 - q\mu(1|\sigma).\}$$

We note that a case for a different approach could be made. For instance, we could ask that we always loose the same fractions of capital while modifying only the reward. Whether these are more *intuitive* is up to reader to decide. That being said, these alternative definitions would have similar problems to those discussed in the last section.

Definition 8. *We will say that a sequence $x \in 2^{\mathbb{N}}$ is μ-simply random if no simple μ-martingale succeeds on x.*

2.3   Simple Randomness Implies LLN

Theorem 2. *If x is μ-simply random then it is Church μ-stochastic.*

This theorem may be reduced to a statement concerning bounded sequences and their infinite products.

Lemma 1. *Let $a_1, a_2, \ldots$ be a sequence of real numbers such that $-1 \leq a_i \leq 1$ for all $i \in \mathbb{N}$. The sequence $a_1, a_2, \ldots$ is Cesàro limitable with the limit 0 if for every rational number q with $-1 < q < 1$ the following holds:*

$$\sup_{n \in \mathbb{N}} \prod_{i=1}^{n} (1 + qa_i) < \infty.$$

Proof. Fix the sequence $a_1, a_2, \ldots$ and assume the antecedent of the implication. We begin by recalling that

$$\sup_{n \in \mathbb{N}} \prod_{i=1}^{n} (1 + qa_i) < \infty$$

is equivalent to

$$\sup_{n \in \mathbb{N}} \sum_{i=1}^{n} \log(1 + qa_i) < \infty,$$

from which it follows that (†):

$$\limsup_{n \in \mathbb{N}} \frac{1}{n} \sum_{i=1}^{n} \log(1 + qa_i) \leq 0.$$

By Taylor expansion (see e.g. [3]), for every x with $|x| < 1$ we have

$$\log(1+x) = x - \frac{x^2}{2} + \frac{x^3}{3} - \cdots$$

and

$$1 - x^{-1} \le \log(x) \le x - 1.$$

Suppose that $q < 0$. We have $q^{-1}\log(1+qa_n) \le a_n$ for all n. Then it follows from (†) that

$$\liminf_{n \in \mathbb{N}} \frac{1}{n} \sum_{i=1}^{n} a_i \ge \liminf_{n \in \mathbb{N}} q^{-1} \frac{1}{n} \sum_{i=1}^{n} \log(1+qa_i) \ge 0.$$

Now, suppose that $q > 0$. Then we have $q^{-1}\log(1+qa_n) \ge \frac{a_i}{1+qa_i}$. Combining it with (†) we get

$$0 \ge \limsup_{n \in \mathbb{N}} \frac{1}{n} q^{-1} \sum_{i=1}^{n} \log(1+qa_i) \ge \limsup_{n \in \mathbb{N}} \frac{1}{n} \sum_{i=1}^{n} \log(a_i - q/(1-q)).$$

With q converging to 0 we get

$$\limsup_{n \in \mathbb{N}} \frac{1}{n} \sum_{i=1}^{n} \log(a_i) \le 0.$$

It remains to observe that Lemma 1 is basically all we need.

Proof (of Theorem 2). Suppose that x is μ-simply random. We show that x is Church μ-random. Without loss of generality, we restrict our attention only to the selection function f which obliviously selects all bits. For every rational number $0 < q < 1$ consider a martingale d_q such that for every $\sigma \in 2^{<\mathbb{N}}$

$$d_q(\sigma 1) = (1 + q\mu(0|\sigma)) d_q(\sigma)$$

and

$$d_q(\sigma 0) = (1 - q\mu(1|\sigma)) d_q(\sigma).$$

Note that if $x_1^i = \sigma 1$ for some $\sigma \in 2^{<\mathbb{N}}$ then $(1 + q\mu(0|\sigma)) = (1 + q\mu(X_i \neq x_i|x_1^{i-1}))$. Similarly, $x_1^i = \sigma 0$ then $(1 - q\mu(1|\sigma)) = (1 - q\mu(X_i \neq x_i|x_1^{i-1}))$. Each martingale d_q is simple. By simple randomness of x, for each of d_q we have

$$\sup_{n \to \infty} d_q(x_1^n) = \sup_{n \to \infty} \prod_{i=1}^{n} (1 + (-1)^{x_i - 1} q\mu(X_i \neq x_i|x_1^{i-1})) < \infty.$$

Note that given a rational $0 < q \le 1$ the following defines a simple martingale as well.

$$d_q(\sigma 1) = (1 - q\mu(0|\sigma)) d_q(\sigma)$$
$$d_q(\sigma 0) = (1 + q\mu(1|\sigma)) d_q(\sigma).$$

So, for every $0 < q \leq 1$ we have

$$\sup_{n \in \mathbb{N}} \prod_{i=1}^{n} (1 + (-1)^{x_i} q\mu(X_i \neq x_i|x_1^{i-1})) < \infty.$$

By Lemma 1, it follows that

$$\lim_{n \to \infty} \frac{1}{n} \sum_{i=1}^{n} (x_i - \mathbb{E}(X_i|x_1^{i-1})) = \lim_{n \to \infty} \frac{1}{n} \sum_{i=1}^{n} (-1)^{x_i-1} \mu(X_i \neq x_i|x_1^{i-1}) = 0.$$

The case of other selection functions is analogous with the one exception. If the selection function f does not select the i-th bit we simply require martingale d_q to abstain from betting on the i-th bit. With that adjustment we repeat the argument with respect to the selected subsequence.

2.4   Failure of the Other Implication

Unfortunately, the other implication fails as witnessed by the following construction.

Theorem 3. *There exists a computable measure μ and a sequence $y \in 2^{\mathbb{N}}$ which is Church μ-stochastic but is not μ-simply random.*

Proof. Suppose that $X = X_1, X_2, \dots$ are mutually independent and for each $n \in \mathbb{N}$ we have

$$\mu(X_n = 1) = 1 - \frac{1}{n+1}.$$

Note that for every infinite sequence of strictly increasing indexes $n_1, n_2, \dots$ the sequence

$$\mu(X_{n_1} = 1), \mu(X_{n_2} = 1), \dots$$

converges to one. Consider the sequence $y = 11\dots$ and note that every infinite subsequence of y has density one. Consequently, y is Church μ-stochastic. However, it is also true that

$$\sup_{n \in \mathbb{N}} \prod_{i=1}^{n} (1 + \mu(X_i = 0)) = \sup_{n \in \mathbb{N}} \prod_{i=1}^{n} \left(1 + \frac{1}{i+1}\right) = \infty.$$

In other words, the following martingale succeds on x:

$$d(\sigma 1) = (1 + \mu(X_i = 0)) d(\sigma).$$

This means that y is not μ-simply random.

Interestingly, the presented counterexample is relatively simple. In particular, the process X consists of independent variables and the construction does not rely on any carefully tailored conditional probabilities. Furthermore, we have the following.

Corollary 1. *There exist a computable measure μ and a sequence $y \in 2^{\mathbb{N}}$ such that:*

1. *y is Church μ-stochastic,*
2. *y is computable,*
3. *y is not a μ-atom.*

Proof. We take the measure μ and $y \in 2^{\mathbb{N}}$ as in the proof of Theorem 3. To show that y is not a μ-atom consider:

$$\prod_{i=1}^{\infty} \mu \left(X_i = 1 \right) = \prod_{i=1}^{\infty} \left(1 - \frac{1}{i+1} \right) = 0.$$

This is quite different from the case of continuous i.i.d. measures—where no Church stochastic sequence is computably enumerable, let alone computable. Indeed, recall that an infinite computably enumerable set A has an infinite computable subset - this subset defines a selection function which selects $11\ldots$ on A.

References

1. Ambos-Spies, K., Mayordomo, E., Wang, Y., Zheng, X.: Resource-bounded balanced genericity, stochasticity and weak randomness. In: Puech, C., Reischuk, R. (eds.) STACS 1996. LNCS, vol. 1046, pp. 61–74. Springer, Heidelberg (1996). https://doi.org/10.1007/3-540-60922-9_6
2. Downey, R.G., Hirschfeldt, D.R.: Algorithmic Randomness and Complexity. Springer Science & Business Media (2010). https://doi.org/10.1007/978-0-387-68441-3
3. Knopp, K.: Theory and application of infinite series. Courier Corporation (1990)
4. Martin-Löf, P.: The definition of random sequences. Inf. Control **9**(6), 602–619 (1966)
5. Mises, R.V.: Grundlagen der wahrscheinlichkeitsrechnung. Mathematische Zeitschrift **5**(1–2), 52–99 (1919)
6. Muchnik, A.A., Semenov, A.L., Uspensky, V.A.: Mathematical metaphysics of randomness. Theor. Comput. Sci. **207**(2), 263–317 (1998)
7. Schnorr, C.P.: A unified approach to the definition of random sequences. Math. Syst. Theor. **5**(3), 246–258 (1971). https://doi.org/10.1007/BF01694181
8. Van Lambalgen, M.: Von Mises' definition of random sequences reconsidered. J. Symbolic Logic **52**(3), 725–755 (1987)
9. Ville, J.: Etude critique de la notion de collectif. Gauthier-Villars Paris (1939)

Péter on Church's Thesis, Constructivity and Computers

Máté Szabó[✉] [iD]

University of Oxford, Oxford, UK
mate.szabo@maths.ox.ac.uk

Abstract. The aim of this paper is to take a look at Péter's talk *Rekursivität und Konstruktivität* delivered at the *Constructivity in Mathematics* Colloquium in 1957, where she challenged Church's Thesis from a constructive point of view. The discussion of her argument and motivations is then connected to her earlier work on recursion theory as well as her later work on theoretical computer science.

Keywords: Rózsa Péter · Church's Thesis · Constructivity

1 Introduction

Rózsa Péter and László Kalmár, lifelong colleagues and friends, were both invited to the famous *Constructivity in Mathematics* Colloquium held in Amsterdam in 1957. In their talks, both of them challenged Church's Thesis, albeit in quite different ways. The aim of this paper is to discuss Péter's less frequently cited contribution *Rekursivität und Konstruktivität* [21]. And to provide more context and background to her argument and to connect it to her earlier and later work in recursion theory and theoretical computer science respectively. Besides her published paper [21], Péter and Kalmár's correspondence [9] will be used to provide such context. Due to the lack of space, attention will be focused on Péter's work and connections to the ideas of other scholars will merely be indicated.

2 Church's Thesis in Péter's Recursive Functions

Péter's most well known scholarly work, *Recursive Functions* [19], famously the first monograph in recursion theory, was first published in 1951. The phrase

I would like to thank Marianna Antonutti-Marfori and Alberto Naibo for inviting me to the HaPoC Special Session on *Church's Thesis in Constructive Mathematics*. I am also indebted to Alberto for his insightful discussions on this topic and his useful comments on earlier versions of this paper. I would also like to thank Kendra Chilson, Katalin Gosztonyi, Wilfried Sieg and Kristóf Szabó for their help and the anonymous reviewers for their suggestions. The writing of this paper was supported by the UK Engineering and Physical Sciences Research Council under grant EP/R03169X/1.

© Springer Nature Switzerland AG 2021
L. De Mol et al. (Eds.): CiE 2021, LNCS 12813, pp. 434–445, 2021.
https://doi.org/10.1007/978-3-030-80049-9_43

'Church's Thesis' does not appear as a label anywhere in the book, as it was popularized by Kleene's *Introduction to Metamathematics* [13] that was published the next year.[1] But, of course, the identification of the notion of 'calculable functions' with the general recursive functions "proposed" by Church is discussed. As it is described below in detail, in general Péter displayed a "noncommittal" attitude towards the Thesis in [19] and in later editions she kept including materials that challenged it.

In §20, *Calculable Functions*, Péter "quote[s] some of the arguments used in attempts to make plausible the identification." However, this chapter is entirely devoted to a detailed description of Turing's machines and their capability to compute every general recursive function. This emphasis is due to Péter's view that the finite calculability of a function is strongly tied to the existence of a repeatable and communicable "mechanical procedure," where the "single steps of the calculation" could be carried out even by a machine "in principle" (p. 225). Péter acknowledges that Turing's machines satisfy these requirements while his analysis makes it plausible that "this interpretation correctly reflects real mathematical activity" (p. 240) and that it gives "the impression that a very general concept of calculability has been captured here." (p. 234).

Church's "proposed" identification is only discussed in the less than a page long 7th section of §21, *History and Applications*. As support for the identification, in addition to what was already given in §20, Péter points to the 'confluence' of the multiple attempts to formalize the "vague concepts of 'calculability', 'constructibility', [and] 'effectibility'" (p. 245). More precisely, she states the equivalence of general recursive functions, functions computable by Turing's machines, Hilbert and Bernays' reckonable functions, and the functions calculable by Markov's algorithms.

At the same time, whenever Péter presents some of the arguments for Church's Thesis, she also remarks that the acceptance of it would lead to "a certain demarcation of the concept of calculability" (p. 240). She finds this problematic, as she strongly believed that mathematics and its methods would develop without an end. As she puts it here: "the future evolution of mathematics may bring about methods of calculation completely unexpected nowadays" (p. 240). For this reason, while she considers the concept of general recursivity a "most welcome generalization," nevertheless she sees it "merely [as] a stage – albeit a very high one" (p. 9). Thus the concept of calculability should not be formally confined based on our current mathematical knowledge, as it is expected to develop further in the future. For this reason Péter concludes: "In my opinion,

---

[1] Already in his ([12], p. 60) Kleene labeled the assertion that "Every effectively calculable function is general recursive" as a 'thesis' that was stated by Church (and implicitly by Turing) but did not use the phrase 'Church's Thesis' anywhere in the paper. The phrase became widely used after it was popularized in Kleene's [13].

no 'final word' is possible here: the concept of calculability cannot be definitely comprised once and for all" (p. 9).[2]

In later editions of *Recursive Functions*, Péter kept adding materials that supported her "non-committal" attitude towards Church's Thesis. In the *Preface of the Second German Edition*, written in 1955, Péter acknowledges that "it has been noted" by the reviewers and readers of the first edition that she was "non-committal on the question of the identification of calculability with general recursivity." And she adds that "It was at that time my intention, however, to set beside one another the possible positions in this question" ([19], p. 9).

Next Péter alludes to her view that based on the endless development of mathematics, the concept of calculability should not be formally confined based on our current knowledge. She claims that this point of view "has lately been very strongly supported by the new results" of Kalmár, which are then discussed in §19.2 and §20.8. The first one is essentially a summary of an early version of Kalmár's *An Argument Against the Plausibility of Church's Thesis* [8]. Here, assuming Church's Thesis while adopting a much less restricted notion of 'effective calculability' than customary, Kalmár constructed a proposition that is intuitively considered to be false, but its falsity cannot be proven by "any correct means." He regarded this "very strange consequence" of the Thesis as an argument against its plausibility (without claiming to have refuted it). Kalmár concluded his paper with his belief that certain mathematical concepts, such as 'effective calculability' or 'provability,' "cannot permit any restriction imposed by an exact mathematical definition" due to the endless development of mathematics, and thus, the possible further development of these concepts in the future.[3] Then Péter discusses Kalmár's [6]. Here, answering a question of Karl Schröter's in the negative, he showed that a system of functional equations (without restrictions on the operations to compute its values) may have a unique solution without the determined function being general recursive.

In his JSL review of the second edition of [19] Robinson focused only on these new additions and felt compelled to proclaim that "the reviewer is still convinced that the concept of general recursive function does provide the proper formalization of the intuitive concept of computable function" ([25], p. 363).

---

[2] Péter also concludes her popular book, *Playing with Infinity* [22], with the same thought. The last chapter explains Gödel's incompleteness and Church's undecidability results and raises the question whether we have "come up against final obstacles?" (p. 264). This is then answered quite forcefully by the very last paragraph of the book: "Future development is sure to enlarge the framework, even if we cannot as yet see how. The eternal lesson is that Mathematics is not something static, closed, but living and developing. Try as we may to constrain it into a closed form, it finds an outlet somewhere and escapes alive" (p. 265).

[3] For a detailed description and analysis of Kalmár's [8] visit Szabó's [33]. In addition, Gosztonyi's [4] discusses Péter and Kalmár's shared views on mathematics and its education within the Hungarian mathematical culture as their broader context, while Máté's [15] examines their philosophical views on mathematics.

3   Recursion and Constructivity

In the summer of 1956, Péter received an invitation from Arend Heyting to the *Constructivity in Mathematics* Colloquium to be held in August of 1957 in Amsterdam (see the *Appendix*). Péter accepted the invitation and her contribution became known as *Rekursivität und Konstruktivität* [21].

The correspondence of Péter and Kalmár [9] reveals that they discussed what they should present at the meeting. In a letter from the 18th of February, 1957, Kalmár recommended to Péter that she either discuss any current work of hers on open problems of recursion theory or, "Since [the Colloquium] is about constructivity, you should discuss why you do not find the concept of general recursive function satisfactory from the standpoint of constructivity (as you usually say, something is fishy about it), or at least why you find the concept of recursion present in the theory of special recursive functions more satisfactory".[4] Péter ended up exploring this latter suggestion.

Before turning to Péter's paper, it is important to note that she did not hold constructivist views. Her views on the foundations of mathematics, like Kalmár's, were most closely aligned with that of the Hilbert school. However, when Péter began research in recursion theory at the very beginning of the 1930s the notions of 'effective' and 'constructive' were used essentially interchangeably in the community, including by Church and Kleene.[5] Thus Péter was not opposed to classical or non-constructivist approaches in mathematics and logic in general, but considered the concept of recursive function to be an attempt to precisely characterize the concept of constructivity.

In the beginning of the paper, Péter asserts that functions defined via special types of recursion (such as primitive recursion, course-of-values recursion, simultaneous and nested recursions, etc.) are obviously finitely calculable, and thus, constructive functions. Then the "Herbrand-Gödel-Kleene" notion of general recursive function is considered. On Péter's account, the main reason to introduce this notion was to precisely formulate the concept of constructivity, and Church's Thesis identifies this very notion with the concept of calculable functions. Then the question is raised whether every 'effectively calculable' function can justifiably be called 'constructive'. Thus Péter challenges the direction of Church's Thesis that is usually considered obvious or unproblematic.

Péter then goes on to argue that Church's Thesis is either non-constructive or it contains a vicious circle. By definition, a function is considered to be general recursive if *there exists* a finite system of equations from which all of its values

---

[4] The correspondence of Péter and Kalmár is in Hungarian; quotes are translated by the present author.

[5] For examples, see Sieg's ([28], pp. 558–559) quoting Church, and Kleene's [11] where he states that "The notion of a recursive function of natural numbers, which is familiar in the special cases associated with primitive recursions, Ackermann-Péter multiple recursions, and others, has received a general formulation from Herbrand and Gödel. The resulting notion is of especial interest, since the intuitive notion of a 'constructive' or 'effectively calculable' function of natural numbers can be identified with it very satisfactorily" (p. 544).

can be calculated in a finite manner. The existential quantifier in the definition can be interpreted both classically and constructively, and the former is (possibly) not satisfactory from a constructivist point of view.[6] Péter was not the only one to point out this issue. Not only did Heyting and Skolem raise this exact issue around the time,[7] but Church ([1], p. 351, fn 10) and Kleene ([13], p. 319) were evidently aware of it too. Church's recommendation for his concerned readers was to "take the existential quantifier [...] in a constructive sense", and added that "[w]hat the criterion of constructiveness shall be is left to the reader".[8] But, according to Péter, this leads to a vicious circle since general recursion was offered as a precise formulation of constructivity,[9] yet the notion of constructivity is alluded to in the interpretation of the existential quantifier in its definition.[10] She then remarks that the same vicious circle appears however one tries to get around it.

Péter's paper was reviewed jointly with articles that challenged or criticized Church's Thesis ([16,17]) and it is usually cited in that context as well. While this assessment is appropriate, especially with Péter's known non-committal attitude towards the Thesis, it misses an aspect of her undertaking, namely that Péter's primary goal in this work was not to undermine the Thesis, but to give a precise account of constructive functions.

Indeed, after the short discussion of Church's Thesis, Péter considers multiple possible definitions to characterize the constructive functions in the second half of the paper. After the presentation of each possible formalization, she points out the "vicious circle" in them. However, these examples are not merely there to strengthen her challenge to Church's Thesis by pointing out the circularity over and over again. These are her genuine (and failed) attempts to give a positive, non-circular characterization of the constructive functions.

Péter's main aim was to provide a formal characterization of constructive functions that includes all the special recursive functions[11] but does not exhaust the general recursive ones. When she recognized the circularity in all of her

---

[6] Kleene brings up a related issue in his [10]: "The definition of general recursive function offers no constructive process for determining when a recursive function is defined. This must be the case, if the definition is to be adequate, since otherwise still more general "recursive" functions could be obtained by the diagonal process" (p. 738).

[7] See Coquand's [2] for a discussion of their relevant writings.

[8] Here Péter remarks that since no "real" general recursive function is known, i.e. one that is general recursive but does not belong to any of the special types of recursive functions, it is not clear what the difference between the two interpretations could actually amount to.

[9] See Sundholm's ([30], pp. 13–14) on Church's view on the Thesis and its relation to constructivism.

[10] Heyting raises a rather similar concern about the circularity involved in the definition of recursive functions from a constructivist point of view in his [5] without referring to Péter's [21].

[11] The term "special recursive functions" is used here loosely, to refer to the collection of those recursive functions that are defined via a specific type of recursion (see above) and are seen as obviously finitely calculable and constructive functions.

attempts, she was discouraged. In a letter to Kalmár, written on the 2nd of July, 1957, not long before the Colloquium, she reported that "sadly it appears to me that the whole idea is bankrupt." Péter ended her paper on the same note with the following, resigned last sentence: "It seems that the concept of constructivity cannot be captured in a non-circular way at all."[12]

Nevertheless Péter did not entirely give up on the idea to provide a precise formal characterization of constructive functions. The next section describes her subsequent attempt after the Colloquium.

4   Péter's Road to Computer Science

Another aspect of Péter's *Rekursivität und Konstruktivität* worth mentioning is its connection to her later work on recursive functions in the field of theoretical computer science. Péter became involved with the field through the active encouragement of Kalmár. As the interests of Kalmár turned towards the applications of logic in cybernetics, automata theory, computer design and even in engineering in general ([14,34]), he started a research seminar devoted to these topics in the spring of 1956 at the University of Szeged.[13] To involve Péter, who was in Budapest, with the work by the members of the seminar, he sent the papers they were reading and open questions to Péter by mail. The specific research problems he posed to Péter were connected both to theoretical interests, such as Victor Shestakov's [27] and Claude Shannon's [26] work on representing relay and switching circuits with Boolean algebras, as well as their use in actual computer design questions that Kalmár planned to undertake in the near future. By the end of the summer, Péter was sending her results frequently and via express mail to Kalmár in order to be part of the "delightful work" of the group.

Péter's recent engagement with the field that we would today call theoretical computer science provides the background for her remark below. While discussing administrative details about their trip with Kalmár to the Colloquium in Amsterdam in a letter on the 14th of January 1957, Péter casually inserted the following comment mid-sentence: "by the way, and I mention this only to you, it seems to me that the notion of definability by calculators[14] is very closely

---

[12] Translated by Tamás Lénárt.

[13] Kalmár's travel report [7] on the *Constructivity in Mathematics* Colloquium also stands witness to these interests of his. In the second page of the report he mentions that: "[After the Colloquium I had] the opportunity to visit the 'Mathematisch Centrum' and take a look at their operating ARMAC electronic calculator. In addition I had scientific discussion with [Jurjen Ferdinand] Koksma, a mathematics professor, and his colleagues; and with [Adriaan] van Wijngaarden, an engineering professor, about the practical applications of the calculator, as well as about the logical machine under construction in Szeged." (Translated by the present author.) In the report, Kalmár mistakenly refers to "A. Koksma." On Kalmár's description of the ARMAC computer as an electronic calculator, see footnote 14. For short descriptions of the ARMAC computer and the logical machine in Szeged, visit [35] and [32], respectively.

[14] At the time computers were referred to as (high speed) electronic or digital calculators in Hungary.

tied to constructivity." This cryptic comment turned out to be rather prescient, as her first paper published after the Colloquium (even before its proceedings appeared), *Graphschemata und Rekursive Funktionen* [20], engaged with programming in yet another attempt to characterize constructive functions.

In the first paragraph of [20] Péter directly refers to her talk at the Colloquium and her "failed attempts" to characterize constructive functions through non-circular definitions. This paper presents another attempt to provide such a characterization that includes the special recursive functions but does not exhaust the general recursive ones. Péter introduces "graphschemata," a graphical representation of flow-charts used in programming, which later became known as "Kalužnin-Péter diagrams."[15] Then she considers a rather restricted type of diagram, "normalschemata," as a possible candidate for the characterization of constructive functions. However, at the end it turns out that normalschemata are unsuited for her purposes as she shows that every partial recursive function is computable by a "normalschema." Hence Péter considered this attempt to be yet another failed one.

5   Computers and Church's Thesis

From the end of the 1950s until her passing in 1977, Péter published several papers on the use and applications of recursion theory in computer science. Her late, lesser known book, *Recursive Functions in Computer Theory* [24], first published in German in 1976 and translated to English in 1981, provides a great introduction and summary of more than a dozen of her papers from this period and deserves greater attention than it has received so far. This section examines how Church's Thesis is discussed in the book in the context of computing machines and programming.

Surprisingly, while Péter showed a non-committal attitude towards Church's Thesis in the earlier works mentioned above, here in [24], she seemingly commits to what we would today call a (particular version of the) physical version of the thesis in the Preface:

> The action of a computer can always be thought of as a process such that in response to given input data, the machine produces certain outputs. Since both the input data and the sequential output of the results can be encoded into natural numbers, it follows that the functioning of the computer can always be considered as the computation value of a numeric function. With the idealization that the contents of the computer store are unlimited,[16] it can be shown that the functions computable by a computer

---

[15] To learn more about Péter's and Kalužnin's work on diagrams in the historical context of automata theory, visit [18]; for a short biography of Kalužnin see [31].

[16] Later on the page Péter adds the following remark: "The above idealization (which will be assumed throughout in what follows) always arises if a general mathematical theory is applied to practical problems. This is often expressed by saying 'the infinite is a useful approximation to the large but finite"' (p. 9).

are identical with the class of "**partial recursive** functions." [emphasis in the original] (p. 9)

The reason to take such a strong stance already in the 'first' page is to justify the approach taken in the book. From this statement, Péter draws the conclusion that "if we study how the computation of partial recursive number-theoretic functions can be programmed, essentially all questions concerning the problems solvable by a computer will be studied" (p. 9). This statement is repeated again later: "Thus if we study the programming problems of the computation of partial recursive functions, this means, in principle, the study of programming of all the machine solvable problems" (p. 63).

There are two noteworthy qualifications in the above statement: the emphasis on the *partiality* of the functions involved and the claim that the identification in this form "can be shown." The latter is especially surprising, as later in the book Péter describes (the classical) Church's Thesis as neither provable nor disprovable mathematically, as it is, "of course, [...] not an exact mathematical proposition" (p. 142).

Péter devotes Chapter 4, *The Recursivity of Everything Computable*, to *showing* that "for every partial recursive function there is a program such that computation with this program yields the value of the function, if it is defined, and goes on forever, without calculating anything if it is not" (p. 63). Péter then uses an idealized assembly language with a simple system of statements. First she shows that every primitive recursive function is machine computable and then claims that "[m]achine computability is also preserved in the application of a μ-operation, not only in the bounded case [...], but in the unbounded case as well" (p. 57). Here a cycle is used for the μ-operation which exits only if the smallest number that was searched for is found and goes on forever otherwise. Then she points to her construction of a universal program for the calculation of partial recursive functions in [23]. Finally, to establish the reverse direction, Péter shows through examples how to encode programs written in assembly language via partial recursive functions.

Thus it seems that the possibility of "showing" the "recursivity of everything [machine] computable" rests on the availability of concrete computers and their known operations and (assembly) languages. On the other hand, Péter does not provide any argument or even raise as a question whether current computers are capable of computing everything that is machine computable in principle (for example, as Robin Gandy did in his [3]). The inclusion of an argument of this sort most likely could not be considered as "an exact mathematical proposition" either.

Interestingly, Péter later eases up on both qualifications in the remarkably short 10th Chapter, *Does Recursivity Mean Restriction?* First she returns to the question of partiality and states that: "Actually, a really *partial* recursive function might not be obtained at all." Here, surprisingly, Péter alludes to programming practices. The problem with the computation of the arguments of a proper partial recursive function is that "one can never know whether the computer has failed to stop because the computation is lengthy, or if it will work on forever, without computing anything." Hence "[o]ne always strives to feed 'reasonable' programs into the computer, whereby [...] the calculation will come to a halt after a (large) finite number of computing steps." After this practical restriction, Péter arrives at the conclusion that "whatever can really be obtained by the use of a computer is general recursive" (p. 141).

According to Péter this conclusion raises the question in the title of the chapter, namely, whether the recursivity of what is computable means "an essential restriction on the abilities of the computer" (p. 141). This question, then, leads her to Church's Thesis, i.e. the statement that "every numeric function is general recursive if its values are computable in a finite number of steps for all arguments" (p. 142). Under the assumption of the Thesis, recursivity not only does not pose any restriction, but it means that "computers, which in principle are capable of computing every general recursive function, yield the most that can be expected according to the present state of our knowledge" (p. 142).

Of course when discussing Church's Thesis, while admitting that "there are many arguments for it," Péter has to mention that there are "some" against it as well. Here she mentions the informal character of the Thesis and points to Kalmár's argument against its plausibility [8]. More importantly, she re-emphasizes the conviction she shared with Kalmár in the endless development of mathematics, namely that "effective calculability is one of those notions the definition of which can never be considered complete in the course of the development of mathematics" (p. 142).

This allusion to possible future developments leads to the last and most interesting remark of the short chapter:

> Let us hope, provided a counter-example to Church's thesis is made known [...that...] the technological means will develop to modify computers to enable them to compute such functions. (p. 142)

Hence, Péter not only believes in the endless development of mathematics and mathematical methods, but in the possibility of advancements in computing technologies as well.[17] This latter belief might explain why Péter did not even attempt to characterize in any way what actual computing machines are capable of in principle. At the same time, it seems to undermine the possibility of "showing" that the machine computable functions are identical with the (partial) general recursive functions, or at least casts into doubt what it could amount to in general.

Thus in the end, it seems that Péter was not philosophically committed to what we would call a physical version of Church's Thesis either. However, it appears that she truly endorsed it under "the present state of our knowledge" and saw it as appropriate justification of the approach taken in the book.

---

[17] This is in stark contrast with Gödel's view that the human mind infinitely surpasses any finite machine for which the "inexhaustibility" of mathematics or the possible future discovery of humanly effective but non-mechanical processes would provide an argument. See Sieg's [29] for a detailed analysis of Gödel's writings on this issue.

6   Conclusion

This short overview of Péter's main discussions of Church's Thesis show a constant and consistent non-committal attitude towards it. She never questioned the usefulness of the concept of general recursive functions within recursion theory and the importance of undecidability results based on Church's Thesis. Nevertheless she persisted in her belief in the endless development of mathematics and that such a development is an argument against the exact formalization and confinement of the concept of calculability once and for all.

Appendix: The Amsterdam Colloquium

In the summer of 1956, Heyting sent Péter the following invitation (found in [9]) to the Amsterdam Colloquium, which was then in the early planning phase.

Dear[18] Madam Péter,

The International Union for Logic, Methodology and Philosophy of the Sciences has charged the Netherlands Society for Logic and Philosophy of Sciences to organize in 1957 at Amsterdam a colloquium on "The different notions of constructivity in mathematics". I hope to organize this colloquium in the summer of 1957, presumably in August. Subventions have been asked from UNESCO and from the Dutch government, by which probably we shall be able to pay a considerable part of the expenses of the participants.

The object of the colloquium will be to study the different notions of constructivity which have been proposed and the relations between them. I shall be please very much if you will participate in it. I join to this letter a list of persons whom I intend to invite.[19] However, this list, as the plan in general is in a preliminary state. I shall be thankful for suggestions w[h]ich you can give me. If the funds are sufficient, I should like to invite a number of young and promising mathematicians as auditors. You will also oblige me by mentioning names which can be considered in this respect, but I beg you to remember that it is by no means sure that such invitations will be possible.

Yours sincerely

A. Heyting

The working title of the Colloquium appears to be more pluralistic at this stage than the final *Constructivity in Mathematics*. Nevertheless, in the *Preface* of the Proceedings, Heyting states in a similar vein that "Several different notions of constructivity were discussed in these lectures" (p. 9).

Based on Péter and Kalmár's correspondence [9] and Kalmár's travel report [7], we know that Péter was among the first 18 logicians invited to the Colloquium, and Kalmár was later invited on her strong recommendation. Altogether 24 logicians received invitations to the event, 21 attended and 19 gave talks,

---

[18] The greeting is stapled over, thus 'dear' is merely an educated guess here.
[19] Sadly this list was not kept in [9].

while about the same number of "promising young mathematicians" were among the audience.

Originally the organizers offered to cover 60% of the costs of the attendees. However, since even the remaining costs were essentially insurmountable for the Hungarian scholars, the costs of Péter, Kalmár, and their student, the set theorist and combinatorist András Hajnal (later frequent Paul Erdős co-author) were almost entirely covered by the organizers.

References

1. Church, A.: An unsolvable problem of elementary number theory. Am. J. of Math. **58**(2), 345–363 (1936)
2. Coquand, T.: Recursive functions and constructive mathematics. In: Dubucs, J., Bourdeau, M. (eds.) Constructivity and Computability in Historical and Philosophical Perspective. LEUS, vol. 34, pp. 159–167. Springer, Dordrecht (2014). https://doi.org/10.1007/978-94-017-9217-2_6
3. Gandy, R.: Church's thesis and principles for mechanisms. In: Barwise, J., Keisler, J., Kunen, K. (eds.) The Kleene Symposium, pp. 123–148. North-Holland, Amsterdam (1980)
4. Gosztonyi, K.: Mathematical culture and mathematics education in hungary in the XXth century. In: Larvor, B. (ed) Mathematical Cultures. Trends in the History of Science, pp. 71–89, Birkhäuser, Basel (2016)
5. Heyting, A.: After thirty years. In: Nagel, E., Suppes, P., Tarski, A. (eds.) CLMPS 1960, pp. 194–197. Stanford University Press, Stanford (1962)
6. Kalmár, L.: Über ein Problem, betreffend die Definition des Begriffes der allgemeinrekursiven Funktion. Zeitschr. f. Math. Logik und Grundlagen d. Math. **1**(2), 93–96 (1955)
7. Kalmár, L.: Travel Report and Other Documents Related to the Constructivity in Mathematics Colloquium. In: Folder MS 2044 at Kalmár's Nachlass at the Klebelsberg Library at the University of Szeged (1956–1957)
8. Kalmár, L.: An argument against the plausibility of church's thesis. In: Heyting, A. (ed) Constructivity in Mathematics, Amsterdam, pp. 72–80, North-Holland, Amsterdam (1959)
9. Kalmár, L., Péter, R.: Correspondence with Rózsa Péter. In: Folder MS 1966 at Kalmár's Nachlass at the Klebelsberg Library at the University of Szeged (1930–1976)
10. Kleene, S.C.: General recursive functions of natural numbers. Math. Annalen **112**(5), 727–742 (1936)
11. Kleene, S.C.: A note on recursive functions. Bull. Amer. Math. Soc. **42**(8), 544–546 (1936)
12. Kleene, S.C.: Recursive predicates and quantifiers. Trans. AMS **53**(1), 41–73 (1943)
13. Kleene, S.C.: Introduction to Metamathematics. North-Holland, Amsterdam (1952)
14. Makay, Á.: The activities of László Kalmár in the world of information technology. Acta Cybernetica **18**(1), 9–14 (2007)
15. Máté, A.: Kalmár and Péter on the Philosophy and Education of Mathematics. Talk delivered at Logic in Hungary (2005). http://phil.elte.hu/mate/KalP%E9tPhil.doc. Accessed 23 Feb 2021

16. Mendelson, E.: On some recent criticism of church's thesis. Notre Dame J. Formal Logic **4**(3), 201–205 (1963)
17. Moschovakis, Y.: Review of four recent papers on church'sthesis. JSL **33**(3), 471–472 (1968)
18. Mosconi, J.: The developments of the concept of machine computability from 1936 to the 1960s. In: Dubucs, J., Bourdeau, M. (eds.) Constructivity and Computability in Historical and Philosophical Perspective. LEUS, vol. 34, pp. 37–56. Springer, Dordrecht (2014). https://doi.org/10.1007/978-94-017-9217-2_2
19. Péter, R.: Recursive Functions. 3rd edn. Akadémiai Kiadó, Budapest and Academic Press, New York (1967). Rekursive Funktionen. 1st edn, Akadémiai Kiadó, Budapest (1951)
20. Péter, R.: Graphschemata und Rekursive Funktionen. Dialectica **12**, 373–393 (1958)
21. Péter, R.: Rekursivität und Konstruktivität. In: Heyting, A. (ed) Constructivity in Mathematics, Amsterdam, pp. 226–233, North-Holland, Amsterdam (1959)
22. Péter, R.: Playing with Infinity. New York: Dover (1976). 1st English edn (1961)
23. Péter, R.: Automatische Programmierung zur Berechnung der Partiell-rekursiven Funktionen. Studia Sci. Math. Hung. **4**, 447–463 (1969)
24. Péter, R.: Recursive Functions in Computer Theory. Akadémiai Kiadó, Budapest and Ellis Horwood Ltd, Chichester (1981). Rekursive Funktionen in der Komputer-Theorie. 1st edn, Akadémiai Kiadó, Budapest (1976)
25. Robinson, R.: Review of [19]. JSL **23**(3), 362–363 (1958)
26. Shannon, C.: A symbolic analysis of relay and switching circuits. Trans. AIEE **57**(12), 713–723 (1938)
27. Shestakov, V.: Algebra of two poles schemata (Algebra of A-schemata). [in Russian] J. Tech. Phys. **11**(6), 532–549 (1941)
28. Sieg, W.: On Computability. In Irvine, A. (ed) Philosophy of Mathematics (Handbook of the Philosophy of Science). North-Holland Publishing Company, Amsterdam, pp. 535–630 (2009)
29. Sieg, W.: Gödel's philosophical challenge (to Turing). In: Copeland, J., Posy, C., Shagrir, O. (eds.) Computability: Turing, Gödel, Church, and Beyond, pp. 183–202. MIT Press, Cambridge (2013)
30. Sundholm, G.: Constructive recursive functions, church's thesis, and Brouwer's theory of the creating subject: afterthoughts on a Parisian joint session. In: Dubucs, J., Bourdeau, M. (eds.) Constructivity and Computability in Historical and Philosophical Perspective. LEUS, vol. 34, pp. 1–35. Springer, Dordrecht (2014). https://doi.org/10.1007/978-94-017-9217-2_1
31. Sushchanskii, V., Lazebnik, F., Ustimenko, V., et al.: Lev Arkad'evich Kalužnin (1914–1990). Acta Appl. Math. **52**, 5–18 (1998)
32. Szabó, M.: The M-3 in Budapest and in Szeged. Proc. IEEE **104**(10), 2062–2069 (2016)
33. Szabó, M.: Kalmár's argument against the plausibility of church's thesis. Hist. Phil. Logic **39**(2), 140–157 (2018)
34. Szabó, M.: László Kalmár and the first university-level programming and computer science training in hungary. In: Leslie, C., Schmitt, M. (eds.) HC 2018. IAICT, vol. 549, pp. 40–68. Springer, Cham (2019). https://doi.org/10.1007/978-3-030-29160-0_3
35. Unsung Heroes in Dutch Computing History: ARMAC, http://www-set.win.tue.nl/UnsungHeroes/machines/armac.html. Accessed 26 Apr 2021

Constructive Mathematics, Church's Thesis, and Free Choice Sequences

D. A. Turner[✉]

University of Kent, Canterbury, UK
D.A.Turner@kent.ac.uk

Abstract. We see the defining properties of constructive mathematics as being the proof interpretation of the logical connectives and the definition of function as rule or method.

We sketch the development of intuitionist type theory as an alternative to set theory. We note that the axiom of choice is constructively valid for types, but not for sets. We see the theory of types, in which proofs are directly algorithmic, as a more natural setting for constructive mathematics than set theories like IZF.

Church's thesis provides an objective definition of effective computability. It cannot be proved mathematically because it is a conjecture about what kinds of mechanisms are physically possible, for which we have scientific evidence but not proof. We consider the idea of free choice sequences and argue that they do not undermine Church's Thesis.

Keywords: Constructive type theory · Church's thesis · Free choice sequence

Introduction

What makes constructive mathematics *constructive*? I believe it is two things: (i) the *proof interpretation* of the logical connectives, and (ii) restoring the older meaning of function as *rule or method* of which it had been stripped by the development of set theory in late 19C. Both steps are due to Brouwer (1908), whose point of departure was the paradoxes of set theory.

Brouwer's *intuitionism* also drew on his intuitions about free choice sequences for conclusions about properties of the continuum. Modern *constructivism* dates from Bishop (1967) whose treatment of the reals is straightforwardly constructive and doesn't make use of free choice sequences. Bridges and Richman (1987) give a thorough technical comparison of Brouwer's intuitionism, Bishop-style constructivism, and a third strand, *Russian constructivism*, due to Markov, which identifies function with recursive function, thus incorporating Church's Thesis.

Bishop and Bridges (1985) develop constructive analysis within the (informal) framework of set theory using intuitionistic logic and without the axiom of choice. A formal system of constructive set theory (CST) along these lines

© Springer Nature Switzerland AG 2021
L. De Mol et al. (Eds.): CiE 2021, LNCS 12813, pp. 446–456, 2021.
https://doi.org/10.1007/978-3-030-80049-9_44

appears in Myhill (1975). The intuitionist set theory (IZF) of Friedman (1973) is similar, for a full discussion of these theories see Beeson (1985), Chap. VIII.

An alternative to CST or IZF as a framework for the formal development of constructive mathematics, is *the intuitionist theory of types* of Per Martin-Löf (1973), and its descendants such as Homotopy Type Theory (Univalent Foundations 2013). These type theories are based on propositions-as-types and differ radically from set theories, whose essential ingredient is some version of the axiom of comprehension.

In the following sections I will cover:

1. Sketch the emergence of propositions-as-types as an alternative to set theory, with a note on the conflicted status of AC (axiom of choice).
2. Church's thesis and constructivity
3. Remarks on free choice sequences.

1   From Frege to Martin-Löf

Propositional Functions

The *Begriffsschrift* of Frege (1879) broke from the analysis, current since Aristotle, of the proposition as comprising subject and predicate. Instead we have an *n*-ary *propositional function* applied to *n* terms

$$P(a_1, \ldots, a_n)$$

one of the a_i might be the grammatical subject, others direct and indirect objects, but in Frege's analysis the terms are all treated in the same way. Each term is a referring (or denoting) expression which has a reference (or denotation). In the case of a mathematical proposition this will be a mathematical entity like a number, a function, etc.

But what is the reference of a complete proposition, that is of a propositional function supplied with its arguments? For Frege it was a truthvalue, either True or False. If P is a complete (or saturated) proposition for which we have a proof we can write the *judgement*

$$\vdash P$$

asserting that P has the value True. This is the sole form of judgement in Frege's system. Note that the judgement is not manifest, that is valid on its face. To be justified in writing it we must have a sequence of valid steps in Frege's system whose last step is $\vdash P$.

Frege's analysis of meaning in terms of reference led to various difficulties leading him (Frege 1892) to introduce a second notion of meaning, *sense*. So for example $\sqrt{16}$ and 4 have the same reference but different senses. But the exact nature of *sense* remained elusive.

Jumping ahead by 90 years, to Howard (1969), we can see in propositions-as-types, an elegant solution to Frege's difficulties. The reference, or denotation, of a proposition is not a truthvalue, but a *type*, namely the type of its proofs. But what is a type?

Types

Russell (1903) in Appendix B "The Doctrine of Types", defines a type as *the range of significance of a propositional function*, that is what we would now call its *domain*. This is different from a set, which is the *extension* of a propositional function, that is the collection of values for which it is true. Types were introduced in an attempt to block self-reference which appear to be at the root of the paradoxes which had been found in set theory, such as the Russell paradox.

Applying the doctrine of types to Frege's analysis of propositions introduces another form of judgement, which we will write

$$a_i :: T_i$$

saying that term a_i has type T_i. Note that this is a manifest judgement, it should be verifiable on its face as a condition of well-formation of the formula in which it stands. To make this judgement we may need context, because in mathematical reasoning we introduce variables, always with their types e.g.,

Let n be a natural number ...

let f be a function in $R \rightarrow R$ ...

So the general form of a typing judgement is

$$\Gamma \vdash a :: T$$

where Γ is a sequence of hypotheses introducing variables with their types. This is a manifest judgement, whose validity can be mechanically checked, as in the type systems of functional programming languages such as Haskell or Agda.

Types Versus Sets

Types and sets are quite different in behaviour

- set membership $e \in S$ is not in general decidable but requires proof; that is $e \in S$ is a *proposition* not a judgement.
- basic operations on sets include union $S \cup T$ and intersection $S \cap T$, which usually make no sense on types; the natural operations on types are cartesian product $A \otimes B$, disjoint sum $A \oplus B$, the function type $A \rightarrow B$ and the dependent versions of product and function types.
- sets are equal iff they have the same elements—this is *the axiom of extensionality*; the situation with types is more complicated—including that in constructive type theories we must distinguish propositional equality from definitional equality. In Homotopy Type Theory, types are propositionally equal when they are *isomorphic*.
- set theory has the *axiom*[1] *of separation*, or *restricted comprehension*: if T is a set and P a property we can form the set $S = \{x \in T \mid P(x)\}$. This effectively erases the distinction between type and set implicit in Russell's doctrine of types. ZF set theory is typeless, or to put it another way there is only one type—everything is a set.

---

[1] Technically an axiom schema.

– the *axiom of choice* of set theory is constructively problematic, as we discuss later, while a choice principle is provable in the main versions of type theory.

Propositions as Types

The standard account of intuitionistic logic is the BHK (for "Brouwer, Heyting, Kolmogorov") or proof interpretation, see for example (Troelstra & van Dalen 1988). Paraphrasing slightly[2] we have

1. A proof of $A \wedge B$ is given by presenting a proof of A and a proof of B.
2. A proof of $A \vee B$ is given by presenting a proof of A or a proof of B *and saying which has been given.*
3. A proof of $A \supset B$ is a rule or method for constructing from any proof of A, a proof of B.
4. Absurdity $\perp$ (contradiction) has no proof; a proof of $\neg A$ is a rule or method for constructing from any proof of A, a proof of contradiction.
5. A proof of $(\forall x : D)A(x)$ is a rule or method which for any $d : D$ constructs a proof of $A(d)$.
6. A proof of $(\exists x : D)A(x)$ is given by providing a $d : D$, and a proof of $A(d)$.

From 2 we see why the law of the excluded middle is rejected. To assert $P \vee \neg P$ in the general case would require a universal decision procedure for mathematical propositions.

Given the above definitions, propositions–as–types (aka the Curry-Howard isomorphism) jumps out, once we are given the idea. We see that a proof of $A \wedge B$ is a pair (a, b) of $A \otimes B$; that a proof of $A \vee B$ is a left or right element of the disjoint union $A \oplus B$; that a proof of $A \supset B$ is a function in $A \to B$; that Absurdity is the empty type. The relation of proof to proposition proved is seen to be the same "::" judgement already met, of a term to its type.

The interpretations of rules 5 & 6 for the quantifiers require, respectively, dependent function and product types. In stating these rules I have the range of quantification, D, a type. Universal quantification over a set, $(\forall x \in S)A(x)$, can be translated as $(\forall x : T)P(x) \supset A(x)$ where P is the defining property of set S in type T. Similarly $(\exists x \in S)A(x)$ can be translated $(\exists x : T)P(x) \wedge A(x)$.

The coincidence of the types of closed terms in typed λ-calculus with the tautologies of intuitionistic implication, and of the terms themselves with natural deduction proofs of these formulae, was first noted in Curry and Feys (1958). Howard (1969) extends this to a lambda calculus with sums and products and full intuitionistic first order logic.

The *Intuitionist Theory of Types* of Martin-Löf (1973) adds equality types, natural numbers and induction, and a hierarchy of universes so that types have types. This is both a functional programming language and a formal system for constructive mathematics in the sense of Bishop (1967)—with fully formal

---

[2] I have used "rule or method for constructing" where Troelstra & van Dalen say "construction for transforming".

proofs. It is strongly normalising and has decidable typing judgement (which is the relation between element and type aka that between proof and theorem).

This has given rise to a number of descendant constructive type theories, including Homotopy Type Theory, and computer systems for developing proofs in constructive mathematics alias functional programs, including Robert Constable's Nuprl at Cornell, Agda (Norell 2007) designed at Chalmers by Caterina Coquand and Ulf Norell, and COQ, developed at INRIA and based on the Calculus of Inductive Constructions, a development from Coquand and Huet (1988).

Mathematicians have a century of highly successful work based on set theory and it is unsurprising that constructive mathematics has to date, from Bishop (1967) on, often been done using constructive versions of set theory. I believe this will change as systems based on type theories like those above continue to develop and become more convenient to use.

It is in any case likely that the publication standard in mathematics (whether classical or constructive) will come to require the submission of fully formal machine checked versions of proofs.

A Note on the Axiom of Choice

The principle of dependent choice

$$(\forall x : A)(\exists y : B)C(x, y) \supset (\exists f : A \to B)(\forall x : A)C(x, f(x))$$

is easily provable in Martin-Löf type theory and its descendents.

The same principle in set theory is an axiom, the *axiom of choice* (AC), whose constructive status is widely regarded as problematic.

Bishop (1967) remarks in his introduction that "choice is implied by the constructive meaning of existence". But in the development of constructive analysis Bishop is working within an (informal) set theory close to Myhill's CST, which does not include AC.

Goodman and Myhill (1978) give a proof that in ZF set theory AC implies the law of the excluded middle, $P \vee \neg P$. So AC is omitted from constructive versions of ZF for good reason.

Goodman & Myhill's proof uses two other axioms of set theory besides AC: the axiom of pairing, that the set $\{a, b\}$ exists for any (not necessarily different) a, b and the axiom of extensionality. Pairing is a special case of the axiom of separation.

The axiom of separation is not part of Martin-Löf type theory—by conflating types and sets it would destroy the decidability of the typing judgement. That the choice principle for types is constructively valid while AC (for sets) is not, is telling us that the latter is a stronger claim.

There have been proposals to add a subtyping construct, analogous to the axiom of separation, to constructive type theory. In the light of Goodman & Myhill's result, this looks suspect—and Thompson (1992) gives arguments why a subtype construction is not needed in type theory.

2   Church's Thesis and Constructivity

In or around 1936, an objective definition of *effective computability* emerged: the general recursive functions of Herbrand, Gödel and Kleene, the λ-definable functions of Church, and those computable by Turing's "logical computing machines" were found to be one and the same class of number theoretic functions, of which further definitions have since been found.

These are the *partial recursive functions*, whose types are $N^k \rightarrow \overline{N}$, with N the type of the natural numbers and $\overline{N} = N \cup \{\bot\}$ where $\bot$ stands for undefined or non-terminating. Without loss of generality we can restrict ourselves to the type $N \rightarrow \overline{N}$; by the use of Gödel numbering, elements of N can be used to represent any finite input or output data. The *recursive functions* are the *total* partial recursive functions, whose type can be written $N \rightarrow N$.

Church's Thesis (jointly due to Church and Turing) is the conjecture that this class identifies what is computable by any realisable mechanism[3] By mechanism we here mean one that takes a symbolic input and produces a symbolic output; a device with analog components is not excluded provided it meets that condition.

The thesis provides an objective basis to computability results that underly computing science. It is shown that such and such is not recursive (aka not computable by a Turing machine etc.), and Church's Thesis allows us to say that the proposed function is *not effectively computable*. The earliest example of this mode of reasoning is Church (1936b) where he argues from the results in Church (1936a) that the decision problem for predicate calculus is *unsolvable*. It is striking that in stating his conclusion Church uses the word "unsolvable" without any qualifying adverb.

What is equally important and emerged together with Church's thesis is the *undecidability of termination*. Given a computation of type $\overline{N}$ there is no generally effective procedure for determining if it is $\bot$. From this other undecidability results are derived, including that for first order predicate logic.

Church's Thesis is Empirical

Gandy (1980) shows that assuming a small number of rather general physical principles—finite (although possibly changing) number of parts, no infinitesimal parts, no instantaneous action at a distance etc.—what can be computed by any machine is recursive.

---

[3] In his analysis of Church's Thesis, Gandy (1980) distinguishes between two claims, which he calls T and M. Theorem T says anything that can be computed by an idealised human being following rules, with an unlimited supply of paper and a pencil, is recursive (alias computable by a Turing machine). Gandy calls it a theorem because, although not formally provable, it is intuitively clear that a Turing machine models an idealised human computer. Thesis M says anything that can be mechanically computed is recursive. This implies T but is stronger. Hodges (2006) offers evidence that Church and Turing held the stronger claim, and this is what I take as Church's Thesis in this paper.

Quantum computers lie outside Gandy's assumptions by the use of quantum entanglement. They make feasible certain computations that would be exponential on a conventional computer, for example factorizing an integer in linear time. But they do not allow the computation of anything that is not recursive. See Rieffel and Polak (2000) for a survey.

Church's thesis reflects the fact that, according to our current understanding, our universe (or at least the part of it which can have causal effects on us) contains no infinities and no infinitesmals, the latter because space, time and matter are all quantised.

It is thus a conjecture of physics, rather like the second law of thermodynamics, for which we have strong scientific evidence but no proof. Deutsch (1997) takes exactly this view.

It should therefore not be built into constructive mathematics. Mathematics should have the ability to describe universes other than our own.

Are There Non-computable Functions?

In classical mathematics non-computable functions flow from the law of the excluded middle and its friend the law of double negation. It is enough to write a *specification* for a function, prove that it is not absurd, and then, as if by magic, we have the function without having provided a method of computing it. In his "Constructivist Manifesto" Bishop[4] sees these as an obscuring fog that must be blown away to reveal a leaner, algorithmic structure.

If we accept the constructive definition of function as *rule or method* then non-computable function is an *oxymoron*, like a round square or a four cornered triangle. See Greenleaf (1992) "Bringing Mathematics Education into the Algorithmic Age".

A non-recursive function is logically possible—because Church's Thesis might be false—but to demonstrate the existence of one you would have to produce a method for computing it that others can understand and use. Since CT is almost certainly true this is not expected to happen. So constructive mathematics is destined to remain in the position that it can neither produce an example of a non-recursive function nor a proof that none exist. I don't see this as a problem, it seems entirely reasonable.

A fallacious argument from cardinality is sometimes advanced to "prove" that non-recursive functions both exist and greatly outnumber the recursive ones:

1. $N \to N$ is uncountable, by diagonalisation (Cantor).
2. The partial recursive functions $N \to \overline{N}$ are recursively enumerable
3. The total recursive functions, as a subset of (2), are countable
4. Therefore "almost all" members of $N \to N$ are non-recursive.

Classically this is unproblematic. Viewed constructively, the argument fails at step 3. An effective enumeration of a set yields an effective enumeration of a

---

[4] Bishop and Bridges (1985) Chap. 1.

subset only if the subset is decidable. The proof of (3) from (2) requires a *totality* test and totality for partial recursive functions is not recursively decidable. The argument is circular; it assumes a non-recursive function to prove that non-recursive functions exist.

The underlying issue is that the classical theory of transfinite cardinals is almost entirely non-constructive. Constructively, a set that is not countable has a *more complex internal structure* than a countable one but that does not necessarily make it "bigger" nor prevent it from being a subset of the latter. The ordering of transfinite sets by cardinality (trichotomy) is equivalent to the axiom of choice, which is not constructively valid.

Note that the total recursive functions constitute a set whose countability we cannot prove from the given facts. This is different from a set whose countability is absurd (leads to contradiction) meaning it is provably *uncountable*. It is the absence of the law of the excluded middle that separates these two situations.

In constructive mathematics the set $N \to N$ is uncountable, by Cantor's diagonalisation proof[5]. This does not imply the existence of non-recursive functions, which is logically independent.

To show this we use the result, established by Bridges and Richman (1987), that Bishop's constructive mathematics, BISH in their terminology, sits in the logical intersection of classical mathematics (CLASS), the recursive mathematics of Markov (RUSS), and Brouwer's intuitionism (INT), although the last is not relevant here. Anything provable in BISH is provable in the other three systems.

Conversely, any proposition that would lead to a contradiction in CLASS, RUSS, or INT, cannot be provable in BISH.

We see straight away that there cannot be a constructive proof that there are non-recursive functions because that would lead to a contradiction in RUSS, which incorporates Church's Thesis as an axiom[6]. Nor can there be a constructive proof that all functions are recursive—that is impossible in CLASS by the cardinality argument.

From which we conclude that the existence of non-recursive functions can be neither proved, nor disproved constructively.

Even in recursive mathematics $N \to N$ is internally (that is recursively) uncountable by the usual diagonalisation proof, despite being externally countable from a classical viewpoint. There are no objective grounds for regarding the classical view as the "correct" one.

In summary: there is no constructive proof that non-recursive functions exist and constructive $N \to N$ is uncountable, regardless of their presence or absence.

Similar results apply concerning the status of non-recursive real numbers, which have been latched onto by some as a possible analog route for evading Church's Thesis (ignoring quantum mechanics).

---

[5] This is a proof by contradiction of non-existence, which is uncontroversial—what is not allowed, intuitionistically, is a proof by contradiction of existence.

[6] See Bridges and Richman (1987) Chap. 3.

In a survey of schemes for "Hypercomputation" by Stannett (2004), we find this statement[7]: "As is well known, a randomly selected real number has a 100% chance of being non-recursive.". An appeal to the, constructively invalid, cardinality argument, it raises another interesting question—how can a point in say, the interval $[0, 1]$ be "randomly selected"? Presumably by an infinite free choice sequence, to which topic we now turn.

3   Free Choice Sequences

Do free choice sequences contradict Church's Thesis?

Bishop is dismissive of Brouwer's use of free choice sequences for analysis, and develops constructive analysis without them. He writes

> In Brouwer's case there seems to have been a nagging suspicion that unless he personally intervened to prevent it, the continuum would turn out to be discrete. He therefore introduced the method of free-choice sequences for constructing the continuum, as a consequence of which the continuum cannot be discrete because it is not well enough defined.[8]

However, this doesn't settle the question of whether free choice sequences exist, and if so, do they give us access to something non-recursive? Concerning existence, we do not need to get entangled in discussions about whether humans have free will and in what sense.

A genuinely random, i.e., not rule-governed, sequence of bits can be generated from quantum uncertainty. It is possible to design a practical piece of equipment that detects atomic transitions or some other event subject to quantum uncertainty to produce a random sequence of numbers in some finite range $[0, k]$, of any desired length. An apparatus of this kind, ERNIE (for Electronic Random Number Indicator Equipment), first employed in 1956, is used to pick monthly winning numbers for UK government premium bonds.

So does ERNIE, by its randomness, give us access to a non-recursive function? Of course not[9]. The free choice sequences generated by ERNIE and his friends are not rule-generated but they are *always finite*. To get something that might be non-recursive we would have to run ERNIE *forever*—and forever is too long to wait. We are mortal as individuals and as a society; any apparatus we build will eventually stop working, our civilisation will eventually come to an end.

A finite sequence of natural numbers, however long, is always computable. Computable means generable by a rule or method. A rule or method is something that can be used in another time or place to get the same results. For a finite sequence you can record it, and replay it when needed or transmit it to another place to be replayed. Also for any finite sequence, there are an infinite number of algorithms that will generate it and that will remain true as it is extended.

---

[7] Teuscher (2004) p. 152.

[8] Bishop and Bridges (1985), p. 9.

[9] It was shown in (De Leeuw et al. 1956) that adding a random number generator to a Turing machine does not enlarge the class of functions it can compute.

Suppose someone resists this argument and insists that we consider a free choice sequence that runs forever and tries to argue that this has probability 1 of being a non-recursive function in $N \to k$ (appealing to the fallacious cardinality argument). My response is that what we are being asked to consider is not a function. To be given a function, in the general case, we have to be given the function *in its entirety*. Some questions can be answered by looking at points, say $f(0)$, $f(17)$, etc., but to prove something about a function in the general case you need the whole function, and a function on N can only be given, constructively, by a rule or method expressed in finite form.

The proper framework for understanding an indefinitely proceeding free choice sequence is as *codata* rather than as a function on N (see e.g., Turner 2004). What we have is a *colist* of digits which we can interrogate to get the next digit and another colist. In the case under consideration we are not given any rule governing the sequence of digits.

To conclude, I do not see free choice sequences as having any impact on the plausibility of Church's Thesis (that every computable function is recursive), because if finite they are recursive and if considered as running indefinitely they are not functions within the constructive meaning of the term.

References

Beeson, M.J.: Foundations of Constructive Mathematics. Springer, Heidelberg (1985). https://doi.org/10.1007/978-3-642-68952-9

Bishop, E.: Foundations of Constructive Analysis. McGraw-Hill (1967). Revised and reissued as Bishop & Bridges (1985)

Bishop, E., Bridges, D.: Constructive Analysis. Springer, Heidelberg (1985). https://doi.org/10.1007/978-3-642-61667-9

Bridges, D., Richman, F.: Varieties of Constructive Mathematics. Cambridge University Press (1987)

Brouwer, L.E.J.: On the unreliability of the logical principles (1908). New translation with introduction by van Atten, M. & Sundholm, G. (2015). https://www.researchgate.net/publication/283448904

Church, A.: An unsolvable problem of elementary number theory. Am. J. Math. **58**, 345–363 (1936)

Church, A.: A note on the entscheidungsproblem. J. Symbolic Logic **1**(1), 40–41 (1936)

Coquand, T., Huet, G.: The calculus of constructions. Inf. Comput. **76**, 95–120 (1988)

Curry, H.B., Feys, R.: Combinatory Logic, vol. I. North-Holland, Amsterdam (1958)

De Leeuw, K., Moore, E.F., Shannon, C.E., Shapiro N.: Computability by probabilistic machines. In: Shannon, C.E., McCarthy, J. (eds.) Automata Studies, pp. 183–212. Princeton University Press (1956)

Deutsch, D.: The Fabric of Reality. The Penguin Press, Allen Lane (1997)

Frege, G.: Begriffsschrift: a formula language, modeled on that of arithmetic, for pure thought (1879). In: van Heijenoort, J. (ed.) From Frege to Gödel – A Source Book in Mathematical Logic 1879–1931. pp. 1–82. Harvard University Press (1967)

Frege, G.: On sense and reference (1892). In: Geech, P., Black, M. (eds.) Translations From the Philosophical Writings of Gottlob Frege, pp. 56–78. Basil Blackwell, Oxford (1966)

Friedman, H.: The consistency of classical set theory relative to a set theory with intuitionistic logic. J. Symbolic Logic **38**(2), 315–319 (1973)

Gandy, R.: Church's thesis and principles for mechanisms. In: Barwise, J., Keisler, H.J., Kunen, K. (eds.), The Kleene Symposium, pp. 123–148. North-Holland, Amsterdam (1980)

Goodman, N.D., Myhill, J.: Choice Implies Excluded Middle. Zeit. Logik und Grundlagen der Math **24**, 461 (1978)

Greenleaf, N.: Bringing mathematics education into the algorithmic age. In: Myers, J.P., O'Donnell, M.J. (eds.) Constructivity in CS 1991. LNCS, vol. 613, pp. 199–217. Springer, Heidelberg (1992). https://doi.org/10.1007/BFb0021092

Hodges, A.: Did Church and Turing have a thesis about machines? In: Olszewski, A., et al. Church's Thesis After 70 Years, pp. 242–252. Ontos Verlag (2006)

Howard, W.A.: The formulae as types notion of construction (original paper manuscript of 1969). In: Hindley, R.J., Seldin, J.P. (eds.) To H. B. Curry: Essays on Combinatory Logic, Lambda Calculus, and Formalism, pp. 479–490. Academic Press (1980)

Martin-Löf, P.: An intuitionist theory of types – predicative part. In: Rose, H.E., Shepherdson, J.C. (eds.) Logic Colloquium 1973, pp. 73–118. North Holland (1975)

Myhill, J.: Constructive set theory. J. Symb. Log. **40**(3), 347–382 (1975)

Norell, U.: Towards a practical programming language based on dependent type theory. Ph.D. Thesis. Chalmers University of Technology (2007)

Rieffel, E., Polak, W.: An introduction to quantum computing for non-physicists. ACM Comput. Surv. **32**(3), 300–335 (2000)

Russell, B.: The Principles of Mathematics. Cambridge University Press (1903)

Stannett, M.: Hypercomputational models. In: Teuscher, pp. 135–157 (2004)

Teuscher, C. (ed.): Alan Turing: Life and Legacy of a Great Thinker. Springer, Heidelberg (2004). https://doi.org/10.1007/978-3-662-05642-4

Thompson, S.: Are subsets necessary in Martin-Löf type theory? In: Myers, J.P., O'Donnell, M.J. (eds.) Constructivity in CS 1991. LNCS, vol. 613, pp. 46–57. Springer, Heidelberg (1992). https://doi.org/10.1007/BFb0021082

Troelstra, A.S., van Dalen, D.: Constructivism in Mathematics. Studies in Logic and the Foundations of Mathematics, vol. 121 & 123. North-Holland, Amsterdam (1988)

Turner, D.A.: Total functional programming. J. Univ. Comput. Sci. **10**(7), 751–768 (2004)

Univalent Foundations Program: Homotopy Type Theory—Univalent Foundations of Mathematics. Institute for Advanced Study, Princeton (2013). https://homotopytypetheory.org/book/

KL-Randomness and Effective Dimension Under Strong Reducibility

Bjørn Kjos-Hanssen$^{(\boxtimes)}$ (ID) and David J. Webb (ID)

University of Hawai'i at Mānoa, Honolulu, HI 96822, USA
{bjoern.kjos-hanssen,dwebb42}@hawaii.edu
http://math.hawaii.edu/wordpress/bjoern/

Abstract. We show that the (truth-table) Medvedev degree KLR of Kolmogorov–Loveland randomness coincides with that of Martin-Löf randomness, MLR, answering a question of Miyabe. Next, an analogue of complex packing dimension is studied which gives rise to a set of weak truth-table Medvedev degrees isomorphic to the Turing degrees.

Keywords: Algorithmic randomness · Effective dimension · Medvedev reducibility

1 Introduction

Computability theory is concerned with the relative computability of reals, and of collections of reals. The latter can be compared by various means, including Medvedev and Muchnik reducibility. Among the central collections considered are those of completions of Peano Arithmetic, Turing complete reals, Cohen generic reals, random reals, and various weakenings of randomness such as reals of positive effective Hausdorff dimension.

Perhaps the most famous open problem in algorithmic randomness [2,9] is whether Kolmogorov–Loveland randomness is equal to Martin-Löf randomness. Here we show that at least they are Medvedev equivalent.

Randomness extraction in computability theory concerns whether reals that are close (in some metric) to randoms can compute random reals. A recent example is [4]. That paper does for Hausdorff dimension what was done for a notion intermediate between packing dimension and Hausdorff dimension in [3]. That intermediate notion, *complex packing dimension*, has a natural dual which we introduce in this article. Whereas our result on KL-randomness is positive, we establish some negative (non-reduction) results for our new *inescapable dimension* and for relativized complex packing dimension (in particular Theorem 14). These results are summarized in Fig. 1.

Let CR, SR, KLR, and MLR be the classes of computably random, Schnorr random, Kolmogorov-Loveland random, and Martin-Löf random reals, respectively. For basic definitions from algorithmic randoness, the reader may consult

B. Kjos-Hanssen—This work was partially supported by a grant from the Simons Foundation (#704836 to Bjørn Kjos-Hanssen).

L. De Mol et al. (Eds.): CiE 2021, LNCS 12813, pp. 457–468, 2021.
https://doi.org/10.1007/978-3-030-80049-9_45

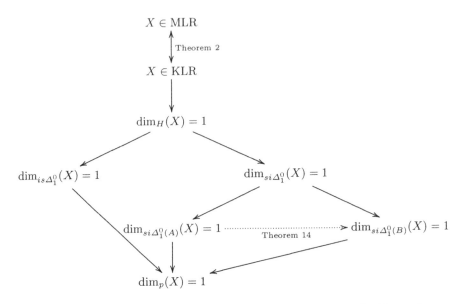

Fig. 1. Truth-table Medvedev degrees of mass problems associated with randomness and dimension. Here $\mathcal{C} \to \mathcal{D}$ means $\mathcal{C} \geq_s \mathcal{D}$, dotted arrow means $\mathcal{C} \not\geq_s \mathcal{D}$ and we assume $A \not\leq_T B$.

two recent monographs [2,9]. Let $\leq_s$ denote the uniform (strong) reducibility of mass problems known as Medvedev reducibility, and let $\leq_w$ denote the non-uniform (weak) version known as Muchnik reducibility. It was shown by Nies, Stephan and Terwijn [10] that CR $\leq_w$ SR. Miyabe [8] obtains an interesting counterpoint by showing as his main theorem that CR $\not\leq_s$ SR.

Theorem 1 ([7]). *Given a KL-random set $A = A_0 \oplus A_1$, at least one of A_0, A_1 is ML-random.*

As a corollary, MLR $\leq_w$ KLR. Miyabe [8] posed one open problem — is MLR $\leq_s$ KLR? — which we answer in Theorem 2.

Let $K(\sigma)$ denote the prefix-free Kolmogorov complexity of a string $\sigma \in 2^{<\omega}$, and let $K_s(\sigma)$ be a computable nonincreasing approximation of $K(\sigma)$ in stages $s \in \omega$. The prefix of A of length n is denoted $A \upharpoonright n$.

Theorem 2. MLR $\leq_s$ KLR.

Proof. Given a KL-random set $A = A_0 \oplus A_1$, we output bits of either A_0 or A_1, switching whenever we notice that the smallest possible randomness deficiency (c such that $\forall n \, (K(A_i \upharpoonright n) \geq n - c)$) increases.

This constant c depends on s and changes at stage $s + 1$ if

$$(\exists n \leq s + 1) \quad K_{s+1}(A_i \upharpoonright n) < n - c_s.$$

By Theorem 1, one of A_0, A_1 is ML-random, hence switching will occur only finitely often. Thus our output will have an infinite tail that is ML-random, and hence be itself ML-random. $\qquad\square$

Inspection of the proof of Theorem 2 shows that we do not need the full power of Turing reductions, but have a truth-table reduction with use $\varphi(n) \leq 2n$.

2 Complex Packing Dimension and Its Analogue

Let $K(\sigma)$ denote the prefix-free Kolmogorov complexity of a string $\sigma \in 2^{<\omega}$. The prefix of A of length n is denoted $A \upharpoonright n$.

Viewed in terms of complexity [1,6], the Hausdorff and packing dimensions are dual to one another:

Definition 1. *Let $A \in 2^\omega$. The effective Hausdorff dimension of A is defined by*

$$\dim_H(A) = \sup_{m \in \mathbb{N}} \inf_{n \geq m} \frac{K(A \upharpoonright n)}{n}.$$

The effective packing dimension of A is

$$\dim_p(A) = \inf_{m \in \mathbb{N}} \sup_{n \geq m} \frac{K(A \upharpoonright n)}{n}.$$

Another notion of dimension was defined in previous work by Kjos-Hanssen and Freer [3], which we review here. Let $\mathfrak{D}$ denote the collection of all infinite Δ_1^0 elements of 2^ω. The complex packing dimension is defined as

Definition 2. $\dim_{cp}(A) = \sup_{N \in \mathfrak{D}} \inf_{n \in N} \dfrac{K(A \upharpoonright n)}{n}.$

This leads naturally to a new notion, the dual of complex packing dimension:

Definition 3. $\dim_i(A) = \inf_{N \in \mathfrak{D}} \sup_{n \in N} \dfrac{K(A \upharpoonright n)}{n}.$

This is the *inescapable* dimension of A, so named because if $\dim_i(A) = \alpha$, every infinite computable collection of prefixes of A must contain prefixes with relative complexity arbitrarily close to α. For such a real, there is no (computable) escape from high complexity prefixes.

As Freer and Kjos-Hanssen show in [3], for any $A \in 2^\omega$,

$$0 \leq \dim_H(A) \leq \dim_{cp}(A) \leq \dim_p(A) \leq 1.$$

The expected analogous result also holds:

Theorem 3. *For any $A \in 2^\omega$, $0 \leq \dim_H(A) \leq \dim_i(A) \leq \dim_p(A) \leq 1$.*

Proof. As the sets $[n, \infty)$ are computable subsets of $\mathbb{N}$, $\dim_i(A) \leq \dim_p(A)$. For the second inequality, notice that for all $m \in \mathbb{N}$ and all $N \in \Delta_1^0$,

$$\inf_{n \in [m,\infty)} \frac{K(A \upharpoonright n)}{n} \leq \inf_{n \in N \cap [m,\infty)} \frac{K(A \upharpoonright n)}{n}$$

$$\leq \sup_{n \in N \cap [m,\infty)} \frac{K(A \upharpoonright n)}{n} \leq \sup_{n \in N} \frac{K(A \upharpoonright n)}{n}.$$

$\square$

Unexpectedly, this is the best one can do. As we will see in the next section, while the Hausdorff dimension of a real is always lower than its packing dimension, any permutation is possible for the complex packing and inescapable dimensions of a real.

3   Incomparability for Inescapable Dimension

We begin with a proof that the inescapable and complex packing dimensions are incomparable in the following sense: $\dim_{cp}(A) \leq \dim_{cp}(B)$ does not imply $\dim_i(A) \leq \dim_i(B)$, nor vice versa. In fact we show a stronger statement:

Theorem 4. *There exist A and B in 2^ω such that $\dim_{cp}(A) < \dim_{cp}(B)$, but $\dim_i(B) < \dim_i(A)$.*

Recall that a real A *meets* a set of strings S if there is some $\sigma \in S$ such that σ is a prefix of A. Moreover, A is weakly 2-generic if for each dense Σ_2^0 set of strings S, A meets S [5].

For a real A, let us write $A[m,n]$ to denote the string $A(m)A(m+1)\ldots A(n-1)$. For two functions $f(n), g(n)$ we write $f(n) \leq^+ g(n)$ to denote $\exists c \forall n\, f(n) \leq g(n) + c$. We write $f(n) = \mathcal{O}(g(n))$ to denote $\exists M \exists n_0 \forall n > n_0\, f(n) \leq Mg(n)$. It will also be useful to have the following theorem of Schnorr at our disposal:

Theorem 5. *A is Martin-Löf random iff $n \leq^+ K(A \upharpoonright n)$.*

Finally, for a real A and $n \in \omega$ we use the indicator function 1_A defined by

$$1_A(n) = \begin{cases} 1 & \text{if } n \in A, \\ 0 & \text{otherwise.} \end{cases}$$

Proof (of Theorem 4). Let A be a weakly 2-generic real, and let R be a Martin-Löf random real. Let $s_k = 2^{k^2}$, $k_n = \max\{k \mid s_k \leq n\}$, $C = (01)^\omega$. Define

$$B(n) = R\,(n - s_{k_n}) \cdot 1_C(k_n).$$

Unpacking this slightly, this is

$$B(n) = \begin{cases} R\,(n - s_k) & \text{if } s_k \leq n < s_{k+1} \text{ for some even } k, \\ 0 & \text{otherwise.} \end{cases}$$

In this proof, let us say that an R-segment is a string of the form $B \upharpoonright [s_{2m}, s_{2m+1})$ for some m, and say that a 0-segment is a string of the form $B \upharpoonright [s_{2m+1}, s_{2m+2})$ for some m. These are named so that a 0-segment consists of zeros, and an R-segment consists of random bits. Notice that by construction, each such segment is much longer than the combined length of all previous segments. This guarantees certain complexity bounds at the segments' right endpoints. For instance, B has high complexity at the end of R-segments: for any even $k \in \mathbb{N}$,

$$s_{k+1} - s_k \leq^+ K\left(B\left[s_k, s_{k+1}\right]\right)$$
$$\leq^+ K(B \upharpoonright s_k) + K(B \upharpoonright s_{k+1}) \leq^+ 2s_k + K(B \upharpoonright s_{k+1}).$$

The first inequality holds by Theorem 5 because $B\left[s_k, s_{k+1}\right] = R \upharpoonright (s_{k+1} - s_k)$. The second (rather weak) inequality holds because from descriptions of $B \upharpoonright s_k$ and $B \upharpoonright s_{k+1}$ we can recover $B[s_k, s_{k+1}]$. Finally, $K(\sigma) \leq^+ 2|\sigma|$ is a property of prefix-free Kolmogorov complexity K. Combining and dividing by s_{k+1} gives

$$s_{k+1} - 3s_k \leq^+ K(B \upharpoonright s_{k+1})$$
$$1 - 3 \cdot 2^{-(2k+1)} \leq \frac{K(B \upharpoonright s_{k+1})}{s_{k+1}} + \mathcal{O}\left(2^{-(k+1)^2}\right) \quad \text{as } k \to \infty. \tag{1}$$

Dually, the right endpoints of 0-segments have low complexity: for odd $k \in \mathbb{N}$,

$$K(B \upharpoonright s_{k+1}) \leq^+ K(B \upharpoonright s_k) + K(B[s_k, s_{k+1}]) \leq^+ 2s_k + 2\log(s_{k+1} - s_k).$$

The first inequality is again the weak bound that $B \upharpoonright s_{k+1}$ can be recovered from descriptions of $B \upharpoonright s_k$ and $B[s_k, s_{k+1}]$. For the second, we apply the $2|\sigma|$ prefix-free complexity bound to $B \upharpoonright s_k$, but also notice that since $B[s_k, s_{k+1}] = 0^{s_{k+1} - s_k}$, it can be recovered effectively from a code for its length. Combining and dividing by s_{k+1}, we have

$$K(B \upharpoonright s_{k+1}) \leq^+ 2s_k + 2(k+1)^2$$
$$\frac{K(B \upharpoonright s_{k+1})}{s_{k+1}} \leq 2^{-(2k+1)} + \mathcal{O}\left(2^{-(k+1)^2}\right) \quad \text{as } k \to \infty. \tag{2}$$

Now we can examine the dimensions of A and B.

Claim 1: $\dim_{cp}(B) = 1$.
Let R_n be the set of right endpoints of R-segments of B, except for the first n of them — that is, $R_n = \{s_{2k+1}\}_{k=n}^{\infty}$. Then the collection of these R_n is a subfamily of $\mathfrak{D}$, so that a supremum over $\mathfrak{D}$ will be at least the supremum over this family.

$$\sup_{N \in \mathfrak{D}} \inf_{n \in N} \frac{K(B \upharpoonright n)}{n} \geq \sup_{n \in \mathbb{N}} \inf_{s \in R_n} \frac{K(B \upharpoonright s)}{s}$$
$$\geq \sup_{n \in \mathbb{N}} \inf_{s \in R_n} 1 - 3 \cdot 2^{-(2s+1)} = \sup_{m \in \mathbb{N}} 1 - 3 \cdot 2^{-(2m+1)} = 1$$

by (1).

Claim 2: $\dim_i(B) = 0$.
Let Z_n be the set of right endpoints of 0-segments of B, except for the first n of them: $Z_n = \{s_{2k}\}_{k=n}^{\infty}$. Similarly to Claim 1, we obtain

$$\inf_{N \in \mathfrak{D}} \sup_{n \in N} \frac{K(B \upharpoonright n)}{n} \leq \inf_{n \in \mathbb{N}} \sup_{s \in Z_n} \frac{K(B \upharpoonright s)}{s}$$
$$\leq \inf_{n \in \mathbb{N}} \sup_{s \in Z_n} 2^{-(2s+1)} = \inf_{m \in \mathbb{N}} 2^{-(2m+1)} = 0$$

by (2).

Claim 3: $\dim_{cp}(A) = 0$.

For each natural k and N in $\mathfrak{D}$, the following sets are dense Σ_1^0:

$$\left\{\sigma \in 2^{<\omega} : |\sigma| \in N \text{ and } (\exists s) \ K_s(\sigma) < |\sigma|/k]\right\}.$$

As A is weakly 2-generic, it meets all of them. Hence

$$\sup_{N \in \mathfrak{D}} \inf_{m \in N} \frac{K(\sigma \restriction m)}{m} = 0.$$

Claim 4: $\dim_i(A) = 1$.

For each natural k and N in $\mathfrak{D}$,

$$\left\{\sigma \in 2^{<\omega} : |\sigma| \in N \text{ and } (\forall s) \ K_s(\sigma) > |\sigma|(1 - 1/k)\right\}$$

is a dense Σ_2^0 set. As A is weakly 2-generic, it meets all of these sets. Hence

$$\inf_{N \in \mathfrak{D}} \sup_{m \in N} \frac{K(A \restriction m)}{m} = 1.$$

$\square$

We say that A is finite-to-one reducible to B if there is a total computable function $f : \omega \to \omega$ such that the preimage of each $n \in \omega$ is finite and for all n, $n \in A \iff f(n) \in B$.

Definition 4. *Let $\mathfrak{B}$ be a class of infinite sets downward closed under finite-to-one reducibility. For $A \in 2^\omega$, we define*

$$\dim_{is\mathfrak{B}}(A) = \inf_{N \in \mathfrak{B}} \sup_{n \in N} \frac{K(A \restriction n)}{n} \quad and \quad \dim_{si\mathfrak{B}}(A) = \sup_{N \in \mathfrak{B}} \inf_{n \in N} \frac{K(A \restriction n)}{n}.$$

Notice that for any oracle X, the classes of infinite sets that are $\Delta_n^0(X), \Sigma_n^0(X)$ or $\Pi_n^0(X)$ are downward closed under finite-to-one reducibility, and so give rise to notions of dimension of this form. We will label these $\mathfrak{D}_n(X)$, $\mathfrak{S}_n(X)$, and $\mathfrak{P}_n(X)$ respectively, leaving off X when X is computable. Interestingly, for fixed n, the first two give the same notion of dimension.

Theorem 6. *For all $A \in 2^\omega$ and $n \in \mathbb{N}$, $\dim_{is\Sigma_n^0}(A) = \dim_{is\Delta_n^0}(A)$.*

Proof. We prove the unrelativized version of the statement, $n = 1$.

[$\leq$] As $\Delta_1^0 \subseteq \Sigma_1^0$, this direction is trivial.

[$\geq$] As every infinite Σ_1^0 set N contains an infinite Δ_1^0 set N', we have

$$\dim_{is\Sigma_1^0}(A) = \inf_{N \in \mathfrak{S}_1} \sup_{n \in N} \frac{K(A \restriction n)}{n} \geq \inf_{N \in \mathfrak{S}_1} \sup_{n \in N'} \frac{K(A \restriction n)}{n}$$

$$\geq \inf_{N \in \mathfrak{D}_1} \sup_{n \in N} \frac{K(A \restriction n)}{n} = \dim_{is\Delta_1^0}(A).$$

$\square$

By a similar analysis, the analogous result for si dimensions is also true.

Theorem 7. *For all $A \in 2^\omega$ and $n \in \mathbb{N}$, $\dim_{si\Sigma_n^0}(A) = \dim_{si\Delta_n^0}(A)$.*

What about the Π_n^0 dimensions? Unlike the Σ_1^0 case, these do not collapse to any other dimension. Two lemmas will be useful in proving this. The first (which was implicit in Claims 1 and 2 of Theorem 3) will allow us to show that an si-dimension of a real is high by demonstrating a sequence that witnesses this. The second is a generalization of the segment technique, forcing a dimension to be 0 by alternating 0- and R-segments in a more intricate way, according to the prescriptions of a certain real. The constructions below proceed by selecting a real that will guarantee that one dimension is 0 while leaving room to find a witnessing sequence for another.

Lemma 1 (Sequence Lemma). *Let $\mathfrak{B}$ be a class of infinite sets downward closed under finite-to-one reducibility, and let $N = \{n_k \mid k \in \omega\} \in \mathfrak{B}$.*

1. *If $\displaystyle\lim_{k\to\infty} \frac{K(X \upharpoonright n_k)}{n_k} = 1$, then $\dim_{si\mathfrak{B}}(X) = 1$.*

2. *If $\displaystyle\lim_{k\to\infty} \frac{K(X \upharpoonright n_k)}{n_k} = 0$, then $\dim_{is\mathfrak{B}}(X) = 0$.*

Proof. We prove (1); (2) is similar.

Form the infinite family of sets $\{N_m\}$ defined by $N_m = \{n_k \mid k \geq m\}$. From the definition of the limit, for any $\varepsilon > 0$ there is an l such that

$$\inf_{N_l} \frac{K(X \upharpoonright n_k)}{n_k} > 1 - \varepsilon.$$

As ε was arbitrary,

$$\sup_m \inf_{N_m} \frac{K(X \upharpoonright n_m)}{n_m} = 1.$$

Thus as $\mathfrak{B}$ is closed under finite-to-one reduction, the N_m form a subfamily of $\mathfrak{B}$, so that $\sup_{N \in \mathfrak{B}} \inf_{n \in N} K(X \upharpoonright n)/n = 1$. $\qquad\square$

Recall that an infinite real A is said to be *immune* to a class $\mathfrak{B}$ if there is no infinite member $B \in \mathfrak{B}$ such that $B \subseteq A$ as sets, or *co-immune* to a class $\mathfrak{B}$ if its complement is immune to $\mathfrak{B}$. We will sometimes refer to these properties as $\mathfrak{B}$-immunity or $\mathfrak{B}$-co-immunity, respectively.

Lemma 2 (Double Segment Lemma). *Let $X_0 \in 2^\omega$ be such that X_0 is $\mathfrak{B}$-immune for a class $\mathfrak{B}$ of infinite sets downward closed under finite-to-one reducibility. Set $X = X_0 \oplus X_0$. Let $s_k = 2^{k^2}$, and $k_n = \max\{odd\ k \mid s_k \leq n\}$. Let A be an arbitrary real and let R be Martin-Löf random.*

1. *If $B = A\,(n - s_{k_n}) \cdot 1_X(k_n)$, then $\dim_{si\mathfrak{B}}(B) = 0$.*
2. *If $B = R\,(n - s_{k_n}) \cdot 1_{\overline{X}}(k_n)$, then $\dim_{is\mathfrak{B}}(B) = 1$.*

Again, we will give a detailed proof of only the $\dim_{si}\mathfrak{B}$ result (though the necessary changes for $\dim_{is}\mathfrak{B}$ are detailed below). Unpacking the definition of B,

$$B(n) = \begin{cases} A\,(n - s_k) & \text{if } k_n \in X \\ 0 & \text{otherwise.} \end{cases}$$

B is here built out of segments of the form $B\,[s_{k_n}, s_{k_n+2}]$ for odd k. Here a segment is a 0-segment if $k_n \notin X$, or an A-segment if $k_n \in X$, which by definition is a prefix of A. These segments are now placed in a more intricate order according to X, with a value n being contained in a 0-segment if $X(k_n) = 0$, and in an A-segment if $X(k_n) = 1$. With some care, this will allow us to leverage the $\mathfrak{B}$-immunity of X_0 to perform the desired complexity calculations.

Specifically, we want to show that for any $N \in \mathfrak{B}$, $\inf_N K(B \upharpoonright n)/n = 0$. It is tempting to place the segments according to X_0 and invoke its $\mathfrak{B}$-immunity to show that for any $N \in \mathfrak{B}$, there are infinitely many $n \in N$ such that n is in a 0-segment, and argue that complexity will be low there. The problem is that we have no control over *where* in the 0-segment n falls. Consider in this case the start of any segment following an A-segment: $n = s_{k_n}$ for $k_n - 1 \in X_0$ and $k_n \in X_0$. We can break A and B into sections to compute

$$K(A \upharpoonright n) \leq^+ K(A \upharpoonright (n - s_{k_n-1})) + K(A[n - s_{k_n-1}, n])$$
$$= K(B[s_{k_n-1}, n]) + K(A[n - s_{k_n-1}, n]) \qquad (k_n - 1 \in X_0)$$
$$\leq^+ K(B \upharpoonright n) + K(B \upharpoonright s_{k_n-1}) + K(A[n - s_{k_n-1}, n])$$
$$K(A \upharpoonright n) \leq^+ K(B \upharpoonright n) + 4s_{k_n-1} \qquad (K(\sigma) \leq^+ 2|\sigma|)$$

Even if n is the start of a 0-segment, if $K(A \upharpoonright n)$ is high, $K(B \upharpoonright n)$ may not be as low as needed for the proof. Our definition of X avoids this problem:

Proof (of Theorem 2). Suppose for the sake of contradiction that for some $N \in \mathfrak{B}$, there are only finitely many $n \in N$ with $k_n, k_n - 1 \in \overline{X}$, i.e., that are in a 0-segment immediately following another 0-segment. Removing these finitely many counterexamples we are left with a set $N' \in \mathfrak{B}$ such that for all $n \in N'$, $\neg[(k_n \notin X) \wedge (k_n - 1 \notin X)]$. As k_n is odd, the definition of X gives that $\lfloor k_n/2 \rfloor \in X_0$. By a finite-to-one reduction from N', the infinite set $\{\lfloor k_n/2 \rfloor\}_{n \in N'}$ is a member of $\mathfrak{B}$ and is contained in X_0 - but $\overline{X_0}$ is immune to such sets.

Instead it must be the case that there are infinitely many $n \in N$ in a 0-segment following a 0-segment, where the complexity is

$$K(B \upharpoonright n) \leq^+ K\left(B \upharpoonright s_{n_{k-1}}\right) + K\left(B\,[s_{n_{k-1}}, n]\right)$$
$$\leq^+ 2 \cdot s_{n_{k-1}} + 2\log\left(n - s_{n_{k-1}}\right).$$

Here the second inequality follows from the usual $2|\sigma|$ bound and the fact that $B\,[s_{n_{k-1}}, n]$ contains only 0s. As $2^{k_n^2} \leq n$, we can divide by n to get

$$\frac{K(B \upharpoonright n)}{n} \leq^+ \frac{2^{k_n^2 - 2k_n}}{2^{k_n^2}} + \frac{2\log(n)}{n} = 2^{-2k_n} + \frac{2\log(n)}{n}.$$

As there are infinitely many of these n, it must be that $\inf_{n \in N} K(B \upharpoonright n)/n = 0$. This holds for every real N with property $\mathfrak{B}$, so taking a supremum gives the result.

The $\dim_{is\mathfrak{B}}$ version concerns reals B constructed in a slightly different way. Here, the same argument now shows there are infinitely many $n \in N$ in an R-segment following an R-segment. At these locations, the complexity $K(B \upharpoonright n)$ can be shown to be high enough that $\sup_N K(B \upharpoonright n)/n = 1$, as desired. □

With these lemmas in hand, we are ready to prove the following theorem:

Theorem 8. *For all natural n there is a set A with $\dim_{si\Pi_n^0}(A) = 1$ and $\dim_{si\Delta_n^0}(A) = 0$.*

Proof. We prove the $n = 1$ case, as the proofs for higher n are analogous.

Let S_0 be a co-c.e. immune set, and let R be Martin-Löf random. Set $S = S_0 \oplus S_0$, and define $k_n = \max\{\text{odd } k \mid 2^{k^2} \leq n\}$. To build A out of 0-segments and R-segments, define $A(n) = R\left(n - 2^{k_n^2}\right) \cdot 1_S(k_n)$.

As S_0 is Π_1^0, so is S. Thus the set of right endpoints of R-segments, $M = \left\{2^{k^2} \mid k \text{ is odd and } k - 1 \in S\right\}$ is also Π_1^0. By construction $\lim_{m \in M} K(A \upharpoonright m)/m = 1$ and thus the Sequence Lemma 1 gives that $\dim_{si\Pi_1^0}(A) = 1$.

As the complement of a simple set is immune, the Double Segment Lemma 2 shows that $\dim_{si\Delta_1^0}(A) = 0$. □

The proof of analogous result for the is-dimensions is similar, using the same S_0 and S, and the real defined by $B(n) = R\left(n - 2^{k_n^2}\right) \cdot 1_{\overline{S}}(k_n)$.

Theorem 9. *For all $n \geq 1$ there exists a set B with $\dim_{is\Pi_n^0}(B) = 1$ and $\dim_{is\Delta_n^0}(B) = 0$.*

It remains to show that the Δ_{n+1}^0 and Π_n^0 dimensions are all distinct. We can use the above lemmas for this, so the only difficulty is finding sets of the appropriate arithmetic complexity with the relevant immunity properties.

Lemma 3. *For all $n \geq 1$, there is an infinite Δ_{n+1}^0 set S that is Π_n^0-immune.*

Proof. We prove the unrelativized version, $n = 1$. Let C be a Δ_2^0 cohesive set that is not co-c.e, i.e., for all e either $W_e \cap C$ or $\overline{W_e} \cap C$ is finite. As $\overline{C}$ is not c.e. it cannot finitely differ from any W_e, so for all e, $W_e \setminus \overline{C} = W_e \cap C$ is infinite. Hence if $\overline{W_e} \subseteq C$, then by cohesiveness, $\overline{W_e} \cap C = \overline{W_e}$ is finite. □

Theorem 10. *For all $n \geq 1$ there exists a set A with $\dim_{si\Delta_{n+1}^0}(A) = 1$ and $\dim_{si\Pi_n^0}(A) = 0$.*

Proof. This is exactly like the proof of Theorem 8, but S_0 is now the Π_1^0-immune set guaranteed by Lemma 3. □

Again, the analogous result for is-dimensions is similar:

Theorem 11. *For all $n \geq 1$ there exists a set B with $\dim_{is\,\Pi_n^0}(B) = 1$ and $\dim_{is\,\Delta_{n+1}^0}(B) = 0$.*

After asking questions about the arithmetic hierarchy, it is natural to turn our attention to the Turing degrees. As the familiar notion of B-immunity for an oracle is exactly $\Delta_1^0(B)$-immunity for a class, we have access to the usual lemmas. We shall embed the Turing degrees into the si-$\Delta_1^0(A)$ dimensions (and dually, is-$\Delta_1^0(A)$). First, a helpful lemma:

Lemma 4 (Immunity Lemma). *If $A \not\leq_T B$, there is an $S \leq_T A$ such that S is B-immune.*

Proof. Let S be the set of finite prefixes of A. If S contains a B-computable infinite subset C, then we can recover A from C, but then $A \leq_T C \leq_T B$.   □

Theorem 12 (si-Δ_1^0 Embedding Theorem). *Let $A, B \in 2^\omega$. Then $A \leq_T B$ iff for all $X \in 2^\omega$, $\dim_{si\,\Delta_1^0(A)}(X) \leq \dim_{si\,\Delta_1^0(B)}(X)$.*

Proof. [$\Rightarrow$] Immediate, as $\Delta_1^0(A) \subseteq \Delta_1^0(B)$.
[$\Leftarrow$] This is again exactly like the proof of Theorem 8, now using the set guaranteed by the Immunity Lemma 4 as S_0.   □

The result for is-dimensions is again similar:

Theorem 13 (si-Δ_1^0 Embedding Theorem). *Let $A, B \in 2^\omega$. Then $A \leq_T B$ iff for all $X \in 2^\omega$, $\dim_{is\,\Delta_1^0(A)}(X) \geq \dim_{is\,\Delta_1^0(B)}(X)$.*

We can push this a little further by considering weak truth table reductions:

Definition 5. *A is weak truth table reducible to B ($A \leq_{wtt} B$) if there exists a computable function f and an oracle machine Φ such that $\Phi^B = A$, and the use of $\Phi^X(n)$ is bounded by $f(n)$ for all n ($\Phi^X(n)$ is not guaranteed to halt).*

Theorem 14. *If $A \not\leq_T B$, then for all wtt-reductions Φ there exists an X such that $\dim_{si\,\Delta_1^0(A)}(X) = 1$ and, either Φ^X is not total or $\dim_{si\,\Delta_1^0(B)}(\Phi^X) = 0$.*

Proof. Let $A \not\leq_T B$, and let Φ be a wtt-reduction. Let f be a computable bound on the use of Φ, and define $g(n) = \max\{f(i) \mid i \leq n\}$, so that $K(\Phi^X \restriction n) \leq^+ K(X \restriction g(n)) + 2\log(n)$. For notational clarity, for the rest of this proof we will denote inequalities that hold up to logarithmic (in n) terms as $\leq^{\log}$.

Next, we define two sequences ℓ_k and λ_k which play the role 2^{k^2} played in previous constructions:

$$\ell_0 = \lambda_0 = 1, \qquad \lambda_k = \lambda_{k-1} + \ell_{k-1}, \qquad \ell_k = \min\left\{2^{n^2} \mid g(\lambda_k) < 2^{n^2}\right\}.$$

These definitions have the useful consequence that $\lim_k \ell_{k-1}/\ell_k = 0$. To see this, suppose $\ell_{k-1} = 2^{(n-1)^2}$. As g is an increasing function, the definitions give

$$\ell_k > g(\lambda_k) \geq \lambda_k = \lambda_{k-1} + \ell_{k-1} \geq \ell_{k-1} = 2^{(n-1)^2}.$$

Hence $\ell_k \geq 2^{n^2}$, so that $\ell_{k-1}/\ell_k \leq 2^{-2n+1}$. As $\ell_k > \ell_{k-1}$ for all k, this ratio can be made arbitrarily small, giving the limit.

A triple recursive join operation is defined by

$$\bigoplus_{i=0}^{2} A_i = \{3k + j \mid k \in A_j, \quad 0 \leq j \leq 2\}, \quad A_0, A_1, A_2 \subseteq \omega.$$

Let $S_0 \leq_T A$ be as guaranteed by Lemma 4, and define $S = \bigoplus_{i=0}^{2} S_0$. Let R be Martin-Löf random, and define $X(n) = R(n - \ell_{k_n}) \cdot 1_S(k_n)$, where $k_n = \max\{k = 2 \pmod 3 \mid \ell_k \leq n\}$. This definition takes an unusual form compared to the previous ones we have seen in order to handle the interplay between λ_k and ℓ_k - specifically the growth rate of $g(n)$. We are effectively "tripling up" bits of S_0 (rather than doubling them as before) to account for the possibility that $g(n)$ grows superexponentially, with the condition that $k = 2 \pmod 3$ replacing the condition that k is odd.

Claim 1: $\dim_{si\Delta_1^0(A)}(X) = 1$.
Proof: As $N = \{\ell_k\}_{k \in S}$ is an A-computable set, by the Sequence Lemma 1 it suffices to show that $\lim_{k \in S} K(X \restriction \ell_k)/\ell_k = 1$. For $\ell_k \in N$,

$$K(X \restriction \ell_k) \geq^{\log} K(X[\ell_{k-1}, \ell_k]) - K(X \restriction \ell_{k-1})$$
$$\geq K(R \restriction (\ell_k - \ell_{k-1})) - 2\ell_{k-1} \qquad \text{(as } k \in S)$$
$$\geq^{\log} \ell_k - \ell_{k-1} - 2\ell_{k-1} \qquad \text{(as } R \text{ is Martin-Löf random)}$$
$$\frac{K(X \restriction \ell_k)}{\ell_k} \geq^{\log} \frac{\ell_k - 3\ell_{k-1}}{\ell_k} = 1 - 3\frac{\ell_{k-1}}{\ell_k}.$$

which gives the desired limit by the above.

Claim 2: $\dim_{si\Delta_1^0(B)}(\Phi^X) = 0$.
Proof: Suppose $N \leq_T B$. For notation, define $a = k_{g(n)}$. By mimicking the proof of Lemma 2, we can use the B-immunity of S to show that there are infinitely many $n \in N$ such that $g(n)$ is in a 0-segment following two 0-segments, i.e., $a - 2, a - 1, a \notin S$. By the definition of X,

$$X[\ell_{a-2}, \ell_{a+1}] = 0^{\ell_{a+1} - \ell_{a-2}}.$$

Suppose the value $X(m)$ is queried in the course of computing $\Phi^X \restriction n$. By the definitions of g, a, and ℓ_k, $m \leq g(n) < \ell_{a+1}$. Hence either $m < \ell_{a-2}$ or $m \in [\ell_{a-2}, \ell_{a+1}]$, so that $X(m) = 0$. Thus to compute $\Phi^X \restriction n$, up to a constant it suffices to know $X \restriction \ell_{a-2}$. Thus

$$K(\Phi^X \restriction n) \leq^+ K(X \restriction \ell_{a-2}) \leq^+ 2\ell_{a-2}$$

As $g(n) > \ell_a$ it must be that $n > \lambda_a$. Dividing by n, we find that

$$\frac{K(\Phi^X \restriction n)}{n} \leq^+ \frac{2\ell_{a-2}}{\lambda_a} < \frac{2\ell_{a-2}}{\lambda_{a-1} + \ell_{a-1}} < \frac{2\ell_{a-2}}{\ell_{a-1}}.$$

As there are infinitely many of these n, it must be that $\inf_{n \in N} K(\Phi^X \restriction n)/n = 0$. This holds for every $N \leq_T B$, so taking a supremum gives the result. □

Remark. We only consider si-dimensions for this theorem, as it is not clear what an appropriate analogue for is-dimensions would be. The natural dual statement for is-dimensions would be that for all reductions Φ there is an X such that $\dim_{is\Delta_1^0(A)}(X) = 0$, and either Φ^X is not total or $\dim_{is\Delta_1^0(B)}(\Phi^X) = 1$. But many reductions use only computably much of their oracle, so that Φ^X is a computable set. This degenerate case is not a problem for the si theorem, as its conclusion requires $\dim_{\Delta_1^0(B)}(\Phi^X) = 0$. But for an is version, it is not even enough to require that Φ^X is not computable - consider the reduction that repeats the nth bit of X $2n - 1$ times, so that n bits of X suffice to compute n^2 bits of Φ^X. Certainly $\Phi^X \equiv_{wtt} X$, so that Φ^X is non-computable iff X is. But

$$\frac{K(\Phi^X \upharpoonright n)}{n} \leq^+ \frac{K(X \upharpoonright \sqrt{n})}{n} \leq^+ \frac{2\sqrt{n}}{n}$$

for all n, so that $\dim_p(\Phi^X) = 0$, and hence all other dimensions are 0 as well.

References

1. Athreya, K.B., Hitchcock, J.M., Lutz, J.H., Mayordomo, E.: Effective strong dimension in algorithmic information and computational complexity. SIAM J. Comput. **37**(3), 671–705 (2007)
2. Downey, R.G., Hirschfeldt, D.R.: Algorithmic Randomness and Complexity Theory and Applications of Computability. Springer, New York (2010). https://doi.org/10.1007/978-0-387-68441-3
3. Freer, C.E., Kjos-Hanssen, B.: Randomness extraction and asymptotic Hamming distance. Log. Methods Comput. Sci. **9**(3), 3:27, 14 (2013)
4. Greenberg, N., Miller, J.S., Shen, A., Westrick, L.B.: Dimension 1 sequences are close to randoms. Theoret. Comput. Sci. **705**, 99–112 (2018)
5. Jockusch, C.: Degrees of generic sets. In: Drake, F.R., Wainer, S.S. (eds.) Recursion Theory: its Generalisations and Applications, pp. 110–139. Cambridge University Press, Cambridge (1980)
6. Mayordomo, E.: A Kolmogorov complexity characterization of constructive Hausdorff dimension. Inf. Process. Lett. **84**(1), 1–3 (2002)
7. Merkle, W., Miller, J.S., Nies, A., Reimann, J., Stephan, F.: Kolmogorov-Loveland randomness and stochasticity. Ann. Pure Appl. Logic **138**(1–3), 183–210 (2006)
8. Miyabe, K.: Muchnik degrees and Medvedev degrees of randomness notions. In: Proceedings of the 14th and 15th Asian Logic Conferences, pp. 108–128. World Sci Publ, Hackensack (2019)
9. Nies, A.: Computability and Randomness. Oxford Logic Guides, vol. 51. Oxford University Press, Oxford (2009)
10. Nies, A., Stephan, F., Terwijn, S.A.: Randomness, relativization and turing degrees. J. Symb. Logic **70**(2), 515–535 (2005)

An Algorithmic Version of Zariski's Lemma

Franziskus Wiesnet[1,2,3(✉)]

[1] Ludwig-Maximilians Universität, Theresienstr. 39, 80333 München, Germany
`wiesnet@mathematik.uni-muenchen.de`
[2] Università degli Studi di Trento, Via Sommarive 14, 38123 Povo, Italy
`franziskus.wiesnet@unitn.it`
[3] Università degli studi di Verona, Strada le Grazie 15, 37134 Verona, Italy

Abstract. Zariski's lemma was formulated and used by Oscar Zariski to prove Hilbert's Nullstellensatz. This article gives an elementary and constructive proof of Zariski's lemma and only uses basics of integral ring extensions under the condition that each field is discrete. After this constructive proof we take a look at the computational side. We give a computational interpretation of Zariski's lemma and use our constructive proof to develop an algorithm which realises the computational interpretation. This is a typical approach in constructive mathematics.

Keywords: Zariski's lemma · Constructive algebra · Computational algebra · Program extraction · Proof mining

1 Introduction

1.1 Historical Background

Presumably the first time Zariski's lemma appeared was in [19]. There Oscar Zariski used it to prove Hilbert's Nullstellensatz. In 1976, John McCabe gave an interesting but not constructive proof [9], which relied on the existence of maximal ideals. In 2020 Daniel Wessel has avoided this maximality argument by using Jacobson radicals [16]. However, the proof still contains a non-constructive moment. To wit, if R is an algebra over a field K and $S \subseteq R$ is a finite subset, then there exists $S_0 \subseteq S$ maximal such that all elements in S_0 are algebraically independent over K. To avoid this, one could use Noether normalization. A constructive proof of Noether normalisation is given in [10] and Zariski's lemma is a corollary of it [5,10,13]. The proofs in [1,2,15,19] are non-constructive but, instead of a maximal algebraically independent subset, they use induction on

I would like to thank the Istituto Nazionale di Alta Matematica "Francesco Severi" for the financial support of my PhD study. Thanks for direct support goes to Daniel Wessel for his ideas and taking a look at the manuscript, my supervisor Peter Schuster for the selection of this topic and support of the publication, and Henri Lombardi and Ihsen Yengui who helped to improve the proof with important comments.

L. De Mol et al. (Eds.): CiE 2021, LNCS 12813, pp. 469–482, 2021.
https://doi.org/10.1007/978-3-030-80049-9_46

the number of generators of the algebra. This will also be part of our constructive proof. The proof in the present paper is a direct and constructive proof of Zariski's lemma. To get this proof, we have analysed the proofs in the sources above and put them together with some new ideas.

1.2   Method of Proof Interpretation

We have considered some non-constructive proofs of Zariski's lemma, analysed them and rebuilt them into a new constructive proof (Sect. 2). This approach was inspired by the methods of *proof mining* [6,7]. Inspired by the methods of the formal *program extraction* from proofs as in [3,14,17], we have turned our constructive proof into algorithms and realisability statements (Sect. 3). But in contrast to formal program extraction, when we speak about "realisability" we do not mean the rigorously defined realisability predicate of program extraction, for example given in [14]. In this paper "realisability" is rather a heuristic notion.

Our approach shows a typical approach in constructive mathematics. Analysing a theorem constructively often goes as follows:

– Formulate a *quite* constructive proof of the theorem.
– Formulate an algorithmic interpretation of the theorem.
– Inspired by the quite constructive proof formulate an algorithm which shall realise the algorithmic interpretation.
– Prove that the algorithm is indeed a realiser of the algorithmic interpretation.

This paper is an example where these steps are done manually on paper and where the formulation of the quite constructive proof is only necessary to get an inspiration for the other steps. As the space in this paper is quite scarce we have to forgo the fourth step. In particular, we do not give proofs in Sect. 3. However, in the example of program extraction from proofs above usually only the quite constructive proof is formulated manually and the other steps are done by the computer. Note that we have written "quite constructive" because sometimes one can bypass a non-constructive moment or it can be included as assumption in the algorithmic version. We also see an example of this in the present paper: since our proof uses case distinction on $x = 0$ or $x \neq 0$ for all x in a ring, we assume that this ring is discrete. However, this is the only computational restriction we have to make.

1.3   Fundamental Notions

Before formulating a proof of Zariski's lemma and the computational interpretation, we define the underlying objects. In Zariski's lemma, we use axioms for rings, field and algebra and their structures. But an algorithm cannot operate on axioms. More specifically: if we state an algorithm about a field, we do not use the field axioms in the algorithm but we use the field structure like $+$, $\cdot$, 0, 1 and so on. Therefore, we first define the underlying structures precisely:

In our setting a *ring structure* $(R, +, \cdot, 0, 1, -, =)$ is a set R equipped with an addition operator $+ : R \times R \to R$, a multiplication operator $\cdot : R \times R \to R$, a zero element $0 \in R$, an unit element $1 \in R$, an additive inverse function $- : R \to R$ and an equality $= \, \subseteq R \times R$. If furthermore $=$ is an equivalence relation and compatible with $+, \cdot, -$, i.e., $=$ is a congruence relation on $(R, +, \cdot, 0, 1, -)$, and the other ring axioms are fulfilled (w.r.t. the equality $=$), R is a *ring*. In our case a ring is always commutative. We call $(K, +, \cdot, 0, 1, -, ^{-1}, =)$ a *field structure* if $(K, +, \cdot, 0, 1, -, =)$ is a ring structure and $^{-1} : K \to K$ is a map. If K is a ring, $xx^{-1} = 1 \vee x = 0$ for all $x \in K$ and $1 \neq 0$, K is a *field*.

Since the notation of $+, \cdot, 0, 1, -, ^{-1}$ and $=$ will not change, we do not mention it and say that R is a ring (structure) or K is a field (structure) and so on. A *homomorphism* $\phi : R \to S$ between two ring structures R and S is a map which preserves the structure in the canonical way.

For a ring structure R we define the *ring structure of polynomials* $R[X]$ with coefficients in R by the well-known construction. For $n \in \mathbb{N}$ we have also the polynomial ring structure in n variables denoted by $R[X_1, \ldots, X_n]$. Obviously, if R is a ring then so is $R[X_1, \ldots, X_n]$.

An *algebra structure* R over a field structure K, or short *K-algebra structure*, is a ring structure together with a map $K \to R$. If R is a ring, K is a field and the map $K \to R$ is a homomorphism, we call R a *K-algebra*. For a K-algebra R and $x_1, \ldots, x_n \in R$ we get an extension $K[X_1, \ldots, X_n] \to R$ of the homomorphism by $X_i \mapsto x_i$. We denote the image by $K[x_1, \ldots, x_n]$, where an element is in the image of a homomorphism if it is equal (w.r.t. $=$) to a value of the homomorphism.

The following definition comes from [8,18]:

Definition 1. *A ring structure R is* discrete *if all its operators are computable. Here $=$ is seen as a Boolean-valued function. A field structure K is* discrete *if it is discrete as ring and $^{-1}$ is computable.*

Here, "computability" means that we can use the operations above freely in our algorithms. In particular, we can use the ring operators arbitrarily, and can distinguish between the cases $x = y$ and $x \neq y$.

We do not specify the underlying theory of computability and how the objects are represented, as there are several possibilities. However, in Sect. 3 we develop an algorithm out of the constructive proof. If one wants this algorithm to be a Turing machine, a discrete structure should be interpreted as a structure where all operators (including $=$) are representable by a Turing machine.

In this article we tacitly assume that each structure be discrete and make case distinctions like $x = 0 \vee x \neq 0$ without explicitly justifying them.

Remark 1. If K is a discrete field structure then the polynomial ring structure $K[X_1, \ldots, X_n]$ is also discrete and for $f \in K[X_1, \ldots, X_n]$ we can decide whether $f \in K$ or $f \notin K$ because $f \in K$ if and only if all non-constant coefficients are zero. Similarly, it is even possible to compute $\deg(f)$ for every $f \in K[X]$.

Let $A \subseteq B$ be a ring extension, i.e., the inclusion $A \to B$ is a homomorphism. An element $x \in B$ is called *integral* over A if there are $a_0, \ldots, a_{k-1} \in A$ such that $x^k + a_{k-1}x^{k-1} + \cdots + a_0 = 0$. The ring extension $A \subseteq B$ is called *integral*, if each $x \in B$ is integral over A.

In our constructive proof we need the following two lemmas. The proofs of them we refer to are also constructive.

Lemma 1. *If $A \subseteq B$ is an integral ring extension and B is a field then A is a field, too.*

Proof. A constructive proof is given in [1, Proposition 5.7].   □

Lemma 2. *Let $A \subseteq B$ be a ring extension. If $x_1, \ldots, x_n \in B$ are integral over A then the ring extension $A \subseteq A[x_1, \ldots, x_n]$ is integral.*

Proof. This follows from Corollary 5.3 of [1].   □

2   A Constructive Proof

In this section we give a new constructive proof of Zariski's lemma. The proof does not use any non-constructive principles (except that the rings be discrete). In the next section we use this proof as basis to create an algorithmic version.

Theorem 1 (Zariski's lemma). *Let K be a field and R an algebra over K which is a field. Suppose that $R = K[x_1, \ldots, x_n]$ for some $x_1, \ldots, x_n \in R$. Then $x_1, \ldots x_n$ are algebraic over K, i.e., there are $f_1, \ldots, f_n \in K[X] \setminus K$ with $f_i(x_i) = 0$ for all i.*

Proof. If $n = 0$, there is nothing to show. We continue by considering the case $n = 1$: if $x_1 = 0$ then $R = K$ and we are done. Otherwise, x_1 is invertible. Since R is a field, there is $p \in K[X] \setminus \{0\}$ with $x_1 p(x_1) = 1$. We set $q := Xp - 1 \in K[X]$. Then $q \neq 0$ because $\deg(Xp) > 0$ and $q(x_1) = 0$.

Next, we consider the case $n = 2$: We show that x_1 is algebraic. The argument for x_2 is analogous. If $x_2 = 0$ we are done by the case $n = 1$ as above. Otherwise, we have $p \in K[X_1, X_2]$ with $p(x_1, x_2)x_2 = 1$. Therefore, $q := Xp(x_1, X) - 1$ is a polynomial in $K[x_1][X]$ with $q(x_2) = 0$ and $q \neq 0$ as its constant coefficient is -1. Let $y \in K[x_1]$ be the leading coefficient of q, which is non-zero by definition. Then $K[x_1, y^{-1}] \subseteq K[x_1, x_2]$ is an integral ring extension by Lemma 2 because x_2 is integral over $K[x_1, y^{-1}]$ witnessed by $y^{-1}q \in K[x_1, y^{-1}][X]$. Therefore, $K[x_1, y^{-1}]$ is a field by Lemma 1.

With this preparation we are now able to construct a non-zero polynomial with root x_1. By $y \in K[x_1]$, there is $f \in K[X]$ such that $f(x_1) = y$. If $f \in K$ then $K[x_1, y^{-1}] = K[x_1]$ and we are done by the case $n = 1$. So, we assume $f \in K[X] \setminus K$. If $1 - f(x_1) = 0$ then x_1 is algebraic over K. Otherwise, $1 - f(x_1)$ is invertible[1] in $K[x_1, y^{-1}]$ and therefore there is $h \in K[X]$ and $N \in \mathbb{N}$ with $(1 - f(x_1))^{-1} = h(x_1)y^{-N} = h(x_1)f(x_1)^{-N}$. So, we have

$$f(x_1)^N - h(x_1)(1 - f(x_1)) = 0.$$

---

[1] The idea to take $1 - f(x_1)$ is based on an idea by Daniel Wessel [16] and an hint by Henri Lombardi. Inspired by [15], the first approach of the author was to take $g(x_1)$ for some irreducible $g \in K[X]$ with $g \nmid f$.

It remains to show that $f^N - h(1 - f) \neq 0$ in $K[X]$. By the binomial theorem there is a $g \in K[X]$ with $f^N = 1 + (1 - f)g$, and so

$$f^N - h(1 - f) = 1 + (1 - f)(g - h).$$

Since f is non-constant, also $1 - f$ is non-constant. Now assume that $f^N - h(1 - f) = 0$ then $g - h = 0$ as otherwise $\deg((1-f)(g-h)) > 0$ and $1+(1-f)(g-h) \neq 0$. But then $0 = 1$, a contradiction.

Finally, we assume $n \geq 2$ and use induction over n. The base case $n = 2$ was done above. For the induction step let $n \geq 3$ be given. Again, we just show that x_1 is algebraic. The arguments for $x_2, \ldots, x_n$ are analogous. Let $L := K(x_1)$ the field of fractions of $K[x_1]$. Since R is a field, we can consider $L \subseteq R$ and therefore $L[x_2, \ldots, x_n] = R$. By induction, each x_i for $i \in \{2, \ldots, n\}$ is algebraic over L. So for each such i, there is a monic polynomial $f_i \in L[X]$ with $f_i(x_i) = 0$. Let v_i be the product of the denominators of all coefficients in f_i and $v := \prod_{i=2}^{n} v_i$. Then all x_i are integral over $K[x_1, v^{-1}]$. Using Lemma 2, $K[x_1, v^{-1}] \subseteq K[x_1, \ldots, x_n]$ is an integral ring extension. By Lemma 1, also $K[x_1, v^{-1}]$ is a field. By the case $n = 2$ it follows that x_1 is algebraic over K. $\qquad\square$

3   Computational Interpretation

The goal of this section is to build an algorithm out of the constructive proof above. One could argue that this is not necessary as a constructive proof provides an algorithm by definition and it is an easy exercise to extract it. However, as we are not using computer support and the proof is not totally formal, there is still some work to do. In particular, we consider the concepts we have used in the proof and give them a computational meaning in the next two definitions.

3.1   Preliminary

We use the following syntactical abbreviations: $\vec{x} := x_1, \ldots, x_n$; $\vec{X} := X_1, \ldots, X_n$; $\vec{y} := y_1, \ldots, y_m$ and $\vec{Y} := Y_1, \ldots, Y_m$. For $n \in \mathbb{N}$ and any $I \in \mathbb{N}^n$ we define $\vec{x}^I := \prod_{i=1}^{n} x_i^{I_i}$ and $\vec{X}^I := \prod_{i=1}^{n} X_i^{I_i}$.

In Zariski's lemma a K-algebra $K[\vec{x}]$ is given. In particular, there is a surjective homomorphism from $K[\vec{X}]$ to $K[\vec{x}]$. It is well-known that the existence of a right-inverse of a surjection in general requires the axiom of choice. That is the reason why we do not use it computationally and we work on the level of the polynomial rings. The following definition is the computational interpretation of $K[\vec{y}] \subseteq K[\vec{x}]$ being a ring extension on the level of polynomials:

Definition 2. *Let K be a field, R be a K-algebra and $\vec{x}, \vec{y} \in R$. We say that $K[\vec{y}] \subseteq K[\vec{x}]$ is a ring extension of K-algebras witnessed by $\vec{h} := h_1, \ldots, h_m \in K[\vec{X}]$ if $h_i(\vec{x}) = y_i$ for all i. In short notation we write $\vec{h}(\vec{x}) = \vec{y}$.*

Similarly, the next definition is the computational interpretation of $K[\vec{x}]$ being a field on the level of polynomials:

Definition 3. *Let a field K, a K-algebra R and $\vec{x} \in R$ be given. A computable function $\iota : K[\vec{X}] \to K[\vec{X}]$ with $f(\vec{x}) = 0 \vee (\iota(f))(\vec{x})f(\vec{x}) = 1$ for all $f \in K[\vec{X}]$ is called* algebraic inverse function *on $K[\vec{x}]$.*

Remark 2. An algebraic inverse function does not have to be compatible with the equality relation of the ring structure $K[\vec{X}]$. From an algebraic inverse function on $K[\vec{x}]$ and a right inverse of a surjection $K[\vec{X}] \to K[\vec{x}]$ we get that $K[\vec{x}]$ is a field. But this is constructively delicate, so in both definitions above we have avoided a direct use of $K[\vec{x}]$ and we also do this in the following algorithms. The occurrence of $K[\vec{x}]$ in the definitions above is just a way of speaking.

Similar to above, "computable" means that we can use the algebraic inverse function freely in our algorithm. For instance, if the algorithm shall be a Turing machine, an algebraic inverse function has to be Turing computable.

In the light of the definitions above: an algorithm which realises Zariski's lemma takes an algebraic inverse function on $K[\vec{x}]$ as input and returns polynomials $f_1, \ldots, f_n \in K[X] \setminus K$ with $f_i(x_i) = 0$ for all $i \in \{1, \ldots, n\}$.

3.2   Some Algorithms for Integral Extensions of Algebras

The following lemma is an algorithmic version, in terms of algebras over a field, of Lemma 2. Given a field K, R be a K-algebra and $\vec{x}, \vec{y} \in R$. As realiser of this lemma we expect an algorithm which takes for each x_i an integral equation in the form $P_i(\vec{y})(x_i) = 0$ for some monic $P_i \in K[\vec{Y}][X]$ and some $f \in K[\vec{X}]$ as input and returns an integral equation of $f(\vec{x})$ as output in the form $Q(\vec{y})(f(\vec{x})) = 0$ for some monic $Q \in K[\vec{Y}][X]$.

Algorithm 1. *Given a field structure K, $f \in K[\vec{X}]$ and $k_i \in \mathbb{N}$, $g_{k_i-1}^{(i)}, \ldots, g_0^{(i)} \in K[\vec{Y}]$ for each $i \in \{1, \ldots, n\}$. We compute $k \in \mathbb{N}$ and $g_{k-1}, \ldots, g_0 \in K[\vec{Y}]$:*

1. *Define $\mathcal{I} := \{I \in \mathbb{N}^n | I_1 < k_i, \ldots, I_n < k_n\}$ and for each $I \in \mathcal{I}$ compute the finite sum $f\vec{X}^I = \sum_{J \in \mathbb{N}^n} f_{IJ}\vec{X}^J$ with $f_{IJ} \in K$.*
2. *For each $I \in \mathcal{I}$ and $i \in \{1, \ldots, n\}$ replace each $X_i^{k_i}$ by $-g_{k_i-1}^{(i)}X^{k_i-1} - \cdots - g_0^{(i)}$ in $\sum_{J \in \mathbb{N}^n} f_{IJ}\vec{X}^J$ one by one until we get a polynomial of the form $\sum_{J \in \mathcal{I}} g_{IJ}\vec{X}^J$ with $g_{IJ} \in K[\vec{Y}]$*
3. *Compute the characteristic polynomial $P \in K[\vec{Y}][X]$ of the matrix $(g_{IJ})_{I,J \in \mathcal{I}}$ as the determinant of the matrix $(\delta_{IJ}X - g_{IJ})_{I,J \in \mathcal{I}}$, where $\delta_{IJ}X := X$ if $I = J$, and $\delta_{IJ}X := 0$ if $I \neq J$.*
4. *Let $P = \sum_{i=0}^{l} g_i X^i$ for some $l \in \mathbb{N}$ and $g_i \in K[\vec{Y}]$. Return $k := \prod_{i=1}^{n} k_i$ and the first k coefficients $g_{k-1}, \ldots, g_0$ of P, where $g_i := 0$ if $i > l$.*

Note that in Step 2 there is no order mention in which each $X_i^{k_i}$ has to be replaced. However, the following lemma is true for any possible order.

Lemma 3. *In the situation of Algorithm 1 we assume that K is a field, R is a K-algebra and $\vec{x}, \vec{y} \in R$ with*

$$x_i^{k_i} + g_{k_i-1}^{(i)}(\vec{y})x_i^{k_i-1} + \cdots + g_0^{(i)}(\vec{y}) = 0 \tag{1}$$

for each $i \in \{1, \ldots, n\}$. Then

$$(f(\vec{x}))^k + g_{k-1}(\vec{y})(f(\vec{x}))^{k-1} + \cdots + g_0(\vec{y}) = 0.$$

The next lemma is an algorithmic version of Lemma 1. In terms of K-algebras and in the light of computational algebra, we want to compute an algebraic inverse function on $K[\vec{y}]$ from an algebraic inverse function on $K[\vec{x}]$ and the integral equations of $\vec{x}$.

Algorithm 2. *Let a field structure K, $\vec{h} := h_1, \ldots, h_m \in K[\vec{X}]$, $\iota : K[\vec{X}] \to K[\vec{X}]$ and $k_i \in \mathbb{N}$, $g_{k_i-1}^{(i)}, \ldots, g_0^{(i)} \in K[\vec{Y}]$ for each $i \in \{1, \ldots, n\}$ be given. We define a map $\tilde{\iota} : K[\vec{Y}] \to K[\vec{Y}]$ as follows:*

1. *Given an input $f \in K[\vec{Y}]$, compute $p := \iota(f(\vec{h})) \in K[\vec{X}]$.*
2. *Apply Algorithm 1 to K, p and k_i, $g_{k_i-1}^{(i)}, \ldots, g_0^{(i)}$ for each $i \in \{1, \ldots, n\}$ to get $k \in \mathbb{N}$ and $g_{k-1}, \ldots, g_0 \in K[\vec{Y}]$.*
3. *Return $-g_{k-1} - g_{k-2}f - \cdots - g_0 f^{k-1}$.*

Lemma 4. *In the situation of Algorithm 2 we assume that K is a field and let a K-algebra R and $\vec{x}, \vec{y} \in R$ be given such that $K[\vec{y}] \subseteq K[\vec{x}]$ is an extension of K-algebras witnessed by $\vec{h}$. Furthermore, we assume that ι is an algebraic inverse function and*

$$x_i^{k_i} + g_{k_i-1}^{(i)}(\vec{y})x_i^{k_i-1} + \cdots + g_0^{(i)}(\vec{y}) = 0$$

for all $i \in \{1, \ldots, n\}$. Then $\tilde{\iota}$ is an algebraic inverse function on $K[\vec{y}]$.

3.3 An Algorithm for Zariski's Lemma

In the following we give an algorithmic version of Zariski's lemma. As in the proof of Theorem 1, we first consider the cases $n = 1$ and $n = 2$. Hence, the next two algorithms construct the polynomials which witness that the generators are algebraic.

Algorithm 3. *Given a discrete field structure K, a discrete K-algebra structure R, $x \in R$ and $\iota : K[X] \to K[X]$, we compute an element $f \in K[X]$ as follows:*

1. *If $x = 0$, return X.*
2. *If $x \neq 0$, return $X\iota(X) - 1$.*

Lemma 5. *In the situation of Algorithm 3 we assume that K is a field, R is a K-algebra, $x \in R$ and ι is an algebraic inverse function on $K[x]$. Then f is non-constant and $f(x) = 0$, i.e., x is algebraic over K.*

Algorithm 4. *Let a discrete field structure K, a discrete K-algebra structure R, two elements $x_1, x_2 \in R$ and $\iota : K[X_1, X_2] \to K[X_1, X_2]$ be given. We compute $f_1, f_2 \in K[X]$ as follows starting with f_1:*

1. *If $x_2 = 0$, we use Algorithm 3 with input K, R, $x_1 \in R$ and $\iota' : K[X] \to K[X]$ defined by $\iota'(p) := \iota(p(X_1))(X, 0)$ and return the output as f_1.*
2. *Otherwise, compute $\iota(X_2)$ and define g as the polynomial which comes from $X_2\iota(X_2) - 1 \in K[X_1, X_2]$ by dropping each coefficient $p \in K[X_1]$ with $p(x_1) = 0$ and let $h \in K[X_1]$ be the leading coefficient of g (and 1 if $g = 0$).*
3. *Apply Algorithm 2 to the input K, $\vec{h} := (X_1, \iota(h))$, ι, $g_0^{(1)} = Y_1$ and $g_{k_2-1}^{(2)}, \ldots, g_0^{(2)} \in K[Y_1, Y_2]$ are the coefficients of $g(Y_1, X)$, except the leading coefficient, multiplied with Y_2. Let $\tilde{\iota} : K[Y_1, Y_2] \to K[Y_1, Y_2]$ be the output of this algorithm.*
4. *If $\deg(h) = 0$, (i.e., $h = h_0$ for some $h_0 \in K$), apply Algorithm 3 to K, R, $x_1 \in R$ and $\iota' : K[X] \to K[X]$ given by $\iota'(p) := \tilde{\iota}(p(Y_1))(X, h_0^{-1})$ and return the output of this algorithm as f_1.*
5. *Otherwise, check if $1 - h(x_1) = 0$. If yes, return $f_1 := 1 - h(X)$.*
6. *If no, compute $\tilde{\iota}(1 - h(Y_1)) = \sum_{i=0}^{N} a_i Y_2^i$ with $a_i \in K[Y_1]$ and $a_N \neq 0$; define*

$$q := \sum_{i=0}^{N} a_i (h(Y_1))^{N-i} \in K[Y_1]$$

and return $f_1 := h(X)^N - (1 - h(X))q(X)$.

Change x_1 and x_2 and repeat the steps above to compute $f_2 \in K[X]$.

Lemma 6. *In the situation of Algorithm 4 we assume that K is a field, R is a K-algebra, ι is an algebraic inverse function on $K[x_1, x_2]$. Then $f_1(x_1) = f_2(x_2) = 0$ and f_1, f_2 are non-constant.*

The next algorithm shows how to compute the field L, which corresponds to the field of fractions of $K[x_1]$ in $K[\vec{x}]$ on the level of polynomials.

Algorithm 5. *Let a discrete field structure K, a discrete K-algebra structure R, $n > 0$ and $\vec{x} \in R$ be given. We define a field structure as follows:*

$$L := \left\{ \frac{f}{g} \,\middle|\, f, g \in K[X], g(x_1) \neq 0 \vee 0 = 1 \right\},$$

$$\frac{f_1}{g_1} = \frac{f_2}{g_2} :\Leftrightarrow f_1(x_1)g_2(x_1) = f_2(x_1)g_1(x_1),$$

$$\frac{f_1}{g_1} + \frac{f_2}{g_2} := \begin{cases} \frac{f_1 g_2 + f_2 g_1}{g_1 g_2} & \text{if } (g_1 g_2)(x_1) \neq 0 \\ \frac{0}{1} & \text{else,} \end{cases} \qquad 0 := \frac{0}{1},$$

$$\frac{f_1}{g_1} \frac{f_2}{g_2} := \begin{cases} \frac{f_1 f_2}{g_1 g_2} & \text{if } (g_1 g_2)(x_1) \neq 0 \\ \frac{0}{1} & \text{else,} \end{cases} \qquad 1 := \frac{1}{1},$$

$$-\frac{f}{g} := \frac{-f}{g}, \qquad \left(\frac{f}{g}\right)^{-1} := \begin{cases} \frac{g}{f} & \text{if } f(x_1) \neq 0 \\ \frac{0}{1} & \text{else} \end{cases}$$

For a given map $\iota : K[\vec{X}] \to K[\vec{X}]$ we define a map $\varphi : L \to R$ by $\frac{f}{g} \mapsto$ $f(x_1)(\iota(g))(x_1)$, which turns R into an L-algebra structure. Furthermore, we define a map $\tilde{\iota} : L[X_2, \ldots, X_n] \to L[X_2, \ldots, X_n]$ as follows:

1. *Given an input $p \in L[X_2, \ldots, X_n]$, it has the presentation*

$$p = \sum_{i_2, \ldots, i_n} \frac{f_{i_2 \ldots i_n}}{g_{i_2 \ldots i_n}} X_2^{i_2} \cdots X_n^{i_n},$$

   for finitely many $f_{i_2 \ldots i_n}, g_{i_2 \ldots i_n} \in K[X]$.
2. *Let $a \in K[X]$ be the product of all these $g_{i_2 \ldots i_n}$, and for $j_2, \ldots, j_n$ let $h_{j_2 \ldots j_n}$ be the product of all these $g_{i_2 \ldots i_n}$ except $g_{j_2 \ldots j_n}$.*
3. *Define $\tilde{f}_{i_2 \ldots i_n} := f_{i_2 \ldots i_n} h_{i_2 \ldots i_n}$ and $\tilde{p} := \sum_{i_2, \ldots, i_n} \tilde{f}_{i_2 \ldots i_n}(X_1) X_2^{i_2} \cdots X_n^{i_n}$; set*

$$\tilde{\iota}(p) := (a(X_1)\iota(\tilde{p})) \left(\frac{X}{1}, X_2, \ldots, X_n \right),$$

   where we consider $b \in K$ also as the element $\frac{b}{1} \in L$.

Because we have to define the algorithm without the ring and field axioms, the definitions of L and the operators are more complex than one might expect.

As already mentioned we cannot define L as $\{ \frac{a}{b} \mid a, b \in K[x_1], b \neq 0 \vee 0 = 1 \}$, which is the field of fractions of $K[x_1]$ if this is an integral domain, because we want to avoid terms like $a \in K[x_1]$, which are constructively delicate. In particular, there is in general no map which takes $a \in K[x_1]$ and returns $f \in K[X]$ with $f(x_1) = a$ without using the axiom of choice. But in the next algorithm we operate on the level of polynomials.

Lemma 7. *In the situation of Algorithm 5 we assume that K is a field, R is a ring, $\vec{x} \in R$ and ι is an algebraic inverse function of $K[\vec{x}]$. Then L is indeed a discrete field, φ turns R into a L-algebra and $\tilde{\iota}$ is an algebraic inverse function on $L[x_2, \ldots, x_n]$.*

With this preparation we now formulate the final algorithm and an algorithm version of Zariski's lemma.

Algorithm 6. *Let K be a discrete field structure, R be a discrete K-algebra structure, $\iota : K[\vec{X}] \to K[\vec{X}]$ be a map and $x_1, \ldots, x_n \in R$. We compute $f_1, \ldots, f_n \in K[X]$ by recursion over n as follows:*

1. *If $n = 0$, return the empty list. If $n = 1$, use Algorithm 3 with input K, R, x_1 and ι and return the output f_1. If $n = 2$, use Algorithm 4 with input K; R; $x_1, x_2 \in R$ and ι, and return the output f_1, f_2.*
2. *Apply Algorithm 5 to K, R, n, $\vec{x}$ and ι and let the field structure L and the map $\iota' : L[X_2, \ldots, X_n] \to L[X_2, \ldots, X_n]$ be the output.*
3. *Apply recursion to L, the L-algebra structure R, ι' and $x_2, \ldots, x_n \in R$ and we get $\tilde{F}_2, \ldots, \tilde{F}_n \in L[X]$.*

4. *For each i we define F_i as $\tilde{F}_i$ divided by its leading coefficient and replacing the leading coefficient by 1 (or $F_i := 1$ if $\tilde{F}_i = 0$). In particular,*

$$F_i = X^{n_i} + \sum_{j=0}^{n_i-1} \frac{a_{ij}}{b_{ij}} X^j$$

for some $a_{ij}, b_{ij} \in K[X]$.

5. *Let $v := \prod_{(k,l)} b_{kl} \in K[X]$, $\tilde{b}_{ij} := \prod_{(k,l) \neq (i,j)} b_{kl}$, and $\tilde{a}_{ij} := \tilde{b}_{ij} a_{ij}$. Define*

$$G_i := \sum_{j=0}^{n_i} \tilde{a}_{ij}(Y_1) Y_2 X^j \in K[Y_1, Y_2, X].$$

6. *Use Algorithm 2 with input K, $\vec{h} := (X_1, \iota(v))$, ι, $k_1 := 1$, $g_0^{(1)} := Y_1$ and for $i \in \{2, \ldots, n\}$ take $k_i := n_i$ and $g_{n_i-1}^{(i)}, \ldots, g_0^{(i)}$ are the non-leading coefficients of G_i. Let $\tilde{\iota}$ be the output.*

7. *Apply Algorithm 4 to the input K, R, $x_1, \iota(v)(x_1) \in R$ and $\tilde{\iota}$, and define $f_1 \in K[X]$ as the output.*

8. *For each $i \in \{2, \ldots, n\}$ exchange x_1 with x_i and repeat the processes starting at Step 2 to get f_i instead of f_1. Then return $f_1, \ldots, f_n$.*

Theorem 2 (Algorithmic version of Zariski's lemma). *In the situation of Algorithm 6 we assume that K is a field, R is a K-algebra, $\vec{x} \in R$ and ι is an algebraic inverse function on $K[\vec{x}]$. Then $f_1(x_1) = \cdots = f_n(x_n) = 0$ and $f_1, \ldots, f_n$ are non-constant.*

4 Summary and Outlook

For $K[x_1, \ldots, x_n]$ and an algebraic inverse function ι on $K[\vec{x}]$ our algorithm computes $f_1, \ldots, f_n$ with $f_i(x_i) = 0$ for all i as follows: The case $n = 0$ is trivial. The case $n = 1$ is given in Lemma 5. The algorithm uses now recursion on n and reduction to the case $n = 2$. The case $n = 2$ itself is considered in Lemma 6. In this lemma the main idea was to find a suitable element u such that $K[x_1, u] \subseteq K[x_1, x_2]$ is an integral extension of K-algebras. By using Lemma 4 we have an algebraic inverse function on $K[x_1, u]$, where Lemma 4 uses Lemma 3. In the case $n \geq 3$, we use Lemma 7 to produce a new field L over which the original algebra is generated by one element less, such that we can use recursion and get $F_2, \ldots, F_n \in L[X]$ with $F_i(x_i) = 0$ for all i. From these F_i's we generate v such that $K[x_1, v] \subseteq K[\vec{x}]$ is an integral extension of K-algebras. Using again Lemma 4 we get an algebraic inverse function on $K[x_1, v]$ and therefore, again by Lemma 6, we get $f_1 \in K[X]$ with $f_1(x_1) = 0$. One now repeats the algorithm where x_1 and x_i are switched for all $i \geq 2$ and get $f_i \in K[X]$ with $f_i(x_i) = 0$.

Using the theory given in [11,12] one can probably formulate an algorithmic version of Hilbert's Nullstellensatz if the underlying field is countable. Another direction in which this paper can be extended is an analysis of the complexity of the algorithm. The algorithm of Sect. 3 as a whole is defined by recursion over the number of generators. In the recursion step (i.e., Algorithm 6) the algorithm with input $x_1, \ldots, x_n$ relies on the algorithm with input $x_1, \ldots, x_{i-1}, x_{i+1}, \ldots, x_n$ for each $i \leq n$. Therefore, the runtime of this algorithm must be at least quadratic in the number of generators.

A   Omitted Proofs

Proof (Lemma 3). We define the $K[\vec{Y}]$-module $M := K[\vec{Y}][\vec{X}]/\langle G_1, \ldots, G_n \rangle$ where $G_i := X^{k_i} + g_{k_i-1}^{(i)} X^{k_i-1} + \cdots + g_0^{(i)}$ for all i, and go through the steps of Algorithm 1: by the definition of M and the process to get the g_{IJ}'s, we have

$$\sum_{J \in \mathbb{N}^n} f_{IJ} \vec{X}^J = \sum_{J \in \mathcal{I}} g_{IJ}(\vec{Y}) \vec{X}^J$$

in M or in other words

$$\sum_{J \in \mathbb{N}^n} f_{IJ} \vec{X}^J - \sum_{J \in \mathcal{I}} g_{IJ}(\vec{Y}) \vec{X}^J \in \langle G_1, \ldots, G_n \rangle$$

seen in $K[\vec{Y}][\vec{X}]$. Note that $(\vec{X}^I)_{I \in \mathcal{I}}$ is a set of generators of M as $K[\vec{Y}]$-module, and multiplication with f corresponds to the matrix $(g_{IJ})_{I, J \in \mathcal{I}}$. Let P be the characteristic polynomial as in the algorithm. By the theorem of Cayley-Hamilton [4], $P(f) = 0$ in M, hence $P(f) \in \langle G_1, \ldots, G_n \rangle$ in $K[\vec{Y}][\vec{X}]$. By (1), we have $G_i(\vec{y}, \vec{x}) = 0$ for all i, and hence $0 = P(f)(\vec{y}, \vec{x}) = (f(\vec{x}))^k + g_{k-1}(\vec{y})(f(\vec{x}))^{k-1} + \cdots + g_0(\vec{y})$. Here we have used the definition of the g_i in the last step, and $\deg(P) = k$ because k is the number of elements in $\mathcal{I}$ which is also the cardinality of the generator $(x^I)_{I \in \mathcal{I}}$.   □

Proof (Lemma 4). Let $f \in K[\vec{Y}]$ with $f(\vec{y}) \neq 0$ be given. Since $\vec{h}$ is a witness that $K[\vec{y}] \subseteq K[\vec{x}]$ is an extension of K-algebras, we have $f(\vec{h}(\vec{x})) = f(\vec{y}) \neq 0$. Let p be given as in Step 1. Then $p(\vec{x})f(\vec{y}) = 1$ because ι is an algebraic inverse function. By Lemma 3 we have

$$(p(\vec{x}))^k + g_{k-1}(\vec{y})(p(\vec{x}))^{k-1} + \cdots + g_0(\vec{y}) = 0.$$

Multiplying this with $(f(\vec{y}))^{k-1}$ and isolating $p(\vec{x})$, we get

$$p(\vec{x}) = (-g_{k-1} - g_{k-2}f - \cdots - g_0 f^{k-1})(\vec{y}) = \tilde{\iota}(f)(\vec{y}).$$

□

The proof of Lemma 5 follows directly by the definition of an algebraic inverse function.

Proof (Lemma 6). It suffices to consider f_1 since the statement with f_2 is proved analogously. We follow the algorithm step by step. If $x_2 = 0$, we use Lemma 5. That ι' is an algebraic inverse function on $K[x_1]$ follows from

$$(\iota(p(X_1)))(x_1, 0)p(x_1) = (\iota(p(X_1))p(X_1))(x_1, x_2) = 1$$

for all $p \in K[X]$ with $p(x_1) \neq 0$.

So, we continue with $x_2 \neq 0$. By definition, $g(x_1, x_2) = x_2 \iota(X_2)(x_1, x_2) - 1 = 0$ and the constant coefficient (as polynomial in X_2) of g is equal to -1.

In the next step it is obvious that $X_1, \iota(h)$ is a witness of $K[x_1, \iota(h)(x_1, x_2)] \subseteq K[x_1, x_2]$ being an extension of K-algebras and that $x_1 - g_0^{(1)}(x_1, \iota(h)(x_1, x_2)) = x_1 - x_1 = 0$. Furthermore, let $g = \sum_{i=0}^{k_2} g_i X_2^i$ for some $g_i \in K[X_1]$ with $g_{k_2} \neq 0$. Then $h = g_{k_2}$ and

$$0 = \iota(h)(x_1, x_2) g(x_1, x_2) = x_2^{k_2} + \sum_{i=0}^{k_2-1} g_i(x_1) \iota(h)(x_1, x_2) x_2^i$$

$$= x_2^{k_2} + \sum_{i=0}^{k_2-1} g_i^{(2)}(x_1, \iota(h)(x_1, x_2)) x_2^i.$$

So, $\tilde{\iota}$ is an algebraic inverse function on $K[x_1, \iota(h)(x_1, x_2)]$ by Lemma 4.

If $\deg(h) = 0$, we have $h = h_0$ and $h_0 \neq 0$ because h is a leading coefficient. Therefore, it follows $\iota(h)(x_1, x_2) = h_0^{-1}$, and we apply Lemma 5 to $K[x_1] = K[x_1, h_0^{-1}]$. To apply this lemma it remains to show that ι' is an algebraic inverse function: if $p \in K[X]$ with $p(x_1) \neq 0$ then

$$(\iota'(p))(x_1) p(x_1) = (\tilde{\iota}(p(Y_1)))(x_1, h_0^{-1}) p(x_1) = (\tilde{\iota}(p(Y_1)) p(Y_1))(x_1, h_0^{-1}) = 1.$$

Now we continue with $\deg(h) \neq 0$, i.e., $\deg(h) > 0$ because $h \neq 0$. If $h(x_1) + 1 = 0$, we have that f_1 is non-constant since $\deg(h) > 0$ and by the case assumption $f(x_1) = 0$.

So let $h(x_1) + 1 \neq 0$. Then

$$q(x_1)(\iota(h)(x_1, x_2))^N = \tilde{\iota}(1 - h(Y_1))(x_1, \iota(h)(x_1, x_2)).$$

Since ι and $\tilde{\iota}$ are algebraic inverse functions and $h \neq 0$ and $1 - h \neq 0$, it follows

$$q(x_1)(1 - h(x_1))) = h(x_1).$$

So for $f_1 := (1 - h(X))q(X) - h(X)^N$ we have $f_1(x_1) = 0$ and $f_1 \neq 0$, similar to the end of the proof of Zariski's lemma.

$\square$

Proof (Lemma 7). L is a discrete field because in the definition of L and its operators we only use the operators of K.

By using the property of an algebraic inverse function, it is also straightforward to check that the map φ is a homomorphism.

It remains to show that $\tilde{\iota}$ is an algebraic inverse function on $L[x_2, \ldots, x_n]$. For this let $p \in L[X_2, \ldots, X_n]$ with $p(x_2, \ldots, x_n) \neq 0$ be given. We take the representation of p, a, $\tilde{f}_{i_2 \ldots i_n}$ and $\tilde{p}$ as defined in the algorithm, and calculate

$$p(x_2 \cdots, x_n) a(x_1) = \sum_{i_2, \cdots, i_n} \varphi\left(\frac{\tilde{f}_{i_2 \cdots i_n}}{1}\right) x_2^{i_2} \cdots x_n^{i_n}$$

$$= \sum_{i_2, \cdots, i_n} \tilde{f}_{i_2 \cdots i_n}(x_1) x_2^{i_2} \cdots x_n^{i_n} = \tilde{p}(x_1, \cdots, x_n).$$

Obviously, $a(x_1) \neq 0$ because it is a product of non-zero factors. Hence, if $p(x_2, \ldots, x_n) \neq 0$, it follows $\tilde{p}(x_1, \ldots x_n) \neq 0$. Since additionally ι is an algebraic inverse function, we have

$$\iota(\tilde{p})(x_1, \ldots, x_n) = (\tilde{p}(x_1, \ldots, x_n))^{-1},$$

and therefore

$$(p(x_2, \ldots, x_n))^{-1} = a(x_1)(\tilde{p}(x_1, \ldots, x_n))^{-1} = a(x_1)(\iota(\tilde{p}))(x_1, \ldots, x_n)$$

$$= (a(X_1)\iota(\tilde{p})) \left(\frac{X}{1}, x_2, \ldots, x_n \right).$$

$\square$

Proof (Algorithmic version of Zariski's Lemma). We use induction on n and consider the algorithm step by step. If $n = 0$, there is nothing to show. If $n = 1$, the statement follows by Lemma 5. If $n = 2$, the statement follows by Lemma 6.

If $n \geq 3$, it suffices to consider $e = 1$. We use Lemma 7 to get that L is a field, R is an L-algebra and ι' is an algebraic inverse function on $L[x_2, \ldots, x_n]$.

We have that $F_2(x_2) = \cdots = F_n(x_n) = 0$ by the induction hypothesis and the fact that F_i is indeed $\tilde{F}_i$ divided by its leading coefficient since L is a field.

Furthermore, $F_i = G_i(x_1, v^{-1}, X)$ as polynomial in $R[X]$ and therefore $0 = F_i(x_i) = G_i(x_1, (v(x_1))^{-1}, x_i)$. So, the non-leading coefficients of G_i (as polynomials in X) witness that x_i is integral over $K[x_1, \iota(v)(x_1)]$ for each $i \in \{2, \ldots, n\}$.

Because of this, the requirements of Lemma 4 are fulfilled and hence $\tilde{\iota}$ is an algebraic inverse function on $K[x_1, \iota(v)(x_1)]$.

Therefore, we get $f_1(x_1) = 0$ and f_1 is non-constant by Lemma 6. $\square$

References

1. Atiyah, M.F., Macdonald, I.G.: Introduction to Commutative Algebra. Addison-Wesley Pub. Co., Boston (1969)
2. Azarang, A.: A simple proof of Zariski's Lemma. Bull. Iran. Math. Soc. **43**(5), 1529–1530 (2017)
3. Berger, U., Miyamoto, K., Schwichtenberg, H., Seisenberger, M.: Minlog - a tool for program extraction supporting algebras and coalgebras. In: Corradini, A., Klin, B., Cîrstea, C. (eds.) CALCO 2011. LNCS, vol. 6859, pp. 393–399. Springer, Heidelberg (2011). https://doi.org/10.1007/978-3-642-22944-2_29
4. Eisenbud, D.: Commutative Algebra: With a View Toward Algebraic Geometry, Graduate Texts in Mathematics, vol. 150. Springer, New York (1995). https://doi.org/10.1007/978-1-4612-5350-1
5. Hulek, K.: Elementare Algebraische Geometrie: Grundlegende Begriffe und Techniken mit zahlreichen Beispielen und Anwendungen. Springer, Heidelberg (2012). https://doi.org/10.1007/978-3-8348-2348-9
6. Kohlenbach, U.: Applied Proof Theory: Proof Interpretations and their Use in Mathematics. Springer, Heidelberg (2008). https://doi.org/10.1007/978-3-540-77533-1

7. Kohlenbach, U.: Proof-theoretic methods in nonlinear analysis. In: Proceedings of the International Congress of Mathematicians, vol. 2, pp. 61–82. World Scientific (2018)
8. Lombardi, H., Quitté, C.: Commutative Algebra: Constructive Methods: Finite Projective Modules, vol. 20. Springer, Dordrecht (2015). https://doi.org/10.1007/978-94-017-9944-7
9. McCabe, J.: A note on Zariski's lemma. Am. Math. Mon. **83**(7), 560–561 (1976)
10. Mines, R., Richman, F., Ruitenburg, W.: A Course in Constructive Algebra. Springer, New York (1988). https://doi.org/10.1007/978-1-4419-8640-5
11. Powell, T., Schuster, P., Wiesnet, F.: An algorithmic approach to the existence of ideal objects in commutative algebra. In: Iemhoff, R., Moortgat, M., de Queiroz, R. (eds.) WoLLIC 2019. LNCS, vol. 11541, pp. 533–549. Springer, Heidelberg (2019). https://doi.org/10.1007/978-3-662-59533-6_32
12. Powell, T., Schuster, P., Wiesnet, F.: A universal algorithm for Krull's theorem. Inf. Comput. (2020, submitted)
13. Reid, M.: Undergraduate Algebraic Geometry. Cambridge University Press, Cambridge (1988)
14. Schwichtenberg, H., Wainer, S.S.: Proofs and Computations. Cambridge University Press, Cambridge (2011)
15. Sharifi, Y.: Zariski's Lemma (2011). https://ysharifi.wordpress.com/tag/zariskis-lemma/. Acccessed 7 Jun 2020
16. Wessel, D.: Making the use of maximal ideals inductive (2021), talk at workshop Reducing complexity in algebra, logic, combinatorics
17. Wiesnet, F.: Introduction to minlog. In: Mainzer, K., Schuster, P., Schwichtenberg, H. (eds.) Proof and Computation, pp. 233–288. World Scientific (2018)
18. Yengui, I.: Constructive Commutative Algebra. Projective Modules Over Polynomial Rings and Dynamical Gröbner Bases. LNM, vol. 2138. Springer, Cham (2015). https://doi.org/10.1007/978-3-319-19494-3
19. Zariski, O.: A new proof of Hilbert's Nullstellensatz. Bull. Am. Math. Soc. **53**(4), 362–368 (1947)

Einstein Meets Turing:
The Computability of Nonlocal Games

Henry Yuen$^{(\boxtimes)}$

Columbia University, New York, NY 10027, USA
hyuen@cs.columbia.edu

Abstract. Quantum entanglement – the phenomenon where distant particles can be correlated in ways that cannot be explained by classical physics – has mystified scientists since the1930s, when quantum theory was beginning to emerge. Investigation into fundamental questions about quantum entanglement has continually propelled seismic shifts in our understanding of nature. Examples include Einstein, Podolsky and Rosen's famous 1935 paper about the incompleteness of quantum mechanics, and John Bell's refutation of EPR's argument, 29 years later, via an experiment to demonstrate the non-classicality of quantum entanglement.

More recently, the field of quantum computing has motivated researchers to study entanglement in information processing contexts. One question of deep interest concerns the computability of *nonlocal games*, which are mathematical abstractions of Bell's experiments. The question is simple: is there an algorithm to compute the optimal winning probability of a quantum game – or at least, approximate it? In this paper, I will discuss a remarkable connection between the complexity of nonlocal games and classes in the arithmetical hierarchy. In particular, different versions of the nonlocal games computability problem neatly line up with the problems of deciding Σ_1^0, Π_1^0, and Π_2^0 sentences, respectively.

Keywords: Nonlocal games · Quantum entanglement · Uncomputability

1 EPR's Dream, Bell's Theorem, and the CHSH Game

In 1935, Einstein, Podolsky and Rosen (who we'll henceforth abbreviate as "EPR") wrote a paper titled "Can Quantum-Mechanical Description of Physical Reality be Considered Complete?", which became one of the most influential papers in the history of physics. The EPR paper was motivated by a fundamental dissatisfaction with quantum mechanics, which at the time was revolutionizing how physicists understood the world at its smallest scales. At the heart of this discontent was the phenomenon of *quantum entanglement*, in which two particles can be separated far away from other but still exhibit "spooky" correlations

© Springer Nature Switzerland AG 2021
L. De Mol et al. (Eds.): CiE 2021, LNCS 12813, pp. 483–493, 2021.
https://doi.org/10.1007/978-3-030-80049-9_47

that defy classical explanation. EPR believed that the *mathematical theory* of quantum mechanics, while accurate in its predictions, must be incomplete. They hoped that quantum theory could be replaced with a completely classical theory of nature that was consistent with the predictions of quantum theory and relativity but free of what EPR thought were apparent paradoxes.

Their paper sparked decades-long debates about the interpretation of quantum mechanics and the validity of EPR's arguments. It wasn't until 1964 when the physicist John Bell came up with a startlingly simple argument for what is known as *Bell's theorem*, which dashed EPR's dream of a classical replacement of quantum theory [1]. I'll present a modern and computer science-friendly formulation of Bell's argument using *nonlocal games*.

Consider a scenario involving three parties; there are two players (who we name Alice and Bob, in the computer science convention) and a referee. The referee chooses two bits $x, y \in \{0, 1\}$ (called "questions") uniformly at random, and sends x to Alice and y to Bob. Alice and Bob are cooperating players on the same team, but during this game they are not allowed to communicate with each other.[1] Instead Alice has to respond with an answer $a \in \{0, 1\}$ and Bob with an answer $b \in \{0, 1\}$ to the referee. The game ends and the players win if $a \oplus b = x \wedge y$. This is known as the CHSH game, named after physicists Clauser, Horne, Shimony and Holt who designed this game as an experimental demonstration of Bell's argument [2].

What strategy should Alice and Bob use to win with highest probability in the CHSH game? This depends on what strategies Alice and Bob are allowed to employ, and this depends on what theory of physics we use to model Alice and Bob's behavior. For example, if we use a *deterministic, classical theory* of physics to model Alice and Bob, then their strategies can be described as a pair of functions $a, b : \{0, 1\} \to \{0, 1\}$; upon receiving question x Alice responds with answer $a(x)$ and upon receiving question y Bob responds with answer $b(y)$.

It is not difficult to see that when Alice and Bob employ deterministic strategies, their maximum probability of success is $3/4$. Furthermore, even if we model Alice and Bob using a probabilistic classical theory – i.e., their answers a, b are not just functions of their respective questions but also some common random variable r – their maximum success probability remains $3/4$. This number is called the *classical value* of the CHSH game.

When Alice and Bob employ a quantum strategy, however, they can win the CHSH game with probability that is *strictly* higher than the classical value. At the beginning of the game, the players share two entangled particles and perform measurements (which depend on their questions) on their own particle, and respond with answers based on the measurement results.

An important thing to note is that the players *cannot* use quantum entanglement to communicate with each other during the game; there is no way for Alice to use entanglement to glean information about Bob's question y, and similarly

---

[1] This non-communication constraint is used to model the situation when Alice and Bob are separated far from each other, and relativity prevents Alice and Bob from instantaneously signaling to each other.

Bob cannot obtain any information about Alice's question x. Instead, quantum entanglement should be viewed as a form of correlations that are stronger than what is possible classically, but still do not allow for instantaneous signaling between two distant parties.

Clauser, Horne, Shimony and Holt showed that there is a quantum entangled strategy for Alice and Bob to produce winning answers with probability $\cos^2(\pi/8) \approx .854$ (this is known as the *quantum value* of the CHSH game). Thus, no classical strategy satisfying the no-communication constraint can match the winning probability of this quantum strategy, and thus there cannot be a classical theory of nature that is compatible with both quantum theory *and* Einstein's theory of relativity, which forbids instantaneous signaling. Furthermore, games like the CHSH game with a separation in their classical and quantum values have been experimentally demonstrated many times over the past 40 years, and the results are unambiguous: Nature is non-classical. These experiments are often called *Bell tests*, in honor of the physicist who conceived of the first such experiment.

Over the past twenty years, computer scientists and mathematicians have also become quite interested in Bell experiments because of their intriguing ties to quantum information theory, theoretical computer science, and pure mathematics. In this paper, we will see how studying Bell tests through the lens of computation and information reveals fascinating connections with computability theory and more.

2   Nonlocal Games and Their Computability

The CHSH game is an example of a *nonlocal game*, which is a general mathematical abstraction of a Bell test. A nonlocal game G consists of a tuple $(\mathcal{X}, \mathcal{Y}, \mathcal{A}, \mathcal{B}, \mu, D)$, where $\mathcal{X}, \mathcal{Y}$ (called *question sets*) and $\mathcal{A}, \mathcal{B}$ (called *answer sets*) are finite sets, μ (called the *question distribution*) is a probability distribution over $\mathcal{X} \times \mathcal{Y}$, and $D : \mathcal{X} \times \mathcal{Y} \times \mathcal{A} \times \mathcal{B} \rightarrow \{0, 1\}$ is a function called a *decision procedure*. The game is played between a referee and two players where the referee first samples a pair (x, y) according to μ and sends x to Alice, y to Bob. They have to respond with answers a and b respectively, and they win if $D(x, y, a, b) = 1$. We assume that Alice and Bob know the question distribution and decision procedure before the game starts, and can choose a *strategy* for answering the questions in order to optimize their probability of winning.

The motivating question of this paper is the following:

What is the complexity of computing the optimal winning probability in a nonlocal game?

To make this question precise, we need to formalize what we mean by optimal winning probability, and this depends on what class of strategies Alice and Bob are allowed to use – for each class, there is an associated complexity question.

A general strategy $\mathscr{S}$ for a game $G = (\mathcal{X}, \mathcal{Y}, \mathcal{A}, \mathcal{B}, \mu, D)$ is a set of conditional probability distributions $\{p_{xy} : \mathcal{A} \times \mathcal{B} \rightarrow \mathbb{R}_+\}_{x \in \mathcal{X}, y \in \mathcal{Y}}$ where $\sum_{a,b} p_{xy}(a, b) = 1$, and the *value* of the strategy is defined to be

$$\omega(G, \mathscr{S}) = \sum_{x,y,a,b} \mu(x,y)\, p_{xy}(a,b)\, D(x,y,a,b).$$

In other words, this describes the probability of winning when Alice and Bob sample winning answers (a, b) from the distribution p_{xy}, when receiving a question pair (x, y) drawn from μ. Let $\mathscr{C}$ denote a class of strategies for a game G. Then the $\mathscr{C}$-*value* of a game G is defined to be

$$\omega_{\mathscr{C}}(G) = \sup_{\mathscr{S} \in \mathscr{C}} \omega(G, \mathscr{S}).$$

In other words, it is the optimal probability of winning the game G using a strategy from the class $\mathscr{C}$. There are three main classes of strategies that we consider in this paper.

First, define the class $\mathscr{C}_c$, which is the class of *classical* strategies $\mathscr{S}$ where there are functions $f : \mathcal{X} \to \mathcal{A}, g : \mathcal{Y} \to \mathcal{B}$ such that $p_{xy}(a, b) = 1$ if and only if $f(x) = a$ and $g(y) = b$. We denote the classical value of a game G by $\omega_c(G)$.

Next, define the class $\mathscr{C}_q$, which is the class of *quantum* strategies. Such a strategy consists of *a state*, which is a unit vector $\psi \in \mathbb{C}^d \otimes \mathbb{C}^d$ where $d > 0$ is some integer; and *measurements*, which for $x, y \in \{0, 1\}$ are sets of positive semidefinite operators $A^x = \{A^x_a\}_{a \in \mathcal{A}}$ and $B^y = \{B^y_b\}_{b \in \mathcal{B}}$ acting on $\mathbb{C}^d$. The operators satisfy the completeness condition $\sum_{a \in \mathcal{A}} A^x_a = \sum_{b \in \mathcal{B}} B^y_b = \mathbb{I}$ for all $x \in \mathcal{X}, y \in \mathcal{Y}$ where $\mathbb{I}$ denotes the identity on $\mathbb{C}^d$. The probability of producing answers (a, b) given questions (x, y) is given by the formula $p_{xy}(a, b) = \psi^*(A^x_a \otimes B^y_b)\psi$. We denote the quantum value of a game by $\omega_q(G)$.

Finally, define the class $\mathscr{C}_{co}$, which is the class of *commuting operator* strategies. This class captures quantum strategies that are *infinite-dimensional* (whereas the quantum strategies of $\mathscr{C}_q$ are by definition finite-dimensional).[2] This generalization of quantum strategies is motivated by quantum field theories, where the natural description of physical phenomena involves infinitely many degrees of freedom; for example the fundamental quantum fields from which elementary particles arise have a degree of freedom for every point in space and time. We denote the commuting operator value of a game by $\omega_{co}(G)$.

We won't really use the specifics of the definitions of the different strategy classes in this paper, but an important point is the following relationship between the different values. For all nonlocal games G, we have

$$\omega_c(G) \leq \omega_q(G) \leq \omega_{co}(G). \tag{1}$$

---

[2] Formally, a commuting operator strategy consists of a unit state ψ defined on a separable Hilbert space $\mathcal{H}$ (which is in general infinite-dimensional), and measurement operators $\{A^x_a\}_{x,a}$ and $\{B^y_b\}$ such that for all x, y, $\sum_a A^x_a = \sum_b B^y_b = \mathbb{I}$ where $\mathbb{I}$ denotes the identity operator on $\mathcal{H}$, and furthermore Alice's and Bob's operators must *commute with each other*: $A^x_a B^y_b = B^y_b A^x_a$ for all x, y, a, b. The probability of producing answers (a, b) given questions (x, y) is given by $\psi^* A^x_a B^y_b \psi$. The essential difference between this model of strategies and the quantum strategies defined above is that (a) the dimension of the Hilbert space may be infinite, and (b) there is not necessarily a tensor product structure in the Hilbert space.

In other words, in any game, quantum strategies can do at least as well as classical strategies, and commuting strategies can do at least as well as quantum strategies. A succinct way to state Bell's theorem is that there exists a nonlocal game G such that $w_c(G) < w_q(G)$; the CHSH game is one example. A fundamental question in the study of nonlocal games and quantum information theory is whether there exists a game for which the second inequality in (1) is strict – this is known as *Tsirelson's problem* [11]. This question was recently resolved in the affirmative: there exists a game G for which $w_q(G) < w_{co}(G)$. The CHSH game is not an example of such a game because $w_q(CHSH) = w_{co}(CHSH)$. We will return to this later.

We can now formalize the complexity question raised earlier. In fact, there are several natural formulations of this question that we can consider, depending on whether one cares about computing the optimal winning probability exactly or approximately, and what class of strategies are allowed. Fix a value type $t \in \{c, q, co\}$. For the exact computation question, to goal is to decide, given a description of a nonlocal game G and a real number $0 \leq \nu \leq 1$, whether $w_t(G) = \nu$. For the approximation question, the goal is to decide whether $w_t(G) = \nu$ or $|w_t(G) - \nu| > \varepsilon$, promised that one is the case. (Here, the error parameter ε can be fixed or provided as part of the input to the problem.) Throughout this paper we will generally fix $\nu = 1$ and $\varepsilon = \frac{1}{2}$ for convenience.

What is the computational complexity of these problems? We start with the classical problem. The question of the complexity of computing and approximating $w_c(G)$ has been central to theoretical computer science: although it may not be obvious when stated this way, the Cook-Levin theorem, which states that boolean satisfiability is NP-complete, is equivalent to the statement that deciding whether $w_c(G) = 1$ is an NP-complete problem. Furthermore, the complexity of approximating $w_c(G)$ to within an additive error of $\frac{1}{2}$ (or any other fixed constant) is *still* NP-complete; this is equivalent to the famous *probabilistically checkable proofs (PCP) theorem*. Thus, even when restricted to considering deterministic classical strategies, computing optimal strategies for nonlocal games is a computationally intractable task.

What about the quantum and commuting operator values of games? The question about the computability of the quantum values of nonlocal games is a relatively recent one; versions of this were first formulated by [3,6,9]. What is striking about these questions is that it is not obvious *a priori* that the quantum and commuting operator values are even *computable*!

I'll discuss the situation with the quantum value first. Recall that it is defined as a supremum over the set of (finite-dimensional) quantum strategies involving an entangled state between the two players and measurements for each of them. This is a very daunting space to optimize over; in particular it is infinite in two ways. First, it is continuous. This is not a critical issue as, for a fixed dimension d, the space of d-dimensional strategies can be discretized and enumerated over in finite time. The more important issue is that there is no *a priori* upper bound on the dimension d needed to come close to the optimal winning probability.

What that means is, even if one were to try to do a naïve brute force search over the space of quantum strategies—e.g., enumerating over a discretization of d-dimensional quantum strategies for increasing $d = 1, 2, 3, \ldots$—it is not immediately clear how to determine whether the best winning probability computed so far in the enumeration is converging to the quantum value. This brute force approach only gives a "semialgorithm" that computes a sequence of values $\alpha_1 \leq \alpha_2 \leq \cdots \leq w_q(G)$ where α_d is the best winning probability amongst a discretization of d-dimensional strategies, and it is only guaranteed that $\alpha_d \to w_q(G)$ in the limit as $d \to \infty$. This shows that the approximation problem of deciding whether $w_q(G) = 1$ or $w_q(G) \leq \frac{1}{2}$ (promised that one is the case) is contained in RE, the class of recursively enumerable problems, because if $w_q(G) > \frac{1}{2}$, then there is a certificate for this fact in the form of a finite-dimensional strategy, and this certificate can be found by the semialgorithm. In other words, you can encode the quantum value approximation problem as an instance of the Halting problem.

The situation is mirrored with the commuting operator value. The space of infinite-dimensional commuting operator strategies is perhaps even more intimidating than the space of finite-dimensional quantum strategies. It turns out that there is a semialgorithm for certifying that $w_{co}(G)$ is *strictly* less than 1: the algorithm computes a sequence of values $\beta_1 \geq \beta_2 \geq \cdots \geq w_{co}(G)$ where β_d is guaranteed to converge to $w_{co}(G)$ as $d \to \infty$.[3] This shows that the problem of deciding whether $w_{co}(G) = 1$ (as well as the problem of approximating $w_{co}(G)$) is contained in coRE, the complement of RE. In other words, you can encode the commuting value problem (both exact and approximation versions) as an instance of the *non*-Halting problem.

These upper bounds were all that were known about the complexity of the quantum and commuting operator values of nonlocal games until very recently. To some, this constituted an embarrassing state of affairs; *surely* there is *some* algorithm for the quantum/commuting-operator value!

A sequence of results have shown that the lack of a computable upper bound on the complexity of nonlocal games is for a good reason: In 2016, Slofstra [12,13] showed that for either $t \in \{q, co\}$ there is no algorithm to decide whether $w_t(G) = 1$. In 2020, Ji, Natarajan, Vidick, Wright and myself [7] showed, via a complexity-theoretic result stated as MIP* = RE, that there is no algorithm to compute an additive approximation $w_q(G) \pm \varepsilon$ for *any* $0 \leq \varepsilon < 1$. Surprisingly, this (un)computability result *also* resolves Tsirelson's problem (i.e., there exists a game G for which $w_q(G) \neq w_{co}(G)$), which in turn resolves *Connes' embedding problem*, an important question in the study of functional analysis that was first raised by Alain Connes in 1976 [4,10]. For more information about the connection between the approximability of the quantum value of nonlocal games and Connes' embedding problem, we refer the reader to Vidick's survey [14].

---

[3] For the curious: β_d is defined to be the smallest number such that the nonnegativity of $\beta_d - w_{co}(G)$ admits a degree-d sum-of-squares polynomial in noncommuting variables. Each β_d can be computed in finite time using the semidefinite programming hierarchies of [6,9].

Furthermore, it is still an open question whether there is an algorithm to approximate $\omega_{co}(G)$, but the current evidence indicates that one is unlikely to exist; for example Coudron and Slofstra have proved lower bounds on the time complexity of approximating $\omega_{co}(G)$ as a function of the approximation quality [5].

While these (non)computability results about the quantum and commuting operator value of nonlocal games present striking statements about the sheer complexity of the space of quantum strategies (both in the finite-dimensional and infinite-dimensional setting), one can even make sharper statements about the *degree* to which these problems are algorithmically unsolvable. In fact, the recent results on the complexity of nonlocal games have uncovered a remarkable correspondence between the complexity of computing the value of nonlocal games and different classes of the *arithmetical hierarchy*.

3   Nonlocal Games and the Arithmetical Hierarchy

We consider sentences S of the form

$$\exists x_1 \forall x_2 \exists x_3 \cdots \phi(x_1, x_2, x_3, \ldots, x_k) \quad \text{or} \quad \forall x_1 \exists x_2 \forall x_3 \cdots \phi(x_1, x_2, x_3, \ldots, x_k)$$

where the variables range over $\{0,1\}^*$, all variables are quantified over, and ϕ is a predicate computable by a Turing machine[4]. In this paper we call these *arithmetical sentences*. Every arithmetical sentence S is either true or false.

Arithmetical sentences are classified according to the number of quantifiers in the sentence, and whether the first quantifier is $\exists$ or $\forall$:

1. Σ_0^0 is the class of arithmetical sentences S of the form $\exists x\, \phi(x)$, and Π_0^0 is the class of arithmetical sentences S of the form $\forall x\, \phi(x)$.
2. For $n \geq 2$, Σ_n^0 is the class of arithmetical sentences of the form $\exists x\, F$ where $F \in \Pi_{n-1}^0$, and Π_n^0 is the class of sentences of the form $\forall x\, F$ where $F \in \Sigma_{n-1}^0$.

The classes $\{\Sigma_n^0, \Pi_n^0\}_n$ form the *arithmetical hierarchy* that is studied in recursion theory and computability theory. For simplicity we abbreviate Σ_n^0 and Π_n^0 as Σ_n and Π_n, respectively.

The set of true arithmetical sentences is undecidable; for example, deciding whether a sentence in Σ_1 is true is equivalent to deciding the Halting problem. It is well-known that the complexity of deciding sentences with k quantifiers is strictly harder than the complexity of deciding sentences with $k-1$ quantifiers; for example, a Turing machine equipped with an oracle to decide Σ_n sentences still cannot decide sentences in Σ_{n+1} or Π_{n+1}.

Fix $0 \leq \varepsilon < 1$ and a value type $t \in \{q, co\}$. Define two sets of nonlocal games $L_t^{yes} := \{G : \omega_t(G) = 1\}$ and $L_{t,\varepsilon}^{no} := \{G : \omega_t(G) < 1 - \varepsilon\}$. (We assume a natural encoding of nonlocal games as binary strings). These two sets are disjoint, and

---

[4] Formally, these are sentences over a first-order language that uses binary strings as its universe, and can encode the behavior of Turing machines.

	$\varepsilon = 0$	$\varepsilon > 0$
$\omega_q(G) \pm \varepsilon$	Π_2 [8]	Σ_1 [7]
$\omega_{co}(G) \pm \varepsilon$	Π_1 [13]	Π_1 (conjectured)

Fig. 1. A characterization of the complexity of computing the value of a nonlocal game in terms of the arithmetical hierarchy, depending on whether the quantum or commuting operator value is being considered, and whether the value is being computed exactly or approximately.

when $\varepsilon = 0$, the union of these two sets is all nonlocal games. These two sets give rise to a decision problem: given a nonlocal game G in the union of these sets, decide whether G is a YES instance or a NO instance.

Figure 1 indicates the following correspondence between deciding L_t^{yes} versus $L_{t,\varepsilon}^{no}$ and deciding whether a sentence S is true (the correspondence is conjectural in the case of $\varepsilon > 0$ and $t = co$):

- (*Sentences to nonlocal games*) There exists a computable map Γ mapping sentences of the type specified by the entry (t, ε) to nonlocal games in $L_t^{yes} \cup L_{t,\varepsilon}^{no}$ such that $\Gamma(S) \in L_t^{yes}$ if and only if S is true.
- (*Nonlocal games to sentences*) There exists a computable map Ξ mapping nonlocal games in $L_t^{yes} \cup L_{t,\varepsilon}^{no}$ to sentences of the type specified by the entry (t, ε) such that $\Xi(G)$ is true if and only if $G \in L_t^{yes}$.

Going from nonlocal games to sentences is straightforward. Fix a nonlocal game G. We consider the different cases.

Let $t = q$ and $\varepsilon > 0$. Define the predicate $\phi(x)$ that is 1 if and only if x is a description of a finite-dimensional strategy $\mathscr{S}$ for G such that $w(G, \mathscr{S}) \geq 1 - \varepsilon$. Here we assume a natural enumeration of a countable net that covers the set of finite-dimensional strategies with arbitrarily fine closeness[5]. It is clear that $\phi(x)$ is computable. If $\omega_q(G) = 1$, then there exists a sequence of finite-dimensional strategies whose value approaches 1, so therefore $\exists x \, \phi(x)$. On the other hand, if $\omega_q(G) < 1 - \varepsilon$, then $\forall x \, \neg\phi(x)$.

Let $t = q$ and $\varepsilon = 0$. Define the predicate $\phi(x, y)$ that is 1 if and only if x is a description of a real number $0 < \delta < 1$ and y is a description of a strategy $\mathscr{S}$ from a canonical net of finite-dimensional strategies (i.e., every finite-dimensional strategy has a sequence of strategies from the net converging to it) such that $w(G, \mathscr{S}) \geq 1 - \delta$. If $\omega_q(G) = 1$, then by definition for all $\delta > 0$ there exists a strategy $\mathscr{S}$ such that $w(G, \mathscr{S}) \geq 1 - \delta$, and furthermore this strategy can be taken from the net, so therefore $\forall x \, \exists y \, \phi(x, y)$. On the other hand, if $\omega_q(G) < 1$, then there exists a $\delta > 0$ such that for all strategies $\mathscr{S}$, the value $w(G, \mathscr{S}) < 1 - \delta$, so $\exists x \, \forall y \, \neg\phi(x, y)$.

---

[5] We assume some natural distance measure between strategies; such as the sum of the ℓ_2 distances between the states and the measurements.

Let $t = co$ and $\varepsilon \geq 0$. Define the predicate $\phi(x)$ that is 1 if and only if x is a natural number d such that there does *not* exist a degree-d noncommutative sum-of-squares certificate, specified with precision d^{-1}, that $\omega_{co}(G) < 1$. The semidefinite programming hierarchies of [6,9] can be used to compute ϕ. If $\omega_{co}(G) = 1$, then $\forall x\, \phi(x)$, whereas if $\omega_{co}(G) < 1$ the completeness of the hierarchy shows that there exists a degree-d certificate (specified with precision d^{-1}) of this fact, for some d, so $\exists x\, \neg\phi(x)$.

It is significantly more difficult to transform sentences into equivalent nonlocal games. Slofstra [13] showed that exactly computing ω_{co} is as hard as deciding Π_1 sentences, and Ji, et al. [7] showed that approximating ω_q is as hard as deciding Σ_1 sentences. In [8], Mousavi, Nezhadi and myself showed that exactly computing ω_q for *three-player games* is as hard as deciding Π_2 sentences, and in upcoming work we show that the same holds for two-player games:

Theorem 1 (Mousavi-Nezhadi-Yuen 2021).   *There exists a computable map from Π_2 sentences S to nonlocal games G such that S is true if and only if $\omega_q(G) = 1$.*

The only remaining piece of the puzzle (which we do not resolve here) is to determine the complexity of approximating ω_{co}, which we conjecture is as hard as deciding Π_1 sentences.

A priori, this close correspondence between nonlocal games and arithmetical sentences seems quite surprising. On one hand, computing the value of a nonlocal game corresponds to a continuous optimization problem over a space of quantum states and quantum measurements, possibly in infinite dimensions. On the other hand, deciding whether a quantified arithmetical sentence is true is a discrete problem in symbolic logic ostensibly having nothing to do with quantum physics. Furthermore, the reader may notice that there are several interesting asymmetries in Fig. 1:

1. If we assume the conjecture, then both exact and approximate computation of the commuting operator value are equivalent to deciding Π_1 sentences, whereas for the quantum value, the complexity splits depending on whether we are considering exact or approximate computation.
2. It is known that there is no computable map Λ from Σ_1 to Π_1 (or vice versa) such that S is true if and only if $\Lambda(S)$ is true[6]. Thus, since approximating the quantum value is equivalent to deciding Σ_1 sentences, the complexities of approximating the quantum value and computing the commuting operator value are *incomparable* in the logic sense, even though both problems are algorithmically unsolvable.
3. Since deciding true Π_2 sentences is strictly harder than deciding Σ_1 or Π_1 sentences, exactly computing the quantum value is strictly harder than approximating the quantum value or exactly computing the commuting operator value. In other words, an oracle that decides whether the quantum value of

---

[6] In the terminology of computability theory, this is stating that the recursively enumerable languages and co-recursively enumerable languages are incomparable sets.

nonlocal games is 1 can be used to approximate the quantum value or compute the commuting operator value of games – but not the other way around.

4 Conclusion

Nonlocal games have become a deeply fruitful research area at the confluence of quantum physics, theoretical computer science, cryptography, and pure mathematics. They have provided a bridge connecting the discrete world of logic and computation to the continuous world of quantum physics, and from them we have learned that the mathematical structure of quantum entanglement is rich enough to capture various phenomena in computability theory. Although I didn't get to the details of how such uncomputability results are proved, they involve sophisticated tools and techniques from quantum information theory, probabilistically checkable proofs, property testing, group theory, and more. All of this suggests that these computability results about nonlocal games is just the start of an exciting research direction with deep connections to many fields.

References

1. Bell, J.S.: On the Einstein-Podolsky-Rosen paradox. Physics **1**, 195 (1964)
2. Clauser, J.F., Horne, M.A., Shimony, A., Holt, R.A.: Proposed experiment to test local hidden-variable theories. Phys. Rev. Lett. **23**(15), 880 (1969)
3. Cleve, R., Hoyer, P., Toner, B., Watrous, J.: Consequences and limits of nonlocal strategies. In: 2004 Proceedings of the 19th IEEE Annual Conference on Computational Complexity, pp. 236–249. IEEE (2004)
4. Connes, A.: Classification of injective factors cases II_1, II_∞, III_λ, $\lambda \neq 1$. Ann. Math. **104**, 73–115 (1976)
5. Coudron, M., Slofstra, W.: Complexity lower bounds for computing the approximately-commuting operator value of non-local games to high precision. arXiv preprint arXiv:1905.11635 (2019)
6. Doherty, A.C., Liang, Y.C., Toner, B., Wehner, S.: The quantum moment problem and bounds on entangled multi-prover games. In: 2008 23rd Annual IEEE Conference on Computational Complexity, pp. 199–210. IEEE (2008)
7. Ji, Z., Natarajan, A., Vidick, T., Wright, J., Yuen, H.: MIP* = RE. arXiv preprint arXiv:2001.04383 (2020)
8. Mousavi, H., Nezhadi, S.S., Yuen, H.: On the complexity of zero gap MIP*. In: Czumaj, A., Dawar, A., Merelli, E. (eds.) 47th International Colloquium on Automata, Languages, and Programming (ICALP 2020). Leibniz International Proceedings in Informatics (LIPIcs), vol. 168, pp. 87:1–87:12 (2020)
9. Navascués, M., Pironio, S., Acín, A.: A convergent hierarchy of semidefinite programs characterizing the set of quantum correlations. New J. Phys. **10**(7), 073013 (2008)
10. Ozawa, N.: About the Connes embedding conjecture, algebraic approaches. Jpn. J. Math. **8**, 147–183 (2013)
11. Scholz, V.B., Werner, R.F.: Tsirelson's problem. arXiv preprint arXiv:0812.4305 (2008)

12. Slofstra, W.: The set of quantum correlations is not closed. Forum Math, Pi **7**, 1–41 (2019). Cambridge University Press
13. Slofstra, W.: Tsirelson's problem and an embedding theorem for groups arising from non-local games. J. Am. Math. Soc. **33**, 1–56 (2020)
14. Vidick, T.: From operator algebras to complexity theory and back. Not. Am. Math. Soc. **66**(10), 1618–1627 (2019)

Computability of Limit Sets
for Two-Dimensional Flows

Daniel S. Graça[1,2] and Ning Zhong[3]

[1] Universidade do Algarve, C. Gambelas, 8005-139 Faro, Portugal
[2] Instituto de Telecomunicações, Lisbon, Portugal
[3] DMS, University of Cincinnati, Cincinnati, OH 45221-0025, USA
zhongn@ucmail.uc.edu

Abstract. A classical theorem of Peixoto qualitatively characterizes, on the two-dimensional unit ball, the limit sets of structurally stable flows defined by ordinary differential equations. Peixoto's density theorem further shows that such flows are typical in the sense that structurally stable systems form an open dense set in the space of all continuously differentiable flows.

In this note, we discuss the problem of explicitly finding the limit sets of structurally stable planar flows.

1 Introduction

In this note, we discuss limit sets of planar C^1 dynamical systems from the viewpoint of computability.

The dynamical systems to be considered are of the type

$$\dot{x} = f(x) \tag{1}$$

where $f : E \to \mathbb{R}^2$ is continuously differentiable (C^1) on E - either an open or a compact subset of $\mathbb{R}^2$ (if E is compact, then f is assumed to be C^1 in some open set containing E), $t \in \mathbb{R}$ is the independent variable, and $\dot{x} = dx/dt$. The solution to the equation with the initial condition $x(0) = x_0$ is a function $\phi(f, x_0)$ of time t that describes the time dependence of x_0 in the phase space, either $\mathbb{R}^2$ or a subset of $\mathbb{R}^2$. The function $\phi(f, \cdot)(\cdot)$ is called the flow defined by the vector field f and $\phi(f, x_0)(\cdot)$ is called a trajectory (or orbit) passing through x_0.

Since the trajectories can be defined for arbitrarily long terms and explicit solution formulas do not exist for most dynamical systems, it becomes necessary and essential to study asymptotic (long term) behaviors of the trajectories. The topic is extensively studied in mathematics and physics. The asymptotic behavior of a dynamical system is captured by its limit sets, which are the states the trajectories approach to or land on as $t \to \pm\infty$. The limit sets are well understood *qualitatively* for C^1 planar dynamical systems: a closed and bounded limit set other than an equilibrium point or a periodic orbit consists of equilibria and solutions connecting them according to the Poincaré-Bendixson Theorem. From the *quantitative* perspective, limit sets of planar flows remain elusive; there are

© Springer Nature Switzerland AG 2021
L. De Mol et al. (Eds.): CiE 2021, LNCS 12813, pp. 494–503, 2021.
https://doi.org/10.1007/978-3-030-80049-9_48

a number of open problems in the field, including Hilbert's 16th problem. The second part of Hilbert's 16th problem asks for the maximum number and relative positions of periodic orbits of planar polynomial (real) vector fields of a given degree. The problem is open even for the simplest nonlinear flows - the quadratic flows.

It is well known that the operator $(f, x_0) \to \phi(f, x_0)(\cdot)$ (as a function of t) is computable (see [3] and references therein). Intuitively, this means that there is an algorithm that plots a polygon curve $p(t)$ on a computer screen satisfying $\max_{-T \le t \le T} \|\phi_t(f, x_0) - p(t)\| \le 2^{-n}$ for every natural number n, every rational number T, and "good enough" information on f and x_0. (It is a convention to write $\phi_t(f, x_0)$ for $\phi(f, x_0)(t)$.) However, the algorithm is local in the sense that it provides little information on asymptotic behaviors of the trajectories. On the other hand, the limit sets are asymptotic and global in nature - global properties are generally more difficult to deal with in classical mathematics and to compute in numerical as well as in computable analysis. It turns out that there are C^1 computable planar flows whose periodic orbits are badly non-computable. This is our first theorem.

Theorem 1. *For any $k \ge 1$, there is a C^k computable function $f : \mathbb{R}^2 \to \mathbb{R}^2$ such that none of the periodic orbits of the C^k planar system $\dot{x} = f(x)$, $x \in \mathbb{R}^2$, is r.e. or co-r.e. as a closed subset of $\mathbb{R}^2$.*

Intuitively, the theorem says that it is impossible to plot any periodic orbit of the flow on a computer screen, not even a good adumbration of it. Then, under what conditions can the qualitatively well-understood limit sets of a planar flow be computable - quantitatively plotted with arbitrarily high magnification? Our second theorem provides an answer.

Theorem 2. *There is an algorithm that locates the positions of equilibrium points and periodic orbits with arbitrarily high precision for any structurally stable C^1 planar vector field defined on the closed unit disk. Moreover, the computation is uniform on the set of all structurally stable planar vector fields.*

Recall that the density theorem of Peixoto [10, Theorem 2] shows that, on two-dimensional compact manifolds, structurally stable systems are "typical" in the sense that such systems form a dense open subset in the set of all C^1 planar systems. Hence, Theorem 2 says that the limit set of a *typical* differential equation (1) defined on the unit disk of $\mathbb{R}^2$, where f is of class C^1, is computable.

2   Preliminaries

In this section, we recall necessary definitions. We begin with definitions related to computable analysis with assumption that the reader is familiar with the classical computable functions from $\mathbb{Z}_1$ to $\mathbb{Z}_2$, where $\mathbb{Z}_i$ is the set of (or the set of tuples of) natural numbers ($\mathbb{N}$), integers ($\mathbb{Z}$), or rational numbers ($\mathbb{Q}$). In computable analysis, informally speaking, an operator $\Phi : X \to Y$ is computable if

(1) elements in X and Y can be encoded by sequences of exact "functions" of finite size (such as rational numbers, polynomials with rational coefficients, polygonal curves with rational corners, etc.) which converge to those elements at a known rate of convergence; such sequences are called names of the corresponding elements; and

(2) there is a computer (a Turing machine, an algorithm or a computer program) that outputs an approximation to $\Phi(x)$ within accuracy 2^{-n} on input of n (accuracy) and (a name of) x.

To execute evaluations in practice, an infinite input datum - a name of x - can be conveniently treated as an interface to a program computing Φ: for every $n \in \mathbb{N}$, the name supplies a good enough (finite size) approximation p of x to the program, the program then performs computations based on inputs n and p, and returns a (finite size) approximation q of $\Phi(x)$ with an error bounded by 2^{-n}. This is often termed as $\Phi(x)$ is computable from (a name of) x. For more details the reader is referred to e.g., [1].

The following is the precise definition.

Definition 1. *1. A name of a real number x is a function $a : \mathbb{N} \to \mathbb{Q}$ such that $|x - a(n)| \leq 2^{-n}$. If the function a is (classically) computable, then x is said to be computable.*

2. Let $f : \mathbb{R}^2 \to \mathbb{R}^2$ be a C^k function, and let $B = \{x \in \mathbb{R}^2 : \|x\| \leq r\}$, where r is a rational number. A C^k-name of f on B is a sequence $\{P_l\}$ of polynomials with rational coefficients such that $d^k(f, P_l) \leq 2^{-l}$, where

$$d^k(f, P_l) = \max_{0 \leq j \leq k} \max_{x \in B} \|D^j f(x) - D^j P_l(x)\|.$$

In particular:

A. f is said to be (C^k-) computable on B if there is a Turing machine (or a computer) that outputs a C^k-name $\{P_l\}$ of f in the following sense: on input l (accuracy), it outputs the rational coefficients of the polynomial P_l.

B. Or, equivalently, f is said to be (C^k-) computable on B if there is an oracle Turing machine such that for any input $l \in \mathbb{N}$ (accuracy) and any name of $x \in B$ given as an oracle, the machine will output the rational vectors $q_0, q_1, \ldots, q_k$ in $\mathbb{R}^2$ such that $\|q_j - D^j f(x)\| \leq 2^{-l}$ for all $0 \leq j \leq k$ (see e.g., [2,9]).

As already mentioned above, an oracle can be conveniently treated as an interface to a program computing f in practice.

We turn now to define computable open and closed subsets of $\mathbb{R}^2$. In $\mathbb{R}^2$, computability can be intuitively visualized by plotting pixels: a subset of $\mathbb{R}^2$ is computable if it can be plotted on the screen of a computer with arbitrarily high magnification. The following definition shares this spirit.

Definition 2. *Let U be an open subset of $\mathbb{R}^2$, and let C be a closed subset contained in B, where B is a closed disk of $\mathbb{R}^2$ centered at the origin with a rational radius.*

1. U is said to be r.e. open if there are computable functions $a : \mathbb{N} \to \mathbb{Q}^2$ and $r : \mathbb{N} \to \mathbb{Q}$ such that $U = \cup_{n=1}^{\infty} B(a(n), r(n))$, where $B(a(n), r(n))$ is the open disk centered at $a(n)$ with radius $r(n)$. In other words U can be filled up by the fattened pixels at a_n - the open disk $B(a_n, r_n)$.

2. C is called co-r.e. closed if $B \setminus C$ is r.e. open in B. C is called r.e. closed if C has a computable dense sequence. C is called computable if it is co-r.e. and r.e. closed. Or, equivalently, if there is a Turing machine that, on input $n \in \mathbb{N}$ (accuracy), outputs finite sequences $r_j \in \mathbb{Q}$ and $a_j \in \mathbb{Q}^2$, $1 \le j \le j(n)$, such that $d_H(C, B \setminus \cup_{j=1}^{j(n)} B(a_j, r_j)) \le 2^{-n}$, where $d_H(\cdot, \cdot)$ denotes the Hausdorff distance between two compact subsets of $\mathbb{R}^2$.

We observe that the r.e. openness is a local property - one pixel at a time - and the plotting does not give any adumbration of the whole picture of U. On the other hand, co-r.e. closeness is a global property: Assume that $C = K \setminus \cup B(a_n, r_n)$. If one plots $B(a_n, r_n)$ as before, then at each step one obtains a set (the portion not covered by the pixels) containing the entire C, an adumbration of C.

Now we turn to define structurally stable planar dynamical systems. Let $K \subseteq \mathbb{R}^2$ be a compact set, and let $f : K \to \mathbb{R}^2$ be a C^1 function. The C^1-norm is used for C^1 functions

$$\|f\|_1 = \max_{x \in K} \|f(x)\| + \max_{x \in K} \|Df(x)\|$$

Definition 3. The system (1), where $f : K \to \mathbb{R}^2$ is of class C^1, is structurally stable if there exists some $\varepsilon > 0$ such that for all $g \in C^1(K)$ satisfying $\|f - g\|_1 \le \varepsilon$, the trajectories (orbits) of

$$y' = g(y) \tag{2}$$

are homeomorphic to the trajectories of (1), i.e., there exists some homeomorphism h such that if γ is a trajectory of (1), then $h(\gamma)$ is a trajectory of (2). Moreover, the homeomorphism h is required to preserve the orientation of trajectories by time.

Intuitively, (1) is structurally stable if the shape of its dynamics is (globally) robust to small perturbations. If K is the unit disk $\mathbb{D} = \{x \in \mathbb{R}^2 : \|x\| \le 1\}$, as in Theorem 2, it is common to assume that the vector field points inwards along the boundary of $\mathbb{D}$. Otherwise, the system may be structurally unstable if the flow is tangent to a point on the boundary. We note that not all planar systems are structurally stable. Explicit examples of structurally unstable systems can be found in e.g., [7, Figure 9.4 in p. 193]. However, due to Peixoto's density theorem, structurally stable systems are typical in the sense that structurally stable systems form an open dense set in the space of all continuously differentiable flows.

3   Proof of Theorem 1

We now proceed with the proof of Theorem 1. We begin by recalling a theorem by Weihrauch (Theorem 4.2.8. [Wei00]): the countable set $\mathbb{R}_c$ of all computable real numbers can be covered by the union of a computable sequence of open intervals, $I_n = (\alpha_n, \beta_n)$, such that the length of $\sum I_n$ is at most 1. Let $A = \mathbb{R} \setminus \bigcup I_n$. Then $A \neq \emptyset$ is co-r.e. closed and none of points in A is computable.

Fix $k \in \mathbb{N}$. Now we construct a C^k computable function $g : \mathbb{R} \to \mathbb{R}$ such that $g(x) = 0$ if and only if $x \in A$. Let $\phi : \mathbb{R} \to \mathbb{R}$ be the computable C^∞ standard bump function

$$\phi(x) = \begin{cases} e^{-\frac{x^2}{1-x^2}} & \text{if } |x| < 1 \\ 0 & \text{otherwise} \end{cases}$$

The function $g : \mathbb{R} \to \mathbb{R}$ is defined by the following formula: for every $x \in \mathbb{R}$,

$$g(x) = \sum_{n=1}^{\infty} \phi\left(\frac{x - \frac{\alpha_n + \beta_n}{2}}{r_n}\right) 2^{-n}$$

where $r_n = \frac{\beta_n - \alpha_n}{2}$. It is readily seen that the function g is C^k computable, $g(x) \geq 0$, and $g(x) = 0$ if and only if $x \in A$. We are now ready to define the desired function $f : \mathbb{R}^2 \to \mathbb{R}^2$: $f(x_1, x_2) = (f_1(x_1, x_2), f_2(x_1, x_2))$, where

$$f_1(x_1, x_2) = -x_2 + x_1(x_1^2 + x_2^2) \cdot g\left(x_1^2 + x_2^2\right)$$

and

$$f_2(x_1, x_2) = x_1 + x_2(x_1^2 + x_2^2) \cdot g\left(x_1^2 + x_2^2\right).$$

It is clear that f is a C^k computable function.

For the following system

$$\dot{x}_1 = f_1(x_1, x_2), \quad \dot{x}_2 = f_2(x_1, x_2)$$

it can be rewritten, in polar coordinates, in the form of

$$\dot{r} = r^3 g(r^2), \quad \dot{\theta} = 1$$

for $r > 0$ with $\dot{r} = 0$ at $r = 0$. It is clear that the system has a unique equilibrium point at the origin of $\mathbb{R}^2$. Since $g(x) \geq 0$ for all $x \in \mathbb{R}$, it follows that the only periodic orbits are circles with center at the origin and radius r satisfying $r > 0$ and $g(r^2) = 0$.

Consider one such circle Γ_0 with center at the origin and radius r_0. Then it follows from the construction of g that r_0^2 is a non-computable real. For any point (x_1, x_2) on Γ_0, if the point is computable, then x_1 and x_2 must be computable reals, which implies that $r_0^2 = x_1^2 + x_2^2$ is a computable number. This is a contradiction. Hence none of the points on Γ_0 is computable, which implies that Γ_0 cannot be r.e. Next we show that Γ_0 is not co-r.e. either. Suppose otherwise Γ_0 was co-r.e. in $\mathbb{R}^2$. Since the intersection of two co-r.e. closed sets is again co-r.e., the set $\Gamma_0 \cap \{(x, 0) : x \geq 0\} = \{(r_0, 0)\}$ is co-r.e.. Then it follows from Theorem 6.2.9 [Wei00] that there is a computable function γ, $\gamma : \mathbb{R}^2 \to \mathbb{R}$, such that $\{(r_0, 0)\} = \gamma^{-1}[\{0\}]$. Since $(r_0, 0)$ is the unique zero of γ, $(r_0, 0)$ is a computable point (Corollary 6.3.9 [Wei00]). This contradicts the fact that r_0^2 is non-computable. Hence Γ_0 is not co-r.e. This completes the proof of Theorem 1.

We mention in passing that the set $\mathcal{P}$ of all periodic orbits of the planar system defined above is co-r.e. closed because it is the set $G^{-1}(0)$, where $G : \mathbb{R}^2 \to \mathbb{R}$, $G(x_1, x_2) = g(x_1^2 + x_2^2)$, is a computable function. Hence, it is possible to plot over-adumbrations of $\mathcal{P}$ with better and better accuracies (although the accuracies are unknown per se since $\mathcal{P}$ is not computable). On the other hand, the relative positions of the periodic orbits are completely in dark - there is no good over- or under-adumbration of any periodic orbit.

4   Main Ideas of the Proof of Theorem 2

One apparent difficulty to plot periodic orbits of the flow in Theorem 1 is that there are too many of them. Can we plot the periodic orbits of a flow if there are only finitely many of them? While the problem is open for C^1 computable planar flows in general, the answer is yes if the planar flow is structurally stable. The structural stability of a planar flow is characterized in terms of its limit set by Peixoto in 1962 in his seminal paper [10]. Let f be a C^1 vector field defined on a compact two-dimensional differentiable manifold $K \subseteq \mathbb{R}^2$. Peixoto showed that if f is structurally stable on K, then, among other characterizations, the number of equilibria (i.e., zeros of f) and of periodic orbits is finite and each is hyperbolic, and there is no trajectories connecting saddle points. Similar results hold if K is a manifold with boundary; in particular, $K = \mathbb{D} = \{x \in \mathbb{R}^2 : \|x\| \leq 1\}$ with the assumption that the vector fields always point inwards along the boundary of $\mathbb{D}$.

We now outline the proof of Theorem 2, which shows that there is an (uniform) algorithm that locates the positions of equilibrium points and periodic orbits with arbitrarily high precision for any structurally stable C^1 planar vector field defined on $\mathbb{D}$.

Remark. The algorithm is a "master" program in the sense that it computes all equilibria and periodic orbits simultaneously when given a structurally stable planar vector field f.

Main Ideas of the Proof. The proof is long and intricate; only a brief outline is presented. The complete proof can be found in the preprints [4,5].

(A) Construct a sub-algorithm to locate the equilibrium points: on input $n \geq 1$ (accuracy) and (a C^1-name of) f, the sub-algorithm outputs a union of mutually disjoint squares such that each square contains exactly one equilibrium and the side-length of a square is at most $1/n$. The construction relies on the fact that there are only finitely many equilibria and each is hyperbolic. The hyperbolicity ensures that the Jacobean of f at each equilibrium is non-zero. By computing f and Df simultaneously on refined square-grids of $\mathbb{D}$, the algorithm will return a desired output after finitely many updates on the square-grids.

More specifically, if s is some square, its corners will have rational coordinates, which ensures all squares are computable. Hence, both $f(s)$ and $Df(s)$ are computable from any given C^1-name of f. If s does not contain any zero of f, then $0 \notin f(s)$ (precisely, it should be $(0,0) \notin f(s)$; 0 is used to denote the origin of either $\mathbb{R}$ or $\mathbb{R}^2$). Consequently, the distance between 0 and $f(s)$, $d(0, f(s)) = \min_{x \in f(s)} d(0, x)$, is greater than 0. Since $f(s)$ is computable (from f), an over-approximation $A_l(s)$ (a polygon with rational corners) of $f(s)$ can be computed with accuracy bounded by 2^{-l} for some $l \geq n$. If $0 \notin f(s)$, then $0 \notin A_l(s)$ for l sufficiently large. Hence, by testing if $0 \notin A_l(s)$ for all squares s and $l = n, n+1, n+2, \ldots$, the algorithm can eventually identify the squares which do not have zeros after finitely many updates on l. The problematic squares are those containing zeros, because whether $d(0, f(s)) = 0$ cannot in general be decided by finitely many approximations - one may not be able to conclude either $0 \in f(s)$ or $0 \notin f(s)$ with (any) current choice of l. To deal with this problem, one makes use of the hypothesis that all zeros are hyperbolic; hence, the Jacobean at each zero is invertible. In other words, f is locally invertible at each of its zeros. This is why the algorithm computes both $f(s)$ and $Df(s)$. After finitely many updates on l, the algorithm arrives at a midway stage that either $d(0, f(s)) > 2^{-l}$ (i.e., s contains no zero of f) or $\|Df(s)\| > 2^{-m}$ for some $m \geq l \geq n$. For the latter case, the algorithm computes - based on an effective version of the inverse function theorem - a polygon (under-)approximation of $f(s)$ as the domain of f^{-1}. Hence, if 0 is in this (under-)approximation of $f(s)$, then s contains a zero of f. A possible problem is that s might have a zero whose image is outside this (under-)approximation of $f(s)$. This problem can be solved by covering s with several overlapping smaller squares and then applying the procedure to all those overlapping smaller squares. By proceeding in this way and by increasing the accuracy $l \geq n$ used in the computations, the algorithm will be able to determine whether or not each square has a zero after finitely many updates on l. If a square contains a zero, the zero is unique because on this square f is invertible. (See [4] for more details).

(B) Construct a sub-algorithm to locate the periodic orbits: It suffices to construct an algorithm that takes as input $(n; f)$ and returns a union A_n of compact subsets, each has polygonal boundaries, such that the Hausdorff distance between A_n and $\mathbf{P}$ - the set of periodic orbits - is less than $1/n$ for every $n \in \mathbb{N}$ and every structurally stable vector field f.

Before constructing this algorithm, it is important to remark that, by Peixoto's theorem (see [10]), the limit set of (1) is formed by hyperbolic equilibrium points and hyperbolic periodic orbits. Hence, each periodic orbit is either attracting or repelling. The hyperbolicity condition ensures (see e.g., [11]) that, for each periodic orbit γ, there is an open set U_γ (a so-called *basin of attraction*) containing γ such that, if the periodic orbit is attracting, then any trajectory entering U_γ will converge to γ exponentially fast as $t \to \infty$ (a similar condition holds for repelling periodic orbits when $t \to -\infty$). Furthermore, if the system (1) has no saddle points, then there exists some time $T_\varepsilon \geq 0$ such that any trajectory starting at a point $x \in \mathbb{D}$ at least ε-distance away from equilibrium point(s) or repelling periodic orbit(s) will be inside the basin of attraction of some attracting periodic orbit γ after time T_ε, i.e., $\phi_t(x) \in U_\gamma$ for all $t \geq T_\varepsilon$. Hence, by computing

$\phi_t(\mathbb{D})$ for increasing (decreasing) values of time, one is able to approach the set of attracting (repelling) periodic orbits as $t \to \infty$ ($t \to -\infty$), as long as the neighborhoods of equilibrium point(s) are avoided (those neighborhoods - squares each containing a unique equilibrium point - have already been identified in step (A)). This can be done with the following steps:

(1) Cover the compact set $\mathbb{D}$ with a finite number of square "pixels;"

(2) Use a rigorous numerical method to compute the (flow) images of all pixels after some time T, and take the union $A_{n,T}$ of the images of all pixels as a first-round candidate for an approximation to **P**. Then pick sets of pixels from $A_{n,T}$ and test whether they are forward or backward time invariant, essentially by testing whether $\phi_t(A) \subseteq A$ for positive or negative t. In this manner, a set $B_{n,T} \subseteq A_{n,T}$ is obtained as a second-round candidate for an (over-) approximation to **P**.

For simplicity and consistency of the algorithm sketched here, $A_{n,T}$ will be used once again to denote $B_{n,T}$. The next step is to see whether $A_{n,T}$ is a "good enough" approximation of **P**.

(3) Test whether $A_{n,T}$ is an over-approximation of **P** within the desired accuracy. If the test is successful, set $A_n = A_{n,T}$ and output A_n. If the test fails, increase time T and use a finer lattice of square pixels when numerically approximating the flow after T. Similar simulations using time $-T$ are run in parallel to find repellers.

Recall that a periodic orbit γ will separate $\mathbb{D}$ into the interior and the exterior according to the Jordan curve theorem. The same can be said to a "good enough" approximation of γ. Hence, $A_{n,T}$ can be separated into connected components; each of the components will have a "doughnut shape." If one is able to identify the interior and exterior of each doughnut, then one can determine the maximum width of each doughnut. The maximum widths provide an upper bound on the error occurred when using $A_{n,T}$ to approximate **P**. To identify the interior and exterior regions delimited by a connected component of $A_{n,T}$, a coloring algorithm is constructed, which works along the following lines. All pixels considered below are disjoint from the component of $A_{n,T}$: (i) pick some pixel and paint this pixel blue; (ii) paint pixels adjacent to blue pixels with the color blue; (iii) when there are no more unpainted pixels which can be painted blue, paint one of the remaining pixels red; (iv) paint unpainted pixels adjacent to red pixels with the color red; (iv) if there are still unpainted pixels, restart the algorithm with a better accuracy (the connected component under consideration has not yet had a doughnut shape). After the successful termination of this (sub-)algorithm, the interior and exterior regions of the considered connected component will correspond to regions of different colors. It can be shown that the main algorithm will eventually halt and that when it does halt, it provides a correct result.

The intricate components of the algorithm are where the saddle points are dealt with, and the search for a time T such that the Hausdorff distance between $A_{n,T}$ and **P** is less than $1/n$ with $(n; f)$ being the input to the algorithm. The problem with a saddle point is that it may take an arbitrarily long time for the flow starting at some point near but not on the stable manifold of the saddle to eventually move away from the saddle. This undesirable behavior is dealt with by transforming the original flow near a saddle to a linear flow using a computable version of Hartman-Grobman's theorem ([6]). The time needed for the linear flow to go through a small neighborhood can be explicitly calculated. (See [5] for details)

We conclude this note with two open questions, which are suggested to us by one of the referees.

– Does there exist a computable function $f : \mathbb{R}^2 \to \mathbb{R}^2$ such that Theorem 1 remains true, where f is either an analytic function or a polynomial?
– It is known that the structurally stable systems defined on $\mathbb{D}$ form an open dense subset of $C^1(\mathbb{D})$. Is this open subset computable? In other words, is it decidable whether a planar system defined on $\mathbb{D}$ is structurally stable?

We remark that the proof of Theorem 1 relies on bump functions. A bump function is usually not analytic at the "foot" of the bump. For a polynomial planar system, it is well-known that such a system can only have finitely many limit cycles [8]. Thus, for polynomial planar systems, the question is: Can the finiteness ensure the computability? Concerning the second question, we note that the open subset of all structurally stable systems defined on $\mathbb{D}$ can be shown to be r.e. open in $C^1(\mathbb{D})$.

Acknowledgments. We thank the referees' helpful suggestions and insightful comments. D. Graça was partially funded by FCT/MCTES through national funds and co-funded EU funds under the project UIDB/50008/2020. This project has received funding from the European Union's Horizon 2020 research and innovation programme under the Marie Skłodowska-Curie grant agreement No. 731143.

References

1. Brattka, V., Hertling, P., Weihrauch, K.: A tutorial on computable analysis. In: Cooper, S.B., Löwe, B., Sorbi, A. (eds.) New Computational Paradigms: Changing Conceptions of What is Computable, pp. 425–491. Springer, New York (2008). https://doi.org/10.1007/978-0-387-68546-5_18
2. Braverman, M., Yampolsky, M.: Non-computable Julia sets. J. Am. Math. Soc. **19**(3), 551–578 (2006)
3. Graça, D.S., Zhong, N., Buescu, J.: Computability, noncomputability, and hyperbolic systems. Appl. Math. Comput. **219**(6), 3039–3054 (2012)
4. Graça, D.S., Zhong, N.: The set of hyperbolic equilibria and of invertible zeros on the unit ball is computable (2020, submitted). https://arxiv.org/abs/2002.08199
5. Graça, D.S., Zhong, N.: Computing the exact number of periodic orbits for planar flows (2021, submitted). http://arxiv.org/abs/2101.07701
6. Graça, D.S., Zhong, N., Dumas, H.S.: The connection between computability of a nonlinear problem and its linearization: the Hartman-Grobman theorem revisited. Theoret. Comput. Sci. **457**(26), 101–110 (2012)

7. Hirsch, M.W., Smale, S., Devaney, R.: Differential Equations, Dynamical Systems, and an Introduction to Chaos. Academic Press (2004)
8. Ilyashenko, Y.: Finiteness Theorems for Limit Cycles, Translations of Mathematical Monographs, vol. 84. American Mathematical Society (1991)
9. Ko, K.I.: Complexity Theory of Real Functions. Birkhäuser (1991)
10. Peixoto, M.: Structural stability on two-dimensional manifolds. Topology **1**, 101–121 (1962)
11. Perko, L.: Differential Equations and Dynamical Systems, 3rd edn. Springer, New York (2001). https://doi.org/10.1007/978-1-4613-0003-8

Author Index

Printed in the United States
by Baker & Taylor Publisher Services